ASTROBIOLOGY
A MULTIDISCIPLINARY APPROACH

Jonathan I. Lunine
The University of Arizona

San Francisco   Boston   New York
Cape Town   Hong Kong   London   Madrid   Mexico City
Montreal   Munich   Paris   Singapore   Sydney   Tokyo   Toronto

Senior Executive Editor: *Adam Black, Ph. D.*
Associate Editor: *Liana Allday*
Senior Marketing Manager: *Christy Lawrence*
Production Supervisor: *Shannon Tozier*
Production Management: *Elm Street Publishing Services, Inc.*
Composition: *Elm Street Publishing Services, Inc.*
Illustrator: *Scientific Illustrators*
Manufacturing Supervisor: *Robert Davis*
Text Designer: *Elm Street Publishing Services, Inc.*
Cover Illustrator: *Blakeley Kim*
Cover Designer: *Stacy Wong*
Text Printer and Binder: *RR Donnelley and Sons, Crawfordsville*
Cover Printer: *Phoenix Book Technology Park*

Library of Congress Cataloging-in-Publishing Data
CIP data is on file at the Library of Congress.

ISBN: 0-8053-8042-6

Copyright © 2005 Pearson Education, Inc., publishing as Addison Wesley, 1301 Sansome St., San Francisco, CA 94111. All rights reserved. Manufactured in the United States of America. This publication is protected by Copyright and permission should be obtained from the publisher prior to any prohibited reproduction, storage in a retrieval system, or transmission in any form or by any means, electronic, mechanical, photocopying, recording, or likewise. To obtain permission(s) to use material from this work, please submit a written request to Pearson Education, Inc., Permissions Department, 1900 E. Lake Ave., Glenview, IL 60025. For information regarding permissions, call (847) 486-2635.

Many of the designations used by manufacturers and sellers to distinguish their products are claimed as trademarks. Where those designations appear in this book, and the publisher was aware of a trademark claim, the designations have been printed in initial caps or all caps.

1 2 3 4 5 6 7 8 9 10 -DOC- 07 06 05 04
www.aw-bc.com/physics

CONTENTS

PREFACE viii

    Introduction—What Is Astrobiology? 1

1   **Historical Background to Astrobiology** 9
- **1.1** Science Retold as History: Buyers Beware! 9
- **1.2** The Astronomical Revolution: Earth as One of Many Planets 11
- **1.3** The Chemical Revolution: The Building Blocks of Matter Revealed and the Vital Spark Removed from Organic Chemistry 18
- **1.4** The Biological Revolution: Life from Life Only 23
- **1.5** The Search for Life's Origins: Life from What, and Where? 28

2   **Essential Concepts I: Some Basic Physics of Forces and Particles** 33
- **2.1** Introduction 33
- **2.2** The Forces of Nature 34
- **2.3** More on the Particles of Nature 50
- **2.4** Quantum Mechanics 53
- **2.5** The Conservation of Mass-Energy 62

3   **Essential Concepts II: The Physics of Chemistry** 66
- **3.1** Introduction 66
- **3.2** Quantum Mechanics and the Electron 66
- **3.3** The Periodic Table 70
- **3.4** Bonding Mechanisms 77
- **3.5** Particular Properties of the Carbon Bond 86
- **3.6** Spectroscopy and the Spectroscopic Signatures of Atomic and Molecular Structure 88
- **3.7** Isotopes and the Stability of the Nucleus 98

4 Necessary Concepts III: The Chemistry of Life 103

4.1 Introduction 103

4.2 The Fundamental Biopolymers, Their Units, and Other Molecules 103

4.3 Informational, Structural, and Transport Systems of the Prokaryotic Cell 116

4.4 Organelles and Their Functions in the Eukaryotic Cell 120

4.5 Energy Storage and Transfer via the Proton and the Electron in Cells 123

4.6 Glycolysis, Respiration, and Fermentation 126

4.7 Photosynthesis and Chemisynthesis 130

4.8 The Genetic Code and the Synthesis of Proteins 133

5 The Cosmic Foundations of the Origins of Life 141

5.1 Introduction 141

5.2 Temporal and Spatial Scales of the Cosmos 141

5.3 Origin of the Primordial Elements Hydrogen and Helium in the Big Bang 154

5.4 Manufacture of the Biogenic Elements in the Shining and Death of Stars 157

5.5 Is the Universe Tuned to Allow Life? 165

6 Planetological Foundations for the Origin of Life 174

6.1 Introduction 174

6.2 Formation of Planets as a Natural Consequence of Star Formation 174

6.3 The Technique of Isotopic Dating 186

6.4 Meteorites: The Age of the Elements and of the Earth 192

6.5 Origin of the Earth's Oceans and Biogenic Elements 197

6.6 Origin of the Moon and the Differentiated Earth 199

6.7 The Earth After Formation and the Geologic Timescale 201

7 The Thermodynamic Foundations of Life 211

7.1 Introduction 211

7.2 Important Concepts of Thermodynamics: Temperature, Energy, Heat, Work, Entropy, and the Laws of Thermodynamics 212

7.3 A Closer Look at Entropy: Maxwell's Demon, Information Theory, and Statistical Mechanics 217

7.4 Thermodynamics of Living Systems 225

7.5 Thermodynamics, Disequilibrium, Chemistry, and the Origin of Life 236

8 Biological Foundations for the Origin of Life 243

8.1 Introduction 243

8.2 Earliest Roles of RNA as Encoder and Catalyst 244

8.3 Formation of DNA 248

8.4 Precursors to RNA 250

8.5 Vesicle Formation and the Precursors to Cells 258

8.6 The Fundamental Role of Containers and Surfaces in Increasing Replicator Complexity 261

8.7 Sizes of Protocells and the RNA World 265

9 From the Origin to the Diversification of Life 274

9.1 Introduction 274

9.2 Complex Behavior in Chemical Systems 276

9.3 Before the Threshold: Evolving Autocatalysis or Life without Replication 282

9.4 Beyond the Threshold: The Evolutionary Diversification of Life 290

9.5 Putting It All Together: A Time-Travel Fantasy 295

10 Extremophiles and the Span of Terrestrial Biotic Environments 305

10.1 Introduction 305

10.2 Extreme Environments 306

10.3 The Implications of Extremophiles for Life Elsewhere in the Solar System 320

10.4 The Significance of Extremophiles in the Earliest History of Life 325

11 Planetary Evolution I: Earth's Evolution as a Habitable Planet 329

11.1 Introduction 329

11.2 Evidence of Habitability Early in Earth's History 330

11.3 Earliest Evidence of the Existence of Life 334

11.4 The Faint Early Sun and a Massive Carbon Dioxide Greenhouse 337

11.5 Carbon-Silicate Cycle, Plate Tectonics, and Weathering 344
11.6 Origin of Granites and the Supercontinent Cycle 348
11.7 The Rise of Oxygen 353
11.8 The Implications of Earth's Habitability for Mars and Venus 356

12 Planetary Evolution II: The History of Mars 365

12.1 Introduction 365
12.2 Mars as a Planet Today 366
12.3 Evidence for Past Epochs of Surface Liquid Water 375
12.4 Mechanism for a Warm Climate on Early Mars 384
12.5 The Drying and Freezing of Mars: Causes and Timing 385
12.6 Martian Crustal and Surface Water Today 387
12.7 The Origin of Martian Water 393
12.8 Future Exploration of Mars 394

13 Planetary Evolution III: The Significance of Europa and Titan 401

13.1 Introduction 401
13.2 Setting: The Outer Solar System 401
13.3 Europa 404
13.4 Titan 416

14 Life Elsewhere I: Direct Detection of Life and Its Remnants Within the Solar System 435

14.1 Introduction 435
14.2 What Are We Searching For? 436
14.3 The Basic Characteristics of Life That Define What We Are Looking For 439
14.4 Modern Techniques to Search for Life on Other Worlds 444
14.5 Case Study in the Search for Life on Mars I: Viking 460
14.6 Case Study in the Search for Life on Mars II: ALH84001 461
14.7 Future Searches for Life on Mars 465
14.8 Searching for Life in the Subsurface of Europa 467
14.9 Exploration of Titan for Clues to the Origin of Life 468
14.10 Coda to the Search for Life 470

15 Life Elsewhere II: The Discovery of Extrasolar Planetary Systems and the Search for Habitable and Inhabited Worlds 474

- 15.1 Introduction 474
- 15.2 The Systematic Discovery of Planets Using Doppler Spectroscopy 475
- 15.3 Astrometry: Another Technique for Indirect Detection of Planets 483
- 15.4 Planets Almost Seen: Detection of Planets by Transits, Phases, and Microlensing 487
- 15.5 Direct Detection 494
- 15.6 Learning About Other Earths 504
- 15.7 The Search for Extraterrestrial Intelligence 508

16 External and Internal Influences in the Evolution of Life 514

- 16.1 Introduction 514
- 16.2 Mechanisms of Evolution 515
- 16.3 On the Origin of Eukaryotes 518
- 16.4 Complexity, Diversity, and Evolution 523
- 16.5 Environmental and Genetic Interactions in the Evolution of Life 526
- 16.6 Snowball Earth and the Cambrian Explosion 530
- 16.7 Impacts, Volcanism, and Major Extinctions 534
- 16.8 Orbital and Spin-Axis Forcing of Climate Variations 539
- 16.9 Variations in the Pace of Biological Evolution on Habitable Planets 541

17 Evolution of Intelligence and the Persistence of Civilization 546

- 17.1 Introduction 546
- 17.2 Ecce *Homo* 546
- 17.3 Human Intelligence as an Evolutionary Specialization and the Biology of Conscious Self-Awareness 554
- 17.4 Climate Change and the Timing of the Development of Human Civilization 560
- 17.5 Future Prospects for the Human Species and Its Civilization 562
- 17.6 Epilogue: Is There Anyone to Talk to Elsewhere in the Cosmos? 567

INDEX 578

PREFACE

The spread, both in width and depth, of the multifarious branches of knowledge during the last hundred odd years has confronted us with a queer dilemma. We feel clearly that we are only now beginning to acquire reliable material for welding together the sum-total of what is known into a whole; but, on the other hand, it has become next to impossible for a single mind fully to command more than a small specialized portion of it. I can see no other escape from this dilemma (lest our true aim be lost for ever) than that some of us should venture to embark on a synthesis of facts and theories, albeit with second-hand and incomplete knowledge of some of them, and at the risk of making fools of themselves.

Erwin Schrödinger (1944). *What Is Life?* Cambridge University Press, Cambridge.

The goal of this book is a comprehensive treatment of the field of astrobiology for upper-level undergraduate students and beginning graduate students. It is to be hoped, as well, that the book will be useful to scientists in related fields who wish to gain a working knowledge of astrobiology. In undertaking the writing of this book, I am painfully aware of the pitfalls articulated 60 years ago by Schrödinger, but I am equally cognizant of the need for such a text. This is not a collection of review chapters by specialists in the various fields currently encompassed by astrobiology; nor is it an introduction so basic that the particulars of the physics, chemistry, and biology are avoided. Instead, this book attempts to be a self-contained treatment of the key issues and questions in astrobiology, written to be equally comprehensible to students coming from physics, chemistry, geology, and biology backgrounds. A conscientious student from any of these fields should be able to pick up this book, read the chapters on the fundamentals (Chapters 2, 3, and 4), and, thus prepared, embark on an integrated study of the origin of the cosmos and planets, the origin and evolution of life, the mechanisms for maintenance of planetary habitability, the search for life elsewhere, and the long-term prospects for technological civilizations on and off the planet.

Students should take care not only to read the text but to at least ponder all the questions and problems at the end of each chapter and try to fully work or write out the answers for a significant subset of them. They involve a significant amount of research of articles in the literature. The World Wide Web is extremely useful in this regard, but great care must be taken to use it to locate valid, refereed, research articles in journals—and not as a source for unpublished and unrefereed tracts that exist on "somebody's" home page. Most university libraries have online search capabilities and electronic subscriptions to many of the most important journals. The Suggested Readings are books and review articles that provide significant coverage, at varying levels, of the topics in a given chapter. The References are a mixture of papers from which I have drawn a particular set of conclusions, framed an important debate, or obtained figures for a given chapter. Selecting a subset of these to read—particularly the most recent ones—will help students gain entry to the professional literature on a wide range of topical areas in astrobiology.

My choice of breadth and depth with which to cover particular topics is, of course, a personal one, but I have been guided by two principles. The first is that a comprehensive treatment of astrobiology ought to tell a story; that is, it must trace a history from the origin of the cosmos to the onset of consciousness. Much of what motivates astrobiologists is, at its core, a desire to work out the details of this evolution in all of its rich complexity. The second principle is that the book must reflect, in a rough way, the overall emphases of research areas in the field of astrobiology as it stands today. In this I risk showing my own biases, but regardless I have tried to strike a reasonable balance that—I hope—mirrors the field as many astrobiologists see it today. To those colleagues whose work I have not mentioned or—worse—misrepresented, I can only offer my apologies and point again to Schrödinger's words reproduced at the beginning of this preface.

To students who intend to embark on a career in astrobiology, I offer just one piece of advice before starting this book: Be sure to become fully expert in one of the basic fields of science underpinning astrobiology. Without it, you will be unable to make the new insights and discoveries required to advance the field; with it, you become a full participant in the search for answers to some of humanity's oldest questions.

Acknowledgments

To write a book such as this requires working in an environment that is both intellectually stimulating and warmly tolerant of a disruptive, personal project. The Lunar and Planetary Laboratory, and the University of Arizona of which it is a part, have been such an environment for all my 20 years here, and I wish to thank my colleagues for making it so. I particularly want to thank Professor Michael Drake, Director of the Laboratory and Head of the Department of Planetary Science, for his strong support and encouragement of this project. My wife, Cynthia, was an active participant in seeing this project come to completion and was crucial in developing the concept of the book. Dr. Adam Black of Addison Wesley conceived of (and sold me) the project for which I am, in the end, grateful; Liana Allday of Addison Wesley kept everything well organized and tolerated my sometimes explosive responses to her queries regarding delivery of each book chapter. I thank our son, Joseph, for not complaining about my sudden disappearances "downstairs" to capture a thought or change the text yet again and for my gross underestimates of when I would be "finished" for the day so as to be free to do other things. Dr. David Morrison, NASA Ames Research Center, read the entire first draft of the book and made detailed criticisms, while at the same time being positive and enthusiastic about the project; thank you, David! Phil Eklund, Sierra Madre Games, Professor Yuk Yung, California Institute of Technology, and Professor Neville Woolf of the University of Arizona have also read large portions of the draft text, and I am grateful for their comments and suggestions. The following referees were essential in improving the book, although all remaining errors are my responsibility: Geoffrey A. Blake, *California Institute of Technology;* Paul S. Braterman, *University of North Texas;* Manfred Cuntz, *University of Texas at Arlington;* Steven Dick, *Historian;* Frank Drake, *SETI Institute;* Ronald Greeley, *Arizona State University;* Karen Kolehmainen, *California State University, San Bernardino;* Geoff Marcy, *University of California;* Cynthia W. Peterson, *University of Connecticut;* Anthony Pitucco, *Pima Community College;* Simon Nicholas Platts, *Rensselaer Polytechnic Institute;* Terry Richardson, *College of Charleston;* Robert T. Rood, *University of Virginia;* John Scalo, *University of Texas;* Mark A. Smith, *University of Arizona;* David Stahl, *University of Washington;* Michael E. Summers, *George Mason University;* Maria Zuber, *Massachusetts Institute of Technology*

Dedication

To Dr. Juan Pérez Mercader, Director, Centro de Astrobiología, Madrid, Spain—colleague and friend—for showing all of us how to create an Institute of Astrobiology from scratch, and for being the consummate astrobiologist.

Jonathan I. Lunine
May 2004

FIGURE CREDITS

1.1 Courtesy of Robert Gendler.
1.2 Courtesy of http://www.michielb/nl/maya/math.html
1.3 From http://starryskies.com/The_sky/events/mars/opposition05.html
1.4 Courtesy of M. J. Crowe (2001). *Theories of the World from Antiquity to the Copernican Revolution, Second Edition,* Dover Books, New York.
1.5 From http://www.go.ednet.ns.ca/~larry/orbits/ellipse.html
1.6 Courtesy of Heinrich Khunrath, Amphitheatrum sapientiae aeternae (Hamburg: s.n., 1595). University of Wisconsin, Special Collections.
1.7 Annalen der Chemie und Pharmacie, VIII, Supplementary Volume for 1872, p. 511.
1.8 Richard Megna/Fundamental Photographs.
1.10 National Library of Medicine.
1.11 Science & Society Picture Library.
1.12 (a) Courtesy of Stanislav Fakan; Alberts et al. (1994). *Molecular Biology of the Cell,* Garland Science/BIOS Scientific Publishers, pp. 352 and 353. (b) From the art of *Molecular Biology of the Cell, Third Edition,* (1995). Garland Publishing, New York.
2.2 National Aeronautics and Space Administration (NASA)/Goddard Institute for Space Studies.
2.4 (a) From http://astrosun2.astro.cornell.edu/academics/courses//astro201/g_lens_sun.htm
2.5 From Phywe Systeme GmbH & Company, KG, Germany/Daedelon Corporation.
2.6 David Taylor/Photo Researchers, Inc.
2.8 Based on Lorraine, F. and Corson, D. (1970). *Electromagnetic Fields and Waves,* W. H. Freeman and Company, San Francisco.
2.10 Based on Lorraine, F. and Corson, D. (1970). *Electromagnetic Fields and Waves,* W. H. Freeman and Company, San Francisco.
2.11 Ocean wave courtesy of http://www.amath.washington.edu/~bernard/kp/waterwaves/panama.jpg. Army Air Corps.
2.15 After Atkins, P.W. (1986). *Physical Chemistry, Third Edition,* W. H. Freeman and Company, New York.
2.16 Based on Feynman, R. P., Leighton, R. B. and Sands, M. (1965). *The Feynman Lectures on Physics: Volumes II and III,* Addison-Wesley, Reading, MA.
2.17 Courtesy of M. F. Crommie and IBM.
3.1 Based on data from Manthey, D. (2002). *Orbital Viewer,* http://www.orbitals.com
3.3 Redrawn from Eisberg, R. and Resnick, R. (1974). *Quantum Physics of Atoms, Molecules, Solids, Nuclei and Particles,* John Wiley and Sons, New York.
3.4 Redrawn from Broecker, W. (1985). *How to Build a Habitable Planet,* Eldigio Press, Lamont-Dougherty Earth Observatory, New York.
3.5 Based on Jacobs, A. W. (1997). *Understanding Organic Reaction Mechanisms.* Cambridge University Press, Cambridge.
3.6 Redrawn Eisberg, R. and Resnick, R. (1974). *Quantum Physics of Atoms, Molecules, Solids, Nuclei and Particles,* John Wiley and Sons, New York.
3.8 Based on Jacobs, A. W. (1997). *Understanding Organic Reaction Mechanisms.* Cambridge University Press, Cambridge.
3.10 Redrawn from Eisberg, R. and Resnick, R. (1974). *Quantum Physics of Atoms, Molecules, Solids, Nuclei and Particles,* John Wiley and Sons, New York.
3.11 Based on figure by Cynthia Lunine in Lunine, J. (1999). *Earth: Evolution of a Habitable World.* Cambridge University Press, Cambridge.
3.12 Redrawn from Eisberg, R. and Resnick, R. (1974). *Quantum Physics of Atoms, Molecules, Solids, Nuclei and Particles,* John Wiley and Sons, New York.
3.13 Redrawn from Eisberg, R. and Resnick, R. (1974). *Quantum Physics of Atoms, Molecules, Solids, Nuclei and Particles,* John Wiley and Sons, New York.
3.14 NOAO/AURA/NSF.
3.15 Courtesy of Athena Coustenis, Obs Paris Meudon.
3.17 Redrawn from Eisberg, R. and Resnick, R. (1974). *Quantum Physics of Atoms, Molecules, Solids, Nuclei and Particles,* John Wiley and Sons, New York.
4.3 Based on http://users.rcn.com/jkimball.ma.ultranet/BiologyPages/A/AminoAcids.html (line drawings) and http://www.nyu.edu/pages/mathmol/library/life/life1.html (ball and stick models).
4.4 Based on Dorin, H. (1971). *Modern Principles of Chemistry,* College Entrance Book Company, New York.
4.5 World Wide Web version courtesy of G. P. Moss, Department of Chemistry, Queen Mary University of London, Mile End Road, London, E1 4NS, U.K.
4.6 Courtesy of Purdue Research Foundation, http://biomedia.bio.purdue.edu/IML/Amino/html/stereoisomers.html
4.7 Redrawn from Brandon, C. and Tooze, J. (1991). *Introduction to Protein Structure,* Garland Publishing, New York, Figures 2.2 c, 2.3.c, and 2.5.c.
4.8 (a) Redrawn from Brandon, C. and Tooze, J. (1991). *Introduction to Protein Structure,* Garland Publishing, New York. (b) Courtesy of William L. Nichols, University of California at San Diego (UCSD), George D. Rose, Johns Hopkins University School of Medicine, Lynn F. Ten Eyck, UCSD and San Diego State College, and Bruno H. Zimm, UCSD.
4.9 Iwamoto et al. "Direct X-ray Observation of a Single Hexagonal Myofibrile of Striated Muscle," *Biophysical Journal,* **83**, 1074–1081.
4.12 Based on Dorin, H. (1971). *Modern Principles of Chemistry,* College Entrance Book Company, New York.
4.13 Based on Morrison, R. T. and Boyd, R. N. (1973). *Organic Chemistry,* Allyn and Bacon, Boston.
4.14 Based on Morrison, R. T. and Boyd, R. N. (1973). *Organic Chemistry,* Allyn and Bacon, Boston.

4.15 Based on Purves, W. K., Sadava, D., Orians, G. H., and Heller, H. C. (2001). *Life: The Science of Biology, Sixth Edition,* W. H. Freeman and Company, San Francisco.

4.17 After Lowenstein, W. R. (1999). *The Touchstone of Life: Molecular Information, Cell Communication, and the Foundations of Life,* Oxford University Press, New York.

4.18 Based on Purves, W. K., Sadava, D., Orians, G. H., and Heller, H. C. (2001). *Life: The Science of Biology, Sixth Edition,* W. H. Freeman and Company, San Francisco.

4.19 Based on Purves, W. K., Sadava, D., Orians, G. H., and Heller, H. C. (2001). *Life: The Science of Biology, Sixth Edition,* W. H. Freeman and Company, San Francisco.

4.20 Based on White, D. (2000). *The Physiology and Biochemistry of Prokaryotes,* Oxford University Press, Oxford, U.K.

4.21 Based on Purves, W. K., Sadava, D., Orians, G. H., and Heller, H. C. (2001). *Life: The Science of Biology, Sixth Edition,* W. H. Freeman and Company, San Francisco.

4.22 Based on Purves, W. K., Sadava, D., Orians, G. H., and Heller, H. C. (2001). *Life: The Science of Biology, Sixth Edition,* W. H. Freeman and Company, San Francisco.

4.23 Based on a figure by Cynthia Lunine in Lunine, J. (1999). *Earth: Evolution of a Habitable World.* Cambridge University Press, Cambridge.

4.24 After White, D. (2000). *The Physiology and Biochemistry of Prokaryotes,* Oxford University Press, Oxford, U.K.

5.1 Based on an original figure by Cynthia Lunine.

5.2 Courtesy of Space Telescope Science Institute and NASA.

5.3 Courtesy of Max Tegmark, from Tegmark, M. and 65 co-authors (2003). "The Three-Dimensional Distribution of Galaxies from the Sloan Digital Sky Survey," *Astrophys, J.,* in press.

5.4 (a) Gregory Brenner/Pacific Analytics. (b) Luciano Miglietta, INAF Arcetri Observatory, Florence, Italy, partner LBT Observatory Project. (c) European Southern Observatory.

5.5 Courtesy of Space Telescope Science Institute and NASA.

5.8 Modified from Cox, A. N. (2000). *Allen's Astrophysical Quantities,* American Institute of Physics Press, New York.

5.10 Based on an original figure by Cynthia Lunine in Lunine, J. (1999). *Earth: Evolution of a Habitable World.* Cambridge University Press, Cambridge.

5.13 Based on an original concept by Wally Broecker, Columbia University.

5.14 Redrawn from Mason, S. F. (1991). *Chemical Evolution: Origin of the Elements, Molecules and Living Systems,* Clarendon Press, Oxford, U.K.

5.15 Courtesy of Naoki Yoshida, Nagoya University, and Lars Hernquist, Harvard University.

6.1 Space Telescope Science Institute and NASA; WFPC2 and NICMOS teams.

6.2 Space Telescope Science Institute and ACS NASA.

6.5 AP/Wide World Photos.

6.6 Redrawn from Morbidelli, A., Chambers, J., Lunine, J. I., Petit, J. M., Robert, F., Valsecchi, G. B., Cyr, K. E. (2000). "Source Regions and Timescales for the Delivery of Water on Earth," *Meteoritics and Planetary Science,* **35,** 1309–1320.

6.7 Redrawn from Lunine, J. (1999). *Earth: Evolution of a Habitable World,* Cambridge University Press, Cambridge.

6.9 Modified from Minster, J-F., Birck, J-L., and Allegre, C. J. (1982). "Absolute Age of Formation of the Chondrites by the ^{87}Rb-^{87}Sr Method," *Nature,* **300,** 414–419.

6.10 Modified from Allegre, C., Manhes, G., and Gopal, C. (1995). "The Age of the Earth," *Geochim. Cosmochim. Acta,* **59,** 1445–1459.

6.11 (a) NASA/Corbis. (b) Consolidated Lunar Atlas; supplement numbers 3 & 4 to the United States Air Force Photographic Lunar Atlas (Kuiper, G. P., Strom, R., Whitaker E., Fontaine, J., and Larson, S. (1967). Lunar and Planetary Laboratory, University of Arizona, Tucson); scanned by Lunar and Planetary Institute, Houston TX. (c) NASA Jet Propulsion Laboratory.

6.12 Courtesy of R. Strom, University of Arizona.

6.13 From Pierazzo, E. and Chyba, C. F. (1999). "Amino Acid Survival in Large Cometary Impacts," *Meteoritics and Planetary Science,* **32,** 909–918 and courtesy of Dr. E. Pierazzo, Planetary Science Institute, Tucson.

6.14 Modified from Lunine, J. (1999). *Earth: Evolution of a Habitable World,* Cambridge University Press, Cambridge.

6.15 Sinclair Stammers/Photo Researchers, Inc.

6.16 (a) Based on an original figure by Cynthia Lunine; (b) Tom Till Photography, Inc.

7.1 Based on an original figure by Cynthia Lunine.

7.2 Based on an original concept by Cynthia Lunine.

7.6 Figure modified from, and courtesy of, N. Woolf, The University of Arizona.

7.8 (a–c) Peter Ruoff, Stavanger University College, Norway.

7.9 Figure adapted from Dutt, A. K., Bhattacharya, D. K., and Eu, B.-C. (1990). "Limit Cycles and Discontinuous Entropy Production Changes in the Reversible Oregonator," *J. Chem. Phys.,* **93,** 7929–7935.

8.1 Based on http://www.cat.cc.md.us/courses/bio141/lecguide/unit1/prostruct/dnareppr/fg18.html

8.2 Courtesy of Cech, T. R. (2000). "Structural Biology: The Ribosome Is a Ribozyme," *Science,* **289,** 878–879.

8.5 Courtesy of Teresa Larsen, The Scripps Research Institute (TSRI).

8.6 Panel (b) adapted from Mason, S. F. (1991). *Chemical Evolution: Origin of the Elements, Molecules and Living Systems,* Clarendon Press, Oxford, U.K.

8.7 After Joyce, G. F. (2002). "Molecular Evolution: Booting up Life," *Nature,* **420,** 278–279.

8.9 From Joyce, G. F. (2002). "The Antiquity of RNA-Based Evolution," *Nature,* **418,** 214–221.

8.10 From Mason, S. F. (1991). *Chemical Evolution: Origin of the Elements, Molecules and Living Systems,* Clarendon Press, Oxford, U.K.

8.11 From Szabó, P., Scheuring, I., Czárán, T., and Szathmáry, E. (2002). "*In silico* Simulations Reveal That Replicators with Limited Dispersal Evolve toward Higher Efficiency and Fidelity," *Nature,* **420,** 340–343.

8.12 Adapted from an original figure by Cynthia Lunine in Lunine, J. (1999). *Earth: Evolution of a Habitable World,* Cambridge University Press, Cambridge.

8.13 Figure from Benner, S. A. (1999). "How Small Can an Organism Be?" In *Size Limits of Very Small Microorganisms: Proceedings of a Workshop,* National Academy Press, Washington D.C., ISBN 0-309-06634-4, pp. 135. (Available through http://www.nas.edu)

9.1 Modified from an original figure by Davis, W. L. and McKay, C. P. (1996). "Origins of Life: A Comparison of Theories and Application to Mars," *Origins Life Evol. Biosph.*, **26,** 61–73.

9.2 Adapted from Baker, G. L. and Golub, J. P. (1990). *Chaotic Dynamics: An Introduction,* Cambridge University Press, Cambridge.

9.3 Belmonte et al. (1997). "Experimental Survey of Spiral Dynamics in the Belousov–Zhabotinsky Reaction," *Journal de Physique II* (France) **7,** 1425–1468.

9.4 From Levy, S. (1992*). Artificial Life: A Report from the Frontier Where Computers Meet Biology,* Vintage Books, New York.

9.5 Based on an original figure in Dyson, F. (1999). *Origins of Life,* Cambridge University Press, Cambridge.

9.6 Based on an original figure in Dyson, F. (1999). *Origins of Life,* Cambridge University Press, Cambridge.

9.7 From Pace, N. (2001). "The Universal Nature of Biochemistry," *Proc. National Academy of Sciences,* **98,** 805–808.

9.8 From Doolittle, W. F. (1999). "Phylogenetic Classification and the Universal Tree," *Science,* **284,** 2124–2128.

9.10 Redrawn from Joyce, G. F. (2002). "The Antiquity of RNA-Based Evolution," *Nature,* **418,** 214–221.

9.11 Adapted from Rasmussen, S., Chen, L., Deamer, D., Krakauer, D. C., Packard, M. H., Stadler, P. F., and Bedau, M. A. (2004). "Transitions from Non-Living to Living Matter," *Science,* **303,** 963–965.

10.1 Lynn Rothschild.

10.2 B. Murton/Southampton Oceanography Centre/Science Photo Library.

10.3 Linda Amaral-Zettler.

10.4 Courtesy UCAR, University of Michigan, adapted from http://www.windows.ucar.edu/ spaceweather/images/how_damage_jpg_image.html

10.5 Michael Daly/USUHS.

10.6 Thomas, D. N. and Dieckmann, G. S. (2002). "Antarctic Sea Ice—A Habitat for Extremophiles," *Science,* **295,** 641–644.

10.7 Adapted from Thomas, D. N. and Dieckmann, G. S. (2002). "Antarctic Sea Ice—A Habitat for Extremophiles," *Science,* **295,** 641–644.

10.8 Courtesy of Freund, F., Dickenson, J. T., and Cash, M. (2002). "Hydrogen in Rocks: An Energy Source for Deep Microbial Communities," *Astrobiology*, **2,** 83–92.

10.9 MOLA Team and NASA.

10.10 Soennke Grossman.

11.1 Figure by Kevin McKeegan, UCLA Earth and Space Science Department. Courtesy of UCLA National Ion Microprobe Facility.

11.3 Modified from a figure in Schopf, J. W. (ed). (1983). *Earth's Earliest Biosphere,* Princeton University Press, Princeton, NJ.

11.4 Figure courtesy of Wes Traub, Harvard University Center for Astrophysics.

11.6 Based on data in Kasting, J. and Toon, O. B. "Climate Evolution on the Terrestrial Planets." In *Origin and Evolution of Planetary and Satellite Atmospheres,* University of Arizona Press, Tucson, pp. 423–450.

11.7 Redrawn from Kasting, J. F. (1991). "CO_2 Condensation and the Climate of Early Mars," *Icarus,* **94,** 1–13.

11.8 David T. Sandwell/Scripps Institute of Oceanography.

11.9 Redrawn from an original figure by Cynthia Lunine, in Lunine, J. (1999). *Earth: Evolution of a Habitable World,* Cambridge University Press, Cambridge.

11.10 Adapted from Nisbet, E. G. and Sleep, N. H. (2001). "The Habitat and Nature of Early Life." *Nature,* **409,** 1083–1091.

11.11 Adapted from Lunine, J. (1999). *Earth: Evolution of a Habitable World,* Cambridge University Press, Cambridge.

11.12 Adapted from Knauth, L. P. (1998). "Salinity History of the Earth's Early Ocean," *Nature,* **395,** 554–555.

11.13 Adapted from an original figure by Kevin Zahnle, NASA Ames Research Center.

11.14 Caltech/Jet Propulsion Laboratory/NASA.

11.16 Kasting, J. F. (1988). "Runaway and Moist Greenhouse Atmospheres and the Evolution of Earth and Venus," *Icarus,* **74,** 472–494.

12.1 Malin Space Science Systems and NASA.

12.2 National Geographic Society, Malin Space Science Systems, NASA.

12.3 Courtesy of Maria Zuber, MIT; (a) NASA GODDARD/David Smith and Gregory Neumann. (b–c) NASA GODDARD/David Smith.

12.4 (a) NASA/Jet Propulsion Laboratory/Cornell University. (b–c) NASA/Jet Propulsion Laboratory/The University of Arizona.

12.5 European Space Agency/DLR/FU Berlin (G. Neukum).

12.6 NASA/Jet Propulsion Laboratory/Malin Space Science Systems.

12.7 NASA/Jet Propulsion Laboratory/Malin Space Science Systems.

12.8 (a) NASA/Jet Propulsion Laboratory/Malin Space Science Systems (b) NASA/Malin Space Science Systems.
12.9 NASA/Jet Propulsion Laboratory/The University of Arizona.
12.10 (a–b), NASA/Jet Propulsion Laboratory. (c) NASA/Jet Propulsion Laboratory/Arizona State University.
12.11 Courtesy of Philip Christensen, Arizona State University and NASA.
12.12 NASA/Jet Propulsion Laboratory/Cornell University.
12.13 Redrawn from Opportunity APX data by permission of Steven Squyres, Cornell University and http://www.mpch-mainz.mpg.de/mpg/english/index.html, Max Planck Institute für Chemie, Germany, NASA/JPL/Cornell/Max Planck Institute.
12.14 NASA/Jet Propulsion Laboratory/Cornell University/USGS.
12.15 NASA/Jet Propulsion Laboratory/Cornell University.
12.16 NASA/Jet Propulsion Laboratory/Cornell University.
12.17 NASA/Jet Propulsion Laboratory.
12.18 NASA/Jet Propulsion Laboratory.
12.19 NASA/Jet Propulsion Laboratory.
13.1 NASA/Jet Propulsion Laboratory.
13.2 NASA/Jet Propulsion Laboratory.
13.3 NASA/Jet Propulsion Laboratory.
13.4 NASA/Jet Propulsion Laboratory.
13.5 NASA/Jet Propulsion Laboratory.
13.6 NASA/Jet Propulsion Laboratory.
13.7 Based on an original figure by Pam Engebretson; the drawings were rendered by Pam Engebretson of Mountain View, CA, from a design by Eric M. DeJong and Zareh Gorjian of NASA/Jet Propulsion Laboratory, Pasadena, CA.
13.8 NASA/Jet Propulsion Laboratory and Galileo NIMS Team.
13.9 NASA/Jet Propulsion Laboratory.
13.10 Courtesy of M. Zuber, MIT and D. Smith, NASA Goddard Spaceflight Center.
13.11 NASA/Jet Propulsion Laboratory.
13.12 From Coustenis, A. et al. in the proceedings of *The Promise of the Herschel Space Observatory*, ed. G. L. Pilbratt et al., European Space Agency Special Publication SP-460, 2001.
13.13 Adapted from R. Lorenz (2000). "The Weather on Titan," *Science,* **290,** 509–513.
13.14 Map from data reported in Smith, P. H. et al. (1996). "Titan's Surface, Revealed by HST Imaging," *Icarus,* **119,** 336–339.
13.15 NASA/Jet Propulsion Laboratory.
13.17 Figure by N. Artemieva, in Artemieva, N. and Lunine, J. I. (2003). "Cratering on Titan: Impact Melt and the Fate of Surface Organics" *Icarus,* **164,** 471–480.
14.3 Redrawn from Rodier, C., Laurent, C., Szopa, C., Sternberg, R., and Raulin, F. (2002). "Chirality and the Origin of Life: In situ Enantiomeric Separation for Future Space Missions," *Chirality,* **14,** 527–532.

14.4 Redrawn from an original concept by Niemann, H. B., Atreya, S. K., Biemann, K., Block, B., Carignan, G., Donahue, T. M., Frost, L., Gautier, D., Harpold, D. M., Hunten, D. M., Israel, G., Lunine, J. I., Mauersberger, K., Owen, T., Raulin, F., Richards, J., and Way, S. (1997). "The Gas Chromatograph Mass Spectrometer Aboard Huygens," In *Huygens: Science, Payload and Mission* pp. 85–107. ESA SP-1177, Noordwijk.
14.5 Based on a figure in Space Studies Board, Committee on the Origin and Evolution of Life. (2002). *Signs of Life: Report of a Workshop on Life Detection,* Washington, D.C., National Academy Press.
14.6 Redrawn from Ueno, Y., Yurimoto, H., Yoshioka, H., Komiya, T., Maruyama, S. (2002). "Ion Microprobe Análysis of Graphite from ca. 3.8 Ga Metasediments, Isua Supracrustal Belt, West Greenland: Relationship between Metamorphism and Carbon Isotopic Composition," *Geochemica et Cosmochemica Acta,* 1257–1268, Fig. 4, Elsevier Press.
14.7 Courtesy of Banfield, J. F., Moreau, J. W., Chan, C. S., Welch, S. A., and Little, B. (2001). "Mineralogical Biosignatures and the Search for Life on Mars," *Astrobiology* **1,** 447–467.
14.8 (a) Courtesy of Banfield, J. F., Moreau, J. W., Chan, C. S., Welch, S. A., and Little, B. (2001). "Mineralogical Biosignatures and the Search for Life on Mars," *Astrobiology* **1,** 447–467. (b) Courtesy of McKay, D. S. et al. (1996). "Search for Past Life on Mars: Possible Relic Biogenic Activity in Martian Meteorite ALH84001," *Science,* **273,** 924–930.
14.9 Redrawn from Kirschvink, Committee on the Origin and Evolution of Life (2002). *Signs of Life: Report of a Workshop on Life Detection,* National Academy Press, Washington, D.C.
14.10 Images provided by Friedman based on work reported in Friedmann, E. I., Wierzchos, J., Ascaso, C., and Winklhofer, M. (2001). "Chains of Magnetite Crystals in the Meteorite ALH84001: Evidence of Biological Origin," *Proceedings of the National Academy of Sciences* **98,** 2176–2181.
14.11 NASA/Jet Propulsion Laboratory.
14.12 Chimera Research Group; original figure by P. Beauchamp, NASA/Jet Propulsion Laboratory.
15.1 (b) Courtesy of Fischer, D. A., Marcy, G. W., Butler, R. P., Vogt, S. S., Henry, G. W., Pourbaix, D., Walp, B., Misch, A. A., Wright, J. T. (2003). "A Planetary Companion to HD40979 and Additional Planets Orbiting HD12661 and HD38529," *Astrophys. J.,* **586,** 1394–1408.
15.3 From *SIM, Space Interferometry Mission: Taking the Measure of the Universe,* NASA Jet Propulsion Laboratory Publication 400-811.
15.4 Modified from an original figure in *SIM, Space Interferometry Mission: Taking the Measure of the Universe,* NASA Jet Propulsion Laboratory Publication 400-811.
15.5 (b) Courtesy of Charbonneau, D., Brown, T. M., Latham, D. W., and Mayor, M. (2000). "Detection of Planetary Transits Across a Sun-Like Star," *Astrophys. J. Letters,* **529,** L-45-48, Fig. 2.

15.6 Space Telescope Science Institute, NASA, ACS and NICMOS Science Teams.

15.8 Figure provided by Wes Traub, original appeared in Des Marais, D. J., Harwit, M., Jucks, K., Kasting, J., Lin, D., Lunine, J. I., Schneider, J., Seager, S., Traub, W., and Woolf, N. (2002). "Remote Sensing of Planetary Properties and Biosignatures on Extrasolar Terrestrial Planets," *Astrobiology,* **2,** 153–181.

15.10 Archive Multiple Mirror Telescope, operated jointly by Harvard University Smithsonian Center for Astrophysics and the University of Arizona.

15.11 Gerard van Belle/MSC/NASA/Jet Propulsion Laboratory.

15.12 Courtesy of Phil Hinz. Observations by the Center for Adaptive Optics, The University of Arizona on the MMT (operated jointly by Harvard University Smithsonian Center for Astrophysics and the University of Arizona).

15.13 Redrafted from Angel, J. R. P. and Woolf, N. (1997). "An Imaging Nulling Interferometer to Study Extrasolar Planets," *Astrophys. J., ***475,** 373–379.

15.14 Image courtesy of Robert H. Brown (Depts. of Planetary Sciences and Astronomy, University of Arizona); appeared in *Failed Stars and Superplanets: A Workshop on Substellar Mass Objects*, National Academy of Sciences (1998).

15.15 Modified from a figure by David Spergel (Princeton University).

15.16 (a) Based on data reported in rossart, P., Rosenqvist, J., Encrenaz, Th., Lellouch, E., Carlson, R. W., Baines, K. H., Weissman, P. R., Smythe, W. D., Kamp, L. W., and Taylor, F. W. (1993). "Earth Global Mosaic Observations with NIMS-Galileo," *Planetary and Space Science,* **41,** 551–561. (b) from Woolf, N. J., Smith, P. S., Traub, W., and Jucks, K. (2002). "The Spectrum of Earthshine: A Pale Blue Dot Observed from the Ground," *Astrophys. J.,* **574,** 430–433.

15.17 Figure by J. Kasting and L. Brown, published in Beichman, C., Woolf, N. J., and Lindensmith, C. A. (1999). *The Terrestrial Planet Finder,* NASA Jet Propulsion Laboratory Publication 99-3, Pasadena, CA, http://tpf.jpl.nasa.gov/

16.1 Modified from Knoll, A. H. (1999). "Paleontology: A New Molecular Window on Early Life," *Science,* **285,** 1025–1026.

16.2 Modified from Gray, M. W., Burger, G., and Lang, B. F. (1999). "Mitochondrial Evolution," *Science,* **283,** 1476–1481.

16.3 Modified from Lynch, M. (2002). "Gene Duplication and Evolution," *Science,* **297,** 945–947.

16.4 Modified from Peixoto, J. P. and Oort, A. H. (1992). *Physics of Climate,* AIP Press, New York, AAAS Science Magazine.

16.5 Courtesy of Solè, R. V. and Newman, M. (2002). "Extinctions and Biodiversity in the Fossil Record." In *Encyclopedia of Global Environmental Change, Volume 2: The Earth System—Biological and Ecological Dimensions of Change* (eds. H. A. Mooney and J. G. Canadell), pp. 297–301, John Wiley and Sons, Chichester, U.K.

16.6 Redrafted from Hyde, W. T., Crowley, T. J., Baum, S. K. and Peltier, W. R. (2000). "Neoproterozoic 'Snowball Earth' Simulations with a Coupled Climate/Ice-sheet Model," *Nature,* **405,** 425–429.

16.7 Courtesy of Hyde, W. T., Crowley, T. J., Baum, S. K. and Peltier, W. R. (2000). "Neoproterozoic 'Snowball Earth' Simulations with a Coupled Climate/Ice-sheet Model," *Nature,* **405,** 425–429, Figure 4b.

16.8 Modified from Knoll, A. H. and Carroll, S. B. (1999). "Early Animal Evolution: Emerging Views from Comparative Biology and Geology," *Science,* **284,** 2129–2137.

16.9 Virgil L. Sharpton/Lunar Planetary Institute.

16.10 (a) Hubble Space Telescope Comet Team/NASA; (b) H. Hammel, MIT/NASA.

17.1 Courtesy of Wood, B. and Collard, M. (1999). "The Human Genus," *Science,* **284,** 65–71.

17.2 (a) Pascal Goetgheluck/Photo Researchers, Inc. (b) John Reader/Photo Researchers, Inc. (c) Pascal Goetgheluck/Photo Researchers, Inc.

17.3 (a) Lester V. Bergman/CORBIS. (b) Thomas F. Fletcher, Ph.D Department of Veterinary Pathobiology, College of Veterinary Medicine, University of Minnesota.

17.5 Redrawn from McManus, J. F. (2004). "Paleoclimate: A Great granddaddy of Ice Cores," *Nature,* **429,** 611–612.

17.6 Redrawn from Flavin, C. et al. (2002). *State of the World 2002,* W. W. Norton Company, New York.

17.7 Redrawn from Flavin, C. et al. (2002). *State of the World 2002,* W. W. Norton Company, New York.

17.8 Drafted from data at http://www.mbendi.co.za/indy/oilg/p0005.htm

17.9 Kobal Collection.

17.10 Roger Angel, Steward Observatory/The University of Arizona.

Introduction—What Is Astrobiology?

Life covers the Earth's surface and permeates the air, water, and even the uppermost solid crust beneath our feet. Life took hold within the first 10 percent of this planet's 4.5-billion-year history and shows up in ways first subtle and then striking through the geologic record. Thus, the history of our planet and the history of life are inseparably intertwined, and it is extremely hard to guess what the course of our planet's evolution might have been had life been absent. As living organisms possessing self-awareness, it is therefore natural that we should ask, Is Earth the only inhabited world in the cosmos, or, instead, is the universe suffused with life in unimaginable variety? Is life now or was it ever present on neighboring planets accessible to direct exploration?

While such questions, in differing guises, have likely been asked continuously since the dawn of human consciousness, only in the past few decades have the tools to answer them become available. Humans can explore space, touching directly or by robotic proxy the worlds of our solar system. Extreme environments on Earth, from the deep oceans to the lakes of the Antarctic dry valleys, have become the haunts of humanity. Laboratory techniques have evolved to the point where most of the major building blocks of life can be manipulated and, in some cases, synthesized from much simpler components. Rocks can be sampled with powerful techniques that allow subtle clues to the past presence of life to be revealed. Humankind's understanding of the physical and chemical mechanisms underlying life are deep enough to conceptualize the detailed pathways by which fundamental particles form the natural elements, and atoms that define those elements combine to make molecules. Tantalizing hints as to how nonliving organic chemistry might have evolved into living systems—that is, how life began long ago on the primitive Earth—exist from modern understanding of physics, chemistry, biology, and geology.

Astrobiology is the multidisciplinary study of the origin, distribution, and evolution (past and future) of life. Its purview is not confined to Earth, even though we have no confirmed samples of life forms nurtured in places other than our Earth. While astrobiology as a name is new, its component disciplines are not—for several decades scientists intent on finding life beyond Earth worked as exobiologists, and those interested in the interaction of life and geology throughout Earth's history were geobiologists. What the name astrobiology connotes is the burgeoning effort in a handful of countries, led initially by the United States, to bring a variety of fields together into a new discipline to collaboratively understand how life began, where it exists today in the cosmos, and where it is going.

The lead role of the National Aeronautics and Space Administration (NASA) in creating a new interdisciplinary field was largely the result of a handful of scientists within the agency who desired, with the support of then–NASA administrator Dan Goldin, to energize the scientific underpinnings of robotic exploration of the solar system by focusing more strongly on the questions of the origin and distribution of life beyond Earth. The creation of astrobiology as a field might have languished were it not for the announcement in 1996 by a team, headed by a NASA scientist, that a meteorite from Mars might hold evidence for nonterrestrial life (a conclusion that, as Chapter 14 outlines, now seems unlikely). This claim, which caught the attention of federal officials all the way up to President Bill Clinton, paved the way for NASA to acquire (through the congressional budget process) additional monies to create a program in astrobiology, and the NASA Astrobiology Institute (NAI), in 1998. NAI serves as the centerpiece of NASA's astrobiology program, involving universities and government laboratories as well as international research teams, in collaborative research on astrobiology. NAI emphasizes the use of electronic technologies to establish more readily a "virtual presence" of widely separated teams at each others' research facilities, and has actively encouraged the development of undergraduate and graduate programs in astrobiology.

As with the rapid creation of any new program or field from existing programs and science disciplines, astrobiology has experienced both intellectual and organizational growing pains. In this text it is appropriate to focus only on the intellectual goals. Readers interested in assessments of the organizational structure of astrobiology at NASA, within other U.S. federal agencies, and in other countries can consult reports prepared by the independent National Research Council, which operates under the aegis of the National Academies of Science, Engineering, and Medicine [*Committee on the Origin and Evolution of Life*, 2003]. Since 1996, NASA has developed two road maps for astrobiology with the participation of many hundreds of scientists from academia, industry, and government. Because astrobiology as a field is so strongly tied to the NASA program, its fundamental goals are aligned with those of the NASA program:

1. How did life begin and evolve?
2. Does life exist elsewhere in the universe?
3. What is life's future on Earth and beyond?

These questions, in turn, motivate a set of more detailed scientific goals and questions, which serve as the starting point for developing interdisciplinary research programs. The scientific goals and questions have been reformulated by various studies and revision of the NASA road map and those of the research programs of other countries, but what follows is a useful organization of goals and questions that devolve from them. The list of questions is by no means complete; after reading the book you should be able to formulate additional ones.

1. **Understand how life arose on the Earth.** The origin of life on Earth represents the starting point for assessing the degree of commonality of life in the cosmos. Where did the raw materials for life on Earth come from? How were the elements manufactured in previous generations of stars, and in what sorts of molecular arrangements were these materials found in the nascent disk of gas and dust from which our planetary system formed? How and from where was this material delivered to the Earth? What were the environments in which life evolved from nonliving chemical sys-

tems? What did the very first organisms look like? To what extent does the current set of genetic relationships among organisms inform us about the nature of the original ancestral forms from which all life today evolved? When did life begin on Earth? Is the origin of life a common part of the processes of planetary formation and evolution?

2. **Determine the general principles governing the organization of matter into living systems.** Complementary to the first goal is the "holy grail" of experimental biologists and chemists who work on life's origins—By just what processes does organic (here meaning carbon-bearing) matter become organized into self-sustaining, living things (i.e., life)? Does the organization proceed within organic chemical systems themselves, or is some type of template provided within mineralogical or other inorganic systems required? What are the temporal and spatial scales, and levels of starting system complexity, required for the origin of life? What sources of energy are required? What other kinds of chemical or physical systems can self-organize into patterns that we might call life? Is liquid water required for life? Is life a phenomenon requiring at least two basic kinds of polymers (molecules comprised of repeated fundamental units)—one devoted to information, the other to structure and catalysis? Indeed, what is the most general but useful definition of life?

3. **Explore how life evolves on the molecular, organismal, and ecosystem levels.** The history of life is recorded both in the rocks of Earth and in the genetic information stored in every organism. From these disparate types of records it is possible to map out an extraordinarily complex history of evolutionary changes in life, punctuated by environmental catastrophes that, in part (sometimes nearly in whole), emptied ecosystems of many fascinating creatures. What are the detailed genetic relationships among the Earth's organisms? How much has the evolution of life been driven by large-scale transfer of genetic molecules among types of microorganisms? How have the mutability and duplicative nature of the genetic material, coupled with environmental changes, driven the evolution of life? Why are there three major domains of life, when did they arise, and have other domains gone extinct? What factors internal and external to life led to the origin of complex cells ("eukaryotes") and to multicellular complex organisms such as plants, animals, and fungi? What is the evolutionary origin of human intelligence, how is it coupled to self-awareness, and has it arisen in other organisms?

4. **Determine how the terrestrial biosphere has co-evolved with the Earth.** The evolution of life on Earth has proceeded over the course of 4 billion years, during which the planet itself has undergone profound changes to its atmosphere, oceans, and geology. The atmosphere has evolved from being rich in carbon dioxide to being rich in oxygen; correspondingly, the oceans have evolved from anoxic to oxygen-rich. The cycling of Earth's outer layer, or crust, has slowed with time, and buoyant masses of granites called continents have come to cover about 40 percent of a surface that initially may have been almost entirely basaltic—and largely submerged. The rate of impacts on the surface of Earth diminished rapidly in the first 10 percent of the history of our planet then declined more slowly with time. When did conditions allow the first organisms and ecosystems to survive? How many times has life

been completely extinguished from the Earth, and when was the last time? How many times and to what extent has the Earth been plunged into a globally frozen state? Has life played more than a reactive role in the changing state of the oceans and atmosphere over time? To what extent has the evolution of life been determined by evolving global conditions and sudden catastrophes versus the changing levels of complexity of the genetic code itself?

5. **Establish limits for life in environments that provide analogues for conditions on other worlds.** Over the past few decades scientists have found life flourishing in environments previously thought to be uninhabitable—from ocean floor vents at high temperatures and extreme pressures to ecosystems buried beneath a kilometer of basaltic rock. Such extreme environments seem in many cases to be the refuge of organisms with very primitive qualities, which—for reasons we discuss in Chapter 9—suggest that they are the modern descendants of very ancient life-forms. What are the most extreme environments of temperature, pressure, salinity, desiccation, radiation, and other parameters under which organisms can survive? What are the limits under which spores can exist and be successfully revived? What are the extremes under which life can form, and what are the extremes within which ecosystems can survive? What are the timescales for dormancy, revival, and survival under various extreme conditions? Can organisms survive impacts from cometary or asteroidal bodies and survive transport on the fragments of ejected crust to a neighboring world?

6. **Determine what makes a planet habitable and how common those worlds are in the universe.** The fundamental requirements for life on Earth are an adequate flow of energy, appropriate sources of carbon and other critical elements, and liquid water. Therefore, the simplest definition of a habitable world would be one that supports these three fundamental requirements. Determining which planets have liquid water today, when and how much liquid water occurred in the past, and whether adequate stores of biogenic elements and energy exist is a daunting problem. In our own solar system, the debates rage on regarding when and where Mars had liquid water and whether it exists today—a quarter century after the valley networks were first revealed (see Chapter 12). Compelling but circumstantial evidence exists for a liquid water ocean under the icy crust of Europa, maintained by tidal heating; if life exists there it is very different from the traditional "habitable zone" defined by orbital distances like that of Earth around Sunlike stars. Do Mars and Jupiter's moon Europa support life now, or has Mars supported life in the past? Does the profoundly cold but organic-rich Titan, the giant moon of Saturn, support some form of life? If not, how far did organic chemistry proceed on this moon? How common is the formation of rocky planets around other stars, like or unlike the Sun? Is the formation of habitable worlds around Sunlike stars a phenomenon restricted to a narrow "galactic habitable zone" within our and other galaxies, or can it occur in many different galactic environments? How common are giant planets, and is the architecture of our own solar system typical? What is the true range of possibilities for habitable zones around stars like and unlike the Sun? And, even for planets possessed of an Earthlike orbit and hence balmy temperatures, how many have the water and organics budget of the Earth? How do Earthlike planets gain their water and organics, and

how many are in planetary system environments that, over billions of years, resemble that of the Earth? Could an Earth-sized body orbiting a star like the Sun at a Marslike distance support life, and for how long?

7. **Determine how to recognize the signature of life on other worlds beyond our solar system.** The detection of planets orbiting stars that shine a billion or more times more brightly than the planets themselves is a daunting task and, as yet, has been done mostly indirectly (as discussed in Chapter 15). Even when large ground- and space-based telescopic systems become capable of teasing the light of a planet out from under the glare of its parent star, the daunting task of determining whether such planets are habitable remains unsolved—even if the biosphere we are looking for is Earthlike. Indeed, distinguishing a planet whose atmosphere contains gases in chemical disequilibrium is not enough because we must know whether the disequilibrium is caused by biological processes or by rapid evolution of the planet's atmosphere itself (see Chapter 11 for the case of Venus). And the cases of planets where primitive life has only a toehold, or on which life-forms very different from Earth's exist, or that have only subsurface life beneath an inhospitable surface like Europa may represent impossible biospheres to detect remotely. How could we detect the effects of biospheres on parent planets many light-years away? How can we separate rapid abiotic planetary evolution from the biochemical modification of surface and atmosphere? To what extent could we detect the signature of life on planets with atmospheres very different from Earth's or planets with no atmospheres at all? Finally, how can the search for signals from extraterrestrial civilizations be designed to provide a useful negative answer in the absence of a history-changing positive result? Should it turn out that humanity's Earth is the singular abode of life in the cosmos, could we ever determine with certainty such a shocking property of the universe—and could we ever truly come to grips with its extraordinary implications?

8. **Determine whether there is (or once was) life elsewhere in our solar system, particularly on Mars, Europa, and Titan.** The very first directed search for life in our own solar system, the Viking expeditions to Mars, yielded negative results regarding the presence of life at the landing sites. Debate still rages on the claimed evidence for biological activity in the Martian meteorite ALH84001 more than five years after the initial exciting announcement. To culture a terrestrial organism is straightforward since we know that it must have DNA and RNA in its cells, with the appropriate five nucleotide bases (one different in RNA compared to DNA, as described in Chapter 4). But the exquisitely sensitive techniques to detect the comfortably familiar, Earthly life-forms would fail should the information-carrying polymers of an extraterrestrial organism be slightly different—and the more general techniques are less sensitive or more open to ambiguity. What are the most robust sets of life-detection experiments to carry to Mars or to Europa and Titan? What types of robotic exploration capabilities are required to find and access sites at which such tools would be useful? Is it best to search for life directly on the surface of a body (e.g., in situ), or to return a sample to Earth? What are the hazards of contaminating a planet with organisms carried from the Earth or of contaminating Earth with a returned sample? How would we recognize life that had an origin separate from Earthly life versus life that had a common origin and was transported to the

target planet on crust blasted off the Earth in an impact (or vice versa)? Indeed, have we missed on Earth the signs of life so primitive that they are always swamped by the life we know—that is, the faint echo of something truly odd lost amidst the cacophony of the three known domains of the phylogenetic tree (see Chapter 8)?

9. **Determine how ecosystems respond to environmental change on timescales relevant to human life on Earth.** Life on Earth today, while still dominated in numbers by the prokaryotes (archea and bacteria), may be profoundly challenged by the vigor and power of human industrial activities. The species extinction rate today—largely caused by human activities—is perhaps a hundred to a thousand times larger than that of the average background seen in the geologic record of the last half billion years (10 percent) of Earth's history. Even humanity's finely tuned agricultural and industrial infrastructure may be challenged in the coming decades by changes we exert on the atmospheric composition, oceanic composition, and continental soil and forest cover. What does the geologic record tell us of the range of climatic extremes and timescales for change over the last half billion years? How does the coupled air-ocean-land system on Earth respond to rapid changes in basic properties? How robust might the mechanisms of human civilization be to sudden and unpredictable changes in the nature of the Earth's climate and ecosystems that sustain us?

10. **Understand the response of terrestrial life to conditions in space or on other planets.** Humanity first left the Earth to touch the surface of another world in 1969. For the astronauts who went to the Moon, even the span of a few days on an alien world was a profoundly moving and lonely experience. While the pace of human expansion beyond Earth slowed after that initial burst of activity, robotic emissaries continue to push outward through the solar system to the realm of nearby interstellar space. Ultimately, humankind might try to colonize the nearby planets and the space around them. To do so requires putting human beings into low-gravity and high-radiation environments—isolated from Earth—for months or years at a time. How does microgravity affect the physiology of life-forms, from bacterial to animal? Is long-term propagation of species under conditions very different from those on Earth possible? Is humankind both technologically and spiritually ready to bud off a small fraction of itself to live in the planetary neighborhood of the Earth—a stone's throw from the home world on cosmic scales yet so remote compared to the bulk of the human experience?

The vast scope of astrobiology, as just defined, may surprise some readers. Its breadth is an outcome of the original vision for the field in which a broad sweep of the space science community involved in biological and medical studies would be involved. In the past few years, a desire to narrow the field a bit to bring it more focus has emerged, and, in particular, goal 10 has become less prominent in recent discussions of the nature of astrobiology. Our focus in this text will be on the vast bulk of goals 1 through 9, leaving out little except for the medical aspects of human exposure to weightlessness and radiation.

Astrobiology is a multidisciplinary endeavor, and scientists who are actively involved in the field have traditional educations in physics, chemistry, geology, biology, medicine, astronomy, and planetary sciences. Scientific disciplines often intersect to create new fields of research, and in the 20th century some of the major cross-fertilizations were physics, astronomy, and chemistry to create astrophysics; geology and chemistry to create geochemistry; geology and physics to create geophysics; and biology and chemistry to create biochemistry. More than just shuffling names, each new field involved scientists bringing novel tools to bear on problems, new ways of looking at long-standing questions, and criticisms from workers in the traditional fields from which the new interdisciplinary science was drawn. Astrophysics developed in the late 19th and early 20th centuries as an outgrowth of the development of spectrometers that could be used to constrain the chemical composition of stars. For the first time, astronomers did more than track the positions and motions of stars in the heavens, they measured what things were made of. Astrophysics was scorned by some traditional practitioners of astronomy, but through its techniques, helium—the second-most abundant element in the universe—was discovered. Today, the application of the modern theoretical and observational tools of astrophysics so suffuses astronomy that it is hard to distinguish between the two.

Likewise, planetary science was the outgrowth of interdisciplinary research between astronomers and geologists, stimulated enormously by the rapid growth of NASA in the late 1950s and early 1960s as the United States looked to the Moon as a target for human exploration. New techniques in studying the signatures by which life leaves its mark on the geologic record, as well as a deepening understanding of how life co-evolved with the Earth, led to the development of geobiology in the 1980s as an academic discipline unto itself. And it can be argued that, together with the fundamental fields of physics, chemistry, and biology, it is cross-disciplinary research among all these fields that has led to astrobiology—stimulated again by new discoveries and new aspirations within the space program of the United States and other countries.

We should not try to go too far with such labels, which can often be confining or serve to set up artificial barriers among researchers (a common phenomenon among academic departments at universities). The intent of this discussion, then, is to illustrate that, to participate in addressing the scientific questions posed today in astrobiology, a researcher must be an expert in at least one of these fields: physics, chemistry, biology, geology, or astronomy. At the same time this same person must have a keen interest in the other disciplines involved in astrobiology. This means that if you are an astronomer, you should also have a thirst to understand aspects of biology, and vice versa. At the very least you must understand the language, the jargon if you will, of the practitioners of the other disciplines that make up astrobiology. Without that understanding, you will be unable to converse with other scientists at a level that would allow meaningful interactions for formulating problems, designing observations or experiments, interpreting data, and elucidating the large-scale meaning of the results of experiments and observations.

This text will help you understand, at a foundational level, much that you should know to successfully work on problems central to astrobiology. Although a text this size cannot hope to cover everything in detail, its careful reading will allow you to usefully understand the literature in all the basic fields underpinning astrobiology. The text also will give you a progress report—at the time of printing—on our understanding of the major scientific issues upon which astrobiology is based. To benefit from the book in both these ways, you must complete many of the problems and thought questions in each chapter. As you think your way through this book (please note the verb!), remember that while the name "astrobiology" is new, and the progress extraordinary, the ultimate questions that we seek to answer are, at their most basic level, as old as humankind itself.

QUESTIONS

1. After reading the book, go back to the 10 scientific goals and augment the questions there with additional ones you might have formulated. Are the goals themselves properly structured, or is there a more logical formulation of these in terms of the intellectual content of astrobiology?
2. Consider the field in which you are majoring in college. Did it have its origin as an interdisciplinary combination of two or more fundamental fields? Write a paragraph on the genesis of the intellectual field (scientific, artistic, humanistic, etc.) that you are pursuing for your major.

SUGGESTED READINGS

This is the book to read if you have had a basic course in one of the fundamental sciences of astrobiology; for example, biology or astronomy. For a more basic introduction read J. Bennett et al., *Astrobiology* (2003, Addison Wesley Longman, San Francisco). An excellent graduate level text, prepared as a series of technical articles, is that of G. Horneck and C. Baumstark-Khan, *Astrobiology: The Quest for the Origin of Life* (2001, Springer, Berlin). The first stop on the Web for astrobiology should be the home page of the NASA Astrobiology Institute at the NASA Ames Research Center, www.nai.arc.nasa.gov/. Here you will find news, links to the NASA road map, various participating institutions around the world, and other astrobiology sites outside of NASA.

REFERENCES

Committee on the Origin and Evolution of Life. (2001). *Signs of Life: Report of a Workshop on Life Detection,* National Academy Press, Washington, D.C

Committee on the Origin and Evolution of Life. (2003). *Life in the Universe: An Examination of United States and International Programs in Astrobiology,* National Academy Press, Washington, D.C.

Morrison, D. (2001). "The NASA Astrobiology Program," *Astrobiology,* **1,** 3–13.

CHAPTER 1

Historical Background to Astrobiology

1.1   Science Retold as History: Buyers Beware!

This historical background offers a sense of how the key scientific revolutions in astrobiology have informed, and in turn been informed by, the prevailing worldviews at the time such revolutions were fomenting. History being characterized by Napoleon as "but a fable agreed upon," it is at some level unavoidable that the reader may be left with the impression of a natural unfolding of the right ideas from the ashes of the wrong. Of course, science isn't done that way, and the traditional pedagogical approach to teaching the scientific method as a cyclical progression from observation through hypothesis through experimentation to revised hypothesis, etc., is highly idealized. Flawed views of the world remain stubbornly in place or rise repeatedly, Phoenixlike, from their own ashes, especially if such views have an appealing cultural or political aspect. Great insights that ignite new and more powerful worldviews seem themselves to rise in a historically complex fashion from a few brilliant minds at some special times characterized by a foment of experiments, observations, rafts of competing (and lesser) ideas, and the ever-present cultural and political background of the day. Far from the stereotypically linear view, an honest look at scientific progress suggests a human epiphenomenon that arises from a tangle of collective experiences and communication—not much different from life itself.

And yet, there is something special about science. Scientists—people who perform scientific research—adhere to an overriding principle that the universe around us can be understood and that such understanding must come from or be supported by observation (be it of a phenomenon in the field or a carefully controlled laboratory experiment). Inspiration is essential but must be grounded before or afterward by the facts. Hence, scientists must be willing to give up their treasured ideas if the facts negate them.

In the second decade of the 20th century, Harvard astronomer Harlow Shapley did battle with Mt. Wilson (California) astronomer Edwin Hubble over whether the so-called spiral nebulae were objects embedded within the massive galaxy of stars (including our Sun) called the Milky Way or were actually much more distant galaxies. Shapley defended the former view. In 1924 Hubble sent Shapley a letter announcing his discovery of a type of variable star in the spiral nebulae, using the most powerful telescope of the time (the 100-inch hooker reflector on Mt. Wilson, California, so-called for the diameter of the primary light-gathering mirror). The discovery clinched the great distance of the spiral nebulae, and hence their identity as distant cousins of the Milky Way Galaxy (Figure 1.1). Shapley, realizing that the hypothesis that he had championed was now invalid, could only remark, "Here is the letter that has destroyed my universe" [*Hetherington,* 1993]. Far from an academic debate, this set the stage for the modern view of the structure of the universe and, ultimately, the concept of the Big Bang (see Chapter 5). Science is successful because it is *fallible*. Progress is possible when properly constructed hypotheses both inform observations and are falsifiable by them. Some science is driven purely by exploration—the ability to go out and examine things with new technologies that have not previously been employed. Planetary exploration is an obvious example, and some aspects of astrobiology are more driven by exploration than by a specifically formulated hypothesis.

In contrast, those who hold their worldviews by faith do so with no fear of confronting observations, which are simply irrelevant. If the world (i.e., universe) that we see around us is inexplicable, because it is the result of creation by a supernatural power that we cannot fathom, then no observation can be useful in defining its nature. There are no hypotheses to falsify; instead, embedded within the canons of any particular religion is the immutable truth. In this fundamental difference of approach between science and faith-based (or religious) worldviews lies the inability to resolve the debates between the two. Of course, there are always scientists who seemingly hold to their favorites theories as a matter of

FIGURE 1.1 Typical spiral galaxy, M81, seen through a ground-based telescope.

faith and religious individuals who seek physical evidence to validate their faith, but the usually failed outcomes of such approaches make them the exceptions that prove the rule.

The remainder of the chapter traces out three revolutions of ideas that are pivotal to the intellectual foundation of the remainder of the book, and of astrobiology itself. Most scientists and historians of science would agree that the first, usually called the Copernican revolution, represents a watershed in the perception of the place of humankind in the larger cosmos. The last two revolutions deal with chemistry and life—and historically were quieter revolutions—but as many would argue, are no less important to the foundations of astrobiology than the first. The twin realizations that life could only arise from other life and that the basic chemistry of life obeyed natural laws, were pivotal in the development of biology as a science. But a piece of this revolution had to be revisited when scientists confronted the origin of planets and life. How did the very first life begin? Both the "vital spark" as a necessary part of living processes, which could be invoked only by special creation beyond the natural laws, and the phenomenon of spontaneous generation of life from nonliving matter, had been contradicted, and something else was needed in their place. As yet, that something else remains poorly understood, and lies at the heart of the question of the commonality of life in the cosmos.

1.2 The Astronomical Revolution: Earth as One of Many Planets

Cosmologies and Cosmogonies of Different Cultures

Cosmology, which describes the structure of the world, and *cosmogony,* the sequence of events of the creation of the cosmos, have historically been expressed through stories that serve to encapsulate for individual cultures their sense of uniqueness, their relationship to the rest of humanity, and foundational experiences in their history. They are interesting in their own right because some of the themes they express reflect common experiences throughout human history. But they also provide a kind of frame of reference from which we can appreciate the first revolution that is the foundation of astrobiology, namely the Copernican model of the structure of the cosmos. The Copernican revolution sparked by this model, wherein humankind's place in the cosmos is based on physical observations rather than cultural considerations, has echoed down through the centuries and continues today with the discovery of over one hundred planets orbiting other stars. This section sketches a few exemplary cosmologies; then focuses on the classical Greeks, whose ideas in this regard form the direct foundation for the Renaissance development of modern cosmology; and finally discusses the Copernican model and its implications.

The book of Genesis in the Judeo-Christian Bible is a set of creation stories from the Middle Eastern crossroads of Europe and Africa. It begins with the separation of the Earth and Heavens by a Supreme Being, a common event in Mesopotamian cosmogonies from at least the second millennium B.C.E. It continues with the sequential creation of plants (not planets!), stars, Sun, Moon, animals, man, and woman, and it ends with the death of Joseph in Egypt. This in turn sets the stage for the seminal wanderings of the Israelites, which include their enslavement and subsequent liberation—classic themes in the collective experience of humankind that mirror the reality of the spread of humankind from a single ancestral home region (Chapter 17).

The Hopi cosmogony, like that of many peoples on the American continent, begins with the impregnation of Mother Earth by Father Sky (in this case, the Sun), leading to life, including humans, that exists in a dimly prehistoric underworld. Humankind falls prey to corruption and evil (a theme as well of the Bible, composed half a world away), and a portion leaves with the help of spirit and animal deities for an Upper World that is reached through a hole in the sky. There, different groups of humanity fan out in different directions to find their own destinies. The cosmology of worlds within worlds and sky portals from one to another illustrates the complexity of human perceptions of the cosmos. The concluding migration of people to different lands reflects again the fundamental theme of wandering, or migration, in the human experience.

The classical Mayan civilization, which flourished in the centuries before the arrival of the European *conquistadores* in the Americas, possessed a remarkable cosmology. The Mayans of this period were careful observers of the sky and discerned the periods of recurrence of various objects in the sky, most notably Venus. On this basis the Mayans constructed accurate calendars and kept several going simultaneously for different purposes—including one based apparently on the orbital period of Venus. The so-called long-count calendar was more accurate than the Julian in terms of remaining synchronized with celestial motion and had a zero point at 3114 B.C.E. Accompanying this precise calendrical system was an efficient number system using base 20 (the number of human fingers and toes), no less practical than the current, almost universal, base 10 system (Figure 1.2). While careful observers of the sky, the Mayans were not interested in constructing physical models of what they saw; instead, they constructed a cosmology based on what they perceived to be the cyclical nature of time. Human history was one cycle embedded within a much larger set of cycles played out on a supernatural scale, and the cycle was predetermined by what happened on the larger scale. Thus, the overall "tone" of Mayan society was perhaps more fatalistic than that of the prevailing Western society today, since the latter is based more heavily on a linear view in which history is an evolving series of human events that do not return to a beginning.

Over the centuries many different cosmologies were invented in China, including some that were echoed in European cosmologies. Early Chinese cosmologies visualized the cosmos as a dome with the Earth in the middle. But very early on, between the first and third centuries C.E., there arose an advanced view of the heavens espoused by Qi Meng, and known as the Xuan Ye school. In this view, the sky is a two-dimensional perception of a three-dimensional space. The analogy was drawn from the fact that mountains look increasingly flat as one moves farther away from them. The universe was visualized as huge, maybe infinite, with enormous space between objects. Objects in the heavens, including the wandering stars, the Sun, and the Moon could move in either direction—they were not fixed together or fastened to the body of the heavens. It was clear to 12th-century philosopher Ma Yongquin that the sky was three-dimensional—the higher you go the farther away you are, with no definite distance between the ground and Heaven.

Deng Mu of the 13th century elaborated on this view, stating that the sky and Earth were huge but, in the whole of everything, were only a grain of rice. He drew the analogy that if all of space were like a tree, then heaven and Earth would be one of its fruits. But a

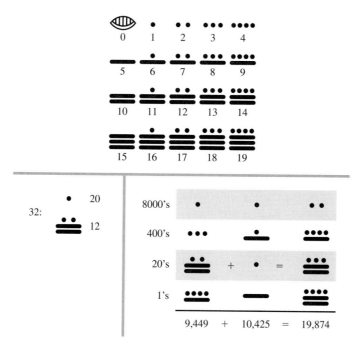

FIGURE 1.2 The symbols for the numbers in the base 20 system of numbers used by the classical Mayans, indexed by the base 10 system in use almost universally today.

tree has many fruits, so Mu concluded that it would be unreasonable to suppose that besides this heaven and this Earth there were no other heavens and Earths. This was a clear statement of the plurality of worlds that was also a common theme in ancient Greek and European Renaissance views of the cosmos.

What most distinguished the Chinese view of the heavens during the great dynasties was the notion that historical events on Earth were tied to, and hence were presaged by, events in the heavens. Hence, the court took careful note of changes in the heavens, including the appearance and disappearance of comets, as well as more extraordinary events such as the exploding stars known today as *supernovae*. While European artists also drew the comings and goings of comets, they missed the 1065 C.E. supernova that created the Crab nebula, well-known and well-imaged by modern astronomy. Chinese observers, in contrast to the Europeans, did not miss this event.

Classical Greek Thought on the Structure of the Cosmos

Classical Greek cosmology focused much more on the content and arrangement of the cosmos and less on how things came to be. Particularly important to the later development of modern cosmology was the empirical, that is, observational, approach taken by a number of classical Greek philosophers to determining the distances to and sizes of various celestial objects. However, the empirical approach was not universally employed, even when it was overlaid with philosophical and aesthetic assumptions in a process called *saving the phenomenon.*

Saving the phenomenon consisted of mechanistically explaining the observed motion of celestial bodies using only geometric tools that were aesthetically acceptable, for example, circles and spheres. The great philosopher Plato, a fourth-century B.C.E. Athenian who was a disciple of Socrates, believed that reality existed in the mind and not in the physical world perceived by the senses and, hence, disdained observation. Consequently, he made little contribution toward cosmological models. But his pupil, Eudoxus, developed sets of simple nested spheres around a common center in an attempt to save the phenomenon. But such a geometry could not cause changes in the distance of objects from the center of the spheres, and hence from Earth (which was positioned at the center). This was a serious problem because the observed cyclical changes in brightness of the planets could not be reproduced.

Aristotle, a pupil of Plato, adopted an approach that seems more modern, namely, trying to understand the underlying laws by which objects move. Aristotle tended to search for mechanisms and can be thought of as having adopted a physicist's point of view (albeit with physics that was later shown to be largely invalid in its assumed laws and causative agents). However, Aristotle did tend to operate in a way different from modern science, via seeking out those ideas from other philosophers that would create a logical framework. Observations were introduced to convince the reader, not to test hypotheses.

For example, Aristotle said the entire universe was contained within the outer surface of the sphere of stars—every point within contained matter of some kind and no vacuum. Beyond the sphere was nothing, including no space. For Aristotle, matter and space went hand in hand; they were intimately related. It was meaningless to talk about what was beyond the stars, in Aristotle's cosmology.

It is fashionable in physics textbooks to use Aristotle as an example of "incorrect" physics, and indeed large numbers of studies have been conducted to point out that students come into physics classes with a preconceived view of mechanics that is "Aristotelian," and wrong. Students, for example, tend to imagine, as did Aristotle, that objects naturally come to rest when, in fact, they will continue to move in a straight line unless a force is imposed upon them. Living on the Earth, which imposes a strong "downward" (really radial) gravitational force and has an atmosphere thick enough for friction to be clearly evident, it is easy to understand how Aristotle arrived at his thoroughly incorrect laws of motion. Hetherington writes, in regard to Aristotle, "Never before, nor since, [in European history], has so pervasive a world view been so thoroughly overturned and replaced in all aspects of human thought." [*Hetherington,* p. 177]. But some of Aristotle's ideas—including those on matter and space—have an interestingly modern ring, and despite the mechanistic defects, we continue to find ourselves strongly influenced by the Aristotelian world view in its broad aspects.

From Ptolemy to Copernicus: Moving the Earth from the Center of Things

Ptolemy was a second-century (C.E.) Greek astronomer, mathematician, and geographer, whose cosmology formed the basis of Arabic and European views of the cosmos until the

16th-century Copernican revolution. His cosmology was founded on the premises that the Earth must sit at the center of the cosmos and that the known planets, the Sun, and the Moon must move in perfect circles around the Earth. The system was a mixture of aesthetically (or philosophically) driven assumptions coupled to precise mathematics, and indeed the Ptolemaic system was remarkably accurate. However, its accuracy came at the expense of complexity because the planets as seen in the sky have paths that include retrograde motion (i.e., against the direction of the Sun and Moon; Figure 1.3). To explain the occasional retrograde motion requires placing the planet on a secondary circle that is centered on the primary circle. The secondary circle, which is called an *epicycle,* moves along the orbit in place of the planet itself; the planet orbits along the epicycle. Provided the speed of the motion is properly adjusted, the epicycle can cause a planet to appear to move backward in the sky for the time that it is on the outer arc of the epicycle (Figure 1.4). However, to explain in detail the motion of the planets, including when and for what duration the retrograde motion occurred, as well as the details of lunar and solar motions, required roughly 40 epicycles in total, as well as some compromises to the aesthetic of circular motion. While the orbits remained circular, Ptolemy displaced the Earth from the center of the orbits of the planets (making what was called an *eccentric*) and displaced the center of the circular orbit from the center (or *equant*) of uniform motion of the planet's main epicycle. Although complex, the system was based on empirical observation of the planets, and in that sense caught the "spirit" of scientific observation that would ultimately topple the Ptolemaic system some 14 centuries after his death.

The observations of 16th-century Danish astronomer Tycho Brahe ultimately brought down the Ptolemaic system, although he could not divorce himself from Earth-centered

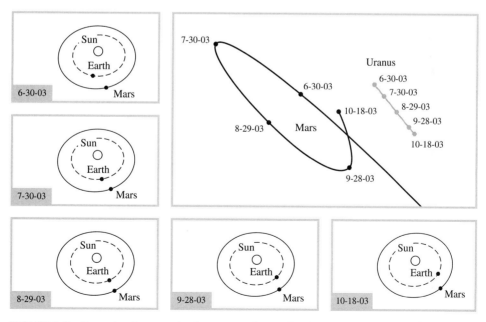

FIGURE 1.3 Example of retrograde planetary motion on the sky: Mars. Small panels show the orbital positions of the planets corresponding to each date.

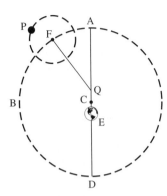

FIGURE 1.4 A planet P orbiting the Earth E in the system of Ptolomy, with the "eccentric" (the orbit center C is displaced from Earth), epicycle (circle around F), and equant (displacement of radius of orbit to Q) needed to account for planet motion.

cosmologies. By the time of Brahe's observations, the model of Nicolas Copernicus, in which the Sun was the center of the cosmos and the Earth moved about it, had been published. Brahe's reluctance to embrace Copernicanism was due at least in part to his own observations and a mistaken assumption about the scale of the universe. Brahe understood the phenomenon of "parallax," whereby objects in the sky appear to shift back and forth as the Earth moves in its orbit around the Sun (See below and Chapter 5). However, he mistakenly assumed that the planets and stars were close enough to the Earth that his instrument would be able to detect the shift. Because he could not, he assumed the Earth stood immobile at the center of the cosmos; in fact his astrolabe simply did not provide the required accuracy.

Brahe's contemporary and sometime-collaborator, Johannes Kepler, adopted the Copernican model and had the insight to assign *elliptical* orbits to the planets; he formulated three fundamental geometric rules about such orbits that properly defined planetary motion. With the ellipse (Figure 1.5) in place as the fundamental shape of planetary orbits, the heliocentric model of the Earth, Sun, and planets assumed a much simpler form that accurately reproduced planetary motions.

The Copernican revolution received a dramatic demonstration of plausibility through the telescopic observations of Galileo Galilei (1564–1642). His maps of the moon revealed a celestial body that, with its craters, mountains, and other features, seemed more geological than ethereal. And his discovery of four "Medician stars" (named after the

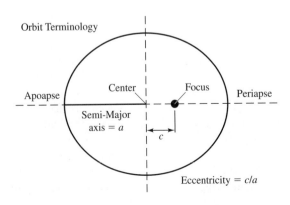

FIGURE 1.5 Structure of an ellipse, the fundamental shape of planetary orbits. Labels define the *semi-major axis*, the *eccentricity*, and the *inclination* of the orbit—fundamental parameters we will use in later chapters.

wealthy Italian benefactor)—the four large moons of Jupiter, now honored as a group with Galileo's name—showed that other planets could have worlds orbiting about them. There was no longer any rationale for seeing the Earth as anything but another planet, one that had by that time been circumnavigated by ship and to which the attribute "global" could now and forever be attached.

The revisionist view of the cosmos went beyond reshuffling the positions of the planets of our own solar system. If the Earth is just one planet out of seven, and life exists on Earth, then might not life exist on other planets? And, why could there not be planets far beyond the classical ones, perhaps orbiting the stars the way Earth and its sister wanderers orbit the Sun? This question was posed, and answered, by Giordano Bruno. Part mystic, part scientist, Bruno was more an acolyte than a contributor to the Copernican revolution, but his remarkably modern expression of the notion that our own solar system is one of many echoes hauntingly through four centuries of time:

> There are countless suns and countless earths all rotating around their suns in exactly the same way as the seven planets of our system. We see only the suns because they are the largest bodies and are luminous, but their planets remain invisible to us because they are smaller and non-luminous.
>
> G. Bruno, *On the Infinite Universe and Worlds*, 1584

Completing the Copernican Revolution

In Chapter 15 we will see that Bruno's speculation about the innumerability of stars is essentially correct, the notion that many such stars should have planets is reasonable based on available data, and his explanation for the invisibility of planets turned out to be correct. Here, however, the reader might puzzle at the depth of Bruno's insight, which seems to include the correct inference that the fixed points of light seen in the sky, the stars, are suns like our own. This would seem to be an enormous leap from the immediate changes wrought by the Copernican revolution, which really just concerned the wandering stars—the planets—and the position of the Earth relative to the Sun. [Bruno's heretical writings led to his conviction by the Catholic Inquisition, and he was burned at the stake in a piazza (il Campo di Fiori) near the center of Rome in 1600 after refusing to recant.]

Indeed, Bruno's insight had its foundation elsewhere—in that of the classical Greeks. Aristarchus (third century B.C.E.), Hipparchus (second century B.C.E.), and then Ptolemy (second century C.E.) were all interested in determining the absolute scale of the cosmos. Ptolemy, long before Brahe, estimated the distance from the Earth to the Moon using the *parallax* effect detailed in Chapter 5. The classical Greeks surely were not the only ones to recognize this effect, but they distinguished themselves by putting parallax to use in determining the distance scale from Earth to Moon. The Copernicans, however, had a more profound application: With the insight that the Earth orbited the Sun came the realization that, as the Earth moved from one side of the Sun to the other along its orbit, the fixed stars should shift in relative apparent position according to their distance from the Earth. The fact that *no* star could be seen to possess a parallax shift led Ptolemy to conclude that the Earth was immobile, but the Copernicans offered a different—and correct—interpretation: The stars are enormously distant compared with the Earth–Moon distance; hence the parallax shift is too small to be seen. Today, measuring stellar parallax is a standard way to determine distances to stars (Chapter 5), but the astronomers of Brahe's era did not have the instruments required to make the measurements with the necessary precision.

Thus, one of the most profound insights of the Copernican revolution arose because of the *failure* to observe a particular effect, which in the context of the model of the Earth orbiting the Sun required that the starry universe be enormously distant from us. And if the stars were so distant, why could they not be Suns like our own? And if they were, then surely they must have planets, as Bruno expressed so clearly. It would take more than four centuries from Bruno's writings to the actual detection of the first planet orbiting a star like the Sun, in 1995. During that time, the notion that there must be many planetary systems like our own rested largely on the philosophical grounds of Copernicanism—that the Earth was not special but was one of many planets orbiting the Sun; hence the solar system in total ought not to be special either. Different models of how the solar system formed, from Descartes's nebular disk of the 17th century to the tidal stripping model of the 19th century (Chapter 6), implied very different conclusions about the ubiquity of other worlds. Indeed, even today we do not know how common planets are, and current techniques are not yet sensitive enough to detect other planets like Earth around other stars. But the fact that giant planets such as Jupiter exist around other stars confirms the long-held Copernican notion that our planetary system is not the only one.

The essence of the Copernican revolution is the switch from the Aristotelian dichotomy between the Earth and sky to the view that the Earth is a part of the heavens (i.e., cosmos). This switch initiated the modern debate on whether life exists elsewhere in the cosmos. On this question, we find ourselves in a situation more akin to that of Galileo and Bruno than to that of the modern astronomers, who almost monthly now announce discoveries of new planets. The fundamental mechanisms of living processes have been determined, and it is possible to use exquisitely sensitive laboratory techniques to observe and manipulate the basic molecules of life (Chapter 4). But we do not yet understand how life came to be on Earth, and there is as yet no indisputable evidence for life—past or present—on other worlds. Nor is it known whether the direct detection of life will happen on a body within our solar system (Chapter 14)—and hence within the next decade—or whether we must search beyond our own planetary system to the stars. In that case, while indirect and disputable evidence might come from advanced telescopes that in the next decade will be deployed in near Earth space (Chapter 15), it could be centuries at least before such life could be directly detected. Hence, we as a human civilization remain very much in the throes of the Copernican revolution, with no immediate end in sight.

1.3 The Chemical Revolution: The Building Blocks of Matter Revealed and the Vital Spark Removed from Organic Chemistry

The classical Greeks had a variety of views regarding the nature of matter, one of which was that there was a limit beyond which we cannot divide matter further. The indivisible and invisible particles were called atoms, and this name has been resurrected for the fundamental units (which are divisible; see Chapter 2) of the chemical elements. Other than the name, and the invisibility to the naked eye, there is no correspondence between the philosophical atoms of the classical Greeks and the physical elemental particles studied today. The classical Greek concept of the elements, articulated by Aristotle and others, was very different from the modern concept of the chemical elements. The classical Greek elements were earth, air, fire, and water, so chosen out of a long pre-Socratic tradition that stretches into prehistory. Each possessed defined "qualities" of dryness or wetness, warmth or cold.

By the time of the alchemists of the European middle ages, the notion of "elements" as having to do with materials such as gold, silver, sulfur, etc., had become more prominent. The endeavor of alchemy was the transformation of one such material into another by the application of certain other substances, appropriate incantations, timings, locations, and so forth. Alchemy was to chemistry as astrology is to astronomy, with one important exception: The goal sought by the alchemists was possible but simply beyond the physical conditions attainable in alchemical practice (Figure 1.6). Nuclear reactions, and radioactive decay, indeed transform one element into another, as described in Chapters 5 and 6.

Modern chemistry, with its focus on empiricism, developed through the 17th and 18th centuries as experimental capabilities improved in Europe. The first truly modern description of chemical principles, indeed arguably the first textbook on chemistry, was that by French chemist Antoine Lavoisier, published in 1789. In this treatise some 23 elements, immutable one to the other, were described, along with the properties of compounds that could be made from these elements and the techniques by which their synthesis was achieved. Particularly important as a laboratory tool was the weighing before and after a set of chemical reactions of both the reactants and the products. By this approach it was

FIGURE 1.6 A classical depiction of a medieval alchemical workshop, engraving by Hienrich Khunrath for his treatise *Amphitheater of Eternal Wisdom* published in 1604.

possible to falsify the claims of earlier experimenters that, for example, water has been transformed into silica.

Debate raged among experimentalists over the nature of the elements and how they bonded to each other. John Dalton (1766–1844) initially postulated that each chemical element was an assemblage of atoms of various different volumes, weights, and propensities for bonding. However, this did not fully satisfy experimental data obtained from measuring the properties of, for example, hydrogen gas, chlorine gas, and the resultant product, hydrogen chloride. Instead, it became evident that the elements themselves were single atoms, characterized by the propensity for bonding with other elements. Initial attempts at describing the nature of bonding held that identical atoms always repelled, so that diatomic molecules such as N_2 (molecular nitrogen) or O_2 (molecular oxygen) could not exist. By this reasoning, which was adhered to by Dalton, water was a *molecule* (combination of atoms) with the formula HO (one hydrogen, one oxygen atom) and ammonia was NH (one nitrogen, one hydrogen atom). However, Dalton's own empirically determined law of partial pressures of gases—and the theory of Berzelius in the early 19th century that allowed attractive electrical forces in bonding—led to the discarding of this view. Berzelius then correctly formulated water to be H_2O (two hydrogens, one oxygen) and ammonia to be NH_3 (three hydrogens, one nitrogen), later confirmed by experiment.

William Prout determined that the densities of a number of different gases made of a single element were roughly whole-number multiples of the density of hydrogen gas, for example, molecular nitrogen being seven times the density of molecular hydrogen gas or H_2. Thus, by the early 19th century, hydrogen was identified as representing the unit building block of the other elements; indeed Sir Humphrey Davy in 1811 proposed that hydrogen was the primordial element from which all others were made. This prescient view came more than a century before the production of elements in stars by fusion reactions (Chapter 5) was understood.

These and other discoveries set the stage for the spectacularly unifying principle of the periodic table of the elements. With the properties of some 67 chemical elements established, Dmitri Mendeleev and J. Lothar Meyer published in 1869 an ordered classification of the elements according to their individual atomic weights and bonding properties (Figure 1.7). A full description of the table is given in Chapter 3. The most important aspect of the table as proposed was that there were gaps in the ordering of the elements, and from these gaps Mendeleev made detailed predictions for the properties of the elements that should fill the gaps. The discovery of elements with the right properties gave tremendous confidence to late-19th-century chemists that they now understood the basic systematics of the chemical elements and, hence, the rules of chemical bonding. This confident command of the properties of the elements and their compounds enabled progress in elucidating the nature of chemical bonding (Chapter 3), though a full understanding had to await the dismemberment of the atom and the development of quantum mechanics in the next half century (Chapter 2).

Although laboratory experiments discovered more than two dozen elements after Lavoisier's work, many more remained undiscovered until the development of quantitative spectroscopy. It had been observed that gases of different elements glowed with different colors when they were allowed to burn. By spreading the component wavelengths of the light using a grating, it was possible to see that the color was caused by preferential emission of light in a characteristic set of narrow bands (Figure 1.8). Indeed, some of the ele-

TABELLE II

REIMEN	GRUPPE I. — R^2O	GRUPPE II. — RO	GRUPPE III. — R^2O^3	GRUPPE IV. RH^4 RO^2	GRUPPE V. RH^3 R^2O^5	GRUPPE VI. RH^2 RO^3	GRUPPE VII. RH R^2O^7	GRUPPE VIII. — RO^4
1	H = 1							
2	Li = 7	Be = 9,4	B = 11	C = 12	N = 14	O = 16	F = 19	
3	Na = 23	Mg = 24	Al = 27,3	Si = 28	P = 31	S = 32	Cl = 35,5	
4	K = 39	Ca = 40	− = 44	Ti = 48	V = 51	Cr = 52	Mn = 55	Fe = 56, Co = 59, Ni = 59, Cu = 63.
5	(Cu = 63)	Zn = 65	− = 68	− = 72	As = 75	Se = 78	Br = 80	
6	Rb = 85	Sr = 87	?Yt = 88	Zr = 90	Nb = 94	Mo = 96	− = 100	Ru = 104, Rh = 104, Pd = 106, Ag = 108.
7	(Ag = 108)	Cd = 112	In = 113	Sn = 118	Sb = 122	Te = 125	J = 127	
8	Cs = 133	Ba = 137	?Di = 138	?Ce = 140	−	−	−	− − − − −
9	(−)	−	−	−	−	−	−	
10	−	−	?Er = 178	?La = 180	To = 182	W = 184	−	Os = 195, Ir = 197, Pt = 198, Au = 199.
11	(Au = 199)	Hg = 200	Tl = 204	Pb = 207	Bi = 208	−	−	
12	−	−	−	Th = 231	−	U = 240	−	− − − − −

FIGURE 1.7 Mendeleev's original periodic table of the elements. The spaces marked with blank lines represent elements that Mendeleev deduced existed but were unknown at the time. Symbols at the top of columns are chemical formulas written in 19th century style (compare with the modern table in Chapter 3).

ments are named for the color, or wavelength, of the most prominent *emission* line in their spectrum: *rub*idium (red), *caes*ium (blue), *thall*ium (green), *ind*ium (indigo).

By spreading sunlight using prisms and gratings, it was possible to identify elements in the Sun through the correspondence of the spectral lines with those obtained in the laboratory. The two yellow "D" lines of the spectrum of sodium were seen in the Sun, and hence sodium was inferred to be present there. The yellow color of sodium street lamps is a result of these lines. In the Sun, these lines appeared dark against the bright *continuum* background of the Sun's spectrum, which we can see naturally in a rainbow when raindrops spread out sunlight into component wavelengths. The dark lines were correctly interpreted to be the result of the elements selectively absorbing light emitted from deeper in the Sun, at the same wavelengths that they emit light when burned in a flame or excited by a electric discharge. Another line in the yellow part of the Sun's spectrum could not be correlated with known elements until 1895, when William Ramsey identified it in the spectrum of gas produced from heating the terrestrial mineral *clevite*. This gas was named *helium*, referring to its discovery first in the Sun (Greek, *helios*), and it was correctly identified as an element in the periodic table.

Using telescopes in the late 19th century, it was possible to take spectra of stars, demonstrating that, in accordance with Bruno's 16th-century speculation, they are simply other Suns scattered through space. The strength of the lines of hydrogen in stars showed that this was the most abundant element in the cosmos, with helium being second. Differences in the spectral appearance and overall color of stars allowed them to be classified. Because stars were self-luminous, and because it had long been observed that cooler objects glowed red while hotter glowed blue, it was clear that blue stars were much hotter than red. By the turn of the 20th century, systematic study of spectra, primarily at Harvard University,

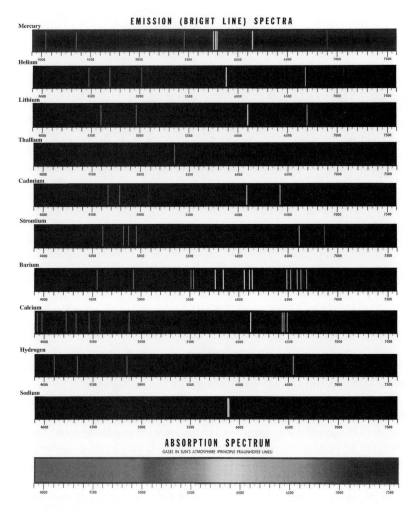

FIGURE 1.8 Spectra of the gases of various elements, showing emission lines (where energy is preferentially emitted) as a function of wavelength in nanometers (10^{-9} meters).

established the range of stellar temperatures and compositions. Stars became classified with a letter and number system that, after several reshufflings, progressed from hot blue "O" stars, through B, A, F, G (our Sun), K, down to the cool, red, and dim M stars. Stars of given temperature were found to vary in their spectra in such a way that some were compact "dwarfs" (of which the Sun is one), and others were bloated "giants." Chapter 5 describes how these observationally-determined conditions relate to the history of stars. Early in the 20th century, laboratory experiments and spectroscopy unified the Copernican and chemical revolutions by making it possible to determine the chemical composition of the countless stars of the observable cosmos.

Organic chemistry, so called because it deals with compounds of atoms that include the element carbon, developed in the 18th and 19th centuries somewhat differently than the

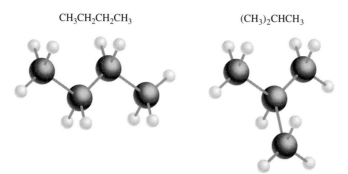

FIGURE 1.9 Isomers of an organic molecule. Butane and 2-methylpropane have the same number of carbon atoms (large circles) and hydrogen atoms (small circles), but the atoms are arranged differently.

rest of chemistry. Because of the way in which carbon bonds with itself and other elements, many different forms (isomers) with the same chemical formula can occur (Figure 1.9). Hence, from a laboratory point of view, organic chemistry looks special because it involves few elements but enormous numbers of compounds. In fact it is not; the chemistry of carbon-bearing compounds follows the same physical principles as do those compounds not containing carbon These principles will be elaborated in Chapter 3 and applied to biochemistry in Chapter 4.

However, prior to the mid-19th century, the majority of practicing chemists believed that organic compounds could only be synthesized by living organisms or, equivalently, that some vital spark associated with life was required for organic chemical synthesis to take place. Indeed, even after laboratory techniques had developed to the point where abiotic (nonbiological) synthesis of organic compounds became possible, thereby falsifying the vital spark notion, the vast array of possible compounds and the complexity of the carbon bond made organic chemistry a separate branch of chemistry, with its own practitioners. This division between organic and inorganic chemistry continues to the present day, a division based not on differences in physical and chemical principles but rather on the practical needs of specialization in kinds of techniques of synthesis and measurement specific to organic compounds.

1.4 The Biological Revolution: Life from Life Only

Throughout human history, spontaneous generation has been taken for granted. Maggots materializing from rotten meat, parasitical worms created from the diseased insides of human beings, geese produced from tree resin upon contact with sea salt, and other impossibilities were all part of the world surrounding people of diverse times and cultures. In the context of observation of the natural world, such claims were not unreasonable given the absence of microscopes until the 17th century, and hence the inability of people to glimpse the invisible processes of germination from single living cells or the division of cells that fundamentally underlies organismal growth.

Likewise, the creation of new human beings could not be understood in the absence of a detailed understanding of what was in a women's egg and a man's semen. As any high school counselor knows, the "how" part of sex comes naturally while the mechanistic "why" (or its consequences) need not be known for the process to be successful. One charming fiction of the pre-microscope occidental world, at least 2,000 years old in origin, is that a miniature human, or *homunculus,* is formed through the mixing of semen with menstrual blood.

By the second half of the 17th century, attitudes among the educated in Europe began to change, as microscopic observations of insects revealed that even these small organisms possessed eggs, and complex internal structures associated with reproduction, making it plausible that new insects arose from the eggs of old. Careful experimental work also revealed that maggots, the initial stage of flies, are generated on top of a fine cloth covering a piece of meat, rather than in the meat itself—and that, indeed, flies laid the eggs that then later hatched as maggots.

The death of macroscopic spontaneous generation as a viable model for the production of animals came at a time prior to detailed mechanistic understanding of the process of sexual reproduction. While 17th-century microscopic studies of human semen revealed the little "animalcules" of sperm swimming around, it wasn't clear what was going on. How could a simple structure like a female egg or a flagellating sperm turn into the pinnacle of creation? Indeed, motivated as much by religious conviction as by plausibility, the notion of an unfolding of a preexisting homunculus became popular in Europe (Figure 1.10). But

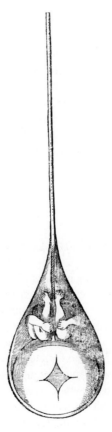

FIGURE 1.10 A homunculus, encased in a human sperm, ready to unfold and grow.

if each human egg or sperm contained a fully developed tiny human ready to grow, then that homunculus should contain an even tinier homunculus, and so on.

To avoid the notion that a natural process could take the formless egg and do with it what God did with a formless pile of dust (make a human being) required from the religious viewpoint that all the generations of humans were preformed. That is, the original humans (Adam and Eve, if you like), contained all the generations of humankind to come, in the form of nested homunculi—the ultimate Russian матрёшка (traditional nested wooden doll). It made little difference to the adherents of this view that a given homunculus could be no larger than the ovum itself, hence only a fraction of a millimeter. Even taking the creationist view that the Earth is only 6,000 years old (a gross underestimate; see Chapters 5 and 6), this requires that some 200 to 300 generations of humans be nested as preformed homunculi, so that you and I were originally $(10^{-3})^{200}$ meters to $(10^{-3})^{300}$ meters $= 10^{-600}$ meters to 10^{-900} meters in length (generously allowing a millimeter if the homunculus were curled around the circumference of the egg). With the background afforded by the Greek atomists (see above), most scientists in the 17th century recognized this to be a total absurdity even in the absence of quantitative information on the size of an atom.

This conundrum would not have to await the 20th century discovery of the mechanisms of heredity for its final resolution. By the mid-18th century, careful examination of the ability of certain animals to regenerate organs or limbs (in some extreme cases, whole new organisms from a part of the original animal) convinced many scientists that growth of a human being from an undifferentiated egg was possible. But this, in turn, helped revive the concept of spontaneous generation—not in macroscopic animals but in the tiny organisms observed in the microscope. These became the new battleground in the debate on whether life could originate from nonlife on a regular, repeatable, basis.

The debate on the spontaneous generation of living organisms came to a head in the middle of the 19th century, and by then it was a practical issue. The ability to treat human diseases in the coming decades hinged on the correct identification of their origin in microorganisms that invaded the human body. The alternative viewpoint, concordant with spontaneous generation, was that disease was caused by microscopic particles or by certain chemicals; as a byproduct, these nonliving agents created microorganisms in the human body under attack. If this debate had not been resolved by the mid-19th century, the ability of modern medicine to prevent or treat infectious diseases might have been delayed many decades beyond the seminal first part of the 20th century.

In the 1850s a French scientist, Fèlix Pouchet, boiled an infusion of hay (i.e., essentially "hay tea") and exposed this supposedly sterile material to an oxygen atmosphere generated artificially from electrolyzing water. Microorganisms appeared from the hay, and under the assumption that the hay really had been sterilized, Pouchet concluded that spontaneous generation had occurred. What Pouchet had in fact determined unknown to him and not revealed until many years later, was that certain kinds of microorganisms can survive boiling. Pouchet's spontaneously generated microorganism was likely a bacterium that formed heat-resistant spores that, upon exposure to oxygen, reverted to viable bacteria. The failure of his experiment was perhaps the first introduction to extremophilic life, that is, life that can survive in extreme environments, which will play an important role in the discussions of Chapter 10 on the span of life-bearing environments on Earth.

In a successful attempt to bring the debate to definitive conclusion, in 1859 the French Academy of Sciences offered a substantial monetary prize to be awarded to a scientist that could produce genuinely new and illuminating experimental results on spontaneous generation. Three years later French chemist Louis Pasteur won the prize and dealt the deathblow to spontaneous generation. His experiments were both clever and careful in procedure (Figure 1.11). Pasteur boiled a mixture of beet juice, urine, and extract of yeast dissolved in water. The flask containing this material was exposed to air only through a long, sinuous glass tube resembling the tight S-shape of a swan's neck. The boiled organic liquid did not ferment (the key macroscopic signature of bacterial microorganisms) while a similarly prepared, but unboiled sample, did. Pasteur speculated that the microorganisms responsible for the fermentation were present on tiny particles of dust in the atmosphere that were unable to work their way through the sinuous flask. To demonstrate this, Pasteur broke off the exposed end of the glass tube and inserted it in the boiled, hence sterile, liquid. This led to fermentation—demonstrating that the exposed end of the flask was contaminated with microorganisms.

Other than the stroke of good luck that Pasteur unknowingly enjoyed in not having to deal with a spore-forming, heat-resistant organism like *Bacillus subtilis* (the undoing of

FIGURE 1.11 Pasteur's setup for the experiment on spontaneous generation.

Pouchet's experiment), the successful experiment was a classic demonstration of sound if not brilliant laboratory science. Pasteur went on to develop the heat treatment that bears his name, which kills microorganisms in milk which therefore could no longer transmit typhoid or tuberculosis. Armed with his laboratory skills and endowed with the knowledge that most diseases certainly were caused by microorganisms, Pasteur developed a veterinary vaccine against *anthrax* and a rabies vaccine for humans.

With the firm resolve that all life—macroscopic or microscopic—comes from life, and with continuing improvements in the ability to manipulate and study individual cells in the laboratory, biological progress in the late 19th century focused on understanding how traits were transmitted from one organism to another. The experiments in pea breeding of the Augustinian monk Gregor Mendel revealed the basic rules of heredity. With a background in physics, Mendel's experiments in the mid-19th century were systematically conducted and allowed him to interpret genetic dominance and segregation of traits mechanistically. His important insight was that traits were transmitted from parent to offspring in discrete packets, or "factors." He inferred that the factors occurred in pairs, each member of a pair coming from one of the two parents of an offspring. Because the parents might have conflicting traits (black hair, blond hair), one of the factors must be dominant—yet the other factor would remain latent in the offspring and could reappear if the appropriate mate were selected.

Mendel's work, published in an obscure journal, was ignored, and other scientists had to rediscover it at the turn of the 20th century. Meanwhile, microscopic studies of cells revealed that the fertilization of a sperm by an egg resulted in the merging of a structure in the cell that came to be called the *nucleus,* and thus the factors of heredity must be located there. Other studies conducted on salamander eggs in the 1880s showed tiny threads inside the dividing cells; some two decades later the connection between chromosomes and heredity was recognized. In 1910 Columbia University scientists T. H. Morgan and A. Sturtevant demonstrated, with experiments on the inheritance of traits in fruit flies, that the genetic factors, now called *genes,* must be arrayed on chromosomes, structures seen in the nucleus of a living cell (Figure 1.12).

By the 1930s scientists applied the techniques of chemistry and physics to biology to allow individual components of the cell to be analyzed and manipulated. H. J. Muller at Texas showed that X-rays could damage genes and alter traits. By the 1930s scientists at the Rockefeller Institute in New York isolated the essential molecule responsible for storing information in the gene, *deoxyribonucleic acid* (DNA). More than a decade of debate on how DNA might store information came to an end in 1953, when James Watson and Francis Crick of Cambridge University determined the double-helix structure of DNA and showed how genetic information could be stored and transferred by that structure. The mechanisms of cellular function, the storage of genetic information and its transfer are described in detail in Chapter 4.

It was less than a century between Pasteur's decisive experiments ruling out spontaneous generation of living organisms and the elucidation of the structure of the molecule that actually carries the genetic information of all life on Earth. This astonishing progress was the culmination of three centuries in which humankind invented and used the microscope and thus was able to examine the inner working of life from individual cells ultimately down to the individual molecules. Yet progress was not only technical. The disposal of the theory of spontaneous generation provided a mental framework, a *paradigm,*

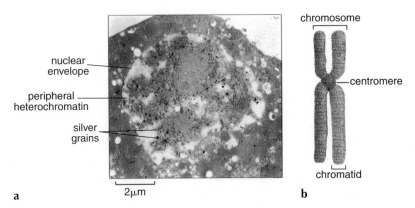

FIGURE 1.12 (a) A portion of the nucleus of the cell, showing the chromosomal material in the form of "chromatin" (scale-a few microns). (b) Structure of an individual chromosome.

within which was the notion that the *instructions* for generating new organisms from old must be contained within the germ cells of parents—rather than believing in fully developed, miniature organisms waiting to unfold. Within this framework, experiments on heredity became important as a means of inferring the rules and source of the inheritance of traits—even though such experiments were not conceptually different from the domesticated breeding of desired traits into plants and animals over the millennia of human history. Thus, both a new worldview and new techniques were required to move to today's understanding of life as a physical phenomenon that obeys natural laws.

1.5 The Search for Life's Origins: Life from What, and Where?

Here, at the opening of the 21st century, the three revolutions described in this chapter have come together in the search for life on worlds beyond the Earth, as well as in the effort to understand how life began on Earth. This is the science of astrobiology. It is tempting to imagine that history has brought us triumphantly, almost teleologically, to the summit of a mighty scientific mountain from which we stand poised to launch ourselves to even higher realms of astrobiological understanding. However, it remains unclear, for at least two reasons, whether this metaphor or the more sobering one at the end of Section 1.2 is in force. In this closing section of this text's historical chapter (indeed, the only chapter that takes a consciously historical approach), we can summarize these reasons. They are covered in more detail in the context of the discussions of Chapters 4, 9, 14, 15, and others as referenced below.

The inaccessibility of other Earths. In our own solar system there are but a handful of objects on which life, or evidence of past life, might manifest itself: Earth, Mars, and Europa. Titan can be added to the list as a potential target on which the chemical steps leading toward life might be preserved, as well as meteorites which might yet preserve very ancient (4.5 billion years old) environments in which liquid water, organic compounds, and rocks mixed. The extraordinary difficulty inherent in exploring Mars, and the possibility that results will be largely negative, provide a natural reason to look beyond the solar system at other stars that might have planets like Earth. While such planets should be detectable within a decade (and crude spectra could reveal the presence of an atmosphere), they lie well beyond the sensitive and invasive in situ techniques for the detection and study of life. There, we are like the astronomers of the 17th and 18th centuries, who could gaze telescopically upon the planets of our solar system but only dream of touching them. (We do somewhat better in possessing the technique of spectroscopy, but such cannot prove inarguably that life exists on a planet orbiting a distant star.) There is no rational timeline for bridging the interstellar distances. Only SETI, if lucky enough to receive a signal from an intelligent form of life elsewhere, would provide an immediate and definitive existence proof (Chapter 15).

The problem of spontaneous generation removed but not eliminated. In the effort to understand how life began on Earth, we are forced to revisit spontaneous generation in a different guise—that of the emergence of life long ago from a set of nonliving and unknown chemical systems. Indeed, at the turn of the 19th century, the grandfather of Charles Darwin, Erasmus, wrote a poem in which the first microscopic bits of life were generated spontaneously beneath the waves of the primordial Earth ocean. This poetic image is no different from the artistic images we see in popular representations of modern scientific theories. The theories themselves add hypothesized physical and chemical processes, usually supported by laboratory experiments that mediated the transformation.

As Iris Fry points out in her book on the history of the search for life's origins, the 18th-century scientists who believed in spontaneous generation tended to be those who also believed (correctly) that organic chemistry did not require a vital spark, and vice versa. Because spontaneous generation negated the view of the special creation of all life by a Supreme Being, it was philosophically akin to the notion that the chemistry of organic compounds was simply like that of all other, nonliving matter. Spontaneous generation of the 18th and 19th centuries did not need to be a violation of physics, even if the specific mechanism was unclear, because it was not the mystical unfolding of homunculi or the creation of geese from trees and salt air. Instead, it was the movement of unspecified, organic fluids through the pores of inorganic rock, according to Lamarck, or other invisible "subtleties" that were postulated to lie just beyond the microscopes of the time. And, as Fry notes, "Materialistically inclined organic chemists . . . saw the synthesis of organic material outside the cell as a demonstration of a possible first step in the spontaneous generation of organisms" [*Fry,* 2000].

I would argue that the models proposed today for the origin of life on Earth carry some of the flavor of the 18th- and 19th-century musings on spontaneous generation. While modern experiments of much greater technical sophistication and much deeper understanding of underlying chemical and physical mechanisms inform such models, there remains an enormous gap between what can be done in the laboratory and what is supposed to have happened on early Earth. Today we substitute for the invisible subtleties of the 18th century an enormity of time to allow the buildup, from simpler precursors, of chemical cycles sufficiently complex to cross some defined (often poorly defined) threshold into the living. Or that time span is used to presume the development of DNA and its simpler cousin RNA (*ribonucleic acid*), usually the latter first, which then coopt the chemistry inside cellular precursors to enable cellular life. Here again, however, the detailed steps by which RNA, synthesized today inside cells, could have been assembled in the absence of life remain unknown. "Time is the greatest innovator," said the 17th-century scientist Francis Bacon; nonetheless to simply invoke time is not a solution to this fundamental question.

Much of what remains unknown of the chemical processes leading to the origin of life lies in the transition to the structured, highly selective stereochemistry of certain classes of asymmetric organic compounds—those that are *chiral*. As discussed in Chapter 4, chiral compounds possess a mirror image that does not superpose on the original, and the source of this asymmetry lies in the way that the carbon atom bonds with other chemical groups, specifically in a tetrahedral bonding arrangement. This issue is raised here because Pasteur pioneered the investigation of chiral organic compounds. While others had long noticed that the large crystals of certain minerals possessed left- and right-handed forms, Pasteur was able to separate (using tweezers and a microscope) the left-and right-handed crystals of the organic precipitate *racemic acid*. He correctly deduced that the source of the asymmetry or "handedness" in the crystal was the structure of the molecule itself, even though he lacked the tools at the time to deduce the structure. Most importantly, he observed that living organisms utilize almost exclusively one or the other handed form of chiral substances, whereas nonbiological organic synthesis produces equal amounts of left- and right-handed forms. Pasteur concluded that the distinction between the living and nonliving was the chiral selectivity of the former. In the 1850s he began searching for a "chiral force" that would have provided the initial separation required for life to begin (see Chapter 8 for a discussion of the present-day search for mechanisms of abiotic chiral separation). By the 1860s Pasteur was publicly espousing the view that spontaneous generation of life from nonlife was impossible and that God had created the first life-forms.

In these discussions I am not imputing that the concept of life arising from nonliving chemistry is faulty. Instead I am drawing a strong parallel between certain 18th- and 19th-century ideas regarding both spontaneous generation and the origin of life and the modern-day theories regarding the latter. The purpose is to justify a cautionary conclusion to this chapter, relying also on the fact that an equivalent situation existed in astronomy for many centuries after Galileo and Bruno. We do not know whether our computational and laboratory techniques have brought us to the immediate threshold of a detailed, step-by-step understanding how life began or whether we stand further from that goal than we might imagine.

QUESTIONS

1. What traditional culture did your ancestors live in? Do you still practice aspects of that culture today? Find some stories about the creation of the world that come from this traditional culture, and compare them to the accounts in this chapter.

2. During the European middle ages the Islamic world preserved much of the knowledge of the ancient Greeks and invented key aspects of mathematics in use today. Do some library research on this contribution, including contributions from the Arab world to the debate between the Ptolemaic and Copernican systems.

3. Compare the 19th-century periodic table of Mendeleev with that of a modern periodic table of the elements. What elements known now were known then? Try to determine what elements would be in the blank spaces in Mendeleev's table.

4. Make a list of technological advances that accompanied or enabled the revolutions described in this chapter, and give a sentence on the role each advance played.

5. Find two or three more examples of organic isomers beyond those in Figure 1.9. What we have called isomers are really "structural isomers," in which the molecules with the same chemical formula assume different structures. Another kind of isomer is the so-called stereoisomer. Consult an organic chemistry textbook to find the definition of stereoisomer and sketch an example.

SUGGESTED READINGS

The classical work on the complexity of the scientific process, reprinted, is that of T. S. Kuhn, *The Structure of Scientific Revolutions* (1996, University of Chicago Press). A recent, somewhat opinionated, but forceful view of the battle between science and superstition in modern society is Carl Sagan, *The Demon Haunted World* (1996, Ballantine Books, New York).

A comprehensive and very readable analysis of cosmogonies and cosmologies from many times and places is *Cosmology: Historical, Literary, Philosophical, Religious, and Scientific Perspectives,* edited by N. S. Hetherington (1993, Garland Publishing, New York). The relationship between the Copernican revolution and the debate on extraterrestrial life is given by Steven Dick in several books, most recently *Life on Other Worlds: The 20th Century Extraterrestrial Life Debate* (2001, Cambridge University Press). Two authoritative and clearly written histories of the efforts to understand the nature of life are Iris Fry, *The Emergence of Life on Earth: Historical and Scientific Overview* (2000, Rutgers University Press) and Michael J. Crowe, *The Extraterrestrial Life Debate, 1750–1900* (1999, Dover). A very comprehensive, though somewhat densely written, review of the history of chemistry is Steven F. Mason, *Chemical Evolution* (1991, Oxford University Press, New York). Daniel Boorstein, *The Discoverers* (1983, Vintage Books, New York) remains a highly readable and broad sweep of the millennia of human discovery and exploration, culminating in the remarkable half-millennium of the Western world's scientific revolution.

REFERENCES

Crowe, M. J. (1999). *The Extraterrestrial Life Debate, 1750–1900,* Dover, New York.

Dick, S. (1982). *Plurality of Worlds,* Cambridge University Press, Cambridge.

Dick, S. (2001). *Life on Other Worlds: The 20th Century Extraterrestrial Life Debate,* Cambridge University Press, Cambridge.

Fry, I. (2000). *The Emergence of Life on Earth,* Rutgers University Press, New Brunswick, NJ.

Hetherington, N. S. (1993). Hubble's Cosmology, in *Cosmology: Historical, Literary, Philosophical, Religious, and Scientific Perspectives,* edited by N. S. Hetherington, pp. 347–369. Garland Publishing, New York.

Lander, E. S., and Weinberg, R. A. (2000). "Pathways of Discovery. Genomics: Journey to the Center of Biology," *Science,* **287,** 1777–1782.

Lederberg, J. (2000). "Pathways of Discovery. Infectious History," *Science,* **288,** 287–293.

Mason, S. F. (1991). *Chemical Evolution: Origin of the Elements, Molecules and Living Systems,* Clarendon Press, Oxford.

Morrison, R. T., and Boyd, R. N. (1973). *Organic Chemistry,* Allyn and Bacon, Boston.

Parezo, N. (1996). Emergence to the Fourth World, in *Paths of Life: American Indians of the Southwest and Northern Mexico,* edited by T. Sheridan and N. Parezo, pp. 240–241, University of Arizona Press, Tucson.

CHAPTER 2

Essential Concepts I
Some Basic Physics of Forces and Particles

2.1 Introduction

This chapter and the next two provide the briefest of overviews of some of the fundamental scientific concepts underpinning astrobiology. To be an effective astrobiologist—one who can evaluate the validity and significance of the work of others as well as formulate and execute research projects of his or her own—a solid understanding of the basic sciences coupled with a trained concentration in one of the core sciences is essential. Physics, chemistry, and biology are the three pillars; astronomy, geology, and planetary sciences represent applied sciences, based on these pillars, whose language and approaches must also be understood. Astrobiology draws on all of these in its discussions.

Our philosophy in introducing these essential concepts is to consider physics in the context of its application to chemistry and chemistry in terms of its role in the biological sciences. An important connection between physics and biology, the thermodynamics of living systems, is covered in Chapter 7. Here, and in the next two chapters, concepts of direct applicability to what we will discuss later in the book are selectively introduced. Your own coursework in one or more of the fundamental sciences will serve as the hook by which you will be able to grab onto and develop an understanding of the concepts discussed here. These chapters are by no means intended to substitute for a thorough and rigorous sequence of university courses in physics, chemistry, or biology.

Physics concerns itself with the behavior of matter in the cosmos, as affected by the fundamental forces and interactions with other particles. Chemistry focuses on the nature of bonding between atoms as determined by the properties and behavior of electrons. This chapter first introduces the fundamental forces, then discusses the particles that are governed by them, and then focuses in on the electron. The mechanisms of chemical bonding and the resulting systematics in chemical systems are discussed in Chapter 3, followed by a detailed discussion of carbon and how it bonds chemically, which plays a crucial role in determining the nature of living systems. The diagnostic expression of atomic and molecular systems in their spectra is the focus of the final part of Chapter 3. The emphasis is on concepts rather than equations or formalism. The reader who comes away with an understanding of the terminology and the scope of concepts associated with the physics of chemical systems will be better prepared for the remainder of the book.

2.2 The Forces of Nature

Forces and Pseudoforces

Objects tend to remain in their existing state of motion—that is, maintain a constant velocity *v*, or speed and direction—unless acted on by a "force." This statement, which is Newton's first law of motion written in his masterwork *Principia*, defines (in a somewhat circular fashion) the concept of force. Another way to define it is to consider the acceleration, **a**, of a particle which is the time derivative of the velocity, $d\mathbf{v}/dt = \mathbf{a}$. (Symbols in bold represent quantities that are vectors, that is, have an implied direction as well as a magnitude.) Then the net force on an object is its mass multiplied by the acceleration. This is Newton's second law. We can also write Newton's second law as

$$\mathbf{F} = m\frac{d\mathbf{v}}{dt} \tag{2.1}$$

Finally, if we define the momentum **p** of an object as *m***v**, for *closed* systems (where the mass contained in the system is held constant), $\mathbf{F} = d\mathbf{p}/dt$. Physically, the consequences of applying a force to a set of particles are varied: for example, curved motion as opposed to straight-line motion; changes in speed; initiation or termination of motion; radiation or absorption of a quantum (*photon*) of light; initiation of a phase change or of a chemical reaction (see Chapter 7).

The concepts of force and acceleration are nonintuitive to us because we live in a world dominated both by frictional slowing (which is the macroscopic manifestation of the electromagnetic force) and by a gravitational force, that of the Earth's, which imposes a constant acceleration of 9.8 m/s^2 (meters per second squared) downward on all objects. As a consequence we grow up with the expectation that objects tend to fall and to slow down and stop unless acted upon (for example, a car rolling on a level plane with the transmission set in neutral). Balls follow curved and not straight trajectories after being thrown or struck by a bat or racket, and gradually slow to a stop on surfaces. A world in which these are perceived to be happening in the absence of forces (because these forces are constantly and pervasively all around us) is called "Aristotelian"; Aristotle referred to these as the "natural" motions of physical systems (Figure 2.1). Not even the environment within Earth-orbiting shuttles or space stations offers a force- or acceleration-free environment. The free-fall trajectory of such craft around the Earth imposes a small net acceleration, the so-called Coriolis acceleration associated with the circular movement around the Earth, and the air inside the vehicle provides frictional resistance to motion. Objects in low Earth orbit, where the tenuous upper atmosphere creates a measurable drag, gradually spiral downward into the atmosphere, all the while experiencing a deceleration that grows as the altitude drops. On the other hand, astronauts conducting operations outside their space-

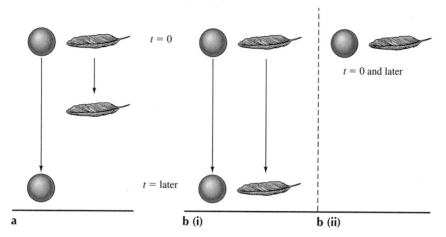

FIGURE 2.1 (a) "Natural" behavior of balls and feathers in the Aristotelian world. (b) Behavior of balls and feathers is surprising even in the Newtonian world; here the behavior is shown in the absence of (i) friction and (ii) gravity.

craft (e.g., to retrieve external film canisters), while returning from the Moon to the Earth in the early 1970s, experienced virtually frictionless, nearly zero-acceleration conditions—the closest any humans have come to the "force-free" conditions that allow Newtonian physics to be experienced firsthand (Figure 2.2).

Thus, forces are, literally, agents of change. Forces can be divided into two types, physical forces and pseudoforces. To understand the difference between the two requires introducing the idea of a reference frame. A *reference frame* is simply the "platform" from which you observe a phenomenon, but it is not merely the location that counts—the velocity and the acceleration of your observation platform must be specified as well. Physical forces act to accelerate masses as viewed in a reference frame—called "inertial"—that is itself not accelerated relative to anything else. Pseudoforces are the manifestation of changes in motion—acceleration—as viewed in a noninertial, or accelerated, frame. They may or may not be the direct result of physical forces.

A simple example of the distinction between the two can be visualized if you are sitting on a spinning carousel (Figure 2.3). Imagine it to be a very large carousel such that you need not see, or even perceive (after a while), your own rotating motion relative to the outside world. The first thing you might notice is the wind on your face. That wind will be present even on a still day because it is actually caused by your face being pushed through the air by virtue of the carousel's motion. But this is merely a case of relative motion, which you would perceive even if your motion were in a straight line (and, if the speed were constant, you would not then be undergoing any acceleration). The circular motion of the carousel, which constitutes acceleration because of the continuous directional change, causes bizarre effects in your local environment. Kneel down and place a marble on the floor of the carousel; unless the floor is sticky, the marble will roll away from you toward the edge, to be dropped off onto the ground. It appears to you that some kind of force is pushing the marble to the edge of the carousel—a "centrifugal" force. This centrifugal force is in fact a pseudoforce, caused by imagining that the frame of reference in

FIGURE 2.2 A nearly—but not completely—acceleration-free environment, Ronald B. Evans travels outside the Apollo 17 command module during a coast from the Moon to the Earth in 1972. Because the spacecraft is roughly at the midpoint of the Earth–Moon distance, the gravitational acceleration he feels from the Earth is 0.97 cm/s^{-2} and from the Moon about 0.01 cm/s^{-2}, vastly less than what is felt on the Earth's surface or in low Earth orbit.

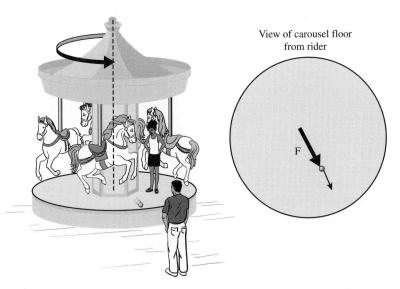

FIGURE 2.3 A spinning carousel contains an observer who, unaware of the carousel's motions, sees marbles rolling in response to an apparent pseudoforce.

which you exist is not itself accelerating in any way whatsoever. But an observer standing on solid ground, beside the carousel, would note that you were rotating (accelerating) and that this was the source of the marble's motion along the floor of the carousel. Your local reference frame—the seat on the carousel—is noninertial. (Although the frame of the observer standing outside the carousel is strictly noninertial, because of the revolving Earth, the spinning of the carousel dominates at the local scale of this experiment. But the "rising" and "setting" Sun, and the spinning-up of storms into spiral shapes, subtly reflect the fact that we sit on a planetary-scale carousel.)

So, what physical forces were in play to create the illusory pseudoforces in the carousel world? A motor of some sort converted stored, or "potential" energy (e.g., gasoline burned in air or the voltage drop of an electric circuit) to drive, by an attached arm, the carousel in a circle. The arm was anchored to the motor by bolts, and this attachment, like the grip of your shoes on the floor, is ultimately rooted in friction caused by electromagnetic force (described later). Your feet, rather than being bolted to the floor, were pressed against the floor of the carousel by gravity. And that is it. All of the macroscopic effects in this scene were due to the forces both of gravity and of electromagnetism. Two other forces shaped the nature of the constituent atoms of your feet, shoes, and the carousel, but they played a much less overt role. The remainder of this section focuses on the nature of the four fundamental physical forces.

Gravity

All matter attracts all other matter toward itself. The force of attraction is called *gravity*, or the *gravitational force*. The gravitational force an observer feels from another object is directly proportional to the amount of matter in the object, and inversely proportional to the square of his or her distance from the object. Hence, a body 2 million kilometers from the Earth feels a force of gravitational attraction one-fourth as strong as would the same body situated just 1 million kilometers from the Earth. The gravitational force is also proportional to the mass of the body on which it acts, so that

$$F_{grav} = \frac{GMm}{r^2} \qquad (2.2)$$

where M is the mass of one body, m the mass of the other, r the straight-line distance between the two, and G is called the *universal gravitational constant*. Its value, determined by laboratory measurements of the force exerted by various masses, is 6.673×10^{-11} Nm^2/kg^2, where a "newton" (N) is a unit of force in the "mks" (meters-kilograms-seconds) system of units, 1 kg $\cdot$ m/s^2. For users of the "cgs" system (centimeters-grams-seconds as fundamental units) the force unit is the dyne, which is 1 g $\cdot$ cm/s^2.

The gravitational force is thus reciprocal; if one body exerts a force on a second body, the second exerts an equal and opposite force on the first. This is Newton's third law of motion (which is traditionally, but less precisely, expressed as "for every action there is an equal and opposite reaction"). However, the acceleration due to gravity is *not* reciprocal; that is the acceleration of the Earth by a smaller body (say, a baseball) is vastly smaller than the acceleration of the small body by the Earth. The gravitational acceleration of the Earth is independent of the mass of the objects it is accelerating. A cannonball and a feather will fall at the same rate and hit the ground at the same time in the absence of the distinguishing effects of air resistance. At the surface of the Earth, the gravitational acceleration is 9.80 m/s². Hence, ignoring air resistance, a body freely falling from a great height reaches a speed of 9.8 m/s at the end of the first second, 19.6 m/s after two seconds, and so on.

To understand the nature of this force requires a deeper understanding of the concept of "mass" as a measure of the amount of material present, its response to a force that accelerates it, and its response to the gravitational force in particular. These last two define the property of the "inertia" of an object. We measure our body mass, typically, by stepping on a scale, which is a platform supported by springs; the more force we exert on the springs the higher the number on the scale. We call this our "weight" and, out of convenience, erroneously equate this to body mass. In fact, however, weight is a force—the mass multiplied by the local acceleration due to gravity. So, if you step on the scale and it reads 70 kg, what is really measured is a force exerted by your body of 70 kg × 9.8 m/s² = 686 newtons of force. (In cgs units this is 7×10^4 g × 980 cm/s² = 6.86×10^7 dynes of force). But since the acceleration due to gravity is practically uniform all over the surface of the Earth, we can conveniently regard this weight as being directly proportional to our body mass. To weigh ourselves on the same scale on the Moon, with about one-sixth the gravitational acceleration at its surface compared with the Earth, we would have to relabel the display to find the same answer for the mass, 70 kg (i.e., there is a new constant of proportionality now). Were we to count the number of protons and neutrons in our body (the heavy particles forming the nuclei of our atoms discussed below) and multiply this number by the laboratory-determined mass of these particles (the two are nearly the same), we should get 70 kg.[1]

The most convenient way to measure the mass m of any object is to apply a constant force generated by, let us say, a spring and then measure the resulting acceleration of the object. Then we take another object, say, one that is much bulkier, and apply the same force. The acceleration of this body will be less. In fact, based on Newton's second law, if the force applied to each of bodies 1 and 2 is the same, the masses are inversely proportional to the accelerations $m_1/m_2 = a_2/a_1$, and the accelerations are easily measured. Then with a standard body defining a "kilogram" mass, we have an internally consistent mass scale. But this is a mass scale based on applying a specified force to an object and then measuring its acceleration. Mass measured in this way is the "inertial mass." Even in the absence of a gravitational field (say, somewhere in interstellar space), the mass of a body would manifest itself through its inertia—its resistance to acceleration when a force is applied. Measuring a body's mass indirectly by determining the force that the body in a known gravitational field applies to, let us say, a spring in a scale, gives us the "gravitational mass."

Are these two masses equivalent? It is straightforward to show that inertial and gravitational masses are proportional to each other. Beginning with Newton more than three centuries ago, scientists have devised intricate and delicate experiments to demonstrate that they are equivalent, and by the mid-20th century it was possible to show experimentally that equal inertial masses always experience equal gravitational forces to better than 1 part in 10 billion in the mass.

---

[1]Measurement of the proton mass was first accomplished by relying on its electric charge and measuring its acceleration in an electric field. The concepts of charge and electric fields will be introduced later.

The equivalence of the inertial and gravitational masses suggested to 20th-century German physicist Albert Einstein that the following two situations are indistinguishable: (1) a room in which there is no acceleration but which is in a uniform gravitational field and (2) a room that is placed in uniform acceleration but is not in a gravitational field (e.g., in interstellar space). Physicists refer to these rooms as reference frames. A person in either of these reference frames will experience *exactly* the same physical phenomena and would be unable to distinguish between the two situations. This *principle of equivalence* is one of the underpinnings of Einstein's theory of gravity, called general relativity. *General relativity* predicts and explains a range of phenomena on cosmic scales, including the correct amount of bending of light away from straight paths caused by the gravitational force of material objects, and was first observed in the distortion of the apparent positions of stars seen against the occulted disk of the Sun during a total solar eclipse in 1919 (Figure 2.4). The bending of light from distant galaxies, by massive foreground galaxies, is easily seen in Hubble Space Telescope images (see Figure 15.6 in Chapter 15). The model of gravity that general relativity provides is a geometric one, in which three-dimensional space is distorted into a fourth dimension by the presence of matter in the same way

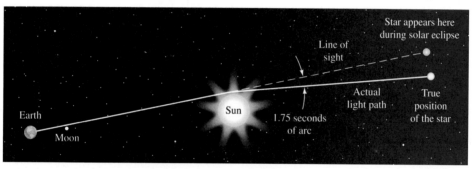

a

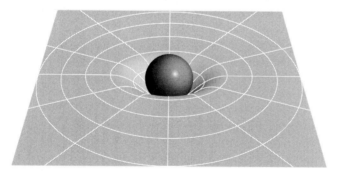

b

FIGURE 2.4 (a) Distortion in the observed position of background stars as gravity from a foreground object causes light to deviate from a straight path, as seen during a total solar eclipse of the Sun. Distances are not to scale. (b) Analogy of the distortion of space–time by objects possessing mass—a ball resting on a thin rubber sheet, which projects the previously two-dimensional sheet into a three dimensional space.

(crudely) that a two-dimensional rubber sheet is distorted into the third dimension by placing an object on it (Figure 2.4).

What is this fourth dimension into which space can be distorted? The fourth dimension is time multiplied by the speed of light to achieve units of a spatial dimension. By visualizing the universe as a four-dimensional construct, with three spacelike dimensions and one timelike dimension, it is possible to treat the movement of matter in a gravitational field as if it were nonaccelerated movement in a curved set of space–time dimensions. Gravity, then, not only bends the path of matter and light; it also causes time to slow down near massive objects. Light moving away from a gravitational field becomes stretched to longer wavelengths, that is, becomes redder in color. These phenomena, verified by experiment, are explained by the general theory of relativity, with the basic postulate of the equivalence of uniform acceleration and a uniform gravitational field and its picture of a four-dimensional space-time distorted by matter.

Later, we discuss the four fundamental forces of nature as being associated with the exchange of transient subatomic particles between material systems (including other subatomic particles). This picture, which fits very well the three forces other than gravity, has been invoked as well for gravity, but with much less success in terms of experimental verification. It is possible that both views are correct—gravity is a space–time distortion associated with matter, and it is also a binding force mediated by an as yet undetected particle (referred to in the scientific literature as the "graviton"). But the former view is far more compelling, and will be essential in our discussion of the structure of the cosmos in Chapter 5.

Electromagnetism

Take two pieces of common household transparent tape, affix the sticky side of one to the back of the other, and then affix the sticky side of the second to a smooth table. Now strip the composite tape quickly off the table, and then separate the two pieces quickly. The separated pieces of tape spontaneously move toward each other and reattach if held close enough. Rubbing your feet on a carpet on a dry day leads to a spark and a brief shock when touching grounded metallic objects such as a light switch or (to its detriment) a computer. Clouds build up in the sky on a summer afternoon, eventually playing host to huge bright bolts of electricity (lightning) that arc within the clouds and between the sky and the ground. A magnetized needle placed in water or suspended from a string points in a preferred direction, which is approximately geographic north. All of these common phenomena illustrate the existence of two forces having apparently nothing to do with gravity; however, we shall see that electricity and magnetism are two manifestations of one and the same fundamental force.

Just as mass is the fundamental property that carries with it the gravitational force, charge is the fundamental property of matter that carries the "electromagnetic" force. Unlike mass, however, it would appear that not all objects carry a net charge, and the net charge on an object seems not to be directly related to its mass. Objects seem capable of being endowed with charge when, for example, they are rubbed against certain other kinds of objects (e.g., a balloon against cat's fur). Further, there are cases in which charged objects repel one another and other cases in which they attract. One historical explanation for these behaviors, promulgated by 18th-century scientist Benjamin Franklin, is that *charge* is a kind of fluid that is conserved in a given system, but one that can be moved by rubbing or other physical manipulations and caused to relocate from one material to another. In this view, if two objects each have excess charge (or if both have a deficit of charge), they will mutually repel, whereas if one material has an excess and the other has a deficit, they will be mutually attracted in order to transfer fluid, hence evening out the charge imbalance. An alternative view is that matter contains two types of charges, which can be named "positive" and "negative," such that like charges repel and unlike attract. (We could argue that perhaps three, four, or even more types of charges in fact exist; an idea discarded in favor of simplicity early on but not disproved until much later. In fact, other fundamental particle properties, but not charge, do come in more than two varieties.)

There was little in the way of experimental evidence to decide between these competing views until the 19th century, when fundamental advances were made in understanding the particle nature of matter. Experiments with apparatuses later applied to the development of television's cathode ray tube demonstrated the existence of discrete, negatively charged particles called "electrons." The experiments suggested that electrons exist in normal matter and could be stripped from any atom, molecule, or even macroscopic object through application of a sufficiently high electrical potential, or voltage. What remained behind was positively charged matter, which later was found to be a combination of positively charged "protons" and uncharged "neutrons." American physicist Robert Millikan measured the value of the smallest single unit of electronic charge in 1911 using charged oil droplets that were dispersed in an electrified medium (Figure 2.5). This very difficult experiment provided a value for the fundamental unit, or quantum, of electric charge, which, based on other laboratory work, was later realized to be the value carried by the particles liberated in the cathode ray tube, namely, the electron. Additional experiments demonstrated that electric charge is conserved, which essentially means it is indestructible. Charges are not arbitrarily created or destroyed, although a negative and a positive charge can combine to form a neutral particle; for example, an electron (e^-) and a proton (p^+) can combine to form the neutron. A complication, which we discuss later, is that protons and neutrons are themselves made of subatomic building blocks that carry positive and negative charges that are a well-defined fraction of the electronic charge.

The force associated with the possession of a net amount of electric charge is enormous. Like gravity, it drops off as the square of the distance between two charges, but the repulsive force between two electrons is of order 10^{42} times stronger than their gravitational attraction. Hence, to avoid all things flying apart in the world in which we live, the number of positively and negatively charged particles must be—and is—exquisitely balanced. And yet, as we see later, the spatial distribution of positively and negatively charged particles—at the submicroscopic scale of atoms—is not uniform, with the positively charged protons clustered in a tiny space (the atomic nucleus) compared with the volume within which the electrons move. It is the repulsion between the electrons both within and between different atoms that provides the strength and stiffness of ordinary solid matter, the incompressibility of the liquids we drink and use, and the pressure and viscosity of the gases that surround our planet.

The electrical force is so intimately tied to the basic properties of matter that we ignore it in everyday life, except in the sudden breakdown of insulating air in a lightning bolt or in the dramatic demonstrations of static discharges with a Van de Graaf generator

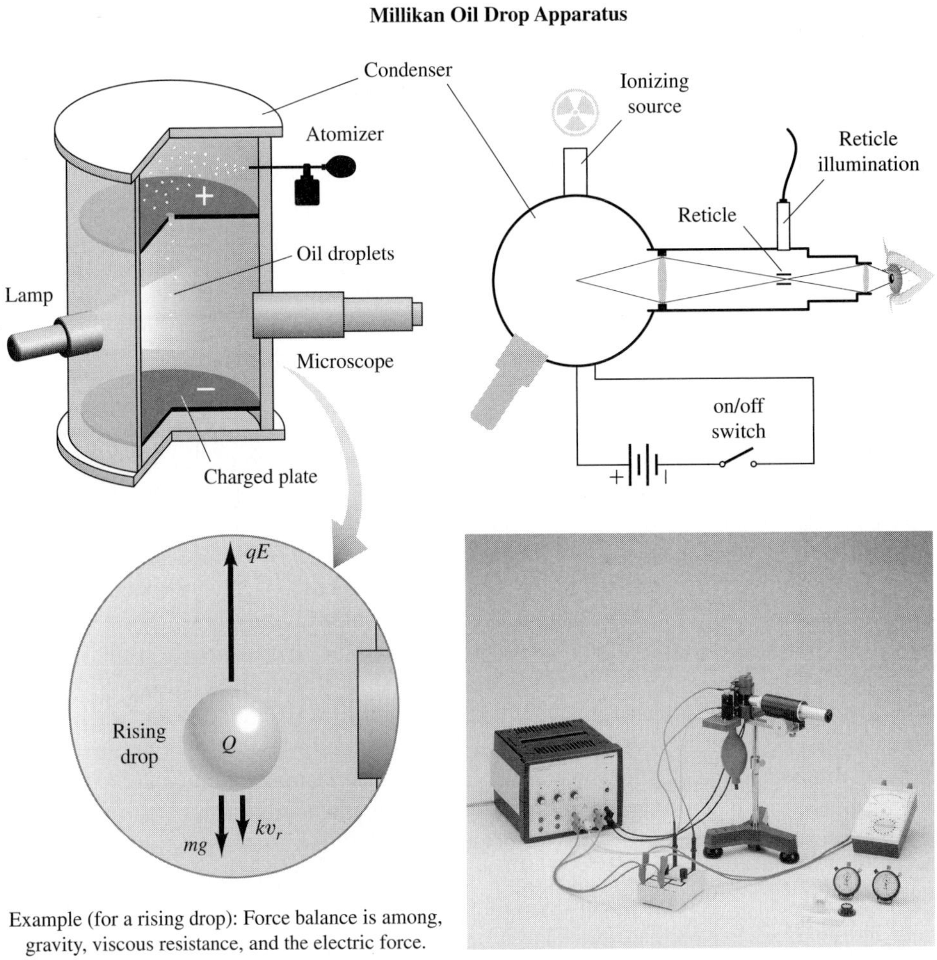

Millikan Oil Drop Apparatus

FIGURE 2.5 A challenging physics experiment of the early 19th century—Millikan's oil drop apparatus designed to measure the elemental value of the electrical charge. Top: Schematic of the apparatus. Lower left: Force balance in a given oil drop. Lower right: Modern example of the experimental setup.

Example (for a rising drop): Force balance is among, gravity, viscous resistance, and the electric force.

(Figure 2.6). While we traditionally come out of our high school science courses thinking of the electron as "bound" to an atom, the situation is more complex. Take, for example, a single water molecule, which is composed of two atoms of hydrogen bonded to one atom of oxygen. All three atoms—oxygen and the two hydrogens—are electrically balanced and neutral to start. But there are certain desirable numbers of electrons, having nothing to do with electrical neutrality, for each of these atoms. For oxygen, it is two more electrons than it possesses in isolation and that guarantee electrical neutrality. Hence, oxygen "shares" a pair of electrons with each of the two hydrogen atoms, and in consequence, the side of the molecule containing oxygen is slightly negatively charged. Likewise, the hydrogen end is

FIGURE 2.6 Static electric discharges emanating from a Van de Graaf generator.

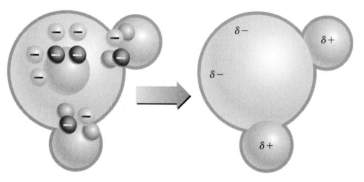

FIGURE 2.7 Schematic of a water molecule in which the origin of the electric dipole moment is shown to be the resultant of four separate dipolar contributions: two "lone pairs" of valence electrons and two polar "covalent" bonding pairs, in which the shared electrons are preferentially drawn toward the oxygen nucleus leading to net negative charges there ($\delta-$) and net positive charges near the hydrogens ($\delta+$). See Chapter 3 for definitions of valence electrons and covalent bonds.

positively charged since the electron "belonging" to the hydrogen changes its mean position to be closer to the oxygen (Figure 2.7). Water is thus a "polar" molecule because it possesses a permanent electric dipole moment associated with this charge asymmetry. In consequence, the liquid and solid phases of water are more stable assemblages than they would otherwise be, and liquid water in particular possesses a very high electrical "dipole moment." Because of this, liquid water is an excellent solvent for ionic substances, which in turn play key roles in the transport of material into and out of living cells through the cell membranes.

Charges moving in a net direction constitute an electric current; such movement is possible through a conducting material. We typically think of metals, in which electrons have extremely high mobility, as carriers of electrical current and other materials, such as silica-based glass, as being insulators in which electrons are bound tightly to the atoms and hence not free to move. The movement of charged particles in the form of a current along a wire, or more generally through free space, reveals the existence of the magnetic force. The ancient Greeks, who collected rocks with the property to exert a force on metals, were the first to record magnetic phenomena in written detail; the site where the rocks were located was in a region called "Magnesia," hence the term magnetism. In fact, however, electricity and magnetism are two aspects of the same fundamental *electromagnetic force*. When an isolated charge exists in vacuum, we can easily draw contour lines representing constant values of the force. That force is then able to act upon another charged particle via its electric component, magnetic component, or both. We can normalize the values of each component (at any point in the surrounding space) by measuring the electric force exerted on a tiny test charge and so derive the value of the "electric field" at that point. Similarly, to determine the magnetic field's value, we measure the magnetic force exerted on a moving tiny test charge (Figure 2.8). Both the electric and magnetic forces and the electric and magnetic fields are vector quantities; they depend on direction.

An electric current generates not only an electric field; it generates also a magnetic field that can pull on a piece of iron in the same way a bar magnet can. The total force **F** exerted by the electric field **E** and magnetic field **B** on a charge Q is

$$\mathbf{F} = Q(\mathbf{E} + \mathbf{v} \times \mathbf{B}) \tag{2.3}$$

where quantities that have both magnitude and direction, vectors, are marked in bold. Notice that the magnetic field can only exert a force if the electric charge it is acting on is moving with a velocity **v** (and hence cutting across magnetic field lines). And in fact, the magnetic field itself is a manifestation of the movement of other electric charges somewhere in the system. The × symbol is a special kind of multiplication, called a cross product, that multiplies the directional components of the two vectors in a definite way and that, in turn, yields a vector quantity. For our purposes it is simply important to realize that the force the magnetic field exerts on the moving charge depends directly on the relative orientation of the magnetic field and on the direction of movement of the charged particle (Figure 2.9).

How can this be? What kind of force depends on the direction the particle is moving, *and* becomes zero when the particle grinds to a halt in the reference frame of the electric field? (The force you feel walking through the running water of a river seems to depend on the direction you are moving but is always applied in a constant direction and does not abate when you stop moving). The answer to this was not understood until the early 20th century when Einstein developed the so-called special theory of relativity. This theory, which preceded the general theory of relativity described earlier, has a fundamental postulate: It is impossible to detect the uniform motion of a frame of reference from observations made entirely within that frame. Uniform motion means that there is no acceleration. The general theory of relativity dealt with accelerating frames of reference and was underpinned by the notion that gravity acting on a nonaccelerating frame is equivalent to an accelerating frame, as far as the observer within the frame is concerned. It is, as noted earlier, a theory of gravity. The special theory of relativity ignores acceleration and deals with frames moving at constant velocity. It has a second postulate, which is that the speed of light in vacuum ($c = 3 \times 10^8$ m/s^2) is the ultimate speed at which anything, including

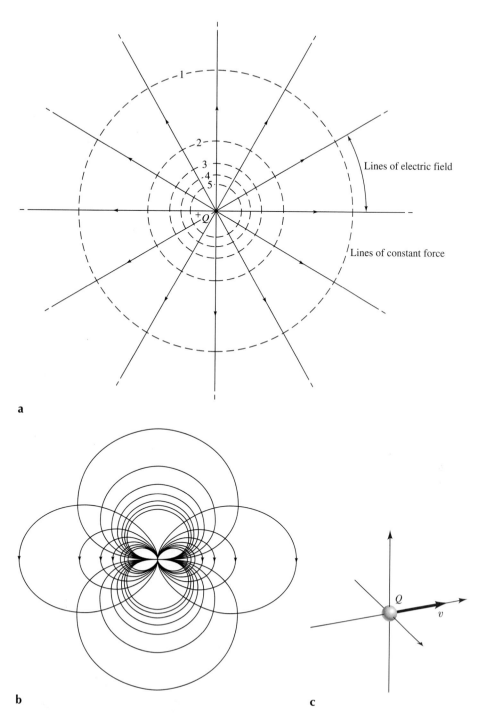

FIGURE 2.8 (a) Electric field around an isolated charge. (b) Electric field around a dipole, which is a positive charge and a negative charge displaced relative to each other. (c) Electric field around a wire of infinite extent carrying an electric current.

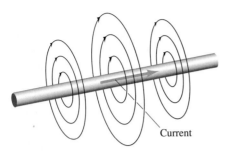

FIGURE 2.9 Lines of force indicating the magnetic field around a wire carrying an electric current (flowing charges).

information, can travel in a reference frame in uniform motion, and that material objects must always move at less than this speed.

These two postulates, both of which have now been verified numerous times through various ingenious experiments, require a change in our expectation as to what happens when we observe activities on moving (but nonaccelerating) platforms, such as a train traveling at constant velocity. If the velocity of the train is **v**, where the magnitude of that vector (which we call its speed) is v along a direction that we call x, then we expect that the position of the train with time is $x = x_0 + vt$, where x_0 is the train's starting position and t is the elapsed time. Now imagine that a passenger on the train is measuring the position of seats along the line of the train's motion and establishes a coordinate system that starts, let us say, at the back of the rear car. Call the distance from the back of the car to a particular seat to be x_2, as measured by the passenger. The same distance, measured relative to a coordinate system fixed to the ground by another observer, is x_1. We expect that $x_2 = x_1 - vt$, the so-called Galilean transformation.

This relationship between x_2 and x_1, however, is not correct, because it implies that the train and its occupants could in fact move faster than the speed of light from the ground observer's point of view. More importantly, when such a rule for the change in position of a moving coordinate system (or a "transformation") is applied to basic physical laws, it implies that these laws act differently on objects in coordinate systems that are in uniform motion than they do on objects in systems that are at rest. This violates the first postulate of special relativity. A transformation that does not violate this postulate, and that accounts for the finite ultimate speed of things, was first introduced by Lorentz and later by Einstein and others at the turn of the 20th century. It turns out to be the correct transformation to satisfy the postulates of special relativity and is given by the following equation:

$$x_2 = \frac{[x_1 - vt]}{\left[1 - \left(\frac{v^2}{c^2}\right)\right]^{\frac{1}{2}}} \qquad (2.4)$$

In a world in which the speed of light is infinite, equation (2.4) would reduce to the Galilean transformation. Because the speed of light c is so large compared with everyday speeds we experience, objects in motion in our low-speed world seem to obey the Galilean transformation, because the v^2/c^2 term is essentially zero in equation (2.4). However, a raft of counterintuitive phenomena occurs when any object is in motion at *any* velocity, and it is most dramatic near the speed of light. As seen from a stationary observer, a moving object seems to contract in the direction of motion—the train seems shorter. A clock on a moving platform seems to run slower as observed by this fixed observer on the ground. (This is directly verifiable by the use of ultra-accurate clocks on high-speed jets, because their precision is sufficient to detect slowing even for sonic speeds, which are a million times smaller than the speed of light.) The mass of an object that is in motion appears to increase to an observer who is at rest. These consequences of the *proper* transformation of coordinate systems [equation (2.4)], according to special relativity, are completely consistent with the basic postulate of the equivalency of inertial frames (those in uniform motion) because the apparent contraction of length, the apparent slowing of time, and the apparent increase of mass of a moving observer are recorded only by the stationary observer outside the frame. The moving observer experiences none of this. And these effects are consistent also with the speed of light being unreachable—a material object moving as speed c would have zero length in the direction of motion, time would cease, and mass would become infinite, according to a stationary observer on the ground.

Returning now to electric and magnetic fields, it is the equivalency of inertial frames from special relativity—and the Lorentz transformations—that correctly predict the magnitude of magnetic and electric fields caused by charges in uniform motion. Imagine a "system" composed of a charged particle and a current-carrying wire; the charged particle is moving uniformly (i.e., with zero acceleration) parallel to the wire. The charged particle experiences a force. If we examine the system from a frame of reference in which the charge is moving, there is found to be both an electric and magnetic force on the charge. If instead we examine the system while moving along with the charge, so that to the observer the charge appears stationary, the force on the charge appears to be purely electric (Figure 2.10).

In summary, electric and magnetic fields (hence, effects) are of differing degrees of importance in any experimental situation, depending on the frame of reference of the observer. *Electricity and magnetism are manifestations of the same fundamental force: the electromagnetic force.* Further, armed with the relationship between magnetic fields and moving reference frames, we can understand several crucial phenomena: First, there is no

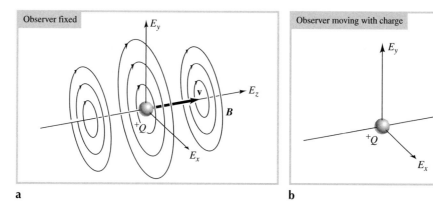

FIGURE 2.10 Electromagnetic fields depend on frame of reference—a charge moving along a fixed wire is affected by electric and magnetic fields if the observer is stationary (a) and by an electric field if the observer is moving with the charge (b).

magnetic equivalent to the point electric charge; that is, there is no such thing as a magnetic monopole. Such a construct, equivalent to a magnetic field around a wire that is shortened down to zero length, would imply that a magnetic field could exist in the absence of the movement of electric charge.

Second, as we discuss later, electrons bound to atoms not only move about the atoms but also have characteristic values of spin angular momentum, or rotational energy. Called *spin*, this is a quantum mechanical property and should not be visualized by simply imagining the electrons to be spinning like tops. In some materials, the correlation between those spins creates a preferred direction that defines—in the rest frame of the observer—a magnetic field. Hence, materials such as magnetite, whose electrons have a preferred spin alignment generate magnetic fields, and hence are "ferromagnetic." "Paramagnetic" and "diamagnetic" materials, on the other hand, only exhibit bulk magnetization when an external magnetic field is imposed.

Finally, planets and moons whose interiors contain electrically conducting fluids, such as molten metals or saline solutions, potentially can generate magnetic fields when such fluids are in motion with a rotational component, which might be induced by a combination of planetary spin and release of interior heat. That the Earth does so, enabling a magnetized compass to point in a fixed direction, is thus a consequence of the fact that moving charges generate magnetic fields.

A situation of fundamental importance concerning the electromagnetic force is the phenomenon of electromagnetic waves, and hence electromagnetic radiation (e.g., light). If an electric current is modulated—turned on and off, or varied repeatedly in strength—then the electric and magnetic fields must oscillate. But they do not only oscillate locally; instead the fields propagate outward from the source of the variation in the form of electromagnetic waves. These waves, in vacuum, travel at the speed of light and consist of alternating electrical and magnetic fields orthogonal to each other. That such waves exist is both experimentally verified and a natural consequence (in the equations developed by the 19th-century physicist J. C. Maxwell that describe the electromagnetic force) of varying the velocity of the charged particles. Indeed, even the acceleration of a lone charged particle will generate such waves. Hence, nonuniform motion of electrons about an atom can cause electromagnetic waves to be emitted or absorbed, as described in Chapter 3. Such waves are defined by the physical distance between peaks of the electrical (or magnetic) component of the wave (the "wavelength") and by the rate at which each peak passes by a given point (the "frequency"). The analogy with a water wave is shown in Figure 2.11. Electromagnetic waves of varying frequency have different interactions with physical matter, including atoms, molecules, and living cells. Certain wavelengths stimulate the retinas of our eyes and produce the phenomenon of vision, in which a rapidly changing spatial map of the incident electromagnetic radiation is perceived. Wavelengths shorter than visible light can enter and damage cells. Longer wavelengths can interact with the molecules in the cells to cause vibration and, hence, the sensation of heat.

Before moving on to the next fundamental force, it is desirable to mention another aspect of electromagnetic waves. Many phenomena observed in the transmission of such waves are consistent not with waves, but with particles. Electromagnetic waves can also be described as "massless" particles called photons, and the wave–particle duality of electromagnetism—that the force is in effect mediated by massless photons—is one of the grand mysteries of the physical world illuminated by quantum mechanics.

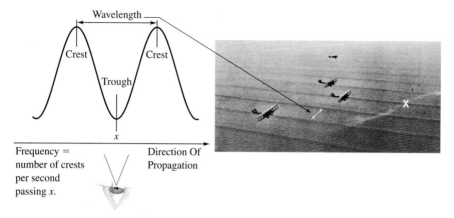

FIGURE 2.11 Wavelength and frequency can be defined for a water wave just as for an electromagnetic wave.

Nuclear Forces

The third and fourth forces are ones with which we have little or no everyday awareness but which nonetheless are crucial to the structure of the atoms of which we are made. As discussed here, the atom's spatial structure is that of a cloud of electrons occupying space around a tiny, compact nucleus. In the nucleus are the positively charged protons and the electrically neutral neutrons. But if the protons are so tightly bound together in a volume much smaller than that of the electrons, why do they not repel each other, based on their like charges? In fact, they do repel each other. Then why does the nucleus not spontaneously fly apart? The attractive force that binds together protons and neutrons, called the *strong nuclear force*, prevents this from happening. The strong force is indeed just that, and over short distances, it overwhelms the mutual repulsion of the protons by the electromagnetic force. However, the nuclear force is a short-range one and falls off extremely rapidly, approximately exponentially, with distance (Figure 2.12).

The relative stability of particular nuclei is determined by the ratio of the number of protons to neutrons, so that nuclei with large numbers of neutrons (relative to the total number of protons) also tend to change "radioactively" into more stable nuclei. In some cases, a neutron will "decay" into an electron and a proton (along with an exotic and very small subatomic particle called the *neutrino*). The force that governs this process is called the *weak nuclear force*. It, too, has a very short range and acts primarily on electrons. Both the strong and weak nuclear forces seem to have their origins in the exchange of exotic subatomic particles that exist only transiently. In the case of the strong nuclear force, the mediating particle is called a "gluon," and evidence for it has been obtained in particle accelerators in which collisions between bits of accelerated matter occur at extremely high velocities. The weak force, in turn, is mediated by a particle called a "massive vector boson," for which some tentative experimental evidence exists. The transient exchange of such particles between protons and neutrons, or between atomic nuclei and electrons, accounts for the short-range attractions observed.

50 Chapter 2 Essential Concepts I

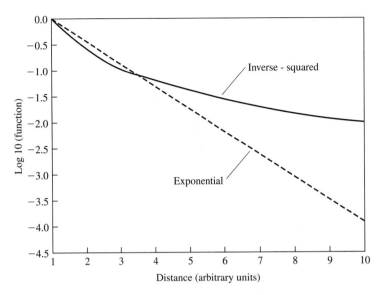

FIGURE 2.12 The exponential falloff of a quantity with distance is much steeper than a falloff inversely proportional to the square of the distance.

Could there not also be an equivalent wave particle that mediates the force of gravity—the so-called graviton introduced earlier? Although gravity seems to be a different kind of force from electromagnetism, the latter is related to the two nuclear forces. At very high energies, achievable in particle accelerators, the behavior and nature of these three forces become the same; that is, they and the particles they affect become "unified." There is a temptation to seek the same for gravity, but it is different from the others: It acts on *any* matter, and on photons, and it is by far the weakest of the four fundamental forces of nature. Nonetheless, the notion of a mediating particle for gravity has been advanced, though its properties would take it out of the experimental reach of modern-day particle accelerators. More promising is the prediction that violent events involving extremely massive bodies, such as the collapse of a star to form an ultradense neutron star or black hole, would generate so-called gravitational waves—ripples in the fabric of space–time. Efforts to detect such gravitational waves are currently under way.

2.3 More on the Particles of Nature

As the preceding section emphasized, the two concepts of forces and particles are intertwined. Normal matter is composed of electrons, protons, and neutrons—subatomic particles that together make up atoms. The ancient Greeks first articulated the concept that matter is composed of discrete, indivisible, elementary particles. Democritus, a fifth-century B.C.E. Greek philosopher, argued that, if the universe was built up of elementary particles, then the possibility existed that it was finite in both extent and complexity and hence understandable. The Roman poet Lucretius (first century B.C.E.) echoed this philosophical argument for the existence of atoms: If matter is made of elemental particles that obey natural laws, then everything in the universe obeyed such laws and the supernatural did not exist.

During the Middle Ages, the practitioners of "alchemy," who sought to transform less valuable substances into gold and silver, were the heirs to the ancient idea of indivisible elemental particles. Ironically, however, to make such a transformation required not the tools of ordinary chemistry that the alchemists employed but the breaking apart and reassembly of the so-called indivisible atoms the ancient Greeks envisioned. The Roman Catholic Church, concerned about the experimental practices of alchemy as a form of black magic, sought to suppress the ideas of atomism because it was seen as encouraging alchemical practices and hence heretical beliefs.

When experimental chemistry developed in the 18th century, evidence for the particulate nature of matter began to accumulate. Many common materials could be shown by quantitative analysis to always consist of irreducible proportions of other (often elemental) substances. Furthermore, when more than one sort of compound could be formed out of two elements, the ratio of the amounts of a particular element in one compound compared with the other could be expressed as small whole numbers. Experiments demonstrated, for example, that oxygen is twice as abundant relative to elemental carbon in carbon dioxide as it is in carbon monoxide. These and other observations led Lavoisier and Dalton in the late 18th to early 19th centuries toward an understanding that the world was indeed comprised of a small number of elemental building blocks, and they also led eventually to the periodic table invented by Mendeleev that displayed and predicted the systematic properties of the known elements. The term *atom* came to be associated with the indivisible elements, and *molecule* to the fixed-proportion ("stochiometric") aggregations of these elements that comprise and account for the enormous variety of materials in the world around us.

Experiments in the late 19th century were able to impart electric charge onto gases, and it became evident that atoms themselves were composites of particles carrying positive and negative charges. To ensure electrical neutrality, experimenters initially thought that the protons and electrons must be mixed uniformly together in the atom. Ernest Rutherford's famous "gold-foil" experiment at the start of the 20th century revealed a very different distribution of these charged particles (Figure 2.13). Rutherford was able to separate a particularly stable, positively charged atomic fragment, which he called an alpha (or α) particle, and focus these through a slit so that they were directed at a thin (10^{-5} centimeters in thickness) foil of gold. He then measured the various directions of scatter of the α particles, which should be repulsed by the positively charged component of the atom. (It would be shown later that the α particle—the nucleus of the helium atom—consists of two protons and two neutrons.)

The gold foil failed to deflect the vast majority of the α particles; a few were deflected over a range of angles. The solid foil thus appeared to consist largely of empty space, with pointlike regions of positive charge. Because the foil was electrically neutral, Rutherford and others inferred that the atoms of gold could not consist of mixtures of negative and positive charges uniformly filling a volume of space defined by the diameter of the atom. Instead, the positive charges (protons) must be concentrated in a very small volume compared with the total physical volume of the atom, which must thus contain the negatively

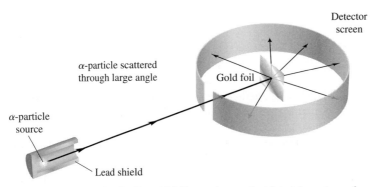

FIGURE 2.13 Rutherford's gold foil experiment elucidated the nature of the atom.

charged electrons. The electrons, in turn, must be much less massive than the protons since they hardly affected the path of the alpha particles; had their mass been comparable with that of the protons, a very different pattern of scattering would have been seen.

The positively charged region, referred to as the atom's *nucleus*, turns out to be tiny indeed. While a typical size for the atom as a whole is an angstrom (Å), (1 Å = 10^{-10} meters), that of the nucleus is 10,000 times smaller. Later experiments measured the mass of the proton to be 1.67×10^{-27} kg, and that of the electron to be about 1,800 times smaller. Other experiments established that the proton could not be the only particle found in the atomic nucleus; an electrically neutral particle (the neutron), just 0.1 percent more massive than the proton, must be present in all elements except the lightest, hydrogen (and even hydrogen, as we shall see, has variants or "isotopes" that contain neutrons).

Because the electron itself is much smaller in extent than the nucleus, the atom is by and large a volume within which the electrons move in particular patterns. If we scale up the nucleus of an atom so as to be represented by a child's marble, the volume within which the electron moves is then equivalent to a football stadium. We must also be careful about trying to define the exact "size" of an electron, since it exhibits the kind of wave–particle duality that was briefly defined earlier for the photon. Hence the cartoon notion of a discrete, negatively charged sphere orbiting a tiny but massive nucleus consisting of a positively charged sphere is so oversimplified as to be *wrong*.

Protons and neutrons are not elemental particles, in the sense of being indivisible. Experiments with ever-higher collisional velocities, and hence energy, confirm a decades-old model that elemental constituents of these particles exist. The so-called quarks come in six varieties, and they carry fractional quantities of charge. Different combinations of quarks form a set of different kinds of so-called hadrons (meaning "heavy ones"), of which the proton and neutron are the most stable and most abundant in nature. The electron itself is elemental and indivisible and is the most stable and abundant member of another class of subatomic particles called leptons (or "light ones" in Greek). There are, in fact, a respectable number of different kinds of subatomic particles, most of which occur in very exotic astrophysical situations or the interiors of accelerators. Their presence in theory and experiment is not just academic, however, because the properties of these particles in turn constrain models of the origin of the forces that shape the universe as we know it. In particular, we already noted that at very high temperatures at least three of the forces unify. And in Chapter 5, we discuss the origin and early evolution of the universe, which involves declining temperatures from an almost inconceivably hot beginning. How forces became separate and distinct from each other is a story whose clues are contained in the properties of some exotic subatomic particles.

Other subatomic particles hold clues to the inner workings of more common astrophysical objects. How our Sun generates energy by nuclear fusion is well understood, and the details of the process can be probed by detecting massless particles called neutrinos that are emitted in the process and, ghostlike, glide unimpeded through the Sun and through the Earth. On Earth, they pass through solid rock, unlike any other particle, and their signature can be filtered from that of more common particles by putting vats of pure elements (such as gallium) that *do* interact with neutrinos in former mines deep under the Earth. Such measurements allow the details of hydrogen fusion in the Sun to be probed on the Earth, and these play a crucial role in constraining the history of the Sun's energy output, which is of primary importance for the existence and maintenance of life on Earth.

2.4 Quantum Mechanics

We now return to electrons to discuss two other aspects of their properties that seem like exotica but are actually fundamental to their role in determining the bonding of atoms to each other by the rearrangements of electron positions around atoms during reactions—what we call chemistry. Both concepts are rooted in so-called quantum mechanics, a theory of the behavior of matter at microscopic scales that has been validated through increasingly sensitive laboratory experiments over nearly a century.

The first has to do with angular momentum, a discussion presaged briefly in the context of the intrinsic magnetism of certain elements and their compounds. *Linear momentum* is a vector quantity equivalent to the mass of a body multiplied by its velocity **v,** while—at least in the macroscopic world—*angular momentum* is defined in terms of motion "around" something else. Planets in orbit around the Sun possess angular momentum; so does a rock tied to a string that is being swung around by hand. (But remember Newtonian physics: Make the Sun disappear or let go of the string and both planet and rock move away from you in a straight line; the rock, of course, is eventually destined to follow a parabolic trajectory because of the pull of the Earth's gravity.) Things that spin on their own axis, like the Earth or a gyroscope, also possess angular momentum. Formally, we find the angular momentum by taking the vector cross product of the velocity and the radius of curvature about which the body is moving.

Quantum mechanical systems, those on the tiny size scale of atoms, also possess angular momentum, but the classical analogies of orbital and spin angular momentum—in spite of the use of these terms by quantum physicists—create somewhat misleading pictures of what is going on. The best way to see how misleading they are is to first visualize a classical scenario and then show how the microscopic "equivalents" are so very different. The classical scenario is a planet orbiting about a star, the subject of Chapter 15, and the exact situation of the Earth and the Sun.

Imagine a moving planet in an orbit (it could be circular or eccentric) around the star. The planet possesses an orbital angular momentum by virtue of its motion about the star, and the orbit itself defines a plane of symmetry that bisects, or cuts, the star in half (in an imaginary fashion). This defines a reference direction for the orbital angular momentum, namely, an axis that is perpendicular to the orbit plane (Figure 2.14). The spin angular momentum of the planet (the star has one too) is defined by the rotation of the planet around an axis drawn through the center (strictly, the center of mass) of the planet. This axis may, in general, cut through the orbital plane of the planet at any angle. Traditionally, 90 degrees minus this angle is called the "obliquity" of the planets' spin. Mercury's spin axis is perpendicular to the orbit plane, defining an obliquity of 0 degrees; that of Uranus is nearly parallel to the orbit plane, with an obliquity of 98 degrees, and Earth's obliquity is a bit over 23 degrees. Now a line drawn perpendicularly through the orbit plane of the planet defines for the spin axis a preferred direction in space; let us call that the *z-direction*. The *z*-component (a scalar) of the vector spin angular momentum of the planet is the *cosine* of the obliquity multiplied by the total spin angular momentum. For Mercury the *z*-component is numerically equal to the total momentum, since $cosine\ 0° = 1$; for Uranus it is nearly zero but is also slightly negative because the axis of spin tips below the plane of the orbit around the Sun. In consequence, Uranus rotates "backward" (or retrograde) relative to its revolution around the Sun.

This classical system defines all the directions we need in order to see why an analogous quantum mechanical system is so different. The quantum system of interest here is an electron and an atomic nucleus (in this case, say, a proton or a proton and neutron) to which it is bound. We focus on the electron, although the proton and neutron as a nucleus have analogous angular momentum states. The fundamental difference with the classical situation is that momentum, and energy, are quantized—that is, they can assume only cer-

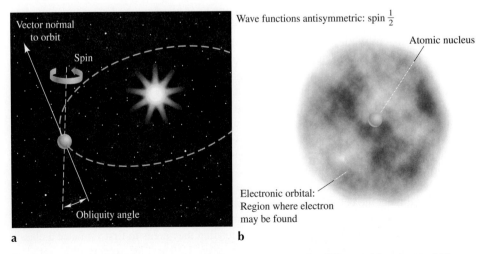

FIGURE 2.14 Classical and quantum angular momenta are quite different: (a) a planet orbiting a star, (b) a lone electron bound to an atomic nucleus. Scale of nucleus is greatly expanded.

tain, discrete values. For the situation we have described, this implies the following properties, which are explored more deeply in Chapter 3:

- While electrons do not follow classical orbits around nuclei, their paths are constrained to be in certain so-called shells that increase in energy as the mean position of the electrons moves progressively farther away from the nucleus. The shells are quantized in their allowable energies and are labeled by *principle quantum number, n.*
- The analogue to the orbital angular momentum of the planet is the *orbital angular momentum quantum number, l,* which defines the shape of the accommodating *orbitals*. However, the electron does not orbit in a steady, progressive, "planetary" path with a predictable position at any given time. Each orbital defines the shape of a three-dimensional region within which the electron is likely to be found, but the specific position and velocity of the electron within that orbital cannot be a priori predicted. This *uncertainty principle* of quantum mechanics is fundamental to the observable behavior of particles on atomic and subatomic scales. The orbital angular momentum quantum number l assumes only integer values constrained to be between zero and $n - 1$. The innermost shell, $n = 1$, can only have $l = 0$; the $n = 2$ shell can only have $l = 0$ or $l = 1$. Chemists abbreviate the $l = 0$ "subshell" the *s* subshell; the $l = 1$ orbital is called the *p* subshell, $l = 3$ the *d* subshell. The subshells are composed of one or more orbitals. An orbital with $l = 0$ describes a spherical volume; $l = 1$ describes three mutually orthogonal dumbbell-shaped orbitals, that is, three dumbbells each aligned along an orthogonal axis (Figure 2.15). Higher values of l display shapes that are progressively more intricate. The *orbital magnetic quantum number*, m_l, indexes the different possible spatial orientations of

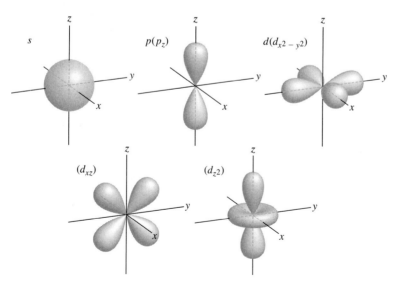

FIGURE 2.15 Shapes of the first three types of atomic orbitals; the shaded surfaces bound regions in which there is a 90 percent probability of finding the electron.

the orbitals. It describes the component of the orbital angular momentum of an electron in the direction of an applied magnetic field (let's identify this with the "z-direction"), and assumes integer values between $-l$ and $+l$. For example, since the p subshell has $l = 1$, there are three orientations corresponding to $m_l = -1, 0, 1$. The next subshell, the d subshell, corresponds to $l = 2$; hence it has five orbital orientations corresponding to $m_l = -2, -1, 0, 1, 2$. Subshells higher than p are progressively more difficult to visualize.

- Another quantum number, s, indexes the property of *spin*. As tempting as it may be to visualize this property as equivalent to the spin of a planet about its axis, there are fundamental differences. The electron is not like a spinning planet with an axis that can be defined relative to a convenient frame of reference, such as the planet's orbital plane. Define any arbitrary direction in space, and the z-component of the angular momentum of the electron relative to that axis will be the same as for any other axis. This sounds like nonsense, but it is not, because the so-called spin is an intrinsic property of the electron—or of whatever other type of subatomic particle we are studying. For a given spin quantum number s, there is a *spin magnetic quantum number*, m_s, analogous to m_l defined above, that can only take on values that are integer increments between s and $-s$. The electron is a "spin$\frac{1}{2}$" particle, that is, $s = \frac{1}{2}$ for the electron, and hence m_s can take on only two possible values: $-\frac{1}{2}$ and $+\frac{1}{2}$. For this reason it is convenient to think of the electron as either "spin-up" or "spin-down" when it is in any atomic or molecular orbital, even if it implies a mental picture of the electron that is not quite accurate.

- The progression of shells with increasing principal quantum number and their corresponding increase in number and diversity of orbitals is at the heart of our understanding of how elements chemically bond because *no more than two electrons—one spin-up and one spin-down—may occupy any individual orbital*. This was first articulated as the exclusion principle by German scientist Wolfgang Pauli in 1925. Particles with half-integer values of s, such as the electron, behave fundamentally differently from those that possess integer spin values. Half-integer spin particles are called "fermions," while those with integer spin are "bosons." Fermions include electrons and individual protons and neutrons. Fermions obey the exclusion principle—two particles may not occupy the same energy state if they also possess the same z-component of the angular momentum. Bosons (e.g., an atomic nucleus with an even number of protons and neutrons), on the other hand, can exist together in the same energy and spin state with no limit on the number of particles. Hence, atomic nuclei with multiple protons and neutrons are possible.

The strange quantum mechanical behaviors that characterize microscopic systems may seem to be arbitrary, even capricious, inventions of the human imagination. But they are, in fact, properties that come directly out of a logically coherent theoretical framework erected in the early 20th century and have been validated by experiment again and again. Indeed, it is the exclusion principle that accounts for essentially all of the variety of different properties that characterize materials from air to metals, from liquid water to living cells.

The view of matter underlying quantum mechanics is a mathematical one, in which particle behavior is governed by wavefunctions. *Wavefunctions* are mathematical constructs that keep track of the phase (position of the peak relative to some reference point) and amplitude of the waves that are associated with all particles, both micro- and macroscopic, and govern how waves interact with each other. Particles are discrete packets of waves, with some finite (rather than zero or infinite) extent, that interact with other waves corresponding either to matter or to forces (e.g., electromagnetic waves). They have the following property: Roughly speaking, the superposition of wavefunctions of two like particles will either cancel each other out, for half-integer spins, or reinforce each other, for integer spins. This corresponds, respectively, to the fermions, which cannot coexist in the same energy state, versus bosons, for which no such restriction holds.

Now just hold on, you might be saying by now. This section is on particles, and yet the author is talking about them as if they were waves. This brings us to an important and "exotic" concept of quantum mechanics, the wave–particle duality that was introduced earlier for the photon. It applies as well to all material particles, those that have mass, but the wavelike nature of matter is best observed in the behavior of particles of lesser mass, such as the electron.

One way to diagnose the presence of waves is through the demonstration of an interference pattern, as shown in the various setups in Figure 2.16. A backstop wall (with an appropriate detector mounted on it) sits behind a barrier screen into which two holes have been drilled, each just a bit larger in size than the diameter of some plastic pellets, which are to be fired at the screen from a pellet gun [Figure 2.16(a)]. Let the operator spray the barrier screen uniformly with bullets. Most will strike the barrier screen and ricochet off. Some will make it through the holes and strike the backstop wall beyond. A smaller number will nick the sides of the holes and be scattered at various angles onto the backstop wall. A separate count for each of the two holes (made with the other one covered) would give impact distributions shown by the two symmetric curves p_1 and p_2 for the two holes. The composite impact distribution p_{12}, observed with both holes open, is a single smooth curve that is approximately the sum of the curves for the two individual holes.

We now repeat the experiment with water waves [Figure 2.16(b)]. We place our barrier with the two holes in the water, with a screen behind that is capable of recording the arrival of waves (characterized by the up and down motion of the water). A wave generator is set up to the left of the barrier, perhaps a bar that moves up and down, and the frequency of motion of the bar is set so that the wavelength (crest to crest) is comparable to or larger than the slit width. Now cover each hole in turn to measure the distribution of wave energy lapping against the detector. We find that the bulk of the wave energy strikes the screen directly behind the open hole, with a lesser tail that extends on each side, just as in the case of the pellets. But when both holes are open, the distribution of the wave energy on the screen looks nothing like the sum of the two separate distributions for each of the single holes [Figure 2.16(b)]. It is, instead, a series of peaks and troughs called an *interference pattern* or *interferogram*. What has happened is that each of the two small holes becomes the source point for a new wave pattern, of identical wavelength to the incident waves, but now emanating as circular wavefronts from the hole. The original wave has been *diffracted* at each hole. The two new waveforms intersect each other and interfere, superposing their wave energy and constructively interfering at some points and destructively interfering at others. The result on the detector is an intensity map of the angular distribution of regions of constructive and destructive interference. The same pattern can be observed using electromagnetic waves, but to do so with visible light requires that the holes

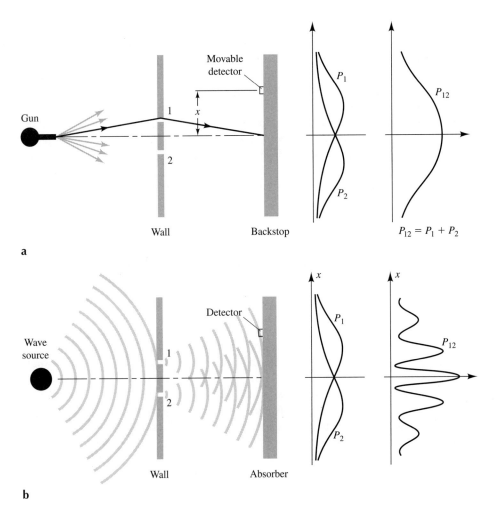

FIGURE 2.16 A two slit interference experiment: (a) with pellets, (b) with water waves, (c) with electrons, and (d) with electrons whose paths are being analyzed through the scattering of light. P is the intensity of the signal at the detector.

be extremely tiny (on the order of the wavelength of visible light, typically 500 billionths of a meter).

Now we return to the electrons [Figure 2.16(c)]. Let's rescale our experiment for these tiny objects, each about 3×10^{-15} m in diameter, which we might expect to behave as the pellets of part (a). (An important caveat is that only recently has it been possible to manipulate electrons with the precision needed to observe directly the results described. The results reported here were well known, however, from many decades of cruder experiments that inferred less directly the same outcome.) To "spray" electrons on the barrier we electrically heat a metal wire (say, tungsten), within which some of the electrons are relatively mobile, surround it with a metal foil, evacuate the air, and apply a voltage between the wire and the surrounding metal enclosure. Electrons will be accelerated off the wire and will strike the inside of the metal enclosure. Let us put a hole in the enclosure so that electrons will shoot out in a well-defined direction, again into a vacuum. Finally, we move our electron gun back and forth to spray the barrier screen uniformly with electrons.

Again, we cover one hole and measure the distribution of electron hits from the other hole on the rear detector screen. (The detector used could be a phosphor screen that emits a point flash and afterglow at each electron-impact site). As shown in Figure 2.16(c), the pattern for each separate hole looks like the pattern for the pellet gun, and for that matter, the pattern for the water waves. Now uncover both holes. Like the pellet gun, we expect that the aggregate distribution will be simply the sum of that for the two separate holes and

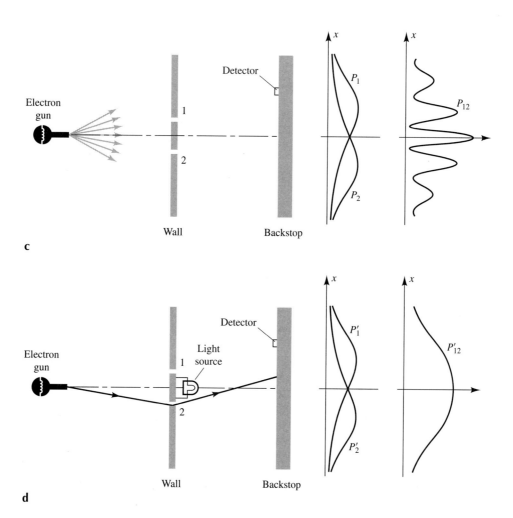

will be a broad smooth curve. But it is not. Instead, the pattern looks precisely like that for the water waves. Perhaps the electrons have interfered with each other as they pass through the holes, in the manner of waves? Perhaps electrons are not particles after all?

We can test this idea by slowing down the electron gun so that it emits just one electron at a time [Figure 2.16(d)]. Each electron is then allowed to travel toward the barrier screen, where it is either reflected away or able to pass through toward the detector. An important point is that we do not attempt to track the electrons in flight; we simply release them slowly by moderating the heating on the wire or the voltage in the box. By doing so, it is clear that what is traveling from the gun to the barrier, and sometimes through to the detector, are particles—or at least lumps of some kind. Yet, the result is still the interference pattern akin to what electromagnetic waves or water waves produce. Hence, the electron is behaving simultaneously as a particle and as a wave. Single electrons fired one at a time (so they do not interact with other electrons) define the interference pattern—a single electron has a wavelike spatial extent that allows it to trace out a portion of the interference pattern, and to interact simultaneously with both holes!

There is no guarantee provided to humankind that nature will look the same at all spatial scales, allowing us to transfer our concepts from the everyday human-scale world to the world of fundamental particles, some 15 orders of magnitude smaller. The ideas of "waves" and "particles" are intuitive for us because we see the waves on a lake and can hold discrete particles of sand in our hands. But these are not fundamental concepts. Both the water and the sand are aggregates of enormous numbers of incredibly small constructs that carry all the properties we perceive in the macroscopic world—mass, electric charge, velocity, etc. But they need not be classifiable in terms of the words we have invented for our own narrow grasp of scale in the cosmos. Indeed, we can rationalize what the interference experiment demonstrated by talking of electrons as "wavepackets" of something with a restricted extent in space. But what is that something, and what does it look like? Modern techniques of imaging electrons and atoms can give us a glimpse into this microscopic universe, revealing both the particle and wavelike nature of a single electron, but still translated into a format that our own eyes and brain can understand (Figure 2.17).

The fact that the electron is not located at a "point" in space but is also not infinite in extent, unlike an expanding electromagnetic wave, is evident from Figure 2.17. How precisely well known is the location of the electron? For material particles, the ultimate precision with which one can determine the position and speed of the wavepacket that is the particle is given by the Heisenberg uncertainty principle of quantum mechanics, $\Delta x \Delta p \geq h/4\pi$. Here Δx is the uncertainty in position in a given direction, Δp the uncertainty in momentum of the particle in that direction, h is Planck's constant, 6.63×10^{-34} joule-seconds (where the joule is the mks unit of force), and π is the ratio of the circumference to the diameter of a sphere, 3.14159.... The more precisely we measure the position of an electron, for example, the more poorly we know its momentum (and hence velocity), and vice versa. Imagining now the wavepacket that is the electron moving around an atom, it is clear that while we can define probabilities of where the electron will be and what its momentum is (which are related to its residency in a particular orbital), we cannot make a motion picture that traces the position of that electron with time. There is a quantum mechanical coarseness to the universe at microscopic levels, defined in terms of position and momentum (or, from the units of h, energy and time), and Planck's constant gives the size of the coarseness. Because h is a small number, well determined by experi-

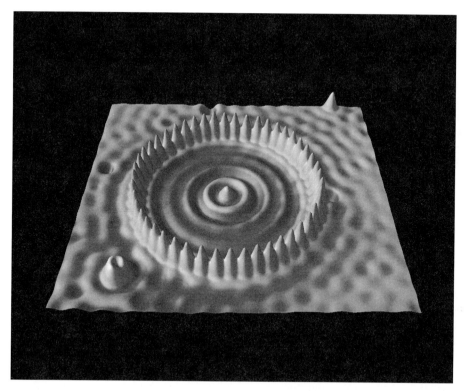

FIGURE 2.17 Image of an electron trapped in a ring of iron atoms. The electron is not the peak in the center; it is the peak plus the pattern of waves that continues beyond the corral, though in much weaker form. The ring of iron atoms acts as a series of slits through which there is some probability that the electron can pass; the interference pattern beyond the slits is the electron interfering with itself as it is diffracted around the slits. The probability that the electron is in a given location is determined by the height of the peaks; it is roughly uniform within the ring and vastly smaller beyond.

ment, the uncertainty principle becomes progressively less important for progressively more massive particles, or for larger populations of particles in a given system. Related to this is the fact that the wavelike properties of, say, a proton or a neutron are much less evident than those of the electron. This is especially important in chemistry, where chemical change is understood in terms of strongly localized (particulate) atomic nuclei, forming the frames of reference for the electronic structures of atoms, ions, and molecules.

It is now tempting to think that we might chase down individual electrons as they find their way through the holes in our interference experiment of Figure 2.16, albeit only coarsely, in accordance with the uncertainty principle. We might try to catch the electrons in the act of extending and contracting as wavepackets interfere with themselves, perhaps by laser light (which is narrow in wavelength and well collimated) scattered off the crests and troughs. The lasers are swept rapidly so that the light cannot miss being scattered off electrons as they move beyond the slits. We confirm, by holding one hole closed, that the resulting pattern is the simple single-slit distribution [Figure 2.16(d)]. But when both holes are open, *no* interference pattern is observed on the detector screen! The interference pattern has been replaced with the pellet-like broad distribution representing the simple sum of the two separate distributions produced by the individual slits. And yet, as soon as we turn off the laser, the pattern on the detector screen rebuilds into the interference pattern [Figure 2.16(c)] associated with the wave. The act of observation has altered the experiment. In quantum mechanical terminology, by allowing photons from the laser to interact with the electron, we have "collapsed" the electronic wavefunction, and for that time the electron behaves perfectly and predictably as a discrete particle.

Wavefunction collapse remains a controversial aspect of quantum mechanics some 80 years after the field was founded by German scientist Ernst Schröedinger. Clearly, as Figure 2.17 shows, we can glimpse the electron in the act of being a wavepacket. But countless experiments also exhibit the results depicted in the thought experiment, that the intrusion of photons into the process, as probe detectors, caused the wavelike electrons to exhibit only particle-like properties. Part of the controversy lies in some of the paradoxes that wavefunction collapse necessarily implies for information transfer, including the notion that the wavefunction collapse of one particle can be detected by another particle immediately, without the delay associated with the transfer of information at the speed of light. Recent experiments seem to suggest that this is the case, although they are preliminary. Is relativity violated if we imagine the wavefunction of a particle spread over such an extent of space that it overlaps with another? This unresolved conundrum is a cousin of the mismatch we encountered earlier between the gravitational force as an intrinsic property of the geometry of the cosmos and the other three forces of nature as being interactions between particles mediated by other transient, or "virtual," particles.

2.5 The Conservation of Mass-Energy

Enough basic physics has been presented at this point to provide conscientious readers with the background required to move forward to consideration of the role of the electron in chemical bonding and the interaction of electrons with electromagnetic radiation to generate atomic (and molecular) electromagnetic spectra. These are the subjects of Chapter 3. The physics discussed here will also help in the discussions of the unfolding of the universe from an opaque soup of subatomic particles to the realm of the galaxies we see today, to the formation of planets, and to the behavior of molecules within cells. To complete this treatment, however, requires at least a mention, however brief, of the fundamental concept of the conservation of mass-energy that will thread its way through most of the chapters to come.

Energy is neither created nor destroyed but only transformed from one kind to another. Energy can be divided into two basic forms: energy of motion (or kinetic energy) and stored (or potential) energy. Kinetic energy is straightforward to visualize; a speeding train and falling brick both possess it. Quantitatively, kinetic energy is $\frac{1}{2}mv^2$, where m is mass and v the scalar magnitude of the vector quantity velocity. Thus, when a car doubles its speed, its kinetic energy quadruples; this is why the destructiveness of automobile accidents increases so dramatically with speed.

Potential energy is stored energy. The storage medium might be certain chemical compounds that tend (under the right conditions) to react in such a way as to release heat or exert pressure on a container; a mixture of oxygen and gasoline is one example relevant to cars, and the phosphate bond in adenosine triphosphate (Chapter 4) is another relevant to cellular processes. Storage of energy can also involve placement of material in a force field that, once a constraint is removed, allows the potential energy to be converted to kinetic energy. A brick suspended above the ground by a string tied to the ceiling possesses potential energy that, after the string is cut, is progressively converted to kinetic energy as the brick accelerates toward the ground (strictly, the brick and the Earth accelerate toward each other). Another form of potential energy is matter itself; mass is equivalent to stored energy and a given mass m can be completely converted to energy according to Einstein's famous formula, $E = mc^2$ where c is the speed of light.

For this reason, a more general statement of the conservation of energy is this: The sum of all forms of an isolated system's energy and the energy equivalent of this system's matter is conserved. When chemical reactions occur, the total mass of the reactants is slightly different from that of the products—the increase or decrease in mass corresponding to a decrease or increase, respectively, in the (electromagnetic) binding energy of the electrons to the atoms. Nuclear reactions also convert some amount of nuclear binding energy into mass, or vice versa, according to the Einstein formula (Chapter 5). Wholesale conversion of matter into energy occurs in the annihilation of matter by antimatter. Antimatter is the equivalent of matter but with all the electric charges reversed. For example, a particle equal in mass to the electron and identical in its spin angular momentum (half, hence, fermionic), but positively charged, is the positron (e^+). It is possible to transiently bond an electron and positron based on the attraction of the charges, but when the two move into close proximity, they annihilate each other with a release of energy given by Einstein's formula. The preponderance of matter in the universe, versus antimatter, seems to be well established observationally and is probably a stochastic outcome of the formation of the cosmos. There would seem to be little or no reason the universe could not have started with a preponderance of antimatter, annihilating the small amount of matter present early on, leading to biochemical systems whose chemistry would have been determined by the positron.

Conservation of energy in terms of potential and kinetic forms is a sufficiently basic concept that little more need be said about it. One subtlety is that, in dissipative systems (e.g., where friction is present), energy tends to cascade down from large-scale, organized motion to random microscopic motion of atoms or molecules. This is a natural consequence of the second law of thermodynamics (Chapter 7), but it can mislead us into thinking that energy has been somehow "lost." Random thermal motions of microscopic particles are a form of kinetic energy and can be measured through the temperature and heat capacity (thermal energy content per unit temperature rise) of a body.

Finally, temperature scales are important in this book. The Fahrenheit scale is commonly used but is awkward scientifically, because 0°F doesn't correspond to a natural physical process such as the freezing of water. The Celsius scale, used in most of the world, is somewhat more sensible. Water freezes at nearly 0°C, and boils at 100°C. A more scientifically useful scale, called the Kelvin scale, will be used here. Kelvin is the Celsius scale displaced by 273 degrees, so that 0 K is absolute zero—the theoretical temperature limit at which systems have no thermal energy within the quantum uncertainty limit. Thus, water boils at 373 K (the degree sign is not used in the Kelvin scale), and the freezing point of water lies at 273 K.

PROBLEMS

1. The quarks that make up protons and neutrons are of two varieties, whimsically called "up" and "down." Knowing that protons are made of two up and one down quark, while neutrons are made of two down and one up quark, work out the fractional charges (magnitude relative to the electron and the sign) of the two quarks.

2. Detail in narrative form or with a flowchart the energy transformations involved in going from thermonuclear fusion in the Sun to your raising of a brick 2 meters above the floor. You may need to skip ahead in Chapters 4 and 5 to identify some of the sources or sinks of energy. Be sure to identify which types of energy are kinetic and which are potential.

3. Centripetal acceleration is just the v^2/R, where v is the velocity of the spinning object and R is the radial length about which the object is rotating. Hence, for the equatorial surface of the Earth, R is just the radius of our planet and v is just the circumference ($2\pi \times R$) divided by the time to make one revolution, 24 hours. (At higher latitudes, you must apply some trigonometry because the circumference over which the motion occurs declines; at the poles, the centripetal acceleration is zero.) Now imagine that you are at a tropical retreat on the equator, and gravity is suddenly turned off. All of the objects at the beach bar are nailed down except your drink. At what speed and in what direction does the glass move for the first second of this "impossible" experiment?

4. Calculate the ratio of the electrostatic force to the gravitational force between two electrons.

5. a. Compare the force of gravity exerted by the Earth to that of the Moon on a person standing at the surface of the Earth.

    b. At what distance between the Earth and the Moon would the two forces be equal?

6. A lightweight and heavyweight body have equal kinetic energies of translation (uniform motion in a single direction). Which body has the larger momentum?

7. By balancing forces, calculate the periods of rotation around the Sun of each of the planets of our solar system using the fact that Earth's period of rotation is one year and we are 150 million kilometers (1.5×10^{11} meters) from the Sun.

8. Imagine you are fixed in an inertial reference frame and a meter-stick moves past you at a uniform speed of 100,000 kilometers per second. Does it appear to you to be a meter in length? How much shorter is it?

9. Imagine you are explaining quantum mechanics to a popular audience. Write down how you might explain the wave–particle duality of matter to such a group and, in particular, what analogies you might use.

10. The hottest air temperature ever measured by thermometer on Earth is 134°F at Death Valley, California. Convert this to Celsius and Kelvin. The surface temperature on Saturn's moon Titan is 95 K; what is this in Celsius and Fahrenheit?

SUGGESTED READINGS

Books on modern physics written for the layman constitute an enduring industry, and as long as you select from a list of credible authors, the choice of reference depends on your literary taste. For a well-written discussion of relativity, and some quantum mechanics, couched in the issue of the nature of time, read P. Coveney and R. Highfield, *The Arrow of Time* (1991, Harper Collins, New York). Likewise, more formal textbooks on physics cater to a range of tastes and capability, and it is difficult to recommend one. I still treasure the two-volume Halliday and Resnick, *Physics I and II*, from the 1960s, but later editions are quite different in style. A. B. Arons, *A Guide to Introductory Physics Teaching* (1991, Wiley and Sons, New York) is a fascinating look at misconceptions students bring to, or acquire in, introductory physics classes. The interference experiment is taken from a longer narrative description by Feynman in R. P. Feynman, R. B. Leighton, and M. Sands, *The Feynman Lectures on Physics, Vol. III, Quantum Mechanics* (1965, Addison Wesley, Reading, MA). It is worth going back to the original lecture in Volume III to pick up some additional details and nuances on the experiment—although because of the age of the book, the caveats on experimental capabilities and limitations are outdated.

REFERENCES

Arons, A. B. (1990). *A Guide to Introductory Physics Teaching*, Wiley and Sons, New York.

Atkins, P. W. (1986). *Physical Chemistry*, W. H. Freeman and Co. New York (third edition).

Cox, A. N. (2000). *Allen's Astrophysical Quantities*, American Institute of Physics Press, New York.

Crommie, M. F., Lutz, C. P. and Eigler, D. M. (1993). "Confinement of Electrons to Quantum Corrals on a Metal Surface," *Sci.* **262,** 218–220.

Feynman, R. P., Leighton, R. B., and Sands, M. (1965). *The Feynman Lectures on Physics: Volumes II and III,* Addison-Wesley, Reading, MA.

Goldstein, H. (1983). *Classical Mechanics*, Addison Wesley, Reading, MA.

Jacobs, A. (1997). *Understanding Organic Reaction Mechanisms*, Cambridge University Press, Cambridge.

Lorraine, F. and Corson, D. (1970). *Electromagnetic Fields and Waves,* W. H. Freeman and Co., San Francisco.

CHAPTER 3

Essential Concepts II
The Physics of Chemistry

3.1 Introduction

Armed with the brief introduction to forces, particles, and quantum mechanics provided in Chapter 2, we are ready to think about the behavior of electrons as they are shared or transferred among atoms, thus creating the enormous variety of substances in the physical universe. We first consider mechanisms of bonding and the resulting systematics when elements combine that is summarized in the "periodic table." We then discuss how electrons that are either free from or bound to atoms and molecules absorb and emit electromagnetic radiation, which is at the crux of the very general and very powerful diagnostic tool called spectroscopy. We close with a summary of how changes in the mass of an element—with no change in its charge state—affect the properties of materials, as a prelude for understanding in Chapter 5 how different mass flavors, or "isotopes" of the elements, can be used to track cosmic and terrestrial history.

The overview of chemistry here sets the stage for subsequent discussions, particularly Chapter 4, on the chemistry of biological processes, and Chapters 5 and 6, in which the origin and evolution of the universe and the early history of the solar system, derived from spectroscopic observations and isotopic measurements, are considered.

3.2 Quantum Mechanics and the Electron

Chapter 2 introduced the concept of energy conservation, and this is our starting point for understanding the behavior of electrons bound to atoms. Because energy is conserved, we can write that the potential, V, plus kinetic energy of a system is constant and that constant is the total energy, E, of the system. The kinetic energy is usually written as $\frac{1}{2}mv^2$, but an alternative form is in terms of momentum $p = mv$, so that the kinetic energy is $p^2/2m$. Then the conservation of energy is

$$\frac{p^2}{2m} + V = E \qquad (3.1)$$

This equation should, in principle, be able to guide us regarding the motion of an electron around the nucleus of an atom, because the electron is bound by the attractive electromagnetic force of the positively charged nucleus and possesses kinetic energy by virtue of its motion around that nucleus. In practice, however, application of equation (3.1) to the electron does not work because there is no provision for the wavelike nature of this subatomic particle. In quantum mechanics, the behavior of a particle is described by a quantity called the *wavefunction*, usually symbolized by ψ, which obeys the following equation (in one dimension characterized by distance x):

$$\left(\frac{-\hbar^2}{2m}\right)\left(\frac{d^2\psi}{dx^2}\right) + V\psi = E\psi \qquad (3.2)$$

Here $d^2\psi/dx^2$ is the second derivative of the wavefunction with respect to the distance, x, and $\hbar$ is Planck's constant divided by 2π. The "denominator" of the derivative is in fact $(dx)^2$ where dx is a differential, that is, infinitesimally small, increment of the linear distance x. This equation looks like the classical one for conservation of energy if we argue that $-\hbar^2/(dx)^2$ has units of momentum squared (mass squared times velocity squared). Because Planck's $\hbar$ has units of energy times time, which is mass times distance squared divided by time, $-\hbar^2/(dx)^2$ has the appropriate units. The subtlety here, however, is that the first term in equation (3.2) denotes a mathematical operation on the wavefunction: Take its derivative twice with respect to distance and then multiply by $(-\hbar^2/2m)$. If we denote $(-\hbar^2/2m)\, d^2/dx^2 + V$ by H, which is called the Hamiltonian, our energy conservation equation becomes $H\psi = E\psi$, which means "perform the operation on the wavefunction specified by the Hamiltonian to obtain the total energy of the system times the wavefunction." Thus, this prescription defines for us the functional form of the wavefunction, because in general just any function wouldn't satisfy the prescription. Equation (3.2) is the *Schrödinger equation*, which in various guises (i.e., one-, two-, or three-dimensional; spherical or Cartesian coordinates; time-dependent or independent; etc.) prescribes the wavefunction in a certain force field (described by the potential energy) and total energy.

So, just what is the wavefunction that we obtain? The wavefunction is a mathematical construct designed to describe the wavelike nature of particles. If we think of an electromagnetic wave, the amplitude of that wave squared is the intensity of energy associated with the wave—that is, the number of photons (elemental particles of light) present. And, indeed, for material particles, the square of the wavefunction is just the probability of finding the particle at any particular point x in space (multiplied by a numerical factor, which is easily normalized out). While the wavefunction itself is hard to visualize physically, the square of it at each point in space tells us where the electron is more or less likely to be—not where it *is*, because the uncertainty principle and the electron's wavelike nature dictate that we cannot precisely specify a position, but a map of probabilities where it might be found.

Now the most interesting property of the wavefunction is that two particles cannot coexist in arbitrarily close regions of space without affecting each other in some way. As the interference experiment described in Chapter 2 shows, the wavelike nature of subatomic particles dictates that the momentum and position of two particles are coupled to each other. If one electron is described by a wavefunction in three coordinates (corresponding to the three spatial coordinates), then the presence of two electrons must be described by a wavefunction with six coordinates (three for each of the two electrons). But the electrons themselves are not labeled—that is, one electron does not have a barcode that distinguishes it from the other electron. Therefore, we should be able to exchange the two electrons and come up with the same physical solution we had before. And we can—with a catch. The physical distribution of positions of the electrons is given by the square of the wavefunction. In interchanging the two electrons in our solution, it would be permissible for the wavefunction to change sign (become negative where it was previously positive); the square of the wavefunction would be unaltered. And for fermions (spin-$\frac{1}{2}$ particles) it is a principle of quantum mechanics defined in Chapter 2 (the Pauli exclusion principle) that the wavefunction *must* change sign under an interchange of the two particles. (For bosons, or integer-spin particles, the wavefunction does not change sign.)

The change-of-sign requirement severely restricts the situations under which electrons can coexist around an atom. We can solve the Schrödinger equation for electrons around atoms to obtain maps of their probability distributions. These distributions are quantized according to energy and angular momentum as defined and described in Chapter 2: The principal quantum number, n, defines a sequence of shells corresponding to progression in energy and can take integer values from 1 upward. The orbital angular momentum quantum number l can take any integer value from 0 to $(n-1)$, and the orbital magnetic quantum number m_l can assume any integer value between $-l$ and l. The spin quantum number, s, for electrons is $\frac{1}{2}$, hence the spin magnetic quantum number, m_s, is either $+\frac{1}{2}$ or $-\frac{1}{2}$. Because these numbers fully define the state of an electron around an atom, *two electrons with the same quantum numbers* will have identical probability distributions given by the square of their wavefunctions. But if we in fact put two electrons into the same set of quantum numbers, the Pauli exclusion principle says that the wavefunction must change sign under an interchange of the two particles, which requires that one electron have $m_s = +\frac{1}{2}$ and the other $m_s = -\frac{1}{2}$. Beyond these two, we cannot add an additional electron because it would have the same spin magnetic quantum number as one of the other two electrons, and the three-particle wavefunction would not change sign under an interchange of those two electrons.

An atomic orbital, defined by the quantum numbers n, l, and m_l, can contain no more than two electrons. This rule, derived from the quantization of energy and angular momentum and the Pauli exclusion principle, defines a set of numbers for filling shells by electrons. When combined with two other rules, it defines the overall pattern of bonding among different atomic elements and creates the concept of *chemical periodicity*. These rules are as follows:

1. Electrons tend to fill the innermost subshell (where a subshell is defined by n, l and innermost denotes the lowest values for these quantum numbers) consistent with the exclusion principle. This rule arises because, for multielectron atoms, the lower values of n and l have smaller energy, and hence electrons will fill these first. (In a single electron atom, namely hydrogen, the energy depends only on n and not on l.)

However, as we shall see, there are important exceptions to this pattern associated with inversions in the dependence of energy on the shell number n.

2. Atoms adopt electron arrangements with the greatest number of unpaired electrons consistent with the first rule, which means that electrons will not pair in the same n, l subshell until other subshells in the same n shell are occupied.

Table 3.1 lists the progression of subshells, according to the quantization rules and the exclusion principle given earlier, while Figure 3.1 shows the probability distributions corresponding to selected orbitals. The general progression of number of electrons per shell can be obtained by memorizing the quantization rules, or by noting that the progression from s, p, d, f corresponds to a progression from a scalar, to a vector, and thence to tensors. A scalar has no direction in space; the spherical shape of the s orbital is completely

TABLE 3.1 The Progression of Electronic Orbitals

n Shell	l [1]	m_l [2]	Electrons to Fill	Total per Shell	Shell Type [3]
1	1 (denoted s)	0	2	2	Scalar
2	1 (s)	0	2		Scalar
	2 (denoted p)	−1, 0, 1	6	8	Vector
3	1 (s)	0	2		Scalar
	2 (p)	−1, 0, 1	6		Vector
	3 (denoted d)	−2, −1, 0, 1, 2	10	18	Tensor (2nd order)
4	1 (s)	0	2		Scalar
	2 (p)	−1, 0, 1	6		Vector
	3 (d)	−2, −1, 0, 1, 2	10		Tensor (2nd order)
	4 (denoted f)	−3, −2, −1, 0, 1, 2, 3	14	32	Tensor (3rd order)
5	1 (s)	0	2		Scalar
	2 (p)	−1, 0, 1	6		Vector
	3 (d)	−2, −1, 0, 1, 2	10		Tensor (2nd order)
	4 (f)	−3, −2, −1, 0, 1, 2, 3	14		Tensor (3rd order)
	5 (denoted g)	−4, −3, −2, −1, 0, 1, 2, 3, 4	18	50	Tensor (4th order)

Notes:
[1] l is the orbital angular momentum quantum number while s, p, d, f are the designations for subshells.
[2] m_l is the magnetic quantum number.
[3] The terms listed mean the following:
  - One lobe (spherically symmetric) can be represented as a scalar quantity.
  - Two lobes can be represented as a vector quantity (lobes are directional along axes).
  - Three or more lobes can be represented as tensor quantities.

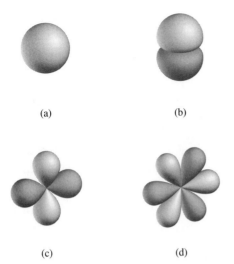

FIGURE 3.1 Selected atomic orbitals. (a) $n = 1, l = 0$ for hydrogen. (b) $n = 2, l = 1, m_l = 1$ for carbon (one of three orthogonal orbitals). (c) $n = 3, l = 3, m_l = -2$ (the five degenerate 3d orbitals) for iron. (d) $n = 5, l = 4$ for uranium-238. The dark and light shading represent the two different phases (signs) of the wavefunction. The surfaces trace out a particular value of the probability density, within which the probability of finding the electron is close to one. Orbitals are not depicted on the same scale.

symmetric (looks the same under any rotation). A vector is indexed by a number indicating direction; the three p orbitals each correspond to an orthogonal direction in space (e.g., p_x, p_y, p_z) and in total hold six electrons (three spin up and three spin down). Two or more numbers index a tensor; the d orbital components each require the specification of two spatial directions (d_{xy}, etc.), and there are five orthogonal combinations, each of which can hold two electrons. The f orbitals require three indexes.

With this model in mind, we can now define what a chemical element is. A *chemical element* (or, more simply, *element*) is defined by the number of protons in the nucleus (the *atomic number*), which is equal to the number of electrons filling the shells when the atom is electrically neutral. As we discuss next, the progressive filling of subshells by electrons creates a predictable periodic structure to the properties of elements that is remarkably simple yet yields the enormous diversity of material properties in the cosmos. Were the electron to behave other than as a fermion—for example, as a boson—the universe would be a completely different place. All electrons would, in the absence of excitation by photons, occupy the 1s subshell—the innermost shell of the electronic structure—regardless of the number of electrons already present. The bonding between atoms to form molecules would be much weaker and less selective. The corresponding diversity in properties of the elements would be vastly smaller and the variety of molecular forms and properties much simpler.

3.3 The Periodic Table

The progression with increasing element number Z in the number of electrons required to fill atomic subshells implies a periodicity in the chemical behavior of the elements. The sharing or transfer of electrons among atoms leads, in some cases, to a decrease in the potential energy of the system, and hence such *bonding* occurs spontaneously (see Chapter 7 for a discussion of spontaneous changes in the context of thermodynamics). But the individual orbital components of the subshells will never accommodate more than two electrons, as codified in the exclusion principle, and hence there are definite patterns of bonding among elements associated with the extent to which their subshells are already filled with electrons.

Hydrogen is an atom with a single proton, hence a single electron, which in its ground state (unexcited by photons) resides in the $1s$ subshell. To indicate that there is a single electron present, let's write $1s^1$ for the hydrogen subshell. Were two hydrogen atoms to associate with each other in such a way as to merge their two s shells and hence share electrons, each would have "filled" its s shell. This $1s^2$ assemblage is called an H_2 molecule.

A hydrogen atom can associate with other elements, too, and the number of hydrogens relative to the other element depends on the electron number of the other element. Fluorine has nine protons, and thus nine electrons, which means it needs just one more to complete its $2p$ shell. We would write its subshell designation as $1s^2 2s^2 2p^5$. But in terms of bonding to other atoms, only one particular shell is active. This is called the "valence" shell; it is usually (but not always) the outermost shell. The inner (core) shells, when filled, exhibit a total probability distribution that is spherically symmetric and has zero orbital angular momentum. This is an extremely stable situation, thus, the filled shells play only an extremely small role in chemical interactions. So for fluorine, just the $2s^2 2p^5$ information is relevant. To complete the shell requires adding an electron, which is supplied by hydrogen. But because hydrogen only has one electron, removing it is energetically expensive, and the "donation" is in the form of sharing in the composite atomic structure, or molecule, of hydrogen fluoride (HF). Oxygen, with eight electrons on the other hand, is $2s^2 2p^4$ and requires two electrons to fill its p subshell; an appropriate composite structure is two hydrogens and an oxygen—H_2O—which we call water. But water is not the only triatomic molecule made with two hydrogens. The element sulfur, with 16 electrons, is $3s^2 3p^4$ and forms the very stable molecule H_2S (hydrogen sulfide) to complete its p subshell.

Some elements do not chemically bond at all or do so with great difficulty. These elements have just the right number of electrons to fully fill their shells. Helium, with two electrons, has a filled $n = 1$ shell; neon with 10 electrons has a filled $n = 2$ shell, argon with 18 electrons has filled the s and p subshells of the $n = 3$ shell, krypton with 36 electrons has filled the s and p subshells of the $n = 4$ shell, and another 18 electrons brings us to xenon with similarly filled s and p subshells of the $n = 5$ shell. Interestingly, xenon does bond with some other elements under certain conditions, indicating that closed shells of high principal quantum number (in essence, far from the nucleus) will interact with electronic orbitals of other elements.

Systematic and repeatable bonding properties in progressively heavier elements were recognized by the late 18th century, and by the late 19th century Russian chemist D. Mendeleev had succeeded in constructing a "periodic table" of the 60 or so elements known at the time. The concept of atomic orbitals and repeating sequences of subshells had not yet been conceived, so the arrangement on the table was empirical, but so accurate that properties of elements could be predicted and, indeed, the existence of yet undetected elements proposed. In this way, a noble gas lighter than neon was suggested years before spectroscopy of the Sun (discussed below) revealed spectral lines not previously seen in the lab, which were eventually ascribed to the new element named helium as discussed in Chapter 1.

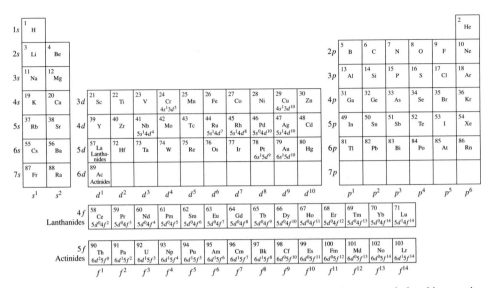

FIGURE 3.2 Periodic table of the elements. Within each box is the element symbol and its atomic number. By each row are the labels for the particular subshells being filled at each column–row intersection, and under each column is the superscripted number of electrons actually present in the valence subshell. Hence d^4 means that four electrons are in the d valence subshell. Where an element is filling a shell other than that given by the column marker, the subshell notation for the shell being filled and the outer shell are indicated. Table 3.3 gives the element names corresponding to the symbols.

A version of the periodic table of the elements is shown in Figure 3.2, in which each of the element names, atomic number, and subshell sequence are given. The overall geometry of the table corresponds to s-subshell filling on the left and p-subshell filling on the right. Completed (closed) shells line the extreme right of the table. Each row represents the filling of orbitals of successive principal quantum numbers. Beyond the row corresponding to $n = 3$, the d shells come into the picture, and the so-called transition elements corresponding to their progressive filling stretch between the s and p columns. All seems regular until we examine the detailed progression of subshell filling in row 4. First, the $3d$ orbitals are filled *after* the $4s$ orbital, but here and there, an electron "slips" from the $4s$ orbital into the $3d$. This offset is repeated in row 5, but a new novelty arises there: An entire row of elements occurs—the "lanthanides," compressed into a single box in the transition elements section, in which the $4f$ shell is filled after the $6s$ orbital. And this pattern is repeated in row 6, in which the lone transition "element" is a box that unfolds into the "actinide" elements, characterized by the filling of the $5f$ orbitals after the $7s$ orbitals.

What is going on here? Is the periodic table only periodic for the s and p subshells? Figure 3.3 and Table 3.2 illuminate what is going on. In general, subshells with lower energy fill before those of higher energy. Were we only to worry about the energy associated with the principal quantum number n, there would be a monotonic progression of energy from the first shell through the seventh. But the energy depends also on the orbital angular momentum number l, except for atoms with a single electron. Examining Figure 3.3, we

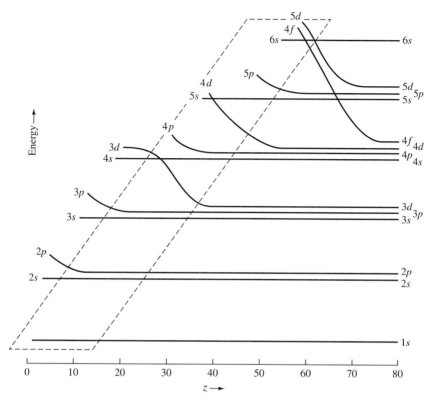

FIGURE 3.3 Ordering of the relative energies of the different subshells as a function of Z, the proton or atomic number. For a given atomic number, shells and orbitals will fill from the bottom up. Diagonal dashed region traces the progression in Table 3.2.

can conclude that for a given value of n, the outer subshell with the lowest l has the lowest energy, and for a given l, the outer subshell with the lowest value of n has the smallest energy. But if the energy variation of l is large, there can be overlaps between a shell with the highest n and a low l, and a high l subshell from the next inner or second inner shell ($n - 1$ or $n - 2$). The ordering also depends on the atomic number Z, so the dependence can get quite complicated. But the essence of the story is easily seen from inspection of Figure 3.3—the difference in energy from one subshell to another declines with increasing n and especially with increasing l. And in lighter (lower Z) atoms, there is a smaller attractive positive electromagnetic force on the electrons; hence, increasing the orbital angular momentum in these must intuitively have a major effect on the energy.

The resulting pattern is a basic template of universal chemical behavior given to us by the properties of the electromagnetic force and the quantum behavior of the electron. Because "chemistry" is mechanistically the interaction between the outermost electrons of different atoms (the inner ones being well shielded by the outer), the lanthanide and actinide elements must have remarkably similar chemical properties within their given series, because the series represents the progressive filling of an inner subshell, with the

TABLE 3.2 Ordering of Atomic Orbitals by Energy[1]

Formal Orbital Sequence	Energy-Based Sequence of Outer Orbital[2]	Maximum Number of Electrons for Energy-Based Sequence	
6d	6d	10	
6p	5f	14	
6s	7s	2	
5g	6p	6	
5f	5d	10	
5d	4f	14	
5p	6s	2	
5s	5p	6	
4f	4d	10	Increasing
4d	5s	2	energy ↑
4p	4p	6	
4s	3d	10	
3d	4s	2	
3p	3p	6	
3s	3s	2	
2p	2p	6	
2s	2s	2	
1s	1s	2	

Notes:
[1] Formal sequence is based on the mathematical progression outlined in Table 3.1. Energy-based sequence is the ordered list of the outermost subshells for progressively increasing Z and follows the diagonal dashed rectangle of Figure 3.3.
[2] The ordering is valid only for electrically neutral and ground-state atoms. For ions, molecules, and excited atomic states, the ordering and orbital energies are different.

outer remaining the same. And, indeed, this is what is striking about these two series—the elements within them behave more like "flavors" of the same element than like different elements. Conversely, the contrast could not be greater than for the progressive changes along row 1 or along row 2 or 3. The element hydrogen is incredibly reactive, bonding in an exothermic fashion with oxygen, sulfur, even with itself, while helium is inert. The difference between hydrogen and helium—the completion of the 1s shell in the latter, with electrical neutrality guaranteed by the nucleus having an extra proton—makes all the difference in the universe in the chemical behavior of these two elements. Reactive fluorine, which readily bonds with hydrogen or lithium, sits one column and one electron from shell completion and the virtually total inertness of neighboring neon. Likewise, the vast difference in chemical bonding behavior of carbon versus nitrogen or oxygen is the result of lacking one or two electrons that the latter two possess, leaving carbon with a half-filled shell that makes it uniquely "multivalent." That is, there are many bonding schemes involving carbon, and this multiplicity is essential for the variety and properties of organic molecules that are fundamental to life. The variety of carbon bonds will be discussed later, but you should here take a glance again at the periodic table, and ponder what a difference a column makes.

As striking as the difference from one column to another is the repetition of this pattern of reactivity from one row to the next. Reactive lithium, like hydrogen, readily donates electrons to the elements in the penultimate column on the right side of the table. Silicon carries some of the same multivalent flavor of carbon, and yet its neighbors phosphorous and sulfur are quite different. Reactive chlorine grabs the lone $3s$ electron of sodium (or the $2s$ electron of lithium, or the lone electron of hydrogen, or the $4s$ electron of potassium) to make a stable compound. Yet, we can add just one electron and proton to obtain argon, a nearly chemically nonreactive, inert, noble gas. Each of the noble gases, despite their vast differences in proton number, behaves in the same fashion, being difficult to react with other elements.

The patterns repeat, but they are not exact duplications from one row to the next. As the shells fill, and progressively higher n shells become the valence shells, the energy differences decline and subshells overlap. The nucleus itself, with more protons, exerts a stronger pull and alters the scale of the shells, but it leaves the outer electrons even more susceptible to removal or electrostatic interaction. These differences lead to variations, some not so subtle, in the chemical behavior of elements in a given column and from one row to the next. Xenon, as noted above, is much more reactive than neon or argon, for example. When we discuss the carbon bond a bit later, it will be important to recognize that many of its bonding properties favorable for life are not shared by silicon, one row beneath. The molecule H_2S is not H_2O, and some of the attractive aspects of liquid water, such as the transport and solvation medium for life, have to do with the particular strength of the bonds between the hydrogen of one molecule and the oxygen of an adjacent one. The strength of these hydrogen bonds differ for H_2S versus H_2O because of the energy level differences between the $2p$ and $3p$ orbitals.

The number of elements that occur naturally in nature is 92. Heavier elements have been produced in accelerators, and some of these were later identified as transient products of violent cosmic explosions such as those of supernovae. The possibility of superheavy elements transcending the $7p$ subshell is limited by the stability of the nucleus, which can split in various ways in the phenomenon called *radioactivity*, discussed later in the chapter.

The atomic number of an element reflects both the number of electrons and the number of protons, but it does not adequately capture the total mass of the element. The nucleus is bound not merely by the strong force acting among the protons; if it were, all elements except hydrogen would be transient because the electrostatic repulsion of the protons would overcome the strong nuclear force. The nucleus is stable because an additional massive particle, exhibiting the strong force but uncharged, is accommodated there as well. As mentioned in Chapter 2, the neutron is just slightly more massive than the proton, 1.675 versus 1.672×10^{-27} kg, and 1,600 times the mass of the electron. The total number of protons and neutrons is, to high accuracy, the mass of the atom and is denoted as the *mass number*.

A given element may come in several flavors distinguished only by the number of neutrons; these flavors are called the *isotopes* of the element. The effect of adding or subtracting a neutron from the nucleus is to change the mass of the atom, but because the neutron is uncharged and the gravitational force is vastly weaker than the electromagnetic force, there should be essentially no change in the bonding properties of the element. Because the presence of the neutron changes the energy state of the nucleus, however, there are subtle changes in the strength of the interactions between atoms when the isotope number of one of the atoms changes. For example, adding a neutron to the single-proton nucleus of the hydrogen atom yields deuterium (D), which is twice as heavy as the most common isotope of hydrogen, called light hydrogen (H) or (in older texts), protium. There is still a single electron occupying the $1s$ shell, and hydrogen is a powerful electron donor. But in HDO, for example, the single-deuterated form of water, the bond strength between D and O is slightly stronger than between H and O, and the bonds between two molecules differ slightly in strength as well. This leads, for example, to a different propensity for HDO to condense as liquid than does H_2O, and thus the variation in the relative abundances of HDO and H_2O in ice cores is used to gauge variations in Earth's climate over time. Even where differences in bond strength are not so important, the different masses of the isotopes can lead to their separations in gravitational fields. As we shall see for Mars in Chapter 12, the loss of water from its atmosphere over time has preferentially left the heavier HDO behind, leading to enrichment in the atmospheric deuterium abundance.

We cannot make isotopes of an element with an arbitrary number of neutrons. As Figure 3.4 shows, with the exception of light hydrogen, all stable isotopes have a number of neutrons at least equal to the atomic number, and most have more. Fewer neutrons mean more electrostatic repulsion among protons; too many lead to the strong nuclear force (which acts over an extremely short range) being overcome, with the consequent breakup of the nucleus. The different ways nuclei split apart or transform lead to different radioactive phenomena, and the predictable nature of the process for a given isotope allows its *decay* (breakup or transformation in the number of neutrons and protons) to serve as a natural chronometer of geologic time.

Among the stable isotopes, abundances as measured in the natural environment of the Earth's crust are used to assign terrestrial abundances to the isotopes. Averaging the isotopic masses weighted by their terrestrial abundances yields average "atomic weights" for the elements. While somewhat arbitrary (because, e.g., abundances in meteorites differ greatly from crustal abundances for most elements), the average atomic weights are an approximate guide to which isotope of an element is most prevalent. For example, on the Earth, deuterium's abundance relative to hydrogen in seawater is 150 parts per million, but analyses of the atmosphere of Jupiter and of the interstellar medium lead to the conclusion that this is six times the primordial abundance of deuterium relative to hydrogen in the material from which the solar system formed. Either way, it is clear that H dominates over D. Readers interested in more detail on the average atomic weight standards are referred to the article by DeLaeter and Heumann (1991).

Finally, the solar system abundances of the elements—painstakingly determined by examining meteorites (Chapter 6) and spectra of the Sun—are given in Table 3.3. They are an important starting point for any discussion of the formation of planets and life. Their abundances reflect the processes by which elements formed in the cosmos—first hydrogen, helium, and lithium in the origin of the cosmos, then the others under conditions so extreme that one element is transformed into another. Such transformation, the dream of the alchemists, makes the Sun shine and defined the opening of the age of the atom for humankind in the mid-20th century. To understand the transformation of elements requires

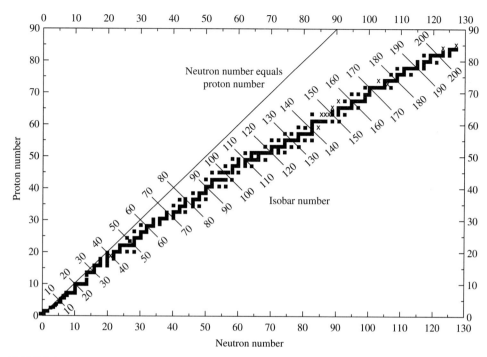

FIGURE 3.4 Distribution of stable atomic nuclei, plotted as the number of neutrons versus the atomic number Z. A theoretical line corresponding to an equal number of protons and neutrons is drawn for reference. Isotopes of different elements that have the same mass number (but, of course, different relative numbers of neutrons and protons) are identified with short diagonal lines and are labeled by their mass number; these are *isobars*. Those isotopes that are not completely stable but in fact decay over long but measurable periods are marked 'x'.

examining the physics underlying the stability of the nucleus, which we defer to later in the chapter after we return to, and more fully consider, chemical bonding.

3.4 Bonding Mechanisms

The bonding of atoms must be consistent with the exclusion principle and with the thermodynamic driver that total Gibbs free energy (Chapter 7), after bonding, be reduced compared with that prior to bonding. These requirements are coupled, as we can see in considering the simplest bond (that between two hydrogen atoms). Each of the atoms is characterized by spherical $1s$ orbitals, and a nucleus vastly smaller in volume—making slightly more precise the analogy of Chapter 2, were the $1s$ shell a football stadium, the nucleus would be a child's marble. The interaction between the two shells is really an interaction between the wavefunctions of the two electrons that obeys the Pauli exclusion principle.

TABLE 3.3 Solar System Abundances of the Elements

Element	Symbol	Atomic Number	Atomic Weight	Log Abundance Meteoritic	Log Abundance Solar
Hydrogen	H	1	1.0079	12.00[a]	12.00
Helium	He	2	4.0026	10.99[a]	10.99[b]
Lithium	Li	3	6.941	3.31	1.16
Beryllium	Be	4	9.0122	1.42	1.15
Boron	B	5	10.811	2.8	2.6[c]
Carbon	C	6	12.011	8.56[a]	8.56
Nitrogen	N	7	14.007	8.05[a]	8.05
Oxygen	O	8	15.999	8.93[a]	8.93
Fluorine	F	9	18.998	4.48	4.56
Neon	Ne	10	20.180	8.09	8.09[d]
Sodium	Na	11	22.990	6.31	6.33
Magnesium	Mg	12	24.305	7.58	7.58
Aluminum	Al	13	26.982	6.48	6.47
Silicon	Si	14	28.086	7.55	7.55
Phosphorus	P	15	30.974	5.57	5.45
Sulphur	S	16	32.066	7.27	7.21
Chlorine	Cl	17	35.453	5.27	5.5
Argon	Ar	18	39.948	6.56[d]	6.56[d]
Potassium	K	19	39.098	5.13	5.12
Calcium	Ca	20	40.078	6.34	6.36
Scandium	Sc	21	44.956	3.09	3.10
Titanium	Ti	22	47.88	4.93	4.99
Vanadium	V	23	50.942	4.02	4.00
Chromium	Cr	24	51.996	5.68	5.67
Manganese	Mn	25	54.938	5.53	5.39
Iron	Fe	26	55.847	7.51	7.54
Cobalt	Co	27	58.933	4.91	4.92
Nickel	Ni	28	58.693	6.25	6.25
Copper	Cu	29	63.546	4.27	4.21
Zinc	Zn	30	65.39	4.65	4.60

				Log Abundance	
Element	Symbol	Atomic Number	Atomic Weight	Meteoritic	Solar
Gallium	Ga	31	69.723	3.13	2.88
Germanium	Ge	32	72.61	3.63	3.41
Arsenic	As	33	74.922	2.37	
Selenium	Se	34	78.96	3.35	
Bromine	Br	35	79.904	2.63	
Krypton	Kr	36	83.80	3.23	
Rubidium	Rb	37	85.468	2.40	2.60
Strontium	Sr	38	87.62	2.93	2.90
Yttrium	Y	39	88.906	2.22	2.24
Zirconium	Zr	40	91.224	2.61	2.60
Niobium	Nb	41	92.906	1.40	1.42
Molybdenum	Mo	42	95.94	1.96	1.92
Technetium	Tc	43	98.906		
Ruthenium	Ru	44	101.07	1.82	1.84
Rhodium	Rh	45	102.91	1.09	1.12
Palladium	Pd	46	106.42	1.70	1.69
Silver	Ag	47	107.87	1.24	0.94[c]
Cadmium	Cd	48	112.41	1.76	1.86
Indium	In	49	114.82	0.82	1.66[c]
Tin	Sn	50	118.71	2.14	2.0
Antimony	Sb	51	121.76	1.04	1.0
Tellurium	Te	52	127.60	2.24	
Iodine	I	53	126.90	1.51	
Xenon	Xe	54	131.29	2.23	
Cesium	Cs	55	132.91	1.12	
Barium	Ba	56	137.33	2.21	2.13
Lanthanum	La	57	138.91	1.20	1.22
Cerium	Ce	58	140.12	1.61	1.55
Praseodymium	Pr	59	140.91	0.78	0.71
Neodymium	Nd	60	144.24	1.47	1.50

(continued)

TABLE 3.3 continued

Element	Symbol	Atomic Number	Atomic Weight	Log Abundance Meteoritic	Log Abundance Solar
Promethium	Pm	61	146.92		
Samarium	Sm	62	150.36	0.97	1.00
Europium	Eu	63	151.96	0.54	0.51
Gadolinium	Gd	64	157.25	1.07	1.12
Terbium	Tb	65	158.93	0.33	−0.1
Dysprosium	Dy	66	162.50	1.15	1.1
Holmium	Ho	67	164.93	0.50	0.26[c]
Erbium	Er	68	167.26	0.95	0.93
Thulium	Tm	69	168.93	0.13	0.00[c]
Ytterbium	Yb	70	170.04	0.95	1.08
Lutetium	Lu	71	174.97	0.12	0.76[c]
Hafnium	Hf	72	178.49	0.73	0.88
Tantalum	Ta	73	180.95	0.13	
Tungsten	W	74	183.85	0.68	1.11[c]
Rhenium	Re	75	186.21	0.27	
Osmium	Os	76	190.2	1.38	1.45
Iridium	Ir	77	192.22	1.37	1.35
Platinum	Pt	78	195.08	1.68	1.8
Gold	Au	79	196.97	0.83	1.01[c]
Mercury	Hg	80	200.59	1.09	
Thallium	Tl	81	204.38	0.82	0.9[c]
Lead	Pb	82	207.2	2.05	1.85
Bismuth	Bi	83	208.98	0.71	
Polonium	Po	84	209.98		
Astatine	At	85	209.99		
Radon	Rn	86	222.02		
Francium	Fr	87	223.02		
Radium	Ra	88	226.03		
Actinium	Ac	89	227.03		
Thorium	Th	90	232.04	0.08	0.12

Element	Symbol	Atomic Number	Atomic Weight	Log Abundance	
				Meteoritic	Solar
Protactinium	Pa	91	231.04		
Uranium	U	92	238.03	−0.49	<−0.45[c]
Neptunium	Np	93	237.05		
Plutonium	Pu	94	239.05		
Americium	Am	95	241.06		
Curium	Cm	96	244.06		
Berkelium	Bk	97	249.08		
Californium	Cf	98	252.08		
Einsteinium	Es	99	252.08		
Fermium	Fm	100	257.10		
Mendelevium	Md	101	258.10		
Nobelium	No	102	259.10		
Lawrencium	Lr	103	262.11		

Notes:
[a] Based on solar data.
[b] Based on stellar observations and solar models.
[c] Uncertain.
[d] Based on other astronomical data.
Source: Cox, A. N. (ed.) 1999. *Allen's Astrophysical Quantities*, AIP Press, New York.

The wavefunctions can add in such a way as to constructively interfere between the two nuclei to produce a higher probability that the electrons lie between the nuclei or destructively to create a region devoid of electrons between the nuclei (Figure 3.5). The former case is called a "bonding" orbital, in which the electron spins are antiparallel to each other, and the latter an "antibonding" orbital, in which the spins are parallel. The bonding orbital has two electrons, and hence a high negative charge density, between the two protons that constitute the hydrogen nuclei. This tends to pull the nuclei together to make a stable molecular assemblage we call H_2. The antibonding orbital can remain empty because both electrons can be accommodated (one spin up, the other spin down) in the bonding orbital. Although the two electrons tend to repel each other in terms of their like charges, the attractive forces between the two electrons and the nuclei are larger, and the system is bound—it is a lower energy state relative to that of two separate hydrogen atoms.

The helium atom, on the other hand, has two electrons. Bringing two helium atoms together so the atomic orbitals interact would produce bonding and antibonding orbitals of

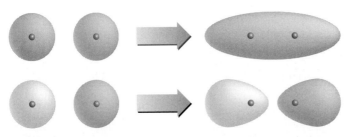

FIGURE 3.5 The *s* orbitals of two hydrogen atoms can combine to reinforce the probability of electrons in between the two nuclei (top) or make the probability essentially zero (bottom) to produce a bonding and antibonding molecular orbital, respectively.

shape similar to that in Figure 3.5, but with four electrons present, two must be accommodated in the antibonding orbital. The electrons in the antibonding orbital cancel out the attractive effect of the two in the bonding orbital, which makes the He_2 molecule energetically neutral or unfavorable with respect to two separate helium atoms. (The cancellation is not perfect—there is a net repulsion at small separations and a weak attraction at large separations.) Thus, He_2 is not observed, and helium is a monatomic (single atom) gas, under most conditions.

The type of bond represented by the hydrogen molecule is called a *covalent* bond. Electrons are shared by the two nuclei, and because the two electrons are now in one common "molecular" orbital, and because the electrons themselves are indistinguishable, we cannot assign a particular electron to a particular nucleus. Covalent bonding in general is directional. Thus, bonding pairs and lone pairs (those not bonded) of valence electrons will repel so as to minimize the electromagnetic repulsive force between them. Any particular atom can interact strongly with only a relatively small number of other atoms; this number is termed the "valency" of the particular atom.

The shapes of molecular orbitals—which are the superpositions of atomic orbitals—can be complex. In some cases, more than just the outermost subshell will participate in the bonding. In the case of oxygen, for example, it is not just the $2p$ subshell that is overlapping with the s orbitals from the two hydrogen atoms. There is actually a further reduction in total energy of the molecule H_2O when the $2s$ electrons of the oxygen atom participate in the bonding. The shape of the orbital is not simply obtained from a superposition of $2p$ from oxygen with the $1s$ of each of the two hydrogen atoms. Instead, the $2p$ and $2s$ orbitals of oxygen *hybridize* into composite molecular orbitals, with a shape that is roughly obtained by mixing the s and p orbitals involved in bonding. In the absence of hybridization, the water molecule would be predicted to have a structure in which the two hydrogens are separated by an angle of 90 degrees, when in fact the angle is 104 degrees, consistent with the hybridization model. In fact, sulfur exhibits this 90-degree geometry because of the much weaker interaction of the outer (valence) electrons compared with that in water.

An important difference between the covalent bonding of molecular hydrogen and of water is that, in the case of the former, the electrons spend most of their time roughly equidistant between the nuclei. This "symmetric" bond, a consequence of the simple nature of the composite molecular orbital, leads to no net charge on either end of the molecule; we say the H_2 molecule is nonpolar. In the case of water, on the other hand, the complex hybridized orbitals are very much asymmetric, the electrons are closer to the oxygen atom, and there is a net positive charge on the hydrogen and negative on the oxygens, as noted in Chapter 2. Water is thus "polar," with a significant charge difference between the positive and negative ends. The asymmetric nature of the bond also permits hydrogen to bond to oxygen atoms of neighboring water molecules. This *hydrogen bond* is much weaker than a standard covalent bond—it changes the energy of each water molecule by less than 0.02 percent—but it is crucial to the structure of water ice and liquid water. Indeed, liquid water has a more regular pattern than many other liquids because of the molecular orientations afforded by hydrogen bonds. And, as we see in Chapter 4, hydrogen bonding is responsible for the establishment of "hydrophobic" and "hydrophilic" ends of biomolecules that are fundamental to the folding of proteins, the shape of DNA, and the formation of cell walls.

Some molecules are formed not through electron sharing but through transfer of electrons from one atom to another; the atoms are then held together by an *ionic bond*. The formation of such a bond is best understood from the point of view of minimizing the energy of the system, although it too must be consistent with the exclusion principle. A simple example is that of sodium chloride, NaCl. Chlorine, which is in the so-called halogen (salt-forming) column, has a strong affinity for an additional electron to fill its $3p$ shell. The sodium atom has a single electron in its $3s$ subshell, and the shielding of that electron from the nucleus of 11 protons (and 12 neutrons) afforded by the $n = 1$ and $n = 2$ shells is huge. Therefore, it does not take much energy to liberate the electron from sodium. Refer to Figure 3.6 to define a useful energy unit, the electron-volt (eV), which is the energy gained by an electron falling through a potential drop of 1 volt; $1 \text{ eV} = 1.6 \times 10^{-19}$ joules. Because of inner shell shielding, the energy required to remove the electron from sodium ("ionize" the sodium) is 5.1 eV, less than half what is required to ionize a hydrogen atom by removing its lone unshielded electron (13.6 eV). By adding that electron to chlorine to complete the $3p$ shell, we reduce the system energy by 3.8 eV. Thus, the net energy penalty for stealing the sodium electron is just $5.1 - 3.8 = 1.3$ eV. But if we were to put the positive sodium ion, Na^+, in the vicinity of the negative chlorine ion, Cl^-, we should get an energy advantage because the oppositely charged ions should attract each other. If that attraction creates an energy reduction for the molecule Na^+Cl^- of more than 1.3 eV, this association would be the preferred state and ought to form spontaneously. The question is whether the two ions would have to be so close that the repulsion of the electron shells around each would negate the energy drop. As Figure 3.6 shows, the total energy of the Na^+ and Cl^- system decreases by virtue of their electrostatic attraction to a minimum of -4.9 eV at a separation of 2.4 Å, before the repulsion of the two electron clouds dramatically increases the system energy at smaller separations. Therefore, the molecule NaCl, which is a positive sodium ion and a negative chlorine ion electrostatically attracted ("ionically" bonded), is lower in energy by $4.9 - 1.3 = 3.6$ eV than the separate neutral atoms and is the preferred configuration of the two elements. This analysis is oversimplified, and a more involved analysis is required to determine whether any particular set of atoms produces a lower energy (more bound) system via ionic bonding than a covalent bond in which the electrons are shared rather than transferred.

Ionic bonds create strongly polar molecules because the transfer of electrons makes positive and negative ends to the molecule. Because the transfer of electrons produces

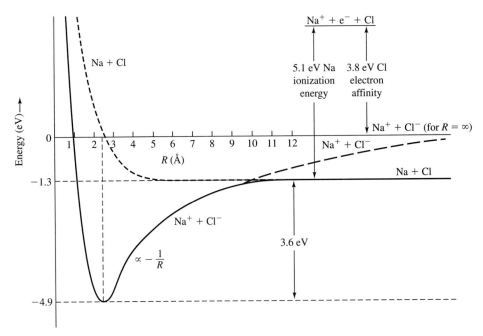

FIGURE 3.6 A graph showing energy (in eV) versus distance of separation between neutral sodium and chlorine atoms, compared with that for the positive sodium ion and negative chlorine ion. A negative energy indicates a net attractive interaction, while positive energy implies repulsive interaction.

ions, each of which has a completed, closed shell, the ionic bond is not directional. This differentiates it from many covalent bonds. However, most chemical bonding usually has both ionic and covalent characteristics—there is a continuum of bond properties between ionic and covalent. Whether a bond will be principally ionic or covalent depends on the affinity of the accepting element for electrons and the ionization potential of the donating electron. Standard chemistry texts contain tables quantifying the electron affinities, ionization potentials, and the related concept of electronegativity for each of the elements.

A final type of bonding important between molecules themselves, as well as between noble gas atoms, involves *inductive* forces. Take the case of, for example, argon. Two argon atoms brought close to each other ought not to have any direct bonding tendency, because their electron shells are filled—hence, spherical—in a time-averaged sense. However, the shells are not spherical at any given point in time, and such distortion leads to a small, and transient, net attractive force. The shape of the energy of interaction between the two atoms is similar to the curve for Na^+ and Cl^-, but the magnitude of the attractive (negative) portion is much smaller. This type of transient force is given the name *van der Waals* attraction, and it is responsible for the stability of liquid and solid forms of the noble gases. Molecules also will undergo van der Waals interactions among each other, the strength of which depends on the ease with which the shape of the electron "cloud" (the shell structure) around the molecule can be altered, and on the total number of electrons contained in the interacting species. The measure of the potential strength of the attractive interaction is called the "polarizability" of the molecule. Many molecules, as noted earlier, have a net polarity, and this allows for stronger van der Waals interactions with nonpolar molecules or noble gases. Because the rate at which collisions occur in a gas depends on the number density—hence, pressure—of the gas, very dense atmospheres made of nonpolar gases may exhibit significant van der Waals interactions, which (as described later) lead to the absorption and emission of electromagnetic radiation (photons). This, for example, leads to a modest warming of the atmosphere of Saturn's moon Titan through the absorption of infrared radiation during collisions among gaseous nitrogen, argon, methane, and hydrogen.

Atoms bound together in a chemically stable configuration (molecules) are not the only composites of atoms. In chemical reactions in natural or laboratory environments, transient molecular fragments exist that lack sufficient numbers of electrons to complete a shell or that have an excess of electrons. Such compounds are *radicals*. While methane, CH_4, is a stable molecule in which the covalent bonding among carbon and four hydrogens fills shells, the radicals CH_2 and CH_3 lack enough electrons to fill the valence shell of carbon. These constructs are not stable and quickly react with other molecules, but their transient existence in a network of chemical reactions is key to understanding the products of such reactions. For example, exposed to ultraviolet sunlight, methane will break apart into CH or CH_2 radicals and eventually form the molecule acetylene, C_2H_2, or CH_3 radicals may form and then combine (with the participation of a third molecule to absorb the kinetic energy of the collision) to form ethane, C_2H_6. Finally, a charged atom, with greater or lesser numbers of electrons than protons, is called an *ion*, examples being Na^+ and Cl^-.

The concept of acids and bases is important in understanding the aqueous (water-based) environment within which cells and organisms exist and that cells themselves regulate within their walls. There are several ways to define acids and bases, and what follows is based on the Lewis theory. An *acid* is a molecule, radical, or ion that potentially can accept additional electrons to fill valence shells, while a *base* is a molecule, radical, or ion that is capable of donating a pair of electrons. These definitions, which suggest some subtleties in the nature of chemical bonding, reflect as well the empirical underpinning of the acid–base concept. For example, ammonia, NH_3, is a molecule in which three hydrogens associate with the nitrogen to create a hybridized sp^3 orbital. The nitrogen in the ammonia molecule has a lone pair of valence electrons in the hybridized orbital that can be donated to an acid. Hence, ammonia is a base. It can combine with the acid BF_3, for example, which lacks two electrons to complete a valence shell, to form a covalently bonded molecule.

The typical example of an acid–base reaction involves the ions H_3O^+ and OH^-, which are, respectively, the hydronium ion and the hydroxyl ion of water. They, of course, will combine to make two molecules of stable and neutral water, $2H_2O$. So why should such constructs exist in an aqueous solution? In pure water, their concentrations are quite small—about 10^{-9} kg/kg of water. But the presence of other substances in water that are strongly acidic or basic will change the concentration of the hydronium and hydroxyl ions, leading to the solution itself becoming acidic or basic. The measure of how acidic or how basic the water solution will be is called the *pH*, and it is the negative of the logarithm (base 10; see Chapter 7) of the *molar concentration* of H_3O^+ in the solution. The molar concentration, in turn, is a traditional way of expressing the concentration in terms of a unit that takes account of the very large number of molecules present in a macroscopic amount of a substance. One *mole* of a material is

6.023×10^{23} molecules. This is a convenient number because it is approximately the inverse of the mass of a proton in grams. So, what is the mass of a mole of atomic hydrogen? It is 1.67×10^{-24} g $\times\, 6.023 \times 10^{23} = 1$ g (0.001 kg)—about the weight of a paper clip. How about a mole of water? Water is two hydrogens and an oxygen—an assemblage that is 10 protons and 8 neutrons. Hence the mass of a mole of water is $18 \times 1.67 \times 10^{-24}$ g $\times\, 6.023 \times 10^{23} = 18$ g or 0.018 kg. Therefore, a mole of a substance is just the atomic mass (or, in the case of molecules, the molecular mass—the sum of the constituent atomic masses).

Now back to pH. The definition of pH requires that the concentration be expressed in moles of hydronium ion per liter. For pure water this is 10^{-7} mole/l. Using our definition above, the pH of pure water is $-(\log_{10} 10^{-7}) = -(-7) = 7$. More acidic solutions have a higher concentration of H_3O^+ and hence a lower pH. More basic solutions have a lower concentration of H_3O^+ and hence a higher pH. The acidity of a water solution determines very strongly its chemical reactivity, and biological systems can be very sensitive to small changes in acidity. Human blood has a pH of 7.5, which is not far from the value for pure water. However, lowering the pH of the blood to 6, which is equivalent to a very mild acid, is usually deadly. On the other hand, organisms have been found in the Rio Tinto river of Spain, which has a pH approaching 1—an exceedingly high acidity. Chapter 10 explores the extremes of habitable environments on the Earth, and one of the key variables defining extreme environments is the acidity of the aqueous medium within which organisms exist. This discussion has suggested that acidity has a tremendous effect on the chemical properties of aqueous solutions.

3.5 Particular Properties of the Carbon Bond

Carbon is composed of six electrons, six protons, and from six to eight neutrons (Figure 3.7). Two isotopes, ^{12}C and ^{13}C, are stable, while the heaviest isotope, ^{14}C, decays (Section 3.7) on timescales of thousands of years into nitrogen. An isolated carbon atom (e.g., in the gas phase) has an electronic structure $1s^2 2s^2 2p^2$. Carbon has a valence number of 4 (i.e., is "tetravalent") and can adopt any one of three possible bonding configurations

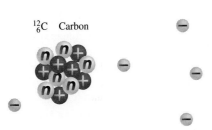

FIGURE 3.7 Schematic cartoon of the mass-12 isotope of carbon, ^{12}C, with enumeration of the number of protons (+), neutrons (n), and electrons (−). Particles and distances are not shown to scale, and the electronic positions cannot be specified, but rather occupy regions defined by the subshells.

involving three different possible hydrid orbitals with tetrahedral (sp^3), trigonal (sp^2), or linear (sp) geometries.

The complexity of the potential bonding combinations associated with carbon is enormous. Carbon can, for example, form a stable molecule with one oxygen atom (carbon monoxide, CO), or two oxygen atoms (carbon dioxide, CO_2), though the former is much more chemically reactive (and indeed, can bind with hemoglobin in much the way molecular oxygen, O_2, does—but much more strongly and with fatal consequences). Carbon can bond with itself in a networked lattice to make graphite or diamond. Because carbon can bond to itself and many other elements, notably hydrogen and the elements important to life (nitrogen, oxygen, sulfur, etc.), the chemistry of carbon—*organic* chemistry—is rich and complicated, with millions of known organic compounds. Chain and cyclical (*ring*) structures associated with carbon's ability to simultaneously bond with itself and other elements leads to a huge range of shapes and sizes of molecules, up through the biopolymers containing many thousands of units.

The rich complexity of organic chemistry cannot be explored in the space of this book, and students preparing themselves for careers in astrobiological research and teaching are urged to pick up an organic chemistry text and at least familiarize themselves with the many classes of organic compounds. By way of introduction, though, we can get a sense of the root of this complexity by following the nature of the carbon bonds in terms of molecular orbital theory.

It is conceptually easiest to begin with a charged radical, the so-called methyl cation, CH_3^+. Because it has only three hydrogens, in contrast to methane, which has four hydrogens, it possesses an open (unhybridized) $2p$ orbital. The configuration of this sp^2 orbital is illustrated in Figure 3.8 as three lobes spaced 120 degrees apart from each other. This configuration geometrically and favorably maximizes the distances between the electrons in each of the orbitals involved in the bonding. When we add hydrogen to make CH_4—methane—an additional bond is created, the last $2p$ orbital is filled, and the hybrid sp^3 orbital is roughly $\frac{1}{4}s$ and $\frac{3}{4}p$. This yields the tetrahedral shape of the methane molecule in

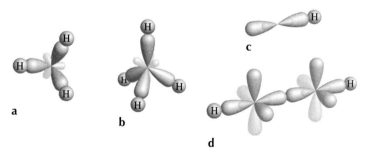

FIGURE 3.8 Diagrams of the orbitals in simple organic molecules. The in-phase (overlapping) lobes are shaded. Hydrogen's $1s$ orbitals are labeled; others are hybrid molecular orbitals centered on carbon: (a) methyl cation. (b) methane. (c) C—H bond in acetylene. (d) acetylene molecule.

Figure 3.8 that is verified through analysis of spectra (discussed in the next section). In neither of these cases should we imagine that it is possible to simply intuit the shapes in detail. But the hybrid approach does allow us to heuristically rationalize the nature of the multivalent carbon bonds and the corresponding molecular structures.

Continuing on, let's move to acetylene, C_2H_2, and first consider each C-H bond. These bonds between carbon and hydrogen are a mixture of the *sp* orbital on the carbon and the *s* orbital on the hydrogen; Figure 3.8 shows the hybrid has a linear characteristic. There remain two unused *sp* orbitals on each carbon, allowing a strong "σ bond" between the two carbon nuclei, in which the two orbitals are facing each other. Finally, two empty *p* orbitals exist in the two carbon atoms, which do not point toward each other but are aligned sidewise. Each pair can form a covalent bond, but this type of bond is a weaker "π bond." In total, then, there are three bonds tying the two carbons together in acetylene—two weak and one strong. The overall shape of the molecule is determined by the C-H bonds and consequent existence of the strong σ bond. Depending on the number and type of other atoms participating in the bonding, carbons may have single, double, or triple bonds with each other.

These few simple cases illustrate the basic principles underlying the versatility of organic compounds, the essence of which is the multivalent nature of carbon and its ability to bond with itself while bonding with other elements. We have reason to use these properties, as well as those of the hydrogen bond, when we discuss the molecules important to biology in Chapter 4. While silicon (Si) is tetravalent like carbon, there are some differences associated with the valence shell being $n = 3$. The Si-Si bond is similar in strength to the C-C bond, but the larger silicon atom is more open to chemical attack than the smaller carbon. For this and other reasons, the number and variety of structures are much less for silicon than for carbon. In effect, the information content of a silicon-based biochemistry is much lower than for a carbon-based biochemistry. Additionally, when silicon bonds with oxygen it forms SiO_2, which under conditions salubrious for life (which are argued later as those in which water is a stable liquid) is a solid (albeit one that is somewhat soluble in water). Carbon, on the other hand, bonds with oxygen to form, most stably, CO_2, a gas that can be extracted from the atmosphere as organic matter or carbonates and then recycled through tectonic processes (Chapter 11). Silicon dioxide—quartz—is immediately sequestered as a mineral and cannot readily be accessed by a silicon-based biochemistry in the way that carbon dioxide can. Silicon cannot be ruled out as a basis for biochemistry, but it seems less suitable than carbon.

3.6 Spectroscopy and the Spectroscopic Signatures of Atomic and Molecular Structure

Careful readers may have already made the connection among the discussion of the electromagnetic force, the photon as the mediating particle of the electromagnetic interaction, the emission or absorption of photons by accelerating electrons, and the presence of potentially varying electromagnetic fields in atoms and molecules associated with the occupancy of orbitals by electrons. Should not electrons in accelerated motion within and across orbitals generate or absorb photons of well-determined energies, which could be diagnostic of the processes occurring within atoms or molecules? How might these be detected and measured? The answer to the first question is yes, and the relevant technique is *spectroscopy*.

In the mid-17th century, British scientist Sir Isaac Newton demonstrated that "white" light from the Sun is made of all the colors of the rainbow, by placing a glass prism in a collimated beam of sunlight and projecting the resulting rainbow on a screen. This effect can be understood by realizing that light of different colors is light of different wavelengths. When passed through a material medium different from the air, the light is slowed down. Bluer light must make a much longer trip through the glass, in terms of number of wavelengths, than does red. Hence, their paths are bent to differing extents, as are the intermediate wavelengths. Indeed, the atmosphere can be thought of (imperfectly) as a prism, with blue light *scattered* (altered in direction) more than red—accounting for the volume of the daytime sky being suffused with a blue light. The intensity of light as a function of wavelength is a *spectrum*, and devices to split the light into its components wavelengths are *spectrometers* or *spectrographs*. These devices rely on the fact that electromagnetic energy is carried in the form of waves through space, and the velocity of these waves (the "speed of light") varies depending on the material medium through which the waves propagate. Our fundamental sense of vision relies on the emission of light (production of photons), absorption of light (destruction of photons), and scattering of light (change in direction of photons) by material media.

We should first consider just what it is that we are looking at when a ray of sunlight is spread into component colors. The Sun appears self-luminous; that is, it emits light of its own accord, a property it has in common with lightning, the hot metal plate of an electric stove, and the flame of a candle. Yet, the integrated color that we perceive varies from one such object to another. Self-luminous objects that are cooler tend to radiate in the red, hotter in the blue. So-called "infrared" night-vision goggles seem to pick up human beings and warm-blooded animals as luminous glowing beings against a darker background, yet to our own eyes we do not glow. Instead we emit light at wavelengths longer than the red—the so-called infrared part of the spectrum. Any object of finite temperature has a radiant spectrum that is centered on a particular wavelength according to its temperature.

The radiant emission of light must involve the emission of energy—but how much? Experiments late in the 19th century showed that light of sufficiently short wavelength could liberate electrons from metal and cause an electric current. The dependence of the production of the electrons on the intensity and wavelength, λ, of light were consistent with light being carried by packets or particles of electromagnetic energy called photons, each of which carries an energy

$$E = \frac{hc}{\lambda} = h\nu \tag{3.3}$$

where h is Planck's constant, c the speed of light in vacuum, and ν the frequency of the light. Furthermore, if electromagnetic radiation is restricted to a box whose walls perfectly absorb and then reemit the radiation (say, a good pizza oven or the inside cavities of a glowing ember), the energy of the electromagnetic waves is restricted to a continuum of

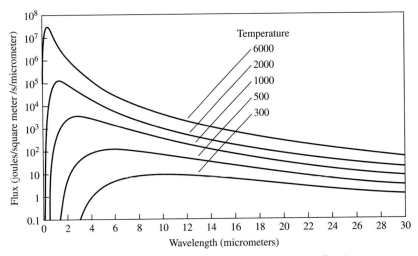

FIGURE 3.9 Brightness of light versus wavelength, expressed as flux (energy per area per time per wavelength) for blackbody radiation at various temperatures ranging from room temperature to that at the surface of the Sun.

values that leads to a distribution of energy as a function of wavelength uniquely defined by the temperature of the box.

This electromagnetic radiation field, or photon flux, is called *blackbody* radiation, and the wavelength dependence follows the so-called Planck distribution (Figure 3.9). The wavelength at which the intensity of the radiation peaks is just inversely proportional to temperature. Although the Planck function is strictly valid only for perfect radiators as defined in the last paragraph, many physical bodies emit radiation that approximately corresponds to blackbody radiation. The Sun, with a surface temperature of 5770 K, emits its peak amount of light at around 5100 Å wavelength, or 5.1×10^{-7} m, or 0.51 μm—between the red and the blue—and our eyes are adapted to seeing this integrated distribution as essentially white light. A body (such as an ember) with one-tenth the surface temperature of the Sun, or 577 K, emits its peak amount of light at a 10 times larger wavelength—5.1 μm. This wavelength lies in the infrared and is beyond the wavelengths to which our retinas are sensitive (but can be felt as heat on our open hands). A body with 10 times the Sun's surface temperature emits at 0.051 μm, which is in the so-called ultraviolet part of the spectrum, again beyond the range of sight.

Figure 3.10 gives the wavelengths, frequencies, and names of the parts of the electromagnetic spectrum. They range from low-energy (10^{-12} eV/photon) and low-frequency radio waves with wavelengths of kilometers to hard X-rays and gamma rays with short wavelengths (10^{-11} m and below) and very high energy (10^5 eV/photon and above). Photons of visible light have energies of a few electron-volts.

It is hard to imagine that the benign, long radio waves picked up by our car antennas and the flesh-penetrating particles called X-rays could be the same electromagnetic phenomena. They are, and the difference is solely in the wavelength and hence frequency—or energy per quantum—of the radiation. Electromagnetic energy of shorter wavelengths is

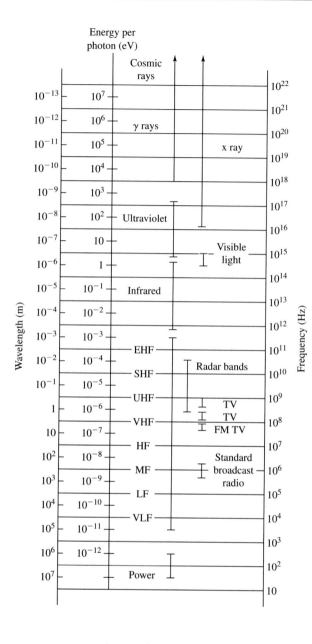

FIGURE 3.10 The electromagnetic spectrum, with regions labeled according to wavelength or photon energy.

easier to visualize as particles because their interaction with matter at our size scale is like that of particles. Electromagnetic energy of long wavelength, macroscopic in the case of radio, interacts with matter at our scale as waves, and so we associate that behavior with a wavelike nature. But at any wavelength, photons behave both as waves and as particles. Photons are emitted by warm bodies in cold surroundings, these bodies cooling down if a sustained source of energy is not available; cold bodies in warm surroundings can absorb photons and increase their temperature. As long as objects are good emitters of photons, that is, are not too different from perfect blackbody radiators, their surface temperatures can be gauged by the wavelength distribution of their emitted light.

But emission is only one part of the story. A white page illuminated by sunlight is not glowing at 5770 K; it is merely reflecting the visible light received from the Sun. Reflection is a special case of scattering in which photons are not absorbed but instead are forced to change direction. The sunlit world around us is visible because of the scattering of radiation; textures are the result of different angular dependencies to scattering, and color is in part the result of different efficiencies of scattering with wavelength. Bodies do absorb some of the light that comes from the Sun, but the expansion of that light through space has so diluted it that even complete absorption of sunlight at Earth leads to a surface temperature less than 300 K (Chapter 11).

Selective scattering of light, to produce the colors that we see, suggests that materials are wavelength-selective in their interactions with photons. This selectivity arises from the electronic structure of their atoms or molecules. Indeed, while scattering over a broad range of wavelengths can tell us something about the nature of materials, it is far from the whole story. Spectral *lines,* and *bands* of spectral lines, map out the patterns within atoms and molecules of their electronic structure and of the shape and mass distribution within molecules. Spectral lines are, for our purposes here, formed in two different processes—emission and absorption (Figure 3.11). *Emission spectra* are bright, discrete lines and bands that stand above the background "continuum" spectrum. They are revealed when self-luminous materials, such as the Sun or a candle flame, are examined with a spectrometer. They can also be produced when atoms or molecules are present in a warm gas and the resulting light is viewed against the background of a colder gas, as happens in some planetary atmospheres. *Absorption spectra* are revealed as the dark lines against a brighter continuum and are formed when light passes through a material medium (gas, liquid, a thin solid) and the materials present selectively absorb or remove photons of particular wavelengths. For the most part, though not entirely, emission and absorption spectra exhibit lines at the same wavelength for a given material, but they differ in the contrast of the line against the continuum background.

The absorption or emission of photons at specific wavelengths in the ultraviolet and optical part of the spectrum are determined by the energy levels of the electron shells within atoms or molecules, and selection rules for the energy changes between levels associated

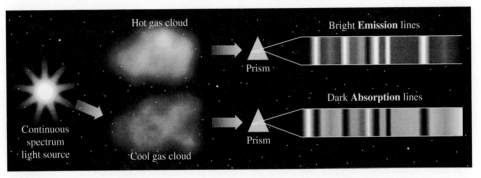

FIGURE 3.11 The distinction between how absorption vs. emission spectra are formed.

with the underlying quantized nature of atomic and subatomic structure. We will not deal with selection rules here, though students who will be pursuing spectroscopic analysis should consult the References for more detailed texts. At longer wavelengths, transitions in energy associated with the vibration of atoms along or across bonds with other atoms in molecules create spectral features, again with restrictions imposed by quantum mechanics. At very long wavelengths—out to the radio part of the spectrum—the rotation of molecules (again quantized) creates series of lines arrayed in wavelength according to the change in energy associated with the rotation. At even longer wavelengths phenomena associated with the spin of subatomic particles can produce spectra when these particles are subjected to a magnetic field—leading (for example) to nuclear magnetic resonance that is in widespread use as a medical diagnostic tool. Figure 3.12 illustrates schematically the embedding in energy levels of electronic, vibrational, and rotational transitions.

It is possible to predict the spectra for simple atoms under restricted conditions. For example, Figure 3.13 on page 95 displays the energy levels for three elements, and the change in energy in going from one level to the next can be converted to the wavelength or frequency of absorption or emission lines. (There is another measure of spectral line position, called *wavenumber*, that readers will see in the literature; it is just the inverse of the wavelength). For example, the energy difference between the $3s$ and $3p$ state in sodium is 2.1 eV. If a gas of sodium were subjected to an electrical current, electrons residing in the $3s$ subshell could gain energy amd hence move up to the $3p$ subshell. In sodium, which has $Z = 11$, the $3p$ subshell is *not* normally filled, except when an electron becomes energetically excited. From equation (3.3) or reference to Figure 3.10, we find that the wavelength corresponding to the transition is 5.9×10^{-7} m, or 0.59 μm—in the yellow part of the optical spectrum. And indeed, most modern streetlamps are composed of bulbs enclosing sodium gas, subjected to an electric discharge, by which they glow over a range of wavelengths, but peaked at 0.59 μm.

Spectral lines are not sharp, infinitesimally thin pointers to a specific wavelength. Even absent environmental effects, the Heisenberg uncertainty principle (Chapter 2) ensures that there is a minimum energy uncertainty and, hence, width to any spectral line. But gases at elevated temperature collide with each other at high speed, distorting in a transient fashion the shapes of the electronic orbitals, and this *collisional* broadening of a spectral line can be substantial—and a measure of the ambient conditions within the gas. High pressure—even at low temperature—can also distort the shapes and energies of the electronic shells and broaden the spectral lines. The temperature of the Earth's surface is increased by the insulating effect of carbon dioxide absorption lines in the atmosphere (Chapter 11); these lines are effective absorbers of infrared radiation because they are broadened by the pressure effects of the dominant gases nitrogen and oxygen. Liquids and solids, in which atoms are close together and affected by the mutual electromagnetic fields they generate, can have different spectra from the pure gases. Finally, small energy changes associated with certain atomic transitions can lead to complex variation within the optical spectrum that we perceive, for example, as the rich colors in minerals.

Despite the complexities associated with thermal and pressure broadening, optical spectra can be used to derive abundances of atoms and molecules in various gaseous and condensed environments, and an example atomic spectrum from the Sun is given in Figure 3.14 (see Color Plate 1). A number of interesting phenomena can occur as electrons are raised to an excited energy level and then "decay" back to a lower energy state. In one

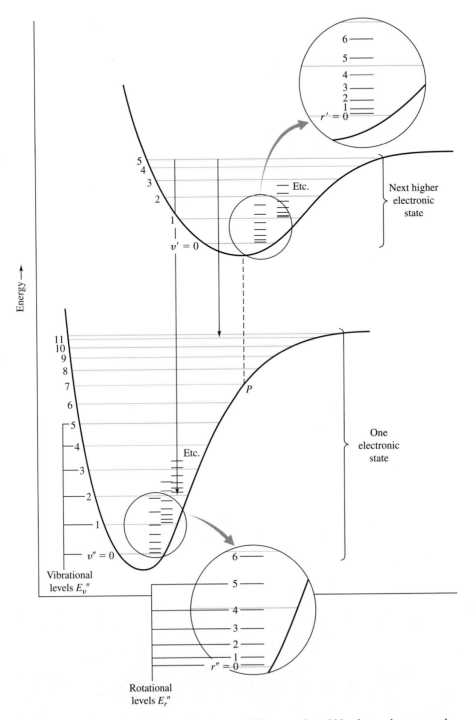

FIGURE 3.12 Vibrational and rotational spectral features arise within electronic states and involve much smaller changes in energy; hence, they occur in the infrared and radio-millimeter part of the spectrum.

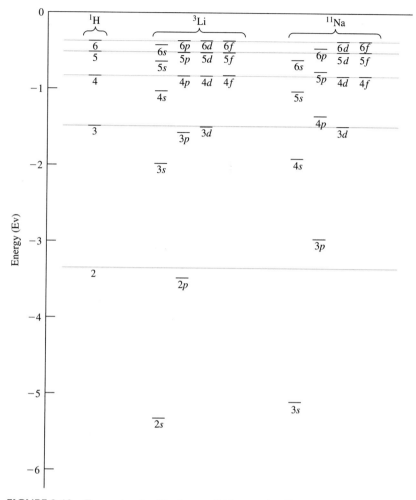

FIGURE 3.13 Energy levels of hydrogen, lithium, and sodium. By comparing the energy difference between two levels to the map of the electromagnetic spectrum in Figure 3.10, we can determine where the spectral lines corresponding to the change in electron energy level occurs.

form of "phosphorescence," for example, the electron is blocked from returning in a single jump to the ground state by quantum mechanical selection rules, and it slowly releases energy through collisions among molecules, leading to a low-energy, long-wavelength glow that persists. *Laser* emission involves pumping an electron to an excited state by putting in energy, and then the molecule to which it belongs emits photons that are contained in a reflective cavity. The photon stimulates other excited photons to emit, until a coherent, very narrow frequency optical beam is produced.

The energy difference between an unbound electron at rest and an electron in the $1s$ shell is 13.6 eV, which from equation (3.3) corresponds to a wavelength of 0.091 μm, or 910 Å, well into the ultraviolet part of the spectrum. The generation of even higher energy, shorter wavelength photons in ionization processes requires digging into the inner shells of elements of higher atomic number, where the energy cost to ionize ("ionization potential") is in the hundreds of electron-volts, on the threshold between the ultraviolet and X-ray part of the spectrum. Ionizing environments are common in the cosmos: lightning on Earth, the interiors of stars such as the Sun, and the outer regions of stars more massive and hotter than the Sun. A cathode-ray tube can send a beam of electrons into the inner shells of heavy atoms, stripping them of inner electrons and generating line spectra in the extreme ultraviolet and X-ray part of the spectrum as well. Simply stopping a beam of electrons on a dense target will emit a continuum of photons in the X-ray region of the spectrum—the electromagnetic emission being a consequence of the deceleration of electrons. Gamma ray generation requires deceleration of very heavy particles, electrons at relativistic velocities, or transitions from one nuclear (i.e., of the nucleus) energy state to another in heavy atoms. Cosmic rays—atomic particles accelerated to large velocities along magnetic lines of force emanating from the Sun or the Milky Way galaxy as a whole—will emit gamma rays when striking the atmosphere or surface of a planet, whose energies (frequencies) reflect the composition of the target material (Chapter 12). Catastrophic astrophysical events (e.g., supernovae explosions to form neutron stars or black holes) generate X-rays and gamma rays, but these are usually absorbed by the intervening interstellar gas and dust.

Moving to the other end of the electromagnetic spectrum, the generation of infrared and radio spectra are the result of lower energy processes that induce molecules to vibrate and rotate. Reference to the hydrocarbon organic molecules in Figure 3.8, and to water in Chapter 2, demonstrates that molecules have different shapes and different bond strengths and masses. Vibration occurs between atoms bonded in a given molecule and between atoms bonded to two different molecules in a solid or liquid. The interaction can be thought of, crudely, as if two masses were at the end of a spring. If the bond strength is thought of as the "force constant" C (related to the stiffness) of a spring, then the classical vibration frequency ν_o is proportional to the square root of $CmM/(m + M)$, where the masses of the two atoms (or molecules) at the ends of the bond are m and M. But because this is a quantum system, the allowed energies of the vibrational modes are quantized, so that $E_{vib} = \left(n_{vib} + \frac{1}{2}\right)h\nu_o$, where n_{vib} can be 0, 1, 2, 3, etc. Typical vibrational energies are 10 to 100 times smaller than those in electronic transitions, which puts the pure vibrational bands in the infrared. (They can, however, occur as small variations in the bands of optical and ultraviolet electronic spectra, as well.) Vibrational spectra can be very complex in multiple-atom (polyatomic) molecules, because many different vibrational modes (corresponding to oscillations in different directions of the same bond and coupling to oscillations of other bonds) can be present.

Rotational transitions are also quantized, and vary in energy because the distribution of mass of rotating molecules can differ greatly from one molecule to another. The energy of rotational transitions is typically 10 times (or more) smaller than for vibrational transitions, and thus such transitions occur in the far infrared and radio parts of the spectrum. Because planets and material in interstellar space are at sufficiently low temperatures that the peak of their photon emission is in the infrared and longer wavelengths, much information on the nature of this material can be derived from the vibration-rotation ("ro-vibrational") and pure rotation spectra. Likewise, as noted earlier, light from Sunlike stars reaching planets in their habitable zones and beyond is diluted enough that, when it interacts with atmospheric or surface material, infrared photons are generated. Selective absorption of infrared radiation by solids and gases provides diagnostic information on composition. Because vibration and rotation are affected by the mass of the atoms at the end of each bond, ro-vibrational spectroscopy is sensitive as well to the isotopic composition of material and provides a key means for identifying low abundance isotopes on remote bodies.

An example of the complexity and beauty of ro-vibrational spectra is given in Figure 3.15, which illustrates an important point: The finer we spread out the light from an object—that is, the higher the resolution in wavelength or frequency—the more detail we will see. However, higher spectral resolution means that a given amount of light is spread out over more wavelength "channels," hence, each channel has less light than for a lower resolution spectrum. Resolution is limited by signal, which means that larger telescopes, more sensitive detectors, or inspection by close-flyby spacecraft (the last for solar system objects only) are desirable when viewing remote objects spectroscopically.

Some molecules are spectroscopically inactive in the infrared because they are symmetric, and hence, the electrons in the orbitals interact very weakly with electromagnetic radiation. Examples include noble gases and molecules such as N_2 and O_2. In the case of the latter two, even ro-vibrational features do not occur, unless the shapes of the molecules are distorted by high-pressure—hence, frequent—collisions. An alternative to examining

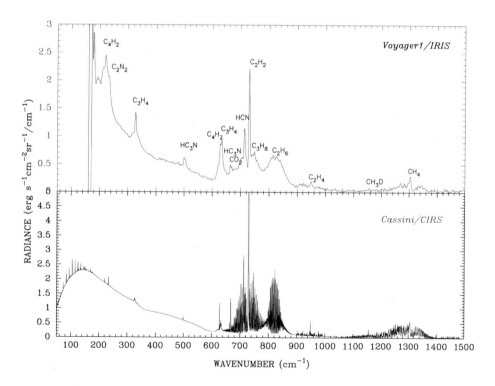

FIGURE 3.15 (Top) Mid-infrared spectrum of hydrocarbons (with chemical formulae labeled) in the atmosphere of Saturn's moon Titan; data from the Voyager spacecraft. (Bottom) theoretical spectrum of Titan at 10 times better wavelength resolution, showing rotational features within the vibrational bands.

the spectra of these types of molecules, at least in the laboratory, is *Raman* spectra. Here, molecules are exposed to intense, narrow-band electromagnetic radiation, and the light is scattered from the molecule inelastically, that is, with a small amount of electromagnetic energy absorbed. This absorption is observed as small shifts in wavelength relative to that of the original light. Weak features corresponding to ro-vibrational transitions, even in symmetric molecules, can be seen in the scattered light.

3.7 Isotopes and the Stability of the Nucleus

Not all nuclei are stable, and it is primarily the heavier nuclei, or those with an excessive or insufficient number of neutrons relative to protons, that break apart or "decay." Four basic types of radioactive decay are observed:

- Alpha decay involves emission from the nucleus of two protons and two neutrons as an α-particle, the stable nucleus of a helium atom. The original atom is left with a reduction of four in atomic mass, and two in atomic number, and is hence converted into a lighter element.
- Beta decay involves conversion of a neutron in the nucleus into an electron and proton. The proton stays behind, and the electron departs from the nucleus. This decay process, mediated by the weak nuclear force, leaves the atomic mass the same but advances the atomic number by one.
- Gamma decay does not alter either the atomic mass or number of the nucleus. A photon at gamma-ray energies is emitted from a nucleus that has (because of a collision or another decay process) been put in an excited state, one in which the configuration of the protons and neutrons is at an elevated energy level. The loss of the photon decreases the energy of the nucleus, but the number of protons and neutrons remains unaltered.
- Fission is the splitting of a massive atomic nucleus into less massive pieces, forming new elements of lower atomic number and mass than the original decaying element. Spontaneous fission involves release of energy as the nucleus splits. Fusion, the opposite process, involves the combining of lighter nuclei to form a heavier one.

Alpha and beta decay occur at predictable rates. While we cannot point to a particular radioactive atom in a box of such atoms and predict when that particular one will decay, the rate at which atoms decay in a large ensemble is readily measurable and a property of the particular isotope (atomic and mass numbers) only. The change dN in the number of radioactive atoms of a particular element is given by $dN = -N\, dt/t_{mean}$, where t is time, and t_{mean} is the so-called mean lifetime of the radioactive isotope; t_{mean} is related to a more intuitive number called the *half-life*. The half-life of a radioactive isotope is the time it takes half of the atoms of that isotope to decay (this assumes a very large sample). The mean lifetime is 1.4 times the half-life. The relationship—that the number of decays depends on the number of that isotope present and the rate of decay intrinsic to the stability of the particular kind of nucleus (i.e., its atomic mass)—makes intuitive physical sense.

By integrating that relation, we find that the number of atoms of the radioactive isotope present at time t is

$$N(t) = N(0) \exp\left(\frac{-t}{t_{mean}}\right) \qquad (3.4)$$

where $N(0)$ is the original number of radioactive atoms present, and exp is the mathematical exponential function (Figure 3.16). (This law does not apply in extreme environments where large amounts of charged subatomic particles are striking nuclei and spalling off particles). We use the law of radioactive decay in Chapter 6 to interpret the record of isotopic abundances in rocks in terms of the ages of those rocks and the processes that occurred on timescales comparable to the half-lives of the radioactive isotopes.

To close this chapter on the physics of the electron (which is chemistry!) and of the nucleus, let us briefly examine the processes of fission and fusion a bit more closely. The stability of protons and neutrons in the nucleus of an atom varies with the number of each in a systematic way that suggests the presence of stable shells analogous to those of the electron. This behavior was noticed in the early 20th century, and it was some time before this behavior could be rationalized in terms of the subnuclear particles moving around in the nucleus under the influence of the strong nuclear attraction and electromagnetic repulsion. For want of space, we must steer clear of this otherwise fascinating chapter in the development of quantum mechanics.

However, the stability of the nucleus as a function of the number of protons and neutrons has a profound consequence beyond the isotopic stability curve of Figure 3.4. The average binding energy of the nucleus per "nucleon" (proton or neutron) becomes greater as the atomic number increases, up to the value of 26—which corresponds to iron. Beyond iron, the binding energy *decreases* with increasing Z (Figure 3.17). Put another way, if we were to find a way to combine four protons—the nuclei of four hydrogen atoms—and

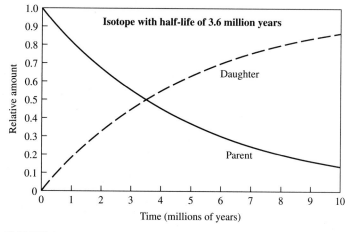

FIGURE 3.16 The radioactive decay law. The number of radioactive ("parent") atoms present is plotted as a function of time for a closed system. The number of stable product ("daughter") atoms is also plotted.

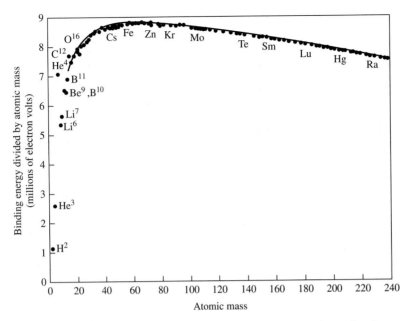

FIGURE 3.17 Binding energy per nucleon as a function of atomic number for the most stable isotope of each element. More negative corresponds to a more tightly bound—hence, stable—nucleus.

convert two of the protons to neutrons, we would have a helium nucleus *plus leftover energy* from the reaction. Or, the conversion of three heliums to carbon should generate excess energy, and so on. This alchemical dream of generating energy from the *fusion* of nuclei ends at iron—fusing nuclei with atomic number beyond iron costs energy. And, for large enough numbers of nucleons, the binding energy per nucleon becomes so small that such large nuclei are unstable, splitting to form lighter elements with a release of energy in the fission process.

Humankind has harnessed both fission and fusion to generate energy, in atomic and hydrogen bombs, respectively, as well as in fission reactors to generate electricity. But the largest nearby *fusion* reactor by far is the interior of the Sun, where hydrogen is converted under tremendous temperatures and pressures to form helium. This conversion is the origin of sunshine, and the details of the process can be monitored through the detection of exotic subatomic products of fusion. The process of fusion appears to have manufactured, in earlier generations of stars, the biogenic (C, N, O, etc.) and planetogenic (Si, Mg, etc.) elements that make up life and the planet Earth it lives upon. This profound link between nuclear fusion in stars and the existence of planets and life is one that illustrates the broad sweep of scientific considerations underpinning astrobiology. In Chapters 5 and 6, armed with the preparation provided here, we return to the details of energy and element generation in stars, and the implications for planets. In Chapter 4 we use the chemistry developed here and the physics described in Chapter 2 to understand the biochemical underpinnings of life itself.

QUESTIONS AND PROBLEMS

1. Pursue further the hypothetical situation raised early in the chapter: What if electrons were bosons rather than fermions? Starting with the text discussion, carry the discussion through as far as you can—even to the point of creating a periodic table of elements for a universe where electrons are bosons and all else behaves as it does in reality.

2. All four of the following gases are held at the same temperature and pressure. Which will have the highest average kinetic energy per mole?
   a. H_2O
   b. O_2
   c. H_2
   d. H_2S
   e. All are the same

3. A lander measures the following atomic composition of surface, expressed as percent number of atoms per total: O, 55 percent. Si, 20 percent. Al, 12 percent. others, 13 percent. What are the fractional contents *by weight* of aluminum, silicon, and oxygen?

4. A potassium-bearing compound emits a spectral line at a frequency of $7.41 \times 10^{14} \text{ s}^{-1}$. What is the corresponding wavelength of light emitted, in meters? What portion of the electromagnetic spectrum does this correspond to?

5. Suppose the Sun's surface temperature were suddenly to drop by half. Leaving aside the fatal effects on Earth's climate, what would be the peak wavelength at which the new "cool" Sun radiates? What would this do to the appearance of typical daylight scenes viewed with our eyes? Consulting Chapter 11, how transparent is Earth's atmosphere to the new peak wavelength of light coming from the Sun, and would the relative atmospheric transparency between the old and new peak wavelengths enhance or buffer the loss by our atmosphere of radiant energy from the Sun?

6. Indicate which of the following does not belong to the indicated family of elements:
   a. argon—noble gas
   b. lanthanum—rare earth element
   c. vanadium—transition element
   d. cerium—actinide element
   e. radon—noble gas

7. "Isoelectronic" means having the same number of electrons. Which of the following is not isoelectronic? (Note: The "+" symbol refers to a charged ion lacking a number of electrons equal to the number of pluses.)
   a. Br_2
   b. ICl
   c. Os^{++++++}
   d. SF_8
   e. GeF_4

8. In Table 3.3 some of the abundances listed under "meteoritic" (pertaining to meteorites) are footnoted as being derived from solar data. What these elements have in common is that they tend to occur in highly volatile form. Can you explain, then, why their abundances are not measured directly in meteorites?

9. What element has the ground-state electronic configuration of $1s^2 2s^2 2p^6 3s^2 3p^2$? (This will turn out to be an element of high abundance in the crust of the Earth). Check the periodic table for help.

10. The copper cation, Cu^{++}, is important in some biological systems. For instance, hemocyanin, a complex of a protein and copper, is used by many invertebrate animals for the transport of oxygen (humans use iron and hemoglobin instead). What is the ground-state electronic configuration of Cu^{++} in hemocyanin? (Note: Only the valence shells are written out, and the nonvalence shells are expressed as the equivalent element having that number of electrons. Hence [He] means two electrons.)

   a. $[Ar]4s^2 3d^8$
   b. $[Ne]4s^2 3d^8$
   c. $[Ar]3d^9$
   d. $[Ar]4s^2 3d^{10} 4p^9$
   e. It cannot be any of the above

SUGGESTED READINGS

As in the previous chapter I leave it to the readers to select the textbooks that best resonate with their particular mental processes in learning more about chemistry. The ones given suit my particular taste. A treatment of physics and chemistry that is well written and draws on the profound insights of its author is F. Shu, *The Physical Universe: An Introduction to Astronomy* (1982, University Science). Readers hungering for more detail on the radiation and quantum aspects of this treatment might refer as well to Shu, *The Physics of Astrophysics* (two-volume set) (1991, University Science).

REFERENCES

Atkins, P. W. (1986). *Physical Chemistry*, W. H. Freeman and Co., New York (3rd edition).

Cox, A. N. (ed.) (1999). *Allen's Astrophysical Quantities*, AIP Press, New York.

DeLaeter, J. R. and Heumann, K. G. (1991). "Atomic Weights of the Elements," 1989, *Pure Appl. Chem.*, **63**, 975–990.

Eisberg, R. and Resnick, R. (1974). *Quantum Physics of Atoms, Molecules, Solids, Nuclei and Particles*, John Wiley and Sons, New York.

Lang, K. R. (1980). *Astrophysical Formulae*, Springer-Verlag, Berlin.

Lunine, J. I. (1998). *Earth: Evolution of a Habitable World*, Cambridge University Press, Cambridge, U.K.

Manthey, D. (2002). *Orbital Viewer*, unpublished manual on http://www.orbitals.com.

CHAPTER 4

Necessary Concepts III
The Chemistry of Life

4.1 Introduction

Perhaps the most remarkable aspect of life is that its underpinnings lie in the electron—through its bonding properties and through its behavior as an electrically charged particle. This statement may seem a gross oversimplification, and yet virtually all of the biochemical substances and phenomena we discuss in this chapter are manifestations of the chemical principles outlined in Chapter 3 and whose building blocks, in turn, are in the physics of Chapter 2. Indeed, it might even be argued that life is a phenomenon of electromagnetism, with the nuclear forces simply serving—as they do for other normal matter—to ensure that there is mass or "bulk" to life and the force of gravity largely a sideshow that has constrained macroscopic morphology but not biochemistry. This chapter introduces the essence of biochemistry in the basic biopolymers of life, the structures and processes in cells, key energy transport, generation and storage mechanisms, and the workings of the genetic code and protein synthesis. One glance at a comprehensive text on biochemistry may convince the reader that attempting a 30-page review is sheer folly. But such a distillation, taken in concert with your favorite (or newly encountered) biochemistry textbooks, will provide the road map by which our discussion of the origin and evolution of life in the following chapters may be better understood.

4.2 The Fundamental Biopolymers, Their Units, and Other Molecules

We know only of one kind of life, that which exists on Earth, and all such life is based on a small set of *polymers* (high molecular weight chains of more fundamental molecular or atomic units). The key "biopolymers" are nucleic acids and proteins. *Nucleic acids* store

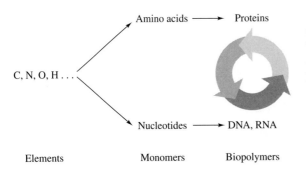

FIGURE 4.1 Symbolic relationship among the biogenic elements, the nucleotides and amino acids, and their polymers DNA, RNA, and proteins.

and transmit genetic information necessary for the structural and functional properties of the basic unit of life, the cell. The two different kinds of nucleic acids used by life are *RNA (ribonucleic acid)* and *DNA (deoxyribonucleic acid)*. *Proteins* are the principal structural polymers of cells, and they are the primary "catalysts" of biochemical processes. By definition, catalysts are molecules that change the speed of chemical reactions without themselves being destroyed. A specific catalyst may be consumed in one reaction, but there is then another subsequent reaction or set of reactions that regenerate the catalyst. Catalytic proteins, or *enzymes*, achieve their catalytic properties through their shapes, which result from intricate folding of the long polymeric chains. Proteins and nucleic acids have a tightly coupled relationship in which DNA provides the information needed to code for proteins, RNA transfers this information and, together with certain proteins, implements the assembly; DNA itself is replicated with the assistance of other proteins. Figure 4.1 symbolizes these relationships.

The molecular units that form the nucleic acid and protein polymers are simple molecules composed of the common elements (in no particular order) carbon, hydrogen, oxygen, nitrogen, sulfur, and phosphorous. These are the *biogenic elements*. They are by no means the only elements used by life on Earth, but they are the most abundant and most crucial in their structural and energetic roles. The building blocks of proteins are *amino acids*; those of DNA and RNA the *nucleic acid bases*.

Formation and Structure of Proteins

Amino acids are compounds containing two organic "functional groups," that is, a set of atoms that bond to other species as a self-contained unit: $-COOH$ and $-NH_2$. The $-COOH$ is called a *carboxyl group* and has a double bond between the carbon and one of the oxygen atoms and a single bond between the carbon and the OH (hydroxyl) group. The carboxyl group sits on one side of the amino acid; what distinguishes this class of carboxylic acids is that on the other side of the molecule is an *amino* group, NH_2. This is simply an ammonia molecule missing a hydrogen atom. Replacing that hydrogen atom—with the same net valence contribution—is a central carbon (called "C_α") bonded on one side to the carboxyl group, on another to a hydrogen atom, on a third to the amino group, and on a final side to any of a variety of groups "R" ranging from hydrogen itself to, (for example), $-CH_2-CH_2-S-CH_3$ (Figure 4.2). This R group comprises the "side chain" of the amino acid. There are hundreds of different kinds of amino acids, all distinguished by a

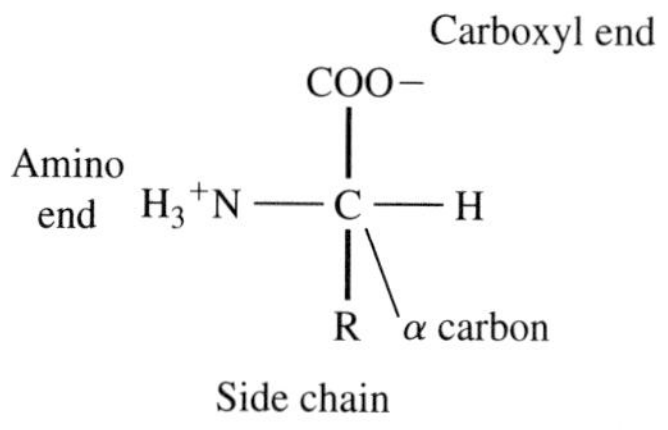

4.2 The Fundamental Biopolymers, Their Units, and Other Molecules

```
                              Carboxyl end
                           COO−
                            |
    Amino           +       |
      end      H₃ N ——— C ——— H
                           |\
                           R  α carbon
                        Side chain
```

FIGURE 4.2 Basic structure of an amino acid. The "R" group is any of 22 different groups, 20 of which are shown in Figure 4.3.

unique R group. Of these, only 20 are commonly used in extant life forms (Figure 4.3, see Color Plates 2–4). Two others, selenocysteine and pyrrolysine, are less common. The former occurs in prokaryotic archaea and bacteria (see Section 4.3) and in animals, including mammals. The latter has been found in only one archeal organism and may occur in bacteria as well. The 20 biochemically important amino acids are conveniently classified according to how the side chain tends to interact with other molecules. The *hydrophobic* class consists of nine amino acids, whose side chains repel water. The *polar* class contains five amino acids, whose side chain tends to interact strongly with water. The *charged* class consists of five amino acids, each of which possesses an R group bearing a residual positive or negative charge. Finally, the small amino acid glycine has a single hydrogen atom comprising its side chain. While hydrophobic, it can exist in a large range of chemical environments and hence is in a class by itself. These distinct classes of chemical behavior, particularly with regard to water, are fundamental to the manner by which the proteins fold and to the stabilities of the folds; subtler differences within each class add substantial additional richness to the protein structures.

That life uses only a fraction of the chemically possible amino acids to build proteins is a remarkable example of the specificity or selectivity of biochemistry relative to the larger range of possible organic chemical compounds. Another is that for all amino acids, except symmetric glycine, biochemical processes employ—with few exceptions—only one of the two asymmetric or "chiral" forms, crudely, left- and right-handed. Each form of a chiral molecule is called an "enantiomer." When an amino acid with a definite handedness reacts chemically with a molecule that is symmetric (or otherwise does not have a particular handedness), the left- and right-handed amino acids have similar properties. Likewise, the chemical properties of an interaction between two left-handed molecules or two right-handed molecules are the same. However, neither of these interactions is the same when a left- and right-handed molecule are interacting with each other. Hence, the handedness of biological molecules such as amino acids or nucleotides plays a role in their functionality. This so-called "enantioselectivity," or exclusive preference for one "handedness" of chiral molecules, is essential to the proper functioning of the proteins. Chirality also plays a role in the sugar backbones of DNA and RNA, which imposes important constraints on the origin of these molecules as discussed in Chapter 8.

The linkage of amino acids to form long polymeric chains is achieved through the so-called dehydration reaction between amino acid functional groups and involves the removal of a water molecule. As shown in Figure 4.4, the amino group of one amino acid is oriented toward the carboxyl group of the other. The hydrogen and hydroxyl groups

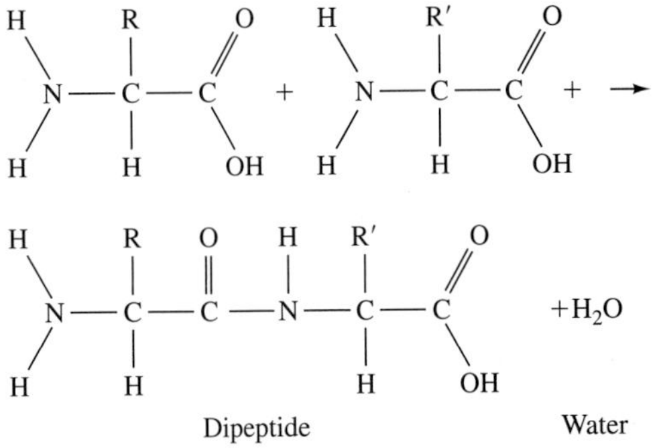

FIGURE 4.4 How two amino acids bond together to form a "peptide linkage."

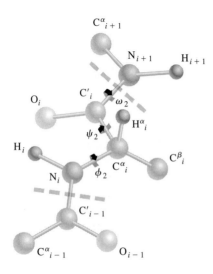

FIGURE 4.5 A section of polypeptide chain representing two peptide units. The limits of a "residue" (each individual amino acid unit) are indicated by dashed lines; portions of three amino acids (indexed as $i - 1$, i, and $i + 1$) are shown. Notations for atoms (for example, C^α) and torsion angles ω, ϕ are indicated.

(H and OH) bond to form water, which is detached from the combined structure, and a bond is formed between the amino nitrogen and the (previously carboxyl) carbon. The combined dual amino acid structure is called a *dipeptide* and the chainlike assembly of large numbers of many hundreds to thousands of amino acids comprises the protein. Because each amino acid has now lost an H and OH group from each respective end, it is chemically convenient to think of the basic units of the chain as running from one C_α atom, to the adjacent C═O double bond, the N—H residue of the amino group, and ending with the next C_α atom. Such a *peptide* unit has a planar shape and well-defined angles between bonds (Figure 4.5); the chain made up of these units is called a *polypeptide*. However, the functionality of the protein is not expressed through the assembly of the amino acids into a polypeptide chain; the resulting structure must then fold, creating the functional shape. Why does the long chainlike polymer fold into a particular form?

The fundamental freedom that the peptide units have is rotation about particular bonds—those involving the C_α carbon, the adjacent carbon, and the adjacent nitrogen; see Figure 4.5. The overall arrangement of rotations in the polypeptide is called the "conformation" of the chain, and these rotations allow the R groups comprising the side chains to assume different angles relative to each other and hence to the surrounding medium,

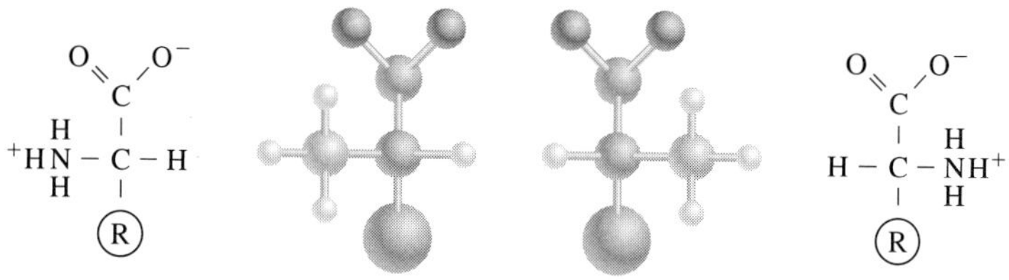

FIGURE 4.6 Asymmetric amino acids are chiral; that is, the mirror image forms cannot be superimposed one atop the other. In the figure an amino acid with a generic R-group is shown, via the chemical formula and a ball-and-stick model, in its two enantiomeric (left- and right-handed) forms.

which will include other parts of the folding main polypeptide chain, other polypeptide chains, or other molecules. The positions of the R groups are well specified by the rotation angles because living processes—as noted earlier—use amino acids of a single orientation or handedness. In fact, this asymmetry in the use of "left-" versus "right-" handed amino acids (Figure 4.6) puts severe restrictions on the combination of rotation angles possible because the side chains repel each other via repulsive electromagnetic interactions if they are brought too close together.

The restricted conformations of the polypeptide chain are one element of the folding rules for proteins. The other is the set of interactions between the side chains themselves, and between the side chains and the environment, associated with the folding. In particular, because proteins function in an aqueous (liquid water) medium, the hydrophobic side chains tend to fold to minimize their contact with water. As a result, the interior of the folded polypeptide chain—the protein—is hydrophobic. A complication arises, however, because the main chain itself is a polar structure and, hence, tends to interact strongly with water molecules. This hydrophilic behavior must be negated to achieve a stable structure with the hydrophilic side chains isolated from water. The NH group can hydrogen bond (Chapter 3) with the C=O group, decreasing the hydrophilic behavior, but forcing the chain to assume one of two primary shapes—either a spiral helix (Figure 4.7, see Color Plate 5) or a set of strands that form a pleated sheet. These helices and sheets form the interior of the protein and are connected to each other along the protein surface by a series of loops of various lengths.

Together, these three common "secondary" structures—helices, sheets, and chains—provide a set of topological tools that constrain the folding of the polypeptides into complex shapes (Figure 4.8, see Color Plates 6–7). A complete protein may result from a single polypeptide chain forming these so-called tertiary structures, or several polypeptides will assemble to form a single protein (the "quaternary structure"). Additional stability of the folded protein structure is attained by attractive interactions among side chains that are brought close to each other by the folding. This is a remarkable consequence of the three-dimensional structure of the protein—peptides that are aligned along the main chain at great (primary sequence) distances from each other may come into proximity and interaction by folding. Because, as we see later, the genetic instructions in DNA for the construction of proteins refer only to the ordering of the amino acids along the main chain, such folded interactions represent a kind of epiphenomenon that result from, but are not directly encoded by, the basic nature of the information carrying nucleic acids. The complexity of the final expressed protein structure, which can include rolls, nooks, mountains, and so on, creates an intricate world of different surfaces on or in which other molecules interact. These other molecules may be other proteins (to form structures and motors), so-called cofactors (e.g., vitamins) that participate in chemical reactions, foreign proteins that must be removed from a cell or organ, DNA itself during the replication process, and so on. Chemistry on surfaces—which can be much more rapid and more specific in the reaction sets than chemistry in a uniform three-dimensional volume—becomes possible on the intricate sculpture of the protein, created when amino acids are strung together along one or several one-dimensional chains.

These higher-order bonding relationships, and the complexity of the three-dimensional structures themselves, make it virtually impossible to predict a priori the structure of a protein. The structure is worked out experimentally, primarily through a technique called *X-ray crystallography* or *X-ray diffraction*. In Chapter 2 we explored how light from a single source, passed through two slits, forms two sources of light that interfere with each other to form a pattern of regular, light and dark, fringes, on a screen behind the slits. A more complex arrangement of slits will produce a more complex pattern, and in principle, there is enough information in the pattern to deduce the arrangement of the slits. Instead of slits, a similar phenomenon is achieved by the scattering of light off a pattern of obstructions; the bending of light around the obstructions is also *diffraction*. The diffraction pattern can be used to deduce the positions of the obstructions, if the wavelength of the light is chosen to match the scale of the system. Hence, electromagnetic radiation at radio wavelengths will diffract around conductors on scales of millimeters to meters. Visible light will diffract around obstructions on scales of a micron. Because the structure of proteins is determined by amino acids with size scales of thousandths of a micron, to use the amino acids as obstructions requires wavelengths of comparable size scale—X-rays. To obtain a clean diffraction pattern that is directly relatable to the positions of the various peptide bonds requires that the protein be isolated—purified—from others and from the aqueous medium.

Proteins can be grown, in different kinds of solvents, as regular crystal arrangements, and X-ray diffraction is performed on these crystals. The large number of repeated examples of the same protein structure, seen in the crystal, generates a definite diffraction pattern that can be interpreted to infer the original structure (Figure 4.9). Multiple wavelengths of X-rays can be employed to resolve structures on different scales and eliminate ambiguities. The growth of protein crystals can be a laborious process, but automation of the crystallization as well as a high-speed computational capability, which allows many different possible structures to be tried as solutions to the diffraction pattern, have reduced the analysis time for the structural determination enormously.

Nuclear magnetic resonance spectroscopy, or NMR, can also resolve the structure of small proteins. Here the protein is placed in a very strong magnetic field, which forces the protons constituting the hydrogen atoms to align themselves in a preferred direction relative to the field (the other nuclei are affected as well, but the strongest signal comes from the hydrogen nuclei). If a variable electromagnetic field is applied, at radio wavelengths, the aligned protons will go into and out of excited states that are well defined and will absorb the radio-frequency energy at a particular set of frequencies for a given protein.

 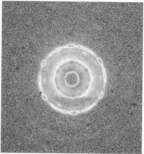 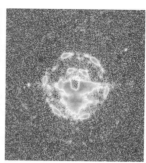

FIGURE 4.9 X-ray diffraction pattern obtained from protein. On the left, a 50-μm-wide beam of X-rays was used to record the diffraction pattern from the fibers of the flight muscles of a bumblebee. The simple ringlike diffraction pattern is a result of the match between the beam size and the width of a muscle fiber. By narrowing the beam to 2 μm wide, the patterns in the center and right-hand panels are obtained; here the structural features of the muscle fiber protein can be diagnosed. Labels indicate the positions of diffraction rings expected for a particular kind of symmetric lattice (hexagonal); the strengths of the corresponding diffraction rings for the muscle fiber then provide an indication of the structural symmetry of the protein.

Because the protons are held themselves in particular electromagnetic environments depending on their positions relative to the other atoms in the peptide structure, the absorbed radio frequencies are discrete and diagnostic of the protein structure. The advantage of NMR over X-ray diffraction is that the protein does not need to be crystallized; the protein simply needs to be in a concentrated solution. However, NMR provides diagnostic spectra only for small proteins.

Formation and Structure of Nucleic Acids

DNA and RNA are very similar polymers with a couple of differences in composition that lead to quite different functions. Their long polymeric chains are assembled from nucleotides. Each nucleotide is composed of three distinct structural components. First is a molecule of *sugar*. Sugar is a subclass of *carbohydrates*, a class of molecules that have the form $(CH_2O)_n$, (n being an integer) and are a principal source of energy for biological processes through their oxidation. The sugar in RNA is ribose; in DNA, it is deoxyribose. Connected to the sugar molecule is a *phosphate* group, which has at its center a phosphorus atom. Phosphate bonds store large amounts of energy, and they will be discussed further in connection with the energy-storing molecule ATP. Together the sugar and phosphate form the backbone of the nucleic acid molecule, alternating one with the other to form a chain.

Positioned along the chain, by connection to the sugar molecules, are the nucleic acid bases—*purines* and *pyrimidines*. These are small molecules composed of—as with the amino acids—C, H, O, and N atoms. The structure of the pyrimidines is that of a hexagon of four carbon and two nitrogen atoms, with hydrogen, nitrogen-hydrogen, oxygen, or methyl (CH$_3$) groups projecting from the ring. Purines have a second ring, pentagonal this time, joined to the first (Figure 4.10). DNA contains the purines guanine and adenine and the pyrimidines cytosine and thymine; in RNA, uracil is present in place of thymine. Although only two of the purines and three of the pyrimidines are used in DNA and RNA, these are members of a much larger class that otherwise are not participants in the fundamental biochemistry of life—illustrating again the molecular selectivity of biochemical processes. Purines and pyrimidines share with carbohydrates, chlorophyll, and hemin—which we will encounter later—the property of being *heterocyclic*, that is, containing rings made up of more than one kind of atom. Life uses heterocyclic compounds in many fundamental processes, and it is again the multivalent versatility of carbon, and especially its propensity to form rings with itself and other atoms, that make these possible. Silicon's bonding properties are sufficiently different that they cannot support silicon-based heterocyclic rings.

The H, NH, O, or CH$_3$ protuberances on the purines and pyrimidines are capable of hydrogen bonding with each other, but they can do so only in very specific ways. In particular, the purine adenine will bond with the pyrimidine thymine, or with the closely related uracil, via two hydrogen bonds. Guanine and cytosine, on the other hand, can and do form three hydrogen bonds since there is an extra electron donor-acceptor pair on those two molecules. Therefore, *the purines and pyrimidines in DNA and RNA pair with each other in specific ways: adenine with thymine or uracil, and guanine with cytosine. The alternative pairings are not stable.* If we abbreviate the purines and pyrimidines by their first let-

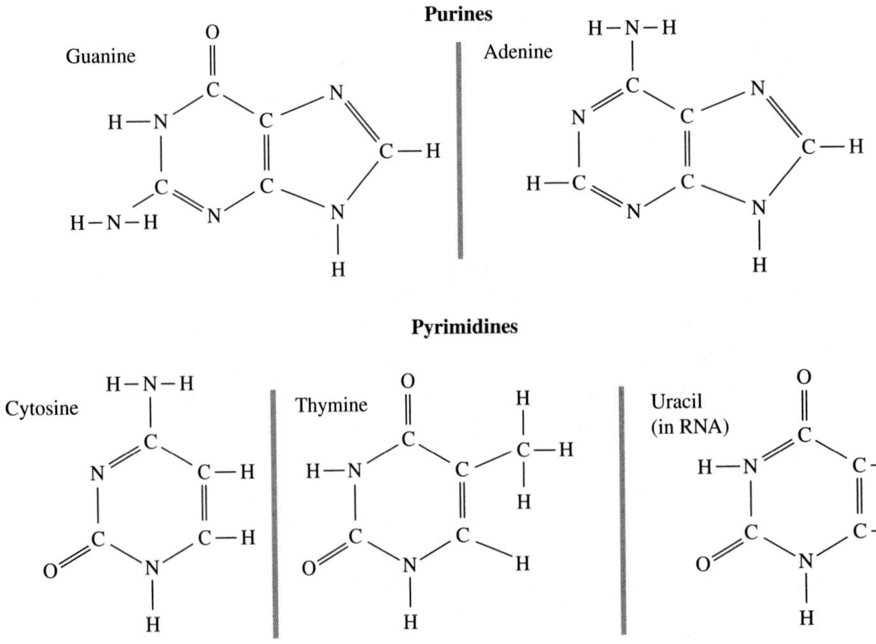

FIGURE 4.10 Schematic structures of the purines and pyrimidines in DNA and RNA.

ters, A (adenine), C (cytosine), G (guanine), T (thymine), and U (uracil), we obtain the abbreviated pairing rules A-T (DNA), A-U (RNA), and G-C (DNA and RNA). We call each of these pairs a *base pair*, and one member of each pair is the complement of the other.

The base pairings lie at the heart of the functionality of the DNA and RNA molecules, but to understand this we must delve further into how the three basic structural components—sugar, phosphate, purine/pyrimidine bases—assemble. If we imagine the sugar-phosphate main chain to be the rail of a ladder, then the bases are the rungs of the ladder—projecting out into space. But attached to the projected rung—in DNA—is its complement. Because it is connected to its own sugar-phosphate backbone, the combined structure is a complete ladder: two sugar-phosphate rails bound by rungs composed of base pairs of purines and pyrimidines. The sugar-phosphate ladder is not stable as a flat chain but instead has a helical shape, like a spiral staircase. For the bases to face and hence bond to each other, the two helical rails must run in opposite directions; hence, the helices are wound around each other with 10 base pairs per turn and a diameter of about 20 Å (Figure 4.11). This structure is stable, particularly in an aqueous environment (as in the interior of a cell), because the interior of the double helix (where the base pairs are located) is hydrophobic. The double helix was discovered through X-ray diffraction studies and first correctly interpreted in the 1950s; the base pair complementarity was deduced earlier through analytic chemistry, which found equal numbers of G-to-C and A-to-T in DNA.

Because the complementary base-pairing of each rung is redundant—if we know one half of the pair, we know the other—the structure of DNA is fully functional as a means of replicating a string of information. The molecule need only tear down the middle, and then each half can rebuild, rung by rung, the same ordering of base pairs as the original maintained. Hence, in each DNA molecule is the capacity to create two daughter DNA molecules. And what is that string of information? We discuss it in detail in Section 4.8, but in

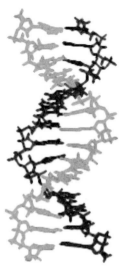

FIGURE 4.11 Helical structure of the DNA molecule, with the bonds of the deoxyribose/phosphate backbone and the nucleotide bases represented as sticks.

short, it is the ordered chain of nucleotides. There is no chemical or functional barrier to any arbitrary ordering of purines and pyrimidines along the chain—AATCTGCATGTC is just as possible as CCTTTGACCCTA, for example. (But once we have AATCTGCATGTC as one half of the base-pairing ladder, we must have TTAGACGTACAG as the other half). And so, if a particular base order has no functional advantage, then why does it arise, and why ought it to be duplicated? Given that cells must divide to survive into succeeding generations, a fair guess is that the information contained in the ordering is something that the cell needs to survive. It is; in fact, the ordering of the nucleic acid bases—taken three rungs at a time, like three-letter words—dictates the order in which amino acids are assembled to make proteins.

RNA, in which the deoxyribose is replaced by ribose, does not form a double helix. Instead, it is a single strand, and hence the base pairs are exposed. As we discuss later, RNA's roles in the cell are more varied than those of DNA, and corresponding to this, its shape can be quite varied. For example, RNA can fold over so that the bases pair with each other, yet the molecule remains a single strand. We might guess—with its exposed base pair rungs—that RNA plays some sort of messenger or transferring role in taking the genetic information stored in DNA and converting it to proteins. Before delving into this role (which is surprisingly intricate), we need to understand more about the structure of cells, which in turn requires considering a few more of the basic biomolecules.

Other Important Biomolecules

Beyond the fundamental biopolymers, a number of other molecules play important roles. Closely related to a nucleotide unit of RNA are adenosine diphosphate (ADP) and adenosine triphosphate (ATP). As shown in Figure 4.12, the three fundamental units of these molecules are the purine adenine, the sugar ribose, and a phosphate component. The ade-

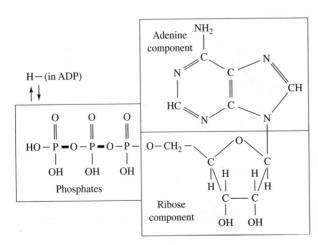

FIGURE 4.12 Structure of the ATP molecule. Heavy lines are the high-energy phosphate bonds.

Adenosine triphosphate (ATP)

nine "residue" differs in form from that of the free base through the substitution of the ribose for the H at the N-H group. Attached to the other end of the ribose is the phosphate group. In ADP this group consists of two phosphate groups connected by a high-energy phosphate bond. This bond releases 31 kilojoules (thousands of joules) of energy per mole of phosphate. ATP by contrast, adds another phosphate group to the two present in ADP; hence, there are two high-energy phosphate bonds in ATP. If a phosphate group is removed from ADP, adenosine monophosphate (AMP) would be left.

The formation of ATP from ADP requires *adding* energy to the molecule to increase the number of phosphate groups from two to three. Removing the additional phosphate group releases that energy, making it available for other reactions. The reason for the inverse relationship between bond formation/breakage and energy addition/release is that the bond is formed, in net terms, by dehydration—H and OH are removed from the molecule to form water. It requires energy to remove the H and OH from the molecule, and the bond left behind "contains" that energy. The added high-energy phosphate bond in ATP is not strictly a source or sink of energy; rather it is the biological equivalent of monetary currency that is passed around the cell to absorb the energy of metabolic processes and make it available for protein production, replication, and cellular repair. Photosynthesis (the oxidation of carbohydrates) allows the cell to "earn" phosphate bonds, which can then be "spent" to construct molecules within the cell—as well as to initiate metabolic processes that require a kick-start to begin.

Carbohydrates and, secondarily, fats and some proteins are broken down and oxidized to provide the principal sources of metabolic energy in cells. This energy is not invented out of whole cloth—the synthesis of the vast majority of carbohydrates involves absorption of sunlight and conversion of carbon dioxide and water through the process of photosynthesis described in Section 4.7. The oxidation of carbohydrates (combination with oxygen or a net loss in the electron number of the carbohydrate) leads to ethyl alcohol and carbon dioxide in the *fermentation process* and then all the way to carbon dioxide and water in *aerobic respiration*, which produces a higher energy yield (more ADP to ATP conversions) than does fermentation.

Fats, while a secondary source of metabolically derived energy, play other important roles of storing energy long term to tide over organisms when ready sources of carbohydrates are unavailable. Fats have several advantages over carbohydrates in regard to storage: Their insolubility in water means they absorb less water and hence weigh less than carbohydrates, and the larger number of metabolically useful C—H bonds in fats as opposed to C—O bonds in carbohydrates means more potential energy is stored. While a typical person has a day's backlog of carbohydrates available for powering cellular processes, about a month's worth is stored in body fat.

Fats are a subset of the larger class of water insoluble, biologically produced organic compounds, called *lipids*. Within this class are the *phospholipids,* which are found in the membranes of all cells and are an essential structural component of those membranes. There they form so-called *bilayers* by virtue of having a polar end to the molecule, which is soluble in water and therefore hydrophilic, and a nonpolar end, which is water insoluble and hence hydrophobic. The polar end is a phosphate, and the nonpolar end is a chain of hydrocarbons (hydrogen and carbon atoms bound together). The phospholipid will array in an opening between two different aqueous media in two rows so that the hydrophobic

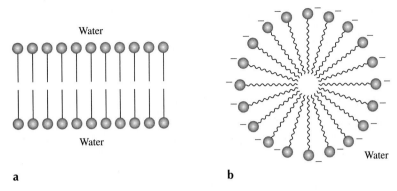

FIGURE 4.13 (a) Arrangement of phospholipids to form a bilayer within a cell membrane. (b) An alternative arrangement, call a micelle.

ends, which are chains of so-called fatty acids, point toward each other and are isolated from the water while the hydrophilic ends are partly dissolved in the water (Figure 4.13). In this way, the bilayers form a barrier that can aid in the selective passage of nutrients, functional molecules, and waste products through the cell membrane. Such bilayers might have played the role of primitive precursors to cells prior to or during the origin of life (Chapter 8). In a uniform aqueous medium, the hydrophobic-hydrophilic interactions lead not to bilayers, but to spherical clusters called *micelles*, which play a key role in the cleaning properties of soap.

Vitamins are nutrients that higher organisms (plants and animals) cannot manufacture for themselves but that are essential to living functions. A primary role of vitamins is as *co-enzymes*, molecules much smaller than the protein enzyme itself, which fit within the functional shape of the protein to enable beneficial interactions with other small molecules to occur. What is and is not a vitamin depends on the organism—most mammals can make ascorbic acid (vitamin C), but primates such as humans cannot. Likewise, many reactions require inorganic minerals such as iron and magnesium that must be ingested in a chemically usable (generally water soluble) form; the metal ions help in catalysis, and are called *co-factors*. That living cells are dependent on their environment for the supply of key components of enzymatic activity is a theme that extends back to the earliest life-forms, which may well have been completely dependent on abiotically produced reduced (hydrogen-rich) organic matter for food (Chapter 9). Other small molecules, *hormones*, also regulate certain enzymatic and protein production functions but are distinguished by being solely cellular products not available from external abiotic sources.

The final class of biochemical molecule is the *porphyrin*. These compounds are distinguished by having metal ions coordinated at their centers, bound by nitrogen atoms contained in heterocyclic rings. The presence of the electronic transitions in metals means these molecules are strongly colored—they absorb and emit light over narrow ranges of optical wavelength. As specialized as such molecules may seem, one of them—"chlorophyll" (Figure 4.14)—is the primary gateway by which the Sun energizes the biosphere; we discuss the process of photosynthesis by which it does so in Section 4.7. "Hemin" is

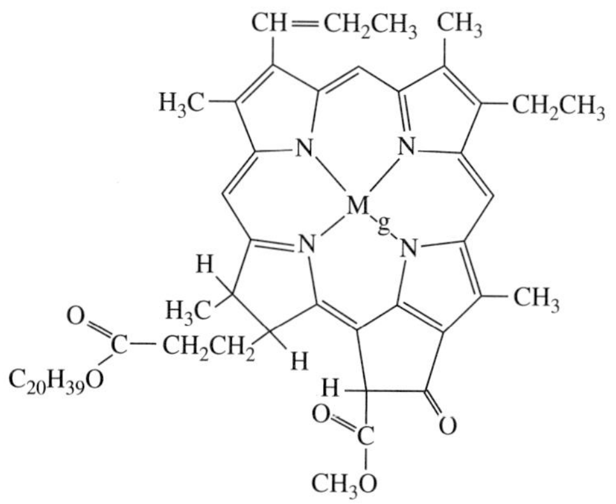

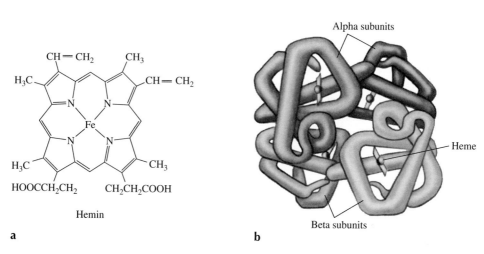

FIGURE 4.14 The structure of chlorophyll A.

FIGURE 4.15 (a) The atomic makeup of hemin. (b) Heme groups (sites of hemin) within the hemoglobin protein.

another porphyrin with iron bound by nitrogen to heterocyclic rings (Figure 4.15). It is, however, not a stand-alone molecule but instead is contained within a protein, which is usually "hemoglobin" (but in most arachnids and some mollusks, e.g., is "hemocyanin"). Hemin's active chemical role is defined by the ability of the iron-bearing ring structure to reversibly bind an oxygen molecule, O_2. Hemoglobin is carried in the bloodstream of animals to transport bound oxygen, acquired in the lungs, to all the cells of the body, where the oxygen is removed and used to oxidize carbohydrates in respiration. When oxygen is bound to the hemin, it gives the hemoglobin protein a red appearance. Chlorophyll, on the other hand, reflects green, indicating that the peak wavelengths at which it absorbs sunlight are red and blue. These complementary processes of aerobic respiration and photosynthesis are global in their impacts, but they reside in and rely on the fundamental unit of living systems, the cell, to which we next turn.

4.3 Informational, Structural, and Transport Systems of the Prokaryotic Cell

Life's biochemical functions are organized into a basic unit called the cell. Even entities that are not cellular, such as viruses (and, to some biologists, that do not qualify as living) must co-opt living cells to replicate. The essential functions of the cell are as follows:

- A dynamic membrane that exhibits fluidlike, flexible motion.
- A set of embedded, membrane proteins that capture energy-bearing molecules, *metabolites*, from the environment, and transport them into the cell.
- A set of enzymes that break down the metabolites and use the breakdown products to construct more membrane, more enzymes, and more genetic material (RNA/DNA).
- A genetic string, RNA coded by DNA, that encodes for the set of structural proteins and enzymes.
- A genetic program, DNA primed by RNA and enzymes, consisting of the set of triggering relations among the sets of instructions for making proteins. The program will cause the cell to grow, duplicate the genetic string DNA, and eventually divide when it has gotten large enough, resulting in two cells that will continue to metabolize, grow, and divide.

The sizes of cells are bounded by physical limitations on the fundamental processes that must go on within them. On the lower end, cells must be able to contain enough DNA to code for the proteins that comprise the organism and its enzymatic functions. As tabulated in Chapter 16, the minimal organism must code for hundreds of proteins, and the volume within which sufficient DNA and RNA (with the proteins needed for replication) can be held corresponds to a diameter of some tenths of a micron (recall that 1 μm = 0.000001 m). Proteins themselves are typically a few thousands to a hundredth of a micron across, comparable to the X-ray wavelengths in the electromagnetic spectrum used for diagnosing their structure. Other biomolecules are comparable or smaller; lipids are typically 0.002 to 0.003 μm in length. Viruses, which pack only a minimal amount of RNA within a protein capsule and contain a drilling device for entering complete cells, are 0.05 to 0.08 μm in size.

Although the theoretical minimally complete organism could approach a virus in size, in practice the smallest cells are 1 μm across. Such cells are *prokaryotes*, organisms that belong either to the Eubacteria ("true" bacterial) or the Archaea domains (Chapter 9). Prokaryotic cells are distinguished by having operational components that for the most part (the exception being photosynthesizing bacteria) are not enclosed by membranes within the cell.

The upper end of the size range for cells is dictated by the surface-area-to-volume ratio of the cell. Because most cells are (very roughly!) spherical, the volume increases as the cell diameter cubed, while the surface area rises as the cell diameter squared. Hence, larger cells have more volume within which to conduct biochemical processes but less surface area over which nutrients can be brought in and waste products expelled. Eventually the available surface area becomes the limiting factor, although it is a complicated affair to determine just what the limiting factor is. Prokaryotic cells range in size up to 10 μm. *Eukaryotic cells*, which contain individual *organelles* enclosed by membranes that compartmentalize cellular function, range from 10 to 100 μm in size. (Certain cells that are highly elongated, like nerve cells, are much larger, but the functions performed by the elongated portion are usually specialized and limited.) Organisms of the eukaryotic domain have overcome the cell size barrier—not with larger cells but through evolutionary specialization of cell types that allows different cells to perform particular functions in a much larger organism. Elaborate networks through which nutrients and oxygen-bearing molecules are transported—by elaborate pumping systems—and delivered to individual eukaryotic cells are required. To bring enough oxygen into the organism to supply all the cells requires specialized organs (lungs and gills) with very large surface areas exposed to the external atmosphere or ocean. Likewise, food must be broken down and digested through an intricate tubular labyrinth, featuring near its climax absorbing cilia that maximize the surface area and resident colonies of bacteria that help complete the breakdown of food stuff. Such are the prices that must be paid for size.

Discussion of the details of eukaryotic cells is deferred to the next section. The essential processes that a cell must undertake to survive are well illustrated by the prokaryotic cell. The essential structures common to all prokaryotic cells are illustrated in Figure 4.16. From the outside in, they are as follows:

- *Glycocalyx.* The outmost shell of the prokaryotic cell may contain an array of protein fibers, as well as fibrous material ("capsule") that may be rigid or flexible. In some bacteria the capsule material, which is largely chains of simple sugars called polysaccharides (another example of which is "starch"), may slough off and give the bacteria a slimy feel (indeed, the material is sometimes called "slime"). Archaea also may have a capsule, and in some archaeal organisms, this serves as the cell wall itself. The capsule allows bacteria to adhere to each other or to other environments, such as (for example) the intricate ecosystem of the human tooth.
- *Cell wall.* The cell wall determines the shape of the cell and protects it from bursting due to the net influx of water associated with the difference in salinity and other

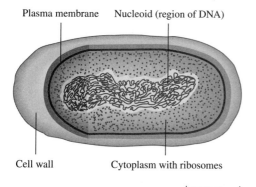

FIGURE 4.16 Schematic cutaway of a prokaryotic (bacterial) cell.

properties of the water within versus without the cell. There are three different architectures for cell walls, corresponding to archaea and two different types of bacteria (labeled gram positive and gram negative according to whether they show a particular color under a microscope after staining). Archaeal cell walls vary significantly in composition from one type of organism to another. The wall of staining (gram-positive bacteria) is quite thick (0.015 to 0.03 μm wide and, hence, comparable to entire viruses). That of the so-called gram-negative bacteria is thinner and consists of the three well-defined layers, the middle one being an aqueous compartment with a variety of proteins and carbohydrates and the inner being the cell membrane. Noteworthy in the middle layer of the gram-negative, and throughout the cell wall of the gram-positive, bacteria is a "glycoprotein" called peptidoglycan, which has sugar chains cross-linked by short peptides. This composite gives bacterial cell walls substantial strength and rigidity. One striking aspect of this structure is that the peptide chains in the peptidoglycan contain not only left-handed alanine (L-alanine), which is the normal amino acid handedness found in life, but right-handed (D-alanine) as well. This is one of the few cases of biological use of the "wrong-handed" amino acid, but it is an important one because it is part of what makes bacterial cell walls durable. Archaeal cell walls lack this interesting chimera. Some of the polysaccharides in bacterial cell walls are toxic to higher animals and are the source of some diseases or poisoning.

- *Membrane.* The cell or "plasma" membrane is the most functionally complex of the structures of the cell. It must move solutes in and out, transport electrons associated with metabolic and photosynthetic processes, establish proton or "electrochemical" gradients in the cell relative to the environment, synthesize ATP (in prokaryotes), anchor the motors for mobility structures such as *flagella*, synthesize lipids and cell wall polymers, secrete proteins, signal to other cells, and sense environmental changes. The bacterial membrane consists overall of the phospholipid bilayer structure described earlier. The membrane contains within it, or is spanned by, many dozens of different kinds of proteins (Figure 4.17). These range from protein complexes that transport electrons, protons, and various solutes to very specialized protein-based systems such as the motor for flagella. The archaeal membrane is composed of different lipids from the bacterial, and less is known of the specific proteins embedded in the archaeal membrane.

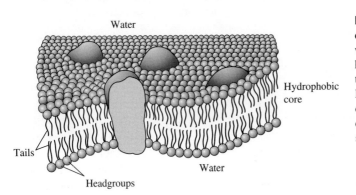

FIGURE 4.17 Drawing of the cell membrane, with the hydrophilic heads and hydrophobic tails organized in a bilayer, and proteins (and complexes of proteins) embedded within and spanning the membrane.

- *Cytoplasm.* The cytoplasm in prokaryotes consists of two components. The *ribosomes* are small granular structures, about the size of a virus, that host the assembly of proteins guided by RNA and using amino acids as raw materials. They and other water-insoluble compounds drift through the aqueous *cytosol*. The cytosol is liquid water that contains dissolved soluble proteins, smaller molecules, and ionic (charged) species. Many of the proteins dissolved in the cytosol are enzymes that catalyze key reactions, such as metabolism of carbohydrates and other energy-yielding molecules, and reactions that synthesize the amino acids as the raw material for protein production. The high concentration of proteins makes the cytosol quite viscous, and a number of subtle chemical interactions and associations among the enzymes may occur. The cytoplasm plays host to a menagerie of other soluble and insoluble components, depending on the type of prokaryote.
- *Nucleoid.* Within the cytoplasm floats the genetic material of the cell, the DNA. Unlike eukaryotic cells, the DNA is unbounded by a membrane, coiled in an amorphous mass called the "nucleoid" approximately in the center of the viscous cytoplasmic environment. Some bacteria may have several such nucleoid regions. Bound to the DNA are the enzymes that replicate DNA and generate the RNA that transfers the genetic code to the ribosomes where protein synthesis occurs.

Other structures comprise parts of cells only in particular prokaryotic organisms. Examples of these include the following:

- *Flagella.* Examples of one type of appendage (others being *sex pili* for linking and exchanging DNA and *fimbriae* used for general adhesion), flagella give bacteria swimming mobility. The flagellum is a semirigid filament that spins like a propeller. At its base is a small spinning motor made of protein and embedded in the cell wall. The force to drive the motor comes, in most bacteria, from protons, which move along an electrical gradient established in the motor, at the expense of energy generated by cellular metabolism. Other bacteria use sodium ions as the charged particles in the electric current. How the current turns the motor is not understood, but most of the parts that correspond to mechanical motors have been identified in these remarkable, nanometer scale, bacterial motors. While single-celled eukaryotes also have filaments that confer mobility, they are completely different in their structure and mode of action.
- *Intracytoplasmic membrane.* Some bacterial cells have specialized membranes that lie just below or are connected to the cellular membrane. Bacteria that grow on methane, for example, have an intracytoplasmic membrane that plays a role in controlling the oxidation of the methane. In cyanobacteria and other photosynthesizing bacteria, the membrane is the site of the photosynthetic reactions that convert electrons excited by light absorption into chemically stored (e.g., ATP) energy.
- *Included bodies.* Within some bacteria lie regions of specialized composition and function that are akin to the organelles of eukaryotes but without the lipid membranes. Examples include gas-filled vesicles, which allow some bacteria to float on the surfaces of bodies of water, and *chlorosomes*, which contain the pigments that

react with sunlight to initiate photosynthesis in bacteria. The latter, of necessity, lie immediately underneath the cytoplasmic membrane where the photosynthesis takes place.

4.4 Organelles and Their Functions in the Eukaryotic Cell

Larger and more structurally complex than prokaryotic cells, eukaryotic cells are really composites of membrane-bound structures called *organelles* whose characteristics suggest a genesis, eons ago, as prokaryotic cells (Chapter 16). The cells of the four kingdoms of the eukaryotic world—plants, animals, fungi, and protists—all possess nuclei wherein lies the cell's primary DNA and possess ribosomes and some other common organelles, but there are also specialized organelles and structures unique to a subset of the kingdoms. Figures 4.18 and 4.19 illustrate the structures of plant and animal cells. We have already discussed the ribosomes and cytoplasm that occur in both prokaryotic and eukaryotic cells, as well as the cell membrane. Although important differences occur in the structure and function of the eukaryotic cellular membrane relative to prokaryotes, the focus in this section is on the organelles as the distinguishing feature of eukaryotes.

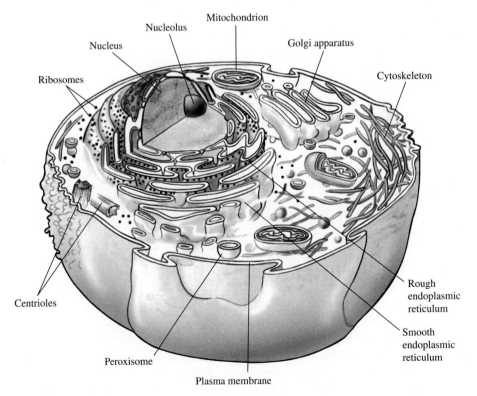

FIGURE 4.18 Representation of the interior of an animal cell.

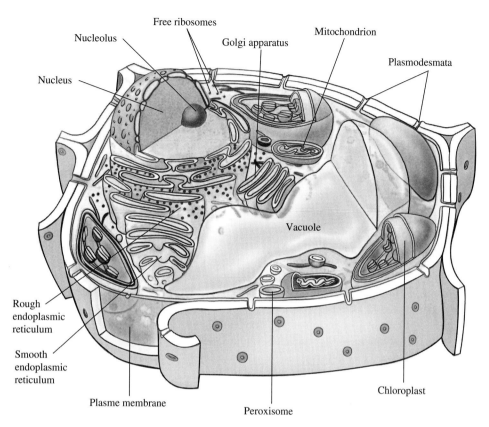

FIGURE 4.19 Representation of the interior of a plant cell.

The *nucleus* of a eukaryotic cell is the site of most—but, importantly, not all—of the DNA in the cell. It is where the DNA duplicates as part of cellular reproduction, where ribosome assembly begins, and where the transferral of genetic information from the DNA to the RNA, for eventual use in protein assembly, occurs. In the nucleus, the DNA is never isolated; it combines with its attendant proteins to form long, thin fibers called *chromatin*. During cell division, the chromatin organizes itself spatially into structures called *chromosomes*.

DNA is also present in the *mitochondria*, which are organelles that occur in all aerobic-respiring eukaryotes. (Indeed, some anaerobic eukaryotes may have had mitochondria but subsequently lost them at some point in their evolutionary development.) And, it is present in the *chloroplasts*. While these give some clues to the origins of these organelles, and, in the case of mitochondria, provide useful ways of relating species one to the other since mitochondrial DNA is inherited only from the female parent, these organelles are most noteworthy in their energy-generating functions. Mitochondria host large enzymes that control the oxidation of food by molecular oxygen to yield ATP. Oxygenic respiration—as we see in Section 4.6—generates an order of magnitude more ATP from ADP (i.e., an order of magnitude more phosphate bonds) than anaerobic metabolic processes such as fermentation. But, as a powerful oxidizer, oxygen is capable of destroying cells, and the food oxidation process must be carefully controlled in terms of rate and location—which the mitochondria do. Chloroplasts are the site of photosynthesis in plant cells. Like mitochondria, chloroplasts have two membranes, and the inner one is folded into a set of closely packed *thylakoid membrane* compartments. Within these is the green chlorophyll pigment and the yellow-orange *carotenoid* pigments that absorb photons and produce excited electrons, which, through a chain of enzyme-mediated steps, produce ATP from ADP. Outside of the membranes is a fluid medium, called *stoma,* where the energy stored in ATP is released and used to convert carbon dioxide and water to carbohydrates. The chloroplast is a member of a more general class of organelles called plastids, others of which provide the bright coloration of flowers.

A variety of other organelles exist, and only a brief mention of each is sufficient here. *Lysosomes* digest food to provide the raw materials for protein synthesis and other processes. Lipids, proteins, nucleic acids, and carbohydrates from consumed organisms are reduced to monomeric units—the raw materials from which the organisms' own proteins will be constructed using the instructions in the nuclear DNA. The eukaryotic cell obtains food by forming a small pocket in the plasma membrane; this pocket contains food materials from the external environment (which could be the ocean surrounding a unicellular protist or the bloodstream of an animal) that are trapped as the pocket pinches off and breaks free of the membrane as a vesicle or *phagosome.* The phagosome moves into the cytoplasm, fuses with the lysosome, and provides the food for digestion. Some undigested products remain in the lysosome, and the expended vesicle fuses with the cell wall to release these wastes. Other organelles store or neutralize hazardous waste products of metabolism; the misnamed *vacuole* (it is not empty) can occupy 90 percent of the volume of the plant cell and plays the roles of storing waste, providing cellular stiffness, and hosting pigments that color some flower petals and fruits.

Beyond the organelles, eukaryotic cells are distinguished by a surprisingly complex transportation and distribution network. The cytoplasm of the eukaryote hosts the *cytoskeleton*, a network of long fibers that maintains the shape of the cell, assists in the mobility of certain cells, and provides surfaces to guide the movement of molecules within the cell. Extending from the cellular nucleus outward is another network of membranes, the *endoplasmic reticulum*. The intricate folding within this structure provides a surface area many times larger than that of the cellular membrane. The endoplasmic reticulum is a key part of the process of moving newly formed proteins to the appropriate part of the cell, where they will perform their structural or enzymatic functions. The sites of protein synthesis, the ribosomes, are attached to the endoplasmic reticulum and feed newly formed proteins to the structure. Within the intricate membranes, proteins fold and gain carbohydrate groups, which begins a tagging process that determines where in the cell the protein will go. The endoplasmic reticulum buds off, like the cell wall, creating vesicles within which proteins move to another structure, the *Golgi apparatus*, where the tagging is made more specific. From there, proteins are moved in vesicles to their final destination.

There is much more to eukaryotic cells than can be described here. In particular, the structures responsible for protist (unicellular) mobility, the filaments assembled from the proteins actin and myosin that interact to create muscle contractions, and the extreme differentiation of eukaryotic cells to perform specific and interdependent tasks are part of a cellular story that is in many ways more intricate and remarkable than the macroscopic physiology of plants and animals. The reader may be left wondering how the lowly cell could acquire or develop such structures through evolutionary processes. In pondering this question, we look forward to Chapter 16 where the evidence is discussed for an originally symbiotic relationship among prokaryotes as the genesis of eukaryotes. The mechanisms, membranes, signaling and tagging systems, and so forth, are likely to be acquired material from eons of prokaryotic experimentation and evolution. Signaling and interaction among prokaryotic cells is itself an intricate chemical dance of electron transfer, proton mobility, and the movement of other ions as well, a subject we shall touch on in Section 4.5.

Eukaryotic evolution likely involved—at an early stage lost in the imperfection of the ancient geologic record—a wide range of possible signaling and transport systems inside symbiotically co-existing prokaryotic cells, and much time was available to weed out the less efficient solutions. What we see and what we are today are the survivors of waves of cellular experimentation made possible by the immensity of time in the billions of years of our planet's history.

4.5 Energy Storage and Transfer via the Proton and the Electron in Cells

Underpinning the dizzying array of structures and functions in the eukaryotic and prokaryotic cells is the fundamental need for useful energy. For the vast majority of the biosphere that energy is sunlight; for a smaller but important subset, it is the chemically reductive composition of the warm seafloor vents. Cells must be able to capture this energy and utilize it—along with raw materials—to sustain the chemical reactions and replication of cellular structures that characterize life. Life is not in equilibrium with anything; organisms are either in a steady state with an environment that provides energy and resources, or they are responding to changes in that environment by altering or adjusting internal cellular processes. The basic principles that describe these processes are physics and chemistry, and we have dealt with the essentials in Chapters 2 and 3. Still to come, in Chapter 7, is a thermodynamic perspective on living processes, which provides a few powerfully general principles by which we can understand the fundamental limitations on life and constraints on its origins. What follows in this section does not require knowledge of thermodynamics, but readers may benefit by turning a few chapters ahead to familiarize themselves with these principles.

Prior to considering specific mechanisms for generating energy in cellular processes, let us consider how the energized cell moves energy around for specific purposes. At the most fundamental level, energy transport and generation in the cell is enabled by the flow of electrons in membranes, and the coupling of this flow to an *ion gradient*. This simply means that organisms move ions across certain membranes to set up gradients (variations) in the electric field and the ion composition itself. For the majority of organisms, the ion is the positively charged proton; among a smaller but important class of organisms, ionized sodium (Na^+) is used. The effect of a gradient is easy to visualize: Imagine beginning with a solution in which electrons and protons are uniformly distributed. The charge, averaged over a volume much larger than the electrons and protons, is zero. We can create a direction to this medium by moving some of the protons to the outside surface of the membrane. Now there is a net positive charge on the outside of the membrane and a net negative charge on the inside; hence, an electric field is created and charges will feel a force. But this is not the only gradient: With more protons on the outside than the inside, even if there were not a repulsive charge, it costs energy to maintain the disparity in protons because there will be a tendency for more protons to move back inward than outward (simply because there are more proteins on the outside). This chemical gradient, when combined with the gradient in the electric charge, becomes an *electrochemical* gradient. The same concept holds when we replace protons with sodium ions.

Figure 4.20 illustrates that, schematically, the proton "pump" can be thought of as maintaining a battery within the membrane of the cell. Energy, from photosynthesis or metabolism of food, is used to pump protons across the membrane, creating an electrochemical gradient. This gradient can be used to perform useful work (Chapter 7), in much the way that a charged battery is available to power an electrical device. The battery must be discharged in a controlled way to perform work, that is, hooked up to an electrical circuit containing motors, semiconductors, resisters, and so forth. (You can also discharge a battery by throwing it into a container of water and short-circuiting it, which will not in general result in the ability to extract work from the discharging battery. Preventing the protons from catastrophically short-circuiting is the lipid membrane itself, which is impermeable to protons.)

Typical voltages maintained by bacterial membranes are 100 to 200 mV, an order of magnitude less than that of small commercial batteries. Proteins embedded in the membrane permit protons to diffuse back to the inner membrane, and this charged particle movement allows work to be done. Some of the cellular functions powered by the proton

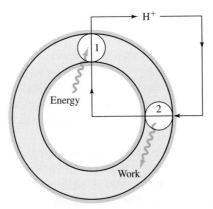

FIGURE 4.20 The proton electrochemical current, shown conceptually for a cellular membrane (either the plasma membrane of a prokaryote or the mitochondrial membrane of a eukaryote). Energy is expended to pump protons (H^+) through the membrane to its outer surface (1). The membrane is thus a kind of battery, and work can be done by allowing protons to flow back through the membrane in a controlled fashion through the use of proton transporters (2).

return flow include maintenance of lower pH in bacteria that live in high pH environments, accumulation of solutes inward of the membrane, turning of the motor that animates the bacterial flagellum, and synthesis of ATP in bacteria and mitochondria. In some membranes, the electrochemical potential at the external surface is converted into a pH gradient (Chapter 3) by pumping negative ions with the protons to maintain electrical neutrality. In place of the electric charge, then, there is an excess of protons that constitute strong electron acceptors, for which the negative ions cannot completely compensate, and hence, the acidity on the exterior of the membrane rises (pH declines).

One protein that channels protons back through the mitochondrial or bacterial membrane is called the *ATP synthase*. This is a complex protein; in mammals, 16 different polypeptides are folded together to comprise the protein, while in the bacterium *E.coli*, eight distinct polypeptides do so. Protons flow through the portion of the protein that is embedded in the membrane and that acts as a channel. As proteins arrive in the portion of the protein inward of the membrane, they exert a force that induces the core of the protein to rotate. This rotation of the core of the enzyme, in the presence of phosphorus and ADP, enables the reaction ADP + (phosphate group) + energy → ATP. Interestingly, the ATP synthase also can extract energy from ATP to be used in other reactions, making ADP and a phosphate group. However, in most circumstances, the ATP is removed quickly from the membrane environment in which the synthase is located, hence keeping the ATP concentration low and tending to drive the reaction toward making more ATP. The cell, as long as it has an adequate food supply, will keep pumping protons across the membrane, sustaining the proton gradient that tends to generate more ATP.

The coupling between energy generated by metabolism and proton pumping is electron transport. The transport is ultimately driven by the energy gained in a series of oxidation reactions, and the most common fuel is the sugar glucose, or $C_6H_{12}O_6$. For brevity's sake, we discuss it only and then permit it to stand as the type example of the extraction of energy by oxidation—that is, a set of chemical reactions in which there is a net transfer of electrons from the reduced species to the oxidizer. The amount of chemical energy in glucose that can be released by oxidation may be measured by burning it in a flame in an atmosphere in which molecular oxygen is present, via the reaction $C_6H_{12}O_6 + 6\,O_2 \rightarrow 6\,CO_2 + 6\,H_2O$ + energy. (See Chapters 3 and 7 for discussions of chemical reactions.) The carbon has been converted to a more oxidized form, carbon dioxide, which is quite chemically nonreactive under almost any circumstances. The hydrogen has been taken up by water, also quite stable. The energy released in making these two very stable molecules is about 686 kcal/mole of glucose. Since each ADP → ATP synthesis requires 12 kcal to generate the high-energy phosphate bond, complete oxidation of a glucose molecule could make 57 ATPs from ADP on a molecule-by-molecule basis. A typical human uses up 10^{25} ATP molecules per day, which would correspond to the complete oxidation of about 2×10^{23} molecules of glucose—or about one-third of a mole of glucose (Chapter 3). Because 1 mole of glucose weighs about 0.2 kgs, this is just under 0.1 kg of glucose per day.

This calculation sets only an upper bound to the energy, expressed in the currency of the ATP molecule, which can be derived from glucose—and hence a lower bound to what must be consumed. Were we to burn glucose by direct exposure to oxygen within a cell, the resulting heat released by the reaction could not be captured by the cell but would instead damage the internal membranes; it would be no more useful than lighting explosive powder within the interior of your body. Metabolic processes have evolved to take the energy available from oxidative reactions in small steps, by utilizing the oxygen along with other oxidants and intermediate products, in long chains of reactions mediated by enzymes. As remarkably efficient as this process is, it provides at best only about half the ATP yield calculated earlier—for eukaryotic cells in which the oxidation occurs in mitochondria. In the absence of oxygenic conditions or the mechanisms (bacterial or mitochondrial) that use it as an oxidizer, the ATP yield is much less—as little as $\frac{1}{18}$ that derived from aerobic respiration.

The stepwise oxidation of glucose can be regarded as the flow of electrons from donors to acceptors. The electron carriers in the processes discussed later include certain lipids, proteins bearing iron and sulfur (which may be a very ancient metabolic pathway), hemin-related molecules, and certain coenzymes. Two important ones that will appear are *nicotinamide adenine dinucleotide (NAD)* and *flavin adenine dinucleotide (FAD)*. Focusing for the moment on NAD, this molecule looks like two AMP molecules attached by an oxygen atom between the two phosphate groups. But the upper "AMP" lacks the adenine double ring, and instead has a single heterocyclic ring. In its oxidized form, this molecule is written NAD^+, and it accepts, from two hydrogen atoms, the two electrons and *one* proton, leaving the other proton behind. Once this has happened, NAD^+ has been reduced—the molecule is then written NADH. In turn, molecular oxygen will accept the extra electrons from two NADH; thus (including the proton we left behind), $2\,NADH + 2\,H^+ + O_2 \rightarrow 2\,NAD^+ + 2\,H_2O + $ energy. The energy from this oxidation of NADH, per mole of NADH, is 218 kilojoules, many times that obtained from ATP conversion to ADP.

Electron transport in cells removes electrons from reduced organic material and transfers them to membrane-bound complexes such as NADH and other carriers through several steps, ultimately to a final, or "terminal," electron acceptor. This flow of electrons in membranes is called *respiration*. In bacteria, the membrane is usually the plasma membrane, and in eukaryotes, it is the mitochondria. The final terminal electron acceptor in aerobic bacteria and mitochondria is oxygen, yielding aerobic respiration; other molecules may be electron receptors in anaerobic respiration. Oxidation of NADH in portions of the cell other than membranes, and with oxidizers other than oxygen, is *fermentation*. During respiration, protons are "translocated" from the interior to the exterior of the membrane, maintaining the proton gradient discussed earlier, which in turn converts ADP to ATP. Thus, the net result of the reaction pathways that "burn" glucose and other reduced organic compounds is the maintenance of the proton pump, the generation of ATP as the mobile currency of phosphate-bond energy, and the conduct of other cellular membrane processes such as flagellal movement.

4.6 Glycolysis, Respiration, and Fermentation

Given the molecular tools for electron transfer through oxidation-reduction reactions described in Section 4.5, the tracing of the basic respiration process is straightforward. Again, we focus on glucose as the type example of a food available for metabolic processing; we mention other metabolisms in Chapter 10. Four basic processes occur to extract and transport electrons from glucose to a terminal acceptor:

1. *Glycolysis*. This process breaks down glucose through a partial oxidation into the molecule pyruvate. NAD^+ becomes NADH, and four ATPs are generated. To initiate the 10-reaction process, however, several reactions occur that require energy, which is donated by the cell in the form of two ATPs. These reactions produce from glucose two molecules of a "sugar phosphate" called glyceraldehydes 3-phosphate, which when oxidized releases 100 kcal/mole of glucose. This enormous release of energy is soaked up by the reduction of two NAD^+ to form NADH. In several more reactions involving the oxidized phosphates, four phosphate groups are transferred to ADP to make ATP, so that in net form glycolysis generates two ATPs and two pyruvate molecules.

2. *Pyruvate oxidation*. While glycolysis occurs within the cell cytoplasm, this next step occurs in the mitochondria or (for bacteria) in the plasma membrane. Pyruvate diffuses into the membrane, where an enormous protein complex (composed in mitochondria of 72 different polypeptide chains) controls a sequence of reactions. First, pyruvate is oxidized to make an *acetyl* group and carbon dioxide. Some of the energy released from the oxidation is used to reduce two NAD^+ to NADH. Bonding the acetyl group to a co-enzyme absorbs additional energy, and this complex is called acetyl CoA.

3. *Citric acid cycle*. The formation of acetyl CoA provides the raw material for a set of reactions that may well be among the most ancient metabolic cycles. The cycle gets its name from the initial reaction in which the two-carbon acetyl CoA reacts with a four-carbon oxaloacetate molecule to make acetic acid. The process is a cycle because, mediated by enzymes, it is maintained essentially in a steady state in cells, with reactants and products entering and leaving (Figure 4.21). The principal inputs to the cycle are acetyl CoA, water, NAD^+, FAD, and other electron carriers. As shown in the energy-level diagram (Figure 4.22), the cycle produces a large amount of energy that is stored in $NADH + H^+$, ATP, and other molecules and yields carbon dioxide.

4. Respiratory *chain*. Examination of Figure 4.22 shows that much of the energy gained in the preceding three cycles is in the form of $NADH + H^+$, which is not the most accessible currency of energy. The electron transport chain discussed in Section 4.5 comes into play here, with a complex hand-off of electrons from NADH to other electron acceptors, and ultimately to the terminal acceptor, which is O_2 in aerobic respiration. The population of protons generated in the production of NADH are pumped across the membrane to generate the electrochemical proton potential, and ATP synthesis proceeds in the membrane powered by the return flow as described earlier. Three molecules of ATP are formed for each pair of electrons transferred along the enzyme mediated chain from $NADH + H^+$ to oxygen. Counting the net number of NADHs generated, along with $FADH_2$ and four ATPs made in glycolysis and the citric acid cycle yields production of 38 ATPs from aerobic respiration. Two ATPs must be used, however, in moving NADH into the membrane of the mitochondria, and hence aerobic respiration has a net yield of 36 ATPs. Hence,

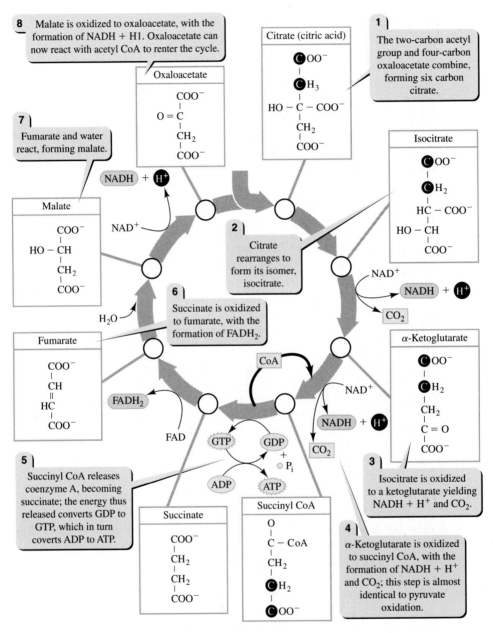

FIGURE 4.21 Schematic of the citric acid cycle.

the analogous reaction to our "burning" of glucose at the beginning of the section is $C_6H_{12}O_6 + 6\,O_2 \rightarrow 6\,CO_2 + 6\,H_2O + 36$ ATP. This is fewer than the 57 ATPs that wholesale burning in oxygen would generate, but it is nonetheless remarkably efficient.

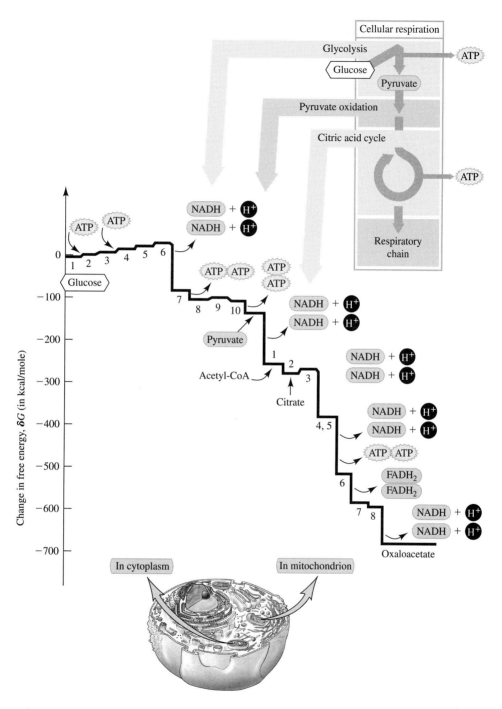

FIGURE 4.22 Energy level diagram of the four processes involved in respiration. A drop in energy means that there is a release of energy in a particular reaction. Numbers identify discrete reactions, but for clarity, the explicit reactions are not labeled. Reactants entering the process are shown by arrows going into the process line; products are labeled by arrows emanating from the line.

In the absence of oxygen, or a suitable substitute for the conduct of anaerobic respiration, the fate of the products of glycolysis is further reaction within the cytoplasm in a process called *fermentation*. Fermentation reduces pyruvate or its immediate products to yield NAD^+ from $NADH + H^+$. The NAD^+ is then available to conduct more glycolysis reactions on glucose. Because glycolysis does generate two ATP molecules, some energy is produced from glycolysis plus fermentation, but this amount is 18 times smaller than what aerobic respiration can provide.

The sites and products of fermentation are diverse. Some yeasts and plant cells undertake alcoholic fermentation, producing ethanol. Muscle cells in animal bodies, when working hard, run an oxygen deficit and produce lactate as part of a "lactic acid" fermentation process. The buildup of this substance eventually limits muscle use, and therefore one must stop exercising and keep panting until enough oxygen is pumped into the muscles to enable aerobic recovery. Those anaerobic bacteria and archaea incapable of respiration conduct fermentation as one of several possible metabolisms, but the energy yield is much too small to maintain eukaryotic life. When a eukaryotic organism finds itself in an anaerobic environment, or its respiratory and pumping systems fail, cellular death is not immediate. Fermentation of remaining glucose goes on in some cells for a short time, but insufficient energy is generated to sustain all but the most minor cellular processes. The powerful electrochemical gymnast—the eukaryotic cell—lives out its final moments reverted to the evolutionarily ancient, feeble, fermenter.

4.7 Photosynthesis and Chemisynthesis

There is not, nor was there ever, an endless supply on Earth of reduced organic compounds to supply energy through metabolic processes. Early in the history of life, organisms may have fed off ambient organic molecules, reduced compared with abundant atmospheric CO_2, that were abiotically produced over the hundreds of millions of years since the formation of the Earth. But this supply would quickly have been exhausted, and today the *organic carbon* is essentially all biologically processed material. Because much of it gets buried, more must be manufactured. The largest source of accessible carbon is CO_2 (buried carbonates are largely inaccessible to the biosphere), and the largest source of hydrogen is water. To reverse the carbohydrate reaction requires usable energy, and life commands that energy from two quarters: the Sun and deep-sea vents. Extracting photon energy from the Sun is *photosynthesis*, and from deep-sea vents *chemisynthesis* (Figure 4.23). The latter process depends on the presence of warm water, which carries reduced gases such as hydrogen sulfide (H_2S), extruded from submarine vents (Chapter 10). The oxidation state contrast between those waters and surrounding cold, sulfide-poor waters provides a *chemical* potential energy difference that can be utilized to make ATP. Photosynthesis, on the other hand, employs porphyrins such as chlorophyll to capture solar energy, and through a chain of chemical reactions, generate carbohydrates from carbon dioxide and water. ATP is also formed during photosynthesis as a direct product. Although some plants have secondary means of getting food, mutants that lack chlorophyll simply starve to death.

But what is crucial about photosynthesis, with respect to the biosphere, is that other organisms can eat the carbohydrates manufactured in plants and other photosynthesizers, thus

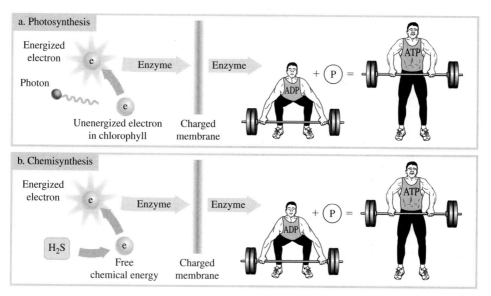

FIGURE 4.23 Schematic comparison of photosynthesis and chemisynthesis. The encircled P represents a phosphate bond. Like a weightlifter storing gravitational potential energy through the suspended barbell, ATP hold chemical potential energy in its phosphate bond.

creating a food chain dependent on photosynthesis. Many photosynthesizers generate oxygen as a byproduct of photosynthesis, and this is the primary source of oxygen for Earth's atmosphere and hence for aerobic respiration. Therefore, the vast majority of the present biosphere of the Earth hinges on photosynthesis, although this was not always the case.

There is not space to describe all of the photosynthetic systems at play on the Earth. Oxygenic *phototrophs* are organisms that generate oxygen during photosynthesis: plants, algae and cyanobacteria. Plants use two slightly different types of chlorophyll molecules, labeled A and B; cyanobacteria employ only the first. These all use H_2O as the principal electron donor, to reduce an ion of NAD called $NADP^+$ to NADPH. The water is oxidized to O_2, which escapes into the atmosphere (but is also used by some plants in a secondary aerobic respiration process). Other bacteria use organic compounds, inorganic sulfur compounds, or hydrogen gas as the source of electrons in photosynthesis. Some of these photosynthetic pathways are likely more ancient than oxygenic photosynthesis, based on the inference that the Earth's atmosphere lacked significant amounts of oxygen for upwards of two billion years (Chapter 11).

All photosynthetic processes use CO_2 as the carbon source with organic carbon a secondary source, except in heliobacteria, which uses organic carbon (reduced carbon) exclusively. We focus on purple sulfur bacteria to outline the steps in photosynthesis by which an electron is excited and transported in a cycle that generates ATP. The action occurs in a "reaction center" embedded in the plasma membrane of the bacteria; in plants the membrane would be that of the chloroplast organelle. The chlorophyll absorbs light at optical wavelengths within the ring structure surrounding the magnesium atom. The absorption of photons raises electrons to an excited energy level. The excited electrons are donated to a receptor molecule called *bacteriopheophytin*, which is essentially chlorophyll from which the magnesium atom is missing. The chlorophyll is oxidized and the receptor molecule is reduced in this electron transfer. The electrons are moved successively to several carrier molecules called *quinones*, which are hydrophobic (and thus mobile) lipids. The electron transfer occurs at a rate such that two electrons, excited by photon absorption in the chlorophyll, are transferred to the same carrier molecule.

The quinone, with two electrons and hence a double negative charge, picks up two protons from the cellular cytoplasm and leaves the reaction center reduced—symbolically, from UQ to UQH_2. The reduced quinone moves to an enzyme complex in the membrane to which it transfers the two electrons. The electrons in turn reduce so-called cytochrome molecules, each of which possesses a "heme" capable of carrying an electron. (Hemes are a class of substances that have heterocyclic rings at the center of which lies iron. Hemin, described in Section 4.2 as the active oxygen-binding group in hemoglobin in the bloodstream of many animals, is a member of this class.) The UQH_2 is oxidized by the removal of the two electrons, and the protons (represented by the H_2) are released and transported to the surface by the enzymes. Each cytochrome molecule accepts a single electron; thus, two cytochromes are reduced in the electron transfer process. These are moved back to the reaction center where they donate the electrons to the oxidized chlorophyll, completing the cycle.

The net result of this cycle is that a proton gradient is set up across the membrane, which can then be used to make ATP. In green sulfur bacteria, $NADP^+$ is the electron acceptor instead of quinone. In cyanobacteria and chloroplasts, importantly, the chlorophyll does not donate an electron: When it loses an energized electron by absorbing light, it becomes a powerful enough oxidizer to grab an electron from water and thus oxidize the water. It is the water that is the electron donor, and it is oxidized to $\frac{1}{2}O_2$ (two water molecules are required for a single O_2). There are also differences in the wavelengths of light that are absorbed. The types of bacteriochlorophyll in purple sulfur bacteria absorb light at a peak wavelength of 0.87 μm, whereas that of chlorophyll A or B peaks at a 0.67 to 0.68-μm wavelength. Thus, the photons absorbed by the latter are 20 per cent more energetic than the former, enabling the chlorophyll to become a strong oxidizer. In bacteria and plants, not all the light is absorbed directly by chlorophyll; additional light-harvesting pigments absorb light at wavelengths other than the chlorophyll peak and funnel the energy to the reaction centers, increasing photosynthetic efficiency.

The production of carbohydrates occurs in a separate cycle (in plants, at night), in which both ATP and NADPH are used to generate carbohydrates. The cycle is called the Calvin cycle and operates in plants, algae, and some bacteria. Adding electrons from NADPH + H^+ reduces three CO_2 molecules. However, NADPH is not a very strong reductant and to make the reaction go requires putting energy in via ATP. The reaction yields a "phosphoglyceraldehyde," which is a complex molecule equivalent from an oxidation point of view to $C_3H_6O_3$. The phosphoglyceraldehyde molecules are then rearranged in a complex series of steps to yield glucose and other sugars. Enzymes control the reactions, a key one in the first stage being ribulose-1, 5-bisphosphate carboxylase, or *rubisco*. Rubisco may be a very ancient enzyme that played a role in the earliest evolution of life.

The energy level diagram for purple sulfur bacteria, Figure 4.24, gives a summary of the photosynthetic process. Absorbed photons raise the potential of the system so that

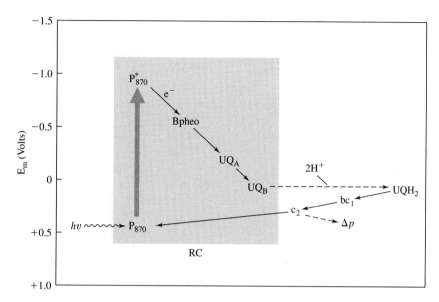

FIGURE 4.24 Energy changes (expressed as electrical potential) associated with the flow of an electron during the photochemical cycle of the purple sulfur bacteria. UQ_a and UQ_b are two quinones through which the electron travels.

work can be done—in particular, establishment of a membrane proton gradient, which in turn makes ATP and reducing agents for other chemical reactions. Paramount among such other reactions is the formation of sugars from water and carbon dioxide, which requires energy, supplied by ATP, and reducing agents for its completion. Through photosynthesis, life uses light to reduce oxidized carbon from previous generations of metabolic consumption of food, so that life may continue as long as the Sun shines.

4.8 The Genetic Code and the Synthesis of Proteins

Given the fundamental importance to cellular processes of the arraying in particular orders of enantiopure amino acids to form proteins, and given the complexity of protein structures, it is remarkable how simple the underlying logic of the genetic code is. The code specifies the order of the amino acids through the order of the nucleic acid bases on strands of DNA. The individual bases do not specify the amino acid sequence; the bases are merely letters in a lexicology of three-letter words, called *codons*, that specify the amino acid sequence. The codons must be capable of uniquely specifying 22 amino acids, and with four different types of bases, this means that the number of letters per word must be at least three. If the number of letters per word were one, there could be only four words: A, C, T, and G. Two-letter words would yield $4 \times 4 = 16$ words, still too small for the number of amino acids. Three-letter words, the codons, yield $4 \times 4 \times 4 = 64$

TABLE 4.1 Specification of the Amino Acids by RNA[1]

Second Base	Uracil (U)	Cytosine (C)	Adenine (A)	Guanine (G)
First Base				
Uracil (U)	UUU phenylalanine	UCU serine	UAU tyrosine	UGU cysteine
	UUC phenylalanine	UCC serine	UAC tyrosine	UGC cysteine
	UUA leucine	UCA serine	UAA stop	UGA stop
	UUG leucine	UCG serine	UAG stop	(UGA selenocysteine)
			(UAG pyrrolysine)[2]	UGG tryptophan
Cytosine (C)	CUU leucine	CCU proline	CAU histidine	CGU arginine
	CUC leucine	CCC proline	CAC histidine	GCG arginine
	CUA leucine	CCA proline	CAA glutamine	GCA arginine
	CUG leucine	CCG proline	CAG glutamine	AGG arginine
Adenine (A)	AUU isoleucine	ACU threonine	AAU aspargine	AGU serine
	AUC isoleucine	ACC threonine	AAC aspargine	AGC serine
	AUA isoleucine	ACA threonine	AAA lysine	AGA arginine
	AUG methionine	ACG threonine	AAG lysine	AGG arginine
	(AUG start)[2]			
Guanine (G)	GUU valine	GCU alanine	GAU aspartic acid	GGU glycine
	GUC valine	GCC alanine	GAC aspartic acid	GGC glycine
	GUA valine	GCA alanine	GAA glutamic acid	GGA glycine
	GUG valine	GCG alanine	GAG glutamic acid	GGG glycine

Notes:
[1] Leftmost column specifies the first base, and the top row the second base, in the codon.
[2] In some organisms.

codons, almost three times the number of amino acids used to build proteins. The redundancy is actually not quite so extensive; there are several codons that code for the initiation and termination of the reading of the DNA chain for a particular protein, which is essential because multiple proteins are coded for by a given strand of DNA.

Table 4.1 lists the amino acids that each codon specifies. The actual protein synthesis requires the mediation of RNA as well. *Messenger RNA* carries the protein-structure information from the DNA, *transfer RNA* attaches to specific amino acids and aligns them based on the messenger RNA sequence, and *ribosomal RNA* (which is located in the ribosomes) receives the ordered amino acid sequence (ferried by the transfer RNA) and then acts as a catalyst to join together the amino acids. Other RNA molecules assist in DNA replication and in the construction of the messenger RNA. This diverse range of roles for a single kind of molecule makes tempting the proposal that, at some time in the distant past, RNA was the central biopolymer (Chapter 8). By contrast, DNA has a very specialized function as a record of the genetic information of the individual organism and (in subsidiary DNA strands) of certain organelles in the cell. This central but more focused role compared with RNA suggests that DNA is a later molecule derived from just a few key changes to RNA.

A length of DNA that carries genetic information, ultimately to be expressed as a single protein, is called a *gene*. The *genome* is the full DNA sequence of an organism. It is similar, but certainly not identical, for all living organisms on Earth that have been examined. The process by which the transfer of information occurs is instructive, particularly from an evolutionary point of view (Figure 4.25). In the *transcription* stage information flows from DNA to RNA. A very large enzyme, the *RNA polymerase,* separates the two strands of the DNA helix and finds the starting point of the sequence to be transcribed. It then links nucleic acid bases floating around within the nucleus (or in the general cytoplasm in the case of prokaryotes) into an RNA chain, messenger RNA (mRNA), that is the base pairing

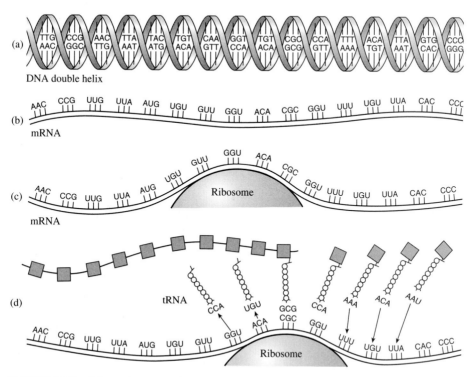

FIGURE 4.25 Schematic process for transcription of gene information and protein construction, beginning with DNA in part (a) of the figure and ending with transcription on a ribosome (shaded sphere) in part (d).

complement (with U instead of T) of the DNA, and then stops the transcription at a "stop" codon (see Table 4.1). The gene may run from 100 to 10,000 nucleotides long; the DNA is pieced back together behind the transcription as the pieces ahead are unzipped, at a rate of about 50 nucleotides per second. The starting point for the reading is not based on a codon, but on a "motif" sequence that may be longer than a codon—in a number of eukaryotic genes, for example, it is TATA. And, because the backbone of DNA involves phosphate groups bonded to ribose groups, there is a definite direction to the molecule defined by the carbon-phosphate positions, and hence a direction to the transcription.

With the transcription completed, the RNA polymerase is not finished with the strand of RNA. It places a special G base with a methyl group attached on one end to define a tail and then performs an editing operation: It removes sequences of nucleotides that do not code for proteins. The origin of these *introns* is unclear, but some may be very ancient sequences related to the construction of RNA in the *absence* of DNA—at a time when RNA alone perhaps encoded life. The splicing of the mRNA is conducted by chimeric pieces of enzymes and RNA (*small nuclear RNA*) that aid recognition of intron sequences.

The mRNA contains all the information needed to make a protein, but it must somehow bring this information together with amino acids. Transfer RNA (tRNA) can fold over on itself, and has an end that holds an amino acid and another end with the nucleotide bases to fit into the mRNA at the correct (complementary) positions. Amino acids are inserted in the appropriate tRNA pieces by a set of some 20 enzymes. The sites at which this RNA–amino acid interaction and assembly occur are the ribosomes—composed of RNA and associated enzymes.

The RNA transcription process is far more error prone than the replication of DNA. RNA production leads to an error of about 1 in every 10^4 to 10^5 nucleotides—still less than one per protein. However, the tRNA coupling to amino acids has an error rate about 10 times worse. In the production of proteins, this is apparently tolerable. But were that rate sustained in DNA replication, the genetic code would be completely scrambled after a much smaller number of generations than characterizes the stable lifetime of species (Chapter 16). Hence, there is a special error-correction mechanism involved in DNA replication. The DNA polymerase enzymes that unstrip DNA, pull in ambient nucleic acid bases, and hence create two new chains that are the complements of the pulled-apart strands to create two duplicate DNA molecules, also perform error correction. They achieve this by testing for proper chemical binding between each complementary nucleic acid base and stripping out those that do not bond correctly. This yields an error rate of one in one billion nucleotides.

DNA replication is the essential core of the division of cells, itself an intricate and beautiful process for which the reader is referred to a standard biology text. But it also encapsulates the fundamental paradox of life raised at the start of this chapter—that DNA replication and DNA transcription to RNA and thence to proteins require the intimate involvement of proteins to work, but proteins require DNA for their specification. The solution to this paradox may involve more diverse roles for RNA early in the history of life, but that discussion is deferred to Chapter 8. Also deferred is a discussion of how unlocking the genetic code and determining gene sequences allow the relationship between organisms to be put on a quantitative basis. With the information at hand—from physics, chemistry, and biology—we are ready to examine how the cosmos evolved from an amorphous mix of matter and energy to one sprinkled with galaxies, stars, planets, and—on at least one—the remarkable physicochemical machines we call life.

QUESTIONS AND PROBLEMS

1. List some of the differences between prokaryotic and eukaryotic cells. Are there any indicators in the structure of the eukaryotic cell for an origin that involved multiple prokaryotic cells?

2. Bacterial cell walls come in two flavors: gram-positive and gram-negative types. The other unicellular cell type, that of the archea, possesses a variety of different kinds of cell walls. Perform a literature search to find the nature of these archeal cell walls, and suggest a rationale for the difference in variety of cell wall structures between the two unicellular domains of life.

3. RNA has received instructions from DNA to make the protein "imaginase," which consists of the following amino acid chain:

    tyrosine—alanine—alanine—methionine—phenylalanine

    a. Write down at least five different nucleic acid base sequences in the RNA that would code for this enzyme. (Use the codons from Table 4.1).

    b. What is the complementary DNA sequence of letters that made this RNA strand?

4. Suppose a form of life began in which only two nucleic acid bases (say, A and U in RNA; A and T in DNA) were used to code for proteins. To make codes for 20 different amino acids the codons would have to be five letters long ($2 \times 2 \times 2 \times 2 \times 2 = 32$ possible codons). In what way might this form of life have been more successful or less successful than the actual life we find on Earth?

5. Imagine a planet with two forms of life—one able to synthesize only left-handed sugars and use only right-handed amino acids, the other able to synthesize only right-handed sugars and use only left-handed amino acids. Could one form usefully eat the other? Could they co-exist? If they could, would their ecosystems be entirely separate or interactive (and if the latter, in what ways?)

6. Perform a literature search to find the reflection spectrum (i.e., the dependence of reflectivity on wavelength) at optical wavelengths of chlorophyll and other photosynthetic pigments. Compare these with a typical spectrum of the Earth's atmosphere. Do the absorptions associated with atmospheric gases obscure the spectral signatures of the photosynthetic pigments? Could photosynthetic pigments be a good indicator of life on another planet orbiting a nearby star, assuming a surface covered with plant life?

7. Compare the absorption spectra of photosynthetic pigments (i.e., the wavelengths at which they best absorb light) to the spectral distribution of light coming from each class of main sequence star—O, B, A, F, G, K, and M (you will have to read Chapter 5 to understand this sequence). You can either calculate the spectrum of each stellar type knowing their photospheric temperature from an astronomy reference book or look in the literature to obtain example optical spectra of each type.

Which photosynthetic pigment best matches the peak wavelength at which the Sun (a G-type star) emits light? What would be the wavelength characteristics and the biological challenges associated with pigments that best match the O- or B-type stars, and likewise that best match the M-dwarfs?

8. Perform a literature search and find out what amino acids have been found in meteorites. Are any of these amino acids also found in life on Earth? What distinguishes the inventory and characteristics of amino acids in meteorites from those found in life on Earth? Finally, have any pyridines or pyrimidines been found in meteorites, and how do they compare with the nucleic acid bases?

9. Other than the 20 fundamental amino acids used by life on Earth, two others, selenocysteine and pyrrolysine, play a limited role in some organisms. Perform a literature search to find out the structures and functions of these less-used, "extra," amino acids.

10. Suppose DNA polymerase left a gap in the copy strand of DNA. How might this be repaired, and could the repair fully recover the lost information? (You can propose a mechanism from the copying process described here, or perform a literature search.)

SUGGESTED READINGS

A well-illustrated, tersely written but encyclopedic treatment of biology at the undergraduate level is W. K. Purves, D. Sadava, G. H. Orians, and H. C. Heller, *Life: The Science of Biology*, (6th ed.) (2001, W. H. Freeman and Co., San Francisco). It is difficult to recommend a specific book for the reader to use in checking and more deeply exploring the organic chemistry; R. T. Morrison, R. N. Boyd, and R. K. Boyd, *Organic Chemistry* (6th edition, 1992, Addison-Wesley), is the classic. A fine upper undergraduate text, clearly illustrated and very systematic, on protein structure is C. Brandon and J. Tooze, *Introduction to Protein Structure* (1991, Garland Publishing, New York). While written at a somewhat more advanced level than the present book, D. White, *The Physiology and Biochemistry of Prokaryotes* (2000, Oxford University Press, Oxford, U.K.) is a clearly expounded and comprehensive introduction to cellular processes, specific to prokaryotes but with applications to eukaryotic organisms. A much shorter and more general treatment is M. Hoppert and F. Mayer, "Prokaryotes," *Am. Sci.*, **58** (6), 518–525. There is not enough space in this book to delve into the subject of intercellular and intracellular communications, which are foundational to the viability of prokaryotic and eukaryotic cells. An entertainingly written and very authoritative treatment of this subject is W. Lowenstein, *The Touchstone of Life* (1999, Oxford University Press, New York).

REFERENCES

Atkins, J. F. and Gesteland, R. (2002). "Biochemistry: The 22nd Amino Acid," *Science*, **296,** 1409–1410.

Ban, N., Nissen, P., Hansen, J., Moore, P. B., and Steitz, T. A. (2000). "The Complete Atomic Structure of the Large Ribosomal Subunit at 2.4 Å Resolution," *Science*, **289,** 905–920.

Baranowski, E., Ruiz-Jarabim, C. M., and Domingo, E. (2001). "Evolution of Cell Recognition by Viruses," *Science*, **292,** 1102–1105.

Bell, A. C., West, A. G., and Felsenfeld, G. (2001). "Insulators and Boundaries: Versatile Regulatory Elements in the Eukaryotic Genome," *Science*, **291,** 447–450.

Berman, H. M., Goodsell, D. S., and Bourne, P. E. (2002). "Protein Structures: From Famine to Feast," *Am. Sci.*, **90,** 350–359.

Bertozzi, C. R. and Kiessling, L. L. (2001). "Chemical Glycobiology," *Science*, **291,** 2357–2364.

Brandon, C. and Tooze, J. (1991). *Introduction to Protein Structure*, Garland Publishing, New York.

Chyba, C. and McDonald, G. D. (1995). "The Origin of Life in the Solar System: Current Issues," *Ann. Rev. Earth Planet. Sci.*, **24,** 215–249.

Compton, R. N. and Pagni, R. M. (2002). "The Chirality of Biomolecules." In *Advances in Atomic, Molecular, and Optical Physics*, **48,** 219–261.

Dorin, H. (1971). *Modern Principles of Chemistry*, College Entrance Book Co., New York.

Fersht, A. (1999). *Structure and Mechanism in Protein Science: A Guide to Enzyme Catalysis and Protein Folding*, W. H. Freeman and Company, New York.

Hayes, B. (1998). "The Invention of the Genetic Code," *Am. Sci.*, **86,** 8–17.

Haynie, D. T. (2001). *Biological Thermodynamics*, Cambridge University Press, Cambridge.

Hoppert, M. and Mayer, F. (1999). "Prokaryotes," *Am. Sci.*, **87** (6), 518–525.

Iwamoto, H., Nishikawa, Y., Wakayama, J., and Fujisawa, T. (2002). "Direct X-Ray Observation of a Single Hexagonal Myofilament Lattice in Native Myofibrils of Striated Muscle," *Biophys. J.* **83,** 1074–1081.

King, J., Haase-Pettingell, C., and Gossard, D. (2002). "Protein Folding and Misfolding," *Am. Sci.* **90** (5), 445–453.

Lander, E. S. and Weinberg, R. A. (2000). "Pathways of Discovery. Genomics: Journey to the Center of Biology," *Science*, **287,** 1777–1782.

Lederberg, J. (2000). "Pathways of Discovery. Infectious History." *Science*, **288,** 287–293.

Levin, B. R. and Antia, R. (2001). "Why We Don't Get Sick: The Within-Host Population Dynamics of Bacterial Infections," *Science*, **292,** 1112–1114.

Lowenstein, W. R. (1999). *The Touchstone of Life: Molecular Information, Cell Communication, and the Foundations of Life*, Oxford University Press, New York.

Lunine, J. I. (1999). *Earth: Evolution of a Habitable World*. Cambridge University Press, Cambridge.

Meyerowitz, E. M. (2002). "Plants Compared to Animals: The Broadest Comparative Study of Development," *Science*, **295,** 1482–1485.

Mistelli, T. (2001). "Protein Dynamics: Implications for Nuclear Architecture and Gene Expression," *Science*, **291,** 843–847.

Morrison, R. T., Boyd, R. N., and Boyd, R. K. (1992). *Organic Chemistry (6th edition)*, Addison-Wesley, San Francisco.

Nurse, P. (2001). "The Incredible Life and Times of Biological Cells," *Science* **289,** 1711–1716.

Purves, W. K., Sadava, D., Orians, G. H., and Heller, H. C. (2001). *Life: The Science of Biology (6th edition)*, W. H. Freeman and Co., San Francisco.

Singer, S. Jr. and Nicholson, G. L. (1972). "The Fluid Mosaic Model of the Structure of Cell Membranes," *Science*, **175,** 720–731.

Soai, K. and Shibata, T. (1999). "Asymmetric Autocatalysis and Biomolecular Chirality." In *Advances in Biochirality*, Palyi, G., Zucchi, C. and Caglioti, L. eds., pp. 125–136, Elsevier, Amsterdam.

Space Studies Board (National Research Council). (1999). *Size Limits of Very Small Microorganisms: Proceedings of a Workshop*, National Academy Press, Washington DC.

Westheimer, F. H. (1987). "Why Nature Chose Phosphates," *Science*, **235,** 1173–1178.

White, D. (2000). *The Physiology and Biochemistry of Prokaryotes*, Oxford University Press, Oxford, U.K.

CHAPTER 5

The Cosmic Foundations of the Origin of Life

5.1 Introduction

The next five chapters take us from the origin of the cosmos to the origin of life. This is a broad sweep of time and space—billions of years and billions of light-years. Yet it is in the physical and chemical evolution of the cosmos that the true linkage between the two seemingly disparate fields embraced in the term "astrobiology" is found, regardless of whether we find signs of life elsewhere in the cosmos. Our existence as carbon-based life-forms on a silicon- and metal-based world owes its genesis to an unfolding of events on a cosmic scale that start, literally, at the beginning of the universe. Tracing this evolutionary inheritance raises some troubling questions about the specific nature of our universe, and its forces and physical laws, that at first sight seem surprisingly tuned to allow life. Deeper exploration of this issue suggests, however, that the tuning is perhaps not as dramatic as some have argued.

This chapter begins the core of the book, which builds on the fundamental physics, chemistry, and biology explained in Chapters 2, 3 and 4, respectively (one more fundamental tool, thermodynamics, will be introduced specifically with reference to life in Chapter 7). Readers may wish to go back and consult some of the concepts and terms introduced previously. From here on in, the "divisions" of science will not seem so clear cut—physics, chemistry, and biology will begin to mix as they in fact do anyway in the natural world.

5.2 Temporal and Spatial Scales of the Cosmos

Olbers's Paradox and the Finite Cosmos

Cosmology is the study of the structure, origin and evolution of the cosmos—or universe—as a whole. Chapter 1 explored some of the vast number of cosmologies promulgated by different cultures around the world. With the scientific revolution, there came the

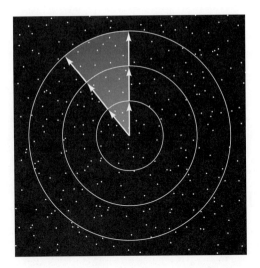

FIGURE 5.1 Olbers's paradox illustrated. If the sky were sprinkled uniformly with stars out to infinity, then any line of sight would intercept a star; because the number of stars increases, and the brightness falls off, as the distance is squared, the brightness of the sky should be uniform.

expectation that the nature of space and its origin might be addressed scientifically. Yet, until the early 20th century cosmology was regarded as too exotic to qualify as a proper field of scientific endeavor. The one fundamental cosmological clue, critics would note half seriously, is that "the sky is dark at night." And yet such a remark is far from simple sarcasm—it does, in fact, point to a fundamental property of the universe.

Were space infinite and stars sprinkled more or less uniformly through space, the sky could not be dark. Looking from Earth at the night sky, the apparent brightness of stars (let us assume they are uniform in intrinsic brightness, for the moment) drops as their distance from the earth squared. But moving outward from the Earth, the surface area upon which the stars are "sprinkled" increases with the distance squared. Hence, the brightness per unit area on the sky remains constant—and the entire night sky ought to have the surface brightness of a typical star (Figure 5.1). Indeed, if the Sun is a typical star, the sky ought to have the uniform brightness of the surface of the Sun—enormously brighter than the actual daytime sky, which is lit by scattered blue light from the Sun's disk. (In fact, most stars are intrinsically dimmer and redder than the Sun, but this detail does not affect the argument, it simply means the night sky would be somewhat redder than what we are used to seeing during the day.) The 17th-century astronomer Kepler, who formulated the laws of orbital motion, first articulated this conundrum and argued that therefore the universe must be finite. But the infinite universe idea was popular from Newton's time onward because of the belief that a finite universe would gravitationally collapse upon itself—another conundrum whose solution also lies at the core of the 21st-century view of cosmology.

Recycled repeatedly through the centuries, the paradox became associated with the name of the 19th-century German astronomer Heinrich Olbers, who argued that a uniform absorbing fluid permeated space and dimmed the light of distant stars. His explanation— itself a retread of one from 80 years before—does not work because the stars would heat up the uniform fluid until it radiated with the temperature (and, hence, color) of the stars. Likewise, the ubiquitous interstellar (and even intergalactic) dust cannot solve the paradox; to be dense enough, it too would contribute significant thermal radiation to the sky. The answer, instead, does lie in finiteness—the finiteness of time, of space, and of the speed of light. As we shall see, the universe—not just the matter within it but the space within which the matter exists—is expanding, and it began this expansion from an arbitrarily small volume roughly 14 billion years ago. Therefore, space is not infinite. And, even if it were, the lifetimes of individual stars are finite, as is the speed at which the light from distant stars reaches us. As we look outward in space, we look backward in time, because the light from increasingly distant stars began its journey at increasingly more ancient times. Because stars have finite lives, we must reach a horizon beyond which the number of stars in existence begins to decline. The darkness of the night sky—foreign, unfortunately, to an increasing number of humans as urban light pollution draws a glowing curtain across the cosmos—tells us that we live in a universe finite in space and time.

Beyond that inference, it required observations with increasingly powerful telescopes to understand just what the content of the cosmos is. By the early 20th century, the notion that our own "galaxy" of stars, the Milky Way Galaxy, could be the total content of the universe gave way to the "island universe" model. In this view, our galaxy is but one of billions of galaxies scattered through the void, each a collection of tens or hundreds of billions of stars and vast clouds of gas and dust (Figure 5.2). By the mid to late 20th century, it was also established that the galaxies themselves, far from being scattered uniformly through space, were largely confined to clusters and "superclusters" (essentially, clusters

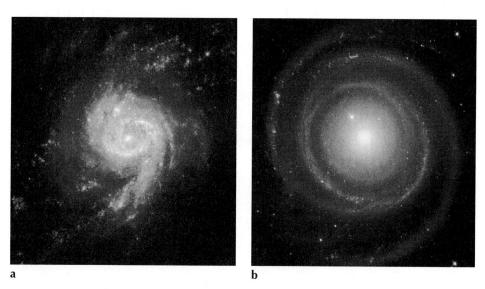

a　　　　　　　　　　　　　　　　b

FIGURE 5.2 Two examples of galaxies imaged by Hubble Space Telescope. (a) NGC 3310, a spiral galaxy with active star formation. (b) NGC 4622. This galaxy is face-on toward Earth and presents a view much like that which our own Milky Way Galaxy would present could we see it from afar. Were this the Milky Way, the Sun would not be at the center but would reside in one of the spiral arms of the galaxy, well away from the center. In both images, a few "large" stars are actually overexposed foreground stars in our own galaxy; like looking at a distant mountain from a crowded parking lot, there will always be foreground objects in the field of view.

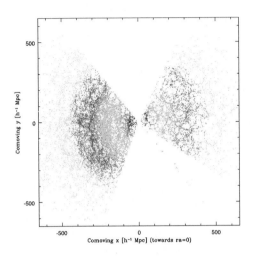

FIGURE 5.3 Distribution on a portion of the sky of more than 67,000 galaxies whose distances were determined using the Sloan Digital Sky Survey results from a network of ground-based telescopes. The horizontal axis is displacement in one direction along the sky plane, while the vertical axis represents distance from Earth. While the figure does not cover the entire sky, the heterogeneous and filamentary nature of the distribution is typical.

of clusters), and the consequent distribution of galaxies on the sky was very much non-uniform (Figure 5.3).

The observations required to draw these conclusions relied on the development of telescopes large enough to identify individual stars in other galaxies and to estimate distances to those galaxies on the basis of the brightness of individual stars (more on this later). Until very recently, larger telescopes did not, in general, serve to increase the sharpness of images because the limiting factor is the turbulence of the atmosphere—not the optical limit defined by the size of the telescope mirror, called the *diffraction limit*. Instead, larger telescopes gather and focus more photons on detectors (100 years ago, photographic plates; today, electronic detectors), enabling more distant and hence dimmer objects to be detected and characterized. *Adaptive optics*, a new technology enabling atmospheric turbulence to be partially corrected, has allowed ground-based telescopes to approach their full resolving power—roughly, the wavelength divided by the mirror diameter (Figure 5.4). Meanwhile, large space-based telescopes, though smaller than the largest ground-based facilities, can peer into deep space unhampered by our atmosphere and see enormous numbers of galaxies stretching off into a distant void (Figure 5.5, see Color Plate 8).

With the "light bucket" capacity of 2- to 5-m-diameter mirrors available by the mid-20th century, spectra of distant galaxies could be taken. Known spectral lines of elements (Chapter 3) are readily identified, but the lines from galaxies usually are shifted, wholesale, in wavelength. More often, the shift is toward longer wavelengths—a red shift—with a minority toward the blue. Furthermore, the dimmer the galaxy, and hence the larger the distance, the more the spectrum is shifted toward the red. This relationship between red shift and distance turns out to be a fundamental observational inference with respect to the overall behavior of the cosmos.

To understand the significance of the red shift, let us return to the phenomenon of waves and consider sound waves, which are compressions and rarefactions in the environment, for example, air, through which the sound propagates. Such waves travel fast—about 1000 km/h in air—but are easily exceeded by a supersonic airplane, and a speeding

FIGURE 5.4 Modern telescopes use very large mirrors and tend to come in pairs (or even quadruples) to combine the light from the mirrors to make an interferometer (Chapter 15). (a) The Keck telescopes in Hawaii, each with a 10-meter-diameter mirror. (b) The structural frame of the Large Binocular Telescope upon completion in Milan, Italy; each of the circular structures is designed to support an 8.4-meter mirror. (c) Three of the four 8.2-meter telescopes comprising the European Space Observatory's Very Large Telescope high in the Chilean Andes.

automobile will move at perhaps 10 percent of the speed of sound. Imagine then a speeding car broadcasting a tone (e.g., car horn) at a constant pitch, or frequency. If the car is moving toward you, the waves will be compressed—the wavelength shortened—by the velocity of the approaching car. Thus, the number of waves per second that move the drum of your ear increases, and the pitch seems to increase. If instead the car is moving away from you, sound waves emanating from the car are stretched out—wavelength lengthened—and thus the number of waves per second flexing your eardrum declines; you hear a lower pitch. This is known as the *Doppler effect*, and it works for light as well because light has a finite velocity in physical media or in vacuum. For velocities much less than that of light, the frequency shift is proportional to the ratio of the velocity—toward or away from the observer—to the speed of light. The shift is toward the red (longer) wavelengths for objects moving away and toward the blue (shorter wavelength) for objects moving toward the observer (Figure 5.6).

Thus, one way to interpret the red shift of the galaxies is to postulate that they are moving away from us at a speed that depends on the extent of the shift of the known spectral

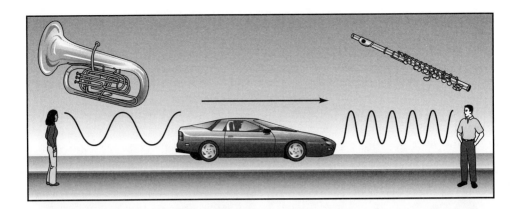

FIGURE 5.6 How the Doppler effect works. Waves appear compressed or extended to an observer moving toward or away, respectively, from the source. On the top, sound waves from a car horn are higher pitched to a listener being approached by the car and lower pitched to a listener from whom the car is receding. By analogy, a celestial object appears bluer to an observer it is approaching and redder to an observer from whom it is receding.

lines toward the red. When this interpretation is applied to the red-shift-versus-distance relationship, it implies that the more distant galaxies are moving more rapidly away from us. However, if the galaxies were simply moving through a fixed background of space, we would come to the troubling conclusion that our galaxy is the center of the universe and all else is moving away from that center. Not only does this seem philosophically unpalatable, but it also leads to fundamental contradictions. For example, the theory of general relativity (Chapter 2) requires that the presence of large masses, such as galaxies and their clusters, distort space so that the path of light is bent. (Indeed, light can be red-shifted merely through the presence of a mass distorting its path, but the relationship between red-shift and distance rules this out as the primary causative agent of galactic red shifts.) Therefore, the geometry of space itself would have to be changing as the galactic clusters fly apart from each other. A self-consistent picture of the expansion of the universe, then, is one in which space itself is expanding, with the galaxies a component of that space, and hence the red shift we measure is that of the general expansion of space and the matter within it.

A very crude analogy is to imagine our universe as the rubber surface of a balloon, with dots placed on the balloon each representing a cluster of galaxies. As the balloon is inflated, the rubber surface expands, and the space between each dot expands. Any dot will see all other dots expanding away from it; there is no special center to the surface of the balloon. This is an apt analogy in two spatial dimensions for our universe, which has three spatial dimensions, even though our universe today is very nearly "flat." (A flat geometry means that space obeys the normal Euclidean geometric rules we learn in school; for example, parallel lines do not cross, and the sum of the angles of a triangle are 180 degrees. In general space could be curved, being closed like a sphere or open somewhat like a saddle, so that these Euclidean rules are violated on large scales.) Like the balloon's surface, the dimensions defining volume in our universe may be only a subset of the total dimensionality of, for want of a better term, a "multiverse" within which our universe is embedded.

The Cosmic Distance Ladder

To quantify the relationship between distance and red shift requires that we set an absolute distance scale. One convenient way to do this is to use the distance that light travels in one year (in a vacuum; its speed is less in material objects). From Chapter 2 the speed of light is 3×10^5 km/s; there are 3.2×10^7 s in every year, so light travels nearly 10^{13} km in a year. Now the Earth-Sun mean distance (one "astronomical unit") is roughly 150 million kms, and hence the annual distance light travels is about 60,000 astronomical units (AU)—more than 1,000 times the mean distance from the Sun to Pluto, the solar system's most distant planet. The distance light travels in a year is called a *light-year*, and it is useful for the remainder of the chapter to visualize it as a distance that corresponds to about 1,000 times the extent of our solar system (that is, from Sun to Pluto).

How can we measure such distances? If stars had clocks, telling us what the "local time" is at the star, we could compare that with Earth-local time to determine how long the light took to travel to Earth. Because stars do not possess clocks, other techniques are required. Several techniques are built one upon the other in a *cosmic distance ladder*. The fundamental step in the cosmic distance ladder is *parallax*, and it is how our dual-eye vision system perceives depth in the world. Hold a pencil close to your face, and alternately blink one eye and then the other (readers with one working eye may shift their faces back and forth while holding the pencil fixed). The pencil appears to move against the background of more distant objects. The effect is also apparent looking out the side window of a car as it is speeding along the road; closer objects seem to move across a field of more distant objects. The origin of this effect is that the angular separation of your eyes (or of the constantly changing position of the car) allows nearer objects to be seen from differing perspectives, relative to more distant background objects. Your brain takes note of the angular shift and uses it to construct a perception of depth—even though each eye by itself has no direct visual clue to depth except size.

Let us imagine constructing a huge dual telescope system—one telescope poised at each end of the Earth's orbit, a distance of some 300 million kms. Blinking the image from one telescope relative to the other should reveal shifts in the positions of the stars in the sky that are proportional to their distance from us. We can do this without investing in two telescopes and a space platform, since the background of stars does not change noticeably over one year (this excludes the orbital motion of the planets of our own solar system). We merely photograph a portion of the sky when the Earth is at one point in its orbit, and then do likewise six months later. By superposing the two images, we can see which stars have shifted the most—they are the closest stars. The geometry is sketched in Figure 5.7. The base of the triangle is the position of the Earth at times spaced by six months—this is roughly 2 AU. The shift in angle of the star we call 2Θ, where Θ is called the "parallax angle." If the parallax angle is small (the triangle is narrow), the distance to the star is $D = 2 \text{ AU}/2\Theta = 1 \text{ AU}/\Theta$. The greater the distance is, the smaller is the angular shift.

Now we must face a small complication with respect to units. To express the distance in physical units (e.g., an AU is 150 million kms), the angle should be in radians. All the way around a clock—360 degrees—is 2π radians, where $\pi = 3.1419\ldots$. But astronomers do not look at the sky in radians, they use degrees. When you stand on a hill at night looking at an uninterrupted horizon (best done from one of the world's great deserts), the sky spans half a complete circle, or 180 degrees. Astronomers divide that into 60 min/deg and 60 s/min "of arc." There are then 3,600 "arcseconds" per degree, and 1.3 million arcseconds in a 360-degree circle. The size of the moon in the sky is 31 minutes of arc, or half a degree. Alpha Centauri C, the very nearest star to our solar system (excluding the Sun about which we orbit) shifts by only 1.5 seconds of arc—all others shift by less. What is its physical distance? From the calculation above, note that 1 second of arc is roughly $\frac{1}{200,000}$ of a radian. Then from the relationship between angle and physical distance, $D = 1.5 \times 10^8 \text{ km} \times 200,000 = 3 \times 10^{13}$ km, or 3 light-years. That is, *a star that has a parallax angle of 1 arc second is about 3 light-years from Earth* (a more precise figure is 3.26 light-years); this is referred to as one "parsec" of distance. The closest star has a parallax angle of 0.77 second of arc (remember the parallax angle is half the observed shift over a year), and so it is $3.26/0.77 = 4.2$ light-years away. All other stars show smaller shifts and are more distant. The closest star to Earth (excluding the Sun) is 4,000 times more distant than the planet Pluto.

The classical Greeks understood the concept of parallax and hoped to measure the distance to the stars by observing such shifts. The 2nd-century B.C.E. Greek astronomer Hipparchos made careful measurements of the sky with this aim in mind. But his effort was rooted in an optimistic expectation that the stars were much closer to the Earth than

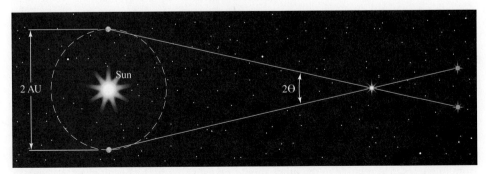

FIGURE 5.7 The geometry of parallax measurements. The Earth is shown in two locations 180 degrees apart in its orbit; the resulting angular shift in the apparent position of the foreground star against the distant stellar background is the parallax shift 2Θ.

they in fact turn out to be. It is hopeless to measure parallax with the unaided eye, and it was not until 1838 that the first parallax was measured. Indeed, atmospheric turbulence limits the sharpness of images to about 1 second of arc at good mountaintop observing sites, so that measurement of parallax requires repeated observations to remove the smearing effects of Earth's atmosphere. From the ground we are limited to parallaxes of 0.01 second of arc or larger. Space-borne telescopes can do better, and the European *Hipparchos* satellite provided parallaxes down to 0.001 second of arc—a milliarcsecond—that in physical distance is roughly 3,000 light-years. Such a vast distance does not get us across even a fraction of our own Milky Way Galaxy.

The careful reader may have asked, somewhere during the preceding discussion, how the AU was itself determined—since it is the underlying unit of physical distance to these measurements. Here again, parallax is employed, along with *Kepler's third law*. The angular motion of a planet under the influence of the gravitational pull of the Sun (Chapter 2) implies that the period of the planet's orbit, squared, is proportional to the cube of the distance of the planet from the Sun. (This holds for circular orbits; for elliptical orbits the relevant distance is the semi-major axis illustrated in Figure 1.5.) Hence, from the time of Kepler (early 17th century) the relative sizes of the orbits of the planets could be determined. The nearby planets, Venus and Mars, are close enough that their parallax shifts from one side of the Earth to the other (i.e., using the diameter of the Earth as a baseline) could be measured. Thus, an accurate determination of the absolute distances of these planets from the Earth—and hence of the astronomical unit—was available by the turn of the 20th century. In the 1960s radar signals were bounced off Venus, and the radio travel time multiplied by the speed of light provided the most accurate foundation for the AU and hence the cosmic distance scale.

To go beyond parallax measurements, that is, thousands of light-years, requires indirect approaches. With thousands of stars whose distances are known by parallax, it is possible to determine the "absolute" brightnesses of these stars by correcting the observed brightness by the distance squared. Conveniently scaling stars to a standard distance, it is possible to classify stars according to their brightness and color. Brightness is just the total number of photons per second a star puts out, this is called the "luminosity" of the star; color (as noted in Chapter 3) can be translated roughly to a surface temperature—or what is often called "effective temperature." A plot of luminosity versus effective temperature (or the observational equivalents of absolute brightness versus color) is called a "Hertzsprung-Russell diagram" after the early 20th-century inventors of this graph. A class of stars called "main-sequence" objects falls on a well-defined line on such a diagram (Figure 5.8). As we see in section 5.4, such stars are stably converting hydrogen to helium by fusion and have very predictable properties as a function of their age. The conversion of hydrogen to helium proceeds more quickly in massive stars, making them hotter, hence more blue, and shorter lived than their lower mass and redder counterparts (the Sun is intermediate in mass, longevity, and color). The spectra of such stars have a recognizable sequence, running from the hot blue stars (called O and B stars) with few atomic lines, to yellow stars (A, F, and G) with many atomic lines, to redder stars (K and M) in which molecules begin to appear (Table 5.1). The absolute properties of stars in the G class can be tied to the Sun since its spectrum in the optical is very typical of a G-type star. Because there are several ways to determine the age of the Sun (Section 5.4 and Chapter 6), astronomers understand the age-dependent properties of such stars.

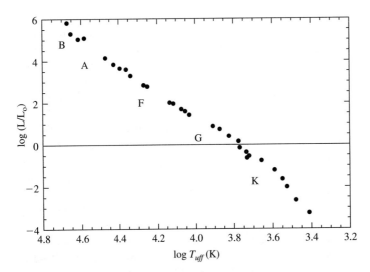

FIGURE 5.8 Hertzsprung-Russell diagram showing the positions of some measured main-sequence stars, with stellar classification letters superposed. The luminosity scale is relative to the Sun; because it is logarithmic, 0 = Sun's luminosity. O stars are too luminous, and M stars too dim, to appear on this graph.

TABLE 5.1 Stellar Classification Sequence

Letter	Criteria
O	Spectra show absorption by doubly ionized helium
B	Absorption by singly ionized helium and atomic hydrogen
A	Strong atomic hydrogen absorption
F	Doubly ionized calcium lines strong; some neutral atomic lines
G	Plenty of atomic lines in the optical
K	Molecular lines of carbon-hydrogen and carbon-nitrogen
M	Many molecules, including water; TiO increases
L	Molecular lines of CO, H_2O, TiO
T	TiO disappears, CH_4 appears

(Note: L and T objects are not true stars in the sense that they are too small to undergo hydrogen fusion to helium)

In consequence, stars like the Sun, and other main sequence stars, can serve as so-called standard candles throughout the galaxy. If the intrinsic brightness as a function of spectral appearance is known, then the actual measured brightness yields the distance—

because the brightness must fall off as the distance to the star squared. The situation is much like that of streetlights. Imagine glancing out an airplane window at night and looking at distant cities lit by streetlights. If you assumed the streetlights were all the same intrinsic brightness, then the apparent brightness of the lights would tell you the relative distances of the lights within a city, and from one city to another. You might even play the game of streetlight colors—blue is a mercury vapor lamp, and yellow is a sodium vapor lamp. If you knew they had different intrinsic brightnesses, you could refine your distances by looking at these two classes of lamps (and, in fact, the sodium lamps come in low- and high-pressure varieties, with again different intrinsic brightnesses). To do this you would also need to correct for two effects: the age of the lamps (they dim with age) and the obscuring effects of smog or cloud.

With stellar standard candles, the age and obscuration complications also must be dealt with. The space between the stars is filled with both hydrogen gas and interstellar dust, and the dust scatters light in such a way as to redden and dim the appearance of distant stars compared with their closer cousins. (You can see this same effect on the setting Sun caused by dust in our own atmosphere.) Correcting the distances for reddening has been a major problem in astronomy, the solution to which has improved as parallax measurements have been extended outward, allowing an absolute correction to be made. Age has subtle effects on stars—because the rate at which hydrogen fusion occurs in the stellar interior changes. One approach is to measure distances using stars in "clusters": Stars within a given cluster are believed to have formed at the same time; detailed spectra of such stars confirm this assumption. Nearby clusters include the Pleiades and the Hyades, and their distances are determined by parallax. More distant clusters, with stars of various colors representing different masses but the same age, can be compared with the nearby standards so that their observed brightness can be translated into distance.

Such a technique works out to distances for which color and brightness of individual stars can be measured. This includes the distant reaches of our own galaxy of 10^{11} stars, the main spiral disk of which turns out to be about 120,000 light-years across, and nearby galaxies out to 10 million light-years distant. Such measurements are helped by a peculiar class of star called a *Cepheid variable*. Variable stars are those whose brightness fluctuates with time; Cepheids are a class of variables in which the rate of variation of the brightness with time correlates inversely with the maximum intrinsic brightness of the star. This empirical rule is tested through the nearby Cepheids, whose distances are known by parallax measurement, and then used to determine absolute brightnesses and hence distances of Cepheids across the galaxy and in other galaxies. While physical explanations for this relationship between maximum brightness and speed of variation have been proposed, the important point is that the observational calibration using nearby Cepheids is *a crucial link* in the cosmic distance ladder. Although Cepheids were used early in the 20th century to determine that galaxies were distant objects, it was not known that in fact there are two distinct classes of Cepheids with different relationships between maximum brightness and frequency of fluctuation; this complication was resolved in the 1950s.

Very distant galaxies, those for which the Cepheids are too dim to be seen, require another rung in the cosmic distance scale. Stars at least several times more massive than the Sun eventually undergo runaway nuclear reactions leading to catastrophic explosions called *supernovae*. While we might imagine that the maximum brightness of the explosion could vary from one star to another, in one class of supernova the maximum brightness is constant. This "Type I" class of supernovae—actually the result of the infall of mass onto a white dwarf star from a close companion star—is recognizable by the rise and fall of the brightness curve during the explosion. Supernovae are very rare events in the Sun's galactic neighborhood (were they not, radiation from the explosions would likely have made life on Earth impossible in the long run); the closest one to us is thousands of light-years away and exists now as a debris field (the "Crab nebula"), the explosion itself was seen on Earth a millennium ago. Useful numbers of such supernova explosions occur for nearby galaxies, where billions of stars at a time can be observed. Because the distances to such galaxies are known through the Cepheid distance scale, the maximum brightness of the supernova explosions can be calibrated for the nearby galaxies; observations of supernovae in more distant galaxies then extend the cosmic distance ladder out to nearly 100 million light-years.

The universe is much larger than than 100 million light-years; the dimmest galaxies, if they have the same intrinsic brightness as closer ones of the same spiral or elliptical shapes (whose distances we know), extend out billions of light-years. How many billions? Here the cosmic distance ladder reaches its pinnacle with the red-shift-versus-distance relationship. Within the "intergalactic neighborhood" encompassing the tens of millions of light-years around our own galaxy, there are blue shifts and red shifts; our galaxy, for example, is falling toward a large mass of material associated with a huge cluster of galaxies in the part of the sky defined by the Virgo cluster. But on scales of hundreds of millions to billions of light-years—still within the realm of the cosmic distance ladder—red-shift seems to increase with distance. We must be careful to interpret this red shift as a general expansion of space–time itself (the four-dimensional reality of general relativity described in Chapter 2) to derive correct distances from the relationship, since the most red-shifted galaxies possess an expansion speed approaching that of light itself—*relativistic* velocities. There is a roughly linear relationship between red-shift and distance, and the slope of that line—given in kilometers per second per million parsecs, is known as the *Hubble constant*. It has values between 60 and 72 km/s/megaparsec (1 megaparsec = 10^6 parsecs) based on determinations made just in the past few years, corresponding to distances from us of 13 to 17 billion light-years.

Galaxies moving at speeds approaching that of light, and hence some 13 to 17 billion light-years away, must define the edge of the universe. There cannot be a component of space–time, containing mass, which travels at or beyond the speed of light. Or, if there were, we could not see it because the information would never reach us. Light from galaxies 13 billion light-years distant began its journey 13 billion years ago. What was happening at the time? If space–time (the universe) is expanding in such a way that things more distant are receding more quickly, then billions of years ago the universe must have been much more compact. Clues to what things were like then came in the 1960s when radio astronomers Penzias and Wilson detected a low microwave hum from their extremely sensitive radio system—a hum that came from all directions and permeated everything. The spectrum of this hum, or microwave background radiation, is as perfect a blackbody as anything in nature—which means the source is a general thermal background of the cosmos, with a temperature based on its blackbody spectrum of 3 K (Figure 5.9).

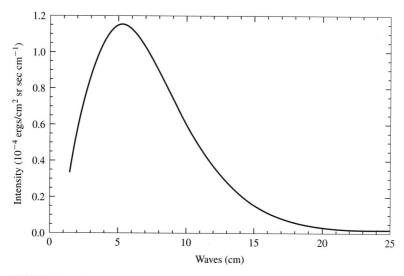

FIGURE 5.9 Measurement of the cosmic background radiation from the COBE satellite, some 30 years after the discovery of the radiation. The COBE data and a theoretical blackbody curve at 3 K are identical at this scale.

What might cause such a hum, permeating all space? If there is no local source, it must come from beyond the realm of any sources we can identify and measure. That is, it comes from beyond the farthest galaxies. In turn, that means it must be a strongly red-shifted signal of something that originally was at much shorter wavelength, much higher temperature. Let us "run the universe" backward to a time, 13 billion years ago or farther, when the scale of the cosmos was much smaller. Having all the matter in the universe in a small package would have meant extremely high temperatures and opaqueness so strong at all wavelengths that light and matter would have interacted on the smallest subatomic scales. In such a milieu of matter and light, electromagnetic energy would have been maintained as an ideal blackbody at a precisely single temperature, until the cosmos expanded enough (around a million years after the beginning) that the universe became transparent—at which point (see Section 5.3) there could be places that were dark and places light. (This part of the cosmological history bears a resemblance to the Judeo-Christian, and other, creation stories in which light is separated from darkness to become distinct properties of the real world.) Once separate, that "horizon" at which light and matter were last coupled in a cosmic soup—a horizon beyond the farthest galaxy—would recede with the rest of the cosmos, its signal progressively red-shifted as space–time stretched out over billions of years.

Given the amount of matter cosmologists estimate to be in the cosmos, and the temperatures at which this great decoupling would have occurred, we arrive at just about the right temperature for this receding echo at an age of the universe somewhere between 13 and 17 billion years. The most recent determination of the age of the cosmos comes from a NASA satellite named the Wilkinson Microwave Anisotropy Probe, which very accurately measured subtle fluctuations in the temperature and other properties of the microwave background radiation. Combined with data from ground-based radio telescopes, these measurements lead to a precise determination of the age of the cosmos—the time since the primordial fireball—of 13.7 billion years, ±200 million years. No more obvious or precise clue to the fact that the universe has expanded from some primordial fireball could be written on the heavens than the 3 K background radiation. Were that signpost of the separation of the light from the darkness not detected in the 1960s, our understanding today of cosmic history would be little better than it was a half century ago.

So, in brief, this universe of four space–time dimensions began some 14 billion years ago in an explosion of an unimaginably dense and hot medium, which then spread outward—space–time and matter with it—cooled, and formed structures that we call stars, organized into galaxies and clusters of galaxies, and around some of those stars planets, at least one of which plays host to life. But were this the only connection to astrobiology, it would be at best a cultural one. There is far more to the story, and it requires that we first delve into the genesis of the primary elements during the cosmic explosion, the Big Bang, and then into the formation of the heavier elements in succeeding generations of stars. Once that part of the story is done, we must return to some of the details of the cosmos around us. Those details—the amazing degree of flatness of space–time, the presence of nonluminous "dark matter," and the observation that the expansion is accelerating at the largest scales, not decelerating—have a remarkable potential implication—that this is the one universe in a multiverse of otherwise compact universes suitable for the formation of macroscopic structures, and life.

5.3 Origin of the Primordial Elements Hydrogen and Helium in the Big Bang

Let us consider the theoretical sequence of events that begins with an explosion creating the cosmos from a point of arbitrarily high density and continues with expansion and cooling. At certain points in the expansion, we will be able to identify key observations that constrain the process, but by and large, the events are based on the theoretical expectations of the behavior of particle systems at various temperatures. Table 5.2 summarizes the key events.

The story begins at 10^{-43} seconds, when the physics of general relativity (and, hence, gravity) begins to dominate the interaction of matter and space. The linear size of the universe at this point can be estimated as the speed of light in vacuum multiplied by the time since the explosion, $ct = 3 \times 10^8$ m/s $\times 10^{-43}$ s $= 3 \times 10^{-35}$ m. In this unimaginably small and hot cosmos, particles do not exist. From that time to 10^{-35} seconds, temperatures are so high (10^{26} K and above), that the other three fundamental forces (electromagnetism, weak and strong nuclear forces) are "unified"; that is, they behave identically and are essentially indistinguishable. Unfortunately, the kinetic energies corresponding to such high temperature are unachievable in any particle accelerator imaginable. However, there are some constraints on that epoch, notably the strong preponderance of matter over anti-

TABLE 5.2 Key Events in Models of the Expansion of the Cosmos

Time (s)	Name of Epoch	Major Events
"Detonation" to 10^{-43}	Planck	All forces were unified
From 10^{-43} to 10^{-35}	Grand unification	Three forces unified
At roughly 10^{-37}	Inflation	Exponential expansion of cosmos
From 10^{-35} to 10^{-6}	Quark	Cosmos made of fundamental "quarks"
At roughly 10^{-11}	Electroweak	Weak and electromag. forces decouple
From 10^{-6} to 10^{-4}	Hadronic	Matter excess over antimatter frozen in
At roughly 1	Neutrino decoupling	Background neutrinos go free
From 10^{-4} to 10	Leptonic	Ratio of protons to neutrons frozen in
At roughly 100	Nucleosynthetic	Light atomic nuclei are formed
From 10 to 10^{11}	Radiation	Scale factor goes as $r^{1/2}$
At roughly 10^{13}	Recombination	Cosmos becomes transparent
From 10^{11} to today	Matter	Formation of galaxies, stars, planets, life

Notes: Hadrons interact by the strong nuclear force (e.g., protons, neutrons); leptons (e.g., electrons) do not.
Source: Adapted from Hawley, J. F. and Holcomb, K. A. (1998) *Foundations of Modern Cosmology*, Oxford University Press, New York.

matter. Nothing in physics requires that an electron be negatively charged; its positively charged counterpart, the positron, can exist and has been observed. Likewise, the negatively charged "antiproton" has been observed. However, contact between matter and antimatter leads to annihilation of both particles, and the initial amount of matter had to be slightly larger than antimatter for the current galaxies that we see to exist. Under current conditions, with distinct electromagnetic and nuclear forces, creation of any particle requires the creation of its antiparticle. But during the grand unification epoch, this symmetry need not have applied: Theory indicates that under these conditions an excess of matter or antimatter would have been permitted. Indeed, the theory is detailed enough to argue that the excess is always matter, not antimatter, leading to the matter-dominated cosmos confirmed observationally in the spectra of both nearby and distant galaxies.

At 10^{-37} seconds, the expansion rate of the universe briefly increased, in an epoch called *inflation*, which we shall discuss in Section 5.5. From 10^{-35} to 10^{-6} seconds, when the temperature fell to a "mere" 10^{11} K, the cosmos was a soup of the elemental building blocks of protons and neutrons, called "quarks," and little else. This epoch is defined by the decoupling of the strong nuclear force from the electromagnetic and weak forces. But the latter two remained unified, and there was no possibility of having stable charged particles exist separately; hence, only their fractionally charged "quark" building blocks could exist. As the electromagnetic and weak forces decoupled from each other late in this epoch (10^{-11} seconds), new fundamental particles appeared. Eventually, at 10^{-6} seconds after the Big Bang (a microsecond!), quarks combined into massive particles, *hadrons*, of which protons and neutrons are examples, with strong nuclear interactions. But at the high temperatures present, neutrons and protons would rapidly convert one to the other. An important event in this stage is that matter and antimatter—hadrons and antihadrons—were distinguished from each other so that collisions between the two led to annihilation of the particles and the creation of photons. Most of the photons in the cosmic background radiation are thought to have formed at this time. The fraction of hadrons that did not annihilate is approximately 10^{-9}, or one billionth, of the original number of hadrons. The survivors constitute the matter in the universe today.

As cooling proceeded, electrons could exist as stable particles, and hence the ratio of protons to neutrons became largely frozen in. By 100 seconds after the Big Bang, the primary stable particles of matter (the electrons, protons, and neutrons) existed in a uniform soup of photon radiation at a temperature of 10^9 K. At this temperature, however, all was not unchanging, because collisions between heavy particles could lead to their fusion into heavier nuclei. Hydrogen—a proton and electron—existed, though the electron always remained unbound in the hot soup. Free neutrons are not stable; they decay after about 1,000 seconds into protons and electrons. However, those neutrons collided with protons to make nuclei of one proton and one neutron, two protons and one neutron, or two neutrons and two protons. The first of these constitutes the heavy but stable isotope of hydrogen, deuterium; the last two are the isotopes of the element helium. Again, all three of these isotopes were fully ionized because it would be some time before the electrons could bind to the nuclei without being knocked off again. In these two configurations, lower in energy than the protons and neutrons flying about separately, neutrons remained stable. The relative stability of the nuclei formed dictated that deuterium would be about 20 parts per million (ppm) the abundance of light hydrogen, and the total helium produced was about 25 percent of the hydrogen abundance.

These numbers, especially the deuterium abundance, depend sensitively on models of the overall density of protons and neutrons in the cosmos at the time. Hence, if the theory is right, this provides an important constraint on the excess of matter over antimatter during the earlier "hadronic" epoch. Measurement of the primordial deuterium abundance is tricky because deuterium is destroyed in stars. The 1995 determination of its abundance in the planet Jupiter by the *Galileo* entry probe was important because the deuterium in Jupiter does not undergo fusion—and hence is a sample of the deuterium abundance at the time of the planet's formation some 4.5 billion years ago when the universe was $\frac{2}{3}$ of its present age.

Beyond hydrogen and helium, only one other element was made in significant abundance in the collisions constituting the "nucleosynthetic epoch" around 100 seconds. Lithium, with three protons and (in stable isotopic form) four neutrons, could be made by colliding a light helium and normal helium, ^{3}He and ^{4}He, along with conversion of a proton to a neutron (and electron) and release of some energy in the form of photons. Because lithium is destroyed by fusion reactions even in the cool outer envelopes of most stars, it is difficult to determine its primordial abundance. However, relatively low-mass stars should retain lithium in their outer layers, and measurements of lithium in very old examples of such stars (in which the lithium might have been preserved from the first generation of star formation) suggest Li/H about 10^{-10}, consistent with theoretical predictions of element production in the nucleosynthetic epoch.

Much beyond 100 seconds, the matter density of the expanding universe declined enough that nucleosynthesis became rare, and then ended, until stars could form and generate their own nuclear reactions. Space was a uniform but cooling soup of photons and matter—though the ghostlike *neutrino* particles, which hardly interact with matter at all and are difficult to detect, had decoupled from this soup at 1 second after the Big Bang. Around 10^{13} seconds, or roughly a million years after formation, when the universe was several times 10^{18} kilometers across (currently the distance from here to the neighboring Andromeda galaxy), radiation decoupling occurred. The temperature of the cosmos dropped to 3,000 K, and electrons began to bind stably to nuclei. Under such conditions, electrons were no longer continually colliding with each other and creating the enormous continuum absorption and emission of photons (Chapter 3) that maintained the coupling between matter and energy in the form of photons. Bound to atoms, electrons absorbed in discrete wavelengths, and photons were largely free to move through the void of space, defining the horizon of the cosmic background radiation that we see—greatly red-shifted—today. From that point onward, despite its compactness, the universe assumed a more familiar form. As expansion and cooling continued, hydrogen and helium gas could form gravitationally bound systems (the precursors of galaxies) and eventually collapse into the bound spheres called stars.

This narrative—in which the characters are bizarre particles moving within an opaque netherworld, rather than the strange creatures and nascent people trapped in the underworlds of the myths of Chapter 1—seems to explain much of what we see in the universe today. It does not succeed in explaining the presence of poorly understood components of matter, called dark matter, and an apparent repulsive force, called dark energy, that we will encounter later in this chapter. But it does explain the existence of the cosmos as one dominated by matter, in which fundamental particles have combined to form three of the fundamental elements, in accord with the properties of these particles and the laws of the forces that govern them as tested in particle accelerators. It leaves us with a cosmos composed of essentially three simple elements that, by themselves, cannot form planets and life. But the natural evolution of the cosmos from then on is, in fact, a story of the progressive formation and dissemination of heavier elements through succeeding generations of stars, elements that are the key constituents of planets and life.

5.4 Manufacture of the Biogenic Elements in the Shining and Death of Stars

> *Stars in your multitudes,*
> *Scarce to be counted,*
> *Changing the chaos*
> *To order and light*
> A. Boublil, C-M. Schönberg, and H. Kretzmer, *Les Misérables*

In the current "matter-dominated" epoch of the cosmos, there are few places where collisions between nuclei occur with sufficient frequency and vigor to allow heavier nuclei to be formed from lighter ones. The binding energy per nucleon (proton or neutron) becomes more negative (more favorable for binding) as the mass of the nucleus increases toward iron, then beyond iron it decreases (Chapter 3). Thus, combining nuclei up to the value of iron, excluding excessively heavy isotopes with too many neutrons, ought to be preferred and lead to release of energy. It does, but it requires overcoming the repulsive interactions of the electron clouds between atoms, and then the repulsive electrostatic interactions between protons in the two nuclei. High temperatures overcome both of these barriers, when atoms are fully ionized and collision speeds sufficient to bring the nuclei in contact with each other.

Such reactions—nuclear fusion—make the Sun shine. Through the 19th to the earliest 20th centuries, after the distance from the Earth to the Sun was determined, various chemical and gravitational sources of energy were proposed to account for the vast production of solar energy. All fell short as the great age of the Earth began to be inferred from the geologic record. The development of models of the atom and the principles of quantum mechanics led in the first decades of the 20th century to an identification of the fusion of hydrogen to helium—the two principal constituents of the Sun—as the Sun's energy source, and later work through the 1950s identified the particular reactions. These are summarized in Figure 5.10.

Two different types of reaction sequences make helium and generate energy in the interiors of the Sun and other stars. The simplest, and most basic, fusion process is called the proton-proton, or *p-p*, chain and requires only hydrogen and helium to be present. The first of the p-p chains (often called p-pI) involves three separate reactions:

1. The first step involves two hydrogen nuclei (protons) colliding to form a deuterium nucleus (one proton and one neutron) and two exotic particles—a positron and a neutrino.
2. Next, the deuterium nucleus collides with another proton. The net result is the release of a photon and the combining of the two nuclei into the light isotope of helium. The detection of ^{3}He emanating from the Sun in the solar wind is evidence of the p-p chain at work.
3. Finally, two ^{3}He nuclei collide together and form the primary isotope of helium, ^{4}He. This nucleus stays intact under conditions in the Sun's interior but undergoes further fusion in stars that are more massive and in later stages of the Sun's evolution. In addition to the helium, two protons are produced.

In net form, four protons have been converted into one helium nucleus, with a release of energy in the form of photons. Photons generated by the reactions diffuse through the Sun, absorbed and emitted by matter along the way. The net direction of energy flow is outward as the fusion reactions occur in the deep interior. At the surface, the energy escapes in the form of a photon distribution appropriate to the photospheric, or surface, temperature of 5,770 K.

The p-pI chain is not the only fusion reaction chain that occurs in stars. Indeed, it is possible for ^{3}He to collide with ^{4}He nucleus to form beryllium, then lithium (p-pII chain), and, in a fraction of reactions, boron (p-pIII chain), but in the end these heavier elements are destroyed in favor of ^{4}He again. Beryllium and lithium act as catalysts in the nuclear reaction; they control the speed of the reaction sequence (in this case by being good targets for electrons and protons) without being consumed in the process.

5.4 Manufacture of the Biogenic Elements in the Shining and Death of Stars

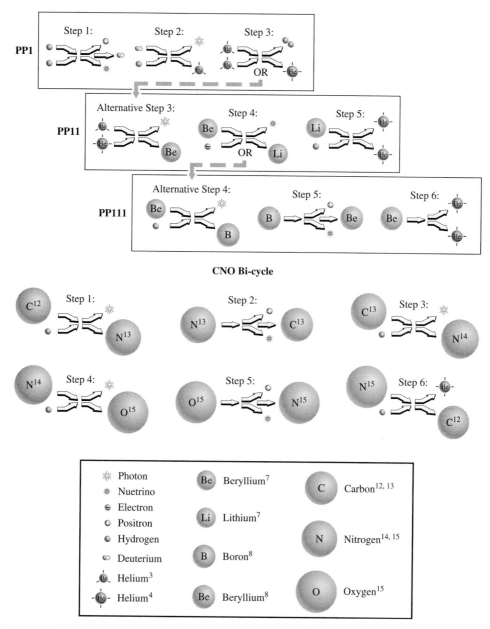

FIGURE 5.10 Steps leading to the production of helium from hydrogen via the proton-proton (p-p) and carbon-nitrogen-oxygen (CNO) nuclear fusion cycles.

The energy liberated per helium nucleus produced is 4×10^{-12} joules, enough to power a 40-watt (which is a 40 J/s) light bulb for only 10^{-13} seconds. However, the sun contains enough hydrogen to produce 10^{56} helium atoms; if all of the hydrogen were so converted, the amount of energy released would be 4×10^{44} joules. The Sun's observed luminosity is 4×10^{26} watts; therefore, the p-p chain could sustain this process for 10^{18} seconds, or 30 billion years. This number is actually an overestimate of the durability of the Sun because (1) much of the hydrogen is too far from the Sun's center to experience high enough temperatures to undergo fusion and (2) the sun's brightness (luminosity, or power) has varied with time. Calculations that are more accurate yield roughly 12 billion years of steady hydrogen fusion for the sun, of which 4.5 billion years have already transpired.

The p-p chain is only one of two cycles converting hydrogen to helium in stars. The CNO cycle requires carbon, nitrogen, and oxygen to be present in the region of nuclear burning. In this cycle, carbon acts as a catalyst to facilitate, through the intermediate creation and destruction of nitrogen and oxygen, the creation of the ^{4}He nucleus from four protons. The sequence is potentially much faster than the p-p chain because, in the latter chain, two hydrogen nuclei (protons) must collide together to initiate the process, and this is inefficient because of the small size of the protons. In the CNO cycle, all collisions are between protons and larger nuclei like the carbon nucleus (six protons and six neutrons). However, there is not much carbon in the center of the Sun, and the CNO cycle is currently less important than the p-p cycle. As fusion proceeds in the Sun and helium builds up, the interior temperature of the Sun will increase; as it does so the CNO cycle will gain in importance and eventually dominate.

Fusion requires very high temperatures to provide nuclei with enough velocity to overcome the repulsive electric force between the protons. In stars the high temperatures are achieved through the enormous pressure associated with the mass of the star: The mutual gravitation attraction of the 2×10^{30} kg of material that comprises the Sun pulls the entire mass into a sphere and creates pressures in the deep interior billions of times the atmospheric pressure on Earth. But most stars are less massive than the Sun. In the interiors of the smallest and most abundant stars, the "M-dwarfs" or "red dwarfs" pressures and mean collision speeds are lower, collisions capable of combining nuclei are much less frequent, and nuclear reactions are much slower than in the Sun. These stars are cooler (hence, appear red rather than yellow), but they are much longer lived because they fuse hydrogen more slowly. Stars more massive than the Sun undergo hydrogen fusion much more rapidly, are much brighter and bluer than the Sun, but are far shorter lived. During the time over which stars undergo hydrogen fusion, their brightness and size change only slowly; this stable epoch of stellar evolution defines the "main sequence" of Figure 5.8.

The lifetime and luminosity of main sequence stars, sorted according to their mass, has important implications for the habitability of orbiting planets and the chance that life will have enough time to evolve into complex forms before these stars go unstable. Figure 5.11 shows the time for stable hydrogen fusion in stars as a function of their mass; this is the main sequence lifetime. Stars several times more massive than our Sun do not last long enough to give complex life a foothold on any planets orbiting about them, if the rate of evolution of life on our planet is a fair guide. This issue, important in the search for life around other stars, is discussed in more detail in Chapters 15 and 17.

Fusion processes in stars deplete deuterium as well as hydrogen. In fact, deuterium can undergo fusion at lower temperatures than can hydrogen. The reaction is simple: It is the

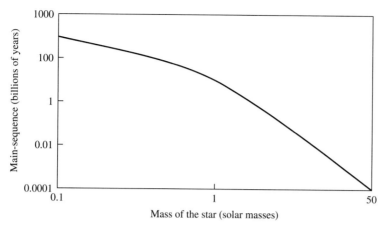

FIGURE 5.11 Lifetime of the hydrogen-fusion, main-sequence stage of stellar evolution, in billions of years, plotted against the star's mass in units of the mass of the Sun.

second step of the p-pI chain in which a deuterium nucleus and hydrogen nucleus collide to form a ^{3}He nucleus with liberation of energy in the form of a photon. The collision of two protons, the first step of the p-pI chain, requires much higher collision velocities and hence limits hydrogen fusion to objects more massive than those that can undergo deuterium burning. The threshold for hydrogen fusion in stars of solar composition is 80 times the mass of Jupiter; for deuterium burning, it is only 13 times Jupiter's mass. Because of the very small abundance of deuterium in the cosmos, 20 ppm relative to hydrogen, deuterium fusion can power a star for a few million years at most, compared with the billions of years the Sun shines by hydrogen fusion.

If the CNO cycle is operative in the Sun, where did the carbon, nitrogen, and oxygen come from? Up to this point, we have discussed processes that yield only the first four or five elements in the periodic table. An important clue to the source of the heavier elements comes from the abundances of the elements in the Sun and primitive solar system material, tabulated in Chapter 3. Displayed as Figure 5.12, the abundance pattern reveals that the elements heavier than hydrogen and helium possess a zigzag abundance pattern. Within the elements up to iron, abundance peaks every two units of element number; that is, the elements with even numbers of protons are more abundant than those with odd number. Element number, of course, does not give us the total atomic mass—the proton plus neutron number. In the lighter elements, the stable and most abundant isotope is that with an equal number of neutrons and protons. Thus, the abundance peaks come every *four* atomic *mass* numbers—via the addition of two protons and two neutrons, which is a helium nucleus. That this is so should not be surprising, because the helium nucleus (what we called an alpha-particle in the discussion of radioactivity in Chapter 3) has an extremely large and negative binding energy. Hence it is very stable.

Nuclear fusion processes can add helium nuclei together, to build heavier elements, but require pressures and hence temperatures in the interiors of stars greater than in all but the most massive main sequence stars. Thus the production of C, N, and O occurs in the

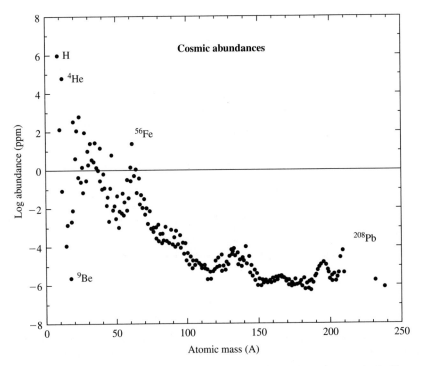

FIGURE 5.12 Relative abundance versus proton number of the elements in the Sun and primitive solar system material.

post-main-sequence phase of stellar evolution. Let us focus on what happens in stars comparable in mass to the Sun (G-type dwarfs); the details differ in other stars according to mass. As hydrogen is depleted in the core of the star, helium "ash" accumulates, and the star becomes both denser and hotter in its interior. Eventually hydrogen fusion becomes impossible because the unburned fuel is so far from the densest, hottest, center of the star. Fusion ceases, and with it the thermal energy it generated—thermal energy that supported the huge mass of the star against gravitational collapse. Collapse occurs, and the conversion of gravitational potential energy into the kinetic energy of collapse (Chapter 2) heats the interior to the point at which fusion is initiated again—this time helium fusion. Helium fusion involves the combination of helium nuclei, with no interconversion of neutrons and protons (because the particles are so stable within the helium nucleus). Two helium nuclei generate beryllium; adding another generates carbon, a fourth generates oxygen, the fifth yields neon, and a sixth produces magnesium.

The ignition pulse of helium fusion causes the star to expand into a *red giant*, with a size much larger than the Sun. Some of the envelope of the star is expelled to the interstellar medium, but by and large the star stays intact. The subsequent evolution of the star is complex and very sensitive to the mass and the original internal composition; the star may oscillate in size and color several times. Side reactions produce elements intermediate in atomic number, such as nitrogen, as well as other stable isotopes of the peak elements (e.g., ^{17}O instead of ^{18}O). A star the mass of the Sun will cease fusion at the stage of oxygen production and collapse finally into a compact object the size of the Earth. This *white dwarf* then slowly cools off, radiating to space.

More massive stars achieve temperatures of billions of degrees in their post-main-sequence interiors, overcoming the progressively greater proton charge barriers of the nuclei to generate silicon, phosphorus, sulfur, magnesium, aluminum, and various other elements and isotopes, up through iron. Beyond iron, fusion costs energy; hence, energy-generating reactions terminate at iron. Stars that have sustained fusion to this level typically began with 10 times the mass of the Sun; they do not end as stable white dwarfs but instead explode as supernovae, expelling the products of their nucleosynthesis into the interstellar void. The remnant collapses as an ultradense, 10-km, agglomeration of neutrons—a *neutron star*. If the compact remnant of the explosion has a mass larger than two or three times that of the Sun, its escape velocity is greater than the speed of light. Because light cannot escape from such objects, the resulting *black hole* is not directly observable. Many weird general relativistic and quantum effects occur at the so-called event horizon, or edge, of this hole in space–time.

The existence of exploding stars provides a straightforward story for the formation of the elements heavier than hydrogen, helium, and lithium. The first generation of stars (for which there are no known examples) must have been nearly pure hydrogen-helium and generated helium via the p-p process. The post-main-sequence evolution of these stars included the nucleosynthetic generation of heavier elements, and the more massive stars of this generation expelled their products into the interstellar medium. Subsequent generations of stars, born in the collapsing fragments of molecular clouds (Chapter 6) containing hydrogen, helium, and the heavier element contaminants, processed yet more hydrogen and helium into heavier elements. The "metallicity" or heavy-element content of the Sun suggests that it is a third-generation star—a so-called Population I star; its more metal-poor progenitor stars were Population II. Because of the longevity of the M dwarfs, many metal-poor Population II stars still exist today and can be studied. The brief lives of the O and B stars imply that many generations of these "flash in the pan" objects have spewed nucleosynthetic products into the environment; astronomical observations of stellar populations suggest, however, that the most massive stars are the least abundant, blunting their contribution to the total heavy-element abundance of the galaxies.

The distribution in the Milky Way Galaxy of Population I and II stars differ. "Metal-poor" Population II stars tend to follow orbits around the center of the Milky Way that cut through the spiral-arm disk of the galaxy, while Population I stars seem to be members of the disk. Many very old stars, metal-poor, exist in *globular* clusters that comprise a *halo* around the galactic disk. This suggests that the early history of galaxies—the initial collapse of gas and then star formation—may not have yielded the spiral disk that is their characteristic signature today. The disk may well have been formed later, but how much later is unclear. Part of the difficulty in resolving this issue is that there are no observable stars that convincingly represent the very first generation (Population III) of star formation. Did this precede or follow the formation of the galaxies themselves? As we see in Chapter 6, evidence from radioactive isotopes in rocks constrains when the formation of elements heavier than helium and lithium began in stars, but where this happened and how the galaxies themselves evolved remains unclear. Galaxies seem to look the same as we peer further back in time with Hubble Space Telescope, until we get to the most distant galaxies. These still are a half billion years younger than the Big Bang; to peer further

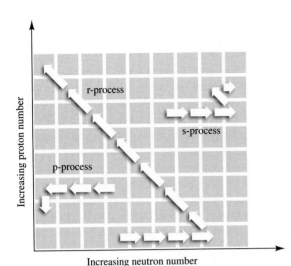

FIGURE 5.13 Schematic "chessboard" of neutron number on the x axis, proton number on the y axis. Neutron addition moves an isotope to the right; neutron decay moves it diagonally to the upper left and transforms the element. Depending on whether neutrons are added slowly (s-process) or rapidly (r-process), the element number may change by one or more protons. The "p" process of neutrons, then protons, "dripping" off superhot nuclei is also shown.

back to the time of galaxy formation will be the purview of Hubble's bigger infrared successor, the James Webb Space Telescope.

But what about the elements heavier than iron? How are they generated? The exotic and energetic environments within and around collapsing massive stars create opportunities for adding neutrons to stable nuclei. While additional neutrons only change the isotope, not the element number, addition of excessive numbers of neutrons lead to the conversion of neutrons to protons and electrons. While the electrons escape, the protons remain in the nucleus and hence the element number changes. Neutron addition and decay are illustrated in the "game board" of Figure 5.13. In the interiors of most stars, but especially those that post-main-sequence become "asymptotic giant branch" (AGB) stars, neutrons are added to nuclei slowly, on timescale of 10 to 10^5 years. After a few such additions, the nucleus becomes unstable and beta decays; a neutron is converted to a proton and electron, with escape of the electron. What was initially formation of a heavier isotope of a given element becomes transformation of the element to one with a higher proton number. This *s-process* can make elements heavier than iron through neutron addition, but it cannot make very-neutron-rich isotopes.

To make very-neutron-rich isotopes, and elements derived therefrom, requires rapid addition of neutrons to overcome the tendency to beta decay until the nucleus is extremely neutron-rich. The *r-process* proceeds until the rate of capture is exceeded by the rate of beta decay, and a cascade of beta decays moves the neutron-heavy element diagonally to the upper left as shown in Figure 5.13. The site of r-process nucleosynthesis is controversial. Neutron-rich environments in supernova explosions can produce some but not all of the elements that require the r-process. Others may be made in this way but are altered or destroyed before they can be expelled to space. Hence, environments that are more exotic have been proposed, including a neutron-rich wind emanating in the first seconds after formation of a neutron star or jets of material driven from disks around neutron stars, in which neutrons are formed by collision of material at very high speed.

Neutron addition is not the only process leading to heavy-element and exotic-isotope production. In extremely high-temperature environments in supernovae, neutrons "drip" off nuclei, leading eventually to a proton-rich nucleus from which protons are expelled. The Milky Way Galaxy itself sports a weak magnetic field, in which bare and hence charged nuclei are accelerated and slam into the ambient interstellar hydrogen gas. In this *l-process*, the heavy nuclei are broken apart or "spalled," leading to additional lithium, beryllium, and boron as well as other elements.

In essence, the violent and high-temperature environments within and around stars, and the acceleration of charged particles in the galactic magnetic field, enable the transformation of elements that the medieval alchemists dreamed of but never commanded enough energy to create (Figure 5.14). Most of the elements and their isotopes—stable and unstable—can be manufactured in abundances corresponding to what is observed in the Sun and solar system via processes that are well understood. However, as noted earlier, the sites of such processes are not fully understood. The explosion of a supernova in one of the nearby Magellanic Clouds, satellite galaxies to our own, in 1987 allowed modern spectroscopic techniques at a range of wavelengths to be trained on the site within days after the observation of the increase in the exploding star's brightness. A number of exotic elements were observed early on and provided a gold mine of data with which to constrain the production of various isotopes—exotic and not—to an extent not possible otherwise. In short, there is little room for arguing with the following statement: *The processes by which the universe is lit—and which grant energy-giving light to life on Earth—are the same processes by which the chemical elements of planets and life were (and are) made.*

5.5 Is the Universe Tuned to Allow Life?

Our story of the origin and evolution of the cosmos would have the nice tidy ending just cited were it not for the following facts:

1. About 90 percent of the cosmos seems to be made of matter that does not emit, scatter, or absorb light.
2. The universe is very nearly flat, rather than curved to some significant extent.
3. Not only is the expansion of the universe not slowing under the mutual gravitational pull of all bodies in the cosmos, it is in fact accelerating at the largest scales.

Let us take each of these in turn, to assemble a picture of the universe that has troubling implications for just what piece of existence the universe represents.

In the mid-20th century, telescopic measurements of the rate at which galaxies spin about their centers suggested that there is a great deal more matter present than can be accounted for by the stars, gas, and dust that can be detected spectroscopically. This extra mass is not concentrated in the center—it seems to surround each galaxy like a halo and alters the rate of rotation of material, as a function of distance from the galactic center, from what a simple self-gravitating mass of the observed stars, gas, and dust would yield. It is as if we were to measure the periods of the orbits of the planets of our solar system

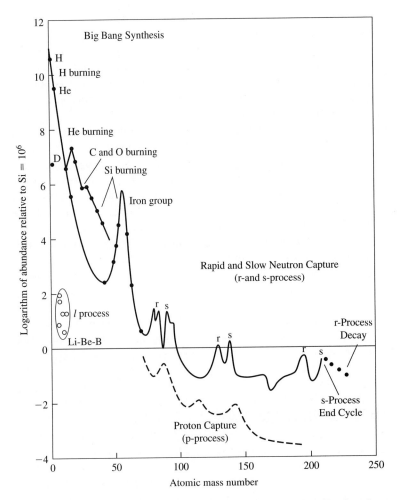

FIGURE 5.14 The alchemist's dream—a map of processes transforming elements. Specific processes, defined in the text, are indicated in the general atomic mass number range in which they predominate.

about the Sun to find that they did not fit the pattern consistent with the Sun being the dominant mass in the solar system. The nature of the "dark matter" remained the purview of speculation, but initially, "brown dwarfs," (objects too small to undergo stable fusion and hence to emit light) were proposed. Various observations of the population of stars, including clever ones that take advantage of the focusing effect of the gravitational field of foreground stars on background objects (Chapter 15), failed to find enough brown dwarfs to account for the galactic missing mass. Extensions of such observations to the gravitational bending of light by clusters of galaxies suggest that *all* such clusters have 5 to 10 times more mass than can be accounted for by stars, gas, and dust. Invoking failed stars or dead stars (white dwarfs, neutron stars, black holes) would require a rate of star formation in galaxies inconsistent with a variety of observations. Ascribing the missing matter to gas would yield a cosmic background radiation much less uniform than is observed.

What, then, is dark matter? No one knows, so exotic and unknown subatomic particles have been proposed, as well as alterations in the properties gleaned from particle accelerators of known subatomic particles. Whatever contributes the missing matter is not electrons, protons, or neutrons; it is a "ghostly" material that possesses mass but interacts with photons in a way vastly weaker than normal matter.

The inventory of normal and dark matter in the cosmos can be used to address another question: Is the universe open or closed? That is, is there enough matter to cause the expansion of the universe to slow, stop, then reverse, or will the expansion go on forever? Put succinctly, is the universe closed or open? This question, debated for decades, now appears to have an answer: The universe is neither—it is in between. There appears to be just enough matter to give the universe a "flat" geometry as described earlier. Such a universe ought to be very uniform in its properties, but detailed measurements of the cosmic background radiation reveal very small fluctuations that are hard to derive in a flat universe derived in a smooth expansion from the Big Bang.

To derive a flat universe with the right spatial fluctuations requires that the early expansion of the universe be very nonuniform. A sudden, rapid "inflation" would take a bumpy universe, initially curved in a closed or open fashion, and render it flat but retain enough nonuniformity to be consistent with what is seen in the cosmic background radiation. (Imagine a small, unpoppable balloon, suddenly inflated to the size of a blimp. To a small bug, the surface curvature would decrease dramatically.) This inflation had to occur before the decoupling of all but the gravitational force, that is, during the grand unified epoch of Table 5.2, and required the "scale factor" to increase by between 10^{40} and 10^{100} to explain the current flatness. This exponential inflation has no physical analog in the post-grand-unified epoch of our current universe, but it is possible in the almost inconceivably small cosmos existing 10^{-37} seconds after the Big Bang. That such an inflation must occur leads to other exotic ideas: that our four-dimensional space–time universe represents only a subset of the dimensions created in the Big Bang, those which successfully inflated; that particles are actually vibrational modes on a space–time topological construct that has 10 or more dimensions (superstring theory); and so on. But, more fundamentally, inflation suggests that the universe was born with a small repulsive tendency in its space–time dimensions, a repulsive tendency that is built into Einstein's general relativity as an adjustable parameter. Einstein himself, before the expansion of the cosmos was detected, set his "cosmological constant" to a nonzero value; he then retracted the concept of such a constant and called it his greatest blunder.

It was, apparently, not a blunder. Very recently, careful measurements of the rate of expansion of the cosmos indicate that on the largest scales the expansion is *accelerating*. This is not a contradiction of the flat universe; indeed, it merely requires a cosmological constant—the fudge factor in the geometry of Einstein's general relativity—that is nonzero and forces a repulsive component to gravity at the largest spatial scales. On the scale of planetary systems and galactic clusters, this repulsive term would be undetectable. Considered another way, the cosmos is suffused with energy—now called *dark energy*—that creates repulsion to space–time on cosmic scales. Because energy and matter can be equated through the formula $E = mc^2$, we can compare directly the amounts of dark energy and matter to obtain the coarse "composition" of the cosmos. To account for the rate of expansion of the cosmos, as well as the various indirect observations of dark matter, 35 percent of the cosmos must be made of dark matter, 3 to 7 percent must be normal ("hadronic") matter, and the remainder of the cosmos would be dark energy.

But what, then, is dark energy? One intriguing possibility is that the universe really is a subset of other dimensions that form a "multiverse." Black holes become connections in the fabric of space–time to the other dimensions, and the presence of other dimensions might be felt in the existence of the dark energy. Currently, the concept is so new that reconciliation of the magnitude of the dark energy required by the observations with that of a cosmological constant consistent with inflation has not been successful, though few still question the telescopic observations leading to the conclusion of an accelerating cosmos.

We seem to be embedded within a cosmology of an expanding universe that both inflated from a substrate containing a greater number of dimensions than four, and that contains unknown and unseen matter and forces vastly more important to cosmic structure than the "normal matter" comprising the 92 natural elements. It would seem the ultimate exercise in Copernicanism to displace life, not just from the center of the cosmos but from the center of the action as well in terms of the most common forms of matter and energy in the cosmos. Yet, we must ask the following question: If our universe is one of many simultaneous or possible outcomes of the original Big Bang (merely the subset of space–time dimensions that inflated to a scale accommodating particles), is the existence of life in our universe a lucky accident of that inflation?

That we are here to ask the question implies that this universe—whether singular or one of many possible sets of universes—evolved in such a way as to permit life to arise. But there are a number of troubling coincidences with respect to the way this cosmos works that have led some to postulate a certain "specialness" to a universe that can accommodate life—under the guise of a loosely defined so-called "anthropic principle." On the broadest scale, the number of dimensions that inflated to form this cosmos is such as to allow stably bound gravitational systems. Were four spatial dimensions to have inflated, the force of gravity (as a characteristic of the geometry of space–time) would increase as a higher power than r^{-2} as two bodies approach each other, rendering impossible the stabilization of planetary orbits. The same problem with the electromagnetic force in four dimensions would render unstable electronic orbits around four-dimensional atoms. A universe with two spatial dimensions, on the other hand, might not allow sufficient complexity in chemistry or structural topologies for something like life to occur.

Moving one step down from dimensionality, we come to the issue of inflation of the cosmos early in its history. In the absence of inflation, the scale of the cosmos would not have accommodated material particles at any time in its history. Some inflation scenarios allow universes to rapidly collapse on themselves, prior to the time when star formation and element production would have allowed life to occur. On the other hand, were inflation not brief, the expansion of the scale of the cosmos might have been such as to render the matter density too small for the formation of elements, even primordial hydrogen and helium. Or even a milder form of excessive inflation might render the cosmos too underdense for the formation of successive generations of stars in which the production of the biogenic elements could take place. It also seems somewhat serendipitous, given the possibilities, that multiple generations of stars have been able to form and die, producing the biogenic elements, at a time when the cosmos is still compact enough that the spatial scales between galaxies are not much larger than the galaxies themselves and hence the whole cosmos can, in essence, be observed (Figure 5.15).

Even in a cosmos "suitably" expanded for life, there is no guarantee that the relative strengths of the different forces will allow the kinds of bound atomic systems that characterize normal matter. This issue is encapsulated in the so-called fine structure constant, which characterizes the inherent strength of the electromagnetic force relative to the fundamental quantum "coarseness" of the cosmos. It is the square of the electronic charge divided by Planck's constant and by the speed of light. The value of this parameter determines how well bound atoms are. A variation in the constant of a factor of 10, for example, would lead to a cosmos in which carbon (as a bound system of electrons, protons, and neutrons) simply could not exist. Other subtler changes in the parameter would alter the rate of fusion reactions and lead to a stellar and element-production evolution very different from what transpired, perhaps in a way militating against the origin of planets and life. Why the parameter possesses its measured value is unclear; specific models of the origin and inflation of the cosmos do not specify it. On the other hand, universes with very different fine-structure constants might not be out of the running for material systems akin to the biogenic elements. It has been pointed out that for values of the fine-structure constant outside of the range permitting carbon, bound systems of boron and hydrogen could exist that would mimic some of the properties of carbon and lead to molecules that—while not precisely like organic molecules—are sufficiently stable that something like life could exist.

Finally, even given the outcome of the creation of this universe with its particular dimensionality and physical forces, the formation of life may be confined to particular times and places within galaxies themselves. In some of the most metal-poor stars in the galaxy,

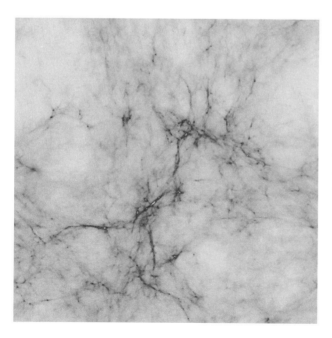

FIGURE 5.15 Computer simulation of the distribution of matter in the early universe. This image shows hydrogen and helium in a volume of space about 100,000 parsecs across at about 200 million years after the Big Bang. The first generation of stars are thought to form within the densest (dark) filaments of gas seen in the simulation.

only 1 of every 100,000 atoms is not hydrogen or helium; in the Sun, it is 1 out of a 1,000. "Metal-poor" (in the astronomical sense of elements heavier than hydrogen and helium) regions of the galaxy—the halo and even the outer portions of the disk—have a mix of elements that may limit the abundance or size of rocky planets. Looking at an even more detailed scale, the nature of our own Earth, with its moving crust that recycles volatiles such as carbon dioxide and water, is partly enabled by a particular mix of heat-producing radiogenic elements (Chapter 11). Some have argued for a "galactic habitability zone," beyond which habitable planets cannot form, and which moves with time over the long-term chemical evolution of the galaxy. The extent to which lesser amounts of specific elements would alter the geologic history (and, hence, habitability) of a planet is not sufficiently well known to define such a zone with confidence. Depending on how strict we believe these requirements to be, life might be possible throughout the disks of spiral galaxies or confined to narrow zones that move outward with time within these disks.

Discussions of the anthropic principle, whether on the galactic or multiverse scale, often spiral into teleological musings about the specialness or purpose of life. Such discussions are fascinating but are not informed by observations of the natural world. That we might well live in a universe that has particular properties salubrious for the genesis of matter and the formation of life does not exclude multiple other possibilities, occurring simultaneously or in other episodes of universe formation, in which something with the complexity of life arises. There is no observational constraint yet on whether our universe is one of endless possibilities, played out on a stage beyond time itself, or whether it is a singular occurrence. Should future observations support the notion that our cosmos is one of a multiverse of evolving dimensions, then even the seemingly infinite sky into which we gaze at night is but one corner of a much larger and varied reality.

QUESTIONS AND PROBLEMS

1. Does the expansion of the cosmos play the principal role in accounting for the darkness of the sky at night? Why or why not? What are the other possible contributing factors, and how important are each?

2. The critical density for closing the universe—that is, for having enough matter to eventually stop the expansion of the cosmos—is of order 10^{-26} kg/m^3. How many hydrogen atoms is this per cubic meter? This seems like a very low number. But suppose the universe contained a galaxy like the Milky Way in every cubic megaparsec of space, on average, and the typical mass of a star is $\frac{1}{10}$ that of our Sun. What would the equivalent average mass density of the cosmos be in units of kilograms per cubic meter?

3. The Big Bang model of the cosmos seemingly creates everything out of nothing in a single moment, while the now-discredited steady-state model gradually produces matter out of nothing, steadily, over time. Is there a scientific or philosophical rationale for preferring one over the other? (Hint: Is it necessary to assign the moment of the Big Bang to a "special point" in time?)

4. Does the apparent preponderance of dark energy *and* dark matter on cosmic scales necessarily militate against conservation of "normal" matter and energy at local

scales? Why or why not? (Hint: Consider the fact that the three fundamental forces other than gravity all conserve energy and are symmetric under a reversal in the direction of time; the same is true for Newtonian gravity but *not* for gravity as described by general relativity).

5. Calculate the parallax angle as seen from Earth for stars that are at the following distances:

   a. 10 light-years

   b. 100 light-years

   c. 50 parsecs

6. Using a general book or encyclopedia of astronomy, make a list of 15 named objects, including planets, stars, galaxies, and galaxy clusters (e.g., Venus, alpha centauri, Andromeda, Virgo cluster). Next to each of these write down the known distances to each using the *same* physical units (parsecs or light-years; the planetary distance will have a very small magnitude in these units). Then, next to these distances, list the technique or techniques used to determine these distances, based on the discussion in the text or other sources.

7. Hubble Space Telescope (HST) can see in optical and near-infrared wavelengths and has a mirror aperture of 2.5 meters. James Webb Space Telescope (JWST), under development, has a mirror diameter of 6.5 meters and can see objects into the mid-infrared. Why are both of these characteristics important for seeing further back into the early history of the universe with JWST versus HST?

8. While stars more massive than the Sun burn hydrogen on the main sequence for a shorter time, those less massive than the Sun are on the main sequence longer. However, the less massive stars (K and M dwarfs) tend to be less stable, producing large flares to a greater extent than the Sun. Speculate on how this trade-off might affect the suitability of planets around K and M dwarfs for long-term evolution of life. (Hint: Think also about the lower luminosity of the less massive stars and the resulting distances at which planets around such stars would be habitable.)

9. How might the presence of intergalactic dust as an agent for reddening and dimming the light from ever-more distant galaxies affect the determination of galactic distances by (a) the assumption that galaxies all have the same brightness and (b) red-shift associated with expansion of the cosmos?

10. Suppose ancient civilizations in distant galaxies sent immensely powerful radio signals out into the cosmos, using well-known frequencies (e.g., that associated with microwave lines of water or hydrogen). How would you use the red-shift associated with cosmic expansion to determine the time since the original signal was sent out by such civilizations?

SUGGESTED READINGS

A fine introduction to modern cosmology, tracing historical roots and clearly explaining the current physics, is that of J. F. Hawley and K. A. Holcomb, *Foundations of Modern Cosmology* (1998, Oxford University Press, New York). This book, though a textbook, is far more readable than the "popular" books written by leading modern cosmologists, which are, unfortunately, sometimes confusing and idiosyncratic. Breaking this mold, because of its spectacular illustrations, is S. Hawking, *The Universe in a Nutshell* (2001, Bantam Books, New York). A more elementary but clearly written introduction to cosmology is provided in the general astronomy textbook, T. P. Snow, *Exporing the Dynamic Universe*, (1999, West Publishing, New York).

REFERENCES

Abel, T., Bryan, G. L., and Norman, M. L. (2002). "The Formation of the First Star in the Universe," *Science*, **295,** 93–98.

Albert, D. Z. (2001). *Time and Chance*, Harvard University Press, Cambridge, MA.

Brumfiel, G. (2003). "Cosmology: Welcome to the Real World," *Science*, **426,** 751.

Cameron, A. G. W. (2001). "Some Properties of r-Process Accretion Disks and Jets," *Astrophys. J.,* **562,** 456–469.

Chiappini, C. (2001). "The Formation and Evolution of the Milky Way," *Am. Sci.* 89(6), 506–515.

Circovic, M. and Bostrom, N. "Cosmological Constant and the Final Anthropic Hypothesis," *Astrophs. Space Sci.*, **274,** 675–687.

Cox, A. N. (2000). *Allen's Astrophysical Quantities*, American Institute of Physics Press, New York.

Cox, P. et al. (eds.). (2000). *The Universe as Seen by ISO*, European Space Agency ESA, Paris.

Davies, P. (1999). *The Fifth Miracle: The Search for the Origin and Meaning of Life*, Simon Schuster, New York.

Geiss, J. (1993). "Primordial Abundance of Hydrogen and Helium Isotopes." In *Origin and Evolution of the Elements*, Prantzos N., Vangione-Flam L., and Casse M., eds., p. 89–106. Cambridge University Press, New York.

Gonzalez, G., Brownlee, D., and Ward, P. (2001). "The Galactic Habitable Zone I. Galactic Chemical Evolution," *Icarus*, **152,** 185–200.

Guthkie, K. S. (1990). *The Last Frontier: Imagining Other Worlds from the Copernican Revolution to Modern Science Fiction*, Cornell University Press, Ithaca.

Hawley, J. F. and Holcomb, K. A. (1998). *Foundations of Modern Cosmology*, Oxford University Press, New York.

Hetherington, N. S., ed. (1993). *Cosmology: Historical, Literary, Philosophical, Religious, and Scientific Perspectives*, Garland Publishing, New York.

Horneck, G. and Baumstark-Khan, C. (2001). *Astrobiology: The Quest for the Conditions of Life*, Springer, Berlin.

Kauffman, S. A. (1995). *At Home in the Universe: The Search for Laws of Self-Organization and Complexity*, Oxford University Press, New York.

Kuhn, T. S. (1957). *The Copernican Revolution: Planetary Astronomy in the Development of Western Thought*, Harvard University Press, Cambridge, MA.

Lineweaver, C. H., Fenner, Y., and Gibson, B.K. (2004). "The Galactic Habitable Zone and the Age Distribution of Complex Life in the Milky Way," *Science,* **303,** 59–62.

Lunine, J. I. (1999). *Earth: Evolution of a Habitable World*, Cambridge University Press, Cambridge, UK.

Mason, S. F. (1991). *Chemical Evolution: Origin of the Elements, Molecules and Living Systems*, Clarendon Press, Oxford.

National Reseach Council. (2002). "Connecting Quarks with the Cosmos: Eleven Science Questions for the New Century," National Academy Press, Washington, DC.

Oberhummer, H., Csótó, A., and Schlattl, H. (2000). "Stellar Production Rates of Carbon and Its Abundance in the Universe," *Science*, **289,** 88–90.

Sackman, I.-J., Boothroyd, A. I., and Kraemer, K. E. (1993). "Our Sun. III. Present and Future," *Astrophys. J.*, **418,** 457–468.

Tegmark, M. et al. (2003). "The Three-Dimensional Distribution of Galaxies from the Sloan Digital Sky Survey," *Astrophys. J.*, in press.

Trimble, V. (1995). "Galactic Chemical Evolution: Implications for the Existence of Habitable Planets." In *Extraterrestrials: Where Are They?*, Zuckerman B. and Hart, M. H., eds., p. 184–189, Cambridge University Press, Cambridge.

Van Dishoeck, E. F. and Blake, G. A. (1998). "Chemical Evolution of Star-Forming Regions," *Ann. Rev. Astron. Astrophys.*, **36,** 317–368.

Ward, P. and Brownlee, D. (2000). *Rare Earth*, Copernicus Books, New York.

Zeilik, M. and Gregory, S. A. (1998). *Introductory Astronomy and Astrophysics*, Saunders College Publishing, Fort Worth.

CHAPTER 6

Planetological Foundations for the Origin of Life

6.1 Introduction

Chapter 5 set the cosmological stage for the formation of planetary systems, with the birth and death of succeeding generations of stars, the production of elements during and at the end of the lives of stars, and the dissemination of that material into interstellar space. This chapter is devoted to an exploration of the mechanisms of star and planet formation and the technique of isotopic dating that puts these events—for our own solar system—in an absolute chronological sequence. The chapter then shifts to the specifics of our own solar system—the interiors and atmospheres of the planets, delivery of water and organics to the Earth, and the earliest environments on the Earth that may have played host to life. The chapter closes with a description of the roles impacts may have played once life began—destroyers of life and transporters of life from one world to another.

6.2 Formation of Planets as a Natural Consequence of Star Formation

Molecular Clouds in the Galaxy

The galaxy is host to so-called molecular clouds, regions of elevated densities (amounts of material per unit volume) of hydrogen, helium, and heavier elements that orbit about the galactic center along with the spiral-armed disk of Population I stars (Chapter 5). The clouds are called "molecular" because hydrogen, which in the background of interstellar space is in the form of atoms, exists as molecules of H_2 in molecular clouds. Some molecular clouds extend for hundreds of light-years and contain enough mass to allow the formation of hundreds of thousands of stars—such are called "giant" molecular clouds. The closest typical molecular clouds, those in the constellation Taurus, are some 400 to 500

6.2 Formation of Planets as a Natural Consequence of Star Formation 175

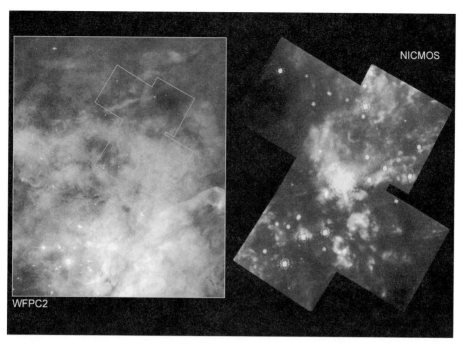

FIGURE 6.1 A portion of the Orion molecular cloud complex, which in total spans 40 to 60 parsecs (hundreds of light-years), and has a mass 200,000 times that of the Sun. The image is shown in visible light (left) and infrared light (right); the latter sees right through the obscuring dust to the newly forming stars embedded within.

light-years distant, and contain thousands of solar masses of gas and dust. The Orion giant molecular cloud complex is about 1,600 light-years distant and at least 10 times more massive (Figure 6.1). The relationship between the smaller molecular clouds and the giant molecular clouds is far from completely understood, but the giant complexes seem to be associated with the spiral arms of the galaxy. The spiral arms themselves are manifestations of a wave pattern set up by the rotation of the galaxy, and the concentration of material in the arms may lead to the convergence of a number of molecular clouds to form a giant complex. Once the spiral arm pattern passes by the giant molecular cloud, the latter may break up again into smaller units.

There is ample evidence that stars form in these mixtures of primordial hydrogen and helium and the regurgitated products of dying stars. The properties of molecular clouds can be sampled through observation in the radio part of the spectrum, where rotational transitions of molecules are present. The spectral lines of these species trace out the density of gas in different parts of the molecular clouds, and the resulting observations suggest very large variations. Some very dense, fairly warm, and compact places called "hot cores" have at their centers pointlike sources that glow strongly in the infrared, suggesting that something self-luminous is heating the dense dust around it, which then is reradiating in the infrared. In other regions, flattish disks of dust and gas surround stars that are clearly detectable in the optical. These "T-Tauri" stars, named after the first such object to be characterized, are otherwise normal stars whose characteristics suggest that they are about to enter the main sequence of stable hydrogen burning. Because the hot cores and T-Tauri stars are associated with molecular clouds, they are among the best indicators that the formation of stars is ongoing there.

Formation of Stars

How does a star form? Simply, it forms by collapse of interstellar gas. But, why does collapse occur? Recall that all matter has an intrinsic gravitational attraction. Dispersed interstellar gas is no exception, but under normal interstellar conditions, the attraction is not sufficient to cause the material to move together.

However, in molecular clouds much more matter is present per cubic centimeter (>100 hydrogen molecules/cm^3) than the average for interstellar space (0.1 hydrogen atoms/cm^3 in the region near the Sun). Because the attractive force increases inversely as the square of the distance between the attracting bits of matter (Chapter 2), higher density means a greater attractive potential. Compression of the gas—perhaps as molecular clouds enter spiral arms and are squeezed by the external interstellar medium or by collisions with each other—further increases the gas density. Finally, molecular clouds contain substantial amounts of dust, which helps shield regions against strong heating from ultraviolet light, and molecules, which emit photons and hence further cool the clouds. By reducing the temperatures of the clumps of gas, collapse becomes possible. (Indeed, in the very first generation of star formation, dust and molecules other than hydrogen would have been absent, and so the process by which stars formed in the early universe may have been quite different.)

To understand how the collapse of gas leads to the formation of a star, it is necessary to focus on just what happens in a collapse and, in particular, on the transfers of energy that occur. A star undergoing hydrogen fusion is a self-gravitating sphere that represents a balance between the inward force of gravity associated with the matter in the star and an outward force associated with the gas pressure. In normal stars, the gas pressure is caused by the repulsive (electrostatic) force associated with the high-speed collisions among atoms or their electronic and nuclear fragments. Remove that pressure and the star would collapse again. Or dump more mass on the star and further compression occurs, which leads to heating and increased outward pressure, establishing a new balance.

And so, compression of the material by the force of gravity causes heating and an outward pressure. Anyone who has operated a manual bicycle pump knows this to be so; in that analogy, though, a muscular force pushing against the pump piston replaces the force of gravity. We can feel the heating of the air trapped inside the pump after exercising the piston repeatedly. In the astrophysical case, the potential energy of the gaseous material in more extended, diffuse form, is being transformed into thermal energy—random kinetic energy of motion—during the collapse of gas to form a star. As material distributed across a region of the molecular cloud (typically, light-years) is drawn inward, moving closer and closer to a common center of condensation, the potential energy decreases and the kinetic energy increases.

The maximum kinetic energy a body may gain by this collapse process is related to the potential energy of the *final*, compressed body by the virial theorem; the kinetic energy is half the potential energy (with a switch in sign because the potential energy of a self-gravitating body is less than zero). The factor of a half arises as follows: The object is assumed to start dispersed over a (formally) infinite space, which would render its potential energy zero; as the object contracts, the potential energy becomes progressively more negative and the maximum energy extractable is, by integration, half the (negative of the) final potential energy. The remaining energy is removed in the form of photons, that is, radiated away. Because most of the kinetic energy ends up in the form of random collisions among molecules, it is expressed as a thermal energy within the star.

We can calculate from the virial theorem an average temperature in the deep interior of the Sun. The mass M of the Sun, 2×10^{30} kg, and its radius R, 7×10^8 m, leads to a potential energy (assuming a uniform sphere) of $3GM^2/5R = 2 \times 10^{41}$ joules. Recall (Chapter 2) that G is the universal gravitational constant that sets the strength of the gravitational force and is 6.67×10^{-11} m^3/kg·s. Then, from the virial theorem the kinetic energy is about 10^{41} joules. The Sun is about 85 percent hydrogen, most in ionized form in the deep interior, so the electrons and protons are separate. Because electrons are so much less massive than protons, to a good approximation the average mass of a particle is a bit more than half that of the proton, or roughly 10^{-27} kg. Then the mass of the Sun divided by this average mass per particle is the number of particles, or roughly 2×10^{57} particles. The kinetic energy per particle then is 5×10^{-17} joules. Now we can find the temperature from the fact that the energy per particle is also $\frac{3}{2}kT$, where k is the Boltzmann constant, 1.38×10^{-23} J/K: $\frac{2}{3} \times 5 \times 10^{-17}$ joule/1.6×10^{-23} J/K, or 2×10^6 K. As it turns out, 2 million Kelvin is the temperature about 70 percent out from the center of the Sun. Models that are more detailed yield a central temperature of 15 million Kelvin. Such temperatures are sufficient to initiate fusion in a clump of gas that is collapsing to form a star, exchanging potential energy for kinetic energy.

Large-scale compression of the molecular clouds leads to "stellar nurseries" within which very massive stars, those of spectral types O and B, are generated. Because these stars rapidly fuse hydrogen into helium, they do not last very long (i.e., millions of years), and hence are almost never seen very far from the molecular cloud regions in which they are born. That they are seen at all is a testament to the powerful heating and ionizing effect the ultraviolet radiation from these stars has on the surrounding gas and dust, effectively evaporating away obscuring material (Figure 6.2). This ablative and evaporative effect probably inhibits further star formation in the surrounding regions over longer timescales, which otherwise might proceed at a pace far larger than is observed as the O and B "association" members become supernovae and send compressive shocks through the clouds.

Stars like the Sun may form in small regions within molecular clouds, light-years across, which are quite dark and dense—enough that they ought to be collapsing to form stars. In some of the densest clumps, hot cores with a pinpoint infrared signature at the center are seen, suggesting that collapse is occurring. The dynamics of the collapse of dense clumps within the cloud are controlled not only by gravity but also by the magnetic component of the electromagnetic force. Our Milky Way galaxy generates, in a poorly understood process, a large-scale magnetic field that threads the molecular clouds. Charged particles are constrained to certain kinds of motion associated with the magnetic lines of force, and, in most of the regions of the molecular cloud, enough charged particles exist that their forced motion constrains the motions of neutral particles through collisions

FIGURE 6.2 The cone nebula, a portion of a molecular cloud in which ultraviolet radiation from nearby new stars is evaporating and blowing away the dust and gas of the cloud.

between the two. In this way, there is a "magnetic pressure" exerted on the molecular cloud material that counteracts collapse. But as clumps become denser, the charged particles—electrons, protons, and other ions—tend to recombine more quickly and become neutral, and the shielding effect of the clump against ultraviolet and particle radiation reduces the efficiency of formation of new ions. Thus, the denser the clump, the fewer the ions and less magnetic pressure. Clearly this is an unstable situation, since the less magnetic pressure, the more the clump will tend to contract and densify, reducing the ion content further. This "slippage" against the outward pressure of the magnetic field is a bit like a person trying to hold onto a steep cliff with his fingernails—as he begins to slip, the static friction of fingernails against the rock declines, and he begins to fall. The result, in the interstellar context, is that sufficiently dense clumps will collapse and not stop until the central portion of the clump becomes hot enough to create a thermal force that balances the inward gravitational force of the collapsing material. Any such protostar more massive than 15 times the mass of Jupiter will initiate deuterium fusion, and those more massive than 75 times the mass of Jupiter will fuse hydrogen and become stable stars.

Planet Formation

Were this the full physical picture, planets would never form—the spherical collapse of a clump to make a star carries with it no additional process that tends to make planets as well. The structure of our own solar system provides important clues to the processes that do make planets. The planetary portion of our system, with the exception of Pluto, lies in a plane with orbits inclined by 7 degrees or less relative to each other. Most of the angular momentum (Chapter 2) of the solar system, namely, the momentum associated with rotational motion of bodies or their orbits about the Sun, is contained in the planets; the Sun possesses only 0.5 percent of the solar system's angular momentum in its spin. Yet, almost 99.9 percent of the *mass* of the solar system is contained in the Sun. There is a chemical gradation as well, with the inner planets being largely rocky and the outer planets a composite of gas giants and their numerous natural satellites, lunar- and smaller-sized bodies made of water ice and rock. Beyond the planets themselves, beginning just beyond 30 AU, lies a belt of rocky and icy debris. Pluto, once considered the ninth planet, really fits in better as the largest known member of this *Kuiper Belt*. Many comets with orbital periods of a few decades or less ("short period" comets) appear to come from this belt. Finally, at vast distances up to hundreds of thousands of astronomical units from the Sun is a loose cloud of icy material, the Oort cloud, from which comets with long orbital periods are derived.

These properties suggest that whatever process formed our planetary system generated a geometry of raw material in the form of a flattened disk, that there was an efficient process of removal of angular momentum from the central part of the disk to the outlying regions, and that the disk was hot close to the center and much cooler further out. The existence of debris belts and clouds suggests that planet formation was truncated after some time, and there was a large amount of material remaining that was subsequently scattered by the gravitational fields of the planets themselves.

The speculation that planets form in disks of material has been made for hundreds of years—since the 18th century French scientist Pierre Laplace first suggested this idea. But in the past decade, first with radio telescopes and then with Hubble Space Telescope and ground-based adaptive optics telescopes, it has been possible to directly image disks—some of which are much larger than the extent of our planetary system, while others have the appropriate 100 AU scale. It is possible, using spectroscopy most recently from Spitzer Space Telescope, to infer the temperature drop with distance on the surface of and within such disks (particularly in the infrared and radio portions of the spectrum), and indeed the temperature typically falls rapidly with increasing distance from the center.

Could such disks be a part of the collapse process that forms stars? The answer, from simple physical reasoning backed by more detailed computer models, seems to be yes. The missing ingredient from the picture given here is rotation of the collapsing clump. Because the Milky Way Galaxy is rotating about its own center, there is an overall angular momentum associated with this system of 100 billion stars. Individual molecular clouds have a sufficiently great extent that some of this rotational motion creates shearing, or relative movement, within the cloud itself. Therefore, clumps possess angular momentum, and they are spinning—but much too slowly to spin out a disk given their extended (light-years across) nascent size. It is when collapse occurs that the rotation increases because of a principle at the core of Newtonian physics (Chapter 2): the conservation of angular momentum. Unless acted upon by an appropriately directed force, the angular momentum of a system is conserved, it can be transferred internally from one portion to another but cannot be lost or added to from outside.

Angular momentum conservation is dramatically illustrated by a spinning ice skater. Because of the nearly frictionless nature of the ice (until the skater digs her blades into the surface), a spin can be sustained for some time without application of additional force by the skater. Often we see a skater with arms outstretched going into a slow spin, the rate of which then increases as the skater pulls in her arms. What has changed in the skater is her *moment of inertia*—essentially, the distribution of her mass. Because the angular momentum depends on the product of the moment of inertia and the angular speed (just the number of revolutions per second), a decrease in one must mean an increase in the other. Thus, returning to the astrophysical situation, as the clump contracts from an extent of light-years to one of hundreds of astronomical units—shrinking in scale by several orders of magnitude—the spin rate increases enormously. There is little to prevent it from doing so, since the rapid densification makes the gas nearly neutral and renders braking by the magnetic field ineffective. The dramatic increase in angular velocity forces some of the material to be flung outward to form a disk.

What happens next to the disk depends on its mass and angular momentum, and it must be modeled through detailed computer simulations because the answer is difficult to divine from direct physical reasoning. If the collapsing clump has a large angular momentum, it breaks directly (or, in some cases, through the brief mediation of a disk) into two clumps, which form a binary star system. If the disk is very massive, portions can themselves become gravitationally unstable and collapse directly into one or several clumps that form companion stars or even giant planets. If both the disk mass and angular momentum are modest, the disk will stabilize and "feed" material with moderate angular momentum to the central star. Such disks may eventually form planets (Figure 6.3). Not all stars possess disks during their formative phases, the remainder being single or binary (sometimes multiple) stars, but astronomical observations suggest that half of the stars from A-type down to M-type do form disks as part of the clump collapse. With the launch of the Spitzer Space Telescope in 2003, a telescope that observes in the middle infrared part of the spectrum, scientists are better able to determine how many stars possess disks capable of forming planets.

Three "torques," all of which serve to alter the distribution of mass and angular momentum, act upon the disk itself. Early in the history of the disk, when it is at its most massive, both gravity and magnetism act on it. Models and observations suggest that the central portion of the collapsing clump, as it heats up, traps a magnetic field from the surrounding molecular cloud that intensifies as it shrinks in size. This, now local, magnetic field can exert torques on the disk if the latter becomes hot as a result of rapid addition of material from the surrounding cloud, since it then ionizes again to a significant extent. Gravity acts because, in its early phases, the disk is relatively massive compared with the central star. Random variations in the mass distribution in the disk, which is rotating about its center, become amplified and can lead to the formation of "spiral density waves," which resemble in miniature galactic spiral arms. These regions of density enhancement exert gravitational forces on other parts of the disk, tending to make the faster inner regions slow down and the outer regions speed up. The net effect of such "torques" is to cause mass to fall toward the center of the disk—where the nascent star is located—but shift angular momentum *outward*.

As the disk ages with time and addition or *accretion* (the term used by astronomers and planetary scientists) of material slows, the disk becomes less massive and the torques from magnetic fields and gravity weaken. Disk evolution does not cease, however, because as the disk tries to shed thermal energy into the colder surroundings, bulk "convective" motions occur: Warm gas rises, radiates photons at the disk surface, then sinks as cold gas.

6.2 Formation of Planets as a Natural Consequence of Star Formation

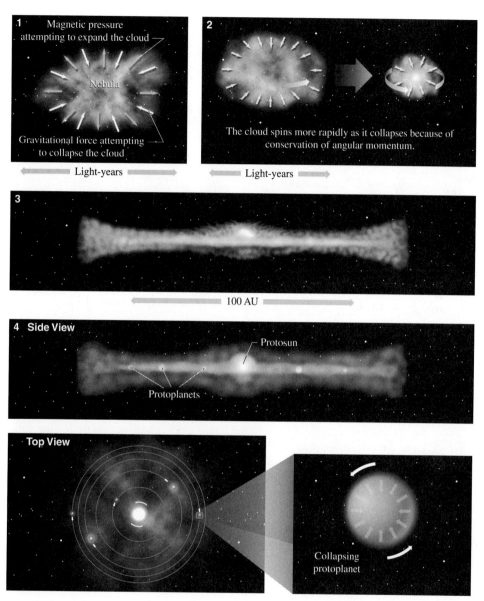

FIGURE 6.3 Stages of contraction of a clump of molecular cloud to form a disk and then planets. The scale changes from one panel to another.

Because the disk is in so-called Keplerian motion, with each element orbiting at a speed appropriate to its distance from the nascent ("proto") star, the rising and falling motions create a frictional drag on the disk. This "viscous" torque does, with less intensity, what the gravitational and magnetic torques did earlier in the history of the disk—it moves matter inward and angular momentum outward. Because the rotation speeds are higher closer

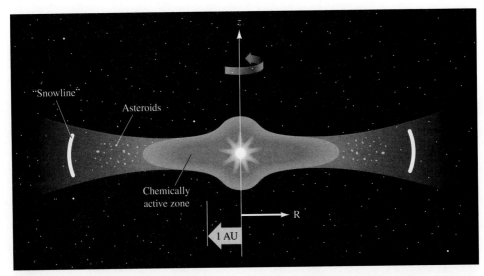

FIGURE 6.4 Schematic edge-on view of the protoplanetary disk that formed our own solar system, at a stage where water ice becomes stable just inside of the present-day orbit of Jupiter.

to the protostar, the viscous motions also tend to heat up the inner part of the disk much more than the outer. Hence, the temperature in the middle—the "midplane"—of the disk declines steeply with radial distance from the protostar.

During this epoch, different molecules in the gas phase begin to condense in the disk at temperatures in accordance with their relative volatilities and abundances. Volatility refers to the tendency of a material to be in the liquid or gas phase as opposed to the solid phase (Chapter 7); abundance refers to the number of atoms or molecules of that substance relative to the primary gas of the disk, hydrogen. "Silicates," for example, the magnesium silicate Mg_2SiO_4, are much less volatile than ices such as water (H_2O). There is also a progression of volatility within each class of compound. Methane (CH_4) is far more volatile than water; Mg_2SiO_4 is more volatile than corundum (Al_2O_3). Thus, where these materials condense out in the disk of gas depends on the local temperature; rock will predominate at distance inward of where water ice predominates, in good correspondence with the composition of solid planetary bodies in our own solar system. But the situation is actually more complicated because material in solid form from the surrounding molecular cloud—micron-sized or larger grains of silicates containing and encased by ices—falls into the accreting disk and may only partially sublimate into the gas phase. Hence, rocky compounds throughout the disk may be an amalgam of molecular cloud silicates and of material condensed out of the disk as the latter cooled; likewise is true for the ices in the outer part of the disk (Figure 6.4).

The Case of Our Solar System

Let us focus now on the case of our own solar system, for which detailed evidence exists regarding the formation of planets. The solid grains formed within, or that had fallen into, the disk have a tendency to stick together because residual charges on their surfaces lead to net attraction. Initially the grains are so small that they are trapped in the gas and are carried with it through its various orbital and turbulent motions. As the grains grow, the gas can no longer carry them, and they "decouple" from it, assuming their own orbits about the proto-Sun. But the grains feel a "wind" associated with their motion relative to the gas (the gas, because it exerts a pressure, actually orbits the proto-Sun at a speed slightly different from the grains). The magnitude of the wind depends on the size of the grains, so the grains themselves tend to move relative to each other in the disk. This enhances the rate at which particles collide with and stick to each other. Solid bodies grow by this accretional process to sizes of tens to hundreds of kilometers, at which point they begin to perturb each other by virtue of their gravitational fields.

The ultimate result of solid body accretion is the growth of planets. In the outer part (>5 AU) of the solar system, where abundant solids were available because of the stability of the ices, lunar- to Earth-sized bodies grew during the time that the gaseous disk (which is also referred to as the "solar nebula" in some of the planetological literature) was still present. Astronomical observations of disks around other stars suggest that disks begin to dissipate on timescales of millions of years and are gone by 10 million years (the timescales are established by comparing the properties of the central star in each disk to that in computer models of stellar evolution). The initial cause of the dissipation is cessation of accretion, but as the star itself ignites hydrogen-fusion, it generates an outflow—a stellar wind—that speeds up the loss of the disk. Computer simulations suggest that solid bodies between lunar- and Earth-mass attract large quantities of gas around them, and this further enhances the capture of solids and of gas. Eventually, giant planets—Jupiter, Saturn, Uranus, and Neptune—are formed, with the last two perhaps being the "latecomers" whose growth was truncated by the dissipation of the gaseous disk. While giant planets might also form by direct collapse of gas in the disk, as noted earlier, there are two arguments in favor of a process that initially involves accretion of solid material. First, the giant planets in our system have an abundance of elements heavier than hydrogen and helium that is much larger than in the Sun. Second, the giant planets have multiple moons, which in the case of Jupiter and Saturn probably required addition of large amounts of rock and ice during formation. (Note that two of the larger moons will be featured as possible abodes of life or of advanced prebiotic chemistry in Chapter 13.)

Within the inner solar system, giant planets either did not form or, through gravitational tugs of war between the planets and the disk, evolved inward until they were consumed by the proto-Sun. As we shall discuss in Chapter 15, many other stars have giant planets, some in very close orbits to the central star. These might have formed in place, by direct collapse, but more likely they migrated inward from orbits that were more distant.

As the giant planets neared their current size, the powerful gravitational influences they exerted on smaller material reshuffled the architecture of the solar system. Leftover small bodies (referred to as "planetesimals" in the planetological literature) were scattered out of the giant planet region into more distant orbits. Some entered the inner solar system, but most were ejected into highly elongated orbits that reached thousands of astronomical units beyond the planetary realm. These bodies, whose orbits were slowly perturbed by the gravitational forces of the surrounding molecular cloud and then later by other nearby clouds and stars as the Sun moved away from its nascent birthplace, today form the Oort cloud. Bodies from this region are occasionally perturbed today by passing stars or molecular

FIGURE 6.5 Comet Hyakutake, a planetesimal left over from giant planet formation.

clouds and redirected into the inner solar system, manifesting themselves as long-period comets such as Hyakutake, Hale-Bopp, and Halley (Figure 6.5). Leftovers just beyond the orbit of Neptune in the Kuiper Belt remained in stable orbits for a while, some growing to significant size (Pluto) and one captured by Neptune (Triton). Gradually, the perturbations of Neptune have cleaned out the Kuiper Belt leaving behind relatively few planetesimals on stable orbits.

In the inner solar system, planetesimals felt the influence of the gravitational fields of the growing giant planets, through increases in the eccentricities and inclinations of their orbits relative to the plane of the disk. This influence became especially important as the gaseous disk was swept away, because there was no longer a damping effect associated with the gas. Collisions between planetesimals led to the rapid growth of lunar- to Mars-sized bodies in the region stretching from the present-day asteroid belt to the realm of the inner planets (Mercury, Venus, Earth, and Mars). Those bodies closest to Jupiter—in the region of the present-day asteroid belt—had especially large orbit eccentricities, and repeatedly intersected the orbits of the terrestrial planets. In this way, Venus and Earth were not built exclusively from the gentle addition of dust and pebble-sized material but through the collision of lunar- to Mars-sized objects—so-called giant impacts. One of these may have formed the Moon, and others may have supplied water to make the oceans of Earth, as discussed in Section 6.5. The clearing out of the inner solar system could have taken upward of 100 million years, by which time there were four terrestrial planets in circular orbits, one with a large natural satellite, and a depleted region with only remnant rocky debris, which we call the asteroid belt (Figure 6.6).

The careful reader may wonder whether the process of heating by conversion of potential to kinetic energy—crucial to bringing the interiors of young stars up to the point of hydrogen fusion—operated on the planets during their accretion. Indeed, it must have, be-

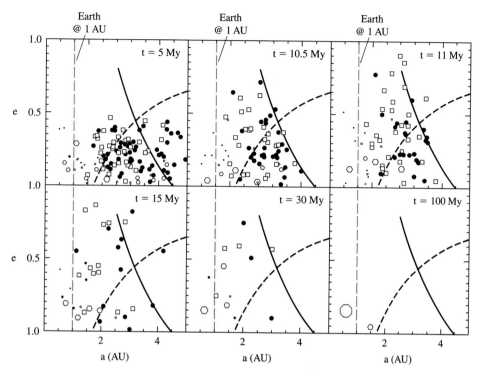

FIGURE 6.6 Simulation of the assembly of the terrestrial planets from lunar- to Mars-sized fragments. Each panel shows the distribution in orbital semimajor axis and eccentricity (Chapter 1), and the series of panels are a progression in time (labeled in millions of years). Planetesimals grow by collision, and their increasing mass is schematically shown by the size of the circles. The dotted line illustrates the realm (to the right of the line) where Jupiter's gravity is so large that rapid ejection of planetesimals occurs. To simplify the calculation, Jupiter's influence is "inserted" into the picture beginning in the second panel. The location of water-bearing planetesimals is shown as being at 2.5 AU and beyond. At the end of this simulation two terrestrial planets are formed.

cause the conversion of a distributed area of dust into a self-gravitating solid planet (and likewise for the disk gas evolving into a giant planet) must obey the virial theorem. For the giant planets, the "virialization" of the potential energy of the disk gas was not enough to generate deuterium fusion. The amount of thermal energy contained within these bodies is substantial, however, and it is possible to measure today, in the thermal infrared, emission of energy from these planets that is over and above that which can be accounted for by absorbed sunlight. In the case of the terrestrial planets, the computation of the heating due to accretion is complicated because it depends on the size of the impactors and the depth to which the energy is deposited. The virial theorem still applies, but much of the energy of accretion may have been lost immediately by radiation of photons from the surface or ejection of hot debris back into space. Most calculations suggest, however, that the Earth became sufficiently hot to have had a fully molten outer layer (essentially, a worldwide sea of lava or magma). Some have suggested that Earth—for a short time—had an atmosphere composed of silicate vapor. The implications of this hot start include a delay of perhaps 100 million years or so between completion of the Earth's accretion and the stabilization of liquid water at the surface (Chapter 16).

Summary of the Planet Formation Story

The story presented here is based on the amalgamation of many different kinds of modeling and observational efforts. Three points stand out. First, the process of spinning out planet-forming disks appears to be a relatively common one among stars similar to the Sun. Second, the assembly of the terrestrial planets in our own system involved a relatively small number of lunar- to Mars-sized bodies, themselves built rapidly from much smaller material. The accretion of these bodies to form the terrestrial planets was a stochastic process, in that small changes to the initial assumptions lead to very different outcomes in terms of planet masses and positions. We might expect that other planetary systems would show wide variation in the architectures of the terrestrial planets. Finally, although the "astronomical" part of the story can be put on an absolute timescale based on the observation of other young stars and comparison with models, the planetary formation epoch requires a different approach to assigning absolute dates of events. We turn to this next.

6.3 The Technique of Isotopic Dating

Radioactive decay occurs on a wide range of timescales in various isotopes of different elements, dependent on the relative stabilities of the nuclei and on the decay processes involved (Chapter 3). Decay itself is a *probabilistic* event, that is, we cannot point to a particular atom and predict at what time in the future it will decay. However, the time for half of a large ensemble of the same radioactive isotope to decay is known extremely accurately, and is called the *half-life*. For isotopes commonly used in dating or those abundant in the crust of the Earth, the half-life is given in Table 6.1. The half-life is *independent* of the environment within which the atoms find themselves; as long as there are a large number of them, the half-life is well defined and precisely measured for each radioactive isotopic species.

As described in Chapter 3, the fundamental property of radioactive decay is that it is proportional to the number of radioactive isotopes of that type present in the sample. That is, if we define as ΔN the number of atoms that decay in a small time interval Δt, it follows that $\Delta N = -N \Delta t / t_{mean}$, where t_{mean} is the characteristic time for the atoms to decay (1.4 times the half-life). It is the time for the radioactive isotope to fall in abundance by a factor of e, the exponential function. In Chapter 3 we obtained the basic decay equation $N(t) = N(0) \exp(-t/t_{mean})$, where $N(t)$ is the number of remaining radioactive atoms at time t and $N(0)$ is the number that were present at some arbitrary time $t = 0$. As the number of radioactive *parents* declines, the number of radioactive products or *daughters* commensurately increases (see Figure 3.16).

Suppose that we can measure, in a sample containing a radioactive isotope, both the number of parent atoms and the number of daughters and that, furthermore, we know (or were

TABLE 6.1 Decay Data for Radioactive Isotopes Common in Dating and in Geologic Processes

Parent	Daughter	Half-Life (years)	Decay Mechanisms
^{238}U	^{206}Pb	4.468×10^9	alpha (8), beta (6)
^{235}U	^{207}Pb	7.04×10^8	alpha (7), beta (4)
^{232}Th	^{208}Pb	1.401×10^{10}	alpha (6), beta (6)
^{87}Rb	^{87}Sr	4.9×10^{10}	beta
^{147}Sm	^{143}Nd	1.060×10^{11}	alpha
^{40}K	^{40}Ca or ^{40}Ar	1.250×10^9	beta, electron capture
^{176}Lu	^{176}Hf	3.730×10^{10}	beta
^{187}Re	^{187}Os	4.200×10^{10}	beta
^{14}C	^{14}N	5,730.	beta

Notes: Alpha decay is emission of one or more helium nuclei (number of emissions in parentheses); beta decay the conversion of neutrons to protons and electrons, with emission of the electron. See Chapter 3 for more details.
Data from Cole and Woolfson (2002).

told) that when the sample was formed (by whatever process) there were no daughter molecules present. We also know t_{mean} for this radioactive material. The number of parent atoms remaining $N(t)$ is simply related to the original number $N(0)$ by $N(t) = N(0) - D$, where D is the number of daughter atoms. Then the radioactive decay equation becomes $N(t) = (N(t) + D) \exp(-t/t_{mean})$. By counting $N(t)$ and D, and by knowing t_{mean}, we can solve for the time since the sample was formed. Because radioactive isotopes occur in rocks within the Earth; in meteorites that are fragments of asteroidal, lunar, and even Martian material; and in lunar samples brought back by the astronauts, it would seem that the decay law provides a powerful tool for the determination of the ages of the samples. However, some important complications must be overcome in radioisotopic dating:

- To count both parent and daughter molecules, we must be assured that the daughter isotopes indeed are from a batch of parents that together form a contained, or "closed," system. Remember that the daughter products are different elements from the parents, which means they may interact chemically with the surrounding rock in a different way. If, for example, the rock had melted sometime during its history, and the daughter isotopes consisted of an element more volatile than the parents, the daughter elements would have migrated through the rock, ending up in a different mineral suite or perhaps leaving altogether. An example of this is the decay of potassium, a moderately volatile element, to argon, a highly volatile gas. Hence, ways must be found to determine whether melting episodes occurred since a rock first crystallized. Indeed, partitioning of the parent and daughter species can occur without melting, and a means must be found to identify such partitioning because it makes the determined age unreliable.

- Often a rock will be formed with daughter isotopes from decays predating the rock. Without knowing how many of the daughter isotopes fall into this category, it is possible to get an age for the rock that is too old, by assuming that all of the daughters were created in the rock after it formed.

- We must define what is meant by "age." Did the rock solidify from a magma, or did it condense from the elements in the cold nebula? Some portions of meteorites appear to have formed in the latter fashion, never having been melted, while other portions clearly were melted. On Earth, where rocks have been cycled through the crust and melted repeatedly over time, the age of an *igneous* (derived from a melt) rock is the time of last melt. But some rocks—the *sedimentary* rocks—are simply lithified fragments of broken up rock deposited in an environment with water (lake, stream, river, ocean). While these may contain radioactive isotopes, the isotopic age is not useful because it reflects the last melt of the igneous portion, which may have been billions of years before the sediment was formed. *Metamorphic* rocks have been partially melted under pressure, and for these repartitioning of the parents and daughters has occurred between some of the mineral compounds—sometimes allowing determination of a reliable age of the metamorphism.

- Some parents transform through a chain of radioactive decays, and the decay equation becomes a series of equations that must be solved in a coupled fashion. The exception to this complication occurs when one of the daughter products has a much longer half-life than the others in the chain; then only the long half-life decay need be considered and the simple equation applies. The decay of ^{238}U to ^{206}Pb can be solved in this way. Another complication is that some decays, notably that of ^{40}K, involve the production of two daughter products (for potassium, these are argon and calcium). To handle this decay properly, the equation must include the mean decay times for both products, yielding a slightly more complex final expression.

The surmounting of just a few of these difficulties can be described here for want of space. The classic way around the ambiguity in the number of preexisting daughter atoms is through the isochron diagram. The case of the transformation of ^{87}Rb to ^{87}Sr, via beta decay, is an important example because the half-life of 4.9×10^{10} years makes this system particularly suited to the dating of events in the early solar system. The problem is that in the meteorites, which are the usual target of the dating, some preexisting ^{87}Sr is usually present. Switching notation, let $N(t)$ be relabeled as $[^{87}\text{Rb}]_{\text{today}}$, $D(t) = [^{87}\text{Sr}]_{\text{today}}$, and likewise for the values at the $t = 0$, defined as the formation of the rock. Then we must find $[^{87}\text{Sr}]_{\text{today}} = [^{87}\text{Sr}]_0 + [^{87}\text{Rb}]_{\text{today}}[\exp(t/t_{\text{mean}}) - 1]$. We know neither the age of the rock, t, nor $^{87}[\text{Sr}]_0$, the amount of the daughter present at the time of the formation of the rock. To work around this ambiguity we may use measurements of the amount of the parent and the daughter from multiple grains in the rock, grains that we hypothesize to be all of the same age (a hypothesis we can check when the analysis is complete).

Because each of the grains will potentially have different amounts of strontium, we normalize the daughter isotope, ^{87}Sr, by the stable lighter strontium isotope ^{86}Sr. Both strontium isotopes behave the same chemically, and the lighter isotope is not produced by a radioactive parent and hence should have the same abundance at $t = 0$ as at present (unless the rock has been altered, as described later). Hence $[^{86}Sr]_{today} = [^{86}Sr]_0$. Notice as well that the half-life of ^{87}Rb is much longer than even the age of the cosmos discerned from the galactic red shift (Chapter 5). So, t/t_{mean} is much less than 1, and under those conditions, the exponential function $[\exp(t/t_{mean}) - 1]$ can be approximated as simply t/t_{mean}. With the normalization to the amount of light strontium and the approximation applied to the exponential, we have the following equation:

$$\frac{[^{87}Sr]_{today}}{[^{86}Sr]_{today}} = \frac{[^{87}Sr]_0}{[^{86}Sr]_0} + \frac{[^{87}Rb]_{today}}{[^{86}Sr]_{today}} \frac{t}{t_{mean}}. \qquad (6.1)$$

This is the equation of a straight line, $y = mx + b$, where y is the ratio of the heavier to lighter strontium today (the "y-intercept"), b is the ratio of heavier to lighter strontium at the time the rock formed, and x is the ratio today of the radioactive rubidium to the lighter isotope of strontium. The slope of the line, m, is the time since the rock formed relative to the characteristic decay time. These relationships are illustrated in Figure 6.7. If the rock has not been altered, measurements from individual grains should fall on a straight line, the slope of which is the age, and the y-intercept is the amount of daughter

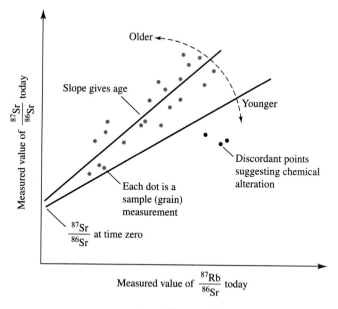

FIGURE 6.7 Schematic plot of $[^{87}Sr]_{today}/[^{86}Sr]_{today}$ versus $[^{87}Rb]_{today}/[^{86}Sr]_{today}$. Straight lines drawn through the points indicate both the age and the amount of ^{87}Sr present in the rock when it formed; a steeper slope implies greater age. Points from grains within the same rock sample that deviate from the straight line suggest thermal or chemical alteration of the rock, making age determination difficult or impossible.

strontium (normalized to the lighter isotope not produced by decay) at the time the rock formed. If the points stray from a straight line, then we must conclude (absent significant measurement error) that heating of the rock subsequent to its formation caused the strontium and rubidium to migrate away from each other. Under such circumstances, we might choose not to attempt to determine the rock's age, depending on the severity of the deviation of the points from a straight line.

A second check on the extent of the alteration of a rock being radioactively dated can be applied to elements with more than one radioactive isotope. Uranium has two radioactive isotopes: ^{238}U decays to ^{206}Pb, while ^{235}U decays to ^{207}Pb. Since the half-lives of these two decays are different, 4.47 billion years for the heavier uranium and 704 million years for the lighter uranium, a plot showing the ratios ^{206}Pb/^{238}U versus ^{207}Pb/^{235}U for different rock ages will be a curved line called a "concordia" curve. Even though both uranium isotopes transform to lead via chains of alpha and beta decays, there is one step for each that has the longest decay time; hence, the concordia curve can be constructed using the simple decay law given above.

The utility of such a curve is evident. Imagine having a rock sample to test for subsequent episodes of melting that might have caused the loss of lead. Were there no such loss, the analysis of all grains within the sample should yield a single point "1" on the concordia curve (Figure 6.8). However, if a single melting event leads to the loss of lead, the analysis of the lead–uranium ratios will yield discordant points away from "1" on the concordia curve, defining a different curve. Because both lead isotopes should behave the same chemically, they will be lost in equal proportions, and a single melting event will lead to a set of discordant points that is a straight line on Figure 6.8. The extension of the straight line upward, to where it crosses the concordia curve, yields the age of the rock. Further, the slope of the straight line gives a measure of the time when the heating event occurred. If multiple heating events have occurred, or complex chemical alteration or contamination, then the discordant points will not form a straight line, and an age determination becomes impossible.

The preceding discussion pertains to the decay of long-lived radioisotopes, with decay half-lives of billions of years, enabling events on cosmological timescales to be dated. Several examples of important dates determined from long-lived isotopes are given in the next section. Before doing so, let us close this section with a discussion of another dating technique, this one involving a radioactive isotope of carbon with a half-life less than 6,000 years, that provides information on the time since a piece of organic matter was part of a living organism. This technique plays an important role in determining climate change over the past few tens of thousands of years (Chapters 11 and 17).

The stable isotope of carbon ^{12}C is one of the more abundant atoms in the cosmos and a foundation for biology on Earth. ^{13}C, or Carbon-13, also stable, is present in all natural carbon-bearing systems but at much lower abundance. The next heavier isotope, ^{14}C, decays with a half-life of 5,730 years (Table 6.1) and is continually produced in the Earth's atmosphere as the most abundant nitrogen isotope, ^{14}N, collides with neutrons produced from an influx of energetic galactic cosmic rays (Chapter 5). The absorption of the neutron leads to ejection of either another neutron or a proton, but primarily the latter. When a proton is ejected, the atomic number decreases by one while the mass remains the same, and hence nitrogen is converted into heavy carbon.

The production by neutron bombardment and then decay leads to a roughly constant abundance of ^{14}C as a minor isotope of carbon in the atmosphere. Carbon-14 combines

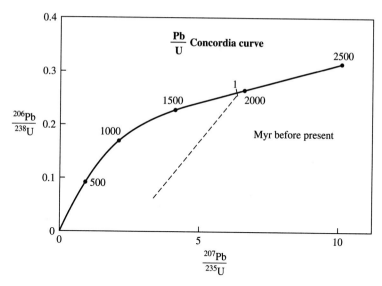

FIGURE 6.8 Uranium-lead concordia diagram plotting ratios of the daughter lead normalized by parent uranium in the sample for the two isotopes of uranium. The concordia curve is the set of points corresponding to a set of ages of the rock from the decay law. An unaltered rock grain will fall on the concordia curve, as shown at point 1; that is the age of the rock. If the rock was subsequently altered, by melting, the measured ratios in various grains will fall below the concordia curve and form the dashed curve. If the latter is a straight line, as shown, then the alteration was a simple discrete melting event, and the line may be extrapolated back up to the concordia curve to find the age. In this example, the rock with an age given by point 1, once altered, would exhibit a straight line that intersects the concordia curve at that point.

with oxygen to make heavy carbon dioxide, $^{14}CO_2$, and then finds its way into plants through photosynthesis and thence through the food chain to the rest of the biological world. Respiration and other biological processes return carbon to the atmosphere so that living organisms and the atmosphere are essentially in continuous exchange on timescales much less than the ^{14}C decay half-life. Hence, the ratio of ^{14}C to ^{12}C in living organisms is a constant and reflects the atmospheric value. (There is a difference in the uptake of light versus heavy carbon dioxide by photosynthesis, but this difference is precisely measured and easily accounted for.)

When an organism dies, exchange stops (technically it slows greatly, as bacterial and abiotic processes move carbon in and out of dead organisms at vastly reduced rates). The ^{14}C within tree trunks, animal bones, and other durable remains of organisms undergo beta decay back to ^{14}N, with release of one electron per decay. Biological materials that are less than roughly 70,000 years old contain enough remaining ^{14}C that the electrons resulting from the decay can be directly counted in the laboratory, thus sensitively measuring the amount of ^{14}C remaining relative to the total carbon (masses 12, 13 and 14) in the sample. By comparing this number with the ratio of ^{14}C to total carbon in the atmosphere (and correcting for the small difference in uptake of the various isotopes of carbon by plant photosynthesis) the age is determined. Unlike the techniques described previously for long-lived isotopes, the daughter ^{14}N is of no help, because it is highly volatile and immediately joins the rest of the ^{14}N that in molecular form constitutes the bulk of the Earth's atmosphere.

The use of ^{14}C to date the age of organic material is subject to the uncertainty of possible variations in the production rate of ^{14}C from ^{14}N in the atmosphere. The manufacture of ^{14}C depends on the cosmic ray flux, and this is known to vary as the strength of the Earth's intrinsic magnetic field changes over time, and as the ionized wind from the Sun—the "solar wind"—fluctuates over time. The ratio of $^{14}CO_2$ to $^{12}CO_2$ in the atmosphere also varies with changes in ocean circulation, which brings up varying amounts of carbon dioxide stored for hundreds of years or longer in deep water. Cross-correlation where possible with independent dating techniques (e.g., counting growth rings in tree trunks), allows the technique to be well calibrated over nearly the past 10,000 years.

Attempts to account for variations in the solar wind, Earth's magnetic field, and ocean circulation lead to the conclusion that dates obtained with ^{14}C are smaller than the actual age of the sample, and this discrepancy increases with age. At 20,000 years before the present, the discrepancy is 3,000 years; it increases modestly for larger ages. Atmospheric nuclear testing through the 1970s has artificially raised the production rate of ^{14}C in the atmosphere so that future users of this dating technique will need to correct for the heavy-carbon pulse of the atomic age.

6.4 Meteorites: The Age of the Elements and of the Earth

With the mathematical tools provided in this chapter it is possible to determine various events associated with the origin of the solar system and the accretion of the planets, as well as earlier cosmic events. Here we focus, by way of example, on two key dates: the appearance of the first solids in the solar system and the initiation of element formation much earlier. Among the most primitive meteorite samples are the *chondrites*, which contain so-called chondrules that appear to be pieces of material melted and cooled over restricted timescales. Most striking about the chondrules is that, for all but the most volatile elements, the bulk composition of the chondrites is very similar to that of the observable portion of the Sun—the so-called photosphere. This suggests that the material out of which the chondrites formed was that out of which the Sun formed as well—the protoplanetary disk and surrounding molecular cloud. Chondrites are divided into classes according to how much of certain elements (such as carbon) are retained, the ratio of rock-forming elements such as magnesium to silicon, and the amount of oxidation the material has undergone. These seem to be related to where these meteorites formed, within the protoplanetary disk.

Because of their resemblance to solar composition, unlike the crusts of any of the planets sampled so far, and their primitive overall appearance (e.g., lack of melting of the matrix holding the chondrules), chondrites may be samples of some of the oldest material in the solar system. Radioisotopic dating using uranium-lead concordia diagrams yield ages for the oldest chondrites of 4.55 billion years. Based on these diagrams and variations in

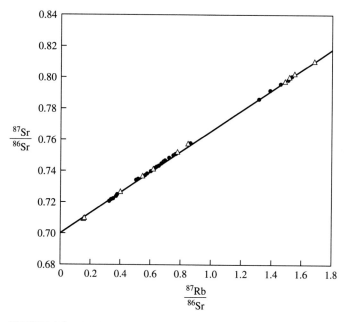

FIGURE 6.9 Rubidium–strontium diagram for a number of different chondritic samples of various types (labeled by different symbols), forming a beautiful straight line with an age of 4.55 ± 0.01 billion years.

the lead abundance, this date appears to be when rocky material was condensing from the protoplanetary disk. Slightly earlier dates for other radioactive isotopes may reflect contamination of planet-forming material with preexisting grains from the molecular cloud. A similar age of solid material is obtained from a rubidium–strontium analysis for a number of meteorites, which form a good straight line (Figure 6.9).

The formation of the Earth's Moon appears to be from a huge impact that gouged out a significant portion of the mass of the Earth. Isotopic dating of lunar rocks and meteorites suggest that this happened about 100 million years after the appearance of the first solids in the solar nebula. This timescale is longer than the time over which core formation occurred in the Earth or its precursor Mars-sized bodies, perhaps 15 to 30 million years, derived from analysis of the hafnium–tungsten radioisotopic system in meteorites and the Moon (Figure 6.10). These timescales seem to be in fair agreement with those derived from the dynamical calculations of the accretion of the terrestrial planets from lunar- to Mars-sized bodies, described in Section 6.2. The oldest rock sample on the Earth is 4.4 billion years old (see Chapter 11) and shows evidence of the presence of liquid water. It must thus postdate significantly the formation of the Earth itself.

The sampling of numerous chondritic and other meteoritic samples, as well as lunar rocks and volatile daughter products of long-lived isotopes in Earth's atmosphere, leads to a consistent picture in which the first solids appeared in the protoplanetary disk 4.55 billion years ago. This initiated a hundred million year period of planetary accretion that,

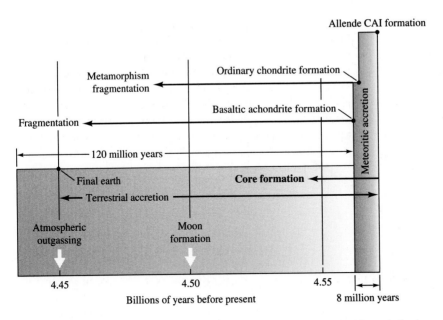

FIGURE 6.10 Various timescales in the accretion of the Earth and the Moon derived from radioisotopic systems.

for the Earth at least, included the formation of the Moon through a giant impact. We move through more of this chronology shortly, but first it is useful to step backward and ask a different question: When were the elements themselves created?

To answer this question requires identifying radioactive parents with long half-lives that are chemically similar to each other so that little or no separation occurred during planet formation. The clear choices from Table 6.1 are the two radioactive isotopes of uranium and that of thorium. While thorium is not chemically identical to uranium, it is of similar volatility. Both uranium and thorium are produced via the nucleosynthetic r-process at known rates (Chapter 5). If we normalize the abundance of ^{235}U and ^{238}U by the abundance of ^{232}Th, we can express the predicted ratios of the three isotopes for ongoing production in the r-process as ^{235}U/^{232}Th = 0.79, ^{238}U/^{232}Th = 0.53. Reference to Table 6.1 indicates that both isotopes of uranium are shorter-lived than the thorium, and the lighter isotope of uranium has a half-life only one-seventh that of the heavier one. Thus, over time, we expect that both isotopes of uranium would have declined relative to thorium, and the lighter isotope of uranium become much more rare than the heavier. Indeed, analysis of carbonaceous chondrites today yields values of ^{235}U/^{232}Th = 0.02, ^{238}U/^{232}Th = 0.3, qualitatively what we expect.

Now we know the age of the carbonaceous chondrites from the rubidium–strontium analysis to be 4.55×10^9 years. If we back up the abundances of the uranium and thorium isotopes from the current value in chondrites to that at their birth, we get ^{235}U/^{232}Th = 0.1, ^{238}U/^{232}Th = 0.4. How much farther must we "back up" in time to get the values predicted in a "fresh" batch (^{238}U/^{232}Th = 0.79, ^{238}U/^{232}Th = 0.53)? Application of the radioactive decay law yields a value of about 2 billion years, *assuming that all of the ura-*

nium and thorium were formed at a single time. Then to get the age of this event we add 4.55 billion years and 2 billion years to get 6.55 billion years. That is, under the hypothesis of a discrete event, element formation occurred (rounding up) 7 billion years ago.

A discrete element formation event is unreasonable. Elements are produced by the r-process in supernovae scattered throughout the galaxy over long periods, and the products from different supernovae are likely mixed quite effectively within a single giant molecular cloud. So, our calculation must be revised to take account of a continuous production of uranium and thorium in supernovae over time. This continuous production means that fresh ^{235}U, ^{238}U, and ^{232}Th injected into clouds slows the decline of the uranium relative to thorium and the decline of ^{235}U relative to ^{238}U, so that our previous timescale is an underestimate. Instead we must write a time-dependent decay equation that allows for injection of a fresh supply of isotopes. Models that do this yield a time from start of element formation to the formation of the solar system about three to four times more than what the discrete model calculates, or about 6 to 8 billion years. This, in turn, implies that element production (at least for the r-process elements) began about 10 to 13 billion years ago.

This result is remarkable for its consistency with the overall picture of cosmic evolution described in this and the last chapter. The universe was formed about 10 to 20 billion years ago, with most recent estimates tending toward the larger figure; element production in the r-process began 10 to 13 billion years ago, and the solar system formed 4.55 billion years ago—with accretion of the Earth complete by perhaps 4.45 billion years ago. What "complete" means is not entirely easy to define. The sweep-up of debris left over from planet formation is clearly etched into the surfaces of the airless or geologically less active bodies of the solar system—in the form of *impact craters*. Impact craters are the holes gouged out in a solid surface by the high-speed impact of another object. The Earth itself has few craters because most have been removed by plate tectonics or erosion (Chapter 11). On the Moon, the southern highlands of Mars, and many of the natural satellites of the giant planets, the record is better preserved (Figure 6.11).

It is clear from this record that the rate of collisions declined very steeply during the first few hundred million years of the solar system's history and has declined more slowly since (Figure 6.12). However, evidence from the spectrum of ages of meteorites suggests that there was a large bombardment event between 3.9 and 4.0 billion years ago, after which the cratering rate declined. This is consistent with the age of the more lightly crated lunar mare, which are places on the Moon where volcanic lavas obliterated the earlier cratering record. Whether the cratering rate declined soon after accretion and then increased again as some additional supply of impactors (possibly due to a large collision in the asteroid belt or elsewhere in the inner solar system) was created, is unclear. The crater density on "old" surfaces (those not subject to geological erasure) is so high that new cratering events simply obliterate older craters.

This record of the violent sweep-up of leftover debris from planet formation has several implications: It is by this process that volatiles (water, organics) might have been added to Earth, and the Moon may have formed (Section 6.5). Frequent violent impacts might have delayed life gaining a foothold on Earth by repeatedly destroying environments in which organic chemical evolution was taking place. Some lucky impacts have ejected rocks on trajectories taking them from one neighboring planet to another, for example Mars to Earth, or vice versa. Direct evidence for this comes from a number of meteorites (Chapter 12),

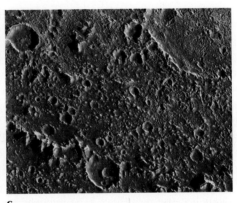

FIGURE 6.11 Cratered surfaces of the solar system: (a) The cratered highlands of the Moon; (b) the lightly cratered mare of the moon; and (c) craters on the Jovian moon Callisto, imaged by the Galileo spacecraft.

whose trapped gas composition pegs them as having originated on Mars. Simulations and impact tests on living microbes suggest that survival during such interplanetary transfers and landings is possible, so Earth and Mars may have been cross-contaminated several times during their histories.

The variation in cratering rate from early to late in the history of the solar system provides a relative chronometer for dating geologic activity. For example, the dried river channels or "valley networks" on Mars occur in areas of heavily cratered terrain, and the channels appear to both carve through some craters and be excavated by others. This implies that their formation occurred, for the most part, during the time of frequent bombardment of the Martian surface. There are other features on the Martian surface, such as apparently flood-generated outflow channels, that occur on much less heavily cratered surfaces and are hardly ever overlain or modified by subsequent impacts; they are almost certainly much younger than the valley networks (Chapter 12). The entire surface of Venus, on which impact craters are sparse and volcanoes abundant, has apparently been resurfaced by massive volcanism at least once during its history, and perhaps more recently than a billion years ago.

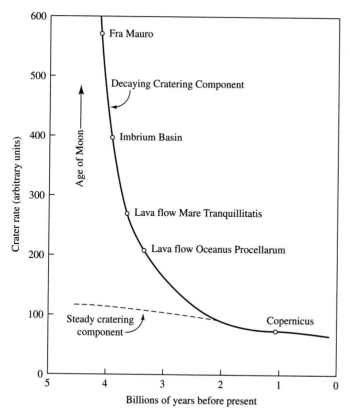

FIGURE 6.12 Schematic change in cratering rate over time for the Moon.

6.5 Origin of the Earth's Oceans and Biogenic Elements

Planet Earth is defined by the presence of water and abundant organic molecules that today are almost entirely incorporated in life. The origin of these materials is a key issue in the formation and early evolution of the Earth because it has implications as well for the supply of water to Mars, as well as for whether planets in Earthlike orbits around other stars will be supplied with water under a broad range of formation conditions.

The amount of water cycling through the upper layers of the Earth (its crust) is about 20 percent relative to the total volume of the Earth's oceans. The deeper interior of the Earth, which we describe briefly in Sections 6.6 and 6.7, might today contain several Earth's oceans worth of water. What is unknown is whether the deep interior, early on, played host to much larger amounts of water, perhaps 10 to 50 times what we have at the surface. The most specific and constraining clue about the nature of Earth's water comes from the measured deuterium-to-hydrogen ratio, HDO/H_2O, in the ocean water—about 150 ppm.

This value, uniform throughout the Earth's surface waters, is known as Standard Mean Ocean Water, or SMOW. It is approximately equal to the average value in carbonaceous chondrites—which contain oxygen and hydrogen bound to mineral grains. It is, however, about half the value measured for the ices in comets. To date, only three comets have been large enough and come close enough to permit this measurement; their value is similar for all three and is about twice the SMOW. This, plus dynamical calculations, seems to mitigate against comets—remnants of icy planetesimals from beyond the 5-AU region in the protoplanetary disk—as the source of Earth's water.

Two alterative sources of water for the Earth are (1) the planetesimals within the 1-AU region itself or (2) planetesimals that today are represented by the carbonaceous chondrites and would have originated in the asteroid belt, beyond about 2 to 3 AU. The attractiveness of the second possibility lies in the apparent decline in water abundance in different chondritic meteorite types that represent material formed progressively closer to the early Sun. Carbonaceous chondrites formed beyond 2 AU have water fractions 1 percent or more by mass. Enstatite chondrites, which on the basis of their oxidation state are thought to have formed closer to the Sun, perhaps within 1.5 to 2 AU, are much dryer; they have about 0.01 percent water by mass. This decline is such that material present at the orbit of the Earth (1 AU) would have been too dry to supply the mass of water we see today at the surface and in the near surface of the Earth—about 0.03 percent of the mass of the Earth. Supporting this view are models of the protoplanetary disk suggesting temperatures at 1 AU during formation of rocky solids too high for the accommodation of large amounts of water in rocky planetesimals.

If water cannot be gained from local sources, it must be derived from ones that are more distant. The simulation of the growth of planets depicted in Figure 6.6, in which bodies in the inner solar system are scattered by the gravitational influence of Jupiter, suggests that material from beyond 2 to 3 AU made its way into the inner solar system, where the Earth and Venus were growing, and in some cases collided with the growing planets. The collision of one or several of these lunar- or Mars-sized impactors with the Earth would have been sufficient to supply the Earth's water inventory, provided that the water was not lost during the collision process—an open issue.

On the other hand, there are problems with this story. If it is assumed that the composition of the material beyond 2 to 3 AU was that of carbonaceous chondrites, then the detailed elemental composition of that material can be compared with that of the modern Earth as derived from deep rock samples. The Earth's composition is vastly different from carbonaceous chondrites, and only a very small fraction of the Earth (one to a few percent) can be made of such material. But exactly how much depends on when the material was added to the growing Earth—if it was early on, the reshuffling of material during the formation of the Earth's core would hide some of the evidence. The dynamical calculations imply that as much as 5 percent of the Earth is derived from the 2- to 3-AU region. Hence, this model is just marginally consistent with our knowledge of the composition of our planet. The alternative—that the region around the forming Earth was cool enough to grow local water-laden planetesimals—implies that there existed a kind of planetesimal not reflected in the known classes of meteorites.

If the Earth had depended on planet-sized bodies in what is now the outer asteroid belt for its water, the implications for habitable worlds around other planets (Chapter 15) is interesting. The history of the accretion of our own terrestrial planets could have been quite different had Jupiter formed in a different location, or not formed at all. The sequence and timing of the formation of planet-sized bodies in the inner solar system could have been vastly different, and perhaps much less or much more water might have been granted Earth. Hence, the architecture of other planetary systems—including the positions of the giant planets in those systems—may be key to the dryness or wetness of Earth-sized planets in orbits appropriate for the stability of liquid water. Alternatively, if the Earth had accreted from local planetesimals containing abundant water, we might expect such a process to be replicated around any other forming star with disk conditions similar to those that obtained in our own solar system.

Whether water came to the Earth locally or from mostly rocky bodies beyond 2 to 3 AU from the Sun, comets would appear to be a minor source (at most 10 percent) of Earth's water. They may, however, have been a primary source of the organic content of the Earth and possibly, as well, of nitrogen. Comets are rich in carbon-bearing molecules, much more so than the carbonaceous chondrites, and were abundant in vast quantities—tens of Earth masses—in the outer solar system during planet formation. Calculations of the impact of comets on the Earth toward the end of our planet's formation suggest not only that more than enough carbon was delivered to explain the Earth's inventory but also that despite the violence of high-velocity cometary impacts, amino acids could have survived in significant quantities (Figure 6.13, see Color Plate 8). Whether nucleic acid bases are present in comets as well, and were delivered to Earth in impacts, remains an open question.

6.6 Origin of the Moon and the Differentiated Earth

One object, shining in the sky from crescent to full each month, is a constant reminder of the impact of planet-sized bodies with the Earth during the epoch of formation. The Moon is striking both in its relatively large mass for a natural satellite (only five moons elsewhere in the solar system are as massive) and its chemical composition. Much poorer in iron than the bulk of the Earth on the basis of comparison of overall density and the shape of its gravitational field, the Moon also has an elemental composition that crudely is reproduceable by heating the outer portion of the Earth to remove volatiles, then reconstituting the remainder as a separate body. The Earth itself, toward the late stages of accretion, became hot enough that the bulk interior melted, causing iron (denser than the surrounding silicate melt) and chemically compatible elements to gravitationally settle into a *core*. This core today, with a liquid outer layer and solid inner layer, is electrically conducting, and the twisting motions of the liquid core as it rotates and convects to remove heat generates a magnetic field (Chapter 2). The Earth's magnetic field reverses polarity stochastically over time, induced by mechanisms not well understood, and the reversal sequence imprinted on surface rocks is an important record of geologic processes—especially the horizontal motions of the Earth's outer layer (its "crust")—via plate tectonics (Chapter 12).

The presence of a differentiated core of iron and other elements (perhaps sulfur, oxygen, even hydrogen) implies that the bulk of the Earth (its "mantle") should be depleted in iron. Reference to Table 6.2 shows that, relative to chondrites, this is indeed the case. Thus, we can suggest that the Moon resembles, crudely, the Earth's mantle in being iron poor. Had the Moon simply accreted out of the same material that formed the Earth, it would have contained much more iron. But to make the Moon from a chunk of the Earth's mantle is no easy dynamical task. Spinning up the Earth to split off a chunk to make the Moon does not put the latter in a stable circular orbit, rather the ejected material leaves the orbit of the Earth. The one plausible dynamical alternative is that the Earth was struck obliquely by a large body, gouging out the mantle and putting some of the material into Earth orbit. Detailed computer simulations of this process require that the impacting body be roughly the mass of Mars and that the impact angle and velocity be arranged so that much of the impactor and ejected material returns to impact the Earth a second time, with a fraction ejected again. Only a small part—less than 10 percent—of the ejected material ends up in a stable orbit and accretes to form the Moon. This dynamical model is consistent with the detailed chemical differences between the Moon and the Earth's mantle, notably the lack of volatile elements in the former.

That a Mars-sized body struck the Earth late in the formation of the planets is consistent with the scenario depicted in Figure 6.6. Nonetheless, there were sufficiently few bodies the size of Mars that the impact should have been a somewhat unusual event. Venus—with no natural satellite—may have suffered collisions with lunar- to Mars-sized bodies, but the conditions were not right to spin off a Moon. The Moon has played a complex role in stabilizing the obliquity of the Earth (Chapter 11), and much marine life beats to the rhythm of the lunar-generated ocean tide. But to what extent the Moon is required for the persistence and evolution of life remains poorly understood; we can readily imagine a Moonless Earth, in a system perhaps bereft of giant planets in the 5-AU region, that supports a biosphere but knows only the weaker tides of its parent star.

TABLE 6.2 Typical Elemental Composition (in percent of total) of Meteorites and Earth Rocks

Element	Chondrites	Earth's Mantle	Basalt (Crustal Rock)	Granite (Crustal Rock)
Oxygen	32.	43.	44.	47.
Iron	29.	7.	10.	3.
Silicon	16.	21.	24.	32.
Magnesium	12.	23.	3.	0.7
Aluminum	1.	2.	8.	8.
Calcium	1.	2.	7.	2.
Sodium	1.	0.5	2.	3.
Potassium	0.1	0.02	0.1	2.
Other	8.	1.	2.	2.

6.7 The Earth After Formation and the Geologic Timescale

With the formation of the Moon, the Earth began to cool down from a molten state. Like slag atop molten metal, the upper layer of the Earth was chemically distinct from the mantle and solidified to form a crust. Today that crust is further chemically differentiated into basalts and granites (Table 6.2), but, at the time of crust solidification, it may have been less distinct from mantle composition. Preferentially sequestered in the upper layers of the Earth were radioactive isotopes of uranium, potassium, and thorium, which adds the heat of radioactive decay to the accretional heat of the Earth that still escapes from the interior today. Detailed analysis of the passage of waves through the Earth generated by sudden crustal shifts (earthquakes) leads to a model for the interior of the Earth sketched in Figure 6.14.

The subsequent history of the Earth, as accretional plus radiogenic heat is lost from the interior, is part of the story of our planet's habitability, described in Chapter 11. The geologic manifestations of this loss of heat include mountain building, volcanism, vertical uplift and subsidence, and horizontal movement. Three different kinds of rock appear in the crust of the Earth. Igneous rock, formed from a melt, may solidify beneath the surface after being "intruded" from below or may be extruded onto the surface as magma. The action of water, and to a lesser extent wind, on rock causes its breakdown into particles ranging from dust to sand to pebbles. This material, ground down and transported by streams, river, and glaciers, is eventually deposited as sedimentary layers in a variety of environments ranging from the ocean floor to lakes. Burial of this material leads to increased pressure and increased heating, leading to lithification to form sedimentary rocks and sometimes to the further modification and even melting of this material to form metamorphic rocks. Some rocks, such as calcium carbonate, precipitate directly out of solution and form intricate limestone caves. Uplift brings buried sediments and metamorphic rocks to the surface in the form of mountains, and subsequent erosion by water or glaciers exposes the more resistant granite cores deep within the mountains, or levels them entirely.

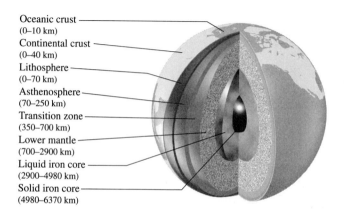

FIGURE 6.14 Schematic cutaway of the interior of the Earth.

An exposed section of rock might (for example) contain pieces of a granitic intrusion embedded in a series of sedimentary layers, overlain by another series of sediments, tilted at a large angle relative to the lower, and then cut through by a columnar basalt "dike." Geology is a "forensic" science in the sense that the clues at hand in such exposed outcroppings of rocks are used to reconstruct a relative sequence of geologic events in a particular region, supported by detailed chemical and isotopic analyses. Only on one other world—the Moon—has such field geology been possible. Elsewhere in the solar system, remote sensing from orbit supplemented for a few bodies by direct chemical analyses has had to suffice to assemble geologic histories.

The determination from exposures of rock of the relative ordering of geologic events provides only limited information on the history of the Earth. It is difficult or impossible from the rocks alone to reliably correlate a sedimentary layer (or sequence) from one part of the world to another, or even from one part of a continent to another. Because of this, filling in the gaps associated with unconformities is often impossible. What is required (other than the absolute timescale afforded by radioisotopic dating, which has been available only in the last half century) is some sort of indexing system for rock layers. Such an indexing system exists in the form of fossils of plants and animals.

Defined broadly, a fossil is physical or chemical evidence in the Earth's rocks of an organism. Chemical evidence in rocks for life is discussed in Chapters 11 and 14. Here we focus on fossils of multicellular organisms as indices of relative geologic time. Fossil re-

FIGURE 6.15 Example of a very well preserved fossil, that of an extinct Cambrian creature known as a trilobite.

Color Plate 1

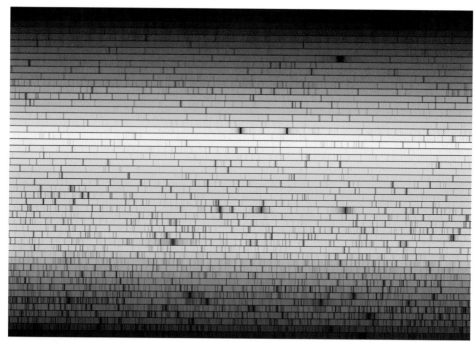

FIGURE 3.14 The Sun's spectrum from the ultraviolet (0.3 μm wavelength) to the near-infrared (1.3 μm). Spectrum is folded into 50 strips; bottom strip (bluest) is the shortest and top strip (reddest) the longest wavelengths. Black features are absorption lines.

Color Plate 2

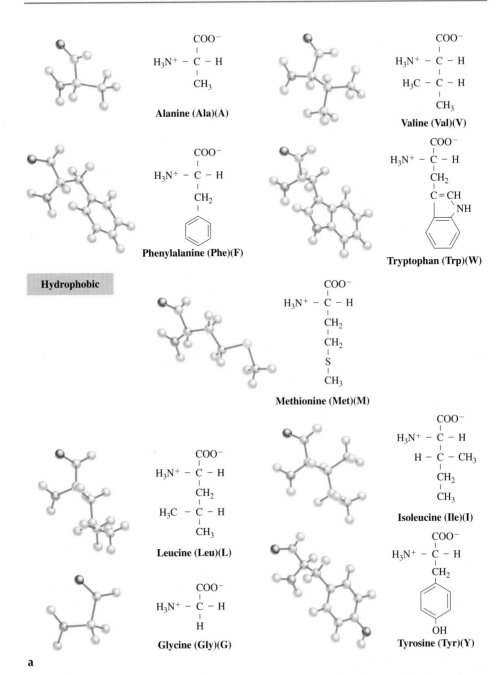

FIGURE 4.3 Structures of 20 of the 22 amino acids used by living organisms on the Earth, organized as (a) hydrophobic (including glycine), (b) polar, and (c) charged. Each is shown in schematic chemical bond notation (single lines and double lines are single and double bonds, respectively), and the entire amino acid is shown in a ball-and-stick three-dimensional representation. One hydrogen atom from H_3N^+ is hidden from view in each panel. The three-letter and single-letter codes for each name are also given. Color code: red-oxygen, white-hydrogen, green-carbon, blue-nitrogen, and yellow-sulfur.

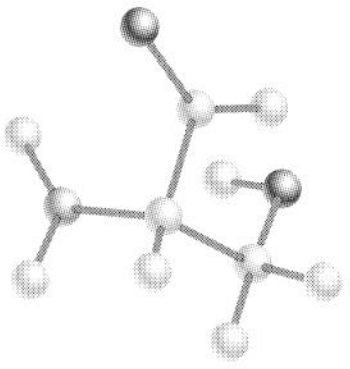

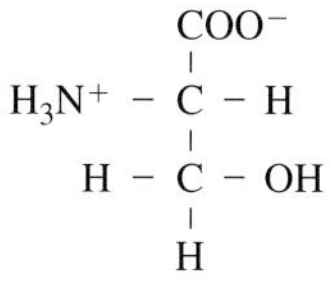

Color Plate 3

Polar

$$\begin{array}{c} COO^- \\ | \\ H_3N^+ - C - H \\ | \\ H - C - OH \\ | \\ H \end{array}$$
Serine (Ser)(S)

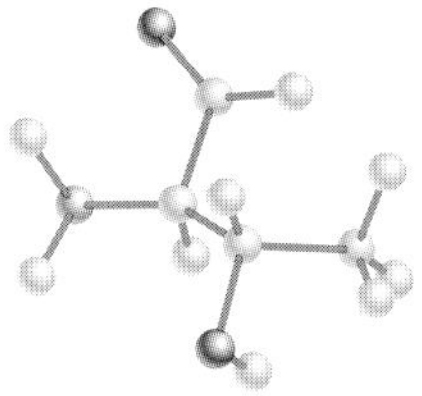

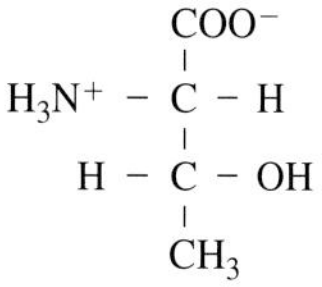

$$\begin{array}{c} COO^- \\ | \\ H_3N^+ - C - H \\ | \\ H - C - OH \\ | \\ CH_3 \end{array}$$
Threonine (Thr)(T)

$$\begin{array}{c} COO^- \\ | \\ H_3N^+ - C - H \\ | \\ CH_2 \\ | \\ SH \end{array}$$
Cysteine (Cys)(C)

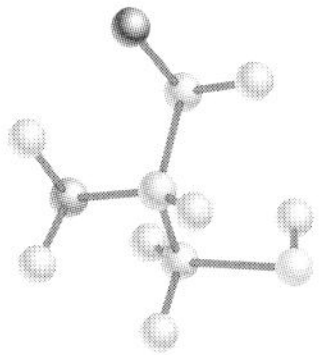

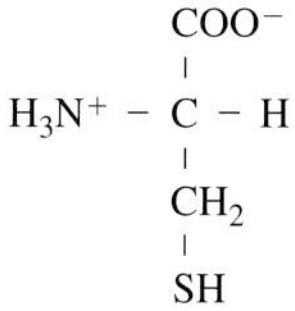

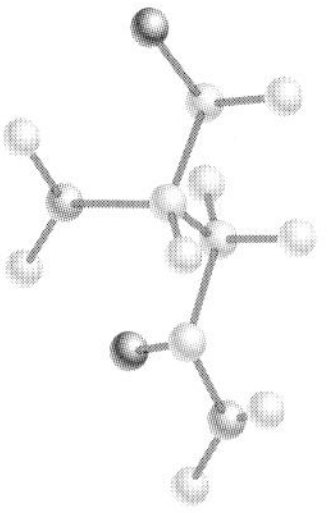

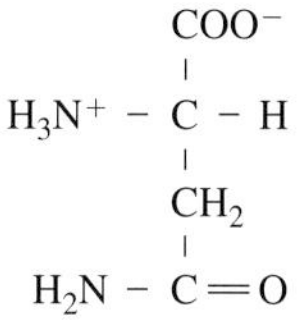

$$\begin{array}{c} COO^- \\ | \\ H_3N^+ - C - H \\ | \\ CH_2 \\ | \\ H_2N - C = O \end{array}$$
Asparagine (Asn)(N)

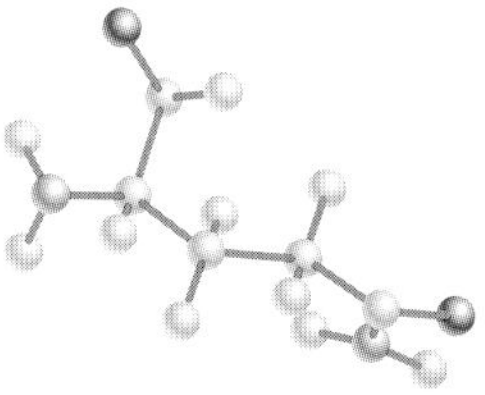

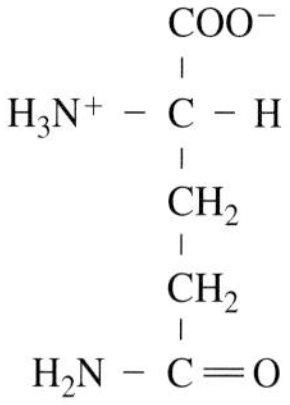

$$\begin{array}{c} COO^- \\ | \\ H_3N^+ - C - H \\ | \\ CH_2 \\ | \\ CH_2 \\ | \\ H_2N - C = O \end{array}$$
Glutamine (Gln)(Q)

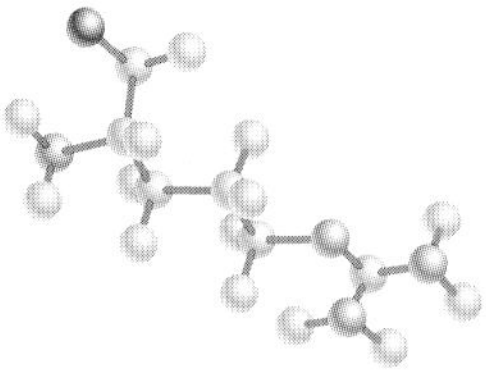

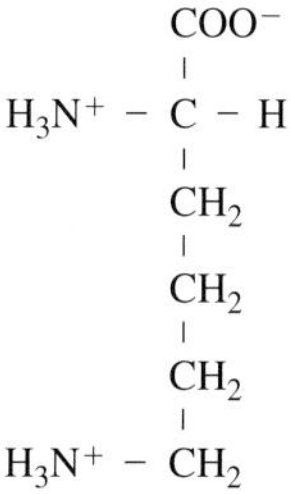

$$\begin{array}{c} COO^- \\ | \\ H_3N^+ - C - H \\ | \\ CH_2 \\ | \\ CH_2 \\ | \\ CH_2 \\ | \\ H_3N^+ - CH_2 \end{array}$$
Lysine (Lys)(K)

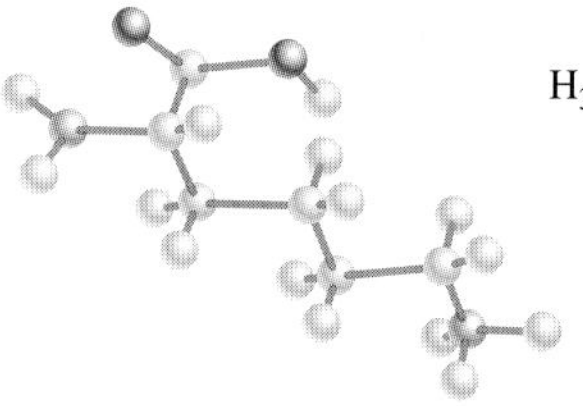

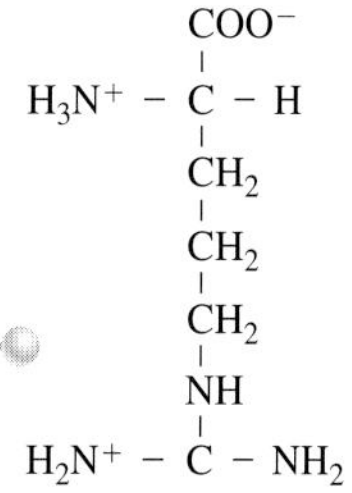

$$\begin{array}{c} COO^- \\ | \\ H_3N^+ - C - H \\ | \\ CH_2 \\ | \\ CH_2 \\ | \\ CH_2 \\ | \\ NH \\ | \\ H_2N^+ - C - NH_2 \end{array}$$
Arginine (Arg)(R)

b

Color Plate 4

Charged

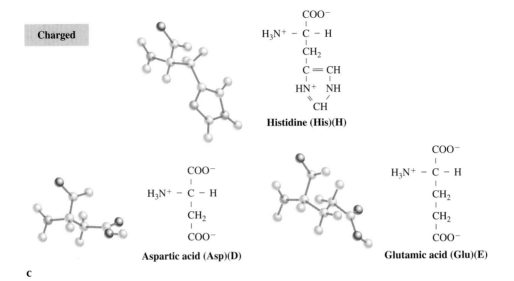

Histidine (His)(H)

Aspartic acid (Asp)(D)

Glutamic acid (Glu)(E)

c

FIGURE 4.3 continued

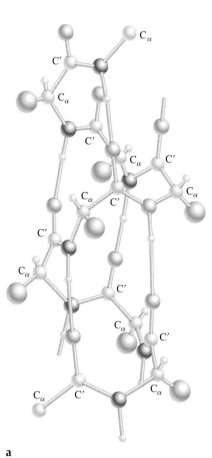
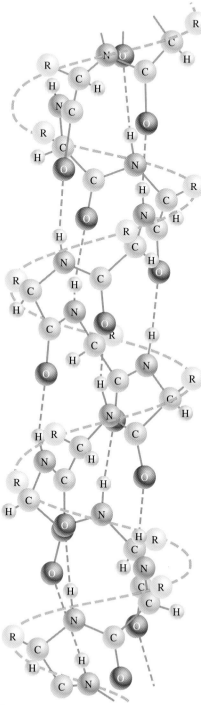

FIGURE 4.7 (a) Examples of a so-called alpha helix, a structure formed by the need to sequester hydrophobic side chains in the interior of the protein while tying up nearby hydrophilic parts of the main chain. Hydrogen bonds that achieve this are shown. (b) Polypeptide chain in the form of a spiral helix, with dashed lines indicating hydrogen bonds.

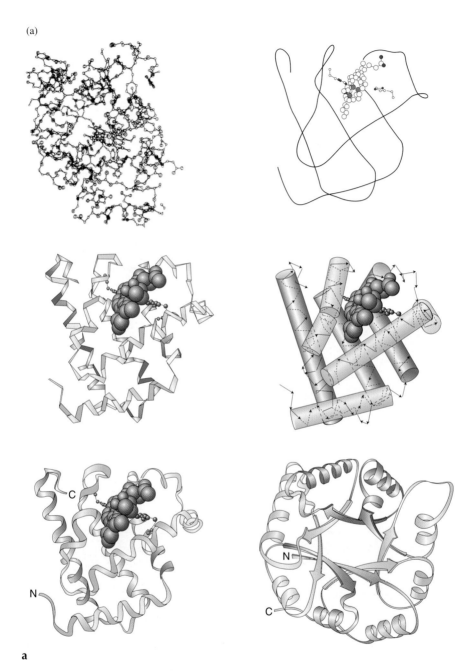

FIGURE 4.8 (a) Structure of the protein myoglobin, which binds oxygen for use in the muscles. Secondary structures—helices, sheets and loops—are shown. (b) An alpha carbon tracing of deoxyhemoglobin from four different views. Hemoglobin is composed of two alpha chains (green and yellow) and two beta chains (light and dark blue). The heme groups, where oxygen binds, are red.

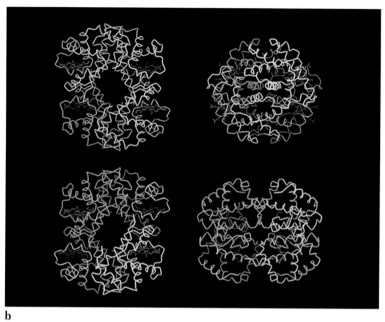

b

FIGURE 5.5 Hubble Space Telescope "ultra-deep-field" image of a small piece of the sky, equivalent in area to 1 percent of the full moon, taken over four months in 2003–2004 with the Advanced Camera for Surveys (ACS) and the Near-Infrared Camera and Multi-Object Spectrometer (NICMOS). The image reveals galaxies more than 13 billion light-years distant, equivalent to going back to the first 10 percent of the history of the universe. Although the part of the sky chosen for the image, just below Orion in the northern hemisphere sky, is fairly devoid of foreground stars in our own galaxy, a few show up in the image. (STScI/NASA).

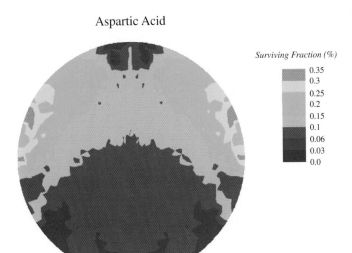

FIGURE 6.13 Results of a computer simulation of a comet striking the Earth. The circular area is a map of the crater showing the fraction of a typical amino acid, aspartic acid, that survives the impact—based on the peak impact pressures.

mains can be actual organic material; petrified remains in which the organic matter has been replaced by minerals such as calcite, quartz, and chert; molds or casts of organisms; prints in lithified mud or sand; or excrement. Petrification of biological structures is the sort of fossil with which most people are familiar, and those of macroscopic animals are most spectacular (Figure 6.15). The formation of such fossils depends on the chemical similarity between carbon and silicon (Chapter 3).

The process of natural fossilization is a chancy and rare event. The processes that result in fossilization vary greatly; only deposition in certain environments can preserve fossils, and the metamorphism of the sedimentary layer containing the fossil can easily destroy it. Hence, the vast majority of individual organisms that lived show no lasting evidence of their existence, their atoms being consumed by other organisms or otherwise dispersed. Enough fossilization occurs, however, to provide a picture of the progressive evolution of life-forms over time (Chapter 16) and hence a relative geologic timescale. Sediments that have been overturned, or "unconformities" where a number of sedimentary layers in a sequence have been removed by erosion (Figure 6.16), can be identified through key fossils.

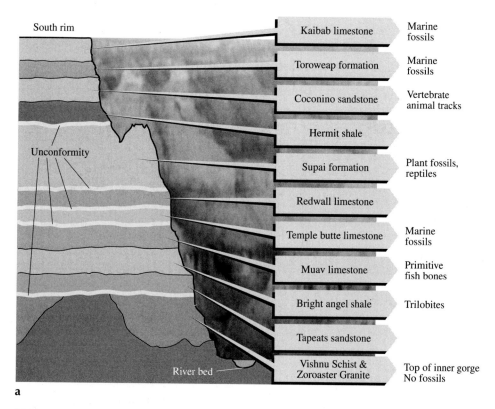

FIGURE 6.16 Sequence of sedimentary layers exposed in the Grand Canyon of the Colorado River, in northern Arizona (a). Layers, including ones missing in "unconformities" (b), are identified through fossil relationships.

b

FIGURE 6.16 continued

Because fossils are not found in igneous rocks, and only rarely in metamorphic rocks, such rocks must be fitted into the time sequence according to their relationship to sedimentary layers that they may overlie, underlie, or intrude. Igneous intrusions, in turn, constrain the absolute ages of sediments because they can be isotopically dated.

Type examples of sediments correlated worldwide by fossils have lent their name to a relative geologic sequence, or geologic "timescale." Based on radioisotopic dating of intruding igneous rocks, this sequence—divided into eras, periods, and epochs (from coarsest to finest divisions)—goes back about 540 million years, which constitutes the "Phanerozoic eon." It is married to a sequence of three earlier "eons" over 4 billion years divided according to major geologic events (e.g., the Archean being traditionally the appearance of the first large-scale rock record on Earth). The sequence constitutes the basic working vocabulary of geologists in discussing the history of the Earth subsequent to its formation, and it is important for astrobiologists to familiarize themselves with the names of at least the eons, eras, and periods (Figure 6.17).

With the tail-off of planetary accretion, the cooling of the Earth's crust, and the appearance of liquid water (which, as discussed in Chapter 11, occurred about 4.3 billion years ago), the next major event in Earth's history is the origin of life. This event, or sequence of events from which life emerges, is shrouded in mystery—and is at the core of the endeavor of astrobiology. The next three chapters describe the thermodynamic and chemical underpinnings of the origin of life and provide ideas for how organic chemical systems evolved to become living things.

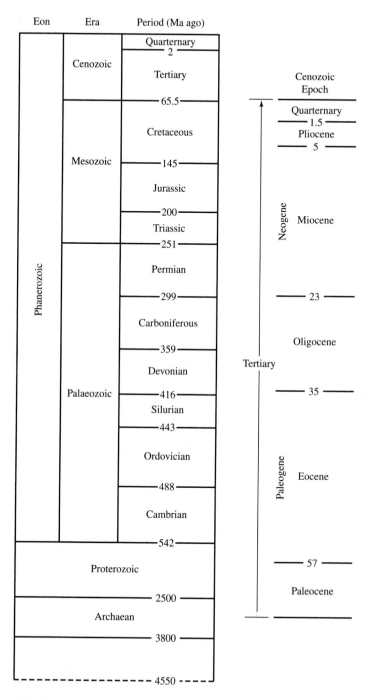

FIGURE 6.17 Geological sequence based, after the Proterozoic, on the fossil record. Period boundaries are updated via Whitfield, J. (2004). "Time Lords," *Nature*, **429**, 124–125.

QUESTIONS AND PROBLEMS

1. a. On the rubidium-strontium graph shown, which line (A or B) represents data from the older rock? (Axes are in arbitrary units.)

   b. If the squares are the data corresponding to line A, and the circles to line B, which line gives a more accurate age?

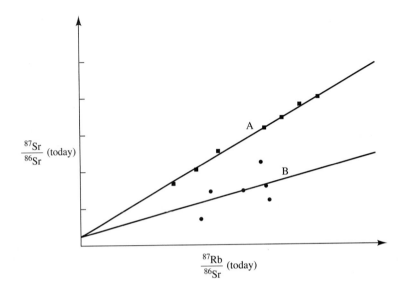

2. Calculate the central temperature of a red dwarf star 1/10 the mass of the Sun, a brown dwarf 1/100 the mass of the Sun, and Jupiter, which is 1/1000 the mass of the Sun. Use an astronomical reference source to obtain the radii. Assume the red dwarf and brown dwarf are mostly ionized hydrogen, whereas Jupiter is an even mix of atomic and molecular hydrogen.

3. Assume the Earth to have the abundances of potassium, uranium, and thorium present in ordinary chondrites. Do a geochemical literature search to find the abundances and heat output of the long-lived radioactive isotopes of these elements in the Earth.

   a. How much heat is produced by these elements today, considering the decay half-lives of these isotopes given in Table 6.1?

   b. How much heat did these produce 4.5 billion years ago?

   c. Calculate the total heat released by these elements over the history of the Earth.

   d. Compare the number you find in part c to the virialized energy of collapse of the Earth. You may make the oversimplification that the Earth is largely iron and silicon to obtain the mass per particle.

4. A sample of wood dated using ^{14}C is found to have been from a tree cut down 9,756 years ago, assuming the atmospheric ratio of ^{14}C to ^{12}C has always been the value we measure today. However, shortly after the analysis it is discovered that the car-

bon in the atmosphere is strongly enriched in ^{14}C due to atmospheric nuclear testing over the last 40 years. Is the sample of wood we analyzed older or younger than we originally thought? Explain why.

5. Calculate the temperature increase of hydrogen gas that has collapsed from a clump of molecular cloud light years across, to something the size and mass of the Sun.

6. Venus has no liquid water remaining on its extremely hot surface, but the water in the atmosphere has been measured to have a D/H ratio two orders of magnitude larger than that in our oceans. Much of this enrichment is thought to reflect that loss of water from Venus early in its history, with the heavy water (containing deuterium) preferentially left behind. Could one use this D/H ratio to further constrain the sources of water for the inner solar system? What else would one need to know, for example, in order to be able to determine the original D/H value for Venusian water?

7. If large bodies in the asteroid belt were the primary source of Earth's water, and comets contributed (let us say) only 10%, then what was the primary source of organics (carbon-bearing molecules) for the Earth? Use tables of the organic abundances in carbonaceous chondrites and comets to help answer this question.

8. ^{26}Al is a short-lived isotope of aluminum that decays to ^{26}Mg with a half-life of 730,000 years. There is evidence that some ^{26}Al was incorporated in rocky bodies early in the formation of planets, and that the source of the isotope was a nearby supernova. Hence, the amount of ^{26}Al may vary dramatically from one forming planetary system to another. Would having more ^{26}Al in a given planetary system lead to wetter or dryer terrestrial planets in that system? Explain your reasoning, using the model for delivery of water to Earth presented in this chapter, or other competing models.

9. Consult a geologic map for the area in which you live. What are the oldest sediments in your region? What are the key fossils that have been found? Describe in words the geologic history of your area as it is known from the rock and fossil record.

10. Do some Web research on our current knowledge of the population of bodies in the Kuiper Belt. In particular, consult the Minor Planet Center at the Harvard-Smithsonian Astrophysical Observatory to determine how many such objects are known today and how this number has grown over the past few years. How much of the sky has yet to be surveyed?

SUGGESTED READINGS

As an introduction to the formation and properties of planets, this chapter should be supplemented by texts devoted wholly to the subject of planetary science and geology. Two readable and recent efforts in planetary science aimed at upper level undergraduates, but with very different approaches, are G. Cole and M. H. Woolfson, *Planetary Science: The Science of Planets around Stars* (2002, Institute of Physics Publishing, Bristol, U.K.) and I. de Pater and J. J. Lissauer, *Planetary Sciences* (2001, Cambridge University Press, Cambridge, U.K.). For geology, Preston Cloud, *Oasis in* Space (1988, W. W. Norton, New York) cannot be topped for its grace and accuracy. A well-written collection of articles on the geology of the Earth is that edited by G. C. Brown, C. J. Hawkesworth, and

R. C. L. Wilson, *Understanding the Earth: A New Synthesis* (1992, Cambridge University Press, Cambridge). To experience excursions with field geologists read John McFee, *Basin and Range* (1981, Farrar, Straus, Giroux, New York). A graduate-level text comprised of excellent chapters on the origin of our own planet is *Origin of the Earth and the Moon*, R. Canup and K. Righter, eds. (2001, University of Arizona Press, Tucson). A series of occasional books at the graduate level, on the origin of stars and planets, is into its fourth volume: *Protostars and Planets IV*, V. Mannings, A. Boss and S. S. Russel, eds. (2000, University of Arizona Press, Tucson).

REFERENCES

Abe, Y. and Matsui, T. (1988). "Evolution of an Impact-Generated H_2O-CO_2 Atmosphere and Formation of a Hot Proto-Ocean on Earth," *J. Atmos. Sci.*, **45**, 3081–3101.

Abe, Y., Drake, M., Ohtani, E., Okuchi, T., and Righter, K. (2001). "Water in the Early Earth." In *Origin of the Earth and the Moon*, K. Righter and R. Canup, eds., p. 1. University of Arizona Press, Tucson.

Allegre, C., Manhes, G., and Gopal, C. (1995). "The Age of the Earth," *Geochim. Cosmochim. Acta*, **59**, 1445–1459.

Artymowicz, P. (1997). "Beta Pictoris: An Early Solar System?" *Ann. Rev. Earth Planet. Sci.*, **25**, 175–219.

Beckwith, S. V. W. and Sargent, A. I. (1993). "The Occurrence and Properties of Disks around Young Stars." In *Protostars and Planets III*, E. H. Levy and J. I. Lunine, eds., pp. 521–541. University of Arizona Press, Tucson.

Briceño, C., Vivas, A. K., Calvet, N., Hartmann, L., Pacheco, R., Herrera, D., Romero, L., Berlind, P., Sánchez, G., Snyder, J. A., and Andrews, P. (2001). "The CIDA-QUEST Large-Scale Survey of Orion OB1: Evidence for Rapid Disk Dissipation in a Dispersed Stellar Population," *Sci.*, **291**, 93–96.

Broecker, W. (1985). *How to Build a Habitable Planet*, Eldigio Press, New York.

Canup, R. and Righter, K. (2000). *Origin of the Earth and the Moon*, University of Arizona Press, Tucson.

Chyba, C. F., Thomas, P. J., Brookshaw, L., and Sagan, C. (1990). "Cometary Delivery of Organic Molecules to the Early Earth," *Sci.*, **249**, 366–373.

Cole, G. H. A., and Woolfson, M. M. (2002). *Planetary Science: The Science of Planets around Stars*. Institute of Physics Publishing, Bristol, U.K.

Delsemme, A. (2001). "An Argument for the Cometary Origin of the Biosphere," *Am. Sci.*, **89** (5), 432–442.

Drake, M. J. and Righter, K. (2002). "What Is the Earth Made Of?" *Nature*, **416**, 39–44.

Dreibus, G. and Wanke, H. (1987). "Volatiles on Earth and Mars: A Comparison," *Icarus*, **71**, 225–240.

Duncan, M. J. and Quinn, T. (1993). "The Long-Term Dynamical Evolution and Stability of the Solar System." In *Protostars and Planets III*, E. H. Levy and J. I. Lunine, eds., pp. 1371–1394. University of Arizona Press, Tucson.

Gaidos, E. (2000). "A Cosmochemical Determinism in the Formation of Earth-like Planets," *Icarus*, **145,** 637–640.

Greenberg, J. M. (1998). "Making a Comet Nucleus," *Astron. Astrophy.*, **330,** 375–380.

Guillot, T. (1999). "A Comparison of the Interiors of Jupiter and Saturn," *Planet. Space Sci.*, **47,** 1175–1182.

Holman, M. J. and Wisdom, J. (1993). "Dynamical Stability in the Outer Solar System and the Delivery of Short Period Comets," *Astron. J.*, **105,** 1987–1999.

Hubbard, W. B., Burrows, A. S., and Lunine, J. I. (2002). "Theory of Giant Planets." *Ann. Rev. Astron. Astrophys.*, **40,** 103–136.

Hunten, D. M., Donahue, T. M., Walker, J. C. G., and Kasting, J. F. (1989). "Escape of Atmospheres and Loss of Water." In *Origin and Evolution of Planetary and Satellite Atmospheres*, S. K. Atreya, J. B. Pollack, and M. S. Matthews, eds., pp. 386–422. University of Arizona Press, Tucson.

Jakosky, B. M. and Phillips, R. J. (2001). "Mars' Volatile and Climate History," *Nature*, **412,** 237–244.

Jewitt, D. (1999). "The Kuiper Belt," *Physi. World*, 37–41.

Kring, D. A. and Cohen, B. A. (2002). "Cataclysmic Bombardment Throughout the Inner Solar System," *J. Geophys. Res.*, **107** (E2), 10.1029/2001JE-001529.

Levy, E. H. and Lunine, J. I. (1993). *Protostars and Planets III*, University of Arizona Press, Tucson.

Lunine, J. I., Engel, S., Rizk, B., and Horanyi, M. (1991). "Sublimation and Reformation of Icy Grains in the Primitive Solar Nebula," *Icarus*, **94,** 333–343.

Malhotra, R. (1993). "The Origin of Pluto's Peculiar Orbit," *Nature*, **365,** 819–821.

Mannings, V., Boss, A., and Russel, S. S. (2000). *Protostars and Planets IV*, University of Arizona Press, Tucson.

Minster, J-F., Birck, J-L., and Allegre, C. J. (1982). "Absolute Age of Formation of the Chondrites by the ^{87}Rb–^{87}Sr Method," *Nature*, **300,** 414–419.

Morbidelli, A., Chambers, J., Lunine, J. I., Petit, J. M., Robert, F., Valsecchi, G. B., Cyr, K. E. (2000). "Source Regions and Timescales for the Delivery of Water on Earth," *Meteoritics Planet Sci.*, **35,** 1309–1320.

Nisini, B. (2000). "Water's Role in Making Stars," *Sci.*, **290,** 1513–1514.

Nuth, J. A., III. (2001). "How Were the Comets Made?" *Amer. Sci.*, **89** (3), 228–235.

Pierazzo, E. and Chyba, C. F. (1999). "Amino Acid Survival in Large Cometary Impacts," *Meteoritics*, **34,** 909–918.

Pudritz, R. (2002). "Clustered Star Formation and the Origin of Stellar Masses," *Sci.*, **295**, 68–75.

Robert, F. (2001). "The Origin of Water on Earth," *Sci.*, **293**, 1056–1058.

van Dishoeck, E. F. and Blake, G. A. (1998). "Chemical Evolution of Star-Forming Regions," *Ann. Revi. Astron. Astrophys.*, **36**, 317–368.

Whitfield, J. (2004). "Time Lords," *Nature,* **429**, 124–125.

Woolum, D. S. and Cassen, P. (1999). "Astronomical Constraints on Nebular Temperatures: Implications for Planetesimal Formation," *Meteoritics Planet. Sci.*, **34**, 897–907.

Zapatero-Osorio, M. R., Rebolo, R., Martin, E. L., Rebolo, R., Barrado y Navascues, D., Bailer-Jones, C. A. L., and Mundt, R. (2000). "Discovery of Young, Isolated Planetary Mass Objects in the σ Orionis Cluster," *Sci.*, **290**, 103–107.

Zharkov, V. N. (1993). "The role of Jupiter in the Formation of Planets." In *Evolution of the Earth and Planets; Geophysical Monograph 74,* pp. 7–17. IUGG Volume 14.

CHAPTER 7

The Thermodynamic Foundations of Life

7.1 Introduction

Thermodynamics is, in its most basic and historical form, the study of heat and its transformations to and from other sources of energy from a *macroscopic point of view*. That is, thermodynamics does not deal with the microscopic mechanisms of energy transfer. Instead, through a remarkably simple set of laws and quantities related to the energy, thermodynamics sets fundamental limits on the behavior of physical phenomena. A later development, *statistical mechanics*, connects thermodynamics to the microscopic world, but it does so by adopting a global (and hence simple) view of the behavior of the enormous number of atoms or molecules that constitute a macroscopic system. As a branch of physics, thermodynamics lacks the glamour of quantum theory and of general relativity. It is not the most favored topic of science journalism, in which field tastes tend toward the exotic. Yet the methods of thermodynamics and its sibling discipline statistical mechanics are powerful and simple, and their applicability—from engineering studies to the exotica of black holes—is remarkably broad. Indeed, thermodynamics, in its extension to systems that are not in equilibrium, is in many respects the physical underpinning of living processes.

How can a mundane subject such as energy transfer underlie the subtle mysteries of living processes? In fact, summarizing thermodynamics as the study of energy and its transformations is a bit like saying that chemistry is the study of electron transfer—words fail to capture the breadth of phenomena covered. In the transfer of energy to and from heat, in the application of temperature differences and potential differences to the production of useful work (accompanied by the production of heat), lie much of the phenomena of the cosmos. In the case of life, potential differences are usually those associated with chemical reactions, and temperature differences are a part of the defining differences between the cellular innards and the external environment. Thus, to truly understand the physical basis of living processes, we must also understand thermodynamics. Indeed, life is not a violation of the laws of thermodynamics (in spite of the claims of some who seek to characterize life as beyond natural laws); rather it is a manifestation of them. And, more profoundly, the origin of life from nonliving processes must at its foundation have been the result of processes working in conformance with these laws.

This chapter attempts to provide students with a brief but useful introduction to the concepts and laws of thermodynamics, and applied them to living processes. Thermodynamics is then used to draw the set of constraints that must have been obeyed by processes that led to life's origin from nonliving systems. This last portion of the chapter is the most speculative, but it is also the most interesting from the point of view of astrobiology, insofar as understanding the origin of life on Earth is a central goal of the field.

7.2 Important Concepts of Thermodynamics: Temperature, Energy, Heat, Work, Entropy, and the Laws of Thermodynamics

Thermodynamics is founded on four empirically determined laws having to do with energy and associated quantities. As is often the case in physics, the laws and the quantities they pertain to intertwine. Hence we introduce the laws in context and not in their numerical order. [I leave it to the reader to create a table in which these laws are neatly arrayed from zeroth (as it is traditionally labeled) to third!] To introduce the fundamental quantity—energy—it is most desirable to do so simultaneously with the *first law of thermodynamics*, which says simply that energy is conserved. In any physical process in the cosmos, of any kind, energy is neither created nor destroyed—it is simply changed into a different form. This concept was introduced in Chapter 5 in connection with fusion reactions in which matter and energy themselves were interchanged.

In what various forms does energy exist? Indeed, the concept of energy itself is not really intuitive, nor can energy be defined in a fully primitive way as we did, for example, for matter in Chapter 2. However, operationally we can think of stored energy, or potential energy, in which the storage may be chemical (locked in bonds of molecules that, upon reaction with other molecules, is released) or the result of the position of a body in a force field (e.g., a ball held above the ground). Kinetic energy, on the other hand, is energy of motion, examples of which include a ball freely falling toward the ground, a runner moving along a race track, or the random motion of a collection of atoms. Energy then, can be transferred from kinetic to potential and back and among the different forms of potential or kinetic energy. The thermodynamic formulation of energy conservation, called the first law, expresses the conservation of energy in a particular way. Let us first defines a *system* as being a particular collection of matter under consideration—an ice cube in a glass, a refrigerator, a muscle cell. Then the first law states that the change in the energy within that system is equal to the energy transferred by work plus the energy transferred by heat. Expressed symbolically, $\Delta E = Q + W$, where ΔE is the difference between the energy of the system before and after, Q is the heat added to the system, and W is the work done on the system.

There is a subtlety with respect to the terms "heat" and "work"; namely, they refer to the transfer of energy. *Heat* is the transfer of energy to or from a system associated with a temperature difference—for example, submerging a cold washcloth into a hot bathtub (or a hot washcloth into a cold bathtub, reversing the direction of energy transfer). *Work* is the transfer of energy to a system by a change in the external conditions that describe the system, and that change is enabled by a force (gravitational, electromagnetic, strong or weak nuclear), for example, moving a block of wood from the floor to a position six feet above the floor, compressing a brick in a vise to increase pressure, or placing a cube of iron in an electrical or magnetic field. Q and W describe not different kinds of energy but two fundamentally different ways of transferring energy to (or from) a system. These two ways are fundamentally different in how they change another quantity of the system, called the *entropy*, defined below. Before doing so, it will help to point out a consequence of this subtlety and give examples of the first law.

There is a *zeroth law of thermodynamics* that says that when a hot and cold body are put in contact with each other, the two will eventually come to the same temperature. Temperature is a quantity that was first measured empirically with liquids that would expand in tubes when warmed. Later, it was found that the temperature is a measure of how fast the molecules or atoms in a substance are moving around, or rotating, or vibrating, or all of the above. Imagine the *heat* in a body as the temperature multiplied by the number of ways (modes) the body stores kinetic energy (number of molecules or number of ways a molecule can turn, vibrate, or otherwise move), multiplied by the amount of energy stored in each of these modes. This is actually a good description of the content of *thermal energy* in the body, but it is not strictly correct to say it is "the heat stored by the body." Heat, as strictly defined, is an energy transfer process, and while this may seem a pedantic distinction, it is not. In the equation given above, ΔE has very different mathematical properties than do Q and W: The first is energy, and the second and third are two different means of energy transfer.

As one example of the first law, let us think first of a bicycle pump. If we define the system as the tire and the pump being used to inflate it, then energy is transferred to the system (by a human pumper) both through work done (the tire inflates as a consequence of the increased pressure) and heat added. The transfer of energy as heat is the result of the tire and the pump increasing in temperature as friction between the parts increases the microscopic thermal motions of the particles of air, rubber, metal, and plastic. Thus, the total increase in energy imparted to the pump and tire is $\Delta E = Q_{\text{friction}} + W_{\text{pumping}}$. The increase in energy of the pump and tire is offset by the chemical energy, stored as adenosine triphosphate (ATP) in the human muscles, being converted through a complex set of chemical reactions into the mechanical motion of sliding the pump piston up and down against the air pressure resistance. Of course, to be a fully accurate account, we must also look at other expenditures: loss of thermal energy from the pump to the outside cold air by heat transfer; increased breathing and hence diaphragm motion of the human pumper, which leads to energy transfer to the environment by work and heat; heat transfer between sweaty hands and a slick pump handle; slick pump sliding on the gravel road.

A second example is the stirring of a vat of liquid. Vigorous stirring by an external agent will heat the vat of liquid as surely as placing a fire underneath it. No work is done by or on the system (excluding the agent doing the stirring) if the stirring does not lead to the expansion of the liquid or its spillage out of the vat. All of the energy transfer is via heat, so that the first law in this case reads $\Delta E = Q$. (In truth, hot liquid does expand a little relative to cold, so in that sense a small amount of work is done.) The energy of the liquid has been raised purely by transfer of heat. The energy is donated by a human or mechanical pot stirrer external to the pot of liquid, and various other energy transfers may occur there—biochemical energy stored in the phosphate bonds of ATP converted to kinetic energy of the stirring arm, some kinetic energy of stirring converted into thermal energy of the hands or stirring spoon associated with rubbing or slippage of hands against the spoon, or so on.[1]

Heat and work differ in how the *entropy* changes during each transfer process. Entropy is a quantity as fundamental to the behavior of physical existence as are energy and matter. Entropy is, from a macroscopic point of view, the heat (the quantity Q), divided by the temperature. Written symbolically, with S as entropy, $S = Q/T$. Strictly speaking, however, as was true of energy as well, we can only refer to differences in the entropy from some reference or starting state, that is, $\Delta S \equiv S - S_{\text{reference}} = Q/T$. This limitation seems inconvenient, but there is a *third law of thermodynamics* that states that as the temperature tends toward absolute zero, the entropy tends to a constant, small value. The small value can be measured or calculated, but for our purposes it is all right to take it as zero. Henceforth $S_{\text{reference}} = 0$ and $S = Q/T$.

With the definition of entropy we are back to the Q quantity again, and it would seem that entropy is simply another way of describing Q, normalized by the temperature at which the transfer is taking place. By making this normalization, entropy becomes endowed with the same property that energy has: Both depend purely on the state of the system and are not dependent on how we arrived at that state. This is completely different from Q and W. Historically, entropy was formulated in this way to make a convenient quantity—quantities that are *state functions* have mathematically special properties that make them convenient to use. For now, this is as much as we need to know about entropy, but later we will see that entropy has a deeper meaning concordant with two related concepts in a system: the amount of disorder possessed by the system and the amount of information content of a system. We will further see that these in turn simply count for us the fraction of accessible *phase space*—the amount of volume, velocity, and other physical parameters—that the system actually occupies.

Unlike energy, entropy is not conserved. But a law of thermodynamics governs its behavior, and this *second law of thermodynamics* states that the entropy of an isolated system increases when that system undergoes a spontaneous change. If the system is somehow in contact with its surrounding environment, then the second law says that the total entropy of the environment plus the system must increase when the system undergoes a spontaneous change. An alternative way to say the same thing, without invoking entropy, is that it is impossible for any process, working in a cycle, to do nothing but extract heat from a reservoir and convert it into the same amount of work. This statement, which dates to the 19th century, demands that a process that takes heat and converts it to work must expend some amount of waste heat. The stipulation that the process is done in a cycle is reasonable; it says that the external parameters of the system must be returned to their original values at the end of the process. Were we to halt in the middle of the cycle, then

---

[1] In each of these cases I have finessed the question of the sign of the Q and W functions by simply stating that work done on a system or heat transferred to a system is positive. If a system does work on something else, then the change in energy of the system is given by $-W$, and likewise for $-Q$, if energy is lost from the system by heat. This is not a universal convention, unfortunately, and so sign the convention must be explicitly defined ahead of time, for readers. *Never* assume it is the same from one textbook to another!

some other agent (e.g., human muscles) must reset the system, and hence additional energy has been added.

The link between the two expressions of the second law of thermodynamics is made if we recognize that the definition of entropy is in terms of Q and that Q and W are only related to each other by the first law. This implies that entropy enters a system with the heat, but does not leave with the work done by the system. Different types of work are freely convertible into each other by the first law, and work can be converted entirely into heat, but by the second law the reverse is not true. Thus, as heat is converted to work, more and more entropy is generated in the system. Some way of removing entropy must be accomplished so that the system can continue to do work without building up an arbitrarily large amount of entropy. The removal of entropy from the system requires that the system eject heat. A classic prototype of an engine is shown in Figure 7.1. A reservoir of some material at high temperature is connected through some set of workings (pistons, etc.) to a reservoir at a lower temperature. Recall that the zeroth law of thermodynamics says that the flow of heat proceeds from high temperature to cold temperature until the two reservoirs eventually equalize in temperature. Some of the energy transfer in the form of heat, Q, can be converted into work, W, by the pistons, etc., in accordance with the first law in the engine. But not all; the second law says that some of the heat must be ejected to the cold temperature reservoir.

Were the engine to take all the heat and convert it to work, it would be 100 percent efficient. So, under the second law, what is the maximum efficiency of such an engine? If the input heat is called Q_{in}, coming from a reservoir at temperature T_{in}, then the entropy input is $S_{in} = Q_{in}/T_{in}$. Likewise that going out is $S_{out} = Q_{out}/T_{out}$. The second law says

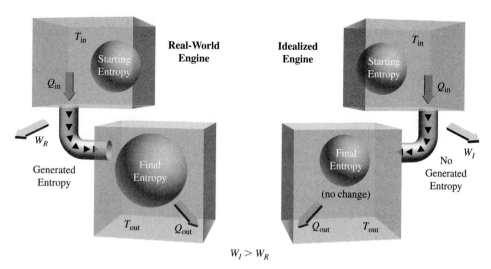

FIGURE 7.1 A schematic engine, with a hot reservoir at a temperature T_{in}, a cold reservoir with temperature T_{out}, that inputs heat Q_{in}, produces work W, and rejects waste heat Q_{out} at the bottom. On the left, the engine is shown as a real-world device; it generates some entropy during the process of turning some of the heat to work. To find the maximum efficiency of any engine, we assume the generated entropy is zero, leading to the idealized or imaginary engine on the right.

that the entropy must increase in the process, or at least it can never decrease. So the efficiency will always be worse than what we would get if the entropy were to remain *constant* in the process—a situation just on the edge of impossibility according to the second law. So our imaginary most efficient machine has $S_{in} = S_{out}$, or $Q_{in}/T_{in} = Q_{out}/T_{out}$, or $Q_{out} = (T_{out}/T_{in})Q_{in}$. But we can also express the work done by the system in terms of the heat because the engine works in a cycle, and hence total energy change is zero. So $E = 0 = Q_{in} + (-W) - Q_{out}$. (Remember the system is doing work, not having work done on it.) Hence $W = Q_{in} - Q_{out}$, and if we substitute for Q_{out} the little expression just derived, we find $W = Q_{in} - (T_{out}/T_{in})Q_{in}$, or $W = [1 - (T_{out}/T_{in})]Q_{in}$. Now the efficiency of an engine is just the ratio of the work done to the amount of heat taken in (perfect efficiency, that is, 100 percent, implies that $W = Q_{in}$). So the limiting efficiency for our imaginary engine, which just barely violates the second law by having the entropy change be zero, has an efficiency $W/Q_{in} = 1 - T_{out}/T_{in}$.

The efficiency given here is called the *Carnot efficiency*, and the idealized engine the *Carnot engine*, after the early 19th-century French military engineer who first outlined the thermodynamics of a steam engine much as was done here (but without mention of entropy per se). Table 7.1 shows that, for typical real-world heat engines, efficiencies cannot get much above 50 percent. These efficiencies are absolute upper limits—the actual efficiencies are lower, usually in the 20 to 40 percent range, because, in all the pieces used to turn the heat into work, substantial amounts of entropy are generated in accordance with the second law.

Cells cannot operate as heat engines because they maintain a nearly *isothermal* (constant temperature) environment; hence their efficiency would be essentially zero. Instead, cells convert the energy gained by sunlight or by food into usable energy and dump waste products of the reaction as well as heat. The conversion of either sunlight or food—the latter with its high-energy chemical bonds—provide much more efficient means of performing work than would the heat engine approach. (An organism operating as a heat engine at only 20 percent efficiency, with its cooling tail always submerged in ice-water at 273 K would require an operating cell temperature of 341 K, or 69°C—destructively far too high above the homeostatic mammalian temperature of 35 to 40°C.) That life can exist on highly efficient sources of energy conversion other than heat-to-work is not a violation of the second law of thermodynamics; the generation of these energy sources elsewhere (e.g., sunlight in the Sun) involves a large production rate of entropy. Indeed, the use of highly efficient sources of energy and energy transfer are the essential characteristic of life, and one that is discussed later in the chapter.

TABLE 7.1 Limiting Carnot Efficiencies for Heat-Driven Engines

Device	High Temperature (K)	Low Temperature (K)	Efficiency = W/Q_{in}	Efficiency (%)
Power plant (steam)	800	300	0.625	63
Home heat pump	400	300	0.25	25
Human cells	310	310	0	0 (see text)

Let us close this section with the one instance in which the entropy does not change in a spontaneous process, and it is, on its face, paradoxical. The second law allows the entropy to remain unchanged if the process is *reversible*, that is, if it can be taken back along the same set of steps without adding some additional energy from another source. In general, real processes are not reversible because energy is transferred in many complex ways as gears turn, boilers heat water, gasoline explodes inside cylinders, and mitochondria burn carbohydrates. These complex energy transfers cannot be undone without the action of an outside agent. A reversible process is one that proceeds in unimaginably tiny steps, each of which has the system close to the *equilibrium* state. What is equilibrium? It is a condition in which a body or a system is unchanging, which means that it is completely isolated, or it is in contact with an environment that is exactly the same in temperature, pressure, or electric field. Equilibrium states are unchanging states—nothing happens. And, as you might have guessed by now, such states do not exist anywhere in the real world. We can imagine situations very close to equilibrium, and it is these situations that allow processes as close to reversible, and hence as close to constant entropy, as possible. But they are never truly so. The limiting case of unchanging entropy in a reversible process, one that "moves" from one unchanging state to another, is intrinsically paradoxical. It illustrates that the thermodynamics of which we have been speaking, or *equilibrium thermodynamics*, provides limits on real physical systems—powerful and important limits, but limits nonetheless. Artfully employed, equilibrium thermodynamics provides results that are often extremely close to what is observed. And its simplicity makes the theory extremely attractive. However, to delve deeper into the physical meaning of entropy requires moving away from macroscopic, equilibrium thermodynamics, to the microscopic world sketched in Chapter 2.

7.3 A Closer Look at Entropy: Maxwell's Demon, Information Theory, and Statistical Mechanics

James Clerk Maxwell was a mid-19th-century Scottish scientist who developed fundamental results in thermodynamics, in *kinetic theory* (which describes the evolution of systems at the microscopic level), and, most importantly, in electricity and magnetism. Maxwell developed a hypothetical system that he thought might violate the second law of thermodynamics. This system, that of *Maxwell's demon*, seems not to violate the second law but to open the door to a much deeper understanding of entropy.

Imagine a box containing molecules of some gas, which can be divided into two compartments by use of a partition (Figure 7.2). The gas in the box is at a constant temperature (*isothermal*); that is, the distribution of speeds of the molecules is well defined and corresponds to a single temperature. Now we invoke a device, Maxwell's demon, to assess the speeds of individual particles, and we slide the partition closed to make "left-hand" and "right-hand" regions. A particle in the left-hand box moving faster than average and moving to the right, will trigger the demon's opening of a trapdoor to allow it into the right-hand side; the door is then closed. A particle in the right-hand box, moving slower than the average for all the molecules, will trigger the opening of the door if it is heading left. The door is then closed so that this particle is trapped on the left-hand side. Every time either

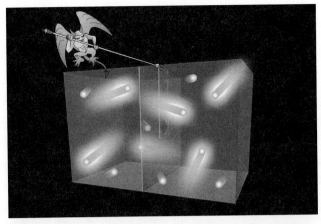
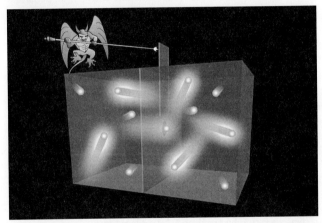
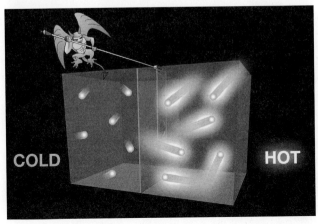

FIGURE 7.2 Maxwell's imagined experiment to violate the second law. (a). Setup of the box with partition and gas of equal temperature in each side. (b) When the demon sees a molecule with higher-than-average speed moving toward the right-hand side, it opens the partition to admit it into the right-hand region. When the demon sees a molecule with lower-than-average speed moving toward the left it opens the partition to admit it into the left-hand region. (c) In the end, the left- and right-hand regions have gases with very different mean speeds, hence temperatures, and work can thus be extracted from this system which was originally at constant temperature.

of these operations is done, the left-hand box acquires a speed distribution with a mean lower than that of the right-hand side. The shapes of the distributions remain like those in Figure 7.3 because collisions among the particles will adjust the distributions. But the door prevents the right and left regions from mixing, only being opened when a slower-than-average molecule from the right heads left, and vice versa for a faster-than-average particle. The result is that the left-hand region cools and the right-hand region heats up; a temperature difference between left and right sides has been created. We can make a heat engine to extract work from the temperature difference between the two regions, and hence we have generated work out of an initially constant temperature box, with no other source of energy. This is a violation of the second law of thermodynamics.

Immediately, careful readers will object that energy has been expended. The demon has been furiously raising and lowering the partition at the appropriate time to move molecules around. The weight of the partition, friction among parts, and movement of the demon all demand energy. So is this not where the work comes from? Maxwell, and later examiners of this thought experiment, argue that, in principle, we can make an operable partition so small and efficient that the energy expended in the raising and lowering approaches zero, and likewise for the demon. In other words, were we clever enough, we could create a

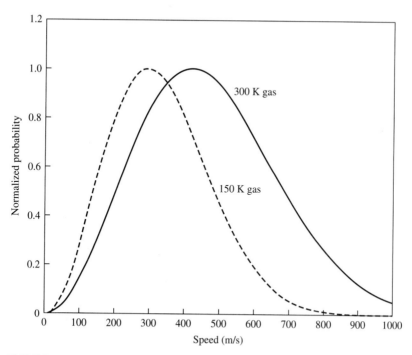

FIGURE 7.3 Graph showing the equilibrium distribution of speeds of nitrogen molecules, in the gas phase, with a temperature of 300 K (solid line) and 150 K (dashed line). The speeds are in meters per second, and the probability is normalized to unity at its peak for each temperature.

substantial temperature difference between two halves of an initially isothermal box with a minimum expenditure of energy.

This apparent violation of the second law was solved a century later, but only bit-by-bit. The solution lies in the demon itself. We might accept that to slide the partition open and closed requires arbitrarily little energy. But the demon must *decide* when to open or close the partition. It must make a measurement, on which basis the partition stays closed or open. Now, this measurement might or might not take energy—a photon has to be scattered by an appropriately moving molecule to trigger the raising of the partition, for example. Perhaps this measurement could be economical enough to avoid expending more energy than is gained with each emigration of an appropriate molecule from one side of the box to the other. What cannot be economized is the information gained or lost by the demon. If we think of it as a computer, with a single register (the most economical), the triggering of the partition is the result of a definite bit being put in the register—say, 0 changing to 1. This, in principle, takes vanishingly small amounts of energy. But then the register must be cleared—and information is lost in the clearing. *This loss of information corresponds to an increase in entropy of the system.*

The insight that loss of information is equivalent to increase in entropy came in the mid-20th century as the statistical mechanical definition of entropy, which is introduced later, was recognized to be equivalent to entropy as a measure of information or uncertainty. There is no finessing this change in the informational entropy. Once the demon's register is cleared, it gains a fundamental unit of information entropy related to the new uncertainty in the register being 0 or 1. Were the register in a definite known state, then the uncertainty is zero. How do we relate the fact that the computer is in one state only to our assignment of the information entropy being zero? Information theory showed that the correct way is the simplest: The information entropy is the natural logarithm of number of states the system might be in. The natural logarithm is a mathematical function that is the inverse of the exponential function "e" graphed in Figure 2.12. The symbol $\ln(x)$ is the natural logarithm of x, where $\ln(1) = 0$, $\ln(10) = 2.3$, $\ln(100) = 4.6$, and $\ln(e) = 1$, where $e = 2.718\ldots$. Then $S_{info} = \ln(1) = 0$. When the register is cleared, instead of the register being a definite value, it might be one of two values. Hence the $S_{info} = \ln(2)$. The increase in information entropy from clearing the register is $\ln(2) - \ln(1) = \ln(2)$.

The reader might endow the computer with two registers rather than one, or a million registers. But eventually these registers fill and must be cleared—and when a million (10^6) registers clear, the information entropy increase is $10^6 \ln(2)$. Stopping the experiment before the registers fill does not work either, because the memory had to be cleared initially anyway. If the memory is not cleared, then whatever sequence of 0's and 1's were there before will create a sequence of partition openings and closings that bear no relation to the trajectories of the gas molecules in the container. So, however we arrange the problem, the movement of a molecule from one side to the other requires an increase in the information entropy of $\ln(2)$.

What we have called the information entropy is, in fact, the thermodynamic entropy with a scaling factor removed. That this is so was demonstrated when the entropy was derived early in the 20th century using the techniques of statistical mechanics. Statistical mechanics, which is not discussed in depth in this book, treats the motion of a system of atoms or molecules not individually, but in averages. And it does this in *ensemble* averages. Imagine a particular system—say, the box full of gas described earlier, *but with the door in the partition closed*—as having a particular set of macroscopic properties, temperature and pressure of the gas in this case. But the macroscopic temperature and pressure are, respectively, the result of the distribution of speeds of the molecules in the gas and the rate at which molecules strike the walls of the box (Figure 7.3). It does not matter which molecule has which speed or which two molecules strike each other at a given time and place—as long as the distribution of speeds and rate of collisions of the system as a whole are known (i.e., the temperature and pressure). So now we can imagine different versions of the same system with particular molecules in different positions or with different speeds in these different versions. *From a macroscopic viewpoint, all these different versions are the same if they have the same distribution of speed and collision rate in the box.* We also must, of course, use the same kind of molecules in each version (e.g., nitrogen) and have them occupy the same total volume (that of the box). What we have done is to create an *ensemble* of model systems each of which has the same temperature and pressure.

Now we can go a step further. Because the motions of the molecules in the box vary with time, there may be brief moments when there are more collisions on one side of the box than the other or when the distribution of speeds is briefly not what is shown in Figure 7.3. These fluctuations must happen in real systems. If we were to have only two or three molecules, the fluctuations in the ensemble of the system would seem huge—one system might have the three molecules each hitting a different side, while another has them all hitting one side. But with 10^{24} molecules in the box (corresponding to, e.g., 3 grams of gas composed of molecular hydrogen, H_2), the fluctuations are much smaller. It does not matter to the macroscopic parameters that one of the ensembles might have an excess of 50 out of 10^{24} molecules in the left corner of the box, while another has an excess of 25 molecules there. Extreme fluctuations—having all the molecules suddenly occupy just the left half of the box without aid of the trapdoor—are so improbable that we expect never to see this. (If we were to spray some perfume into a corner of a room, the expectation is that it will spread throughout the whole room and never collect itself back into the corner without the work of an external agent.)

In the statistical mechanics of macroscopic systems made up of large numbers of particles, the probability of finding a system in its most probable state is extremely sharply peaked around that state. This will be the equilibrium state of the system—the state of maximum entropy. In it, externally measured quantities such as temperature and pressure are well defined and do not change unless an external agent does work on, or adds heat to, the system. If the system were changed somehow, it would reach a new equilibrium, but the manner in which each of the systems in the ensemble did so might be very different. Such a change in conditions gives us a clue as to how to actually count the number of possible states of the system.

Suppose again we have our box, but now the door is closed, and all of the gas is in the left-hand portion. We have a definite volume that the gas occupies. Now we open the door, and, for an instant, the gas remains in the left-hand portion but has twice the volume available to occupy. Of course it will quickly fill the whole box, with a concomitant change in internal parameters (temperature and/or pressure). The second law tells us that this spontaneous change must be accompanied by an increase in entropy. So, by increasing the number of states the system can be in (in this case, the possible volume in our local corner of the universe that the gas molecules can be in), we have increased the entropy.

Might the entropy be, in reality, a measure of the number of possible states the system can find itself in, consistent with the particular parameters (temperature, pressure) of the system? Indeed it is.

Let us calculate the increase in entropy when a gas is allowed to expand from half the volume of the box to the full volume, by opening the door. We isolate the box from the outside world. Hence, in opening the door, no change in the internal energy of the gas is achieved. So in the first law, $\Delta E = 0$, hence $0 = Q + W$. Now Q is $T\Delta S$, and ΔS is what we wish to calculate. What is the work done? Well, the work is the displacement against a particular force. The force is the pressure P, the displacement is the increase in volume ΔV in the gas going from filling half the box to filling all of it. So the work done by the gas is $P\Delta V$, and since the first law was defined in terms of the work done *on* the gas we put a negative sign on it, $-P\Delta V$ (see footnote 1). Then $0 = -P\Delta V + T\Delta S$, or doing a bit of rearranging $\Delta S = \Delta V P/T$. We can go farther, if we assume our gas to obey the perfect gas law that we learned in high school and introductory college science classes. This says that P/T is just proportional to $1/V$, the proportionality being the Boltzmann constant k. So $\Delta S = k\Delta V/V$. Now to do this calculation right, the "deltas" must be very small, infinitesimal in fact, and to get the bulk change in entropy we must perform the summing operation called integration. Let us rewrite the equation above as $dS = k\,dV/V$, where the "d" symbol indicates an infinitesimally small amount of the quantity. When we integrate the dS equation, which just means that we are finding the area below the curve defined by $dS = k\,dV/V$ (see Figure 7.4), we obtain the entropy increase $S_{expand} = k\ln(V_{end}/V_{start})$. The logarithmic function appears in calculus when we integrate the delta or *differential* of a particular quantity divided by the quantity itself, as we did for the decay equation in Chapter 3. Because the volume change is a factor of 2, we can write $S_{expand} = k\ln(2)$.

The entropy as a measure of the number of states accessible to a system looks exactly like the information entropy except for the Boltzmann factor k. But this constant k is just that—a constant—and could not have been intuited the way we introduced the information

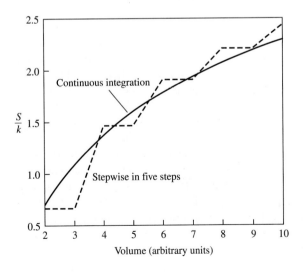

FIGURE 7.4 Plot of the entropy (divided by Boltzmann's constant, k) as a function of the volume (plotted in arbitrary units). The solid line is the result we obtain from integrating $dS/k = dV/V$ to obtain $S/k = \log V/V_o$, where V_o is set equal to 1. The broken line is obtained by simply adding up $\Delta V/V$ from the starting point $V_o = 1$; five steps were used. The broken line is a pretty good approximation of the smooth solid line, illustrating that the operation of integration in calculus is an "adding-up" process along a curve given by the *integrand*, the function to be integrated.

entropy. So we suspect that the information entropy, multiplied by k, really is the same as the statistical mechanical/thermodynamic entropy counting the number of states. And we can show this is true if readers will allow that the second law does indeed hold. Let us return to our Maxwell's demon and its box, and let us focus now on a single molecule bouncing around. The action of the demon is to take a molecule that might be rattling around either portion of the box and putting it on one definite side or the other—the molecule has been restricted to half the original volume it could have roamed. From our calculation above, this means that the entropy has declined by $k \ln(2)$.

However, for the demon to perform this function, it must again have made a definite decision and then cleared its memory for the next action. This implied an increase in information entropy of $\ln(2)$ multiplied by some constant. Because no work has been done by an external agent, the total entropy of the system could not have decreased—it must have either remained the same or increased (again we assume the demon has no energy transferable as work to contribute to the system). For the total entropy change to be zero, the proportionality constant in front of the information entropy must be Boltzmann's constant k—that is, the information entropy *is* the entropy of thermodynamics and statistical mechanics. *These entropies are one and the same.*

Although the correspondence just identified seemed to rely on the gas in the box being a perfect gas, more elaborate considerations show that the conclusion is a general one. And it is a conclusion that not only illuminates what happens to boxes equipped with partitions and tiny demons. It is a conclusion that informs the entire working behavior of physical existence as we know it—the irreversibility of all things, the arrow of time. Everything in our existence is tied to, and informed by, a seemingly inexorable progression of time. All things have a definite direction to them. Exhaust comes out of the back of a car, and gasoline is depleted as the car drives forward; the exhaust does not go back in and make gasoline as the car rolls in reverse. The things we own and treasure, and our own friends and family, age inexorably over time, become chipped and worn, fade with the days. The reverse never happens. Although movie film and videotape can be run backward, the resulting effect never seems to quite correspond with what we see in reality. Yet the basic laws of mechanics (be they classical or relativistic), and the comparable laws governing electromagnetism and nuclear forces, do not have an explicit dependence on the direction of time. A film of a high diver splashing into the pool, run backward, still obeys the laws of classical mechanics, but in real life we will never see such a diver levitate herself out of a shrinking splash of water to descend gently in a supernatural fashion back onto the high platform.

The arrow of time remained a puzzle even through the late 19th century when the concept of entropy as a measure of heat was developed. By the mid-20th century, the full implications of entropy as a measure of the number of states accessible to a system began to be realized. A system in equilibrium, which is the macroscopic state of maximum entropy, will not move in any direction—there is no arrow of time in this unchanging ultimate stasis. But all physical systems interact with each other, and each interaction represents, at some level, the equivalent of our proverbial trapdoor being opened. More and more microscopic states become accessible to physical systems over time, which means that the entropy increases. Every spontaneous process increases the number of available states. Each process that imposes a restriction on the number of states of a system decreases that system's entropy. But to accomplish such an imposition requires an agent that

TABLE 7.2 Entropy per Energy Content of Various Systems

Energy Source	Entropy Generated per Unit of Energy Released (sunlight = 1)
Reactions converting one element to another (hydrogen fusion reactions in the Sun)	10^{-6}
Internal thermal energy in the Sun	10^{-3}
Sunlight received at the Earth	1
Typical chemical reaction (bonding of elements)	1 to 10
Heat dumped from a heat engine	100

Source: Adapted from Haynie, 2001.

expends energy, and that expense—however close to reversible it might be—is never quite reversible and so its own entropy increases.

Finally, heat and entropy are two sides of the same coin. Because the sum of work and heat are conserved for a given energy change (first law), processes that produce a large amount of entropy for a given amount of energy released are less able to do work than those that generate less entropy for the same amount of energy. Table 7.2 illustrates the broad range of useful work that can be obtained from different energy sources.

It is often stated that the entropy also expresses the degree of order of a system. In the sense that an ordered system is restricted to many fewer possible microscopic states than a disordered one, this is indeed true. Likewise, as the information content of a system increases, the number of accessible states decreases. A line of decipherable English text in this book requires the restriction of the ordering of letters to make decipherable words and of words to make meaningful sentences. But there are vastly more ways to type letters, such as qqq 111111111111111115uuuuuuuu, with such random (or pseudo-random) assemblages conveying much less information.[2] Entropy is both a measure of disorder and a measure of information content, and both are aspects of the general definition of entropy, which is the number of microscopic states accessible to a system.

The unfolding or evolution of the universe with time involves a gradual increase in the number of accessible states to all of the matter in the cosmos, with the consequent increase in entropy. This entropic increase seems exactly opposite to the apparent complexification of the structure of the universe with time that we explored in Chapters 5 and 6, from primordial gas to galaxies, stars, planets, and then (at least on Earth) life, with its continuing evolution into different forms. But indeed, throughout all of this the total entropy content of the universe is increasing—stars fuse hydrogen to form helium, generating energy and entropy, and that hydrogen is no longer available to generate additional useful energy once fused into helium. Stars pump heavy elements into the interstellar medium, and these be-

---

[2] Allowing my son's pet rat to run on the keyboard generated this particular example.

come planets, which are themselves illuminated by the light of new generations of stars that further consume hydrogen. Nonetheless, self-organizing systems do arise, and in isolation they seem like nothing else but direct violations of the second law. They are not, because their formation requires the consumption of large amounts of energy generated elsewhere, energy that may have a very low entropy-to-energy ratio but that was made available nonetheless with the concomitant production of entropy. The mistake is in looking at such systems in isolation.

Life, for example, is a system open to the flow of both energy and matter, and its continued existence depends on the entropy-generating processes of stellar fusion and (at the seafloor vents) emission of hot volcanic gases into the oceans. The large-scale system is generating entropy; that part of it we call life is self-organizing and self-complexifying through the transformation of energy and material resources. And, as energy and matter flow through the living systems of the Earth, entropy is generated there as well. The next section examines a few example processes in living cells that illustrate the adherence of life to the laws of thermodynamics.

7.4 Thermodynamics of Living Systems

To apply thermodynamics to processes inside cells requires introducing yet another concept, that of *free energy*, and specifically, the *Gibbs free energy*, G. The free energy might be better referred to as the available energy for work. It is a way of combining the parameters of a given system—temperature, entropy, pressure, volume—and adding or subtracting these from the system's own internal energy so as to provide a measure of how much work the system can do. The need for the Gibbs free energy, named after the 19th-century American scientist J. Willard Gibbs, arises because the differential form of the first law, with the work expressed in terms of the change in volume and the heat as the change in entropy, is

$$dE = TdS - PdV \qquad (7.1)$$

Now this equation implies that the two controlling variables in the problem are the entropy and the volume. These two variables are inconvenient for biological systems because nearly all cellular processes operate at constant temperature and pressure. Further, even in nonbiological systems the first law equation (7.1) is difficult to work with because, although the change in entropy of the system can be measured straightforwardly, that of the environment cannot.

Therefore, it is convenient to express the first law in terms of differentials of pressure and temperature instead of entropy and volume, and the resulting expression will specify the change in Gibbs free energy. To do so requires some algebra, which is useful to write out because it shows that the relationship between E and G is straightforward. We want dG to have two terms, like E, but with differentials dT and dP. To accomplish this, let $G = E + PV - TS$. Then $dG = d(E + PV - TS) = dE + PdV + VdP - TdS - SdT$ (the terms obey the algebraic "chain rule" $d\{xy\} = xdy + ydx$). But equation (7.1) says that $dE = TdS - PdV$. If we plug equation (7.1) into our equation for dG, we find

$dG = TdS + PdV - PdV + VdP - TdS - SdT$. Four of the terms cancel each other out, hence

$$dG = VdP - SdT \qquad (7.2)$$

This is exactly what we desire: a function that behaves like energy but for which the variables that can be independently adjusted (or held fixed) are pressure and temperature. Now, for a system in equilibrium, neither the pressure nor the temperature change, so by definition the change in the Gibbs free energy is zero. In this sense again it behaves both like the energy and the entropy for which $dE = dS = 0$ at equilibrium, but this means different things for the energy versus the entropy. The entropy of a system at equilibrium will take on its maximum value, while the energy is a minimum. Which way does the Gibbs free energy go? Looking at equation (7.2), you might guess it must be the opposite of the entropy, and indeed *in a system's equilibrium state the Gibbs free energy is minimized*. But another important property of the Gibbs free energy is that it represents the maximum amount of work that can be done at constant temperature and pressure by a process, taking the system from a nonequilibrium to an equilibrium state. It is therefore a measure of the limit to which cellular processes can do useful work—hence sustain life—as opposed to simply releasing energy in the form of heat.

A nonbiological example of the application of the Gibbs free energy is the melting of water. The conversion from one phase to another (in which a *phase* of a substance is a region of uniform properties) occurs because the Gibbs free energies of the two phases have different dependencies on temperature and pressure. Let us leave the pressure alone and focus on the effect of temperature—in effect just looking at what happens, for example, to a frozen patch of ice as the temperature warms up during the day. If the pressure is fixed, and we also assume that the entropy does not depend much on temperature near the freezing point of water, then G is proportional to $-ST$. Now because entropy is a measure of disorder, we might guess that solid water (ice) has a smaller entropy than liquid water, since the former exhibits a well-ordered crystalline structure, and this is correct.[3] Therefore, as shown in Figure 7.5, the Gibbs free energy of liquid water will fall more steeply than that of ice as the temperature rises, because the proportionality constant of the dependence on temperature, which is the entropy, is greater in the liquid. Hence, even though the solid is the most stable phase at low temperature (its Gibbs energy is lower than that of the liquid), the Gibbs energy of the liquid will cross and then become less than that of the solid with rising temperature. This occurs at 0°C (273 K), the melting point of water.

Adding antifreeze to water will depress the freezing point, largely for one reason: By having two components in the liquid phase, the entropy of the liquid is increased (the number of accessible states is larger). Because most antifreeze substances—glycol, ammonia—do not dissolve as fully in ice as in liquid, the same is not true for the solid phase. Hence the entropy of the liquid phase is increased, the Gibbs free energy of the water in the liquid phase is lower, and for this reason it crosses that of the solid at a lower temperature than for pure water (again, see Figure 7.5).

---

[3] Of the liquid phases of most common materials, water is the most structured because of the way the hydrogen atoms from one water molecule attract the oxygen on another. In spite of this, the crystalline form of water indeed has the greater ordering and, hence, lower entropy.

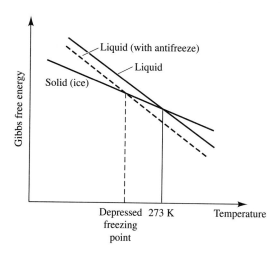

FIGURE 7.5 Sketch of the Gibbs free energy of liquid and solid water as a function of temperature. The melting point of water is where the two cross. If the liquid contains a second substance, the entropy is increased and the Gibbs energy is lowered (dashed line).

With the Gibbs free energy as our determinant of the equilibrium, or preferred, state, we have the power to go beyond the melting of ice to chemical reactions—and, therefore, biochemical reactions. What happens in a chemical reaction? Photosynthesis is not a single chemical reaction, but the long chain of events can be summarized as $CO_2 + H_2O + h\nu \rightarrow CH_2O + O_2$, where carbon dioxide is CO_2, water is H_2O, molecular oxygen is O_2, the unit sugar is CH_2O, and $h\nu$ is the energy absorbed from a photon of sunlight (Planck's constant times frequency, yielding units of energy). In a chemical reaction individual elements are moved from one molecule to another. Analogously, if one were to have two components in a liquid, as in our previous example, different amounts of the two might move into solid phases as the temperature is lowered. To handle these complications in the context of thermodynamics, we extend the definition of Gibbs free energy by adding a term, thus

$$dG = VdP - SdT + \mu dN \qquad (7.3)$$

The new term consists of N, which is the number of particles present (atoms, molecules, whatever) and μ is the *chemical potential* of whatever substance we are considering. In spite of the fact that μ appears with S and P as whole quantities rather than infinitesimals, it is more akin to the temperature and pressure (T and P) because it does not depend on the size of the system. It is a quantity that measures the chemical pressure for a molecule or atom to move out of one phase into another. Notice that it exists in a positive term in the Gibbs energy expression. The more molecules of a given substance in a particular phase, the higher the Gibbs energy of the phase. Hence the tendency is greater for a phase transformation or ejection of some of the particles into another phase, which lowers the Gibbs energy. The chemical potential is the factor that says how great a change in G is effected by a change in N.

Unlike T and P, which we can sense with our bodies, there are no senses that directly measure μ. Hence, it seems an arcane concept. But a mundane analogy can be drawn, that of a crowd in a room. Pack in ever more people—increase N, that is—and eventually some will want to leave. How big does N have to be before some people move into a different room? Well, if the majority of people are attractively sexy in the way they look or smell, you might get a very high value of N to remain in the room—μ is low. But if the crowd is predominantly, well, suffering from serious cases of deodorant failure, you might imagine the room would clear before N gets very large—μ is large in this case. Molecules or atoms do not use deodorant, but their ability to remain in a liquid solution or be part of some molecule depends on how well they "fit" in terms of size, bonding strength, and so forth.

Now you might object that equation (7.4) does not contain more than one μ—so how do we deal with multiple components? The answer is by extending the equation to as many different components as are present:

$$dG = VdP - SdT + \mu_1 dN_1 + \mu_2 dN_2 + \mu_3 dN_3 + \mu_4 dN_4 + \mu_5 dN_5 + \cdots \text{etc} \quad (7.4)$$

where each of the components is labeled by subscripts. Then the condition for equilibrium of a system again requires minimizing the Gibbs free energy, which means making dG equal to 0. To do this requires that the chemical potential for each of our components 1, 2, 3, 4, 5—separately—be equal across the various phases or compounds in which they exist. For water freezing in the presence of dissolved ammonia, it means that the chemical potential of water in the liquid equal the chemical potential of the water in the solid phase (ice), and likewise for the ammonia. The amount of each component in each of the phases is adjusted, as well as temperature and pressure, to accomplish this.

In chemical reactions, where we can write out how much of this or that molecule or atom goes into each compound, this leads to a simple result. For example, to convert nitrogen (N_2) to ammonia (NH_3) in the presence of hydrogen (H_2) and in the absence of oxygen, a reaction such as one nitrogen molecule plus three hydrogen molecules transformed into two ammonia molecules is plausible (but in fact schematic only):

$$N_2 + 3H_2 \rightarrow 2NH_3 \quad (7.5)$$

The conditions on the chemical potentials then would combine into a single equation that reads

$$\mu_{\text{nitrogen}} + 3\mu_{\text{hydrogen}} - 2\mu_{\text{ammonia}} = 0 \quad (7.6)$$

If we know the chemical potentials at some particular temperature and pressure, we can work out how much nitrogen and hydrogen we can force into the ammonia.

How can we figure out the chemical potentials? For perfect gases and simple solutions, the chemical potential of a substance is simply proportional to the logarithm of the concentration of that substance in the liquid solution or the gas. That is, μ_{ammonia} is proportional to $RT \ln[\text{ammonia}]$, where [ammonia] is meant to be the concentration of the ammonia in the gas or solution, T is temperature, and the constant R is 8.3 J/mole/°C. There are two caveats: First is that we must always add a "standard-state" chemical potential to this expression. Like entropy, chemical potentials are computable only as differences from some standard point. Unlike entropy, we cannot cheat and invoke a law to say that the standard state is at zero temperature and the chemical potential there is zero. So someone has to measure standard-state chemical potentials for us, and this is, by agreement, done under the easiest conditions possible: room temperature and pressure (for substances unstable under those conditions, measurements are done at other T, P and extrapolated). Let us call this standard state μ_o and worry about it no further. Second, not all gases are perfect or solutions ideal. So, in fact, we must account for this with another factor, γ, which is peculiar to each component under the given conditions of T, P and the abundance and type of other substances. It cannot be reliably conjured up from theory and is usually experimentally determined. Then

$$\mu_{ammonia} = \mu_{o,ammonia} + RT \ln \gamma \tag{7.7}$$

but we will also neglect γ for our discussion.

Thus, with an ideal version of the chemical potential being the logarithm of the concentration of the substance in gas or solution, we can take equation (7.6) and rewrite it as

$$RT \ln\left(\frac{[\text{nitrogen}][\text{hydrogen}]^3}{[\text{ammonia}]^2}\right) = 0 \tag{7.8}$$

or, for a general chemical reaction where we just symbolize components as A, B, C, D, and their relative amounts as a, b, c, d:

$$RT \ln\left(\frac{[A]^a[B]^b}{[C]^c[D]^d}\right) = 0$$

The relative amounts end up as powers because of the way a logarithm works: $2 \ln(n)$ can also be written as $\ln(n^2)$.

The quantity in parentheses, $([A]^a[B]^b/[C]^c[D]^d)$, is called the *equilibrium constant* of the reaction, and is usually written K_{eq}. It is in fact just the ratio of the speed of the chemical reaction going forward to that of the same reaction going backward (the *reverse* reaction). It is defined *at equilibrium*—that hypothetical point of no change. But it crucially tells us how much of the different components we actually will have in the soup when reactions do grind to a halt as equilibrium is achieved. For this reason it is a key concept in chemistry and biochemistry.

One can easily tell how temperature and pressure will affect equation (7.5); that is, what must be done to the pressure to make more ammonia and have less hydrogen and nitrogen. The Gibbs free energy, which goes up when the pressure goes up, must be minimized. The multiplicative factor for the pressure is the volume. The more molecules there are, the higher the volume. So if we crank up the pressure in a gas, the Gibbs free energy will go up unless there is a change of state that the system can achieve to at least reduce the effect of the pressure rise. Taking four particles (one nitrogen and three hydrogen molecules) and transforming them into two particles (two ammonia molecules) can most dramatically reduce the volume. Thus, as the pressure increases, more ammonia will be made at the expense of hydrogen and nitrogen. But when four particles are converted into two, the number of states in the system is reduced, and hence so is the entropy. Reducing the entropy raises the Gibbs free energy—against the principle that systems evolve to minimize G. Therefore, a reaction that drops the entropy is only preferred at low temperatures, that is, where the ST term is low to begin with. Hence at a given pressure, to get more ammonia, one must lower the temperature.

Let us return to the formulation we have painstakingly developed for chemical reactions and apply it briefly to three cases in biochemistry. (For an elaboration of these applications and many more, see the excellent book on thermodynamics in biology by Haynie, listed in the References.) Recall too, as we go along, that the results of our formulation—chemical potentials and equilibrium constants—are merely a way of conveniently expressing two fundamental laws: that energy is conserved and transferred through two distinct kinds of processes, heat and work, and that an isolated system reduces its ability to perform work when spontaneous changes occur.

The first case is that of ATP, adenosine triphosphate, which was introduced in Chapter 4. This molecule has a broad range of roles in energizing cellular metabolic processes that would not occur spontaneously, permitting muscle contraction and transport of ions against a concentration gradient. All of these processes involve generating useful work, or moving systems *away* from equilibrium (toward higher Gibbs free energies). To accomplish these things, ATP must release energy, and it does so through a chemical reaction that liberates a phosphate group:

$$ATP + H_2O \rightarrow ADP + HPO_4 \qquad (7.9)$$

In this equation ADP is adenosine diphosphate (one less phosphate bond than ATP), and HPO_4 is the phosphate group (this group may be attached typically to another molecule such as an alcohol, which is not shown here). To understand how much energy is released we can analyze this reaction by writing the chemical potentials as in the model equation (7.7):

$$\mu = \mu_o + RT \ln\frac{[ADP][HPO_4]}{[ATP]} \qquad (7.10)$$

The water that reacted with the ATP can be left out of equation (7.10) because its abundance and that of alcohols or other compounds to which the phosphate is attached are adjusted by other reactions in the cell. Chemical reactions and membrane pressure are tightly controlled in cells so that temperature and pressure remain close to constant. So, the chemical potentials directly reflect the Gibbs free energy changes in the reactions; hence, from equation (7.10),

$$\Delta G = \Delta G_o + RT \ln\left(\frac{[ADP][HPO_4]}{[ATP]}\right) \qquad (7.11)$$

which expresses the energy released in the reaction as a function of the amounts of the reactants and products involved. We must measure the standard-state Gibbs free energy change; it is $-31,000$ J/mole. The negative sign indicates that energy is released. The term "mole" refers to a standard number of molecules—6×10^{23} (Avogadro's number), which is more conveniently macroscopic than writing the energy change per individual molecule. In most cells ATP and ADP are about equal in abundance to each other and the phosphorous, at concentrations such that $\ln([ADP][HPO_4]/[ATP]) = -4.6$. Then at typical cellular temperature the RT term in equation (7.11) is $-11,000$ J/mole (8 J/mole/°C $\times$ 298 K $\times$ -4.6). Hence the total free energy released to the cell is $-(31,000 + 11,000) = -42,000$ J/mole. This is a substantial amount of energy, but in skeletal muscle, ATP is 10 times more abundant than ADP, the log term is even more negative at -7, and $\Delta G = -(31,000 + 17,000) = -48,000$ J/mole. Where does all of this energy come from? In general, it must come from an energy source with comparable or lower entropy per energy value; otherwise, the cellular energy transfer system would be hopelessly inefficient. And, indeed, it comes from sunlight—through the production of carbohydrates, through efficient electron transfer in plants, and then through metabolism of the resulting carbohydrates to "charge up" ADP (and its even more phosphate-deprived cousin, adenosine monophosphate, or AMP) to ATP.

The second case is the folding of proteins, which is crucial to the enzymatic action of these polymers of amino acids. One way to understand the conditions under which proteins will fold or unfold is to apply the thermodynamic edifice we have constructed. Consider the folding and unfolding to be essentially equivalent to a chemical reaction: Folded → Unfolded or Unfolded → Folded. Then we can write the equilibrium constant K_{eq} = [Unfolded/Folded]. What determines the value of the equilibrium constant? Interestingly, in describing the folding process as one involving an equilibrium constant, we implicitly assume that it is determined by some relative values among Gibbs free energies. And remember that G acts like an energy, which means (looking back toward the beginning of the chapter) that ΔG (like ΔE) depends only on the starting and final conditions—not on the path from start to finish. That is, the proteins fold into a particular conformation not because of any particular process by which a folding event occurs or the condition of the protein and its environment during a folding or unfolding event—but because of its intrinsic structure. That intrinsic structure is simply the sequence of amino acids that make up the protein (Chapter 4). This view of protein folding is generally accepted and is fundamental in understanding the chain of relationships from the DNA base sequence to protein function in cells.

Finally, let us briefly look at another "reaction," that of the conversion of double-stranded to single-stranded DNA. Here we can summarize the conversion of double- to single-stranded as Double → Single and vice versa as Single → Double. Then the equilibrium constant is K_{eq} = [Single]/[Double]. However, the stability of the doubled DNA depends on the stability of each of the base pairs, so that $K_{eq} = K_{A-T}^n K_{G-C}^m$ where n, m express the number of each type of base pair. (See Chapter 4 for a discussion of the four nucleic acid bases in DNA.) Interestingly, the equilibrium constant for $G–C$ is much larger than for $A–T$, and hence it is the former that determines the stability of the double-stranded DNA. Chemically, this is a result of the larger number of stabilizing hydrogen bonds that $G–C$ provides versus $A–T$. Hence, the more $G–C$ pairs, the higher the temperature at which DNA thermodynamically splits into single strands. For 20 percent of the pairs being $G–C$, that temperature is 80°C; for 80 percent of the pairs being $G–C$, it is 100°C.

Now, it would be a disaster for the role of DNA as an information-bearing molecule if its replication or transcription processes, which involve unzipping the two strands, were in fact thermodynamically driven. The propensity for replication or transcription, and the stability of the double helix, would become a sensitive function of the genetic information carried by the molecule. Proteins that might be very useful for cellular function could be selected against if their coding resulted in, for example, an excessively low number of $G–C$ pairs relative to the average for the ensemble of proteins synthesized by a particular cell. Hence, for the cellular information replication and transcription mechanism to work, temperatures in the cell must be well below the values where thermodynamic conversion of double-strand to single-strand DNA happens. Indeed this is the case, and the enzyme-driven unzipping of DNA is not a thermodynamically driven process. Were cells to be heat engines, relying on temperature differences to sustain cellular processes, the calculation earlier in the chapter shows that cellular temperatures would have to be much closer to or in the range of those that thermodynamically convert double-strand to single-strand DNA. Again, the use of low entropic chemical processes to generate free energy differences to power cellular functions is critical to the viability of the life as we know it, in which DNA is a stable repository of the genetic code.

Let us close this section with three broad applications of thermodynamics to life on the planetary scale. The first deals with the way life, versus the inanimate planet itself, uses sunlight. It is evident that the process of sunlight reaching the surface of a planet and warming it generates entropy. Imagine an ensemble of photons of visible light being absorbed by the surface. The frequency distribution of these photons corresponds to that at the photosphere of the Sun, where the blackbody temperature is roughly 5770 K (Chapter 3). The absorption of these photons in the ground, or in the water, leads to these abiotic materials increasing in temperature and hence reemitting the absorbed radiation as a spectrum of infrared photons. The infrared photons are much lower in frequency (longer in wavelength) than the absorbed photons; the corresponding surface temperature is 285 K (this includes re-radiation back down from the atmosphere in the so-called greenhouse effect—see Chapter 11). The peak wavelength of reemission is about 20 times longer than the 0.6 μm of sunlight, and in consequence the average energy of each photon is 20 times less than in the original sunlight. Hence, to conserve the incoming energy, the outgoing energy stream must consist of more than an order of magnitude more photons in number than streamed in. Looking at entropy as a measure of the number of available states, by increasing the number of energy-carrying particles (photons), the number of available states has increased greatly, and hence so has the entropy. We can use the heat added to the system in this way in a heat engine, which is essentially what our planet's weather and ocean circulation patterns are. But the amount of entropy generated per unit energy is large—(see the heat engine entry in Table 7.2).

Earth, however, unlike the other bodies in the solar system, has a surface covered to a large extent by green plants and algae. The chlorophyll and related dyes in the exposed cells of these plants also absorb sunlight, but they do not passively generate heat. Instead, photosynthesis is initiated when the sunlight is used to liberate electrons from the chlorophyll and, through a series of reaction steps, creates a gradient in the amount of protons from one side of a membrane (in the chloroplast) to the other. As described in Chapter 4, the protons are an energy source for charging up the cell through the manufacture of ATP from ADP, and for making carbohydrates. From a thermodynamic point of view, the protons flow along a gradient of chemical potential, from high to low, and can do work. Although some heating of the plant material results from the absorption of sunlight, it is far less than the ground or water suffers and is largely at wavelengths where the chlorophyll does not efficiently absorb. Indeed, there is a strong cutoff in wavelength at which chlorophyll no longer absorbs, and this "red edge" (Figure 7.6) is an important adaptation (or fortuitous feature) of green plants that prevents excessive heating. Thus, green plants convert sunlight into chemical energy, which carries a much greater potential for doing work (an entropy per energy 10 times smaller) than a heat engine whose high-temperature reservoir is heated by the same amount of incoming sunlight. This efficient use of solar energy has triggered a long chain of events beginning with the global spread of photosynthesis and formation of an oxygen-rich atmosphere, culminating in the spread of eukaryotic forms of life including land plants and animals and, ultimately, humans. Even those forms of life that do not employ or couple into photosynthesis use mechanisms of energy transfer that do not involve heat engines. For example, chemisynthesis at mid-ocean ridges relies on chemical potential gradients associated with gases coming from the vents—far more

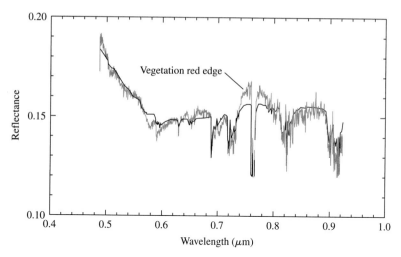

FIGURE 7.6 Spectrum of terrestrial vegetation showing the sudden increase in reflectivity in the red part of the spectrum. The reflectivity scale is linear; perfect absorption (black) is 0, while perfect reflection (white) is unity. A smoothed version of the spectrum (superposed) still shows the jump.

efficient than trying to extract energy via a heat engine relying on the temperature difference between warm and cold seawater.

The high thermodynamic efficiency of cellular processes may not be the only reason for the diversification of life on Earth, which brings us to the second application. Harold Morowitz of George Mason University has pointed out an intriguing aspect of Earth's modern biosphere with regard to its energy density and diversity. Recall from Chapter 4 that the diversity of fundamentally important molecules such as proteins and the nucleic acids relies on the ability to construct very long chain polymers. This requires that certain bonds be formed between the monomers that comprise the units of the polymer chain. However, it is also possible to terminate polymer chains with groups that do not allow the continued lengthening of the polymer. This can be a distinct chemical group, (e.g., the hydroxyl radical OH), or it can be the wrong-handed version of a monomer that goes into a chiral molecule (e.g.), a left-handed ribose attached to an otherwise right-handed RNA molecule) (Chapter 4). Either way, thermodynamics provides a guide to the preference for chain termination versus chain building. When the temperature is high, high-entropy systems are preferred by the definition of the Gibbs free energy; because long-chain polymers have a low entropy relative to a soup of unattached monomers, chain termination to achieve very short (or no) polymers is preferred at high temperature. At very low temperature, although low-entropy systems win out, the type of chemical bond must also be considered. Bonding with groups that tend to terminate chains is preferred to bonding with chain-extending organic groups, so that at very low temperatures short chains are also preferred. The predominance of bonds that extend the chain, and hence the potential for long polymers in a given organic chemical system, peaks at intermediate temperatures (Figure 7.7).

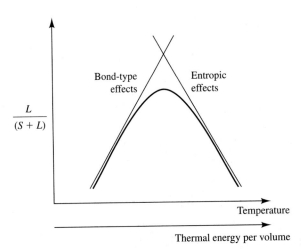

FIGURE 7.7 Sketch of the dependence of the ratio of chain-extending bonds (L) to the sum of chain-extending and chain-terminating (S) bonds in a biochemical system, versus the temperature and, equivalently, the thermal energy per unit volume. At low and high temperatures, short chains are preferred.

To relate this result to biological systems, recall that the temperature is related to the amount of thermal energy in a system per unit volume. (Temperature is not a function of system size; a temperature of 37°C means the same thing in a pea as it does in an elephant.) Low temperature means low thermal energy per unit volume; high temperature means high thermal energy per unit volume. We can calculate the mean energy density of the Earth's organisms, by computing what the energy density is for cells of all different types (from archaea to eukaryotes). When Morowitz did this for polymers relevant to biological systems, he found that the mean energy density of the Earth's biosphere was very close to the value for which polymer-lengthening bonding is maximally preferred—the peak of the curve in Figure 7.7.

The significance of this lies in the fact that biological diversity relies on being able to build very long polymer chains. Longer DNA molecules carry more genetic instructions than do shorter DNA molecules. Longer chains of amino acids have the potential to fold into more varieties of proteins than do shorter ones. Hence, an environment in which the mean energy density is such as to maximize the chain length in biological polymers enables the greatest variety of key biopolymers and, hence, the greatest extent of molecular diversity in the biosphere itself. If diversity provides a higher assurance of survival of organisms in an external environment that changes unpredictably (Chapter 16), then this concordance of the overall cellular energy density with that which is needed to maximize diversity might be an evolutionary, rather than a fortuitous, outcome.

There are a number of further considerations and caveats that must be attached to this conclusion:

1. While the calculation of the variation in the ratio of chain-lengthening to chain-shortening bonding versus thermal energy is well determined, the same is not necessarily the case for the determination of the mean energy density of the biosphere. Further, we must be careful in identifying the thermal energy density (which relates directly to temperature) to total energy density of a system that includes stored

chemical energy. Finally, the significance of an average over many systems, some of which are interacting with each other while others are not, is reduced if there are large variations in the energy density from cell type to cell type, organism to organism, or ecosystem to ecosystem.

2. There are many examples in science of seeming concordances that turn out in the end to be coincidence. And, indeed, there are many concordances whose significance remains unresolved. Like many "just so" situations in which two ostensibly unrelated values agree with each other, the potential profundity of this concordance cries out for additional tests of its true significance—however, it is unclear how this might be done.

3. It would be of extreme interest to determine how late in the history of life the concordance identified by Morowitz was achieved and how sensitive it is to the kinds of organisms that inhabit the planet. Was this concordance also in play early in the history of the Earth, when only primitive members of the domain Archaea, and possibly Bacteria, existed? If not, is this concordance a uniquely eukaryotic phenomenon? What feedbacks would lead to such an optimal situation being achieved; that is, what are the evolutionary pressures pushing biological diversity at the molecular level?

The final application leads us to a powerful means for detecting the past presence of life, even when the only remnants of an organism are literally atoms, namely, that of signature isotopic ratios left behind by the metabolic processes of life. The strength of chemical bonds are affected not only by the electron configuration of the atoms but also by their mass. For example, each atom in a diatomic molecule feels a force that depends on the mass of the other, even though the force is electromagnetic, because the vibration frequency of the molecule is affected by mass. Because the heavier isotope tends to exert a stronger force, it is harder to dissociate a molecule with the heavier isotopes of a given atomic species. Therefore, the rates of the reactions involved in breaking the molecule are slower for the heavier isotope. Because so many biological processes are rate dependent, slower reaction rates for isotopically heavy molecules mean that these are preferentially excluded from biochemical processes. While this is a kinetic, not thermodynamic effect, there is also an effect of isotopic mass on the equilibrium constants of the chemical reactions. Where the kinetic and isotopic effects work in the same direction, substantial differences in the isotopic ratios of certain elements can build up between biological systems and the surrounding environment. These can persist as concentrations of isotopic anomalies in phases of minerals that retain elemental or chemical traces of past biological activity, long after the organism itself has died and morphologically disappeared from the rock record.

The best known and perhaps most reliable isotopic signature of life is the enrichment of light carbon (^{12}C) relative to the stable heavy isotope ^{14}C in photosynthetically generated organic molecules. Iron isotopes have also been used as indicators of past biological activity (Chapter 14). Because such tracers can persist for billions of years, they may be crucially important for the detection of past biological activity on planets such as Mars, where life may have been eradicated billions of years ago. The generality of biochemical processes as being "lazy," in the sense of preferring to use molecules with weaker chemical bonds and hence faster reaction rates, suggests that the isotopic systematics found in terrestrial life might be general to organic life elsewhere. However, there remain complications in applying this property to life detection, which are explored in more detail in Chapter 14. For now, it is simply fitting to reflect on the fact that the most fundamental aspects of kinetics and thermodynamics in chemical systems enable life to leave behind traces of its biochemical activities aeons after they have ceased.

7.5 Thermodynamics, Disequilibrium, Chemistry, and the Origin of Life

The examples in Section 7.4 failed to go into any detail on nonequilibrium processes in cells. Indeed, the speeds at which chemical reactions occur are not just important in determining equilibrium constants, but they are crucial in a host of processes in which equilibrium is neither achieved nor desirable. The whole purpose of enzymes is to selectively speed up some chemical reactions, and not others, so as to achieve a very specific and selective set of products vastly different from those that would be found in a chemical reactor allowed to achieve equilibrium. Enzymes are not the only agent of ensuring the cell's constituents do not come to equilibrium—the continued flux of reactants into the cell and waste products out, carefully mediated by cellular membranes, ensures that vast suites of chemical reactions *never* come to equilibrium. And this is essential to sustaining a self-organizing, self-regulating system, which is cellular life reduced to its core characteristic.

Where, then, does thermodynamics come into play? Well, thermodynamics is certainly discussed in the examples of specific processes covered earlier and in many more covered elsewhere (again, the reader is urged to consult Haynie). But it also comes into play in the disequilibrium state that is viable life. In the mid-20th century, physicists around the world, notably the Belgian theorist I. Prigogine, extended the field of thermodynamics into the realm of nonequilibrium and irreversibility. Although this physics lacks the conceptual simplicity that makes equilibrium thermodynamics so powerful, a number of important specific results were obtained, particularly with regard to the rate of entropy generation as a function of energy input into systems held away from equilibrium. Inherent in these results, and confirmed in numerical simulations on computers, is that *complicated systems that are held away from equilibrium and have access to sufficiently large amounts of free energy exhibit self-organizing, self-complexifying properties.*

Complicated systems are those with many different degrees of freedom—many component parts, many different chemical reactants, intricate morphologies. To be held away from equilibrium means having a large temperature contrast across the system, for example, or being supplied with chemical reactants (and having products removed) at a sufficiently high rate. Large amounts of free energy here means in the Gibbs sense—with the potential to do large amounts of work, that is, with low entropy per unit energy. Abiotic systems that exhibit self-organization include, for example, thunderstorms, which are born from small-scale turbulent motions of air strongly heated by strong temperature gradients between ground and air and which have an additional source of energy in the form of condensation of water. The condensation releases heat and changes the local water abundance of the air, creating a buoyancy difference between dry- and wet-air parcels. As long as the

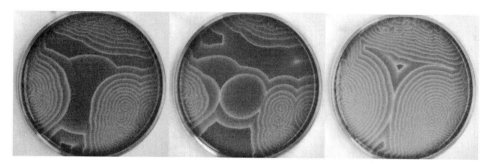

FIGURE 7.8 An example of the Belousov-Zhabotinsky reaction, in which distinct spiral patterns are created corresponding to spatial variations in the products generated by the reaction itself.

energy of sunlight is available to drive the system, the potential is present (and often realized) for large-scale columns or cells of cloud to form, in which strong electrical potentials are generated and well-organized patterns of rainfall occur. Indeed, groups of thunderstorm cells can organize into larger-scale weather systems—*mesoscale convective complexes*—that exhibit self-organized behavior over hundreds of kilometers. Once the Sun sets, the source of free energy is removed, and these complexes disintegrate on a timescale of hours.

Chemical reactions can exhibit self-organizing behavior. The so-called Belousov-Zhabotinsky (BZ) reactions can exhibit remarkable patterns delineated by colors as long as reactants are continuously supplied so that the system does not too closely approach equilibrium (Figure 7.8). In one version, in which citric acid, potassium bromate, sulfuric acid, and an iron compound acting as catalyst are combined, the iron compound alternates between red and blue as it goes from a more reduced to more oxidized state. Analysis of the reactions indicates that more than 30 different species are created, some of which serve as key catalytic intermediates that are regenerated in a process called *autocatalysis*. The spatial distribution of the color patterns changes as the addition of reactants is altered; in some versions, exposure to light can alter the degree of self-organization. The party is over when the amount of product is so large that addition of enough reactants to overwhelm the detritus becomes impractical.

It is possible to succinctly describe the BZ system with a set of five reactions among five compounds: (1) $A + Y \leftrightarrow X + P$; (2) $X + Y \leftrightarrow 2P$; (3) $A + X \leftrightarrow 2X + 2Z$; (4) $2X \leftrightarrow A + P$; (5) $2Z \leftrightarrow fY$. Here A and X are reagents (respectively, BrO_3^- and $HOBr$ for a real system), Y (Br^-) an intermediate product, Z (Ce^{++++}) a catalyst, and P ($HOBr$) the final product. The amount of Y manufactured at the end depends on the rates (kinetics) of the reactions and this is symbolized by the factor f. The reactions may move forward or back, depending on the amount of reagent relative to product.

Why is it important to be able to list the reactions themselves? The answer lies in the basic first law of thermodynamics [equation (7.1)] which we can extend by including the chemical potential term [defined in equation (7.3)] as $TdS = dE + pdV - \Sigma_i \mu_i dN_i$. The symbol Σ_i indicates we sum over all five different components in the BZ system. Then the rate of change in entropy, dS/dt, if we assume the system energy and volume are held constant, is just depends on $\Sigma_i \mu_i dN_i/dt$. (Holding the energy and volume constant is a convenience for this argument but does not affect its generality). Here t is time.

In the simplest model of chemical reactions (which is still fairly good), the chemical potentials for each of the species in a given reaction j is proportional to $ln(\Lambda_j^+/\Lambda_j^-)$, where Λ_j is the rate of the reaction and the + or − superscript indicates the forward or reverse reaction [forward reaction is that read from left-to-right in (i) through (v) above]. The change in the amount of each component in the reactions just depends on the difference in the forward versus reverse reaction rate, $\Lambda_j^+ - \Lambda_j^-$. Hence the rate of entropy change is

$$\frac{dS}{dt} = k \Sigma_j (\Lambda_j^+ - \Lambda_j^-) \ln\left(\frac{\Lambda_j^+}{\Lambda_j^-}\right) \tag{7.12}$$

where we sum over the five reactions listed earlier. k is just Boltsmann's constant, and is of course positive.

Equation (7.12) contains a key message: The rate of entropy change is always positive no matter which way each individual reaction goes. If Λ_j^+ is less than Λ_j^-, both $(\Lambda_j^+ - \Lambda_j^-)$ and $(\Lambda_j^+/\Lambda_j^-)$ are negative and hence the product is positive; if Λ_j^+ is greater than Λ_j^- then the product of the two terms is also positive. Hence, despite the presence of organized spiral structures in the chemical reactor hosting the BZ solution, the entropy of the system increases over time, as the second law of thermodynamics prescribes. The specific dependence of the entropy change for the BZ reaction on the reagents and product is complicated, as shown in Figure 7.9. In particular, for a particular range in the amount of product the system exhibits some unpredictability in behavior, called chaotic behavior, which is discussed further in Chapter 9. But the rate of entropy production in the BZ system remains positive—entropy increases with time.

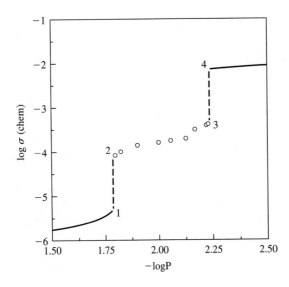

FIGURE 7.9 Production of entropy versus the abundance of HOBr in the BZ reactions. As the concentration of HOBr increases, the system entropy jumps discontinuously from point 1 to point 2 and then from point 3 to point 4. Between points 2 and 3 the system oscillates in time, exhibiting spiral patterns of varying form and definition.

In none of these examples is the second law violated. Entropy is generated at a high rate in the systems described and, indeed, in any system that exhibits self-organizing behavior. But if the rate of free energy addition is large enough, the systems behave as low entropy systems because the entropy per unit free energy is low. Life is no different. Its access to solar energy, and the strong chemical gradients at deep sea vents, ensure a large supply of energy that is captured and transduced through chemical processes. Entropy generation occurs at a high rate, but so does the access to free energy. The cost to the cosmos, if you will, is high—the Sun has generated enormous amounts of entropy over its 4.5-billion-year history, but life on Earth gets the associated sunlight almost for free. On the molecular level, we can see where the entropy generation is occurring; that is, we can identify where the irreversibility occurs in key biochemical reactions. ATP conversion to ADP not only provides energy for other cellular processes; some of the energy yield in every ATP $\rightarrow$ ADP reaction is released as heat. Virtually all of the innumerable catalyzed chemical reactions in cells occur well away from equilibrium, are thus irreversible, and generate entropy. Yet, as a complicated physical system with an abundant source of free energy and mechanisms for obtaining abundant reactants and removing products, the cell is the quintessential example of a self-organizing, self-sustaining system.

In the coming chapters we will examine models for the origin of life. Some of these models are amenable to analyses that devolve from the considerations of entropy, irreversibility, and self-organization detailed earlier. You, the reader, should take away from this chapter, when considering those that follow, a few key principles:

1. Life obeys the laws of thermodynamics. Indeed, biochemical processes can be analyzed from the point of view of thermodynamics, even though the vast majority of these processes do not operate close to equilibrium nor do they serve to bring products and reactants to equilibrium. The cost of these processes is the generation of entropy at a substantial rate; the cost is offset through access to abundant sources of free energy (sunlight and resulting carbohydrate products, chemical gradients at seafloor vents).

2. The selective and well-controlled chemical processes that the cell undertakes allow it to exist in steady state even though it is not in thermodynamic equilibrium. An analogy again is the atmosphere of the Earth, which is not in equilibrium with the sunlight received. Energy flows outward through the atmosphere continuously, and that energy flow generates entropy, which does not happen in an equilibrium state. Likewise for life. The difference in the case of life is an ability to compensate for changes in the environment, so as to maintain the integrity of the cellular structure and the chemical reaction networks occurring within.

3. The origin of life from nonliving processes faced two fundamental difficulties. First, the extent to which chemical synthesis of polymers had to evolve from relatively random, nonspecific, poorly catalyzed systems to something approaching the selective, specific, exquisitely catalytically controlled biochemistry of even the simplest cell seems daunting. Everything from selection of a single-handedness in amino acids and ribose to the evolution of folding proteins as catalysts requires a degree of self-organizing unachieved in laboratory or computer simulations. Second, these processes had to occur in systems that were both poorly *autopoetic* and poorly

homeostatic. Here autopoesis is the ability of a system to regenerate itself and homeostasis the ability to maintain a steady state in the face of changing environmental conditions. The more time is invoked as the key ingredient in evolution from nonliving organic chemistry to the cell ("Time is the greatest innovator," said 17th century British philosopher Francis Bacon), the more imperative is the need to explain how evolving systems avoided being reset by environmental change.

QUESTIONS AND PROBLEMS

1. Equations (7.1) and (7.2) express the first law of thermodynamics in terms of the independent variable pairs (entropy, volume) and (temperature, pressure), respectively. Following a procedure similar to that which led to the Gibbs free energy of equation (7.2), derive a first-law expression for the independent variable set (entropy, pressure). The corresponding free energy is called the enthalpy. Now derive the first-law expression for the independent variable set (temperature, volume). This free energy is called the Helmholtz free energy.

2. When we build a beautiful arch bridge or arrange chess pieces in their proper starting positions on a chessboard, we seem to increase order in the cosmos. Why does this not violate the second law of thermodynamics?

3. Which of the following statements is true only for the change in internal energy (ΔE), the change in Gibbs free energy (ΔG), is true for both ΔE and ΔG, is false for both ΔE and ΔG?

    a. It depends on the pathway during a change of state, that is, depends on the functional relationship between temperature and pressure in going from the initial to final state.

    b. It includes only the energy that—at constant temperature and constant pressure—can perform work.

    c. If a process occurs spontaneously at constant entropy and volume, it must be negative.

    d. It is a state function.

    e. It can predict which phase in a system is likely to be the preferred state at a given temperature and pressure.

4. Construct an argument for or against the notion that life could survive on ice rather than liquid water by comparing the free energy available from the reaction converting ATP to ADP to that required to melt ice. You may also need the amount of ATP available versus the amount of water that must be melted.

5. The simplest expression for the entropy added to a system by mixing of two ideal gases (for example) is $S = k[-x \ln x - (1 - x)\ln(1 - x)]$, where x is the fraction in the mixture of one gas, $(1 - x)$ is the fraction in the mixture of the other gas, and k is the Boltzmann constant. For what value of x does this expression reach a maximum? (You can either graph it or use calculus to find the maximum.) In real mixtures, the entropy of mixing is not always maximal at this value, in fact it usually is not. Speculate why this is so.

6. An inch of rainfall over a medium-sized city of 150 km^2 is equivalent to 10^{10} liters of water. What is the total energy release in the condensation of that water to make clouds and rain? (You must look up the density and appropriate latent heat of water in a meteorological or thermodynamics text.) Calculate as well the gravitational potential energy released when this amount of water fell as rain from a height, let us say, of 2 km. By what factor is the heat released by condensation larger than the release of gravitational potential energy?

7. In an adiabatic process, no energy transfer as heat occurs between the system and its surroundings. For such a process, what is the form of the first law of thermodynamics, mathematically and in words?

8. Which of the following is a state function? Explain your answer.

   a. Q

   b. W

   c. $Q + W$

   d. $Q - W$

9. Which of the following hands in the card game "poker" has the lowest entropy? Which has the highest entropy? Explain why. (Consult a book of games to determine the ranking of the card hands, and the basis for the rankings.)

   a. A♠ A♣ A♦ A♥ J♥

   b. 5♥ 10♥ J♥ 6♥ 8♥

   c. A♣ K♣ Q♣ J♣ 10♣

   d. 2♥ 2♣ 2♦ 10♣ 10♦

   e. A♣ 2♦ 3♦ 4♦ 5♥

10. a. Describe the chain of energy exchanges by which we can argue that our muscles are powered by hydrogen fusion.

    b. The photosynthetic energy conversion process by which sunlight, CO_2 and H_2O are converted to biomass is not 100% efficient. Explain why this must be so based on the first and second laws of thermodynamics, and identify what happens to the sunlight *not* converted into biomass in the photosynthetic process.

SUGGESTED READINGS

A basic undergraduate introduction to thermodynamics can be found in many undergraduate general physics and chemistry textbooks. A well-written treatment of thermodynamics at the upper undergraduate level is C. Kittel and H. Kroemer, *Thermal Physics* (1980, W. H. Freeman and Co., San Francisco). The thermodynamics of living systems is well covered for undergraduate students in D. T. Haynie, *Biological Thermodynamics* (2001, Cambridge University Press, Cambridge, U.K.). A detailed but quite readable treatise on information theory and entropy as applied to cellular communications is contained in W. R. Lowenstein, *The Touchstone of Life* (1999, Oxford University Press, Oxford, U.K.). Application of thermodynamics and statistical physics to the origin of life is well covered in the engaging, semipopular book by P. Coveney and R. Highfield, *The Arrow of Time*

(1990, Flamingo Press, London, U.K.). For those with advanced preparation in physics, the field of nonequilibrium thermodynamics may be tackled with the help of B. C. Eu, *Kinetic Theory and Irreversible Thermodynamics* (1985, John Wiley and Sons, New York). Finally, although it is excessively autobiographical, Stuart A. Kauffman, *Investigations* (2000, Oxford University Press, Oxford, U.K.) is essential reading for key insights on this subject by one of its central players. This book updates and extends his 1995 work published at Oxford, *At Home in the Universe*.

REFERENCES

Dutt, A. K., Bhattacharya, D. K., and Eu, B. C. (1990). "Limit Cycles and Discontinuous Entropy Production Changes in the Reversible Oregonator," *J. Chem. Phys.*, **93**, 7929–7935.

Callen, H. B. (1985). *Thermodynamics and an Introduction to Thermostatistics*, Academic Press, New York.

Coveney, P. and Highfield, R. (1990). *The Arrow of Time*, Flamingo Press, London, U.K.

Fersht, A. (1999). *Structure and Mechanism in Protein Science: A Guide to Enzyme Catalysis and Protein Folding*, W. H. Freeman and Company, New York.

Haynie, D. T. (2001). *Biological Thermodynamics*, Cambridge University Press, Cambridge.

Lieb, E. H. and Yngvason, J. (2000). "A Fresh Look at Entropy and the Second Law of Thermodynamics," *Phys. Today,* **53** (4), 32–37.

Loewenstein, W. R. (1999). *The Touchstone of Life: Molecular Information, Cell Communication, and the Foundations of Life*, Oxford University Press, New York.

Shinbrot, T. and Muzzio, F. J. (2001). "Noise to Order," *Nature*, **410**, 251–258.

CHAPTER 8

Biological Foundations for the Origin of Life

8.1 Introduction

How did life begin on Earth? This question is a central one to astrobiology, and yet, paradoxically, it may never be answered in a straightforward way. Indeed, the question may be posed incorrectly because it implies a discrete threshold across which living systems were separated from nonliving. But perhaps there is no such threshold. Life might be considered a kind of epiphenomenon of the sum total of selective sets of chemical reactions, catalytically controlled and guided by templating molecules and contained within a series of structures constructed by these self-same reactions. Strip away progressive elements of the biology (which has been done in the laboratory as we shall see in Section 8.6) and eventually the organism becomes nonviable—but we are left with no perspective on what were the original forward pathways by which viability became a meaningful, identifiable property. Like human self-awareness (Chapter 17), we "know" life when we see it, but we cannot imagine dissecting it in a way that reveals any sort of continuum from the nonliving to the living.

This chapter discusses particular aspects—with an emphasis on particular—of biology that seem fundamental to life and that can be "regressed" in such a way as to illuminate some of the possible steps that led to living organisms as we know them today on the Earth. By undertaking such a dissection, no comprehensive model for the origin of life is achieved . . . but some perspectives on what might have been required are gained. Chapter 9 then attempts a synthesis of potential pathways to the origin of life, using the considerations presented here as a foundation.

A word of caution is in order here to the reader; namely, that *Chapters 8 and 9 are the most speculative chapters in the book*. Chapter 8 begins with a straightforward discussion of the roles of DNA and RNA in storing genetic information and encoding the production of proteins. However, we then move into discussion of some ideas in the literature as to how the present dual nucleic-acid system of life—DNA and RNA—might have arisen from simpler precursor systems. While many of these ideas are based on laboratory experiments, much remains poorly understood about the early roles of RNA and possible precursor molecules. Chapter 9 brings us squarely into the question of how a self-organizing system such as life might have arisen, by looking at the general physical aspects of such systems and then creating a *story* of life's origins based on the considerations in both chapters.

8.2 Earliest Roles of RNA as Encoder and Catalyst

While we tend to think of DNA and proteins as the two quintessential biopolymers, the "yin" and "yang" of life representing, respectively, information and action, RNA plays far more diverse, yet central, roles. As in the transcription of DNA described in Chapter 4, the replication of DNA requires a large enzyme—in this case a DNA "polymerase"—to control the reaction. However, all the enzyme can do is to attach new nucleotides onto one end (the 3′ end as labeled in Figure 8.1) of the existing nucleotide—it cannot initiate the synthesis of DNA. Therefore, also required is a complex of RNA and enzymes called a *primase*, which has the ability to assemble a short strand of RNA with base pairing complementary to the DNA being replicated. This RNA *primer* (see Chapter 14 for more detailed discussion of DNA primers in the context of life detection) provides a structure by which free nucleoside units can be assembled to begin the replication of the original DNA strand. While some DNA replication mechanisms have evolved to use a DNA primer instead, most use RNA. There is an *absolute* requirement for an RNA or DNA primer in the replication of DNA but, as we shall see in Section 8.3, not for the replication of RNA.

RNA is, as described in Chapter 4, the functional molecule that translates and transports the genetic code to the sites of protein synthesis, the ribosomes. But RNA does more than that in protein synthesis. RNA resident in the ribosomes (ribosomal RNA, or rRNA) has at least three different functions:

1. It assists in catalysis of the assembly reactions.
2. It is crucial to the correct positioning of the messenger RNA (mRNA) and transfer RNA (tRNA) that bring the genetic information and the amino acids to the protein assembly site.
3. It folds into structures on which the ribosomal proteins are assembled.

Thus, while the DNA is the repository of the information on amino acid ordering from which proteins are built, all of the active steps in the construction involve RNA (Figure 8.2, see Color Plate 9). Indeed, the active sites of assembly of amino acids are largely devoid of proteins that could act as enzymes. RNA is now known to be the primary catalyst within the ribosome.

RNA participates in a wide range of cellular processes other than protein assembly. Small strands of RNA (snRNA) are critical to formation of mRNA by the editing out of the heterogeneous nuclear RNA (hnRNA)—the sequences of bases corresponding to the

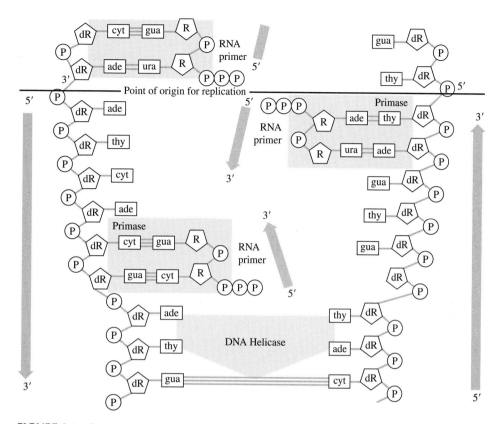

FIGURE 8.1 Schematic of the replication of DNA, focusing on the region where the two complementary strands of DNA have been pulled apart and each is being replicated. Part of that process requires the creation of short strands of RNA. The nucleic acid bases are labeled with the first three letters of their names (gua = guanine, etc.), ribose is R, deoxyribose is dR, P is the high energy phosphate bond, and the 3′ and 5′ indicate the particular carbon bonds in the standard nomenclature given in Chapter 3. The DNA helicase, powered by hydrolysis of nucleotide bases, unwinds the DNA strand; the primase is an RNA-enzyme complex that has the ability to assemble a short strand of RNA with base pairing complementary to the DNA being replicated.

base pairs in eukaryotic (and some prokaryotic) DNA that carry no genetic information ("introns"). Many enzymes use RNA as co-factors crucial to the control of various biochemical reactions. And, as noted earlier, RNA can act as a catalyst ("ribozyme") in and of itself. Catalysts in general stabilize key transition states between reactants and products—states that are otherwise not accessible or too transient to be accomplished absent the catalyst. Protein enzymes do this by adding or removing protons during reactions, thus orienting reactants on a surface so they are in an optimal position to react—and, as well, create interactions that bind reactants in an *unfavorable* way that is relieved by otherwise inaccessible transition states. Thus, the free energy that goes into the construction of the enzymatic catalysts yields structure—that is, organization—that enables otherwise thermodynamically unfavorable states in a reaction to be accomplished. Ribozymes function in much the same way, and indeed some fold into three-dimensional, tertiary structures (Chapter 4) but lack the complexity of shape and diversity of functional groups to be as effective as enzymes. In only a few cases are ribozymes able to achieve rates of key chemical steps in reactions they catalyze comparable to those in enzymes.

While a number of ribozymes are known, only one example of a catalytic DNA molecule has been found—a DNA molecule that can cut ("cleave") RNA. What makes RNA different from DNA is the properties of the sugar ribose versus deoxyribose as the RNA versus DNA backbone. The depths of the grooves defined by the backbone and the nucleic acid bases, the tilt of the bases, and other factors make RNA a better molecule for catalytic activity. Conversely, however, DNA is very robust, with the double helical structure readily reassembled after denaturation (Chapter 4), and with strands extending for several thousands to millions of nucleic acid bases. Lengthy strands of RNA are generated only by direct transcription from DNA and then must be actively edited down to make an intron-free messenger of the genetic code. Like DNA, hnRNA and mRNA are never found free in cells but are bound at each end by proteins. DNA, relative to RNA, is optimized for long-term storage of large amounts of genetic information, but it is, on the other hand, a poor catalyst based on the limited experiments conducted so far.

A self-sufficient living system must be able to encode information, catalyze its own replication, and catalyze other reactions necessary for its maintenance. Neither DNA nor proteins separately can perform all three of these functions; given the divergent nature of the fundamental building blocks of these two classes of polymers (nucleic acids bases versus amino acids), it is attractive to postulate a more versatile precursor molecule. RNA qualifies on the basis of its capabilities observed in the laboratory, and it is now widely assumed that RNA preceded DNA in living organisms. Whether RNA preceded proteins or always encoded the production of at least simple peptides is less clear. But RNA's catalytic abilities suggest that it could have been an important player in, if not the foundation of, a simple metabolism. Indeed, some experiments have shown that some RNA sequences can catalyze RNA replication itself.

Further suggestive of the fundamental nature of RNA is that the energy storage molecule ATP is obtained from AMP by adding two phosphate groups (see Chapter 4). AMP is produced as a byproduct of the ATP–ADP energy charging–discharging cycle in the cell (Figure 8.3). The adenine nucleotides of RNA synthesis and energy metabolism are basically the same, differing only in the number of phosphates present.

This correspondence between one of the four fundamental building blocks of RNA and the energy-storing molecule ATP is striking in its thriftiness, given the very different functions of RNA and the ATP–ADP cycle. Might we imagine, as has been suggested by some authors, that the energy-carrying function of ATP preceded the origin of RNA and that the buildup of AMP in primitive cells led to their polymerization and the formation of a primitive precursor to RNA? If so, how did other nucleotide bases get incorporated into the cycle to produce a genuine information-coding molecule? Indeed, while the role of RNA as catalyst and encoder is straightforward to argue, the formation of RNA abiotically is little easier than that of DNA. It requires the dominant production of ribose over other sugars and the selection of ribose with a preferential handedness or chirality (see Section 8.4). It also might imply that several different RNA structures be constructed, since researchers

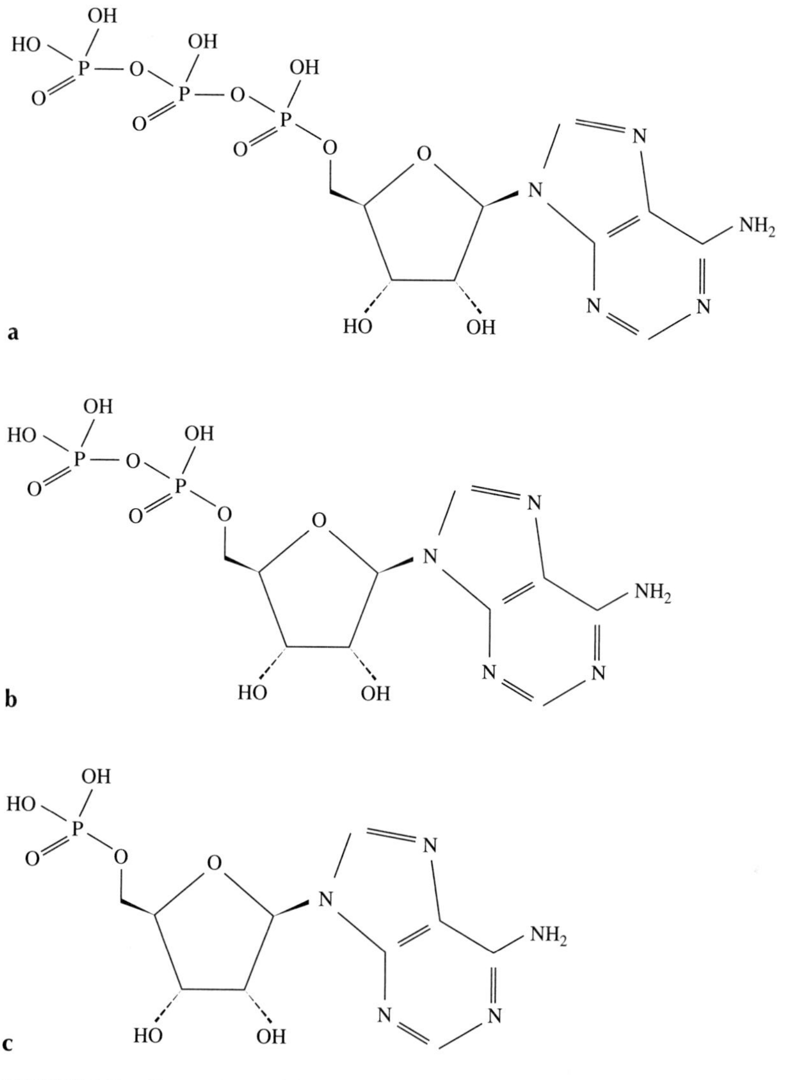

FIGURE 8.3 Comparison of the structures of (a) ATP, (b) ADP, and (c) AMP. Carbon bonds are shown single or double, with either C or CH at the unlabeled vertices, depending on the valence. Solid lines with expanding width indicate bonds coming out of the paper; alternating black and white indicates bonds going into the paper.

have yet to find a particular RNA molecule that functions effectively as both encoder and catalyst. While this last would not seem a problem if any RNA molecule can be synthesized in a natural abiotic system, it is a severe problem if a random accident must be invoked in RNA synthesis. If an RNA encoder is the product of that random accident, it is unclear how additional copies are catalyzed, and if the RNA catalyst is formed instead, then there is no encoder to replicate it—unless a precursor–encoding molecule (see Section 8.4) is invoked.

While these problems cast doubt on the notion of RNA being abiotically produced as the primitive cell's workhorse molecule, it does not detract from the notion that RNA preceded DNA or even proteins. Indeed, apart from the problem of producing RNA, it seems well suited as the central molecule in a primitive biochemistry that was the precursor to the modern DNA–RNA-protein triad. RNA molecules are amenable to the exploration of alternative configurations. They have just four highly interchangeable subunits (the bases), and quite distinct *conformations* (structures distinguished by the degree of rotation around chemical bonds) are achievable with few changes or mutations. The resulting secondary structures (Chapter 4) are simple and modular. Most sequences are soluble in water. For these reasons, once RNA was available it would have been straightforward for these molecules to be incorporated in catalytic cycles of the sort to be described in Chapter 9. In these cycles error rates in the production of new molecules were higher than today, yet the very presence of such errors demanded the system be forgiving enough that it could be sustained in the absence of very rigid, complex encoders such as DNA. The RNA "world," while not necessarily the first form of biochemistry on Earth, would be the only precursor that has left a trace in modern biology; its faint echoes are heard in the diverse roles that RNA plays in the cell today.

8.3 Formation of DNA

If biochemistry using RNA as the fundamental polymer was not only feasible but also innovative, why then was it replaced by a two-polymer system? Ironically, RNA's advantages as a changeable molecule and its disadvantages as an inefficient catalyst probably conspired to extinguish the RNA world as it was outcompeted by the DNA–protein duality. As noted in Section 8.2, ribozymes are, with few exceptions, vanquished in their catalytic performance by proteins, both in terms of speed and the complexity of the reaction sequences they can catalyze. Yet, proteins require a lengthy, complex genetic sequence for their construction—one that short strands of RNA cannot code for.

We can imagine the progressive incorporation with time of increasing lengths of peptide chains in an RNA-based cell to assist (as co-factors or replacement catalysts) in metabolic and replicative processes, but with a limit on the complexity of the peptides that could be generated. Because RNA is at the heart of the ribosomal apparatus for aligning and connecting amino acids together to make long peptide chains, the assembly of very long peptides and hence complex proteins is not limited by the absence of DNA; it is the coding of a complex protein, corresponding to thousands or more bases, that is at issue. Synthesis of the more specialized alternative to RNA, namely, DNA with the deoxyribose sugar in place of ribose, would have permitted instruction sets with vastly larger numbers of bases. This, in turn, would have allowed for the construction of peptides of sufficient complexity to exhibit vastly better and more complex catalytic capability than was possible with RNA alone. But synthesis of DNA requires proteins acting as enzymes; a plausible story for DNA's synthesis in their absence has yet to be developed. Undoubtedly countless "inventions" of DNA by primitive RNA-based life failed because of a lack of appropriate catalytic machinery.

How might the successful invention of DNA have proceeded within an RNA-based cell? RNA can make DNA from its own nucleotide sequence *ab initio* if the enzyme "reverse transcriptase" is present, along with the appropriate raw materials (including the pyrimidine thymine in place of RNA's uracil). As is the case for other "retroviruses," the HIV virus responsible for acquired immune deficiency syndrome (AIDS) has a genome composed of RNA and, upon entering the human cell, generates the corresponding DNA sequence using the enzyme reverse transcriptase. While viruses in general are too exquisitely engineered to be relics of the RNA world, instead representing obligate parasites of the DNA-based cell, the reverse transcription process is instructive in illustrating what is required to make DNA from RNA. Reverse transcriptase does two things: It (1) transcribes RNA into DNA and (2) degrades (removes) the RNA from the resulting DNA–RNA pair, leaving behind the DNA. As shown in Figure 8.4, the RNA sequence has a tail of adenine nucleotides. To initiate the reaction, a corresponding polymer of thymine nucleotides with deoxyribose as the backbone, available in the cellular soup of nucleotides or supplied as a primer in laboratory reverse transcriptase reactions, adheres to the "polyadenylate" tail. Then the reverse transcriptase enzyme catalyzes the addition of the corresponding free deoxy-nucleotides adenine, guanine, cytosine, and thymine (dATP, dGTP, dCTP, and dTTP) to the end of the primer in the order corresponding to the base pairing defined by the RNA template. It then catalyzes the removal of the RNA strand, leaving behind a single DNA strand that contains the base pair sequence of the original RNA.

Because retroviruses perform these reactions, we might expect that this process could be a model for the invention of DNA in an RNA-based cell. But the formation of the reverse transcription enzyme, with its lengthy peptide chains and complex tertiary structure (Figure 8.5, see Color Plate 10), represents a potential stumbling block. Retroviruses use the transcription machinery of the cell to express their genetic sequences once converted to DNA, including the production of the reverse transcriptase enzyme. Could an RNA-based cell support an RNA genetic code of sufficient length and stability to generate a reverse transcriptase enzyme, without first requiring the presence of that enzyme to create

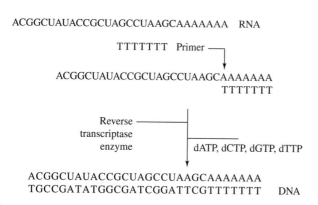

FIGURE 8.4 Schematic of the copying of RNA into DNA by reverse transcriptase. See text for details of the process.

the equivalent DNA? The longest ribozyme observed contains 200 nucleotides and synthesizes RNA molecules up to 14 nucleotides long. The DNA polymerase enzyme, of which the reverse transcriptase is a variant, is of order 10^3 amino acids in extent and hence is encoded by a DNA with thousands of nucleotides. While this seems to be the same sort of chicken-and-egg problem encountered before in the protein-DNA duality, the way out might be to invoke simpler catalysts to build a DNA strand on the RNA template. Indeed, the complexity of proteins and the lengthiness of the DNA that encodes them are in part a consequence of the two-polymer duality, which demands both replication machinery involving proteins and machinery for synthesis of the proteins themselves. In principle, a much smaller genome is required if the genome and the replicator are the same kind of molecule.

The two alpha helices in the reverse transcriptase enzyme serve as a clamp on the thymine primer attached to the RNA, aligning it toward the active site of the enzyme and enabling the rest of the DNA chain to be constructed. This is necessary to overcome the thermodynamically unfavorable entropy decrease associated with taking a random soup of free nucleotides and ordering them in specific locations along the RNA template. (This unfavorable entropy decrease occurs as well in the duplication of a strand of DNA, but once bound together, the hydrogen bonding between the two strands stabilizes the double-helix structure of the DNA). Any simpler catalyst present in an RNA-based cell must have functioned to overcome the same entropy decrement that the reverse transcriptase more efficiently overcomes today. The identity of this catalyst—and whether it was a peptide or nucleic acid—is probably lost in time, as are those of many other primitive catalysts whose functions were superseded by more effective and specific enzymes. It would be of keen interest to the origin of life problem to identify in the laboratory relatively simple catalysts that, however inefficiently, aid in the transcription of RNA into DNA. But the problem of the catalysis of the formation of DNA from RNA is arguably much less difficult than the origin of RNA itself.

8.4 Precursors to RNA

Could short RNA strands have formed abiotically, or was a simpler precursor-encoding molecule required? Given the existence of phosphate-bonded nucleotides such as ATP, ADP, and AMP, whose abiotic formation is straightforward, it would seem as if the essential ingredients for the formation of RNA were present in the abiotic world. Two fundamental problems stand in the way of a simple pathway to the construction of RNA from the monomeric nucleotide units. First, ribose is but one of only a large range of sugars that could have been formed abiotically in the early Earth environment. While the specific distribution depends on the ambient environment and the starting ingredients, one popular route is the abiotic production of sugars from formaldehyde, H_2CO. Formaldehyde is readily made abiotically and is observed, for example, in interstellar space. The formose reaction (Figure 8.6) normally produces a host of different sugars, including several with five-member rings as ribose, unless one biases the reaction by starting with glycoaldehyde phosphate. In that case, the dominant five-member sugar produced is a ribose diphosphate. While this chemically resolves the issue of what is required to preferentially generate

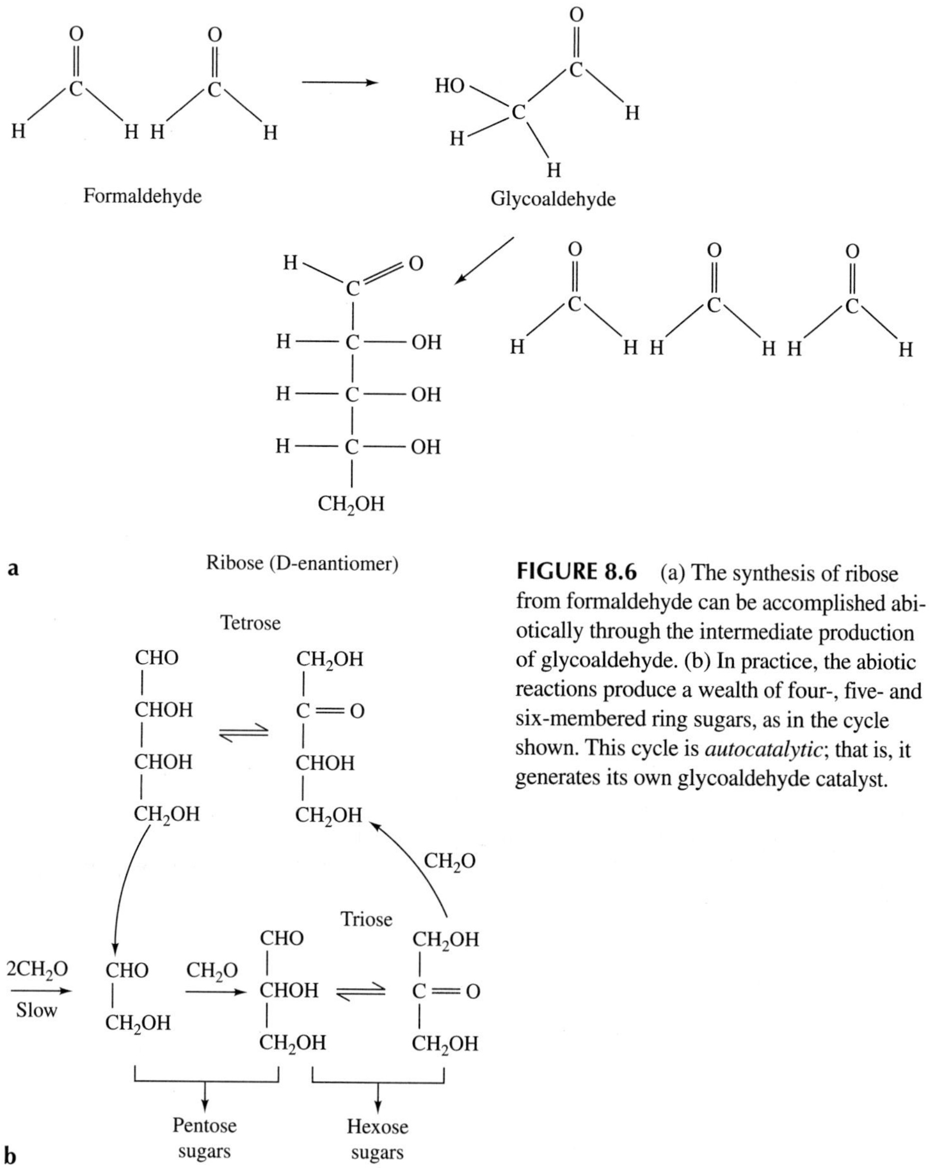

FIGURE 8.6 (a) The synthesis of ribose from formaldehyde can be accomplished abiotically through the intermediate production of glycoaldehyde. (b) In practice, the abiotic reactions produce a wealth of four-, five- and six-membered ring sugars, as in the cycle shown. This cycle is *autocatalytic*; that is, it generates its own glycoaldehyde catalyst.

ribose, there is no particular reason abiotic chemistry on Earth would proceed along this route in the absence of catalysts or templates forcing such a preference.

Another problem in the synthesis of ribose for RNA is the issue of chirality: All of the sugar molecules in the backbone of a given RNA or DNA molecule are the same handedness—the same *enantiomer* (Chapter 4). This is usually but not always D-ribose ("right-handed') for RNA and D-deoxyribose for DNA. More important than which enantiomer is

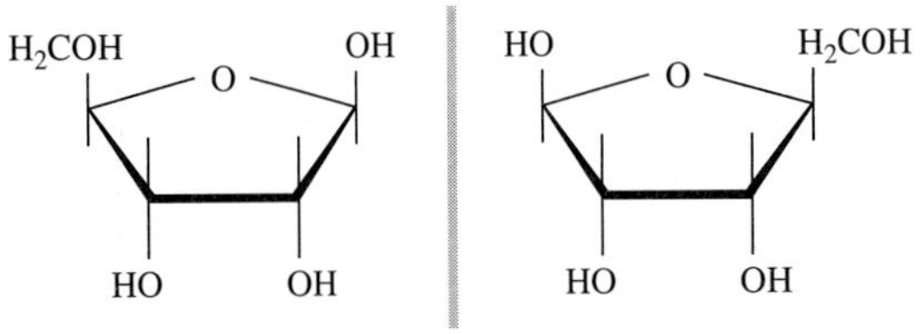

FIGURE 8.8 D-ribose (left) versus L-ribose sugar (right); the structures are identical but are mirror images of each other.

present is that it is always one or the other: the polymer is enantiopure. (Likewise, with few exceptions, biological systems use only left-handed amino acids to synthesize proteins.) We can see why this is necessary from the structure in Figure 8.7 (see Color Plate 10): If a mirror image of the D-ribose (Figure 8.8) were inserted into the positions of the corresponding nucleic acid bases, the phosphate bonds would have to be flipped in that unit, altering the structure of the polymer and the position of the active site for base pairing. Indeed, laboratory experiments indicate that the presence of even a small amount of the opposite enantiomer truncates the polymerization process. How such purification was accomplished in the prebiotic environment is unclear because the chemical properties of the two enantiomers of the ribose are identical. Abiotic production through the formose or other reactions would have led to a racemic mixture, that is, one with equal amounts of the two enantiomers.

Various ideas have been proposed for overcoming the racemic nature of abiotic production of sugars, generally by invoking an intrinsic or imposed asymmetry in the amount of left- versus right-handed forms. An intrinsic asymmetry between the two forms exists at the elemental particle level, associated with the interaction between the weak nuclear force and the electromagnetic force (Chapter 2). This imposes a small chiral asymmetry to atoms in their ground state, an asymmetry that increases as $(Z)^6$, where Z is the atomic number. For D-ribose versus L-ribose, the difference in energy is about 10^{-14} J/mole of ribose, extremely small, for example, compared with the energies stored in chemical bonds in the sugar, typically of the order of kilojoules per mole. This difference, however small, favors the D-sugars over the L-sugars, and the L-amino acids over the D-amino acids, which are the preferred enantiomers in biochemistry. It has been argued that under certain conditions, amplification of this excess to macroscopically significant levels might occur spontaneously in abiotic environments of sufficient size and complexity, but to date no experimental detection of such amplification has been found. Other schemes for enantio-amplification have focused on environmental asymmetries, such as the possibility of asymmetry in the sense of circularly polarized light in interstellar or planetary environments that would lead to preferential photochemical production of one or the other enantiomer. More plausible is the possibility that adsorption of chiral molecules on mineral surfaces could provide some enantioselectivity that would subsequently be amplified; modest selectivity has been demonstrated for the adsorption of chiral amino acids on calcite. The difficulty of assessing the possibilities for amplification of small excesses is that laboratory spatial and temporal scales are limited, and natural environments in which such effects might be observed are limited and difficult to access (e.g., Saturn's moon, Titan, discussed in Chapter 13).

The selection of certain sugars and of one or the other stereochemical orientation (chirality) are particular examples of the need to reduce the entropy—or increase the information content—of the abiotic system as it progresses toward life. In the case of chiral selectivity, for example, the entropy penalty per mole of ribose (let us say) in selecting one or the other handedness is at least $\Delta S = -R \ln 2$, where R is the gas constant and ln the natural logarithm (base e). This comes directly from the thermodynamic result that the entropy is related to the number of accessible states of the system. One way to reduce the entropy is to impose constraints on the system associated with templates or with a reduction in the number of degrees of freedom. Formation of molecules on mineral surfaces does both: It provides a particular set of preferred sites for chemistry and forces the chemistry to occur on two-dimensional surfaces rather than in a three-dimensional space. Many schemes for catalyzing or controlling a variety of prebiotic reactions on mineral surfaces have been proposed, but no direct experimental demonstration of abiotic synthesis of an RNA molecule, for example, has been achieved. Despite the reduction in the degrees of freedom, there remain an enormous number of possible reaction sets and outcomes.

An alternative to the direct abiotic synthesis of RNA is the abiotic synthesis of precursor molecules that are in some way easier to form than RNA. The two examples in Figure 8.9, both known to occur only in laboratory synthesis, provide separate potential solutions to the problems of sugar selectivity and enantiopurification outlined earlier. Threose nucleic acid—TNA—is similar to RNA, but threose is one of only two sugars with a four-member ring. Further, the positions where phosphate bonds can attach to the sugar ring to

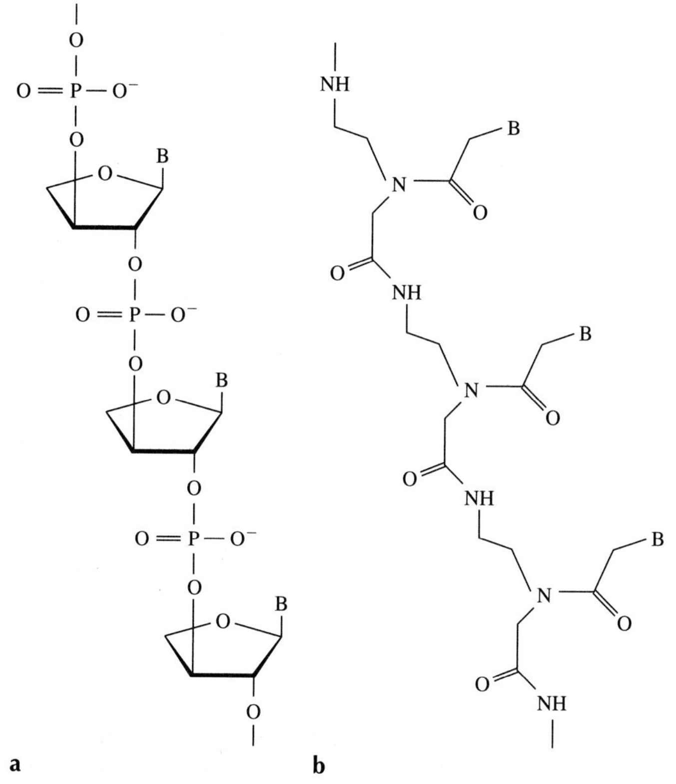

FIGURE 8.9 Two examples of possible precursors to RNA: (a) threose nucleic acid and (b) peptide nucleic acid. In both cases "B" refers to a nucleotide base.

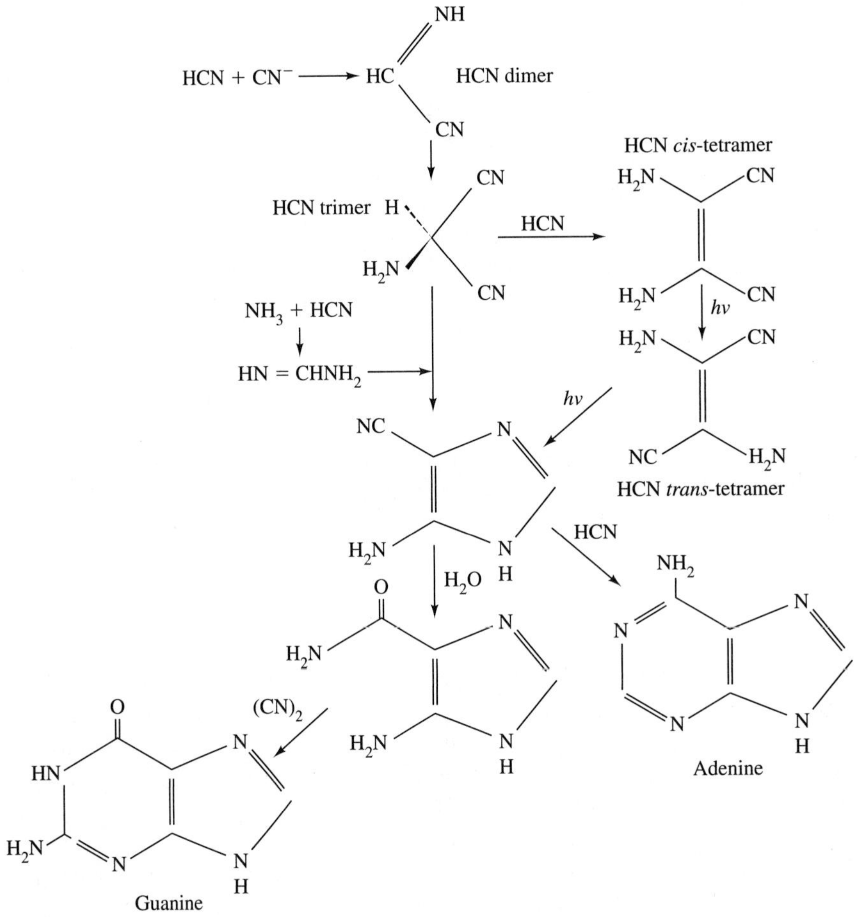

FIGURE 8.10 Possible pathways for the abiotic synthesis of purines (a) and pyrimidines (b). The purine synthesis uses hydrogen cyanide (HCN, observed in molecular clouds in interstellar space), ammonia, and water in a reducing environment in which ultraviolet light is available. HCN might also be manufactured in a prebiotic terrestrial environment as it is on Saturn's moon Titan. The pyrimidine synthesis uses cyanoacetylene (HC_3N), which is also seen in interstellar clouds, and cyanogens $(CN)_2$ in aqueous (water-based) solution. Single lines indicate single bonds, two lines double, three lines triple. Also, "cis" and "trans" refer to the differing conformations or orientations of the HCN around the central double bond. As before, unlabelled vertices are bonds with C or CH groups.

form the polymer are fewer for TNA than for RNA. For these reasons, the structure of TNA is much simpler than that of RNA, and the number of possible substitutions for the threose much less than for ribose. TNA also has the attractive property that it undergoes bonding between the nucleic acid base pairs following the same rules as for RNA. And it does so not only to another TNA polymer but to RNA as well—hence, genetic information can be transferred between the two.

[Chemical reaction scheme showing synthesis of uracil and cytosine from HC≡C—C≡CN via NCO⁻ addition, producing intermediates with CH, NCO, and cyanoacetaldehyde-type groups, yielding Uracil and Cytosine with release of CO₂.]

b

Peptide nucleic acid (PNA) is another polymer in which aminoethylglycine and methylene-carbonyl groups stand in for the sugar and phosphate bonds in the nucleic acid backbone. PNA's prebiotic advantage is that the backbone of the polymer is not chiral, so no enantioselectivity is required to synthesize it. Like TNA, PNA can bond across the base pairs not only with itself but also with RNA and presumably TNA. In neither case (TNA or PNA) have the monomer units been synthesized in a simulated prebiotic environment. However, aminoethylglycine is a product of nitrogen, methane, ammonia, and water mixtures subjected to a spark discharge, an experiment first performed by Miller and Urey well over a half-century ago. The pyrimidines and purines that form nucleic acid bases for attachment to any of the backbones described earlier in this section are readily synthesized abiotically in relatively reducing environments exposed to ultraviolet light or spark discharges (Figure 8.10). Such environments need not be the same ones in which the nucleic acids themselves are subsequently synthesized. Indeed, the purines and pyrimidines need not necessarily be formed on the Earth, since a fraction of the monomer units such as amino acids or nucleic acid bases carried in meteorites survive large impacts on the Earth. This is an important point, because the early Earth's atmosphere may not have been sufficiently reducing to readily synthesize such molecules or string them together as polymers.

PNA does not appear to have the same propensity for replicative fidelity as does RNA, and the catalytic properties of both PNA and TNA remain to be examined in detail. But we could imagine a sequence of steps in which PNA was abiotically synthesized and was the replication centerpiece of a primitive "proto-biological" system in which catalysts of various types were encoded for production by PNA. As discussed in the next chapter, chemical systems that generate their own catalysts—*autocatalytic cycles*—might have proceeded initially without any templating molecules whatsoever. But those systems that do possess such templates might more faithfully have regenerated important catalysts or supported more elaborate autocatalytic cycles. If PNA were the first templating molecule, it might have supported the production of amino acid–based catalysts—a primitive precursor to the protein-based catalysts of today, or the production of sugar-based catalysts such as RNA and TNA. In the former situation, PNA might have been both replicator and catalyst. In the latter case, the primitive living system would be a kind of mirror image of today's, having a peptide-based replicator and sugar-based catalysts. PNA's attractiveness as a templating molecule prior to the era of enatiopure sugars comes from laboratory results suggesting that, while not itself based on a chiral backbone, PNA does inhibit the insertion of the wrong enantiomer when assembling ribonucleotides to make RNA. Thus, PNA could have been the vehicle for template-driven enatiopurification.

The next stage of template-directed life might have come when the sophistication of the catalysts began to approach the limit of the replicative fidelity of PNA. All replicating polymers have error rates associated with the replication, in which the sequence of the nucleic acid bases is not faithfully copied from one molecule to another. In modern biological systems, mutations are rare and usually do nothing because the likelihood that they occur in a piece of the genetic code that specifies a crucial protein is small. If an expressable mutation does occur, usually (but not always) it is fatal. Too much mutation and the information contained in the genome would leak away (the system entropy would increase) to the point where replication itself becomes impossible—a situation known as "error catastrophe." The larger the number of genes in an organism, the lower the error rate must be (thus, bacteria can tolerate a higher error rate than can eukaryotes).

The champion of replicative fidelity is DNA because it is copied with the aid of DNA polymerase, which reduces the error rate from 1 in 10^4 nucleotides (roughly that of RNA) to as few as 1 in 10^9 nucleotides (Chapter 4). A crude rule of thumb for the maximum error tolerance permitted in a system where the copying is done in a linear fashion (nucleotide by nucleotide) is that an order of one error per genome length is permissible. Apparently, life sits on the edge of this permissible error rate: Eukaryotic cells have DNA with error rates of 1 in 10^9 and genome lengths of order 10^9 nucleotides. Bacteria, with $\frac{1}{1000}$ the genome size, have an enzymatic copying system a thousand times less reliable. While avoiding much larger error rates is intuitively sensible if cells are to faithfully generate the required proteins generation after generation, the apparent tendency to be "on the edge" of such a threshold may have to do with the potential advantages innovation offers. Too reliable a genetic system would inhibit evolution of cellular structure and function, and indeed would inhibit the growth of genetic code by gene duplication (Chapter 16); such nonevolving systems would be outcompeted in a Darwinian sense by more mutable systems that developed better metabolisms, cellular transport systems, mobility, or a raft of other possible novel adaptations.

Suppose PNA's error replication rate exceeded significantly—by one to two orders of magnitude—the 1 in 10^4 that characterizes RNA. The length of genetic instructions would have been limited to an order of 100 bases, and the number of codons (Chapter 4) a factor of two or three smaller still (allowing for the possibility that codon length in PNA was two bases rather than the three in use today). Alternatively, the chemical stability of a PNA polymer might have limited the length as well. In either case, living systems using such a replicating system would have reached what the German physical chemist Manfred Eigen called the "information crisis," wherein catalysts requiring larger numbers of codons could not have been synthesized. Living systems reaching this crisis point would have been limited in the sophistication of the catalysts they could generate or be forced to move off a template-guided biochemistry to (or back to) autocatalytic cycles whose stability was determined entirely by the rules of a metabolism-only approach described in the next chapter. But PNAs could construct TNA and RNA and might have preferentially synthesized the former because the number of competing sugars (with four-member rings) is far less than the number of competitors to ribose. Because PNA prefers one enantiomer in assembling the sugar-based nucleic acid polymer, it would seem an attractive, if not essential, precursor to the TNA or RNA world.

How long would the "PNA world" have gone on before yielding to the TNA or RNA world? If PNA served as both catalyst and replicator, or synthesized peptide-based catalysts closely related to it, then the advantage in moving to a more faithful sugar-based replicator would have been outweighed for some time by the need to have separate catalysts and replicators—and, hence, a more elaborate genome. If PNA instructed the production of sugar-based catalysts (TNA or RNA), then the PNA era might have been extremely short—this replicator preceding TNA or RNA only because no chiral selectivity, and hence no entropy decrement associated with the imposition of enantiopurity, was required.

Suppose, then, that the distinctiveness of threose as a four-carbon sugar made its synthesis easier; the next era in replication might then have been the TNA world. Why don't cells use TNA today, in place of RNA? Either the error rate or the catalytic properties of TNA must have been inferior to RNA. Laboratory studies on TNA are far less extensive than on RNA because the latter is of biological and medical interest whereas the former is not (no viruses exist that use TNA). Clearly, the determination of the catalytic power and replicative fidelity of TNA are of keen interest to the origin of life question. TNA might have encoded for ribose-based catalysts if, relative to what could previously be coded for, they were more versatile and powerful. Then such organisms—TNA based in their replication and ribose based in their metabolism—would have selected ribose out of the prebiotic sugar forest, even driving its synthesis, until it became the dominant sugar. At this point, even if the TNA and RNA error rates were comparable, a distinct advantage in having the same molecule be replicator and catalyst would have given RNA-based organisms the edge over their TNA-based cousins. The templated production of RNA from TNA—which is facile and might have occurred incidentally or in catalyst production since the dawn of the TNA world—would have led to the takeover by RNA as the templating molecule. Alternatively, if RNA replication simply possesses a lower error rate than that of TNA, takeover might have occurred in an information crisis akin to that speculated earlier to have engendered the earlier PNA–TNA transition.

The specific progression sketched here is highly speculative; it is intended to be illustrative of how a progression of ever more specific and sophisticated replicators might have led to the current situation. It is also simplified and linear in time (first one nucleic acid, then the other, and so on) to make the exposition clear. However, the progressive development of ever more sophisticated replicators must have been a much more complicated and messy affair, with perhaps other nucleic acid–based replicators in play as well. PNA and TNA are attractive because they avoid two of the more daunting problems in the *ab initio* synthesis of ribonucleic acid (i.e., synthesis in the absence of an existing nucleic acid and primers): enantiopurification and selection of the sugar ribose out of a forest of other sugars. However, the abiotic synthesis of neither PNA nor TNA has been achieved without the overt intervention of the experimenter—that is, in a way that could convincingly have occurred in a prebiotic environment. But the laboratory studies do show convincingly that once one kind of self-replicating polymer was developed, template-driven production of potentially more sophisticated polymers was possible.

What was prior to PNA or its achiral cousins? Chapter 9 discusses a form of life that undergoes metabolism with no replicating molecule to template and build the catalysts that sustain the system. Perhaps such systems were based on peptides and related compounds, and, in a subset of such systems, a templating polymer (PNA or likewise) was created in sufficient numbers that it took over the system and directed the production of a subset of the system's catalysts. Presumably, additional nontemplated catalytic molecules continued to be created repeatedly in autocatalytic cycles within the system, but the disadvantages of relying on catalysts that could not be produced in a directed fashion led to these being eliminated from the system, in favor of those PNA could template. Gerald Joyce of the University of California, San Diego, notes that this kind of transition led to the birth of Darwinian evolution, in which those systems that survived—by virtue of their more robust catalytic set or more reliable templating—perpetuated their genetic codes at the expense of others. And this, he argues, is the birth of life, on the basis of the definition that life is a chemical system that undergoes Darwinian evolution. In this view, life has a definite origin, though that origin likely occurred many times and in many places.

Alternatively PNA might not have been the first self-replicating polymer, even if it represents the first such polymer to undergo replication on a "residue by residue" (base-pair by base-pair) basis, which is a particular approach to self-replication. More generally, self-replication merely requires that rate of production of new copies of the molecule, by whatever means, exceed the rate of destruction. This opens up the possible chemistries of self-replicating, templating molecules to a much broader realm—even to the limit of arguing that the charge structures on mineral surfaces such as clays would serve as a template for replicating molecules. Such systems would not necessarily undergo Darwinian evolution, particularly if templating molecules were not confined to containers (primitive cells) or the fidelity of replication strongly depended on ambient, fluctuating conditions. But if such systems were intermediate between metabolism-only, autocatalytic systems explored in the next chapter and the Darwinian systems guided by nucleic acid templates, then the origin of life might not have constituted a threshold at all. Instead, it might have proceeded in a gradual fashion, as a sloppy free-for-all in which self-replicating polymers gave way eventually to the more orderly residue-by-residue replication of the nucleic acids.

8.5 Vesicle Formation and the Precursors to Cells

Life as we know it is organized around cells, which function as dynamic containers holding the genetic code, the protein assembly apparatus, and membranes within which energy generating and storage functions are conducted; in conjunction with cell walls, the flow of material and energy inward and outward is controlled (Chapter 4). The abiotic formation of simple containers within which chemical reactions may be isolated from the external environment has been demonstrated in the laboratory. *Amphiphilic vesicles* are one such class of structures that might possess some of the properties that would qualify them as protocells. Amphiphiles are the name for a general class of compounds (of which phospholipids are one example) in which one end of the molecule is polar and the other nonpolar. The nonpolar end is hydrophobic, so that it is repelled by water, whereas the polar end is hydrophilic. As in Figure 4.13, which depicted a lipid (fatty) membrane, a two-layer ("bilayer") assemblage will have the hydrophobic ends of each layer pointed inward toward each other, with the hydrophilic ends extending into the solution. While the figure depicted a short section of such a bilayer as a flat sheet, the minimum energy configuration of such an assemblage in water is that of a closed "vesicle," in which the hydrophilic ends extend outward toward the environment in the outer layer and toward the interior in the inner layer. Soap bubbles are a common, if transient, example of such amphiphilic containers.

While biologically generated lipids form bilayer vesicles, they do not inform us directly about the formation of such amphiphilic enclosures in the prebiotic world. Long-chain hydrocarbons—organic molecules composed of carbon and hydrogen—are manufactured in a variety of abiotic environments ranging from interstellar clouds to the upper atmospheres of the giant planets and Saturn's moon Titan. Hydrocarbons can react chemically with phosphate-based minerals to produce amphiphilic compounds such as phospholipids. For these to be the raw materials for protocells requires a source of hydrocarbons on the Earth.

Early models of the Earth's atmosphere postulated a reducing atmosphere, the inspiration for the Miller-Urey experiments of the mid-20th century in which water, methane, and ammonia were reacted using a spark discharge to produce amino acids. However, a strongly reducing atmosphere is not consistent with the initial volatile budget with which the Earth was likely endowed (Chapter 6). Nor would it be consistent with the need for the Earth to have a massive carbon dioxide atmosphere early in its history to sustain liquid water in the face of the faint early Sun, or with the presence of massive amounts of inorganic carbon in the present crust of the Earth and the atmosphere of Venus (Chapter 11). The more likely case of weakly reducing or mildly oxidizing atmosphere militates against the production of significant amounts of hydrocarbons within the surface-atmosphere system of Earth, as occurs today on Saturn's moon Titan (Chapter 13).

However, delivery of hydrocarbons by comets or asteroidal fragments, with survival of a nonnegligible fraction of this material indicated by impact simulations, is a viable alternative source, as quantified by E. Pierazzo (Planetary Science Institute) and C. Chyba (SETI Institute). Indeed, it is possible to imagine a very early Earth environment in which hydrocarbons, including methane, were delivered in abundance by impacts so that they provided a part of the greenhouse warming required for liquid water to be sustained at Earth's surface and thereby maintained a weakly reducing atmospheric oxidation state. Later, carbon dioxide may have taken over in the atmosphere, by which time primitive organisms might have tolerated the relatively more oxidizing conditions (but still much less oxidizing than the current, aerobic, atmosphere). The reader can see that the jury is still out on the oxidation state of the early Earth's atmosphere.

Some work has been done on the tendency of vesicles to grow when additional, raw, membrane-forming material is added, eventually splitting to form smaller and more stable vesicles. Certainly, this must reflect the general property that, as the surface area increases, the surface tension (surface energy) of the vesicle changes in such a way that the energy advantage associated with the formation of a spherical system from a planar one declines. This in turn encourages the formation of two smaller vesicles. While of some interest in resembling the division of one cell into two, this process may be less significant than the simple existence of the enclosure itself. And, as Princeton physicist Freeman Dyson points out, there is an enormous difference in the significance of reproduction (the division of one cell into two) and replication (in which a precise copy of a molecule or molecular system is made). Life's special property is its reliability of replication; a vesicle that splits after growing is merely reproducing.

The advantage of having an "inside" and an "outside" is that gradients can be set up between the two; these gradients might be compositional, chemical (pH, acidity), electrical, thermal or other. Such gradients may relate simply to the presence of a finite diffusion rate of material and energy into and out of the vesicle so that chemical reactions within the vesicle lead to changes in temperature and composition that are maintained against the external environment by the finite rate of diffusion across the membrane. The membrane itself may provide a site for surface chemistry particularly suited to the production of certain compounds. It may also absorb certain pigments that could capture light energy and convert it into an electrical potential for doing work across the membrane into the vesicle.

On the other hand, the German chemist G. Wächtershäuser has pointed out that membranes behaving like lipids allow water through but are impermeable to hydrophilic organic molecules—making such vesicles more like coffins than sites of primitive organic chemistry. Vesicle proponents disagree, arguing that uncharged amino acids diffuse reasonably quickly through the membrane. And RNA has been shown in the laboratory to affect the permeability of membranes to transport of ions. Furthermore, in living cells various molecular structures breach the bilayer to create channels where exchange of matter and energy is accelerated (see Figure 4.17). The controversy illustrates that the extent to which abiotic amphiphilic vesicles and potential chemical systems contained within remain poorly understood, despite two decades of laboratory experiments.

If not protocells, then where were the sites of prebiotic autocatalytic chemistry or the protobiological development and utilization of replicators? An alternative is the restricted dimensionality and enhanced catalytic action of surface chemistry. With regard to the former, organic molecules can, in principle, be inserted into the surface of a microcrystal in a way that restricts their lateral mobility. For a replicating molecule such as RNA or its precursors or successors, this might provide the equivalent of the clamping capability that is key to the replicase or transferase enzymes. Given this, University of California chemist L. Orgel has speculated that a nucleotide polymer attached to the surface of a suitable mineral surface could organize a complementary sequence. Such a system might quickly saturate, that is, be unable to add more than a few polymers, if the original nucleotide polymer remained attached to the surface. But Orgel goes further to suggest that the phosphate groups on the complementary nucleotide strand might then seed the formation of a new mineral crystal. In this way, the growing surface of the crystal functions as the equivalent of a ribosomal site, a replicase enzyme, and the two-dimensional version of a host cell.

Wächtershäuser examined the potential role mineral surfaces would play in providing a source of usable energy for organic reactions and a redox environment suitable for the synthesis of organic molecules. In the anaerobic conditions present at the surface of the Earth during its early history (Chapter 11), the mineral pyrite (FeS_2) was likely to be stable, and perhaps abundant in the more reducing conditions around submarine vents. At such vents the reaction of a less oxidized iron–sulfur compound, FeS, with hydrogen sulfide (exothermic; see Chapter 3), would lead to the production of pyrite; the reaction can be written $FeS + H_2S \rightarrow FeS_2 + H_2$. As the reaction indicates, the formation of pyrite involves the release of hydrogen. Both the free hydrogen and the heat released might provide a net amount of free energy (Chapter 7) for the synthesis of organic molecules on the pyrite surface itself. Residual positive electrical charges on the surface of the pyrite would force carbon-bearing compounds such as hydroxyl carbon dioxide (HCO_3^-), and the organic compounds formed from them, to adhere to the surface. Thus, the pyrite plays two roles that echo the roles of enzymes in the DNA world: (1) accelerating or enabling chemical reactions through enabling chemical states that would not otherwise exist and (2) forcing the reactants into restricted surface sites that place them in mutual contact with each other.

Other substrates for protobiological reactions have been proposed, including "oil slicks" of hydrocarbons brought to the Earth in comets and floating atop primordial oceans. Diverse, transient surfaces composed of a wide range of different materials may have served as sites for interesting organic chemistry, but eventually enclosed and semipermeable structures—cells—won the day.

8.6 The Fundamental Role of Containers and Surfaces in Increasing Replicator Complexity

Postulates regarding particular chemical reactions in specific environments can be tested by laboratory experiments or experimental data regarding reaction rates. But life today is not a single chemical reaction or a small suite of reactions—it is a much more complicated set of autocatalytic cycles regulated by the templated production of very specific suites of enzymes. How can we hope to understand the level of complexity possible on mineral surfaces or in protocellular systems such as vesicles? The chemistry over brief periods can be simulated in the lab, but the long-term evolution of such systems is beyond the reach of laboratory experiments. Nowhere in natural environments on present-day Earth should we expect to find such systems ongoing. Modern DNA-based life is ubiquitous so that it has control of all of the organic carbon accessible to sampling on Earth (and, as well, the aerobic conditions in atmosphere and ocean would quickly oxidize any organic compounds not under control of modern metabolic processes). Evidence for such protobiology might be preserved on other planetary surfaces, notably Titan, but detailed exploration that would find such systems is many years away (Chapter 14).

Computer simulations represent the only means by which the evolution over long timescales of complex chemical systems can be explored at present. And such simulations have already indicated crucial differences between the evolution of such systems in a large-scale organic soup versus a compartmentalized environment afforded by mineral surfaces or vesicles. One recent such effort is that of Hungarian mathematician P. Szabó and colleagues. They focus on replicator-catalyst systems intended to simulate the essential features of an RNA world. At issue is the extent to which such a system can evolve ever-longer genomes with replicative accuracy that keeps pace with the increasing genome length, according to the rule of thumb discussed in Chapter 4: The error rate per nucleotide is the inverse of the total number of nucleotides in the genome.

Absent any other effects, a lengthening genome using a given base-pairing system and nucleotide backbone would have a linearly increasing number of errors as the genome length increases, or worse if the increasing length leads to a longer total replication time with a concomitant increase in the possible opportunities to introduce errors. Counteracting this unhappy situation is the fact that the genome-bearing molecule in this simulation is also the replicase "enzyme" (or, more precisely, ribozyme or "peptozyme" if, as is intended, the simulation refers to RNA or simpler nucleic acid precursors like TNA or PNA). The longer the replicase, the potentially more effective it could be at catalyzing an accurate replication since a longer polymer has a potentially more complex tertiary structure.

The computer simulation allows interaction among a population of evolving molecules, each of which possesses the capability for templating and ribozymic activity. The world within which these activities occur is a grid of 16,000 cells, and each of them can be occupied by one or zero polymers. A population of monomers, raw material for the growth process, is fed into each grid in such a way that the monomeric abundance is constant from one grid to another. Polymers in neighboring cells can interact with each other according to the rule that one of the molecules would be a template for its own replication while the other would catalyze the reaction. Within a given polymer, different sequences code for three different traits:

1. The speed of the replicase in performing a replication of a template.
2. The accuracy of the replication accomplished by the replicase activity.
3. The affinity of a given molecule to act as a template.

Mutations are permitted that change the length of the genome of the particular sequence of the bases so that different molecules over time would express each of these traits to greater or lesser degree. As sequences increase in length, however, their viability as replicase or template depends on evermore efficient replication. Hence, the system has a number of possible outcomes. Replicase catalytic function and templating efficiency might improve at rates sufficient to produce ever-larger genomes, the system could stagnate at a particular mean genome length, or genomes could degrade into ever-shorter and less effective catalysts and templates.

In fact, the system tends to evolve toward the production of two types of molecules:

1. Large, complex replicators with high fidelity and strong replicase catalytic activity. These "altruists," as Szabó and colleagues referred to them, serve as templates and replicase catalysts that increase the mean size of the molecules as the evolution wears on; eventually a steady-state size is reached.
2. A class of short molecules with very high quality as templates but very poor replicase function. These "parasites" contribute nothing to the growth of the mean size of the molecules in the system; they simply use the replicase function of the larger molecules to make additional copies of their own short selves.

The parasites remain a limited fraction of the population because the simulation specifies that only local interactions among molecules should lead to replication, and diffusion of molecules from the site of their production is very limited. This immobility of molecules ensures that there are regions in which the altruists tend to be clustered and where they repeatedly undergo replicator–template reactions—encouraging growth and discouraging extensive contact with the parasites. When the restriction on diffusion across the model world is relaxed after the simulation builds up a population of large polymer altruists, a startling change occurs. There is no clustering of the altruists; the parasites dominate the replication activity and inhibit the growth of ever-larger altruists. Because of this, there are too few replicase molecules with sufficient activity and fidelity to sustain an evolution toward larger sizes. Template quality is the last property to decline, which means the parasites are sustained even as they drag the system down toward a functionless soup of ever-smaller polymers. Eventually the polymer population goes extinct in favor of small molecules that do not exhibit replicative or templating function (Figure 8.11, see Color Plate 11).

The criterion for the changeover in the system behavior from growth of altruists to destruction by parasites is that the rate of diffusion of molecules across cells exceeds the rate of birth of new molecules formed from the replicative-templating activity. The analogy to the real world is clear—autocatalytic evolution hosted on restricted sites or in confined spaces imposed by low diffusion across grids corresponds to protobiology occurring in favorable sites on mineral surfaces, or in the confined spaces defined by the interiors of vesicles. Evolution in a high-diffusion environment is the equivalent of chemistry in a vast aqueous soup, where molecules formed in one place may be wafted significant distances before another replicative event takes place.

By imposing fixed sites on the simulation, a potentially key source of entropy—random diffusion of molecules—is removed from the system. Hence, the net tendency of the simulation is not so surprising—evolution toward increasing levels of self-organization in an imposed low-entropy environment, evolution toward randomness in a high-entropy situation. However, the details of the outcome are also tied intimately to the assumptions made by Szabó and colleagues. The grid (the "world") is two dimensional for computational convenience; this removes degrees of freedom that might, even in a low-entropy environment, contribute to an increase in disorder. The template-copying rules differ from the complementary Watson-Crick base pairing known to occur in DNA, RNA, TNA, and PNA, in which A pairs with U and G with C (using the base abbreviations introduced in Chapter 4). Instead, again for computational efficiency, each of the four functional monomers pairs only with its own type: A with A, and so on. Because each of the four base monomers carries a different characteristic (A, replicase speed; B, replicase fidelity; C, template activity; D, neutral), changing the rules from replication of the identical monomer to a Watson-Crick complementary pairing could dramatically change the evolution. For example, a polymer that evolves to become an excellent replicase could then give birth to a molecule with high propensity for being a template. Finally, the "large" polymers are not very large—they are typically several dozen monomers in length, orders of magnitude below the minimum genome size for real organisms.

These and other limitations of the computer simulation relative to a real biological or protobiological system only illustrate the potential extraordinary complexity of the latter relative to the former. But it is to be hoped (and, unfortunately, that is the strongest statement we can make at this point) that such simulations do map out the general behavior of systems that might have been precursors to life.

Indeed, laboratory experiments on RNA since the 1970s have demonstrated aspects of the evolution simulated (much later) by Szabó and colleagues. German biochemist

M. Eigen demonstrated the ability to grow an RNA molecule from the monomeric units catalyzed by a biologically produced polymerase enzyme, in the absence of a templating RNA molecule. L. Orgel (UC San Diego), conversely, demonstrated that nucleotide monomers could polymerize to form RNA using another RNA as a template but with no catalyst other than zinc ions, that is, with no polymerase present. Such experiments illustrate the tendency of nucleic acids to template and to replicate, essential to the growth of the altruists though they prove nothing for abiotically generated molecules. Required was the presence of biologically produced molecules in the soup, either replicator or template, but this simply reflects the inability of a laboratory experiment to handle the large number of molecules, the complexity of surfaces or compartments, and the extended time that a "real-world equivalent" of the Szabó and colleagues computer simulation would entail. The system entropy must be artificially decreased, substantially, to see self-organization "happen"—and this was done by employing molecules whose high degree of order is owed to their biological origin. More recent directed evolution experiments refine these results by showing how RNA can catalyze the construction of ever more powerful ribozymes—RNA strands as catalysts—but still require that we begin with RNA.

S. Spiegelman and colleagues at the University of Illinois demonstrated the opposite evolution—decreasing the order of the system—in dramatic experiments performed in the 1960s. Recall that the RNA genome of retroviruses is replicated within the host cell using replicase enzymes that are manufactured by the ribosomes of the host cell (accomplished through the reverse transcriptase enzyme that generates DNA with the viral genome which the host cell incorrectly recognizes and transcribes as its own). Leaving aside its parasitic nature in utilizing the transcription and ribosomal apparatus of the host cell, the virus is akin to the altruistic polymers of the numerical experiments in sustaining a complex genome generation after generation.

Spiegelman and colleagues took a virus (the so-called $Q\beta$ virus whose host is the *E. coli* bacterium) and stripped it of what few proteins it possessed (including its protective protein coat) so that it was a naked strand of RNA, 4,500 nucleotide units long, in a test tube. Into the tube were added the replicase enzyme specific to the virus and abundant monomeric nucleotide units. Thus, the complete genome was present with the raw materials and the enzyme needed to undergo replication, with no need to invade a host bacterial cell and co-opt its replicative machinery. Because all of the functions of the virus except replication itself had been removed, a number of mutant strains of the RNA began to appear and propagate that would otherwise have been eliminated in the normal demands of the viral life cycle. By altering the environment in terms of the amount of replicase versus monomer units, temperature, or presence of enzymes that tended to degrade the RNA, Spiegelman and colleagues could alter the mix of mutants that persisted. Most interesting was the imposition of a test tube environment in which rapid replication of the genome was encouraged. Shorted strands of RNA could replicate faster, so the system evolved with time toward shorter and shorter mean lengths of RNA. From the original 4,500-unit polymers, the experiment yielded a soup of RNA strands no longer than 220 units, comprising the site that attached to the replicase to initiate replication. None of the other portions of the genome—involving host recognition, protein coats, and so on—were required; in fact, they only got in the way of rapid replication. The end state was a genome completely dependent on the supply of enzyme and raw materials—the ultimate parasite—that excluded from the test tube world longer genomes with greater functionality.

Although the starting materials were derived entirely from biological processes, the progression toward less elaborate genomes with lower information content should be a general property of any system in which constraints have been removed and, hence, the overall entropy of the system increased. In the case of the $Q\beta$ genome, those constraints were the processes by which the genome as a virus must invade the host cell and interact with its machinery—and the constraints were removed by providing the essential replicative ingredients in a test tube. In the case of the Szabó et al. computer experiments, the constraint removed is the restriction in volume (really, area) accessible to the polymers. In the limit of very rapid diffusion from one cell to another, the area accessible to a given polymer increased from 1 to 16,000 cells; hence the increase in entropy must equal or exceed $k \ln(1.6 \times 10^4)$ per polymer.

Dyson has noted that the final size of the bottom-up and top-down experiments of Eigen and Spiegelman, respectively, yielded RNA strands of about the same size—100 to 300 bases. In the $Q\beta$ virus, the three RNA "genes" have approximately 1,500 bases each, so the Eigen and Spiegelman strands are of order $\frac{1}{10}$ the size of each gene. It is intriguing that an RNA strand of order 100 bases represents an order-of-magnitude dividing line between a completely nonfunctional fragment of order 10 bases that cannot even serve as a recognition site for the replicase and that of a single, functional RNA gene of order 10^3 bases.

8.7 Sizes of Protocells and the RNA World

The essence of the arguments presented in Section 8.6 is that cells are compartments that reduce the entropy of the system by creating partitions in the world between an "inside" and an "outside," restricting or specifying the positions of polymers. Whatever other functions cells have acquired, including energy and material capture and transport, the entropy-reducing function would seem fundamental. But why cells, when the planet is filled with mineral surfaces that can do the same? Almost certainly countless autocatalytic organic chemical systems utilized mineral surfaces as the two-dimensional equivalents of protocells, but such surfaces have disadvantages relative to the protocell. The interior of the protocell can be packed with membranes that possess an enormously larger surface area than that of a simple spheroidal container, in the same way that a gas chromatographic column (Chapter 14) or the linings of our lungs possess surface areas far in excess of that defined geometrically by their total enclosed volumes. Protocells can divide and—as long as the supply of nutrients is sufficient—provide new spaces for chemistry when the old surfaces become saturated. And protocells can be mobile, which may be advantageous relative to the fixed sites of minerals when the local nutrient pool is expended or destroyed by environmental changes.

How large might protocells have been? Experiments with lipid vesicles, as a proxy for abiotic amphiphilic vesicles, suggest a size of 0.1 μm or less, limited by stability of the membrane. This is one-tenth the typical size of a prokaryotic cell (Figure 8.12) and leads then to several questions regarding the relationship of such entities to the modern cell. What is the capability within such vesicles for housing and replicating a genome of sufficient length for the persistence of a self-contained organism? Are there organisms on the

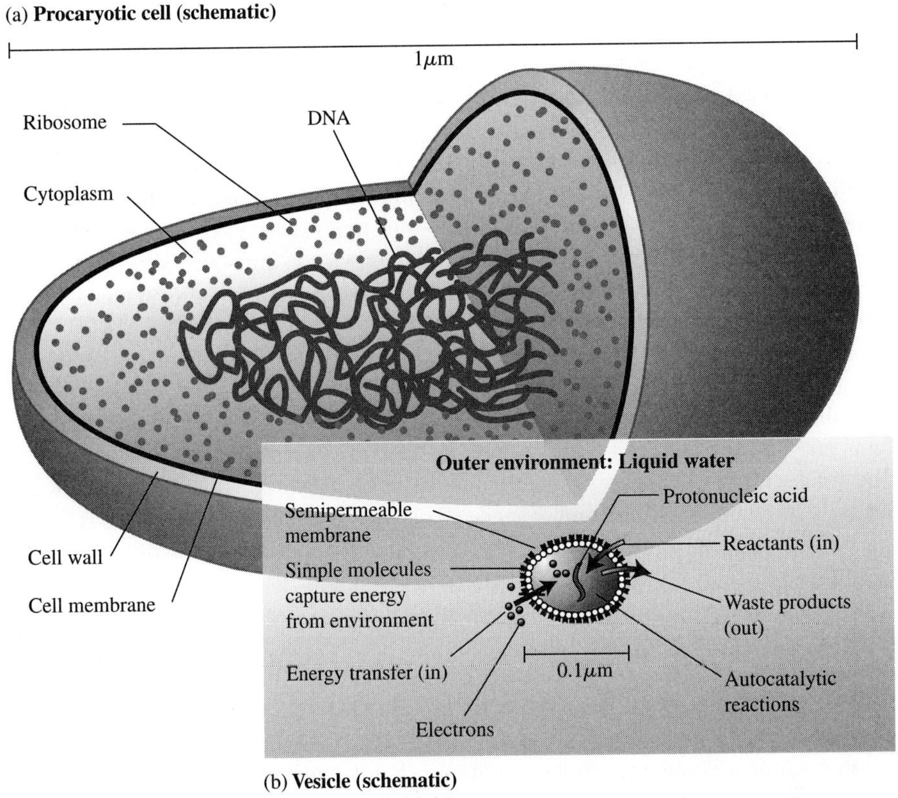

(a) Procaryotic cell (schematic)

$1\,\mu m$

Ribosome
DNA
Cytoplasm
Cell wall
Cell membrane

Outer environment: Liquid water

Semipermeable membrane
Simple molecules capture energy from environment
Energy transfer (in)
Electrons
Protonucleic acid
Reactants (in)
Waste products (out)
Autocatalytic reactions

$0.1\,\mu m$

(b) Vesicle (schematic)

FIGURE 8.12 The size relationship is indicated between (a) a typical prokaryotic cell and (inset b) an amphiphilic vesicle (a putative protocell) of the sort synthesized in the laboratory. Protocell size is shown relative to the prokaryote. Possible structures within the protocellular vesicle are sketched.

Earth today with cell diameters comparable to those of lipid vesicles? If not, is the size disparity between vesicles and prokaryotes suggestive of an origin of true cells completely unrelated to that of vesicles (perhaps coming directly from chemistry on mineral surfaces)? Or were the first true cells enclosures around multiple vesicles that reflected a symbiosis akin to the later prokaryotic symbioses that led to the much-larger eukaryotes (Chapter 16), both events constituting the solution to a crisis of internal space?

Claims by R. Folk and others of evidence for prokaryotic life-forms much smaller than typical bacteria, referred to as nannobacteria or nanobacteria (the spelling varies in the literature), would suggest that minimum prokaryotic cell size may approach the maximum size of vesicles. Bacterial cells of different types are found in the 0.3- to 0.5-μm-diameter range, and there are reports of cultures from human and cow blood with cell diameters down to 0.2 μm. Circumstantial evidence based on the passage of some organisms through very fine filters suggests forms with linear extent as small as 0.05 μm. (Viruses, of course typically exhibit diameters of 0.05 μm, but the fact that these are not cells argues against the relevance of these entities to the present discussion.)

A minimum cell size may be theoretically estimated, for modern prokaryotes, by assuming a minimalist cell with a 500-gene genome (Chapter 16), one ribosome, plus one mRNA (messenger RNA) and a set of 20 tRNA (transfer RNA) molecules per protein type in the cell, and a number of proteins (10^2 to 10^3) that would give the cell a minimal self-sufficient functionality. The cell membrane plus wall is assumed 0.01 to 0.02 μm thick. The result, for reasonable assumptions about the packing capability of the DNA-based genome, is a cell with a diameter between 0.2 and 0.4 μm across, corresponding to a volume larger by an order of magnitude than our putative protocellular vesicles. Such a minimum size also calls into question the existence of the nannobacteria that are claimed to have diameters several times smaller than the calculated minimum.

To countenance vesicles, then, requires that we rid our protocell of the bulkiest machinery—the ribosomes. And to get rid of ribosomes requires almost certainly moving from the dual-polymer system of DNA and proteins to the single-polymer replicator–catalyst system of RNA. A reasonably sized vesicle, roughly 0.1 μm in diameter, could house of order 10^3 ribozymes each containing about 50 to 100 nucleotides and a small genome of about 50,000 bases, or 30 to 50 RNA genes, according to calculations performed by J. Szostak of the Howard Hughes Medical Institute. This is a genome 10 times longer than that of the $Q\beta$ virus but 20 times shorter than the minimum free-living bacterial genome. However, as pointed out earlier, a single-polymer life-form requires, in principle, a smaller genome to perform the same set of functions as a dual-polymer DNA-protein life-form, albeit not as well.

Hence, it is possible that RNA-based life-forms with genomes and catalysts enclosed in amphiphilic vesicles could have existed as free-living organisms, feeding off organic matter in the environment and deriving thermal control or additional energy from chromophores attached to the vesicle surface. Because ATP and ADP are directly related chemically to the AMP base of RNA, energy storage and transport via an ATP system can also be envisioned for such vesicles. An example minimum metabolism for such an RNA-based protocell has been invented by Florida biochemist S. Benner and is illustrated in Figure 8.13. It allows for nucleic acid synthesis, incorporation of carbon dioxide, and production of carbohydrates—relying on roughly 50 different kinds of catalysts, none of which are proteins.

There might even be enough room in such vesicles for membrane-based compartments, which would allow RNA-based genes to remain close to their ribozymic products, and perhaps to metabolic energy sources, in the absence of the transport systems seen in modern cells. Like the need for vesicles themselves, the compartments would provide the localization and entropy decrease that would encourage the growth of more sophisticated genomic systems, systems that eventually required the abandonment of the amphiphilic vesicles in favor of the first true cells. This process might have proceeded gradually, as colonies of vesicles built lipid-based cellular membranes around themselves or small parasitic vesicles invaded the newly formed membranes around larger vesicles, ending up in symbiotic relationships ranging from unfavorable to the larger vesicle (the prototype of the virus-bacterium relationship) to beneficial (the prototype of the much later prokaryotic symbioses leading to eukaryotic cells).

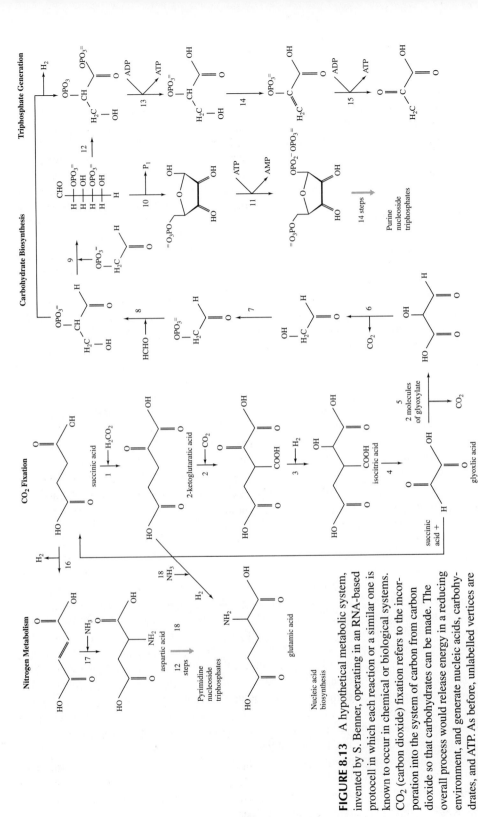

FIGURE 8.13 A hypothetical metabolic system, invented by S. Benner, operating in an RNA-based protocell in which each reaction or a similar one is known to occur in chemical or biological systems. CO_2 (carbon dioxide) fixation refers to the incorporation into the system of carbon from carbon dioxide so that carbohydrates can be made. The overall process would release energy in a reducing environment, and generate nucleic acids, carbohydrates, and ATP. As before, unlabelled vertices are bonds with C or CH groups.

Indeed, it is possible that yet smaller vesicles housed metabolizing systems without functioning genomes at all, a form of life that Chapter 9 explores in more detail. If indeed such "nondirected" metabolisms preceded the RNA world, we might imagine eventual changes in the nature of such systems as first PNA and then TNA were synthesized and incorporated into vesicles, co-opting the chemistry and inventing Darwinian evolution. Protocells hosting nondirected metabolisms might have persisted for some time after the invention of Darwinian evolution, perhaps interacting with PNA- or TNA-directed protocells as diseases or parasites that used the primitive genetic machinery to spin off new nondirected metabolisms.

Why, then, does RNA exist? If TNA were a reasonable Watson-Crick replicator and templating molecule, could not life have utilized TNA, perhaps as a subsidiary templating molecule in much the way RNA is used, up to the present? Suppose laboratory studies show that TNA is not that much worse a replicator than RNA? Suppose ribozymes could not outcompete other catalysts in the system so that there was no catalysis-based pressure to synthesize RNA? What else might have mitigated in favor of the synthesis of RNA, given the difficulty of plucking the ribose out of the sugar forest? The abiotic synthesis of AMP, and then ADP and ATP from AMP (both of which appear straightforward from laboratory studies), would have led (perhaps inevitably) to the incorporation of the ADP–ATP energy storage system in protocellular metabolisms. In those protocells employing nondirected metabolisms, this would have created a rich supply of ribose and adenine, but the jump to RNA all at once might have been too difficult. In those protocells already using PNA or TNA as replicators, all the requisite nucleic acid bases were present. There the buildup of ribose associated with the ADP–ATP cycle might have led to the facile templating of RNA via TNA or PNA—and eventually to the takeover of the templating role by RNA just by virtue of the abundance of ribose. RNA might then have co-opted the catalytic role from other molecules, even if ribozymes were not necessarily better catalysts, because it created the advantage of a single molecule as replicator and catalyst in vesicle-based protocells increasingly squeezed for internal volume.

Today the only trace of an ancient life different from our own is the presence of RNA alongside DNA, and the many roles—from viral parasite to faithful scribe and messenger—it plays in modern life. Yet, we do not know when the RNA–DNA transition occurred in the history of the Earth. We would like to imagine that it was early, but there is nothing to rule out the persistence of RNA-based life for many hundreds of millions of years after the formation of the Earth. And why could it not have persisted a billion years into the history of life? Suddenly, we turn around and remind ourselves that there is no direct evidence—geologic or genomic—that RNA-based life existed, ever. It is just so tempting an idea; but then, to explain the origin of the RNA world, we must populate earlier epochs with yet other replicators, simpler but absent from modern biochemistry, in the hope that we will find one so simple that its abiotic synthesis can be finally understood. Now the simple model has become a complex story. Perhaps the answer will be found in the lab, in the output of powerfully fast computers, or in the preserved organic detritus on other worlds in our planetary system.

Or perhaps the answer is already there—in the extraordinary power and persistence of the DNA-protein duality, which has carried life across billions of years and along paths of metabolic and morphological evolution remarkably diverse, perhaps there is a message that RNA is, and has always been, just a helper.

QUESTIONS AND PROBLEMS

1. Assume comets to be mostly water ice with 1 percent hydrocarbons and nitriles (by number) relative to the ice. Typical size of a comet (the nucleus) is 1 km, and the ice porosity (ratio empty space to total volume) should be taken as 50 percent. For hydrocarbon survival rates of (a) 1 percent and (b) 10 percent on impact with Earth, how many comets would be required to generate a hydrocarbon "slick" a meter deep on the surface of the prebiotic Earth's oceans (assume 90 percent coverage of Earth's surface by its oceans early on)?

2. Calculate the ratio of surface area to volume for a typical prokaryotic cell, a nannobacterium, and a prebiotic vesicle relative to the same ratio for a eukaryotic cell. Assume all cells are spherical. What might be the challenges of having high surface area? What might be the advantages? Referring back to Chapter 4, how do large cells compensate for having such small surface areas (relative to their volumes) across which nutrients and wastes can be transferred?

3. Do a literature search to determine if other alternatives to RNA beyond TNA and PNA have been proposed (you will probably want to focus on the more recent literature). Are these alternatives based on chiral or achiral molecules? What are the advantages and disadvantages of the various polymers as information-carrying systems, and as catalysts?

4. DNA and RNA have rather different stabilities in alkaline environments. Quantify these differences using information in a biochemistry textbook or literature review article. How might these differing stabilities have helped determine the differing roles these two biopolymers play in life?

5. Nannobacteria remain a rather controversial subject. Find papers in the literature that claim evidence for very tiny bacteria, as well as papers that may refute such claims. What are the challenges in identifying or culturing (growing large numbers of) such ultrasmall organisms?

6. Because some viruses contain RNA only and require host DNA to replicate, might viruses be relics of a pre-DNA time, when only RNA existed? Try to find arguments in the literature for and against such a notion.

7. A "DNA world" is one with only DNA as the replicating molecule and no RNA, TNA, PNA, and so on. Is such a world plausible, based on your understanding of the properties of DNA? Explain.

8. Information content of a system is based on a binary choice—either something is or is not what we seek. For DNA, each nucleotide contains two bits of information: It is either a purine or pyrimidine, and then it is either a cytosine or a thymine or a guanine or an adenine. (This assumes no bias in the ordering of nucleotides along the DNA chain.) Hence, at two bits of information per nucleotide, what is the genetic information content (in bits) of a genome that is 4,000,000 bases long (typical of a bacterium)? If each position (base) on the DNA polymer can have one of four letters, what is the total number of sequences possible? How does this compare with the total number of types of bacteria thought to exist today?

9. As described in Chapter 7, the information definition of entropy implies that one bit of information costs us $k \ln 2$ of entropy, where ln is the natural logarithm. Thus, the

gain of one bit of information requires the expenditure of $kT \ln 2$ units of work. (This arises because, at constant volume, $\Delta E = T \Delta S$ from the first law presented in Chapter 7, and is a minimum because not all of the internal energy change can be converted to work.) What is this number in joules at room temperature? What is the minimum energy required to construct the genome of a typical bacterium, given the genome length cited in problem 8? The heat released by oxidation of glucose to CO_2 and H_2O is 15,000 J/g of glucose. How many grams of glucose are required, then, to synthesize the bacterial genome? (This is, of course, a gross minimum, since not all the energy released as heat can be converted into usable metabolic energy.) How does this compare with the weight of a bacterial cell?

10. What do the considerations in problems 8 and 9 regarding the energy cost of information imply with regard to "economies of scale"—an ultrasmall cell with a minimal genome versus a large cell (eukaryotic) with a large genome? (It may be convenient to consult Table 16.1, which lists the number of genes in various organisms, and to make a crude rule of thumb approximation that each gene contains of order 10^3 base pairs.)

SUGGESTED READINGS

Iris Fry, *The Emergence of Life on Earth* (2000, Rutgers University Press, New Brunswide, NJ) an account of efforts to understand the origin of life, is authoritative and well written. It does not entirely supplant Stephen F. Mason, *Chemical Evolution: Origins of the Elements, Molecules and Living Systems* (1991, Clarendon Press, Oxford), because the latter is more technical and hence provides interesting details on prebiotic chemistry. The July 11, 2002, issue of *Nature* (vol. 418) contains a series of articles (see References) by Joyce, Cech, and others on RNA chemistry and the possible nature of RNA-based life. The NRC's Space Studies Board hosted a workshop summarized in the excellent report *Size Limits of Very Small Microorganisms* (1999, National Academy Press, Washington D.C.) that includes speculations on the details of RNA-based chemistry and the possible size range of protocells that might have predated the DNA era. Harold Morowitz, *The Origins of Cellular Life* (1992, Yale University Press, New Haven, CT), treats the origin of life with a focus on the development of the first cells. Much more recently, Geoffrey Zubay, *Origins of Life on the Earth and in the Cosmos*, 2nd ed., (2000, McGraw-Hill, New York), also focuses on the chemistry of life's origin. Freeman Dyson's *Origins of Life* (1999, Cambridge University Press, Cambridge), which is featured prominently in the next chapter, takes a broad, physicist's view of the metabolism-first versus replication-first approaches to life's origin.

REFERENCES

Avetisov, V. A. and Goldanskii, V. I. (1996). "Mirror Symmetry Breaking at the Molecular Level," *Proc. Nat. Acad. Sci. USA*, **93**, 11435–11442.

Benner, S. A. (1999). "How Small Can an Organism Be?" In *Size Limits of Very Small Microorganisms: Proceedings of a Workshop*, National Academy Press, Washington D.C., p. 135. (Available through www.nas.edu)

Bonner, W. A. (1991). "The Origin and Amplification of Biomolecular Chirality," *Orig. Life Evol. Biosph.*, **21,** 51–111.

Cech, T. R. (1993). "The Efficiency and Versatility of Catalytic RNA: Implications for an RNA World," *Gene*, **135,** 33–36.

Cech, T. R. (2000). "Structural Biology: The Ribosome Is a Ribozyme," *Sci.*, **289,** 878–879.

Chyba, C. F. and McDonald, G. D. (1995). "The Origin of Life in the Solar System: Current Issues," *Ann. Rev. Earth Planet. Sci.*, **23,** 215–249.

Cody, G., Boctor, N. Z., Filley, T. R., Hazen, R. M., Scott, J. H., Sharma, A., and Yoder, H. S., Jr. (2000). "Primordial Carbonylated Iron-Sulfur Compounds and the Synthesis of Pyruvate," *Sci.*, **289,** 1337–1340.

Deamer, D.W. (1997). "The First Living Systems: A Bioenergetic Perspective," *Microbiol. Molec. Biol. Rev.*, **61**(2), 239–261

Doudna, J. A. and Cech, T. R. (2002). "The Chemical Repertoire of Natural Ribozymes," *Nature*, **418,** 222–228.

Dyson, F. (1999). *Origins of Life*, Cambridge University Press, Cambridge.

Fersht, A. (1999). *Structure and Mechanism in Protein Science: A Guide to Enzyme Catalysis and Protein Folding*, W. H. Freeman and Company, New York.

Folk, R. L. (1993). "SEM Imagining of Bacteria and Nanno Bacteria in Carbonate Sediments and Rocks," *J. Sedimen. Petrol.*, **63,** 990–999.

Fry, I. (2000). *The Emergence of Life on Earth*, Rutgers University Press, New Brunswick, NJ.

Hayes, B. (1998). "The Invention of the Genetic Code," *Am. Sci.*, **86,** 8–17.

Haynie, D. T. (2001). *Biological Thermodynamics*, Cambridge University Press, Cambridge.

Horneck, G., Rettberg, P., Reitz, G., Wehner, J., Eschweiler, U., Strauch, K., Panitz, C., Starke, V., and Baumstark-Khan, C. (2001). "Protection of Bacterial Spores in Space, a Contribution to the Discussion on Panspermia," *Orig. Life Evol. Biosp.*, **31,** 527–547.

Johnston, W. K., Unrau, P. J., Lawrence, M. S., Glasner, M. E., and Bartel, D. P. (2001). "RNA-Catalyzed RNA Polymerization: Accurate and General RNA-Templated Primer Extension," *Sci.*, **292,** 1319–1325.

Joyce, G. F. (1989). "RNA Evolution and the Origins of Life," *Nature*, **338,** 217–224.

Joyce, G. F. (2000). "RNA Structure: Ribozyme Evolution at the Crossroads," *Sci.*, **289,** 401–402.

Joyce, G. F. (2002a). "The Antiquity of RNA-Based Evolution," *Nature*, **418,** 214–221.

Joyce, G. F. (2002b). "Molecular Evolution: Booting Up Life," *Nature*, **420,** 278–279.

Lunine, J. I. (1999). *Earth: Evolution of a Habitable World*, Cambridge University Press, Cambridge.

Mason, S. F. (1991). *Chemical Evolution: Origin of the Elements, Molecules and Living Systems*, Clarendon Press, Oxford.

Melosh, H. J. (1988). "The Rocky Road to Panspermia," *Nature*, **332,** 687–689.

Miller, S. L. and Urey, H. C. (1959). "Organic Compound Synthesis on the Primitive Earth," *Sci.*, **130,** 245–247.

Morowitz, H. (1992). *Beginnings of Cellular Life*, Yale University Press, New Haven, CT.

Nilson, P. R. (2002). "Possible Impact of a Primordial Oil Slick on Atmospheric and Chemical Evolution," *Orig. Life Evol. Biosph.*, **32,** 247–253.

Orgel, L. E. (1986). "Did Template-Directed Nucleation Precede Molecular Replication?" *Orig. Life*, **17,** 28–34.

Pierazzo, E. and Chyba, C. F. (1999). "Amino Acid Survival in Large Cometary Impacts," *Meteor.*, **34,** 909–918.

Shapiro, R. (1988). "Prebiotic Ribose Synthesis: A Critical Analysis," *Orig. Life Evol. Biosph.*, **18,** 71–85.

Sowerby, S. and Petersen, G. B. (2002). "Life Before RNA,"*Astrobiol.*, **2,** 231–239.

Szabó, P., Scheuring, I., Czárán, T., and Szathmáry, E. (2002). "*In Silico* Simulations Reveal That Replicators with Limited Dispersal Evolve Toward Higher Efficiency and Fidelity," *Nature*, **420,** 340–343.

Wächtershäuser, G. (1992). "Groundworks for an Evolutionary Biochemistry: The Iron-Sulfur World," *Prog. Biophys. Molec. Biol.*, **58,** 85–201.

White, D. (1995). *The Physiology and Biochemistry of Prokaryotes*, Oxford University Press, New York.

Khvorova, A., Kwak, Y-G., Tamkun, M., Majerfield, I. and Yarus, M. (1999). "RNA's That Bind and Change the Permeability of Phospholipid Membranes," *Proc. Nat. Acad. Sci. USA*, **96,** 10649–10654.

CHAPTER 9

From the Origin to the Diversification of Life

9.1 Introduction

Theories of the origin of life fall into two categories (Figure 9.1). In the extraterrestrial, or *panspermia*, model, Earth was seeded by life from elsewhere in the cosmos. For completeness, we can divide the seeding into "random" (the result of an impact of a body already containing spores of viable organisms) and "directed" (the deliberate intervention of an intelligent or Supreme Being to initiate life on Earth). The random panspermia model relies on the conclusion from analysis of meteorites and from computer simulations that portions of meteorites impacting the Earth are subject to only modest pressure and temperatures in the atmospheric entry and shock upon hitting the surface, so that at least spores might remain viable. And both panspermia models sidestep the question of how, mechanistically, life develops from nonliving matter.

Models in which life forms on the Earth itself from nonliving matter (the terrestrial origins in Figure 9.1) must deal with the dilemma of, as the biologist Philip Morrison expressed it, how to "separate the quick from the dead." And, indeed, even this does not quite express the enormous gulf of organization and structure between living and nonliving systems, since a dead organism retains, for some time, a record of the highly organized state that allowed it once to be alive. The "inorganic origins," discussed briefly in the last chapter, imposes order on organic systems by hosting the chemistry on mineral substrates, which have catalytic and ordering properties that originate in the structure of the crystals. The "organic origins" approach postulates that self-organization of the chemical systems occurred in organic molecules from the beginning, perhaps with the early participation of self-forming enclosures in the form of amphiphilic vesicles. Of course, both inorganic and organic origins may have played a role, the mineral surfaces opening the door to particular kinds of polymers that, later incorporated in vesicles, served as templators and replicators that further ordered organic chemical systems that had reached a plateau of complexity on their own. And once such systems were established and able to replicate their particular chemical habits through templating polymers, Darwinian selection took hold and enabled

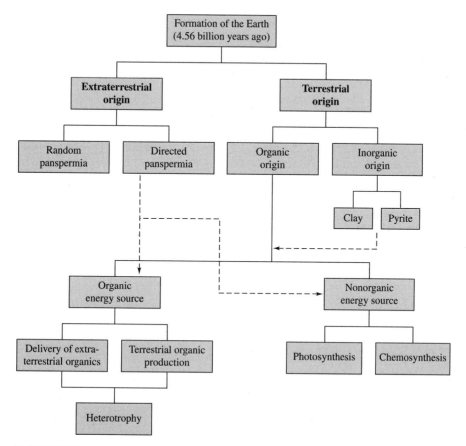

FIGURE 9.1 Flowchart illustrating relationships among different models for the origin of life on Earth. Below the terrestrial origin category are theories divided according to the role of organic structures versus mineral surfaces. Below these are the origins of metabolism in terms of energy sources and sources of organics. Dashed lines indicate possible connections later in the formation process; inorganically catalyzed or organized life must have transferred to cellular (organic) containers. Life supplied through panspermia would have undergone metabolic evolution in terms of using novel energy sources but could have done so before or after delivery to the Earth.

a competitive diversification of metabolic types, from organisms that consumed abiotically generated organic molecules to the first chemisynthetic and photosynthetic organisms.

In Chapter 8, biochemical constraints were discussed on the development of two polymer life-forms—those with catalysts and replicating molecules coming from different classes of organic monomers. But the dynamics of chemical systems that might evolve into self-replicating living systems were not explicitly explored, and for good reason—the problem is beyond a complete solution either in the laboratory, at least at present. Insight into the boundary conditions under which chemical systems might evolve into living systems can be explored computationally, and one such model is described in this chapter.

Many biologists and chemists are skeptical of such models, arguing that they do not exhibit a high fidelity of simulating real chemical cycles, with all of the intricacies of the action of catalysts and assembly of monomers that would be required to truly simulate a living or protoliving system. Nonetheless, such models may, if nothing else, allow deeper insight into the different ways in which living versus nonliving chemical systems explore the phase space of polymer length, monomer diversity, and overall system size.

Once life formed, it diversified at some undetermined point in time into three different cellular types, or domains, of life: the bacteria, archaea, and eukaryotes. Why there are three domains, and not two or five, might simply have to do with the geologic time available for life on Earth and the rate of change of organisms or might reflect a fundamental limitation in the variety of fundamentally different cellular forms achievable with the DNA/RNA–protein duality. Regardless, elucidating the manner in which living organisms are related to each other depends on the analysis of gene sequences and the creation of topologies that realistically describe the similarities and differences of such sequences. *Phylogeny*, at the heart of understanding the nature of life today as having a universal ancestor, will be described later in the chapter. Included in this is a discussion of the last universal common ancestor, commonly abbreviated as LUCA, which might be found at the root of the phylogenetic tree but which—if it existed at all—probably represents an evolutionary point far removed from the origin of life itself.

Some may question the juxtaposition in the same chapter of the exploration of life's origin—particularly through the dangerous choice of a single model—and the exploration of the phylogenetic tree of extant organisms. Certainly the latter prepares us for the discussion of extremophiles in Chapter 10. But it also provides a flavor for the difficulties of trying to go forward with origins models and backward with phylogenetic exploration of the most primitive organisms we find today—the chasm in the middle is enormous and daunting. Finally, it gave me, as the author, a chance, midway through this book, to have just a little fun, by painting a fantastical "what if" story about witnessing the origin of life.

9.2 Complex Behavior in Chemical Systems

The notion that the laws of nature have embedded within them the capacity for generating not only regularity and structure but intricate and unpredictable behavior has evolved over the centuries after Kepler and Newton quantified the workings of the clocklike cosmos. The 18th-century German philosopher Immanual Kant wrote that "God has put a secret art into the forces of nature so as to enable it to fashion itself out of chaos." But by the late-20th century, the idea of chaos as the simple opposite of order had to be replaced by a different view. A particular set of so-called dissipative nonlinear equations describing the dynamics of a complicated system (e.g., a chemical one) can give rise to ordered or complicated behavior depending on the choice of parameters that describe the system. Furthermore, complex physical systems can exhibit the property of self-organization that at first glance seems—but is not—a violation of the second law of thermodynamics requiring a net increase of entropy during spontaneous changes. Examples of self-organization in physical systems have been given in Chapter 7.

9.2 Complex Behavior in Chemical Systems

Chemical reactions can exhibit self-organizing behavior. The simplest chemical reaction, in a formal sense, is substance A becoming substance B, or

$$A \to B \ (k_f)$$
$$B \to A \ (k_r)$$

where k_f, k_r are the rate constants (Chapter 7) of the two reactions. For this system the concentrations of A and B change with time as

$$\frac{d[A]}{dt} = k_r[B] - k_f[A] \tag{9.1}$$

$$\frac{d[B]}{dt} = k_f[A] - k_r[B] \tag{9.2}$$

where $[X]$ is the concentration of the substance X. This simple system, as written, does not exhibit chaotic behavior; the abundances of A and B change in simple, linear ways as we alter the values of the rate constants.

It does not require too much of an increase in laboratory sophistication to find a reaction that is capable of exhibiting complicated and unpredictable behavior. Imagine a vessel with input ports for the reactants, output ports for the products, and a device for stirring the material within. Further, suppose we have a reaction involving three substances

$$A + B \to C$$

for which the rate constant moving to the right (production of C) is k_f and that to the left is k_r. Let the flow rate into the vessel be controllable and labeled Ψ; there is also an exit port that ensures no net buildup in the total amount of material in the vessel. Then the rate equations for this reaction are

$$\frac{d[A]}{dt} = k_r[C] - k_f[A][B] - \Psi \left([A] - [A_0]\right) \tag{9.3}$$

$$\frac{d[B]}{dt} = k_r[C] - k_f[A][B] - \Psi \left([B] - [B_0]\right) \tag{9.4}$$

$$\frac{d[C]}{dt} = k_f[A][B] - k_r[C] - \Psi \left([C] - [C_0]\right) \tag{9.5}$$

Here $[X_0]$ is the concentration of substance X at the entry port; the equation can be simplified further by having no product C_0 at the entry port ($C_0 = 0$).

If Ψ is set equal to zero, equivalent to no flow of reactants through the vessel as the reactions proceed, then the set of three equations (9.3), (9.4), and (9.5) resemble the simple two-equation set (9.1) and (9.2). The reactions proceed to equilibrium, at which point $d[A]/dt = d[B]/dt = d[C]/dt = 0$. We can then compute the equilibrium constant from the reaction rates, as described in Chapter 7, to determine the amounts of the three components at equilibrium, and that becomes the end of the story. Likewise, if we were to set Ψ to an arbitrarily large value, the equations would become $d[A]/dt = -\Psi([A] - [A_0])$,

$d[B]/dt = -\Psi([B] - [B_0])$, $d[C]/dt = -\Psi[C]$. The solutions to these equations are of the form $[A] = [A_0] + \exp(-\Psi(t - t_0))$, t_0 some arbitrary starting time, and equivalently for B and C (with $C_0 = 0$). The equations imply that material flows through the reactor so fast that it does not react at all, and the long-term (steady-state) concentrations of A and B are simply those at the input port, while for the product C the steady-state concentration is zero.

The interesting case is the intermediate one, in which material flows into the container at a speed that permits reactions to occur. Then the full potential of equations (9.3), (9.4), and (9.5) for complex behavior comes into play. The system of products and reactants exhibits various behaviors in terms of the abundance with time of the reactants (A, B) and the product (C) within the chemically active portion of the vessel. These range from a simple periodic swing in abundances, akin to a pendulum swinging back and forth (Problem 1), to so-called chaos, in which the abundances of chemicals in space and time is extremely complex and—at some level—unpredictable (Figure 9.2).

A real-world example of this kind of complexity is exhibited in the Belousov-Zhabotinsky reaction, in which potassium bromate is used to oxidize citric acid and sulfuric acid. To make the reaction go, a catalyst that donates metal ions is added, usually a reactive iron compound. The solution changes color depending on the state of the iron in solution. But the color change is not uniform; the mixture can exhibit spiral patterns defined by the color differences, and these patterns change in time. Depending on the rates of introduction of the various reactants and the catalyst into the solution, the color changes may oscillate in a periodic fashion or form intricate and stable patterns (Figure 9.3).

What is striking about this result is that a map of the system, akin to Figure 9.2 for the simplified chemical reaction set of equations 9.3, 9.4, and 9.5, exhibits the characteristic pattern diagnostic of a kind of unpredictability physicists call "dynamical chaos," and yet the system also displays patterns of structure that are self-organizing and self-dissipating. These two aspects of the system are *not* contradictory; they are in fact two manifestations of the same characteristic property of complex dynamical systems. When held far from equilibrium via a source of energy (including abundant chemical reactants), complex systems can exhibit self-organization and an unpredictable—chaotic—evolutionary path.

While beautiful patterns in a chemical soup are fascinating, they are not directly relevant to the origin of life. They simply illustrate a principle, that of self-organization, which is also manifest in a more relevant chemical context—the progressive production over time of increasing numbers of catalytic species within a reaction network. Such autocatalytic chemical systems exhibit complex behavior and may well have been essential to the origin of life. If a chemical system, consisting of a set of reactants, products, and the reactions that convert one to the other, has the property that the amount of a product produced at a given time depends on the amount of the product there initially, then that system is nonlinear, as discussed earlier. But going further, if the product of the reaction actually *catalyzes* its own production—a restrictive case of the nonlinear system—then the system is called *autocatalytic*. A conceptual example of an autocatalytic chemical scheme is given in Figure 9.4 (see page 281). The quintessential autocatalytic systems seen today in nature are biochemical processes where enzymes catalyze their own production in a complex sequence of steps.

The growth of ever-more complex cycles sustained by autocatalysis has been invoked by several authors as the means to grow and then select prebiotically promising cycles out

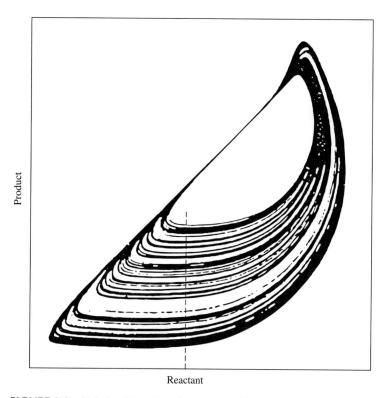

FIGURE 9.2 Relationship of the abundances of the reactants and products is shown for the system described by equations (9.3), (9.4), and (9.5). Here the flow rate into the vessel is such as to cause chaotic behavior of the system. The intricate expanse of dark regions, representing allowed abundances of reactants and products, are not simple curves; instead the system can shift into an enormous number of different states so that even if a portion of the dark area were expanded it would still not resolve into simple curves. Under such conditions, a given amount of product (dashed line) can be generated by any of a number of different amounts of reactants (measured in some combined way to allow the graph to be only two dimensional).

of the daunting number of possible chemical reactions that organic systems can undergo—the vast majority of which lead to nothing but a set of products from a given reaction. The autocatalytic cycles or networks obey all the rules of chemical reactions: consume the reactants without resupply or remove any required energy source and the cycles cease. But they are special in the sense of synthesizing over time increasing varieties of catalysts, and in some cases selecting particularly effective catalysts. Thus, such systems can, under certain conditions, exhibit increasing levels of self-organization as catalysts generated by intermediate reactions act on other reactions in the network. While some laboratory examples of autocatalytic systems exist, the limitations of time and reactor volume (as well as perhaps the limited complexity of surfaces within a reactor) make it difficult to explore the full complexity of autocatalytic cycles suggested by computer simulations. But in suitable

FIGURE 9.3 Spiral patterns demarcating differences in abundances of products and catalyst in the Belousov-Zhabotinsky reaction. The different panels reflect different relative abundances of the reactants; in the final panel, an incipient breakup of the regular pattern is seen.

abiotic environments, such systems ought to exist; they use the free energy available to work themselves toward a state of higher organization at the expense of that energy, the raw materials in the environment, and the overall generation of entropy in the system-plus-environment. While lacking the ability to faithfully replicate themselves for lack of sophisticated templating molecules, such abiotic systems might once have—however imperfectly—rehearsed what life does today with the free energy and nutrients extracted from the terrestrial environment and from the Sun.

Abiotic chemical systems can reduce their entropy (at the cost of a larger entropy generation in the environment) in a number of ways. They can exist on ordered mineral sur-

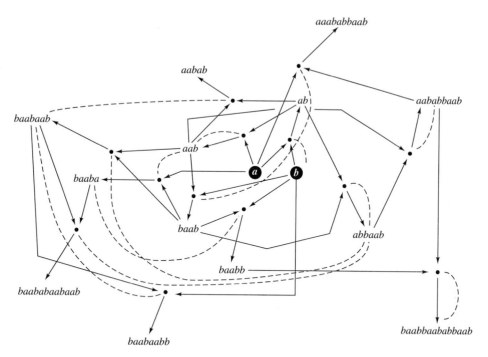

FIGURE 9.4 Schematic example of autocatalytic cycles that together form a self-contained network, where reactions will continue as long as reagents (reactants) are supplied. The catalysts that control the chemical reaction rates are generated by the reactions themselves. Two monomers, *a* and *b* (circled), are injected into the system. Chemical reactions are shown with black dots. Some reactions are catalyzed and are connected by dashed lines to the relevant polymer. While the system shown has been created on a computer, abiotic autocatalytic cycles have been demonstrated in the laboratory.

faces on which the possible range of chemistry is restricted and certain reactions catalyzed (Chapter 8). There is also the natural tendency of complicated systems (i.e., with large numbers of reactants and possible products) to exhibit self-organization if they are held far from equilibrium through availability of abundant free energy (Chapter 7) and reactants. Are these enough to make, eventually, living entities? The answer remains unknown. But an appealingly simple, albeit heuristic, model that exhibits the essential behaviors expected of such chemical systems—developed in the 1980s by the Princeton astrophysicist Freeman Dyson—suggests the constraints under which such systems must operate for such a transition to occur. The detailed discussion of this model in Section 9.3 is not necessarily an endorsement of it as a high-fidelity simulation of how life began, and many other models exist in the literature. I chose to highlight this model because of its clarity and simplicity, which enable it to illuminate those aspects of the behavior of physical systems that allow life to form from abiotic chemical systems—as apparently it did, once upon a time.

9.3 Before the Threshold: Evolving Autocatalysis or Life without Replication

We wish to understand the conditions under which a chemical system will evolve from a state of higher entropy and hence less information (less order) to a state with lower entropy and hence more information. The system ought not have a templating and replicating molecule to remember the right catalysts to synthesize, because such a molecule would already confer on the system a state with a very high degree of order. Instead, we expect the system to evolve from the very simplest kinds of chemical systems we can imagine, for example, those that produced the organic phases in meteorites. Some workers have called this approach "metabolism first" on the basis that the chemistry that the model represents is sufficiently complex and self-organizing that it qualifies as a kind of primitive biological metabolism. However, this is a matter of definition.

Rather than delineate a dichotomy between metabolism first and replication first, let us simply use the model to ask how far a chemical system might self-organize and what its properties are. If abiotic chemical systems can evolve in a self-organizing way, then the prebiotic Earth might have been replete with all sorts of organic and inorganic chemical systems, confined by mineral surfaces, amphiphilic vesicles, or other containers. We can imagine these feeding off of each other and their products, some systems more ephemeral than others, some using and then throwing away templating molecules or surfaces, in a tapestry that would be impossible to order in a temporal sense.

To visualize this system mathematically, Dyson imagines this prebiotic world as an abstract space of different populations of chemical systems. We can visualize each point in this space to represent a set of molecules that can react with each other, confined in some way akin to the chemical system of Section 9.2. Let us call these confined systems, despite the evident prejudicing toward the biological organization of life, *cells*. To keep the model mathematically tractable, the diffusion of molecules into and out of a "cell" is ignored, even though it is assumed implicitly that such diffusion is maintaining an adequate supply of reactants and removal of products. The molecules within the cell change by reacting with each other, and by doing so, change from one state to another.

We seek to understand under what conditions this change represents an increase, versus a decrease, in order. Further, it is necessary to understand whether the new state persists for some time (i.e., is quasi-stationary) and how other similar states might evolve toward it. This requires identifying regions called "attractors" in the phase space that describe the state of the chemical system. And this, in turn, requires that Dyson make some specific assumptions about the chemistry that permit a quantitative model to be constructed.

The first assumption is that within each cell—each point in our space of populations of molecules—there are a number of sites N on which to do the chemistry. These sites might be favorable places within the inner surface of the vesicle membrane, or on the surface of a suitable mineral, upon which the chemistry can take place. Individual molecules, which we will now call monomers, are the building blocks of polymers. We assume that polymers are made when monomers attach to adjacent sites. The chemistry that is assumed is very simple—attachment of a monomer to a site (*adsorption*) or removal from the site by the inverse of adsorption (*desorption*). The sites (there are "N" of them) are always in contact with the soup of monomers in the cell, but attachment onto a particular one of the N sites requires that certain conditions be met, in particular, the intercession of catalysts. Cells do not interact with each other; this is consistent with the other simplification stated, namely, ignoring the movement of molecules into and out of the cell.

This depiction of chemistry is extremely simple, indeed too simple to provide a system that can shift from a less to more organized state or vice versa. We can make additional assumptions on the nature of the monomers to rectify this, namely, that there are two types of monomers, "active" and "inactive." The active monomers are defined as those that at a particular adsorption site can combine with their neighbors—also active monomers—to make a catalyst that can then preferentially insert particular other monomers at other sites to make more catalyst. This is the essence of the autocatalytic nature of Dyson's system and a very crude analogue to metabolism. The fact that this must be specifically inserted as an assumption is not a weakness of the model. In the real world, molecules have different shapes, valence states, and surface charges that cause them to adhere to surface sites differently and to combine selectively with other molecules. In the organic chemical world these range in complexity from very simple molecules such as acetylene (C_2H_2) catalyzing the formation of ethane (C_2H_6) in the stratosphere of Titan (Chapter 13) to the enormously complex and involuted structures of proteins that catalyze cellular reactions.

In Dyson's model, the action of the catalysts is to accelerate the adsorption of the active monomer species onto a surface site. If the total number of chemical species is a, then only one of the species is assumed active; the other $a - 1$ are inactive. An empty site will adsorb an inactive monomer with a probability p, while the active monomer is adsorbed with a greater probability $f(x)p$. The function f depends on the number of sites that already contain the active monomer species and represents the action of the catalyst. To compare the effect of these model catalysts with real ones requires that we go further in defining the catalytic function. The catalysts in this model discriminate between the adsorption of an active versus an inactive monomer by creating a difference in the energy (let us call it ΔU) that is required to insert a monomer in an adsorption site. Refer back now to Chapter 7, equations (7.9) or (7.10). These concern the energy change (as either chemical potential or Gibbs free energy) in a chemical reaction—the particular one there being the ADP $\Leftrightarrow$ ATP transformation. The process to cause an adsorption of the correct monomer species can be written the same way: ΔU is of the form $RT \ln b$. The symbol b is a discrimination factor; it expresses the action of the catalyst in putting an active monomer into a single adsorption site in preference to an inactive monomer. Thus, b is just $f(1)$ for a single adsorption site: $f(1) = b$. For two sites containing an active monomer, $f(2) = b^2$; for x active sites $f(x) = b^x$. Once a site is filled, to empty it (desorb a molecule) is assumed to cost the same amount of energy for an active or for an inactive monomer. To simplify the model, assume that, at any given time, the vast majority of adsorption sites are filled with either active or inactive monomers; there are few empty sites.

Now we are nearly done; there are some other rather technical assumptions that Dyson makes that are important to the model but that we will not discuss here. What we have are a number of different kinds of monomers, a. In an abiotic hydrocarbon system with $a = 4$ we might have, for example, methane, ethane, acetylene, and molecular hydrogen. In the environment of the RNA and DNA backbones we have $a = 5$, with the different monomer types being the five nucleic acid bases cytosine, adenine, thymine, uracil, and guanine. We have also defined the discriminating capability b of the catalyst. In the RNA–DNA system, this would be the ability of enzymes, using other DNA and RNA molecules as templates, to insert the nucleic acid bases at the right positions and in the right order on new nucleic acid copies. In the case of the hydrocarbon system, we might imagine chemistry on the surface of water ice for which polymerized forms of acetylene might act to discriminate among the monomers in building more polymers. In the former case (RNA–DNA), the enzyme-based catalysts must be very powerful to work; b_{enzyme} in such a biological system is as high as 10^4. In the hydrocarbon-on-ice system, the properties of the hydrocarbon acetylene suggest it would be a relatively poor catalyst; perhaps $b_{hydrocarbon}$ is between 1 (no catalytic action at all) and 10.

What are the characteristics of this system as we change the three key parameters (the number of types of monomer species a, the catalytic efficiency b, and the number of sites N)? If we imagine allowing our system to evolve with time, let each time step be defined by one opportunity for each of the adsorption sites to fill or empty. Let the population of adsorbed monomers after a time step be the "daughter" population and that in the previous time step the "parent" population. If the daughter population has the same fraction of sites occupied by active monomers, then the cell is assumed to be in a quasistationary state, in which x (the number of active sites) does not change. But in general, it is highly unlikely that in a single cell the value of x will not change. And here is where the assumption of a larger population of cells—points in the multidimensional space we introduced at the beginning—comes into play.

By evolving a population of cells over time—remember, they do not interact with each other—we can form the average of x over the whole population of cells. We then may define the quasistationary state as one in which the average value of x over all the cells remains unchanged from one time step to another, to within some suitable tolerance.[1] Because the cells do not interact, why is this at all useful? There are two answers. First, we might try to make the model more intricate by allowing some interaction between cells in terms of exchange of molecules, something that certainly occurred on the prebiotic Earth. Second, we may assume that the average behavior of a population of cells represents the average behavior *in time* of a single cell. This assumption plays a fundamental role in treating the behavior of many microscopic systems, but it is actually difficult to prove without the detailed numerical simulations that the assumption is intended to obviate.

The model as described yields three different quasistationary points for our model cell for a given set of values of N, a, and b. The three different quasistationary points correspond to different values of x. The quasistationary point with the intermediate value of x (let us call it x_m) is unstable in the sense that if the parent population were to have a value of $x > x_m$, the daughter population would have an even larger value of x. And, if the parent has $x < x_m$, the daughter population would have a smaller value of x. Thus, the daughter population evolves away from the unstable quasistationary point. The other two quasistationary points are stable—the daughter population tends to approach the quasistationary point relative to the position of the parent population. Because these two points

---

[1] In Dyson's model the quasistationary states are referred to as equilibrium states, which mathematically is correct terminology. However, these states are not in equilibrium in the strict thermodynamic sense—they do not represent states of maximum entropy—and so I have chosen a different term to avoid confusion. Quasistationary is the appropriate term when considering the behavior of these points in phase space.

are well spaced—by virtue of having an unstable quasistationary point in between—they define two very different situations (Figure 9.5). The high-x point has many more sites occupied by the active monomer than does the low-x point, and therefore the former represents a state of greater order—lower entropy—than does the latter. Dyson calls these points, respectively, alive and dead, by analogy with the situation in real (biological) cells, but it might be better to avoid forcing the comparison with biology and refer to them as organized and disorganized states.

The existence of these states is not in itself interesting, unless real chemical systems were actually to transition between them. This is clearly what we want for the disorganized state, since the ability of a chemical system to self-organize, through growth of autocatalytic networks, for example, would seem a necessity for the origin of life. But even a jump downward to the disorganized state is significant, from our point of view, in the sense that a system that remains in the organized state—in the context of this model—has no place further to go on the organizational scale. Organized stasis may be no less prohibitive for the origin of life than disorganized stasis. Thus, we seek to find the conditions—in a, b, and N—under which cells may jump with significant probability between the organized and disorganized stationary states, as a proxy for the condition that chemical systems

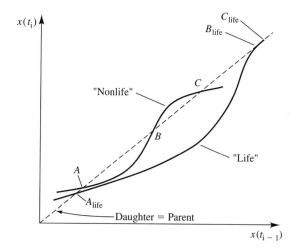

FIGURE 9.5 Schematic of the existence of three quasistationary points implied by the Dyson model. Plotted is $x(t_i)$—the fraction of sites occupied by active monomers at a particular time step t_i—versus the same for the previous time step $x(t_{i-1})$. In effect this is the fraction of sites occupied by active monomers in the daughter cell versus that in the parent. (All of these are assumed to be in an average sense, derived by looking at the behavior of a large number of noninteracting cells.) If the parent and daughter states were the same for any value of x, the curve would be the straight dashed line with slope unit labeled Daughter = Parent. Instead, the Dyson model—with appropriate choices of a, b, and N—may yield the solid curve labeled Nonlife. Stationary points are indicated by the intersections with the dashed curve at A, B, and C. A and C are stable while B is not. Relative to that curve, one for a different set of parameters more appropriate to life is depicted by the curve labeled Life.

continue to change and complexify. (I am aware that the model has no real provision for growth to ever-more organized states, so instability must substitute for this feature which is actually very difficult to simulate in a computer.)

We want the time it takes for a cell to transition from the ordered to the disordered state and vice versa be as small as possible. To set a timescale, let us normalize that time by the average time required for desorption of a monomer from a site in the cell—this is now the number of desorptions that occur in a cell before such a transition occurs and has the simple form $\exp(\Delta V \cdot N)$. Here N is the number of adsorption sites as before, and ΔV is the potential barrier in phase space that the system must cross to make the transition. The equation indicates that we wish this barrier to be as small as possible for the transition rate from ordered to disordered and back to be high. (If $\Delta V = 0$, for example, then the system would make the transition every time a monomer desorbed; however, the model as described does not allow the potential barrier to become exactly zero.) Also, N must not be too large.

The expressions required to compute ΔV are somewhat involved and require delving deeper into statistical physics than is appropriate here. From the point of view of the language of phase space we developed earlier, the two stationary points are places in phase space called attractors, to which the cell will tend to evolve. We desire those attractors to be far enough from each other in phase space that they represent significant differences in the extent of organization; but they cannot be so deep or so widely separated that the system will never jump from one to the other. These conditions depend on the choices of a, b, and N in a way that is illustrated in Figure 9.6.

Furthermore, the choice of a and b determines how interesting the system will be in terms of excessive simplicity or excessive error rates. For a very poor catalyst (b low) operating on a large number of different kinds of monomers (a large), disorganization reigns supreme, and the probability that there will be enough active monomers in adjacent sites to build catalytic polymers is very small—transitions to a more ordered state are rare. If we regard the insertion of inactive monomers into adsorption sites as the chemical equivalent of an error (because it prevents the formation of additional catalytic polymer), then the error rates are simply too high in the low-b/high-a cases for self-organization to take hold.

A highly efficient catalyst operating on only a few different kinds of monomers leads to very low error rates, but the number of sites filled with active monomers is so large that the system simply remains in this state for arbitrarily large time-steps. This simple chemical system will function indefinitely as long as additional material and energy are supplied, but it will never fluctuate between order and disorder in a way that might lead to the introduction of additional complexity into the mix—nor does it have the complexity to be of interest to the origin of life. We can imagine that many systems with this property might have existed on the prebiotic Earth, with fates ranging from starvation to extinction by environmental fluctuations to consumption as food by more sophisticated chemical systems.

For small values of a, b only two quasistationary states exist, and the system does not exhibit the property of switching between an ordered and disordered state. For large values of N, the transition time is too long to be of interest. Dyson, experimenting numerically with intermediate values of a, b, and N, finds systems that exhibit quasistable states that are reasonably well separated but for which the transition rate is large have values of $a \sim 10$, $b \sim 10^2$, and $N \sim 10^4$. All these numbers are smaller than in biological cells, but not by enormous amounts. Life uses roughly 20 amino acids as its building blocks,

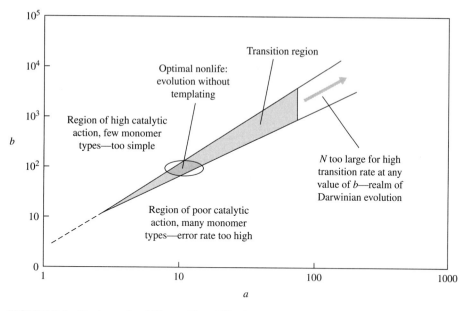

FIGURE 9.6 Regions of stability and instability in a phase space defined by the values of a and b, defined in the text. For high a and low b, the system is stable in a high-entropy, low-order state. For high b and low a, the system is stable in a low-entropy state of high order but great simplicity relative to the biochemistry of the living cell. At intermediate values of a, b, a region of rapid transition between the ordered and disordered states exists, and in which the value of a is sufficiently high to be of interest as a model for a chemical system on the road to life. At low values of a, b, this transition region does not exist. At very high values of a, the number of adsorption sites N must be so large that the transition rate is too low to be of interest.

when, in fact, hundreds are chemically possible. The model makes it possible to understand why life selected 20 out of hundreds—too many flavors of building blocks too easily lead to unwanted chemical reactions, and hence disorder in the absence of templating molecules, and put excessive demands on templating molecules in terms of variety of codons required. Many fewer amino acids may lead to nicely stable, ordered, chemical systems, but with insufficient variety to synthesize the range of enzymes needed to maintain metabolism and enable replication of the templators DNA/RNA or to build complex structures of cells from membranes to ribosomes.

The catalytic power expressed by b is two orders of magnitude below the corresponding value for enzymes, and this too is reasonable—enzymes are complex three-dimensional structures whose synthesis requires detailed templates. On the other hand, the value of b is comfortably high enough that it disqualifies many molecules from acting as catalysts in the real world version of this model—prebiotic chemical systems had to bootstrap their way toward the use of relatively efficient catalysts. The value of N is smaller than the number of monomers in our minimalist cell of Chapter 8, which might be on the order of 10^6, but innovation in such a cell is driven by Darwinian evolution associated with the

DNA and not by dynamical instability. Abiotic vesicles, perhaps 10 times smaller in volume than the smallest viable living cell, might have operated with N close to the value prescribed by Dyson's model. Presumably, the number and size of enzymatic proteins grew with the advent of templating molecules such as RNA and its precursors, which completely redefined the rules relative to the model illustrated here.

This lengthy description of Dyson's model required some stamina on the part of the reader. But it is worth thinking about, in spite of its lack of incorporation of specific chemical bonding rules, its reliance on adsorption to take the place of chemistry, and the deliberate neglect of interaction between the nodes that we have called cells and that constitute his ensemble of tiny chemical systems. The important lessons that can be derived from the model include the following:

- *Chemical systems of sufficient complexity can undergo transitions to more ordered states.* By assuming the presence of relatively efficient catalysts that can be built from a certain class of smaller, monomeric units and a sufficient number of monomeric types and adsorption sites, we can construct a system that has a probability of decreasing its entropy (increasing its order) enough to be of interest to the real world. The numerical probabilities for such transitions calculated from Dyson's model imply that one requires of order 10^{10} systems for interesting transition rates (here 1 per 1,000 time-steps) to occur. On the prebiotic Earth, it is not difficult to imagine far more than 10 billion vesicles, or their surface mineral equivalents; such numbers represent an infinitesimal amount of the organic material available on the early Earth. And this result is entirely in accord with the second law of thermodynamics; it is, in fact, just a particular example of the self-organizing behavior of dynamical systems with sufficient degrees of freedom held far from equilibrium. In our case, the degrees of freedom are defined by appropriate values of a and b; the assumed ability to make catalysts, to obtain fresh material, and to avoid "clogging" sites via desorption are simply an abstract way of saying that free energy and raw materials are available from an environment to which unusable products are flushed.

- *The required catalytic power and number of fundamental building blocks required for self-organizing behavior are reasonable.* Dyson's model, which is reasonably general, does not place strong demands on the action required of the catalytic molecules. They are no challenge for enzymes or even ribozymes, yet they must be respectable enough as catalysts that not just any molecule will do. Indeed, peptide chains assembled without templates, or sugar-based catalysts without the refinement of ribozymes, might possess the catalytic power suggested by Dyson's model. The number of fundamental monomeric types required for self-organization—10—is less—but not much less than what living cells use today. If such self-organizing systems were the precursor to template-directed life, then we have a natural explanation based solely on the general behavior of dynamical systems as to why life utilizes only a small fraction of the total number of amino acids—and, for that matter, nucleotide bases—that are abiotically produced in nature.

- *There is a natural cutoff in the size and complexity of the system beyond which non-template-driven self-organization ceases.* Dyson's model naturally explains why cellular life without genetic templating does not exist. The structures of even the simplest living cells require for their formation, replication, and sustenance an army of compounds that—while vastly smaller than what nature could produce—are too numerous to appear through progressive self-organization from less sophisticated systems. But—and this is crucial—the degree of complexity below which such nontemplated evolution could occur is only two orders of magnitude below that of the simplest living cell. Had it been 10 or 100 orders of magnitude less (for example), then the plausibility of a causal chain from nontemplated, self-organizing, autocatalytic systems to life would be in grave doubt.

An optimal system size of 10^3 to 10^4 monomers allows us to imagine plausible precursors to cellular life, such as amphiphilic vesicles (or suitable mineral surfaces) hosting self-organizing autocatalytic chemistries sophisticated enough to be the ancestors of cellular life, yet without genomic polymers. Subsequent to this appeared precursors to RNA that (i) functioned as primitive genomes; (ii) could be synthesized in a system with the complexity level $a \sim 10$, $b \sim 10^2$, $N \sim 10^4$; and (iii) put an end to the era in which precursors to cellular life functioned without genomes. However imperfect these molecules might have been, by directing the synthesis of the better catalysts, they broke the rules by which the dynamically evolving chemical systems in Dyson's model functioned. Hence, they broke the bottlenecks on catalytic specificity and system complexity.

RNA arose later, we argue based on Chapter 8, because it is too complex a molecule to have been synthesized prior to the template-driven chemistry that succeeded the dynamically evolving autocatalytic world. It arose in the context of cells that evolved by Darwinian evolution acting through templates with an order of magnitude lower information capacity than RNA itself and via catalysts with an order of magnitude less specificity than those of the RNA world.

We close this section by emphasizing again that the origin of life model outlined here considers only a toy chemical system, stripped of its details to the point where it is abstractly mathematical. Yet it quantifies in an intuitively understandable way the level of complexity that might be expected, in a general sense, from chemical systems evolving without benefit of sophisticated templating and replicating molecules. It also suggests that the complexity and sophistication of life were set not by the particular choices of the number of nucleic acid bases or amino acids but by general properties of dynamical systems that are expressions of fundamental physical laws and the dimensionality of space–time in the universe—and not with the details of organic chemistry per se. Through the invention of templating molecules, prebiotic autocatalytic systems bootstrapped upward in complexity several orders of magnitude to become living cells, but the base level from which they bootstrapped was set, in essence, by the Big Bang and cosmic inflation that determined the physical laws and geometry of the cosmos (Chapter 5).

As sophisticated as DNA and its enzymatic handlers are, they existed in cells at least as primitive as the most primitive life-forms on Earth today. We have nothing fundamentally new (on the scale of the invention of RNA or DNA itself) to show for the billions of years over which vast quantities of nucleotides have been replicated and manipulated and recreated. On the other hand, this stasis in complexity also means that we might be able to trace the evolution of life over billions of years of time through the record contained in DNA and RNA, to which we now turn.

9.4 Beyond the Threshold: The Evolutionary Diversification of Life

Phylogeny is defined broadly as the study of the evolution of life. After centuries of classification based on the appearance and behavior of organisms—virtually useless for the prokaryotes—modern phylogeny is molecular in nature and based on the genomes of organisms. By comparing sequences of DNA or RNA among organisms, it is possible to determine relationships among organisms. These relationships are relative, not absolute, in time, although attempts have been made to use calculated or observed mutation rates to fix the absolute times in geologic history of major evolutionary branchings (Chapter 16).

The basic visualization and organizational tool of phylogenetic analysis is the phylogenetic tree. One detailed version of the so-called universal tree of life, encompassing all organisms, is given in Figure 9.7; a more schematic, but more informative version for the discussion that follows is shown in Figure 9.8. In Figure 9.7, time (in the sense of the evolution of organisms) moves forward from the center radially outward along the branches; in Figure 9.8 it moves from bottom to top. Where two lines converge to a point, two organism types diverge from a common ancestor. This divergence may be at the species level, for the fine structure of the tree that concerns the appearance of individual species of organisms, or it may be at the kingdom or domain level in the broader, schematic view of the entire evolution of life. In most phylogenetic trees, the distance between organisms along the branches of the tree expresses the genetic distance between organisms. Whether an organism is located to the left or the right of another usually has no significance; what matters are relative distances one from the other, and the relationships expressed by the branching topology.

Many different parts of the genome (and different genomes expressed, for example, in mitochondrial DNA or in RNA) have been used in attempts to gauge relationships among organisms. Most phylogenetic-tree research concerns relationships among a limited number of closely related organisms, and thus cover a small part of a single domain. To construct the universal tree of life (i.e., the relationship among all organisms in the three domains of life), the genome of choice has been the so-called small subunit ribosomal RNA (SSU rRNA). This material is abundant in organisms, it is coded for in the eukaryotic organelles as well as in prokaryotes, and it has portions that evolve quickly while other parts evolve slowly. Thus, we can look at the detailed relationships between recently diverged organism groups as well as ancient bifurcations in domains and kingdoms. SSU rRNA plays essential roles in the assembly of proteins as discussed in Chapter 4 and has been an integral part of cells perhaps all the way to the beginnings of life. Its central roles in cellular function make it among the least likely parts of the cellular genetic code to be tossed from one organism to another in a process called *lateral gene transfer*, which obscures or even invalidates parts of the tree and which we shall discuss later in this section.

Genomic analysis of extant organisms using SSU rRNA has established, through the pioneering work of Carl Woese, Norm Pace, and others, the existence of three major domains of life. The two prokaryotic domains are bacteria and archaea, while the eukarya form a separate third domain. Out of the three domains radiate hundreds of kingdoms and finer divisions. Figure 9.8 implies that the eukaryotes and the archaea represent a second branching of the domains and that the initial branching was between bacteria and the

9.4 Beyond the Threshold: The Evolutionary Diversification of Life

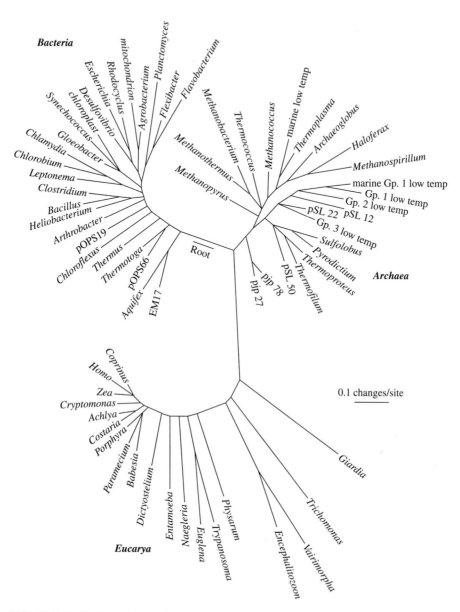

FIGURE 9.7 Phylogenetic tree based on analysis of ribosomal RNA sequences. Various organisms for which the rRNA has been analyzed are labeled, and a scale bar is included to show the extent of the genome change as a function of distance along the branches.

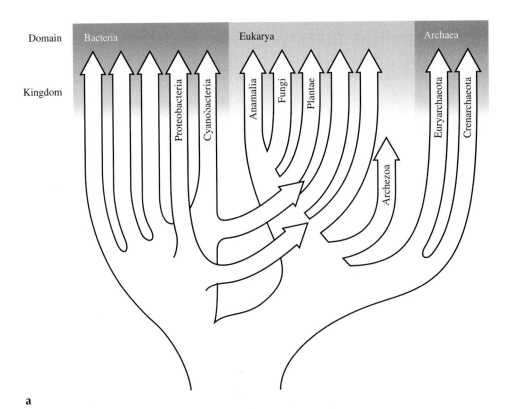

a

FIGURE 9.8 (a) The standard model of domain relationships based on molecular phylogenetic analysis. The smaller branches within the domains are those of the kingdoms, for which the order of the branching remains in dispute. Diagonal arrows suggest the symbiosis of originally independent organisms to form mitochondria and chloroplasts within eukaryotic cells. (b) A speculative tree suggestive of extensive lateral gene transfer and other symbiotic events throughout the evolution of life would make phylogenetic interpretation more difficult or even—in some extreme views—not useful in characterizing the most ancient evolutionary branchings.

common ancestor of the eukaryotes and archaea. Because the eukaryotic cell appears from a physiological point of view to be a product of a symbiotic relationship (Chapter 16), it might be argued that the common ancestor at the branch point of the archaea and eukaryotes was itself an archaeal organism. Indeed, many versions of the tree show the archaea more deeply rooted (closer to the common origin point) than are the eukarya. Thermophilic (heat-loving) organisms appear to populate the deepest branches of the phylogenetic tree, suggesting that they are in an evolutionary sense closest to the origin of life.

However, this successive branching of domains is not universally accepted, in large part because of problems in using rRNA to establish the detailed relationships among the three domains. rRNA appears to come from a single ancestral gene in the so-called LUCA, or last universal common ancestor, mentioned at the beginning of the chapter. Because LUCA is presumed to be the very root of the tree, rRNA cannot be used to distinguish the

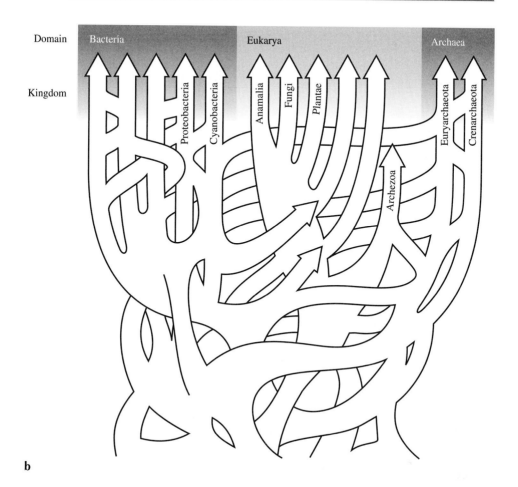

b

order in which the domains appear. In its place certain protein coding DNA genes have been used among a much smaller number of families of organisms than have been subjected to the rRNA analysis, leaving (according to some researchers) the later bifurcation of archaea and eukarya in doubt.

Indeed, this problem is just one of many associated with reconstructing the detailed topology of the phylogenetic tree—that is, how the branches and subbranches are organized. Because mutations are likely to be, by and large, a random process, which includes gene alteration and gene doubling among other processes (Chapter 16), knowing the number of differences between the same portions of the genetic code of a suite of organisms does not uniquely specify the phylogenetic relationship among them. Often, trees with different patterns of branches among which the organisms are placed will satisfy the genomic analysis. Several different reconstruction algorithms are used. It is generally assumed that the two types of organisms that share the greatest number of base-pair orders in a sequence will be the most closely related. But how these organisms relate to others that must be placed prior to a branch point requires additional information on the rate and nature of the mutations that often is not available. Hence, ambiguities in the details of the branching are usually present.

A second, much more severe problem is that of lateral gene transfer. Many enzymes involved in aspects of the metabolisms of eukaryotes are of bacterial, not archaeal, origin in spite of the closer phylogenetic relationship of the latter to the eukarya. This suggests that genes have been directly transferred from one prokaryotic organism to another and that some of the transferred genes are active in the recipient in controlling important cellular processes. Detailed analysis of bacteria and archaea suggest that such lateral transfers have continued throughout the history of life. For example, after the divergence of the bacterium *Escherichia coli* from the lineage of the *Salmonella* bacteria, estimated on the basis of mutation rates to have been 100 million years ago, hundreds of lateral gene transfer events into *E. coli* appear to have occurred from the manner in which individual codons (Chapter 4) are expressed. If this interpretation is correct, much that distinguishes *E. coli* from one type of *Salmonella*, *S. entereca*, was acquired not by mutation and evolution but through direct transfer of genes from other types of microbes.

The importance of lateral gene transfer in obscuring (or even invalidating) the universal tree of life remains controversial among workers in this field. One view is that dramatized in Figure 9.8b in which much of the evolutionary branchings inferred among the domains and kingdoms have in fact been short circuited through gene transfer. Such extensive lateral gene transfer, along with other effects such as gene duplication (Chapter 16), might require that we consider the archaea and bacteria as a sort of common emergent property of the evolution of the genome, in which the only definable units are the genes themselves. We could trace the history of life, in this view, only in terms of the history of genes: Which genes traveled together for given amounts of time through various genomes, and which "jumped ship" at various times in their history? The organisms that carry the genes, in this view, are an "epiphenomenon" that might retain certain stable lineages for some time but whose precise evolutionary relationships to each other cannot be defined.

What, then, of the very root of the tree of life? Does life have a single origin? The commonality of the genetic code among all life has been used to argue that all the branches of the tree arose from a LUCA. But this need not have been a single organism—a set of closely related organisms could have provided parts of the genome we study today, and indeed we cannot rule out small contributions via gene transfer among a much larger set of less closely related ancient ancestors. LUCA, or the group it symbolizes, almost certainly was part of a larger population of diverse kinds of organisms, with metabolisms probably less sophisticated but no less diverse than today's and perhaps fundamental genomic differences as well.

The tree of life may, in fact, be more like its namesake than this description suggests: The tree is the top half of the structure, and below ground exists a far more complex network of roots representing all sorts of experiments in life that do not exist today. If all but one "type" of life became extinct, then we are the result of an evolutionary bottleneck that appears to be a single root of the tree of life—LUCA—but in fact is not. All but the bottleneck lineages are gone because the former were outcompeted by the genetic and metabolic apparatus possessed by the LUCA group, for which they became merely foodstuff.

(Indeed, such a process could have happened multiple times in the root system of the phylogenetic tree, a point that is recognized in the adjective "last" attached to the universal common ancestor acronym.) How long the extermination of all other life-forms required, and how much transfer of genetic machinery to our ancestral lineage took place during that time, will likely never be known. With molecular phylogenetic analysis, we can ascertain that we share a common origin with the bacteria and archaea, but how that origin came about and when it occurred may be beyond the reach of this technique.

9.5 Putting It All Together: A Time-Travel Fantasy

Having approached the origin of life moving forward in time with the models of this and the previous chapter and working backward through the phylogenetic tree, it should be evident to the reader that there remains a huge gap between the two. Even were there a sensibly definable LUCA (which, from the preceding discussion, a reader should consider skeptically), there is no consensus on the temporal gap between that ancestor and the earlier "event" of the origin of life—whether it was a billion years, millions of years, or even less. Indeed, even the notion of defining a single origin of life is suspect. Darwinian evolution initiated by the appearance of templating and replicating molecular polymers is a convenient point to call the origin of life, but the first such templator/replicator might have been quite different from DNA and RNA. We might instead mark life's origin at the appearance of RNA, or DNA. And what do we call the putative self-organizing chemical factories that preceded the replicators, whose evolution was not Darwinian but was driven instead by the physical laws governing complex dynamical systems: Life, non-life, or protolife?

Figure 9.9 is an attempt to put together one possible chain of events, without reference to time, in the march from random organic chemistry on the early Earth (or even in space) up through the appearance of bacteria. It is very much a personal view, picking and choosing from ideas in the literature to come up with one possible solution to the extraordinary problem of how life began. There are many other novel ideas that have been offered; for example, rather than linear polymers as precursors to RNA as the carrier of genetic information, perhaps purines could have formed two-dimensional templates that would have coded for specific peptide sequences, as in the suggestion of two New Zealand biologists, S. Sowerby and G. Peterson.

The steps shown in the model are sequential only in the sense that those preceding seem to be required for, or are helpful to, the steps following. Along the way, increases in the efficacy of catalytic molecules, in the size and complexity of the self-contained chemical systems, and in the appearance of novel templates mark watersheds in the changing tapestry leading toward life as we know it today. While the timescale for the entire process is, as noted, unknown, it certainly was shorter than for the later evolution of DNA-based life from bacteria through multicelled eukaryotic plants and animals (Figure 9.10). Only when DNA and protein-based enzymes came into the cellular picture did the fundamental innovations end, choked off by a form of life so efficient, so sophisticated, that everything else became merely food for it.

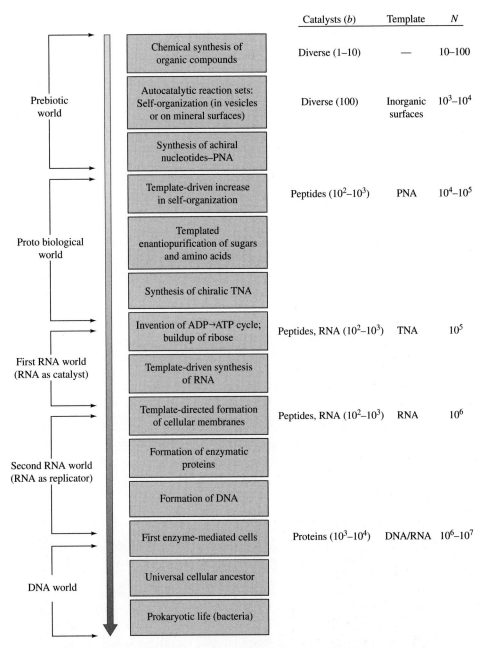

FIGURE 9.9 One possible sequence of events in the development of life from nonliving but self-organizing chemical systems—a pathway to the "origin" of life—is shown. Major events in this history are labeled as "worlds" on the left, in conformance with current astrobiological usage. A more detailed sequence of steps is given in the boxes, with time running downward on the chart. The efficiency of catalysts b (see text), the identity of the information-carrying molecule or inorganic structure (the "template"), and the number of monomers N (see text) per chemical system are estimated in order of magnitude form for several of these steps.

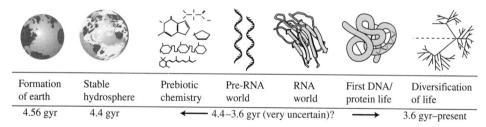

Formation of earth	Stable hydrosphere	Prebiotic chemistry	Pre-RNA world	RNA world	First DNA/ protein life	Diversification of life
4.56 gyr	4.4 gyr	←	4.4–3.6 gyr (very uncertain)?	→		3.6 gyr–present

FIGURE 9.10 Possible timeline for key steps in the origin and evolution of life. The pre-RNA world is a generic term for chemistry more complex than prebiotic but, in the model of this chapter, extends from the metabolism-first systems through PNA- and TNA-based systems.

Suppose we could go back in time to revisit those steps. We might choose to make journeys to five different points in time on the early Earth, corresponding to the different "worlds" of life of which our "DNA world" is the last (and which stands as the superdomain to the three domains of life today). We would equip ourselves with breathing gear against the anoxic atmosphere and advanced laboratory tools to identify and analyze the tiny organic systems we sample. We would take ever-so-much care to avoid leaving behind DNA and RNA samples from ourselves and the bacteria we carry—lest we end up in that paradoxical infinite loop common to most time-travel stories.

The Prebiotic World

Stepping onto the planet soon after the crust cooled sufficiently to allow liquid water to exist in certain places for extended periods of time, we see a world heaved by the frequent impacts of debris left over from planet formation and by the high heat flow of an Earth that was nothing but rocky debris some 50 million years earlier. Many crater rims poke above a global ocean that is gradually deepening with the outgassing of crustal water. Some craters host inland seas where, on the shores, organic deposits are available to be sampled. These are a mixture of unaltered extraterrestrial organics and material processed further by the action of liquid water and the presence of crustal heat sources. Within the extraterrestrial organics are a wide suite of amino acids; nucleic acid bases; sugars of four-, five-, and six-member rings; and compounds that host phosphate bonds, including some AMP, ADP, and ATP. Many chiral molecules (e.g., some amino acids) are present but with virtually equal amounts of left- and right-handed forms—no process of enantioselection seems to be present. At the edges of the seas are a number of hot springs and surrounding them are patches of organics subjected to hydration and dehydration reactions, in which small molecules act as weak catalysts and a few polymers are assembled occasionally, some reaching chain lengths of many dozens of monomers. Suddenly, with a rumble, a geyser discharges at the site of our sampling, and the organic compounds are boiled, disrupted, and mechanically dispersed.

At another site, however, the organics are not simply exposed to the open air. Some of the samples contain amphiphilic vesicles, some 10 nm across, in which organic molecules have been sequestered. In a number of these, the organic chemistry seems different from that of the surroundings. Only a subset of the amino acids found in the meteorite samples are contained in the vesicles, and polymer chains of amino acids extending out to hundreds of monomers have been constructed in a few of the vesicles. Some of these peptides, as well as other shorter polymers composed of other monomers, seem to function as catalysts considerably better than the molecules outside the vesicles. A few of the larger vesicles contain many thousands of amino acid monomers, but the number of distinct monomer types is only a dozen or so; some vesicles are warmer than their surroundings through the attachment on their surfaces of pigmenting molecules that allow greater absorption of sunlight. Other vesicles are relatively boring, with poor catalysts and a hodgepodge of random organic monomers clouding up the microscopic soup. But vesicles are not the only organic chemical factories in the region. Nearby on some clay, purines and pyrimidines have attached themselves preferentially to certain surfaces relative to other organic molecules, yielding a surface rich in nucleic acid bases. Elsewhere in the region, pyrite seems to host a primitive metabolism that is catalyzed by the iron and energized by the redox contrast with the surrounding minerals and water lapping across the rocks.

The Protobiological World

We return much later to the same region of the Earth. Many of our sample sites are now submerged beneath the global ocean; others have been wiped clean by impacts or volcanism. A few sites remained relatively stable, though, and the organic deposits now seem more heavily dominated by vesicles than had been the case before. Those vesicles located close to or on top of the clays whose surfaces hosted a rich extract of nucleic acid bases seem to be different from the others. To our surprise, we find under a microscope a slow but continual growth and splitting of vesicles, with material from the original being roughly divided between the two daughters. Analysis of the interior soup of tens of thousands of monomers reveals peptide chains of hundreds of members, as before, but the variety of peptides present seem fewer than before—it appears that those particularly suited to catalysis have been preferentially selected. More interesting is another molecule that has appeared in the dividing vesicles—a PNA with an aminoethylglycine and methylenecarbonyl backbone attached to nucleic acid bases. One PNA is seen Watson-Crick-pairing with another PNA; others are adhering to amino acids that themselves are attaching to the ends of nearby peptide chains. The PNA examined here uses just adenine and thymine; another in a larger vesicle is using four bases (Figure 9.11).

In many of the vesicles containing PNA, there is a strong excess of left- versus right-handed amino acids, and some vesicles contain chiral sugars inherited from extraterrestrial AMP in which the right-handed form has been preferentially extracted. A very few vesicles have essentially only left-handed amino acids; in these, peptide-based chains have been constructed with the help of PNA in such a way as to yield surprisingly specific catalytic action. But the real surprise comes from a foray to a site some 100 km away that seems particularly salubrious from the point of view of access to liquid water, clay minerals, and modest but not devastating amounts of geothermal heat. Here the vesicles reproduce more quickly than their cousins across the bay, are larger and more colorful, and have a characteristic and distinct pH and redox state different from their environment. Some

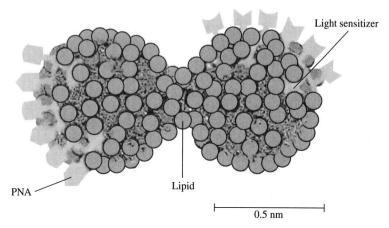

FIGURE 9.11 Conceptual model of a dividing protocell, just 5 nm (5 billionths of a meter across), and hence 20 times smaller than nanobacteria described in Chapter 8. The spherical clusters are micelles (Chapter 4) made of a carboxylic acid, which is composed of a functional group (e.g., H or CH_3) connected to COOH. Carboxylic acids are easily made when hydrocarbons are exposed to liquid water. The light sensitizer is a simple molecule that acts as a pigment to absorb sunlight and transfer the energy, as electrons or simply heat, through the agglomerate of micelles comprising the protocell. PNA acts as both the genome and helps relay electrons through the system, hence playing a simple metabolic role.

appear to tolerate immersion in ocean water, and the tides appear to carry a fraction of them off into the open ocean. Examining the vesicles, we find PNA along with another nucleic acid based on the chiral sugar threose—TNA. Uniformly employing four nucleic acid bases (adenine, guanine, cytosine, and uracil), the TNA with the participation of PNA seems to be making a much more specific set of peptide chains that undergo simple folding and hence have primitive enzymatic function. Nonetheless, as in the PNA-only vesicles, there is surprising variation in the inventory of molecules from one vesicle to another, and acceleration cultivation starting with one TNA in our mobile lab leads to many variants both in the TNA and in the metabolic reactants and products. Energy storage in some vesicles is performed via the ADP–ATP cycle, and in many of the vesicles, a detritus of ribose in the form of AMP seems to be a steady-state feature of their interiors.

The First RNA World

We return some hundreds of thousands of centuries later to find that our TNA site was destroyed in an impact. Further, the Earth has lost its hydrocarbon shield against ultraviolet radiation, and a weak ozone shield is insufficient to restore a UV-poor environment at the surface. Most of the vesicle sites have disappeared, and we must look for new ones in environments shielded from the Sun. The pigmented solar-energy vesicles are gone. Many vesicles are now found in aqueous environments adjacent to vents that generate reduced gases and, hence, a redox contrast with the surrounding water. Sampling these, we find two surprises. First, most of the containers are now lipid membranes similar to those found in modern bacterial cells, but without the protein-based channel; these containers in fact approach the size of the smallest modern cells. Some much smaller vesicles are still to be found, but in at least one case, such a vesicle was observed being absorbed by the larger lipid cellular membrane. Second, within these containers that we are sufficiently impressed with to call cells, PNA is absent, TNA is the replicator, and a second sugar-based nucleic acid has appeared: RNA. The RNA is in short strands and appears to play dual roles: one to assemble peptide chains based on the instructions in the TNA and the other as a catalyst for membrane-based processes and for the replication of TNA. A few of the RNA catalysts outperform the peptides by one to two orders of magnitude. These cells show remarkable regularity in the types of monomers and concentrations present, and cultivation of a TNA-RNA cell produces many generations without mutations. However, the cells still lack fully functioning, folded peptide structures that could be called enzymes, and for this reason, both the growth rate and viability of these cells are far inferior to modern cells.

The Second RNA World

With some sense of relief we find that our vent site has remained intact after of order ten thousand centuries, and the cell populations in the surrounding water look quite healthy—far more abundant and in a greater variety of redox niches than before. Examining these cells we find that, in many, TNA is gone and RNA has taken its place. These cells have RNA as the information-storing molecule, as the information-transferring molecule, and as a catalyst to build peptides. Other cells in the same environment have RNA and TNA, but the roles of information encoder and messenger have been switched. In a few cells the RNA has actually synthesized PNA-like structures that seem to play a messenger role as well. Indeed, the overall impression is that of a kind of messy cellular exuberance made possible by the lengthier and more robust information-storage chains of RNA compared with those of PNA and TNA.

But all is not entirely rosy in the second RNA world. We witness, in a sample of ocean water, a small TNA-based vesicle being ingested by an RNA-based cell. After a short time, the cell bursts and releases several TNA-based vesicles into the surrounding environment. Other cells ingest these and then burst, and for a while, it appears we are witnessing the beginning of the disease-driven demise of the RNA-based cellular colony. But, as quickly as the attack began, it tapers off, as many of the cells in the colony seemed unaffected by the ingestion of the TNA virus. We culture these cells and extracted what we had hoped to find—a peptide with a tertiary folding structure, an enzyme, that recognized the vesicle coat around the TNA and neutralized the virus. Encouraged in this way, we look further to a little gulley on nearby seafloor partly isolated from the main colony. Here are cells where RNA is present as replicator and catalyst, but also present are about a half dozen enzymatic proteins along with DNA. In these cells the DNA seems to code for the enzymatic proteins that sustain it, and RNA mediates the process in the messenger role. But RNA is still replicated as the cell divides and is responsible for the synthesis of other peptide chains. It is an awkward situation, one that likely will not last long.

The DNA World

We come back hundreds of millions of years later, and we cannot find our biologically prodigious seafloor vent. But it doesn't matter because we roam all over Earth, and wherever there is liquid water and protection from the solar UV flux, there are cells. When we collect and analyze the cells, we find they contain DNA as the repository of genetic information, RNA as the messenger, and proteins as the structure and catalyst. The proteins now number 625—the original five each spawned five, and so forth, in four evolutionary events—but in a few cells we find more. Most remarkable is that genomic analysis of the DNA suggests that all of the widespread cellular life we now find originated from a small group of these cells. While the invention of DNA undoubtedly happened in many places where RNA could synthesize a basic set of proteins, one set of cells in particular passed though an evolutionary bottleneck and inherited the planet. And, of course, we witness plenty of roaming RNA strands, surrounded by protein coats, inserting themselves into DNA cells and forcing the cellular machinery to serve a more ancient master. But we cannot determine from the short strips of RNA whether these retroviruses are relicts of the second RNA world or rogue pieces out of the contemporary genetic apparatus we are exploring.

None of the cells we find have the genome of a modern organism, but we can dimly perceive the root of the phylogenetic tree that eventually will include us. Here on our final stop before returning to the modern world, we arrived in time to see the cells that might qualify as LUCA, or the ancestors of LUCA, but it is much too soon to see any branching into modern domains. We are disappointed to not see a familiar "face" in the form of an *E. coli* or *M. genitalium*. And detailed analyses of the genetic makeup of these organisms in our mobile lab suggests that lateral gene transfer among cells is enormously important at this stage—making any sort of phylogenetic analyses of these cells almost impossible. To keep our sanity amidst this ocean of primitive cells and rivers of freely flowing genes, we permit ourselves a brief trip to the surface of the ocean to gaze at the skies above. There we see an enormous Moon, clearly much closer than it is today and yet to be fully adorned with the dark "mare" lava flows so familiar to us. More than 4 billion years hence, some of the descendants of these simple cells we study will stand on that cratered surface and look back at the world where they were born.

QUESTIONS AND PROBLEMS

1. A deeper perspective on the representation of a chemical system in Figure 9.2 can be gained by considering a simple pendulum. Fixed to a point on a beam, the pendulum is a weight on a rod that swings back and forth and achieves its maximum velocity as it reaches the lowest point in its swing (when its rod is vertical). It stops at the two upper limits of its traverse in each cycle. Let the point at which the pendulum rod hangs down vertically be zero, so that positions to the right are positive and those to the left negative. Likewise, velocity toward the right is positive and toward the left negative. Make a graph of velocity versus position of the pendulum during a complete cycle back and forth in its swing, ignoring the slowing of the pendulum by friction. You may plot linear velocity versus linear position (defined by marking with a pen the surface below the weight) or angular velocity versus angular position, in

which the latter is the angle of the rod (positive or negative) from the vertical. The shape of the resulting graph—a simple oval—will be the same. This graph is a plot of the behavior of the pendulum in a phase space of velocity and position. It is analogous to Figure 9.2, which describes the behavior of a chemical system in a phase space defined by abundances of reactants and products.

2. Now consider that friction acts to make the swing of the pendulum ever smaller, until finally the rod and weight come to rest. What is the shape of the path on the graph of velocity versus position?

3. Suppose the weight on the end of the pendulum were magnetized, and you set up several magnets along the tabletop above which the pendulum swings. With appropriate positioning of the magnets and choice of strength, you would have a "driven" pendulum. What do you suppose the phase space appearance of the pendulum would be in this case? Look up the answer in a recent physics text or introductory book on chaos—the plot is nearly as intricate as that of Figure 9.2!

4. Dyson's model defines monomers and catalysts without including any detailed chemical information, for example, on how certain monomers would form polymers or how catalysts would act. How would you incorporate at least simple chemistry in Dyson's model?

5. Quite a number of origin of life models have been advanced in the physics literature, based on reducing living systems to the bare essentials and trying then to understand how such systems might arise from nonliving chemistry. Perform a search of recent literature to find some models (or nonquantitative scenarios) for the origin of life. Pick the models that are most and least compelling to you, and explain your choices.

6. One of the ambiguities that arise in constructing phylogenetic trees is the topology of the resulting tree—often several topologies are possible that each imply different relationships among organisms. The method of constructing trees most accepted is called cladistics. One authoritative reference to the technique is that of the University of California Berkeley website, http://www.ucmp.berkeley.edu/clad/clad4.html. Go to the site to appreciate some of the ambiguities and complexities of determining relationships among organisms. (Should that website become inactive, try a google search under "cladistics" or "phylogenetic tree.")

7. The article by Boucher and Doolittle (2002) referenced at the end of this chapter concerns the placement in the universal tree of life of a newly discovered organism (and cites the more detailed original report upon which it is based). Read this article (and, if you wish, the more detailed report by Huber et al. it cites) with a copy of the universal tree of life in hand (e.g., Figure 9.7), so that you can follow the controversy over the phylogenetic placement of the species. What, in your view, constitutes the biggest uncertainty or weakness in the analysis? The article refers to the organism's genome as one "begging to be sequenced." What is the difference between the sequencing of the entire genome and the sequencing of the ribosomal RNA (rRNA)?

8. How might you use rRNA and other standard phylogenetic tools to test the hypothesis that there was once a fourth domain of life that was extinguished? In what way does the ability to test the hypothesis depend upon whether the putative fourth domain existed before or after the other major domain branchings on the universal tree?

9. The notion of bottlenecks in biological evolution is a common one and is the analogy for the LUCA bottleneck discussed in the chapter. Do a literature search to find a well-documented example of an evolutionary bottleneck for some higher organism (i.e., plant or animal). Based on your reading, in what way is this a useful or misleading analogy for the LUCA bottleneck?

10. How might the origin of life story be modified if RNA were "delivered" to the Earth by a meteorite impact, as a contaminant from another planet (e.g., Mars)? Rework either Figure 9.9 or the narrative story at the end of chapter given the assumption that RNA simply becomes available, rather than having to be invented, in the prebiotic epoch.

SUGGESTED READINGS

Freeman Dyson explains his heuristic model for self-sustaining metabolism without replication in *Origins of Life*, 2nd edition (1999, Cambridge University Press, Cambridge). G. L. Baker and J. P. Gollub, *Chaotic Dynamics* (1990, Cambridge University Press, Cambridge) gives an admirably concise and clear treatment of nonlinear dynamics. More general discussions of self-organization, chaos, and autocatalysis can be found in Steven Levy, *Artificial Life* (1992, Vintage Books, New York). For an intriguing and entertaining exploration of self-organization (but see the caveat in the Chapter 7 Suggested Readings), go to Stuart Kauffman, *Investigations* (2000, Oxford University Press, Oxford). W. Ford Doolittle, "Phylogenetic Classification and the Universal Tree," *Science*, **284**, 2124–2128, provides a clearly written review of phylogenetic classification. I borrowed the notion that the ancestral protein types were five in number, and that the protein content in the universal ancestor of life was $5 \times 5 \times 5 \times 5 = 625$, from an unpublished poster paper by Monica Riley and colleagues at the Marine Biological Laboratory in Woods Hole, Massachusetts.

REFERENCES

Baker, G. L. and Gollub, J. P. (1990). *Chaotic Dynamics: An Introduction*, Cambridge University Press, Cambridge.

Baranowski, E., Ruiz-Jarabim, C. M., and Domingo, E. (2001). "Evolution of Cell Recognition by Viruses," *Sci.*, **292**, 1102–1105.

Belmonte, A. L., Ouyang, Q., and Flesselles, J.-M. (1997). "Experimental Survey of Spiral Dynamics in the Belousov-Zhabotinsky Reaction," *J. Phys. II France*, **7**, 1425–1468.

Boucher, Y. and Doolittle, W. F. (2002). "Something New Under the Sea," *Nature*, **417**, 27–28.

Davies, P. (1999). *The Fifth Miracle: The Search for the Origin and Meaning of Life*, Simon and Schuster, New York.

Davis, W. L. and McKay, C. P. (1996). "Origins of Life: A Comparison of Theories and Application to Mars," *Orig. Life Evol. Biosph.*, **26**, 61–73.

Doolittle, W. F. (1999). "Phylogenetic Classification and the Universal Tree," *Sci.*, **284,** 2124–2128.

Drossel, B. (2001). "Biological Evolution and Statistical Physics," *Adv. Phys.*, **50,** 209–295.

Dyson, F. (1999). *Origins of Life*, Cambridge University Press, Cambridge.

Ehrenfreund, P., Bernstein, M. P., Dworkin, J. P., Sandford, S. A., and Allamandola, L. J. (2001). "The Photostability of Amino Acids in Space," *Astrophys. Jo. (Lett.)*, **550,** L95–L99.

Joyce, G. F. (2002). "The Antiquity of RNA-Based Evolution," *Nature*, **418,** 214–221.

Lander, E. S. and Weinberg, R. A. (2000). "Pathways of Discovery. Genomics: Journey to the Center of Biology," *Sci.*, **287,** 1777–1782.

Lederberg, J. (2000). "Pathways of Discovery. Infectious History," *Sci.*, **288,** 287–293.

Levy, S. (1992). *Artificial Life: A Report from the Frontier where Computers Meet Biology*, Vintage Books, New York.

Loewenstein, W. R. (1999). *The Touchstone of Life: Molecular Information, Cell Communication, and the Foundations of Life*, Oxford University Press, New York.

Pace, N. (2001). "The Universal Nature of Biochemistry," *Proc. Nati. Acad. Sci.*, **98,** 805–808.

Perez-Mercader, J. (2002). "Scaling Phenomena and the Emergence of Complexity in Astrobiology." In *Astrobiology: The Quest for the Conditions of Life*, G. Horneck and C. Baumstark-Khan, eds., 337–360, Springer, Berlin.

Rasmussen, S., Chen, L., Deamer, D., Krakauer, D.C., Packard, M.H., Stadler, P.F., and Bedau, M.A. (2004). "Transitions from Non-Living to Living Matter," *Sci.*, **303,** 963–965.

Reysenbach, A-L. and Shock, E. (2002). "Merging Genomes with Geochemistry in Hydrothermal Ecosystems," *Sci.*, **296,** 1077–1082.

Ruvkun, G. (2001). "Glimpses of a Tiny RNA World," *Sci.*, **294,** 797–799.

Sagan, C. and Chyba, C. (1997). "The Early Faint Sun Paradox: Organic Shielding of Ultraviolet-Labile Greenhouse Gases," *Sci.*, **276,** 1217–1221.

Sowerby, S. J. and Petersen, G. B. (2002). "Life before RNA," *Astrobiol.*, **2,** 232–239.

Xiong, J., Fischer, W. M., Inoue, K., Nakahara, M., and Bauer, C. E. (2000). "Molecular Evidence for the Early Evolution of Photosynthesis," *Sci.*, **289,** 1703–1705.

White, D. (2000). *The Physiology and Biochemistry of Prokaryotes*, Oxford University Press, Oxford, U.K.

Zainetdinov, R. I. (2000). "Entropy Dynamics Associated with Self-Organization." In *Paradigms of Complexity*, Novak, ed., 229–243, World Scientific, Singapore.

CHAPTER 10

Extremophiles and the Span of Terrestrial Biotic Environments

10.1 Introduction

Once upon a time, all of life was extremophilic—that is, under extreme environmental conditions, or living "on the edge." The Earth itself, changing rapidly as its removal of accretional heat proceeded in the Hadean and Archean eons of geologic time at vastly larger rates than later in Earth's history, may have been abundant in hot springs and volcanic vents, which likely were highly variable and hence fickle abodes for primitive life. Surface climates, buoyed by greenhouse gases against a young Sun 30 percent fainter than today (Chapter 11), may have fluctuated between very warm and freezing conditions. Impact rates declined precipitously but were still significantly larger than those today for about the first 10 percent of life's history, and large portions of the habitable Earth were wiped out on timescales of millions to tens of millions of years. Ocean salinities and oxidation states probably fluctuated with amplitude and frequency unknown today except in restricted regions of the modern world. Further, the coupled dance of photosynthesis and oxygenic respiration had not yet come to be, and the phylogenetic tree suggests that many different metabolic mechanisms existed to take advantage, simply, of what was there in the environment.

Today, research into extremophiles has important implications for biology and other sciences. Many of these organisms, despite their high degree of evolutionary adaptation, provide insights into the history of life on Earth unavailable from *mesophilic* organisms like us that live in environments of intermediate temperature, pressure, salinity, and acidity. They provide guidance on the range of conditions on other worlds in the solar system (and beyond) that might play host to life—and provide cautionary examples to those whose job it is to sterilize spacecraft bound for the astrobiological investigation of our solar system. Extremophiles have also spawned a huge biotechnology industry devoted to the analysis and extraction of proteins responsible for extremophilic survival—for application in a range of industries from medical to cleaning.

This chapter is written as an overview of the range of extremophilic environments and the biochemical mechanisms for survival there. In reading it, the reader should not forget that this aspect of astrobiology is distinguished by its dependence on fieldwork in very difficult and sometimes remote environments on our own planet, work sometimes done by robotic proxies of scientists. In this respect, astrobiologists who seek extremophiles are rehearsing a process that someday, in some vastly more remote and difficult corner of our solar system, could yield the living entity whose existence answers astrobiology's motivating question of whether we are alone in the universe.

10.2 Extreme Environments

Various kinds of known extreme environments are listed in Table 10.1. The values of the various limits mean different things to different organisms. In some cases, organisms flourish under the particular conditions listed; in other cases, these are the limits of viability.

TABLE 10.1 Kinds of Extremophiles

Environmental Parameter	Type of Behavior	Parameter Definition	Example Organisms
Temperature	Hyperthermophile	Growth >80°C	*Pyrolobus fumarii* (113°C)
	Thermophile	Growth 60–80°C	*Synechoccus lividis*
	Psychrophile	<15°C	Psychrobacteria, some insects
Radiation	Atomophile	Up to 6,000 rad/hr; 15 Mrad tot.	*Deinococcus radiodurans*
Pressure	Pizeophiles	Up to gigapascals	*Sh. oneidensis, E. coli*
Desiccation (dryness)	Xerophile	Anhydrobiotic	*Artemia salina*, fungi, etc.
Salinity	Halophile	High salt (2–5 Molar NaCl)	Halobacteriaceae, *D. salina*
pH (acidity/alkalinity)	Alkaliphile	pH > 9	Natronobacterium, *B. firmus*
	Acidophile	pH < 0	*Cyanadium caldarium* (pH 0)
Oxygen pressure	Anaerobe	O_2 intolerant	*Methanococcus jannaschii*
	Microaerophile	Tolerates low O_2	*Clostridium*
Chemical extremes	Gases	Pure CO_2 atmosphere	*Cyanidium caldarium*
	Metals	High metal concentrations	*Ferroplasmic acidarmanus*

Source: Modified from Rothschild and Mancinelli, 2001.

This ambiguity, which the reader can resolve by going to the list of references at the end of the chapter, is normally resolved by defining extremophiles as organisms that flourish under or require particular conditions. However, it is normal in this field to fold in organisms that tolerate, but do not necessarily seek to exist in, the extreme cited in the table. Further, organisms may not tolerate particular conditions in all of the possible stages of life; bacterial spores are far more resistant to certain extremes than the active organism (indeed, this is the formal definition of *spore*).

Extremophiles exist in all three domains of life: archaea, bacteria, and eukarya. There are no known hyperthermophilic eukaryotes (able to endure extremely high temperatures), but some eukaryotes can deal with low temperatures and extremes of pH, dryness, and pressure. Indeed, some would argue that from an evolutionary perspective, any organism capable of existing under aerobic conditions is an extremophile; that is, aerobic organisms are aerophilic. It is certainly true that were we able to visit the Archean Earth (Chapters 9, 11), aerobic environments would be rare. Tolerance and use of oxygen required very particular adaptations that enabled complex organisms to evolve (Chapter 16). But from the point of view of the search for life elsewhere in the solar system, Earth's aerobic atmosphere and mild environment salubrious for liquid water are the least useful guides (Chapters 12, 13, 14). So, in this chapter we apply an intuitive definition of extreme in exploring the various types of extremophiles.

High-Temperature Environments

The first scientific description of hyperthermophiles (organisms adapted to very high temperatures) came in the late-19th and early-20th centuries from the hot springs of Yellowstone National Park in northwestern Wyoming. There, B. Davis and W. Setchell identified microbes that were viable at temperatures approaching 90°C. Today Yellowstone remains the type-locality for the study of organisms at high temperatures and unusual pH values (Figure 10.1). Vast hydrothermal systems exist within the Earth's crust (most of which do not manifest themselves at the surface) and at the seafloor. While extreme temperatures (above 80°C, the threshold between hyperthermophily and just "plain" thermophily) may be found in such systems, often hot springs are overall much cooler; at submarine vents, mixing with cooler water may provide a safe environment for thermophiles that are not hyperthermophilic.

Although organic reactions proceed more rapidly at higher temperatures, the intolerance of most organisms for extreme heat derives from the nature of polypeptide folding. Proteins are not very much more stable than the unfolded polypeptide structures; the energy required for unfolding is about 4 to 15 kcal/mole of protein. (A kilocalorie, or kcal, which is equivalent to the consumer's "calorie," equals 4,184 joules.) Compare this, for example, with the latent heats of melting and vaporization of water, 1.5 kcal/mole and 10 kcal/mole of water, and it is not much different. That proteins are not that much more stable than the "denatured" polypeptides should not be surprising—it is essential to the practicality of protein folding. Were a large amount of thermal energy to be released in the folding process, as if a grenade were being detonated, it would create severe thermal problems in the cell. Likewise, the ability to easily denature proteins without a hugely prohibitive thermal input is essential to the digestion of food by organisms.

FIGURE 10.1 Congress Pool in Yellowstone National Park. The water in the pool is at an average temperature of 80°C and is extremely acidic (pH 3), yet there flourishes an archeal organism, *Sulfolobus acidocaldarius*. Because it exists in an environment with two different kinds of extremes, this hardy organism is a "polyextremophile."

In addition to proteins, nucleic acids are also readily denatured—that is, they lose their helical or other shapes above certain threshold temperatures. Above approximately 100°C, many proteins along with DNA and RNA denature, terminating viable cellular life. Other effects occur around this threshold as well; the fluidity of membranes is increased so much that cells cannot control the input or output of molecules. (That 100°C is the normal boiling point of water under a sea level atmosphere is not significant here; what is important is that this corresponds to a thermal content of the medium that encourages denaturing of the essential polymers.) Temperatures above 75°C are a problem for many photosynthesizers because chlorophyll degrades and becomes nonfunctional under such conditions. A general problem associated with high temperatures for aquatic organisms that rely on oxygen or carbon dioxide is that solubility of these gases drops significantly as temperature rises. Fish, for example, expire above 40°C for this reason.

And yet, organisms seem to exist at and above these temperatures, the champion being *Pyrobolus fumarii* with a maximum growth temperature of 113°C. The bacteria and the archaea are especially adapted to high temperatures; no known eukaryote can survive above 60°C. Essential to the survival of organisms at high temperature is their possession of enzymes—proteins—that do survive, by virtue of their particular structure, at high temperature. Higher-order chain structures, with more bonds between polypeptide chains to provide sturdiness, make the protein more resistant to unfolding, as does decreasing the length of loops in the structure and other chemical features of certain proteins. The enzyme amylopullunase does not denature even at temperatures of 140°C or above.

The problem of DNA denaturing is solved in some organisms through the ingestion or production of salts such as KCl and $MgCl_2$, which enhance the stability of the DNA chain by "screening" (partially canceling out) the negative charges of the phosphate groups in the nucleic acid backbone. Interestingly, the nucleic acid base pairing G-C is more stable at high temperatures than A-T (or A-U), because G-C has an extra hydrogen bond compared with the others (Chapter 4). Among prokaryotes, this difference seems less important in the DNA than it does in the ribosomal RNA, a perhaps obvious result since reliance on only one of the two base pairings in DNA would drastically reduce the potential variety of amino acid orderings available to make proteins. With respect to membranes, heat-tolerant organisms have a different proportion of saturated (no double or triple bonds with carbon; Chapter 3) versus unsaturated fats, which optimizes membrane stability at high temperature.

The propensity for high levels of heat tolerance to be present in the prokaryotes may be at least partially a legacy of the origin and early existence of life. While the question of just how ancient is the archaeal domain remains a complex and controversial one, it is clear that the earliest life was prokaryotic and not eukaryotic. Prior to the development of complex photosynthetic mechanisms for generating organic from inorganic carbon, and prior to oxygenic metabolisms developing in response to an atmosphere growing increasingly oxygen rich, metabolisms relied on other oxidation-reduction chemical systems. Two examples are methanogenesis among the archaea, which in summary is $CO_2 + 4\,H_2 \rightarrow CH_4 + 2\,H_2O$ (in aqueous solution) and sulfur reduction among bacteria and archaea, in summary $S + H_2 \rightarrow H_2S$, again in aqueous solution. These reactions require a supply of reactants, and gases coming from the Earth's crust and deeper, exuded at submarine vents, provide the reductive power to make methane and hydrogen sulfide (Figure 10.2).

The concentration of the reactants must be sufficient to initiate the reactions, so prokaryotes could not stand off too far from the typically 300 to 400°C vent waters to obtain the nutrients. Reduction does not occur in the absence of enzymatic control, even in the hot waters of the vents; reduction of carbon dioxide to methane occurs very slowly compared with the dynamical exudation and dispersal of the gases in seawater, even at 500°C. Hence, only a small subset of the possible chemical reactions associated with vents are appropriate sources of metabolic energy, but in the absence of (efficient) photosynthesis on the early Earth, these hot regions may have been where the action was. And if the interpretation of chert data by P. Knauth at Arizona State University and colleagues is correct, much of the early Earth may have looked like a combination of Yellowstone hot springs and midocean submarine vents.

High-Salinity Environments

A few gulps of high-saline water such as that in the Dead Sea between Israel and Jordan can be sickening or even deadly to human beings, but life overall lives in environments

FIGURE 10.2 Saracen's Head, a black smoker on the ocean floor of the Atlantic.

ranging from nearly pure water to water so salty that the salts are saturated (on the verge of precipitating out). An environment with a particular salt content has two implications for an organism. One is the *osmotic pressure*, which refers to the pressure required to prevent water from flowing across a cell membrane when there is a difference in the dissolved content of salt within and without the cell. This pressure is essentially related to the chemical potential difference (Chapter 7) between the two different solutions—salt water inside and outside the cell. The greater is the difference in salt concentration, the greater is the osmotic pressure. The second implication is with respect to the *ion concentration*; the ability to maintain an appropriate charge balance at the cell membrane and to transport molecules across the membrane depends on the ion content of the water.

Seawater has a salt concentration of only 3.5 percent, yet we cannot drink it to rehydrate our cells. The effect of excessive salt in the water we drink is in fact to desiccate the cells, to draw water out through the membrane, eventually causing the enzymes and DNA—which rely on being in an aqueous medium to function—to denature or break. Membranes faced with external salt concentrations vastly different from their interiors may rupture. Yet, some organisms exist in hypersaline environments such as salt flats, evaporation ponds, closed basin lakes and seas, and deep-sea basins with high salt content. Some organisms are found at salinities 10 times that of the ocean, that is, approaching 35 percent. These include bacteria, archaea, and eukaryotes.

The strategy for surviving in hypersaline environments is generally the use of ions within the cell to reduce the osmotic pressure and prevent desiccation and bursting. These ions may be as simple as potassium, K^+ in halobacteria, to sugars and other small biologically synthesized molecules. Often these dissolved solutes are repelled by proteins and hence accumulate away from them. This in turn forces water to surround the proteins and stabilize them, even in a high-salt environment that would encourage desiccation.

How important high-salinity environments may have been on the early Earth is unclear. The majority of Earth's inventory of water was probably delivered well before life formed (Chapter 6), and we have almost no indication of the extent to which particular water-filled basins—proto-oceans or seas—dried up and refilled once life began. The early oceans may have fluctuated greatly in their saline content, challenging the survival of early marine prokaryotes. The process of desiccation and consequent formation of salt flats is so common geologically that, undoubtedly, organisms have been exposed to high-salt environments throughout Earth's history, even if the fluvial transport systems that salinate the oceans today had yet to fully develop in the Hadean. High salinity is of particular interest for environments on Mars and Europa, to be discussed in Section 10.3.

Highly Acidic or Alkaline Environments

The extent to which a water solution acts like an acid or a base, expressed as pH (Chapter 3), is crucial to the stability of proteins as folded structures, as well as transport of ions through the cells. Proteins in acidic solutions (low pH) will denature; indeed, one technique for denaturing proteins in fish in the absence of heat is to place it in a very acid solution such as lime juice (the popular Mexican dish *ceviche*). Fish and cyanobacteria die at a pH less than 4, where 8.2 is the pH of the standard ocean environment. Some eukaryotes can survive in extremely acidic waters with a pH approaching zero, as do some algae and archaea. Among the most interesting places to find acidophilic organisms are in old mines in contact with groundwater or rivers that flow through abandoned mines and mine tailings such as the Rio Tinto in Spain (Figure 10.3, see Color Plate 11).

At the Iron Mountain mine in northern California, metal sulfide minerals are oxidized, leading to the formation of acid mine drainage. Waters within the sediments there vary in pH from 0 to 1, with temperatures up to 40°C. Acidophilic microorganisms do much of the oxidation, and surprisingly 85 percent of the microbial community in the mine drainage is a single type of archaea, *Ferroplasma acidarmanus*. The organism accelerates the rate of oxidation of the mineral pyrite, FeS_2 at the mine site, contributing to a human-generated mineral extraction process responsible for half of the sulfate deposited in the world's oceans.

Highly alkaline environments pose a different challenge for organisms; there the protons needed to drive electrochemical gradients at membranes, where ATP is made, are scarce. Nonetheless, prokaryotes and eukaryotes both are able to maintain cellular processes in external environments that are very alkaline, that is, with a pH of 11. In both the acidic and alkaline cases, maintaining a significant pH gradient between the inside and the outside of the cell ensures survival. Unlike the maintenance of a salinity contrast, osmotic pressure and membrane bursting is not an issue. But energy cost is; to maintain an internal pH close to that of nonacidophilic or alkaliphlic organisms requires ways of transporting protons into or out of the cell membrane via additional transport mechanisms or charging of cell walls. Some archaea, such as that at the Iron Mountain mine, maintain the pH contrast in the absence of cell walls.

The ability to live at the extremes of pH, while an adaptation that happens to allow some organisms to thrive in human-generated mining operations, is likely a very ancient adaptation. The early anoxic atmosphere would have made minerals such as pyrite and uraninite (Chapter 11) stable, whereas today they are not. Abiotic nitrogen reduction, perhaps still going on for some period of time after life formed, could have occurred via the conversion of FeS to pyrite in the presence of hydrogen sulfide (from deep-sea vents). This class of reactions, whether biologically or abiotically mediated, could have led to significant pH variations, particularly in small, shallow seas. Whether organisms such as the archaeal *F. acidarmanus* represent deep-rooted adaptations to such early conditions or much later adaptations unfortunately remains unresolved.

High-Radiation Environments

The surface of the Earth is well protected from lethal particle radiation associated with cosmic rays. Also, ultraviolet (UV) radiation is severely diminished by a layer of ozone that has probably existed—but fluctuated—over at least 2 billion years of Earth's history. Without such protection, surface-dwelling organisms would sustain photon and particle radiation damage to their cells, the most severe of which is damage to the structure of the DNA (Figure 10.4). While enzymes exist to repair DNA (Chapter 4), if the rate of influx of damaging particle radiation is high enough, it will exceed the ability of the enzymes to make repairs quickly enough, and the genetic code will be corrupted, leading to the cell's inability to replicate, or to its death. In macroscopic, multicellular organisms, the physiological effects of radiation exposure and cell damage are complex, ranging (in humans) from vision-robbing cataracts to sterility, cancer, and, in sudden large-scale exposures, digestive and renal failure followed by death.

It is the complexity of the deposition of radiation into living tissue in macroscopic organisms, or colonies of multicelled organisms, that makes determination of the effects of

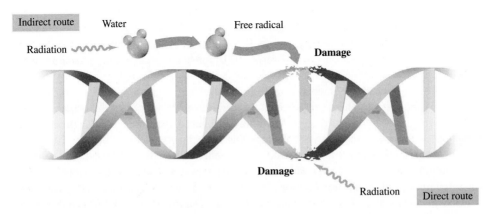

FIGURE 10.4 Schematic of how radiation damages DNA. Direct impingement can lead to alteration of the base bonding, or absorption of the radiation by water can produce a free radical (OH), which then attacks the base pair.

radiation somewhat complex. For a given amount of energy absorbed per mass of tissue, energetic particles cause more damage than photons, and particles with higher atomic mass do more damage than particles of lower atomic mass. Hence, physicists use derived physical units (e.g., the sievert) that attempt to account for both rate of energy input and absorption by tissues. However, we will stay with physical units because the derived units are of little use for microorganisms. The basic unit of radiation dose is the rad: 100 rads equals one joule of radiation input into an organism per kilogram of cells. In these units, the background dose of cosmic radiation (mostly cosmic rays) in the terrestrial environment is about 0.5 rad/yr. An immediately (hours) lethal dose for humans, administered in a single burst, is in the range 2,000 to 5,000 rads.

The range of radiation tolerance in terrestrial organisms is quite large. Organisms with smaller genomes tend to be more resistant; the surface area of DNA present declines steeply with declining number of genes (see Table 16.1 for a discussion of genome size). Organisms in environments with plenty of nutrients can repair DNA faster, presumably by generating more enzymes, than can starving microbes. Finally, both lower temperature and lower water content increase resistance to radiation, presumably through the effect both have on the generation of charged free radicals in the cell.

Deinococcus radiodurans, a bacterium, has the astounding capacity to absorb 6,000 rad/hr of radiation without any evidence of induced mutation or effect on its growth rate (see Figure 10.5). It can sustain a single burst of between 1.5 million rads (1.5 Mrads) at 0°C to 4 Mrads at −70°C without dying, and after 24 hours the genome is apparently fully repaired. Its radiation resistance extends to different types of particle radiation as well as ultraviolet exposure. Further, *D. radiodurans* shows unusual resistance to many chemical toxins and to desiccation. Hence, it is at home in hazardous waste dumps that include toxic chemicals, heavy metals, and radioactive materials. The organism's ability to weather such hazards is apparently connected to an enhanced capability to repair its DNA—vastly better than that of human cells, which die at radiation doses three orders of magnitude less. This capability is in part due to the excessive number of copies of its genome possessed by each bacterium—4 to 10, stacked one atop the other. However, the specific enhancements in the repair mechanisms are not yet well understood nor is their evolutionary origin. While exposure to ultraviolet radiation may have been more extensive over long periods of time earlier in the Earth's history, particle (cosmic-ray) radiation at the surface may not have been much different in a time-averaged sense than it is today. Potential sources of excess cosmic rays include transient astrophysical events such as nearby supernovae (which would be difficult to adapt to from an evolutionary standpoint) or transient fluctuations in the Earth's protective (but inconstant) magnetic field.

The ability of organisms to withstand the ultraviolet fluxes potentially incident on the surface of the Earth prior to the rise of oxygen and the consequent establishment of an ozone shield remains one of the continuing puzzles in astrobiology. Because an ozone shield can be maintained for oxygen levels in the lower atmosphere only 1 percent that of the present-day value, it is possible that the shield has been there for more than 3 billion years if the rise of oxygen occurred in the Archean, as endorsed by H. Ohmoto of Pennsylvania State University (Chapter 11). Prior to that time, the Earth's atmosphere might have been reducing enough to allow small amounts of methane and ammonia to remain stable in the atmosphere and screen out damaging solar ultraviolet radiation. More problematic is the situation assuming that significant levels of oxygen were not present in the atmosphere

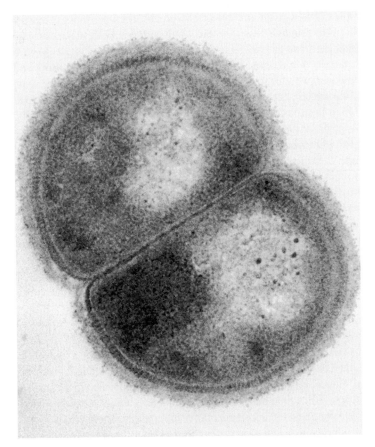

FIGURE 10.5 *Deinococcus radiodurans*, the Olympian of radiation-resistant organisms. The organism is a couple of microns in extent.

prior to 2.5 billion years ago. Given too that it is unlikely that very reducing conditions remained in place for 2 billion years after the formation of the Earth, the late oxygen scenario could imply a long period in which the surface of the Earth had large amounts of ultraviolet radiation. This would have made land areas, and the immediate ocean surface, lethal to organisms without special protective capabilities, such as mutual shielding in colonies (e.g., bacterial stromatolites). Photosynthesizers could still have existed at sufficiently shallow depths beneath the ocean surface to capture the long-wavelength visible light while being protected by a thin layer of water from shorter-wavelength ultraviolet light. Lack of an ozone shield might have spurred evolutionary development of resistance to particle radiation, but we lack direct evidence for significant exposure near the Earth's surface to such radiation for extended time periods in Earth's history.

Regardless of their origin, the existence of radiation-resistant microorganisms poses both a promise and a challenge for the search for life elsewhere. The promise is that high-radiation environments might harbor life, while the threat is that the normal uses of ionizing radiation to sterilize spacecraft and prevent forward contamination of other worlds might not work on all organisms. These issues important to astrobiological space exploration are discussed further in Section 10.3.

Cold and/or Desiccated Environments

On Earth, cold and desiccation often go together. The dry valleys of Antarctica, surrounded by a continent covered in water ice, are among the driest places on the Earth, their snowfall limited by mountain-storm shadow effects and the extreme cold of high altitude and polar latitude that limits storm moisture. The Atacoma desert of South America is extraordinarily dry even by desert standards, again exacerbated by altitude and by the interception of the moist easterly trade winds by the Andes Mountains. The effects of desiccation are similar to those in organisms exposed to high salinity and, to a lesser extent (only because it encourages dehydration), high temperature. In other places, such as in glacial ice and the sea ice near the poles, cold and moisture are paired.

Cold has effects that are in some aspects complementary to heat. Lowering the temperature decreases the fluidity of the cell membrane, so to compensate, the ratio of unsaturated to saturated hydrocarbons must be increased—opposite to the response to high temperature. Low temperatures exacerbate the effect of the activation energy (Chapter 7) on reactions catalyzed by enzymes, so psychrophiles (cold lovers) generate proteins that are very well matched with substrates on which the enzymatic activity occurs or is aided. Likewise, protein flexibility declines at low temperature, so cold-adapted organisms often rely on enzymes with folds and shapes that promote less rigidity.

Cold creates a special challenge for cells when the temperature becomes low enough for freezing to begin. Freezing causes a large expansion in the water phase of almost 10 percent, sufficient to rupture cells if no means of removing water is available. Even if rupturing does not occur, the dramatic increase in viscosity and decline in mobility relative to liquid water means that a frozen cytoplasm cannot play the role of transporting molecules around the cell. One defense against freezing is to delay its onset—freezing temperature within and around cells is rarely 0°C in natural environments because the presence of salts and other solutes depresses the freezing point of the solution. Certain proteins produced by psychrophilic cells can lower the freezing point by 10 to 20°C. Another strategy for protection against internal freezing is to permit the external environment to freeze—the resulting change in thermal conductivity insulates the cell against internal freezing. An ultimate defense is to form spores or cysts and try to outlast the cold period; the evidence that some organisms might be able to form spores for geologic lengths of time and then become viable again suggests this might be a relevant mechanism for putative biota over some periods of Martian history.

The diversity and wealth of low-temperature ecosystems is remarkable. Antarctic sea ice covers 20 million km^2 of ocean in a typical winter and plays host to a food chain that begins with single-celled eukaryotes through krill and on to whales. The krill stock exceeds in mass by a factor of three the total mass of human beings on the planet, and in winter, these animals feed primarily on algae that exist in the sea ice. The presence of microscopic diatoms in sea ice can be of such extent that they visibly discolor the ice, in much the same way that algae discolor continental glacier ice (Figure 10.6). Sea ice is not

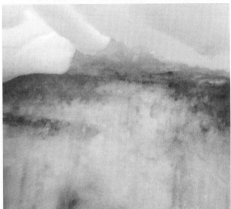

FIGURE 10.6 Diatoms living in Antarctic sea ice create the dark shading, which is brown in color.

very porous, and, as a result, substantial gradients occur from one part of an iceberg to another in salinity, pH values, organic matter, and so on. Organisms have a range of environments to choose from, the base of the ice being the most mesic in terms of temperature and salinity but highest in amount of trapped salty water (Figure 10.7). Because of the small amount of liquid water available in the ice, organisms themselves dramatically affect the balance of pH, concentration of organics, and ammonia. Specialized low-temperature enzymes occur in many of the sea ice organisms, while in other cases changes in the amount of key enzymes compensates for cold. For example, the key enzyme in plant photosynthesis, rubisco (Chapter 4), declines in catalytic efficiency at low temperature; because there is no real substitute, psychrophilic microalgae compensate by increasing the enzyme's concentration in their cells (at the cost of the added energy required to make it).

The origin of cold adaptation must go very deep into the history of life on Earth, if the interpretation of the geologic record as indicating widespread glaciation at times in the Proterozoic is correct. The diverse psychrophilic biological communities seen today argue

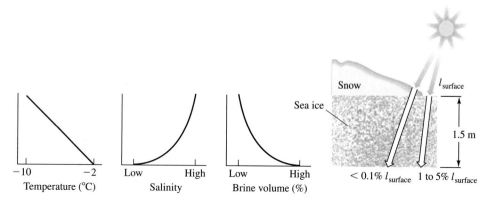

FIGURE 10.7 The environment within an ice floe can have dramatic variations in temperature, ice salinity, coexisting salty liquid water, and sunlight.

against the notion that vast areas of sea ice extending almost to the equator—the current interpretation of the geology—would have wiped out life on Earth. Of course, whales and krill did not exist in the Proterozoic, and perhaps early eukaryotes found the going tough on "Snowball Earth," but prokaryotes likely were well adapted and flourished if today's psychrophilic bacteria are any guide.

Subsurface Rock Environments

Perhaps the most astonishing outcome of extremophile research is the discovery of microorganisms living in rocks, down to great depths in the Earth's crust. Subsurface microbial communities, known for some time to exist in deep mines, have also been found to depths of kilometers in igneous rocks. Some recent estimates put the mass of the subsurface, "endolithic" biosphere as exceeding the mass of all other life-forms on Earth. Although this is a controversial estimate, if it is true, then the search for life in the crust of Mars is really—on a total mass basis—very much a search for a biosphere "like Earth's," though undoubtedly more limited in abundance and variety.

Endolithic life faces several challenges. Temperature increases with depth in the rocks because heat flow is outward from a hot interior; hence, such organisms must be thermo- or hyperthermophilic. In granitic rocks, radiogenic elements are high in abundance; hence, radiation tolerance may be essential to survival. The low abundance of nutrients, which we discuss further later in this section, poses its own challenges as well as exacerbates the problem of responding to internal cellular damage, such as from radiation, by DNA repair. A final challenge is common to organisms that reside at depth in the ocean basins, namely, resistance to pressure. At 1 kilometer below the continental surface, the pressure exerted by the weight of the overlying layers, the hydrostatic pressure, is $1/3 \rho g h = 10^8$ dyne/cm^2 = 100 atm. Here ρ is the rock density, g is the gravitational acceleration, and h is the distance beneath the surface. Because the density of rock is several times that of liquid water, a given distance through rock creates a larger hydrostatic pressure relative to that through water.

Given that organisms exist at depths of several kilometers, we might imagine some of the deepest colonies experience pressures of order 1,000 atm. The potential effects of high pressure include decreased membrane fluidity, as the lipid molecules are pushed closer to each other. As with cold, increasing the amount of unsaturated lipids can offset this effect. While moderate pressures actually stabilize enzymes, a threshold pressure is reached at which the three-dimensional structure of the DNA and the proteins are overcome and the molecules distort to the point of nonfunctionality. Experiments in the laboratory subjected bacteria, including *E. coli*, to hydrostatic pressures of 10 kbars (10,000 atm) in a diamond anvil cell, which applies pressure to a small volume within which a sample (in this case, bacteria in water) is located. Such pressures are so extreme that phases of solid ice form that are not present under normal conditions.

Evidence that bacteria remained viable at high pressure came from measurement of oxidation of HCO_2^- as an assumed measure of metabolism. While some have questioned these results, the recovery of viable bacteria at the end of the experiment suggests that bacteria will recover from, and may even conduct metabolic processes when subjected to, pressures of thousands of atmospheres. Such results are potentially of great importance for Mars, where the only remaining environments for stable liquid water could be well below the surface at high pressures.

Beyond the ability to survive high pressure, endoliths must be able to find something to metabolize. A number of endolithic colonies of microorganisms seem to be severely limited by the availability of nutrients. At deep-sea vents, *chemoautotrophic* prokaryotes sit at the base of the food chain and use the reduced state of the vent water to derive biologically useful energy and nutrients. In rock, the same thermodynamic free energy could be obtained from inorganic sources if a usefully mobile source of reduced substances dissolved in water were available. The iron oxide FeO will react with water, if the latter is available, by the reaction $H_2O + 2\ FeO \rightarrow H_2 + Fe_2O_3$. This liberates molecular hydrogen, which can then be used to sustain biological processes through the oxidation of hydrogen back to H_2O, to H_2S, or other compounds.

However, the attack of iron oxide by water to make hydrogen is a slow process, becoming fast at temperatures approaching the limit for hyperthermophilic organisms, and is often limited by the formation of silica "gels" that coat the reaction surfaces. For these reasons, and because of the small volume of water that may be circulating through igneous minerals in the deep crust, microbial communities living off hydrogen could be quite limited. An alternative source of hydrogen, though, is hydroxyl ions present in some igneous and metamorphic rocks. These anhydrous rocks originally formed in the presence of water and therefore carry the hydroxyl ion OH^-. Laboratory experiments at the SETI Institute, NASA Ames Research Center and Washington State University suggest that an oxidation-reduction reaction can occur under some conditions in the crust that release hydrogen and bind the remaining oxygen to the silicon (Figure 10.8).

To apply the laboratory experiments to the crust of the Earth, we must consider what is going on at depths below those of the microbial communities themselves. Rocks cooling through 400°C, based on the experiments, will release hydrogen to the surrounding environment. Because this temperature is the ambient value for a depth of roughly 20 km, it is

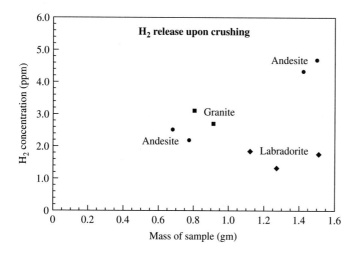

FIGURE 10.8 Results of a laboratory experiment by Freund et al. (2002), in which crushed rocks ranging in silica content from continental granites to volcanic andesite release hydrogen. Shown is the concentration of hydrogen released, in parts per million, as a function of the mass of the rock sample for various rock types.

unlikely that organisms exist there to take immediate advantage of the hydrogen. However, the solubility of the hydrogen in the rock matrix declines with decreasing temperature, and, as hydrogen moves upward in the rock (because of its buoyancy), it experiences cooler conditions. Thus, the net effect is to create a kind of "wind" of molecular hydrogen that passes upward into the cooler pores closer to the surface where the prokaryotes exist.

The amount of hydrogen available for deep endolithic communities is limited, because eventually such communities would draw the hydrogen abundance down to the point where the residuum is in equilibrium with the water in the pore spaces and cannot easily be brought out of solution. Because the total mass of such communities is unknown, it is difficult to assess whether the hydrogen liberated from deeper down is sufficient to sustain colonies in steady-state, or whether such colonies die off regionally, to be reestablished later from others. However, the possibility that past episodes of mineral hydration in the crust ultimately provide food for deep microbial communities is intriguing, particularly in considering the case for Mars (Section 10.3).

Use of hydrogen to sustain biological processes is a chemically simple and hence ancient strategy that may well date back to the dawn of life, at a time when the atmosphere was far more reducing than it is today. Particularly very early on, when frequent impacts disrupted aqueous environments, which had to then form anew, the crust may have been dominated by igneous rock in contact with water—at temperatures often much higher than obtained today at the surface of the Earth. Thus, the production of hydrogen from igneous rocks may have occurred at a much more prodigious rate than today, leading to a dominance of the microbial communities that relied on it. Existence in the rocks, which lent some protection from the effects of geologic and atmospheric cataclysms, could have been the norm of the time. It is an existence wholly different from that which we regard as the "normal" biosphere, yet it may be more common even today than we imagine. If endolithic existence is the norm in the evolution of habitable planets, it will render more difficult the search for life, both in situ on Mars (Chapter 14) and remotely in the spectra of the atmospheres of terrestrial planets orbiting other stars (Chapter 15).

Other Extremes

The preceding discussions in this section have not fully exhausted the kinds of extremes that life may experience on or off the Earth. The exposure of nonmitochondrial organisms to oxidizing conditions represents another kind of extremophilic existence, which, from an evolutionary perspective, has been rendered almost irrelevant by the advent of aerobic respiration. Chemical extremes such as environments with high levels of metal have been covered incidentally in the mentioning of organisms that "love" toxic waste dumps. Strong electric fields, pure CO_2 atmosphere, and life in high levels of organic solvents are other possible extreme conditions to which life on Earth has adapted.

Absence of gravity is not the norm for all but the briefest periods on Earth, but organisms carried into space on impact fragments and delivered to neighboring planets (Earth to Mars or vice versa) experience a lack of gravity for long periods of time. Should such organisms survive on the recipient planet, they must grow in a gravitational field different from that on their home world. To date, effects of microgravity on timescales of months or years have been shown to generate some adverse physiological changes, but mostly in higher organisms. The effect of permanent displacement to a different value of the gravitational acceleration may have no deleterious effect at all, but there are virtually no data available to test such a statement.

10.3 The Implications of Extremophiles for Life Elsewhere in the Solar System

Mars

The history of Mars described in Chapter 12 is one of progressive loss of atmosphere, cooling, and drying of the surface to the point today that liquid water is not continuously available at the surface. However, the geologic evidence for stable, surface liquid water in the past is quite strong (Figure 10.9), and Mars Odyssey gamma-ray mapping of the subsurface indicates that water ice remains in the crust. Early in the history of Mars, life might well have flourished in locations containing water stable on an annually averaged basis. Because there is no detailed information on the nature of such environments, at best we can argue that life could—but need not—have been much like prokaryotic life on Earth. Geologic evidence suggests that liquid water stability on the surface was lost prior to a few billion years ago, so we might posit that time was not available for the development of eukaryotes, or at least aerobically driven, complex multicellular creatures.

The progressive loss of atmosphere was likely experienced across the surface as both increased desiccation during parts of the year and increased levels of cold. In addition to the strategies for surviving desiccation in viable form, described earlier, spore formation for long periods may have dramatically stretched out the timetable for surface life on Mars. Seasonal freezing of standing bodies of water might have been a constant fact of life on Mars even in its most equable climatic stage, but as time progressed, fluctuations in climate leading to ever longer periods of dry, cold conditions became the norm. Suggestions that spores could remain viable for millions of years remain speculative, but it may have

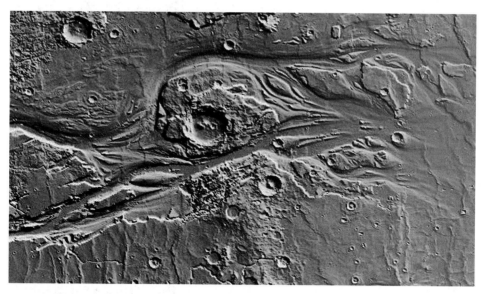

FIGURE 10.9 Outflow channel system on Mars, Kasei Vallis, where large amounts of liquid water scoured out islands and other erosional features billions of years ago. The image was constructed using the Mars Orbiter Laser altimeter on the Mars Global Surveyor orbiter.

become essential for surface and near-surface life on Mars. Psychrophilic bacteria (Figure 10.10) should have been the last surface survivors as mean temperatures over time drifted to well below 0°C, and indeed, in some models annually averaged surface temperatures on Mars never exceeded this value.

Eventually, as the atmosphere thinned enough, other hazards would have challenged surface organisms. Initially cosmic rays, and then ultraviolet photons from the Sun, would create radiation hazards that eventually might be insurmountable. The current particle radiation dose, 0.5 rad/yr at the Martian surface, could be survivable by spore-forming bacteria if the Martian climate were to fluctuate on timescales of millions of years between thick- and thin-atmosphere periods. During the thick-atmosphere epochs, provided liquid water was available, organisms could come out of the dormant spore state to repair DNA and procreate and then become dormant again. The geologic evidence is difficult to interpret in terms of the frequency with which the climate fluctuated and how late in Martian history this might have occurred, but within the last billion years, Mars has almost certainly become cold and dry with sufficient continuity that the spore-forming strategy would fail. An additional problem is that photochemistry in the thin Martian atmosphere produces radical species such as OH that react chemically with the surface. The net result is a powerfully oxidizing soil at the surface that would likely destroy cellular membranes, perhaps even the protective surfaces of spores.

Eventually, then, the only life surviving on Mars would be underground. The gamma-ray spectrometer maps of hydrogen, as a proxy for subsurface water, show vast deposits at mid to high latitudes in the Southern Hemisphere, and more restricted areas in the

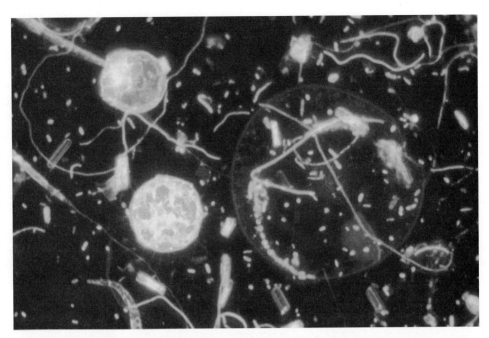

FIGURE 10.10 Psychrophilic bacteria, shown here as micron-sized dots and rods accompanied by much larger algae, have an internal makeup that might be indicative of the kinds of survival mechanisms required for life as Mars became cold and dry early in its history.

Northern Hemisphere (Chapter 12). Although the hydrogen detected is within a meter of the surface, and probably maps out deposits of ice, liquid water could exist deeper in the crust and perhaps might be somewhat more widespread in latitude. Although much water may have been lost to space over the history of Mars (ejected by impacts), the gamma-ray data suggest a good deal has been retained. Organisms living deep in the crust would require a food source, and one possibility is hydrogen produced from OH in minerals that had been in contact with water much earlier in Martian history.

Thus, it is possible to imagine a biota continuing on Mars but well below the surface and hence inaccessible to anticipated capabilities for drilling and extraction of samples by robotic landers. The only hope for accessing living organisms is that a few locations, perhaps near or in the volcanic plateaus, have liquid water and hence biota close to the surface, but no evidence for the existence of such regions has yet been found. Those organisms living deep in the crust of Mars would be living, in effect, on borrowed time. Unlike the Earth, where liquid water and oceanic crust are recycled by plate tectonics (Chapter 11) and where sunlight provides a source of energy that will last billions of years, Mars offers no means for recharging either liquid water or energy sources. Photochemistry at the surface is no longer possible, given the harsh environment, and any hydrogen available to deep crustal organisms will eventually be depleted. It is sobering to think that the exploration of Mars is the exploration of a dying (or already dead planet), but that is the harsh reality of the outcome of planetary evolution on our small, red neighbor.

Europa

The radiation environment at the surface of Jupiter's moon Europa is too intense for any known terrestrial organisms. A megarad of particle radiation strikes the surface each hour, defeating even the repair mechanisms of *D. radiodurans*. However, the strong evidence for liquid water beneath the crust of Europa (Chapter 13) suggests that life might well flourish there. What kind of life and the total biomass depends on whether places exist where the crust is thin enough to admit sunlight. The extensive fracture systems might, speculatively, include places where the ice is only meters thick, and there, sunlight scattered through the ice from above might sustain some photosynthesis. Algae living at the base of terrestrial sea ice floes a meter or so thick are capable of photosynthesis at light levels of 0.1 to 1 percent of the value at the surface, using modified photosynthetic pigments that work at low light levels. Photosynthesizing organisms existing beneath thin portions of the Europan crust would live in an environment of ice and liquid, perhaps somewhat akin to that in the ice floes of the Earth's polar regions. Given that such thin regions might well be transitory, the challenge of moving from one such region to another could be daunting and might simply depend on random movement of organisms by tidal currents.

If, on the other hand, the ice crust of Europa is uniformly thicker than tens of meters, photosynthesis is not possible. Sources of energy for life might exist at the base of the liquid water ocean, at the surface of the rocky mantle, some 100 to 200 km below the surface. Europa's small value of the gravitational acceleration, one-sixth that of Earth, implies that the pressure at the ocean base is a modest kilobar or so. Based on the laboratory experiments discussed earlier, such a pressure does not seem to rule out viable organisms, though there may be a substantial difference between short-term survival in a high-pressure lab experiment and continual existence at such pressures. Other possible sources of energy at shallower levels in the ocean—such as tidal currents—and maximum biomass estimates are given in Chapter 13.

Interplanetary Transport and Contamination

The survival of organisms during impact of asteroids and comets on planetary surfaces has been studied using numerical models that track the time history of pressures and temperatures in both the impacting body and the target material. Much of the material ejected from a forming impact crater suffers pressures of millions of atmospheres, high enough to destroy life, but a significant fraction experiences lower temperatures and pressures that permits the survival of amino acids (Chapter 6) and perhaps organisms as well. In particular, rocks at the surface of an impact site (but not struck directly by the impactor) are not compressed to the same extent and are instead propelled out of the forming crater without being crushed. Organisms must survive an acceleration that amounts to tens of thousands of "g's" (one g being the force of gravity), which is impossible for macroscopic organisms such as animals but possible for microbes. Further, while some rocks would be heated by the impact to high temperatures, others remain relatively cool.

Once ejected into space, organisms hitching a ride on impact debris will be subjected to vacuum, weightlessness, extremes of temperature, cosmic rays, and, for those close to the surface of the rock, ultraviolet radiation from the Sun. Microbes can survive exposure to vacuum, so long as outgassing of solutes in their cytoplasm does not cause the cell wall to burst. Experimental exposure of *Bacillus subtilis* to vacuum in Earth orbit for months resulted in a survival rate of 2 percent; of those exposed to vacuum, cosmic rays, and ultraviolet light for that period, about 0.01 percent survived. Most of the mortality was the result of exposure to ultraviolet light, so that organisms well beneath the upper microns of ejected rocks will have a much higher survival rate. Computer simulations indicated that much of the ejected material has diameters of meters, affording plenty of protection from ultraviolet radiation. Spore-forming bacteria, in particular, could perhaps survive space conditions in such a rocky enclosure for millions of years; exposure to cold actually appears to increase the survival rate for spores.

Once ejected from Earth, spore-bearing rocks may eventually encounter another planet. Close passes by Jupiter almost certainly result in ejection of the rock from the solar system, and the timescale for capture of such rocks by other stellar systems—about 100 million years—is likely too long for the spore to remain viable. Much of the rest of the debris will have orbits around the Sun altered in a stochastic fashion by the gravitational pull of the planets. Meteorites derived from the crust of Mars (Chapter 12) by impacts have a small but significant chance of eventually impacting Earth. During their time in interplanetary space, radioactive isotopes are produced in the near-surface layers of these rocky bodies by impact of cosmic rays. Analysis of the parent–daughter abundances of the isotopes (Chapter 6) suggest travel times from Mars to Earth of hundreds of thousands to tens of millions of years. Because of the much larger gravity of the Earth than of Mars, the probability of bacteria-bearing crust that is not severely shocked being transported off Earth by impact and onto Mars is much lower—but not zero.

Once a carrier rock intercepts the atmosphere of, say, Mars, the organisms inside must survive the entry and impact. Temperatures deep within a rock meters across will remain modest despite the heating of the surface by the friction of entry. Survival at impact depends on whether the rock shatters in the atmosphere—pieces that do so will be slowed further by friction and fall relatively gently, while very large pieces may hit at speeds leading to lethal decelerations.

It is thus possible to imagine that extremophiles—microbes capable of withstanding some degree of cold, radiation, pressure shock, and so on—have been transported from Earth to Mars and vice versa over the age of the solar system. We thus cannot regard these two planets as biologically isolated. Indeed, even transport of impact debris to Europa is possible though highly unlikely, but impact of the carrier on the surface at very high velocity—engendered by Jupiter's powerful gravitational field—makes survival there questionable. What concerns planners of spacecraft missions to Europa and Mars, though, is that if survivable interplanetary trips after natural impacts can be contemplated, then surely organisms left on incompletely sterilized spacecraft might survive deliberate interplanetary journeys. For example, were *D. radiodurans* to find its way onto a spacecraft intended to orbit Europa, it could survive the trip there.

What if, in maneuvering to achieve orbit, the spacecraft hit Europa? The impact speeds could be far less than a direct impact from solar orbit, leading to the worst possible outcome—penetration of the ice crust by the spacecraft but survival of some organisms. These, introduced into the watery world beneath the Europan crust, probably could not find food and would die, but this outcome cannot be guaranteed. In searching for life on

Europa, we could well slip up and introduce our own form of life to Jupiter's watery moon. Indeed, the situation here may be worse than for Mars; the upper meters of Mars contain only frozen water and may be uninhabitable, and injection of organisms into a putative habitable realm a kilometer under the Martian surface would be extraordinarily difficult to accomplish. Europa's habitable environment could be meters below the surface in same places. It was concern over possible contamination of Europa by radiation-resistant terrestrial organisms that led NASA to conclude the Galileo orbiter mission in 2003 with a deliberate hypersonic (and fully sterilizing) burnup in the Jovian atmosphere before the spacecraft ran out of maneuvering gas and became uncontrollable in orbit.

It would be deeply and tragically ironic if our efforts to search for life elsewhere in the solar system ended with the inadvertent colonization of other worlds by the "toughest" and most exotic organisms on Earth, sent as accidental emissaries by the latest form of life—multicelled intelligent eukaryotes—to appear on the scene. But such a possibility after 4 billion years of the evolution of life on Earth might be an accomplished fact; after all, natural impact processes have probably cross fertilized Earth and Mars.

10.4 The Significance of Extremophiles in the Earliest History of Life

The organisms most deeply rooted on the phylogenetic tree of Chapter 9 are essentially all thermophiles. These organisms are remarkable in their metabolic diversity, ranging from fully anaerobic to aerobic. Heterotrophs, those that use existing organic carbon for their metabolism (in some cases reducing iron as well), as well as chemolithoautotrophs that rely on inorganic energy sources, are represented in the panoply of thermophiles, among others. That thermophiles come in so many metabolic flavors could simply be a consequence of the diversity of different chemical energy sources at deep-sea vents, or an indication that thermophiles were a dominant and widespread type of organism early in the evolution of life. Do they speak to a time when the habitable environment of the Earth was mostly hot springs and black smokers, or are they just the survivors of a much more varied biosphere dominated by other kinds of organisms that did not survive in the evolutionary record? The answer to this question may depend in part on more extensive exploration of the endolithic biosphere as well as the deep-sea vent environment to discover yet unknown prokaryotes and a determination of the positions of those organisms in the overall phylogenetic tree.

Regardless of what specific organisms dominated, it is clear that the earliest history of life on Earth was played out by characters and environments different from aerobic eukaryotes and yet tied to us by the commonality of the fundamental biopolymers and the need for liquid water to maintain cellular cytoplasm. It is only in the long-term evolution of Earth that we see the development of complex eukaryotic life that is tuned to a relatively narrow set of environmental conditions, and much of this tuning, with respect to the aerobic atmosphere, was at least partly a result of earlier biological processes that generated large amounts of oxygen. The continuous habitability of the Earth over 4 billion years of the planet's evolving geology and atmospheric composition, as well as changing conditions external to our planet, is fundamentally a story of the maintenance of stable, liquid water environments, a problem to which we turn next.

QUESTIONS AND PROBLEMS

1. a. Calculate the hydrostatic pressure that an organism is subjected to if it were to exist in the deepest ocean trench (Marianas Trench), using the average depth of the trench. The hydrostatic pressure is obtained by recognizing that the pressure is the force per unit area resulting from the weight of the water bearing down on the trench, and the density of water is the mass of water per unit volume.

   b. Now recognize that the pressure is also the energy per unit volume. Compare this energy with the energy associated with the unfolding of proteins in a cell. You might have to determine the number of moles of protein in a cell and the volume of a typical cell (assume prokaryotic).

2. Referring to problem 1, why might a large, uniform hydrostatic pressure not be a problem for a cell? Might there be sources of pressure gradients, or time variation in pressure, that would pose a bigger problem? (Hint—because of concentration differences between the inside of bacterial cells and the environment, a considerable outward "turgor" pressure exists.)

3. What are the so-called bends from which divers suffer, and what is the causative agent? Why does a rapid decrease in pressure cause them, and why would a very slow decrease be much less of a problem?

4. At what wavelength is a photon energetic enough to break a bond in a nucleic acid base? To what portion of the electromagnetic spectrum does this correspond? You might have to look up the bond strength for a typical covalent bond in an organic polymer.

5. For the answer to problem 4, what is the ratio of the amount of sunlight that comes out at the wavelength you found to the amount of energy that comes out at the blackbody peak for the sun, per unit wavelength? (Use the Planck function to calculate the result; do not just look it up.)

6. Using a three-dimensional illustration or description of the prokaryotic cell, calculate the probability that a cosmic-ray particle or penetrating ultraviolet radiation will strike a DNA base pair versus other portions of the cell where such a strike might be a less profoundly damaging event.

7. In addition to the unfolding of proteins and other biopolymeric structures, high temperatures will accelerate many chemical reactions among organic molecules. Either through a literature search or your own thinking, identify some positive and negative effects of speeding up reaction rates in cells.

8. Obtain a detailed phylogenetic tree of life and evaluate the following hypothesis: Some thermophiles and psychrophiles are so deeply rooted in the tree that they must have their origins close to that of life itself, in the unstable and hence strongly

fluctuating (from cold to hot) environments on the early Earth. Can you falsify this hypothesis?

9. Is forward contamination of Saturn's moon Titan (Chapter 13) a significant concern? Why or why not?

10. The very low pH Rio Tinto has been studied extensively. Select a research paper on extremophiles existing in this river in Spain, and summarize the findings on the types of organisms and their survival strategies.

SUGGESTED READINGS

For a well-written and up-to-date review of the field of extremophile research there is none better than Lynn Rothschild and Rocco Mancinelli, "Life in Extreme Environments," *Nature*, **409**, 1092–1101 (2001). A book-length description of the Archeal domain, the phyologenetic home of most extremophiles, is J. L. Howland, *The Surprising Archaea: Discovering Another Domain of Life* (2000, Oxford University Press, Oxford, U.K.). There is even a journal devoted entirely to extremophiles, *Extremophiles: Life under Extreme Conditions*, published by Springer-Verlag.

REFERENCES

Daly, M. J. (2002). "*Deinococcus radiodurans* as an Analogue to Extremophile Organisms That May Have Survived on Mars." In *The Quarantine and Certification of Martian Samples* (Space Studies Board, Ed.), 67–69, National Academy Press, Washington D.C.

David, P. (1999). *The Fifth Miracle*. Simon and Schuster, New York.

Edwards, K. J., Bond, P. L., Gihiring, T. M., and Banfield, J. F. (2000). "An Archeal Iron-Oxidizing Extreme Acidophile Important in Acid Mine Drainage," *Sci.*, **287**, 1796–1799.

Fersht, A. 1999. *Structure and Mechanism on Protein Science*, W. H. Freeman and Co., New York.

Freund, F., Dickenson, J. T., and Cash, M. (2002). "Hydrogen in Rocks: An Energy Source for Deep Microbial Communities," *Astrobiol.*, **2**, 83–92.

Levin, L. A. (2002). "Deep-Ocean Life Where Oxygen Is Scarce," *Am. Sci.*, **90**, no. 5, 436–444.

Marais, D. J. D. and Walter, M. R. (1999). "Astrobiology: Exploring the Origins, Evolution, and Distribution of Life in the Universe," *Ann. Rev. Ecol. System.*, **30**, 397–420.

Melosh, H. J. (1988). "The Rocky Road to Panspermia," *Nature*, **332**, 687.

Pennisi, E. (2000). "AAAS Meeting: Going Deep for an Unearthly Microbe," *Sci.*, **287**, 1580–1583.

Reysenbach, A. L. and Shock, E. (2002). "Merging Genomes with Geochemistry in Hydrothermal Ecosystems," *Sci.*, **296**, 1077–1082.

Rothschild, L. J. and Mancinelli, R. L. (2001). "Life in Extreme Environments," *Nature*, **409**, 1092–1101.

Schoonen, M. A. A. and Xu, Y. (2001). "Nitrogen Reduction under Hydrothermal Vent Conditions: Implications for the Prebiotic Synthesis of C-H-O-N Compounds," *Astrobiol.*, **1**, 133–142.

Schulze-Makuch, D. and Irwin, L. N. (2001). "Alternative Energy Sources Could Support Life on Europa," *EOS*, **82**, 150.

Sharma, A., Scott, J. H., Cody, G. D., Fogel, M. L., Hazen, R. M., Hemley, R. J., and Huntress, W. T. (2002). "Microbial Activity at Gigapascal Pressures," *Sci.*, **295**, 1514–1516.

Thomas, D. N. and Dieckmann, G. S. (2002). "Antarctic Sea Ice—A Habitat for Extremophiles," *Sci.*, **295**, 641–644.

Van Dover, C. L. (2000). *The Ecology of Deep-Sea Hydrothermal Vents*, Princeton University Press, Princeton.

Westall, F., Steele, A., Toporski, J., Walsh, M., Allen, C., Guidry, S., McKay, D., Gibson, E., and Chafetz, H. (2000). "Polymeric Substances and Biofilms as Biomarkers in Terrestrial Materials: Implications for Extraterrestrial Samples," *J. Geophys. Res.*, **105**, 24511–24527.

CHAPTER 11

Planetary Evolution I
Earth's Evolution as a Habitable Planet

11.1 Introduction

The next three chapters cover the properties and evolution of worlds in the solar system that are of interest to astrobiology. One of these worlds, Earth, is known to harbor life; the first half of its history—essentially up to the rise of oxygen and of complex cells—is the subject of this chapter. Mars, the planet that since the 19th century has been the favored choice as the "other" locale for life in the solar system, will be treated in Chapter 12. Finally, Europa and Titan are explored in Chapter 13. The strong but still indirect evidence for a liquid water ocean beneath the icy crust of Europa has focused attention on the possible existence of life there. Titan, almost twice as far from the Sun as Europa has a surface temperature far too low for liquid water. However, thick deposits of organics produced in the atmosphere may cover the surface, and the possibility of slow but continuing abiotic evolution of the organics—at times, perhaps, in contact with liquid water—makes this giant moon of Saturn an intriguing place for exploring the origin of life.

Of these four worlds, only Earth is known to harbor life. Earth, in fact, is teeming with life, from the upper crust of the solid planet to the base of the stratosphere. In the oceans, on the continents, and in the lower atmosphere, life is so abundant that it is a pervasive driver of many of the processes that shape the crust and atmosphere of the Earth. From the aerobic chemistry of surface waters to the profound effect on crustal mineralogy of the oxygen-rich atmosphere, from the long-term sequestration of carbonates on the ocean floor to the change in carbon dioxide abundance over glacial cycles, life is there. For life to play this dominant role, the Earth had to have been continuously habitable over large regions of its surface for billions of years. Within this salubrious venue, life evolved the capability to efficiently capture sunlight via photosynthesis, which led in turn to the global oxidation of the Earth's atmosphere and ocean. This, perhaps the first, major "pollution" of the atmosphere by life as recorded in the geologic record could well have snuffed out cells whose fundamental biopolymers were not protected against destruction by free oxygen. Remarkably, however, cells capable of surviving and utilizing aerobic conditions evolved, ultimately to become plants and animals.

This chapter covers the period in Earth's history during which our planet diverged dramatically in its atmospheric composition and tectonic "style" from that of its neighbors, Venus and Mars. It covers the processes that made the Earth habitable and those that continue to cycle volatiles into and out of the biosphere on timescales ranging from billions down to hundreds of millions of years. Chapter 16 covers the catastrophic events that clear ecosystems, thereby encouraging the diversification and spread of new organisms, as well as cyclical climate processes that operate on timescales from tens of millions of years downward. This division allows us to pick up the search for life in the solar system in Chapters 12, 13, and 14—which is really a search for primitive life and the precursors to life—at a point in history at which life was exclusively unicellular on our planet. It is also a relevant break point for the discussion of the search for life on extrasolar planets, because the rise of photosynthesis and oxygen on a world like the Earth potentially makes it spectroscopically distinguishable from non-life-bearing worlds. That is the subject of Chapter 15.

11.2 Evidence of Habitability Early in Earth's History

The oldest whole rock samples on the Earth are approximately 4 billion years old, are from northern Canada, and are composites of basaltic rocks typical of oceanic crust as well as less iron and magnesium-rich rocks that are closer to (but not quite) the composition of continental material. The rocks have been metamorphosed but also contain pebbles that appear to be sedimentary in origin, that is, debris transported by water and then subsequently lithified (hardened to make rock). Hence, liquid water was present on the Earth's surface some 4 billion years ago.

Chemical evidence from "zircons" suggests liquid water in the upper crust of the Earth, if not the surface, some 300 million years prior to that. Zircon (zirconium silicate, $ZrSiO_4$), a mineral that can be gem quality by virtue of its hardness, is found primarily, but not exclusively, in volcanic or intrusive rocks of granitic composition. Zircons are of interest in Earth history because they can be readily dated using the uranium–lead (U–Pb) radioisotopic system (Chapter 6), they are resistant to melting during metamorphism in the crust, and they are robust against mechanical destruction during sediment transport. Furthermore, oxygen is tightly bound in zircons, and this lack of mobility means that the oxygen isotope ratio in these minerals is hardly altered over time. For these reasons, a group of ancient zircons found in western Australia, which have U–Pb dates between 3.9 and 4.3 billion years before today, are of keen interest.

A team from the University of California–Los Angeles and the University of Colorado, led by S. Mojzsis, found the zircons in a complex of hills on a very ancient and stable piece of the Australian crust. The rocks bearing the zircons consisted of "gneisses," rocks that have been heavily metamorphosed (Chapter 6), with compositions ranging from that of granites to "tonalites." (*Tonalite* is a kind of diorite, a rock type that is more magnesium and iron rich—or mafic—than granite). The zircons that exceed 4 billion years in age are found in a conglomerate of quartz pebbles. None of the rocks within which the zircons were found seem to be the original rocks (the "host rocks") within which the zircons were formed; the tough zircons outlasted their host rocks and were deposited (perhaps repeatedly) in younger sediments, which later were metamorphosed. Hence, we cannot use the rocks surrounding these minerals to determine the environment within which they formed; instead, the zircons themselves must be analyzed.

The ratio of the stable oxygen isotopes ^{18}O and ^{16}O in the zircons provides an indication of the mineral's formation environment. This ratio is well determined for the mantle of the Earth, and eruption of basalts from the mantle, with varying ages, show little change in the oxygen isotope ratio. Crustal rocks, on the other hand, show a wide variation in the oxygen isotope ratio; in particular, crustal material that is repeatedly recycled and heated in the presence of water to form rocks of granitic or similar composition have significantly higher values of the oxygen isotope ratio.

Analysis of the zircons required measuring the isotopes on a scale smaller than the grains themselves, and this is accomplished through the use of an ion microprobe mass spectrometer (Figure 11.1), which can fire a narrow beam of ions at a small spot, tens of

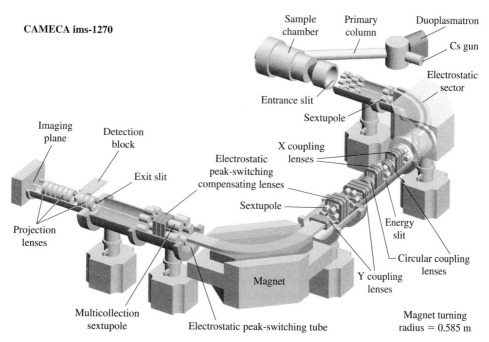

FIGURE 11.1 Diagram of the Cameca ims 1270 ion microprobe at University of California, showing at upper right the sample chamber; the cesium gun that is the source of ions; and then the series of electromagnetic "lenses" that direct the ions, bend them to differing angles in the magnetic field at the bottom of the figure, and then project them at different angles according to their mass-to-charge ratio. The size of the instrument can be gauged from the bend in the magnet at the bottom, which defines a radius of curvature of 0.5 meter. This device, used to image regions of a rock tens of microns in size and analyze the elemental and isotopic composition within such regions, is an example of a sophisticated type of mass spectrometer used in astrobiological research today.

microns in size, thus extracting ionized elements from the sample for analysis by mass spectrometry.

Measurements of younger zircons, still embedded within their host rocks, yield a well-defined relationship between the oxygen isotopic ratio in zircons and that in the host rock. In the case of the ancient Australian zircons, the measured oxygen isotopic ratios imply corresponding values in the now-vanished host rocks significantly higher than for mantle rock and consistent with crustal rocks that had interacted with liquid water. This does not necessarily mean that the host rocks had interacted with water at the Earth's surface; more likely the rocks were subsurface in hydrothermal systems where the intimate and repeated contact of hot water with minerals in the crust raised the oxygen isotopic ratio. This elevated ratio is now reflected in the zircons—all that is left of that ancient environment. How deep the liquid water might have been cannot be determined, but it was definitely within the crust rather than deep within the mantle, where a different oxygen isotope ratio would have been obtained.

Approximately 1 percent of the zircons in the sample were dated to 4.3 billion years before the present time, and they imply that liquid water was at least circulating within the crust, if not at the surface, at that very early time. Life might thus have been possible then, although the still-high rate of impactors should have repeatedly erased such early forays into biochemistry.

A different record of oxygen isotopes, those in cherts, has the potential to constrain surface temperature back to 3.5 billion years ago, that is, into the Archean eon (see Chapter 6 for geologic timescale nomenclature). Cherts are a sedimentary form of silica (SiO_2) in either very small crystal (fine-grained) or amorphous (glassy) form. Cherts form in a wide range of environments, precipitating directly out of rivers or ocean waters or forming from rocks that are subjected to mild increases in temperature and pressure. Biogenic chert is the most commonly found form in which radiolarians and diatoms secrete silica and then lithify in the form of very fine grained, microcrystalline, or glassy quartz.

The key to cherts as an indicator of temperature is again that the oxygen isotopic content of the chert bears a relationship to that of the environment in which it is precipitated. If precipitation occurs in an ocean environment, then the ratio $^{18}O/^{16}O$ of the silica decreases with increasing temperature in a manner that has been determined in the laboratory. Hence, cherts may record the ambient water temperature during their formation. Further, somewhat akin to zircons, cherts tend to be well preserved in the sedimentary record over billions of years.

Because cherts form in so many different kinds of sedimentary environments, however, it is not straightforward to use cherts as a "geothermometer." In some environments, the oxygen isotope ratio may be altered in ways that have nothing to do with the ambient water temperature. Geochemist Paul Knauth and colleagues at Arizona State University have argued that cherts can be used to determine ancient ocean temperatures in spite of these difficulties. They point out that processes during or after formation that alter the $^{18}O/^{16}O$ ratio will primarily tend to lower it relative to the value determined by the precipitation of the silica from the water. Thus, for a collection of cherts of a given age, the cherts with the highest $^{18}O/^{16}O$ values should be the least altered and most nearly reflect equilibration with crustal waters during formation. Therefore, they provide a guide, albeit imperfect, to the temperature of the Earth's surface at the time they formed.

11.2 Evidence of Habitability Early in Earth's History

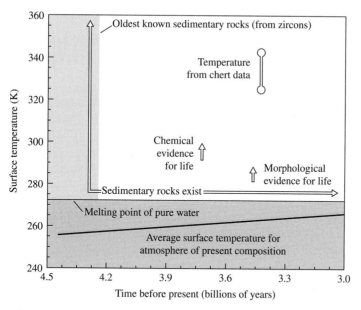

FIGURE 11.2 Summary of evidence for liquid water, and life, on the Earth through time. The temperature constraints from oxygen isotopes in cherts are given schematically based on the data reported by Knauth (1992). Ages of zircons indicating liquid water in the crust are shown. Episodes of deep glaciation are indicated, but it is not known whether the entire Earth's oceans were frozen over. Chemical and morphological evidence for life at specific, early times in the Earth's history are marked on the figure.

A temperature history of the Earth based on the oxygen isotope record in cherts is shown in Figure 11.2, along with other evidence for the presence of liquid water (or ice) on the Earth discussed elsewhere in this chapter. The chert data must be approached with some caution for several reasons—in addition to the uncertainty described in the preceding paragraph:

1. While they provide a measure of the temperature of crustal water at the point of chert deposition, we do not know—especially for the older cherts—how that might relate to global mean surface temperature.
2. The oceanic value of $^{18}O/^{16}O$ is altered by the formation of polar cap and glacial ice because water containing the lighter isotope of oxygen tends to evaporate more readily. Hence, during times when the Earth might have been heavily glaciated, the oceanic $^{18}O/^{16}O$ value would have been higher than today, and cherts would have acquired a correspondingly higher value of the ratio for a given oceanic temperature. Thus, in such epochs the chert data could be underestimating the oceanic temperatures.

3. There is some evidence that the isotopic ratio $^{18}O/^{16}O$ was smaller in the oceans prior to 3.4 billion years ago than it is today so that, overall, the ancient chert isotopic record overestimates the ocean temperature.

These uncertainties are reflected in the temperature range shown in the figure.

Reading the temperatures from the upper envelope of the $^{18}O/^{16}O$ isotopic values in the chert, and considering the uncertainties, we see a general decrease of ocean temperature over the Earth's history. Average surface temperatures recorded in the oldest cherts, dated to be just a billion years younger than the planet's formation, are substantially higher than today's mean value. However, the rarity of suitable deposits containing cherts make the temporal record spotty, so that (for example) we cannot identify the signature of profound ice ages (Snowball Earth) in the chert data. If the data are taken literally, then the Archean Earth's oceans were suitable only for the most heat-tolerant bacterial forms. A step back from committing to the detailed derivation of temperatures from the chert would be to say that they support the view that liquid water was stable on the surface of the Earth billions of years ago. Even this is surprising in view of the history of the luminosity of the Sun, which has gradually increased over time, a subject examined in detail in Section 11.4.

11.3 Earliest Evidence of the Existence of Life

While liquid water is generally regarded as a prerequisite for life, it does not follow that life must have appeared on the Earth soon after the introduction of liquid water. If the origin of life required stringently stable or special conditions for a very long period, the successful establishment of life could have taken a long time, comparable perhaps to "geologic" timescales. Or, even if life got started early, it is conceivable that it took hundreds of millions of years (or longer) for life to become widespread enough that it would show up in the rock record. And it cannot be ruled out that life might have begun elsewhere and was delivered to Earth, for example, from Mars, at some arbitrary point in the early history of the Earth. Complicating the determination of just when life first was on the Earth is that the oldest rocks are precisely those that are least abundant because geologic processes have had so much time to modify, destroy, or bury them.

The earliest hints of life come not from fossilized remains of organisms but instead from chemical remnants of biological processes in rocks. Many fundamental biochemical processes involving the extraction of carbon from an abiotic reservoir preferentially take up the light stable isotope of carbon, ^{12}C, rather than ^{13}C. For example, the rubisco enzyme (Chapter 4) plays a key role in capturing carbon from carbon dioxide in the atmosphere and preferentially removes the ^{12}C from the CO_2 because of the smaller carbon–oxygen bond strength in the lighter isotope. Hence remnants of organic carbon, meaning in this case carbon that has been processed through cells, should be enriched in $^{12}C/^{13}C$ relative to contemporaneous atmospheric CO_2, while "inorganic" carbonate deposits that precipitate from the carbon dioxide left behind should be enriched in the heavier carbon isotope. This pattern is seen, with some variation in the specific value of the enrichment, in deposits of organic and inorganic carbon over a wide range of ages (Figure 11.3).

The oldest isotopic trace of life comes from Akilia Island in West Greenland. Rock samples there are dated to be 3.85 billion years old and are a kind of sediment called a "banded iron formation" (commonly abbreviated as BIF), which is a series of iron-rich layers interspersed with chert. The origin of the banding is not well understood, though; since the solubility of iron in water varies strongly with the oxygen content, it may reflect

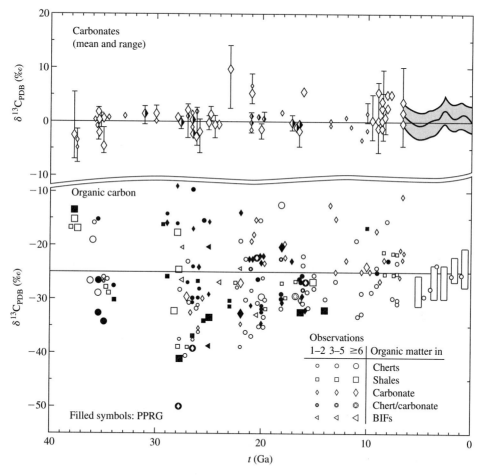

FIGURE 11.3 Plot of the relative enrichment or depletion in the ratio of ^{13}C to ^{12}C in different reservoirs of carbon as a function of time (Ga = billions of years ago). In deposits of biogenic origin, the light isotope is enriched and the heavy isotope depleted; the reverse is true in deposits such as carbonates and the sea floor in which carbon—while ending up in shells of organisms—has never participated in cellular metabolic pathways. Although the notion that certain nonbiological processes might also be capable of enriching the light isotope cannot be ruled out, the consistency of the enrichment in recent deposits suggest that it is a reliable signature of life in ancient deposits as well.

fluctuations in the oxygen content of the water as oxygen-producing processes (including photosynthesis) competed with sinks for oxygen in a time-varying fashion (see Section 11.7). Metamorphic gneisses cut through the sedimentary formation, suggesting they are younger than the banded iron, and zircons within the gneiss provide the radioisotopic (uranium-thorium) age of 3.85 billion years.

Within the sedimentary rock (i.e., the banded iron), carbon deposits are found in the form of graphite that are strongly depleted in ^{13}C relative to inorganic carbon; indeed, they fit very well the pattern of light isotope enrichment seen in later carbon deposits that is demonstrably the result of processing through enzymatic (i.e., biologically mediated) chemistry. Further, inclusions of hydrated (water-bearing) minerals in the Isua rocks show even stronger ^{13}C depletion. The analysis of the carbon within the banded iron formation was possible only recently (the mid-1990s) because the graphite deposits are extremely small (20 μm in diameter) and hence require an ion microprobe for their identification and analysis.

This result pushes the presence of life on Earth back to the Archean-Hadean boundary, almost 4 billion years ago. At this time, the impact-cratering rate was at least an order of magnitude larger than today, and the huge impact that generated the Imbrium basin on the surface of the Earth's moon had yet to occur (Chapter 6). Much debate surrounds the issue of whether the isotopic enrichment was the result of enzymatic processes in a cell of more or less modern construction, or a set of precursor processes that mimicked or preceded true life. That such isotopic enrichment is found in what is an extremely rare, extremely ancient, sediment, argues that the enrichment process—whatever it was—must have been relatively widespread.

Alternative interpretations of the ancient West Greenland rocks have been offered by C. Fedo and others (see references), namely, that they are igneous rather than sedimentary and the carbon isotopic ratios do not reflect ancient biological activity in those samples. This controversy continues as of the time of this writing and may not be resolved quickly. Indeed, its resolution may require the identification of other sedimentary rocks of comparable or greater age, containing isotopic or other evidence of life, to establish a Hadean eon origin of life.

Later evidence for life comes in the form of fossil shapes in sedimentary rocks. Some rocks around 3.5 billion years in age appear to contain so-called stromatolites, which are the layered remnants of biological activity (such as mineral uptake and precipitation) in bacterial colonies, biofilms in which layers of biologically generated material cover mineral surfaces, and tenuous filaments and other structures that might be biological in origin. Enriched light carbon is also seen in reduced or organic carbon phases, while enriched heavy carbon is measured in carbonate phases. These various lines of evidence argue that microbial life was becoming abundant in the Archean.

However, even here the interpretation is subject to disagreement. One sample, a 3.5 billion year old rock from Australia, shows microscopic features that appear to be the remains of blue-green algae, according to a 1993 analysis. That work has been challenged. The geologic context of the sediments was originally interpreted as being a shallow sea, consistent with the need to get sunlight to the photosynthesizing algae. However, reanalysis of the rocks has led to the conclusion that they were part of a deep-sea hydrothermal system, a revision that is not in dispute. What is disputed, though, is whether shapes previously identified as algae are in fact from an organism—which now must be a deep-sea microbe—or simply represent deposits of organic matter from an as yet undetermined source.

Rocks more recent than 3 billion years ago, that is, from the late Archean onward, show abundant evidence for life, including widespread stromatolites and a range of biomarkers. While a limited number of individual fossils may fall into disrepute—a 1.6-billion-year-old "animal fossil" now appears to be mud cracks—there is no question that a continuous record of life exists. Thus, at least 3.5 billion, and maybe 3.9 or 4.0 billion years ago, marks the start of a long record of life on planet Earth—a record that must be understood in the context of processes that maintain habitability.

11.4 The Faint Early Sun and a Massive Carbon Dioxide Greenhouse

Greenhouse Theory

Sunlight warms the Earth and (with the exception of the environments around undersea volcanoes or within the Earth's crust where geothermal energy dominates) makes the planet habitable. Within this simple statement lie a host of complicated issues. The most important is that sunlight, by itself, would produce a mean global temperature on the Earth of 256 K—a number that approximates what would be measured in the mid-latitudes were the Earth without atmosphere and spinning rapidly enough that day/night temperature differences would be erased. The temperature is readily derived by equating the solar energy absorbed by the Earth to the infrared energy reradiated:

$$4\pi R^2 \sigma T^4 = \pi R^2 (1 - A) F(1 \text{ AU}) \tag{11.1}$$

where R is the Earth's radius, A its reflectivity or "albedo," T the temperature we wish to calculate, and $F(1 \text{ AU})$ the solar flux at 1 AU, 1370 W/m^2. The so-called Stefan-Boltzmann constant, $\sigma = 5.67 \times 10^{-8}$ W/m$^2 \cdot$ K^{-4}, expresses the relationship between temperature and radiant flux for a perfect (blackbody) radiator. The numerical constants reflect the fact that the Earth appears as a disk from the point of view of the photons streaming from the Sun but radiates thermal energy outward like a sphere. For a reflectivity of 0.33, close to the present value, $T = 256$ K. Such a temperature is well below the freezing point for the oceans, even in the presence of salts. The Earth, in such an instance, would be uninhabitable.

It is the atmosphere's ability to retain heat that makes the surface of the Earth warm enough to support liquid water. Like the glass comprising the windows of an automobile or greenhouse, our atmosphere is more transparent at optical than at infrared wavelengths. Hence, sunlight streams in to the base of the atmosphere much more easily than infrared radiation can leak out. The lack of transparency of our atmosphere in the infrared part of the spectrum is the result of gases—primarily carbon dioxide, water vapor, and methane—that absorb photons over a range of infrared wavelengths (Figure 11.4).

The surface of the Earth is the primary absorber of photons of optical wavelength streaming in from the Sun (clouds and the lowermost atmosphere absorb a bit, as well).

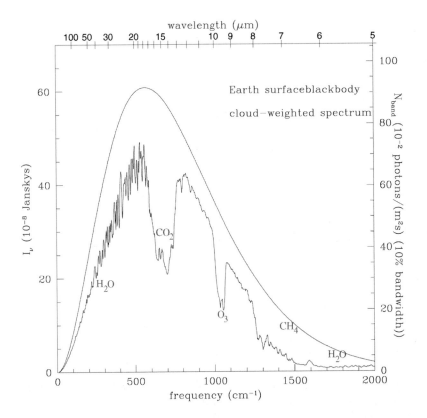

FIGURE 11.4 Spectrum of the Earth's atmosphere is plotted as intensity of light (arbitrary, linear units) versus wavelength in microns (10^{-6} meters). The spectrum shows absorption at near-infrared and infrared wavelengths caused by water vapor, carbon dioxide, and methane.

The optical photons are absorbed by the ground, which then radiates the energy back to space largely in the form of infrared photons. The production of infrared photons from the absorption of optical photons is an inevitable consequence of the dilution of the spatial density of photons streaming away from the Sun and the behavior of bodies in thermal equilibrium. The temperature of the "photosphere" or surface of the Sun is about 5770 K. A blackbody (Chapter 3) radiating at this temperature produces photons with a distribution in wavelength that peaks in the optical. The photosphere is a reasonable but not perfect approximation of a blackbody radiator.

As photons stream away from the Sun into space, they spread outward in spherical shells of ever-increasing area; hence, the number of photons per unit area is reduced. By the time the photons reach the Earth, their numbers are diluted by the ratio of the surface area of the Sun to the surface area of the sphere encompassing the orbital radius of the Earth, a factor of 45,000. The photon flux (the spatial density of the photons times the speed of light) is therefore no longer consistent with the photospheric temperature of the Sun, but the *shape* of the spectrum of the photons—the relative number of photons versus wavelength—is unaltered.

The absorption by atoms and molecules converts the photons into thermal energy within the material, expressed through vibration and rotation of the molecules, and the information on the original wavelength distribution—the spectrum of the photons—is lost. The ground, warmed by the conversion of sunlight into thermal energy, generates new photons with a wavelength distribution consistent with the amount of energy per unit area that it absorbed. This new distribution of photons also approximates a blackbody, but one that peaks in the infrared—at a wavelength of roughly 10 μm.

This redistribution of photons from the optical to the infrared is fully consistent with the second law of thermodynamics (Chapter 7). The number of photons emitted (created) is actually much larger than the number absorbed (destroyed) by the ground, because each photon carries an amount of energy inversely proportional to the wavelength. With the new spectrum peaking at longer wavelength—in the infrared—more photons per second must be emitted than were absorbed in order to ensure energy balance. Each photon also represents an amount of entropy, and the entropy in this reradiation process has increased. The process is irreversible even though energy has been (as it always must be) conserved.

This conversion of optical to infrared photons results in an increase in temperature of the lowermost atmosphere relative to what is obtained with equation (11.1). The mechanism is best understood by turning briefly to the analogous situation of an automobile sitting in the Sun with its windows rolled up tight. On a typical spring day in the desert (say, 30°C air temperature) the air temperature inside the car might be 50°C or more. Why? Because the infrared photons are impeded from escaping the car by absorption in the glass—in contrast to the free inward streaming of optical photons—energy balance associated with the incoming and outgoing photons cannot be satisfied if the internal air temperature equals that of the outside environment. As infrared photons are absorbed and then remitted by the glass both in the outward and inward directions, the rate of energy flow out of the car is less than that coming in. Thus, the inside temperature must rise in response to this imbalance, and in turn this rise increases the rate of energy emission outward. The temperature will stabilize at a value that balances the net outward and inward emission of energy by photons.

The atmosphere of our planet behaves much the same: The much greater absorption of infrared photons relative to optical photons impedes the outward flow of energy relative to incoming sunlight (Figure 11.5). This raises the atmospheric temperature in each layer of the atmosphere until the energy flux in outgoing thermal photons equals the energy flux incoming in solar photons. Because the air is densest near the surface, decreasing with altitude, the air nearest the surface experiences the greatest elevation in temperature. This "greenhouse" effect, so named because the warmth inside a greenhouse is retained by the same process (and was so named when greenhouses were more common than automobiles), imposes a temperature "profile" in the atmosphere that declines with altitude.

A car differs from the atmosphere in that the latter has such large extent that, when greenhouse heating steepens the temperature decline with altitude enough, bulk motion of air parcels begins and reduces the temperature contrast. This convection does not eliminate the greenhouse effect—it just moderates it. (Indeed, convection also occurs inside

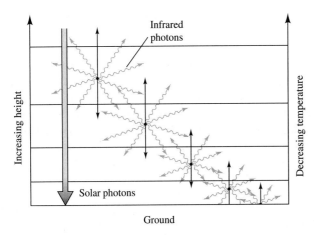

FIGURE 11.5 Schematic of the atmosphere shows the inward streaming of optical photons and the absorption and reemission of outgoing infrared photons.

large greenhouses but because they are capped by a roof it hardly moderates the temperature.) We notice convection when the rising air generates clouds: The release of heat during condensation of water makes the upwelling air parcels even more buoyant relative to the surrounding air and often generates tall columns of cumulonimbus or thunderhead clouds. But even on a dry day, invisible up and down movement of the air, like water in a heated pot, serves to adjust the temperature. Today, Earth's mean global temperature, enhanced by the greenhouse effect and moderated by convection, is approximately 288 K.

Greenhouse Gases on the Early Earth

All well and good—our atmosphere has a modest greenhouse effect that allows liquid water to exist at the surface. There is, however, a serious complication in this story with respect to the habitability of the Archean Earth. Our planet's source of photons, the Sun, did not have a constant luminosity over time. The Sun fluctuates on various timescales, but overall the Sun's luminosity must gradually increase with time—for reasons fundamental to nuclear physics. The source of the Sun's energy, as discussed in Chapter 5, is the conversion of hydrogen to helium by fusion under the enormous pressures and temperatures in the solar interior. The rate of energy production depends steeply on temperature—higher temperature means more protons are colliding at velocities above which the electrostatic repulsive force is overcome so that the attraction associated with the strong nuclear force initiates fusion. As hydrogen is consumed over time in this manner, helium takes its place. The helium "ash" that builds up from the deep interior outward takes up less space than the individual protons of hydrogen. Thus, the interior becomes more compact. This compaction reduces the potential energy of the self-gravitating sphere that is the star (Chapter 5), and the reduction in potential energy must be compensated by an increase in the kinetic energy of the material (Chapter 6)—that is, the temperature of the interior increases. The increase in temperature in turn increases the reaction rates and hence the luminosity.

Through detailed observation of stars such as the Sun coupled with modeling, the predicted gradual increase of luminosity (and size, as the added thermal energy puffs up the star) has been well established. Roughly 4 billion years ago, the Sun's luminosity was 70 percent of the present value; the temperature of an airless Earth in equilibrium with sunlight would then be, not 256 K, but [scaled using equation (11.1)] 256 K $\times$ $0.7^{1/4}$ = 234 K, or 22 K cooler. The surface temperature in the presence of an atmosphere *identical to that at present* but with a Sun 30 percent fainter must be calculated using a greenhouse model, and is only 255 K, a 33 K temperature drop from the current global mean surface temperature. (The greater temperature contrast between the two greenhouse surface temperatures, versus that between the two "airless body" temperatures, reflects the feedback between temperature and amount of water vapor in the atmosphere.) The trend over time as the Sun warmed would keep the Earth glaciated until just a billion years before present.

Such low temperatures over much of Earth's history seem to contradict everything from the geologic record discussed earlier—the zircons, the cherts, the metasedimentary rocks altered by water, the evidence for life. Could Earth have been hot in a few places that were haven for life, with the remainder ice covered? Some have endorsed such a fire-and-ice vision of the early Archean Earth, with impacts and active volcanism sustaining the warm areas. But it is hard to imagine finding the substantial amount of evidence for liquid water in the few samples of Archean rock available, unless most of the Earth was capable of supporting liquid water (or we were very unlucky in having only Archean samples from the putative "hot zones"). Instead—and this is the standard view among astrobiologists—perhaps the atmosphere had a higher burden of greenhouse gases in the Archean, leading to a more substantial surface warming that largely compensated for the weak output of the "faint early Sun."

Because Venus exhibits a carbon dioxide atmosphere of order 10^5 times more massive than the amount of carbon dioxide in our own atmosphere (maintaining the Venusian surface temperature above the melting point of lead—see Section 11.8), it is natural to imagine additional carbon dioxide as the source of an enhanced Archean greenhouse on Earth. Indeed, the amount of carbon dioxide locked in carbonates in the crust of the Earth is, by most estimates, comparable to or slightly less than the mass of carbon dioxide in the Venusian atmosphere. So, if we image that much or all of this was present in the early Archean atmosphere, is it then possible to sustain on Earth—some 30 percent farther from the Sun than is Venus—a mean surface temperature sufficient for liquid water?

The answer from modeling seems to be a qualified yes—qualified because evidence for strongly elevated levels of CO_2 is lacking. An atmosphere with a carbon dioxide pressure between a few tenths of a bar and a bar would have kept global temperatures above the freezing point of water. To sustain global conditions as warm as suggested by the chert data would have required 10 bars roughly, of carbon dioxide. Because the current atmospheric pressure of nitrogen is about 0.8 bar, this is equivalent to an atmospheric composition in which carbon dioxide is the largest constituent—a far cry from the 370 ppm mixing ratio of carbon dioxide today. None of these numbers is unreasonable from the point of view of the potential amount of carbon dioxide available to the atmosphere during the Archean eon. To sustain clement conditions over longer periods then requires a gradual decrease in carbon dioxide along the lines of Figure 11.6, with the most probable mechanism being carbonate formation. The figure suggests that by approximately 2.5 billion years ago, CO_2 in the atmosphere would have comprised 10 percent by number of the atmosphere, some 300 times the current value (abundant enough to be fatally toxic to higher animals, which

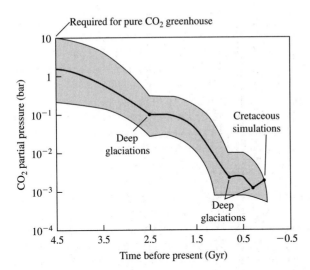

FIGURE 11.6 Schematic possible history of carbon dioxide decline in the Earth's atmosphere. Cretaceous simulations are computer models of the climate during the Cretaceous period, which ended 65 million years ago.

were not present at the time). The oldest paleosol samples, that is, lithified remnants of finely divided solid particles that are sensitive to the chemical state of the atmosphere, come from Canada and date at between 2.2 and 2.8 billion years before present. Were carbon dioxide so abundant, indeed even three times less abundant, iron in the soil would have reacted with it to form siderite ($FeCO_3$), an iron carbonate. No evidence for such carbonates is seen in the soil, which puts the carbon dioxide abundance at 3 percent or less of the total atmosphere some 2.5 billion years ago. Given the luminosity of the Sun at the time (80 to 85 percent present), an atmospheric composition with 3 percent CO_2 would have put the Earth on the edge of catastrophic glaciation, consistent perhaps with the geologic evidence that Earth underwent periods of deep glaciation in the distant past (Chapter 16).

An alternative that potentially reduces the amount of carbon dioxide required to warm the Earth is to posit the presence of additional greenhouse gases such as ammonia and methane. Neither is chemically stable in the Earth's atmosphere today because they are oxidized by molecular oxygen, but the large-scale biogenic production of methane maintains that gas at a level of 1.7 ppm by number in the present atmosphere. Conceivably, a cohort of methane-producing bacteria, existing at a time when the oxygen abundance was much lower than it is today, could have raised the methane concentration to values of order 100 ppm, sufficient to allow a factor of 10 reduction in the amount of carbon dioxide required on the early Earth to maintain a given surface temperature. Whether such high concentrations of methane could have been sustained 2.5 billion years ago, the period when the paleosol record sets strict limits on the CO_2 abundance, depends on the oxygen abundance at the time, another controversial issue covered later in Section 11.7.

Ammonia is a very effective greenhouse gas because it absorbs photons over a broad range of wavelengths. Ammonia, however, is exceedingly unstable to dissociation from solar UV photons high in the atmosphere and would have thus required protection in order to play more than a transient role in warming the early atmosphere. In an anoxic atmosphere, which likely was the situation early in the Archean, few plausible ultraviolet shields exist except perhaps organic hazes, which are found today in the atmosphere of Saturn's moon Titan (Chapter 13). The generation of such hazes, however, would have required the presence of very substantial amounts of methane, which as argued in the previous paragraph was problematic in terms of chemical stability. Other possible shields against destruction of ammonia by ultraviolet radiation include sulfur dioxide (SO_2) and hydrogen sulfide (H_2S).

Much of the impetus for invoking alternative greenhouse gases came originally from the situation on early Mars, which was more severe than on the Earth. As detailed in Chapter 12, there is abundant evidence that early Mars was wetter and, hence, warmer than the Mars of today. Because Mars is 50 percent farther from the Sun than is the Earth, it receives on average over a factor of two less sunlight than does our home planet. Hence, a given surface temperature must be sustained on Mars by a larger burden of greenhouse gases than is required on Earth. Mars's tenuous atmosphere today is almost entirely carbon dioxide, so it is natural to assume that the large greenhouse effect required for an early clement Mars was provided by a massive carbon dioxide atmosphere.

We cannot, however, pump arbitrarily large amounts of carbon dioxide into the atmosphere of Mars (or the Earth) because at some point the atmosphere becomes saturated in carbon dioxide, meaning that "dry ice" (the traditional name for solid carbon dioxide) clouds form (Figure 11.7). Once this occurs, computing the warming associated with carbon dioxide becomes much more complicated and uncertain. Like water clouds in Earth's atmosphere, dry ice clouds may provide net cooling or warming. The bright clouds tend to reflect much of the solar radiation that would otherwise reach and be absorbed by the lower atmosphere and surface; hence, they cool. On the other hand, dry ice clouds also

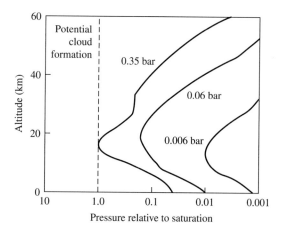

FIGURE 11.7 Carbon dioxide cloud formation on Mars. The figure is based on theoretical calculation of temperature in the atmosphere of early Mars. Each curve traces the ratio of the total pressure of carbon dioxide, divided by the saturation vapor pressure of carbon dioxide, for a given surface pressure of CO_2 (labeled). Because all curves are calculated for the same input of solar radiation (that of the faint early Sun), higher values of CO_2 imply larger temperatures, but they also imply that the ratio of total to saturation pressure moves closer to the value 1, at which cloud formation is assumed to occur (in reality, cloud formation requires a ratio slightly higher than 1). Once dry ice clouds form, the calculation of temperature in the atmosphere requires that the cooling and warming effects of clouds be included.

absorb infrared photons, principally around 15 μm in wavelength, and in this way contribute to the infrared blanketing that warms the surface and lower atmosphere. Detailed modeling of the dry ice clouds suggest that the particle sizes are comparable to thermal infrared wavelengths, and this favors heating by absorption of infrared photons rather than cooling by reflection and scattering of sunlight. Nonetheless, the difficulty climatologists have in reliably predicting the effect of water clouds on the present-day climate of the Earth provides a cautionary note on models for CO_2 clouds on Mars.

Were early Mars to have had methane as well as ammonia, neither of which condenses as readily as carbon dioxide, the dilemma might have been avoided as the added gaseous infrared blanketing made the atmosphere too warm for dry ice clouds to form. The original sources of these gases on Mars are unknown, but if small cometary bodies supplied much of Mars's volatiles, it is possible that they were sources of methane and ammonia as well. The problems of dissociation of ammonia and the relative instability of methane against oxidation apply for early Mars as well as for the Earth. The discovery by orbiting spectrometers of sulfur-bearing compounds on the Martian surface suggests that sulfur in various molecular forms might have been a greenhouse gas on Mars in the past. Ultimately, finding additional constraints on the carbon dioxide abundance and temperature in the Archean geologic record in the form of paleosols, carbonates, cherts, and glacial morphologies will help tie down the initial inventory and then decline of carbon dioxide on Earth.

11.5 Carbon-Silicate Cycle, Plate Tectonics, and Weathering

The loss of carbon dioxide via the production of stable carbonate sediments is a fundamental process of planetary evolution that operates on the Earth and may have operated on Venus and Mars at different times in the past. As Table 11.1 shows, most of the carbon on

TABLE 11.1 Inventory of Earth's Carbon

Reservoir	Amount in 10^9 Tons
Atmosphere	720
Oceans	38,400
Shallow inorganic carbon	670
Deep inorganic carbon	36,730
Sedimentary carbonates	>60,000,000
Kerogens (cooked organics)	15,000,000
Living biomass in biosphere	600–1,000
Dead biomass in biosphere	1,200
Fossil fuels (oil, coal, gas, etc.)	4,130

Adapted from Falkowski et al. (2000).

Earth is stored in the form of carbonates and kerogen compounds that are "cooked" by metamorphosis as sediments are buried. Although carbonate formation can take place in the absence of water, it is much faster when water is present. Carbon dioxide dissolves in rainwater and produces a weakly acidic solution, which attacks the rocks upon which the rain falls. A variety of products are generated as the acidic rainwater reacts with the rock, including silicon dioxide (SiO_2), ions of calcium and bicarbonate (HCO_3^-), and others. Weathering products that are soluble in water are carried downstream with the sediment load, ultimately to find their way to the sea. There, conversion of the calcium and bicarbonate ions to calcium carbonate ($CaCO_3$) is mediated principally by shell-forming organisms. (Biological activity serves primarily to accelerate the precipitation of carbonates in waters whose pH—see Chapter 3—is appropriate for such reactions irrespective of the presence of organisms.)

At current weathering rates and with the present mass of biota in the oceans, the removal time of the current mass of carbon that cycles through the biosphere, including atmospheric carbon dioxide and surface organic carbon, is less than a million years—a small fraction of the 4.5-billion-year history of the Earth. Evidently, there is a source of carbon that balances this removal; otherwise, essentially all life would have disappeared a long time ago. The crustal carbonates constitute the largest potential source, but it is then necessary to ask by what means the carbonates are transformed back into carbon dioxide.

The mechanism is plate tectonics. The Earth's crust is divided into a number of large and small plates, which are generally buoyant relative to the underlying mantle. Two major types of crust exist: (1) the more mafic, and relatively dense, basaltic oceanic crust and (2) the less dense, granitic continental crust. (Intermediate crustal types exist as well, such as andesites at continental margins, but they are less abundant than the basalts and the granites.) The crust is the upper portion of a thick, rigid layer, called the lithosphere, that rides atop the warmer, solid-but-plastic asthenosphere, which comprises the bulk of the mantle. (The terms *crust* and *mantle* distinguish chemical differences between layers; *lithosphere* and *asthenosphere* distinguish different mechanical properties.)

The general convective motion of the Earth's plastic mantle is a response to heating in the interior from several sources: radioactive decay of isotopes (principally those of potassium, uranium, and thorium), as well as remnant energy left over from accretion of the Earth and formation of its core. The convective motion translates into upwelling and melting of mantle to produce basalts that are extruded at mid-ocean ridges in the oceanic crust, such as that in the mid-Atlantic (Figure 11.8). Because this constitutes production of new crust on a planet of constant surface area, there must be a corresponding loss of crust elsewhere, and this occurs at subduction zones where one plate is subsumed underneath another. Continental crust is buoyant relative to oceanic crust; hence, many subduction zones are located along the edges of continents (e.g., Japan on the eastern boundary of Asia), and there lithosphere bearing oceanic crust is forced under the continent.

Pulled by negative buoyancy and pushed by the expansion at the ridges, lubricated by water within the crust, the subducted plates slide diagonally downward until they are softened and partially melted, then chemically altered and returned in part to the surface in andesitic volcanoes or further processed at the roots of continents. This conveyor belt pattern is outlined in the linear zones of greatest earthquake and volcanic activity that follow ridges and subduction zones and is apparent as well in the youthful age of oceanic crust (of order 100 million years or less) compared with the broad span of ages on continents, which

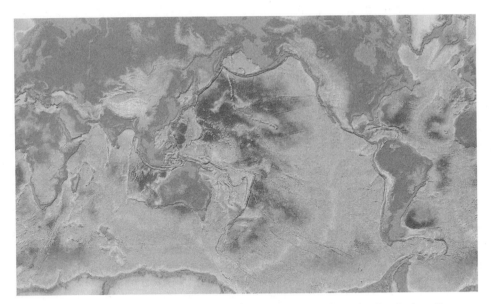

FIGURE 11.8 Map of ridges and subduction zones on the ocean floor showing the long linear features characteristic of plate tectonics on the Earth.

incorporate stable platforms billions of years old. A third type of motion in which plates move horizontally along their margins takes up the distortion of the crust associated with motion of the plates along a sphere; the transform faults along which they move are evident in the ocean floor and along continental margins such as the California portion of the Pacific Coast of North America.

As the ocean floor is recycled back into the asthenosphere at subduction zones, the carbonate sediments are carried with it. Much of the ocean floor material carried downward in subduction zone trenches is melted at temperatures well above 1000 K. The calcium carbonate ($CaCO_3$) reacts with silicates and the water (H_2O) at these high temperatures to make calcium silicates ($CaSiO_3$) and CO_2 gas. The gas makes its way back to the surface—not at the mid-ocean ridges but in the subduction zone regions where andesitic volcanism is occurring. The andesitic volcanoes release the carbon dioxide made from carbonates, resupplying the atmosphere with this important greenhouse gas (and releasing water back to the surface-atmosphere system as well).

While this description is highly schematic (many other mineral reactants and gaseous and condensed products are involved), it captures the most important feature of plate tectonics—namely, the recycling of the important volatiles carbon dioxide and water (Figure 11.9). The burial and then release of carbon dioxide in the weathering tectonic cycle is limited by the time it takes ocean floor (once formed at mid-ocean ridges) to be subducted. At typical plate tectonic spreading rates measured today, the cycle takes about 60 million years; in other words, any given carbon atom in atmospheric carbon dioxide that has been trapped as a carbonate will be subducted and released again as carbon dioxide gas on a timescale that is typically tens of millions of years, up to a hundred million years or more in some cases.

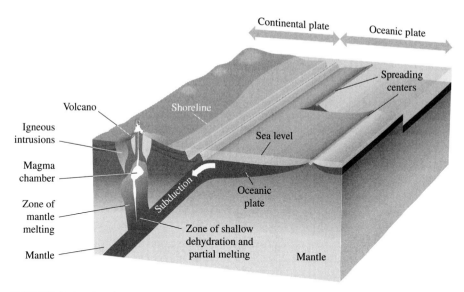

FIGURE 11.9 Cycling of volatiles associated with plate tectonics on Earth.

The carbon-silicate weathering cycle as defined by the plate tectonic timescale is sufficiently short relative to the age of the Earth that it might provide a persistent stabilizing influence on Earth's climate. When the Earth's climate is unusually warm relative to some average value, rainfall and hence rates of erosion increase. This, in turn, leads to an increased sequestration of carbon dioxide in carbonates and thus a lowering of the atmospheric carbon dioxide abundance. All else being equal, the removal of this greenhouse gas should reduce the global mean temperature, that is, cool the climate. But a cooler climate is generally characterized by reduced rainfall and hence lowered erosion rates—which, in turn, slows the loss of atmospheric carbon dioxide by carbonate production. Assuming that the rate of carbonate recycling by plate tectonics has not changed, the carbon dioxide abundance in the atmosphere should then increase as volcanoes continue to release the greenhouse gas into the atmosphere. The atmospheric rise in CO_2 then allows the atmosphere to warm, increasing erosion rates and loss of CO_2.

The dependence of CO_2 loss on rainfall rates and, hence, global surface temperature constitutes a negative feedback loop, in which changing the conditions in one direction tends to cause the system to move back in the other direction, and is characteristic of physical systems that are stable. The negative feedback loop between CO_2 abundance and rainfall-driven erosion provides at least a partial explanation as to why the Earth's climate has remained in the range allowing liquid oceans over geologic time.

Life itself has played a role in altering carbon loss rates: The development of soil-forming microorganisms some 3 billion years ago and the dominance some hundred of millions of years ago of plants both accelerated the trapping of carbon dioxide in soils and hence led to net decreases in carbon dioxide levels in the atmosphere. The development of calcareous plankton about 140 million years ago shifted most of the deposition of carbonate to the deep ocean from shallow continental shelf environments, hastening the transport of carbonates to subduction zones and, hence, increasing the rate at which carbon is cycled into the mantle.

11.6 Origin of Granites and the Supercontinent Cycle

The carbon-silicate cycle generates a negative feedback loop in the absence of variations in the plate tectonic recycling speed. Geologic evidence and computer modeling both indicate potentially large variations in the rate of recycling of oceanic crust. Some of the variations are cyclic and are thought to be associated with the assembly and breakup of so-called supercontinents—single landmasses formed by the merging of all existing continents. Such supercontinents could not have been a driving feature of the first billion years of plate tectonics because continents themselves were much smaller.

During the Hadean and the early Archean, heat flow rates were much higher because the amount of heat from accretion still in the planet was larger then than now and the amount of undecayed radioactive isotopes was larger. Higher heat flow probably meant higher recycling rates. If the modern situation of roughly comparable amounts of continental and oceanic crust had been obtained, collisions between continents to form larger continents, followed by their breakup, would have occurred frequently.

However, the geologic record suggests that there was much less continental landmass in the early Archean compared with today. The amount of ancient continental crust (older than 3 billion years) is a small fraction of the total stable continental landmass, and destruction by burial and metamorphism cannot be invoked to remove large quantities of ancient continental crust. The absence in ancient sediments of an abundance pattern of trace elements characteristic of granites suggests that truly granitic continental landmasses were relatively rare prior to 3 billion years ago, though buoyant platforms of composition intermediate between basalt and granite likely existed then as precursors to modern continents (Figure 11.10).

How granites form remains an enduring mystery left over from the geophysical and geochemical revolution of the 1960s in which the existence of plate tectonics was inferred from a broad range of geologic evidence. From the context of astrobiology, it is not an arcane debate because the existence of large platforms of well-watered land enhanced erosion and loss of carbon dioxide, allowed land plant and animal ecosystems to develop, and changed the nature of the Earth's climate response to orbital forcing and other perturbations (Chapter 16).

Plate tectonics provides a straightforward way of understanding the formation of basalts and its evolution to rocks intermediate in composition, such as andesites. The mantle of the Earth is not liquid; under mantle temperature and pressure conditions, the bulk of the rock is a solid material that flows plastically. However, beneath mid-ocean ridges, the mantle rock is warmer than average, hence is buoyant and rises toward the surface. As it does so, the rock experiences progressively lower pressure, and it enters a regime where a thermodynamically stable liquid phase becomes possible—even though the temperature decreases as the pressure decreases. This phenomenon of "pressure melting" is simply a consequence of the fact that the melting temperature of many materials decreases with de-

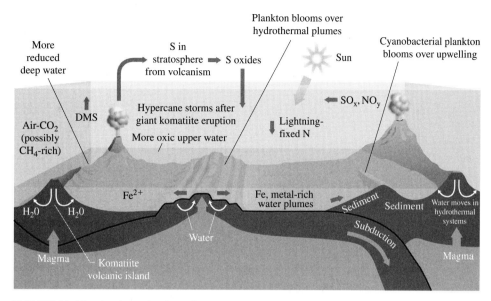

FIGURE 11.10 A schematic view of the possible cycling of volatiles during the Archean showing plate tectonics in a regime of high heat flow and little continental material. "Hypercane" storms are putative extreme-intensity storms generate by large-scale volcanic eruptions.

creasing pressure (a well-known exception being the normal low-pressure phase of water ice). The entire rock does not melt, and the melt has a composition corresponding to basalt—a bit more iron and much less magnesium than in the mantle rock. The molten basalt rises more rapidly than the surrounding rock and is extruded at the mid-ocean ridge volcanic systems. (The combination of fresh minerals, reduced gases, and water heated by the molten rock make the mid-ocean ridge vent systems a rich place for extremophiles as discussed in Chapter 10.)

The formation of basalt is made possible because of a singular, very ancient, stage of chemical differentiation of the Earth. The formation of the iron core, and consequent partitioning of different elements between the core and mantle, led ultimately to a mantle depleted in iron and which, when melted at low pressure, produces basalt. The basalt itself will be remelted tens of millions of years after its extrusion at the mid-ocean ridge, when it has cooled sufficiently to attain a density allowing its subduction at the margin of a continent or island arc such as Japan. By this time, the basalt has become suffused with water as it traveled from ridge to trench. As the slab of wet basalt moves downward into the mantle, it heats up, eventually to temperatures at which the water is liberated. The dehydrated rock becomes denser, aiding subduction of the slab down to depths of 600 km determined by measurements of deep earthquakes in subduction zones. Meanwhile, the liberated water moves upward into surrounding rock, a mixture of basalt and mantle, and contacts the mineral grains. This, in turn, lowers the melting point of the surrounding rock, and a partial melt forms at very different pressure and temperature than the basaltic partial melt at the mid-ocean ridges. The melt in the subduction zone will make a volcanic rock called an

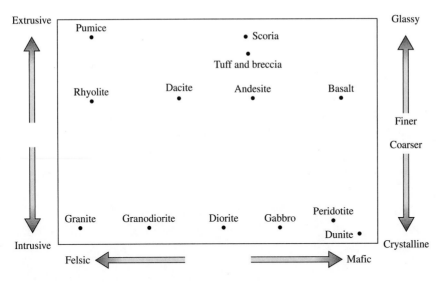

FIGURE 11.11 Schematic relationships between different rock types show those that are felsic (silica rich) to the left and those more mafic (iron–magnesium rich) to the right. Rocks on the top are extrusive, exuded as lava from volcanic vents and solidified with a very fine crystalline or glassy texture; rocks on the bottom are intrusive, slowly solidified below ground leading to large crystals.

"andesite," impoverished in magnesium and iron (in other words, less mafic) relative to the basalt. It rises to the surface and is erupted in volcanoes or intruded as magma into pre-existing rock.

Rocks of andesitic composition (which are called dioritic or gabboric if they are intruded into existing rock rather than extruded as volcanics) can and do form stable island arcs that resist subduction. But they are also the end of the line with respect to major reprocessing of the rock composition toward the silica-rich granites, for the very reason that they do not easily subduct again (Figure 11.11). The formation of granites must happen in a different venue, and it is the rarity of granite as a geologic rock form, on Earth or elsewhere on the rocky planets of the inner solar system, that makes its origin somewhat difficult to divine.

Less than 0.2 percent of the volume of the Earth has been transformed over its history into the *upper* granitic crust of the continents. The word "upper" is important here. Samples of rocks ("granulites") from deep below the surfaces of continents and earthquake data that give information on the speed and hence density of rocks deep below the surface suggest that the bases of continents have a composition that might be obtained if we were to extract material from a mix of basaltic and andesitic rocks. These bases are much deeper below the surface than is the bottom of the oceanic crust—50 km versus 10 km—leading to a very different set of temperature and pressure conditions for forming partial melts. Melting here may yield rock of granitic composition, but not in a single step.

Making granites may be a slow, multistep process, involving contact and partial melting of rocks varying in composition from basalt to andesite at the base of the continents.

11.6 Origin of Granites and the Supercontinent Cycle

The large continental platforms are less dense than the ocean crust, thus are buoyant, and they impede the outward flow of heat beneath them. As a result, temperatures within the base of a continent are sufficient to partially melt the mixture of rocks present, and movement of the melt through overlying layers eventually creates magmas of granitic composition that solidify before they reach the surface and then are exposed by uplift and erosion. But the details remain poorly understood.

The notion that a continent is needed to make a continent suggests that the buildup of granitic continents was initially slow, relying first on the assembly of buoyant protocontinents of andesitic composition. As more of the surface area of the Earth became covered with continents, the occurrence of basal melting of continents and production of new granites increased. It has been suggested that, around 3 billion years ago, enough area was covered by continents that massive basal melting occurred, accelerating the further production of granites and of continents. Declining heat flow and the blocking effect of buoyant continental crust altered the Archean tectonic style of small plates in favor of the large plate form of tectonics present today.

Until that time, smaller amounts of continental mass exposed above sea level meant relatively low efficiency of silicate weathering, and hence lower loss rates of CO_2 from the atmosphere than at present. More rapid recycling of crust in response to the higher heat flows at the time would have led to more rapid return of carbon dioxide to the atmosphere, with less stored as carbonate sediments on the seafloor. The net effect could have been to sustain large carbon dioxide abundances in the atmosphere through much of the Archean.

With the conversion somewhere in the Archean to a style of plate tectonics in which a small number of large plates move relative to each other, a new cycle of tectonics took shape. About every 500 million years or so, most or all continents collide with each other, in an inevitable result of the long-term random movement of a small number of large plates, to form a single supercontinent surrounded by a single global ocean (Figure 11.12).

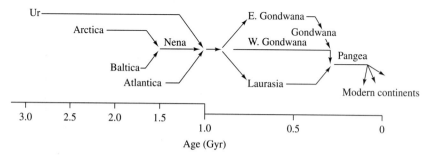

FIGURE 11.12 One possible history of the development of continents on the Earth. The original continental material Ur is small compared with today's continental area, but additional continents form, merge, break up, until, at about 800 million years ago, the oldest discernible supercontinent in the geologic record forms. This then breaks up and reaccretes into the supercontinent Pangea, which, in turn, breaks up to form the modern continents. Previous episodes of supercontinents may have occurred as well, on roughly 500-million-year cycle times, back into the Archean.

The end of a supercontinent comes after hundreds of millions of years during which hot plumes increase in number and intensity under the massive insulating blanket of the supercontinent and act to break it apart, initiating widespread volcanism along the fracture margins, forming new plates, and initiating an era of enhanced seafloor spreading that distributes continents around the globe. Collisions between the new continents eventually occur, leading to compression and the building of large mountain chains and plateaus, such as the Himalayas between the plates carrying Asia and India.

The effects on the abundance of carbon dioxide of the formation and then disruption of a supercontinent may be profound and complex. The clustering of the continents into a single mass reduces the area on the ocean floor where subduction of carbonates occurs and slows the return of CO_2 to the atmosphere via andesitic volcanism. The cessation of collisions between continents terminates large-scale mountain building, and, over time, erosion reduces the mean height of continental land. This, in turn, lowers rainfall rates and reduces the loss by weathering of atmospheric carbon dioxide. Volcanism associated with the early stages of breakup of a supercontinent may enhance the volcanic injection of carbon dioxide and other infrared-absorbing gases into the atmosphere.

Subsequent to the breakup of a supercontinent into smaller pieces, continental collisions lead to the buildup of high mountain ranges, which are sites of increased rainfall, erosion, and (hence) loss of carbon dioxide. Such changes in the topography and meteorology of the Earth may have led to the Snowball Earth catastrophes over the history of the Earth. The most recent supercontinent breakup was followed by a gradual cooling trend in the climate that may be associated with the scrubbing of CO_2 from the atmosphere as the massive Tibetan Plateau and Himalaya Mountains formed.

The supercontinent cycle probably also plays a role in modulating the burial and release of organic carbon. Unlike organic material deposited on the seafloor—which is recycled by tectonics and volcanism within 100 million years, along with carbonates to make carbon dioxide, water, and other simple molecules—organic matter deposited on continents will recycle slowly, if at all. Table 11.1 shows that a large amount of carbon may be stored as kerogen, which is a form of organic carbon trapped in sedimentary rocks such as shales and then cooked as the sediments are buried and metamorphosed. Because organic carbon reacts with and soaks up oxygen, the large-scale burial of organic carbon on continental platforms beginning some time after 3 billion years ago could have represented the removal of an important sink for molecular oxygen. This may have contributed to the rise of O_2 discussed in the next section.

Times of stable supercontinents, in which activity is minimal and burial of organic carbon dominates, might represent times when growth of oxygen in the atmosphere was maximal. Conversely, the large-scale exposure of reduced carbon through increased erosion or continental volcanism associated with supercontinent formation by collision, and then with enhanced volcanism during the early stages of supercontinent breakup, might temporarily reduce the abundance of molecular oxygen in the atmosphere. Complicating matters is the hint from recent laboratory studies that some microbes actually consume kerogen from shales (prior to metamorphism), accelerating the conversion of organics to carbon dioxide and soaking up oxygen. These effects, geological and biological, combined to produce fluctuations in an oxygen abundance that began increasing substantially sometime during the first half of the Earth's history.

11.7 The Rise of Oxygen

The development of photosynthesis changed Earth's atmosphere and oceans into a reservoir of enormous oxidizing power, which mitochondrial-based respirative metabolisms have taken advantage of for at least 2 billion years. It also led to the oxidation of crustal rocks and a dramatic change in the geologic record. The rise of oxygen did not occur immediately after the development of photosynthesis because strong sinks for oxygen exist on our planet and would have overwhelmed the production by photosynthesis. This overall picture is agreed upon by the majority of astrobiologists. What is hotly debated, however, is the specific time in the history of the Earth at which the atmosphere and the ocean went from anoxic to oxidizing and, then, aerobic (i.e., capable of supporting mitochondrial-based respiration). Molecular oxygen was abundant in the Earth's atmosphere at least by 2.2 billion years ago, and maybe earlier.

Prior to the onset of photosynthesis, molecular oxygen would have been produced at a low rate by photochemistry of water in the stratosphere. However, the rate of production was overwhelmed by the rate at which oxygen was taken up by reduced minerals and by reduced gases at the surface of the Earth. Rapid recycling during the Archean provided mineral and gaseous sinks at a sufficient rate to keep oxygen levels at least 3, and maybe 10, orders of magnitude below the current value. In contrast, photosynthesis today operates in an environment in which crustal sinks of oxygen are largely saturated, and the primarily loss of oxygen is via respiration and decay of organic matter (Table 11.2). Absent any biological sources or sinks, oxidative weathering of rocks would remove the current inventory of atmospheric and oceanic oxygen on a timescale of roughly 4 million years.

There are two issues framing the debate on this crucial aspect of Earth's history. The first centers on just when the atmosphere actually did become oxygen rich. While the majority of astrobiologists argue on the basis of geologic evidence that this occurred between 2.2 billion and 2.4 billion years ago, a few (e.g., H. Ohmoto of Pennsylvania State University)

TABLE 11.2 Production and Loss of Oxygen Today

Process	Production (+) or Loss (−) (kg/yr)
Photosynthesis	$+10^{14}$
Respiration and decay	-10^{14}
Fossil fuel consumption (burning + deforestation)	-10^{12}
Burial of organic carbon	$+10^{11}$
Recycling of buried carbon	-10^{11}
Weathering of rock	-10^{11}
Oxidation of volcanic gases	-10^{10}
Photochemical production of oxygen	$+10^{8}$

have argued for a much earlier onset. The second issue concerns the timing of the origin and spread of oxygenic photosynthesis, which as noted might be an evolutionary invention as old as 3.5 billion years. If we take the oldest occurrence of fossil evidence for cyanobacteria (3.5 billion years) and combine it with the commonly accepted estimates for the onset of an oxygen-rich atmosphere, there is a gap of 1 billion years. This gap might indicate that the photosynthetic production of oxygen could not keep up with the geological sinks of oxygen present at the time, a possibility we return to later in this section.

The evidence for an anoxic Hadean and Archean Earth comes primarily from the presence in the ancient rock record of minerals that would be unstable in the presence of an oxygen-rich atmosphere, notably uraninite (UO_2) and pyrite (FeS_2). After about 2.2 billion years ago, the occurrence of "detrital" uraninite and pyrite in the rock record is extremely rare. (Detrital minerals are those deposited in sediments without ever having been dissolved in water during weathering; only this type of deposition can constrain the oxygen abundance). Beginning about 2 billion years ago, so-called redbeds appear in the sedimentary record. The red color forms as iron is weathered out of the rock and reacts with oxygen; substantial amounts of oxygen, but still smaller than what is present today, are required to generate the oxidized iron. Occurring over a broad span of Earth history from 3.5 billion (3.85 billion if the ancient West Greenland samples are included) to about 1.8 billion years ago are the banded iron formations, or BIFs, which consist of interbedded (alternating) layers of chert and minerals containing up to 30 percent iron.

BIFs are abundant—they are the source of 90 percent of the world's commercial iron supply—but their origin is a mystery. Because they are sedimentary rocks, BIFs reflect alternating deposition in water environments of iron-poor and iron-rich material, but the specific mechanisms are debated, and there are probably multiple ones. Varying oxygen content in the water can lead to the alternation because iron is much more soluble in water that is oxygen poor versus oxygen rich. In oxygen-poor water, iron is in the ferrous form, FeO. Adding oxygen leads to reaction with the ferrous iron to produce Fe_2O_3, ferric iron, which is much less soluble and precipitates out of the water. The oxygen fluctuations may be due to seasonal oscillations in the population of photosynthesizing organisms or wind-driven annual upwelling of water with a differing oxygen content or other causes set against the backdrop of a generally oxygen-poor ocean and atmosphere. What is known is that today's oceans and—with few exceptions—lakes are too rich in oxygen to support the kind of fluctuations required to form BIFs. Therefore, the existence of banded iron sediments, regardless of how they specifically formed, points to a time when oxygen levels in the deep oceans were much lower than at present.

Finally, analysis of the pattern of abundances of the sulfur isotopes ^{34}S, ^{33}S, and ^{32}S in rocks formed more recently than 2.3 billion years ago shows a simple fractionation associated solely with the differences in mass among the three. Older rocks show deviations in this mass trend for sulfur, and the deviations have been ascribed to the effect of atmospheric photochemical reactions involving sulfur dioxide. Sulfur dioxide should be so rare in an oxygen-rich atmosphere (it is replaced by sulfuric acid) that, if the photochemical mechanism is correct, then the Earth's atmosphere must have been quite oxygen poor prior to 2.3 billion years ago.

Proponents of an earlier origin for an oxygen-rich Earth atmosphere generally argue that localized or particular types of environments would have been reducing even in the presence of a global atmosphere-ocean rich in oxygen and that the geologic evidence is reflecting those environments. Hydrothermal vent systems pump out reducing gases that yield an anoxic environment even today. The deep oceans could have been entirely anoxic for some time after the growth of an oxidizing or aerobic atmosphere and upper ocean. On this basis, they argue that the banded iron formations reflect anoxic environments at the base of the ocean; the pyrite could be formed by the action of bacteria-reducing sulfates, and the uraninite could be of hydrothermal origin. Further, paleosols dated back to 2.9 billion years ago show the characteristics, according to some analyses, of soils formed in the presence of substantial amounts of oxygen.

Tied to the debate regarding the earliest onset of abundant atmospheric and oceanic oxygen is the issue of how the Earth could have sustained an oxygen-poor atmosphere once photosynthesis became widespread. The simplest model of the response of the Earth's sinks of oxygen to the dramatically increased production of photosynthesis is that they saturate relatively quickly on a geologic timescale, and the rise of oxygen is rapid. To contemplate a substantial gap of hundreds of millions of years between the occurrence of cyanobacteria and the rise of oxygen requires that the geological sinks be large enough to suppress oxygen over very long timescales. Then these sinks must become ineffective at some juncture prior to 2.2 billion years ago, but not necessarily in response to a rise in the oxygen production by photosynthesis. In effect, the crust of the Earth must transition from reducing to oxidizing in the absence of a direct biological stimulus.

Changes in the mantle oxidation state leading to this change in the efficacy of the crustal sinks constitute one possibility, though, absent a mechanism or evidence for it, perhaps a speculative one. Another possibility is that the Archean greenhouse effect was aided by a substantial abundance of methane in the atmosphere; perhaps hundreds of parts per million—an amount that is stable in a sufficiently anoxic environment. The loss of that methane might have played a role in oxidizing the crust in a paradoxical fashion: Under the influence of ultraviolet light, methane actually speeds up the rate at which the atmosphere becomes oxidizing. Ultraviolet photons in the upper atmosphere break the carbon–hydrogen bond of methane to yield free hydrogen atoms, the same photochemical reaction that is the starting point of hydrocarbon chemistry on Saturn's moon Titan (Chapter 13). The free hydrogen atoms attack water so that $H + H_2O \rightarrow H_2 + OH$. The hydrogen molecules escape from the Earth's upper atmosphere (as they do on Titan), and the OH fragments—chemical radicals—give rise to free oxygen. This oxygen would be available to progressively oxidize the crust, atmosphere, and oceans of the Earth.

At issue is why the loss of hydrogen and concomitant additional net production of oxygen would have occurred relatively late in the Archean or in the early Proterozoic rather than much earlier. A clever suggestion by a group at NASA Ames Research Center and the University of Washington, led by David Catling, is tied to the increased production of organic matter by photosynthesis. The buildup of biomass because of increasing photosynthetic productivity would have led to increased scavenging of dead organisms by anaerobic, methanogenic bacteria. The burst of methane production would have compensated for the increasing loss of CO_2 by photosynthesis—which in the Archean might have otherwise catastrophically cooled the Earth's surface—and led to the increased production of oxygen (with loss of hydrogen) from water, hence oxidation of the crust, and the rise of atmospheric and oceanic oxygen (Figure 11.13). As with any model, a number of questions are then raised. To what extent was the rise of continental crust key to providing platforms for the expansive areas of detritus of organic carbon on which methanogenic bacteria could

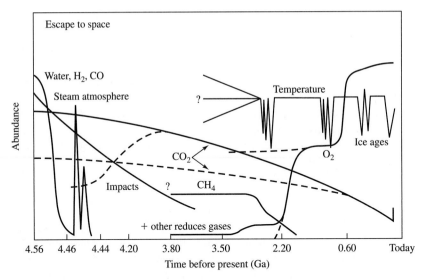

FIGURE 11.13 History of Earth's atmospheric composition, based on a view in which the rise of photosynthesis generates a large biomass, upon which methanogenic organisms feed to generate methane—eventually creating additional oxygen though methane photolysis, which saturates the oxygen sinks.

operate? What were the specific chemical mechanisms for oxidizing continental crust at relatively low temperature? What was the timing of the enhanced methane production and photochemistry relative to the record of pyrite, uraninite, and BIFs?

As additional geologic data are analyzed, especially subtle isotopic clues, it is likely that the timing and even the mechanisms of the rise of oxygen will be better specified. To do so is essential to understanding the history of the habitability of the Earth as a planet and the rise of both photosynthesis and—much later—complex life on Earth. This history will provide the foundation some day for interpreting the spectral signatures of molecular oxygen and photosynthesis that might be seen in extrasolar Earthlike planets as well as provide a perspective on the frequency with which habitable planets evolve to host complex, and eventually intelligent, life (Chapters 15 and 17).

11.8 The Implications of Earth's Habitability for Mars and Venus

Among the three rocky, so-called terrestrial, planets with atmospheres, Earth is unique in having stable liquid water, an oxygen-rich atmosphere, and life. The oxygen-rich atmosphere is a direct result of the long-term oxidation of the Earth's crust by widespread photosynthesis, along with ancillary biological and geological processes (discussed earlier)

11.8 The Implications of Earth's Habitability for Mars and Venus

that aided in the saturation of crustal oxygen sinks. The prerequisite for this situation is long-term habitability of the planet, of order hundreds of millions or billions of years, and so it is this property of the Earth—the stability of liquid water and consequent habitability—that is fundamental. Neither Venus nor Mars maintained long-term habitability, and while Mars may have sustained liquid water over large areas for as much as a billion years, Venus may never have been habitable.

Venus is nearly but not quite identical to the Earth in basic physical properties. It is just 5 percent smaller in diameter and density than the Earth. The density difference is even less significant when we consider that the more massive Earth is compressed by its own mass to a greater extent than is Venus; the intrinsic density of the material making up the two planets differs only by 2 percent. Venus has no large natural satellite, a point to which we return later in this section, and rotates very slowly in a retrograde (opposite to the Earth) direction. Its current atmosphere is 90 bars of carbon dioxide, equivalent to or greater than the entire budget of carbon locked in the Earth's ocean-atmosphere-crustal system as organics, carbonates, and kerogen. Although Venus is 30 percent closer to the Sun than is Earth, much less sunlight is received at its surface because of a massive stratospheric layer of sulfur aerosols, which form a uniform, bright cloud. Despite the much smaller amount of sunlight, the enormity of the carbon dioxide atmosphere blocks infrared radiation so effectively that the surface temperature of the planet is more than 700 K, too hot for liquid water to exist at any pressure and above the melting point of lead.

The surface is a rolling plain with some highlands, mostly volcanic but perhaps with a protocontinental shield in one location and with broad areas of recently extruded basaltic volcanic flows (Figure 11.14). There is no evidence for plate tectonics, as we know it on the Earth, although some remnant trenches that may have supported limited subduction have been identified. There is little in the way of geochemical data for Venus because of the great difficulty of making surface measurements; however, the lack of bimodal topography that characterizes Earth—flat seafloors and relatively uniform continental elevations—suggests the absence of large areas of buoyant rock types on Venus. Small domes seen on the surface may be more silica rich than basalt, but this is an inference based solely on their appearance interpreted to indicate buoyant pieces of the crust.

How did Venus arrive at such a state? The question seems readily answered—it is closer to the Sun, and hence got overcooked—but such a facile answer applies only if Venus very recently achieved its current state. Because the surface seems geologically youthful, with an absence of heavily cratered terrain suggesting an age of 1 billion years, it is difficult to completely rule out the notion that Venus lost habitability late in the history of the solar system. If Venus was never habitable, or lost its liquid water early on, then the issue is more delicate. At its position 0.7 AU from the Sun, Venus receives nearly twice as much sunlight at the top of its atmosphere as does Earth. Four billion years ago, with the Sun only 70 percent as luminous as it is today, Venus would have received merely 40 percent more sunlight *back then* than the Earth receives *today*. Is 40 percent enough to make the difference with respect to stable liquid water?

The answer may well be yes. Our own oceans are prevented from evaporating into the upper stratosphere by the steep temperature profile determined by convection and by the greenhouse effect. The mean minimum temperature at the interface between the troposphere and stratosphere of the Earth—the so-called tropopause—is roughly 190 K

FIGURE 11.14 Physiography of Venus from the Magellan radar mission. Although some long fractures are seen, the global, scale relationships among trenches and ridges as seen on Earth are absent here. Squares with no detail indicate parts of the planet missing topographic data.

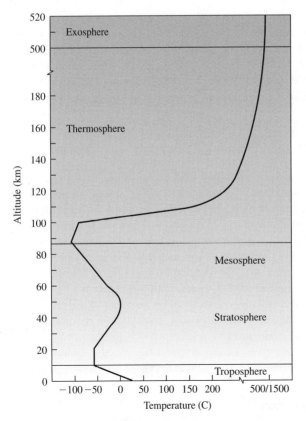

FIGURE 11.15 Typical temperature profile for the atmosphere of the Earth shows the names of several layers that correspond to changes in the slope (gradient of temperature with altitude). The transition between the thermosphere and exosphere is defined by the point—the exobase—at which the air is so tenuous that molecules and atoms are free to escape without colliding with each other.

(Figure 11.15). This is so low that water is almost completely frozen out of the atmosphere at that altitude.

Hence the amount of water vapor that makes it into the stratosphere, where it can then diffuse further up into the so-called mesosphere and where it is destroyed (photolyzed) by ultraviolet radiation, is only a very tiny fraction of the water contained within the hydrosphere (ocean, crust, and troposphere) of the Earth. The mixing ratio of water in the lower stratosphere—2 to 3 ppm—is three orders of magnitude less than that at the surface of the Earth. (Thunderstorms can and do mechanically carry ice crystals into the stratosphere, but their net effect, surprisingly, is to dry out the lower stratosphere—because, on average, the temperatures inside the cumulonimbus anvils are lower than ambient.) Preventing photolysis is the key to retaining the water that makes our planet habitable—once water is photolyzed, the hydrogen escapes, and oxygen is left. The oxygen goes into oxidizing the crust—not making more water because free hydrogen is a rarity.

Imagine heating the system by making the Sun brighter. The surface temperature goes up as does the temperature at each altitude up to the tropopause. The amount of water that can pass through to the stratosphere is limited by the vapor pressure of water. A 40 percent increase in the amount of sunlight could raise the tropopause temperature, determined by radiative equilibrium, from 190 K to 200 K, which, in turn, would increase by a factor of five the amount of water in the stratosphere. But there is a positive feedback—more water in the whole troposphere, and lower stratosphere, increases the infrared absorption and hence, through the greenhouse effect, amplifies the temperature rise, which, in turn, allows more water into the stratosphere. (Figure 11.16). The metaphorical cork is popped from the bottle, and large amounts of water stream into the stratosphere to be photolyzed. Through this "moist convective runaway," a model developed by J. Kasting of Penn State University, the Earth could lose its water—once the threshold solar flux is reached—in some tens or hundreds of millions of years. Kasting calculates that this effect could be initiated once the Sun's luminosity is 10 percent brighter than today—40 percent

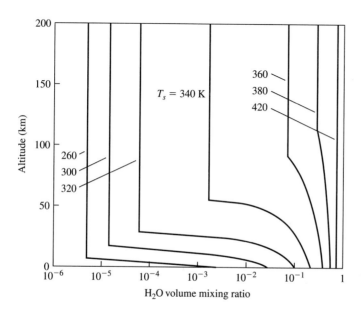

FIGURE 11.16 Moist convective runaway in action for the Earth. Plotted are schematic profiles of water vapor in the atmosphere as a function of altitude, tied to temperature profiles indexed by the surface temperature marked beside each curve. As the surface temperature goes up in response to increased solar luminosity, the amount of water that can be admitted into the stratosphere increases exponentially.

isn't necessary. Models of the evolution of the Sun predict brightness 10 percent higher than today in slightly more than 1 billion years from now.

Venus experienced 40 percent higher solar luminosity 4 billion years ago than we experience today, and so, all else being equal, the most convective runaway model predicts Venus would have lost all of its water then. The loss of water might be recorded in the very high deuterium-to-hydrogen ratio in the remnant water seen in the atmosphere today (about 150 times that measured in Earth's oceans), as heavy (deuterium-bearing) water (HDO) was lost less efficiently than H_2O. Accompanying the runaway heating would have been the outgassing of organic and inorganic carbon into the atmosphere, with the oxygen from the water serving to make CO_2 from more reduced carbon species. Once in the atmosphere, CO_2 could not, for the most part, return to the crust as carbonates without liquid water. And in the absence of liquid water, production of andesitic protocontinents from basalt would have been difficult or impossible, and subduction of crust would have been impeded by the greater stiffness of completely anhydrous rock. Venus irreversibly lost the environment salubrious for liquid water and, with that, the chance for plate tectonics, continents, and life.

Mars, on the other hand, began its planetary evolution with water that coursed across the surface and resided perhaps in low-lying basins, as Chapter 12 describes in detail. The much smaller size of Mars relative to the Earth, however, ensured much more rapid cooling of the red planet and development of a very thick lithosphere—probably too thick for plate tectonics. (Indeed, even the small difference in mass between Venus and Earth means that the lithosphere of the former is twice that of the latter—even with water, plate tectonics on Venus might have been difficult.) Consequently, recycling of surface carbonates into the mantle and regurgitation in the form of carbon dioxide was impossible on Mars. Lower mass also means a relative ease of escape from Mars's atmosphere of gases liberated or heated by impact. It is thus relatively straightforward to understand why Mars lost its volatile envelope that made it habitable.

Much more difficult to guess is what might have happened had a Venus- or Earth-sized planet been formed at the orbit of Mars. Could it have retained its volatiles through tectonic recycling and been habitable through to the present day? The evidence for profound glaciations on the Earth over its history, discussed in Chapter 16, hint that an Earth at 1.5 AU might have been too cold. Would a body 30 percent more massive than the Earth be enough—50 percent? 100 percent? Simulations are ongoing at NASA's Jet Propulsion Laboratory and elsewhere to address this question, but our failure a priori to predict the nature of Mars and Venus does not bode well for accuracy in such predictive ventures. And the test will be years away—when spaceborne telescopes now in development detect extrasolar rocky planets comparable in size to the Earth (Chapter 15).

This chapter closes with a note about Earth's large moon, which may play a role in the planet's climate stability. The presence of the Moon keeps the Earth's axial tilt at close to 23 degrees. The mechanism by which the Moon "protects" the Earth can be understood as follows. The gravitational pull of the Sun and the Moon on the equatorial bulge of the Earth causes the spin axis of our planet to precess (make a slow circle around the sky) and nutate (nod back and forth). The period of the precession is determined by the gravitational effect of the Moon and is about 26,000 years. While long compared with the annual swing of the Earth around the Sun, it is much shorter than it would be were the Moon to be absent. In the hypothetical case of a moonless Earth, the longer precession period would allow the axial tilt of the Earth to be affected much more profoundly by another periodic gravitational pull, that of the solar system's largest planet, Jupiter. With the Earth's axis oriented in the right direction, Jupiter's pull would, over a period of time, create a strengthening "secular perturbation" that would eventually force the axial tilt to vary chaotically (Chapter 9) over a range of timescales. This in fact is happening to Mars, which suffers swings in its axial tilt ranging from 10 to 60 degrees on timescales ranging from hundreds of thousands to millions of years.

Were a moonless Earth to suffer such extreme swings, the effects on its climate would be hard to predict. This situation is not entirely an academic one: If our solar system's general architecture is typical, the question of climate stability on a moonless Earth merits some consideration for the question of the commonality of habitable worlds around other stars. The formation of the Earth's moon was probably a relatively rare event (Chapter 6). However, the same models that predict wild swings for Mars suggest that a moonless Earth with a faster spin rate—a day-length shorter than ours by a factor of two—would precess with a period short enough to retain a stable axial tilt in the presence of Jupiter. And indeed, it is the tidal pull of the Moon that has progressively lengthened the day from a much shorter value billions of years ago. For the Earth we know, other periodic and more subtle changes in its orbital and spin state *do* show up as climate perturbations in the geologic record and have played a role in the evolution of life. We turn to these perturbations, along with catastrophic ones such as volcanism and impacts, in Chapter 16.

QUESTIONS AND PROBLEMS

1. Using equation (11.1), plot the "greenhouse-free" surface temperature of
    a. A planet at the orbit of the Earth but with the albedo of Venus.
    b. A planet at the orbit of Venus but with the albedo of the Earth.
    c. A planet with the albedo of the Earth in orbit around an M-dwarf of the age of the Sun.
    d. The Earth were it completely covered with ice.

   You may need to look up some numbers in astronomical and geological reference books.

2. One simple expression for the enhancement of the surface temperature associated with the greenhouse effect ignores the detailed variation of absorption of photons as a function of wavelength but considers instead the infrared absorption to be characterized by a single number. Then the surface temperature is $T = T_{eff}(0.5 + 0.75\,\tau)^{1/4}$, where τ is the "optical depth" in the infrared—a product of the infrared absorption, the air density, and a typical distance through the air called a scale height. The "effective temperature," T_{eff}, can be assumed to be the temperature derived in equation (11.1). Rather than calculate τ from its definition in terms of density and so forth, use the discussion in the chapter to derive the present-day value of τ at the base of the Earth's atmosphere. Then derive the value of τ required to maintain today's surface temperature when the Sun was 30 percent fainter than today. Given the amount of CO_2 cited in the chapter as required for the enhanced greenhouse effect during the time of the faint early Sun, how does τ depend on CO_2 abundance in

the atmosphere? In what way might this estimate be in error? (Hint: Test how far off this estimate is by calculating the temperature change if you double CO_2 from the present value; numerical models imply that this would increase the global mean temperature by about 1°C.)

3. Use Table 11.1 to calculate the partial pressure of carbon dioxide in the atmosphere of the Earth, assuming all of the fossil fuels stored in the crust to be converted to CO_2 by burning. (The partial pressure of carbon dioxide in the present atmosphere is its abundance, 370 ppm, multiplied by the total pressure, 1 bar.) Now calculate the pressure of carbon dioxide if all carbonates were converted to carbon dioxide. In each case, use the results of problem 2 to crudely estimate the greenhouse temperature rise associated with each case.

4. Show with a simple calculation why planets that are smaller cool off (lose their internal heat) more quickly than planets that are larger—even if they have the same composition.

5. If Mars cooled off more quickly than the Earth, what might be the thicknesses of the crusts of the two planets relative to each other and to the size of each planet? How would this make plate tectonics difficult or impossible on Mars? (Expand on the discussion in the text by going to the literature.)

6. Venus, being nearly the same size as the Earth, ought to have had a similar thermal history. However, water may have been absent from Venus's crust since early in its history. Could Venus have formed the various rock types produced in the crustal recycling process on Earth? Which ones would be absent? Water is not only an agent that changes the melting point and composition of rock; it lubricates rock as well. Describe how the absence of water might militate against plate tectonics on Venus and—by consulting the literature—describe how Venus does get rid of its internal heat, again expanding on the discussion in the chapter.

7. A simple model of the oxygen balance might posit that the rate of loss of molecular oxygen in the atmosphere is proportional to the amount of oxygen present, while the gain in oxygen is not dependent on how much is present already. Write a simple equation that captures this model, and use Table 11.2 to show what happens to the oxygen abundance as you add sources or sinks of oxygen. That is, what does a graph of the oxygen abundance with time look like if you add a source, or add a sink to the equation?

8. Do searches of the planetary literature to investigate further the interpretation of the large deuterium-to-hydrogen ratio in the atmosphere of Venus. What are the estimates for the amount of water lost early in the history of Venus?

9. Where are there anoxic bodies of water on the Earth today? How do they remain oxygen poor in spite of the oxygen-rich atmosphere and ocean we possess today?

10. The use of cherts as an indicator of Earth's surface temperature has been challenged because the technique cannot reproduce the fluctuations of surface temperature during the Phanerozoic eon—the last half-billion years of Earth's history. Can you offer a possible explanation of this failure that still preserves the usefulness of cherts as a measure of more ancient surface temperatures?

SUGGESTED READINGS

Some of the topics covered in this chapter are treated at a more basic level in J. Lunine, *Earth: Evolution of a Habitable World* (1999, Cambridge University Press, Cambridge). A popular account of Earth's habitability and the possibility of its rarity in the cosmos is that of P. Ward and D. Brownlee, *Rare Earth* (2000, Copernicus Books, New York). A classic textbook on Earth's geology is that of F. Press and R. Sevier, *Understanding Earth* (1997, W. H. Freeman and Co., San Francisco). A well-written and comprehensive treatment of Venus, written at a popular level, is D. H. Grinspoon, *Venus Revealed: A New Look Below the Clouds of Our Mysterious Twin Planet* (1998, Perseus Press, New York).

REFERENCES

Catling, D. C., Zahnle, K. J., and McKay, C. P. (2001). "Biogenic Methane, Hydrogen Escape, and the Irreversible Oxidation of Early Earth," *Sci.*, **293**, 839–843.

Falkowski, P., Scholes, R. J., Boyle, E., Canadell, J., Canfield, D., Elser, J., Gruber, N., Hibbard, K., Hogberg, P., Linder, S., Mackenzie, F. T., Moore III, B., Pedersen, T., Rosenthal, Y., Seitzinger, S., Smetacek, V., and Steffen, W. (2000). "The Global Carbon Cycle: A Test of Our Knowledge of Earth as a System," *Sci.*, **290**, 291–296.

Fedo, C. M. and Whitehouse, C. J. (2002). "Metasomatic Origin of Quartz-Pyroxene Rock, Akilia, Greenland, and Implications for Earth's Earliest Life," *Sci.*, **296**, 1448–1452.

Forget, F. and Pierrehumbert, R. T. (1997). "Warming Early Mars with Carbon Dioxide Clouds That Scatter Infrared Radiation," *Sci.*, **278**, 1273–1276.

Kasting, J. F. (1988). "Runaway and Moist Greenhouse Atmospheres and the Evolution of Earth and Venus," *Icarus*, **74**, 472–494.

Kasting, J. F. (1991). "CO_2 Condensation and the Climate of Early Mars," *Icarus*, **94**, 1–13.

Kasting, J. F. (2001). "The Rise of Atmospheric Oxygen," *Sci.*, **293**, 819–820.

Kasting, J. F. and Siefert, J. L. (2002). "Life and the Evolution of Earth's Atmosphere," *Sci.*, **296**, 1066–1068.

Kerr, R. A. (2002). "Reversals Reveal Pitfalls in Spotting Ancient and E.T. Life," *Sci.*, **296**, 1384–1385.

Knauth, L. P. (1998). "Salinity History of the Earth's Early Ocean," *Nature*, **395**, 554–555.

Mojzsis, S. J., Harrison, T. M., and Pidgeon, R. T. (2002). "Oxygen-Isotope Evidence from Ancient Zircons for Liquid Water at the Earth's Surface 4,300 Myr Ago," *Nature*, **409**, 178–181.

Mojzsis, S. J., Arrhenius, G., McKeegan, K. D., Harrison, T. M., Nutman, A. P., and Friend, C. R. L. (1996). "Evidence for Life on Earth before 3,800 Million Years Ago," *Nature*, **384**, 55–59.

Nisbet, E. G. and Sleep, N. H. (2001). "The Habitat and Nature of Early Life," *Nature*, **409**, 1083–1091.

Ohmoto, H. (1996). "Evidence in Pre-2.2Ga Paleosols for the Early Evolution of Atmospheric Oxygen and Terrestrial Biota," *Geol.*, **24**, 1135–1138.

Ohmoto, H. (1997). "When Did the Earth's Atmosphere Become Oxic?" *Geochem. News*, 12–27.

Pavlov, A. A., Brown, L. L., and Kasting, J. F. (2001). "UV Shielding of NH_3 and O_2 by Organic Hazes in the Archean Atmosphere," *J. Geophys. Res.*, **106**, 23267–23287.

Pavlov, A. A. and Kasting, J. F. (2002). "Mass-Independent Fractionation of Sulfur Isotopes in Archean Sediments: Strong Evidence for an Anoxic Archean Atmosphere," *Astrobiol.*, **2**, 27–41.

Petford, N., Cruden, A. R., McCaffrey, K. J. W., and Vigneresse, J.-L. (2000). "Granite Magma Formation, Transport and Emplacement in the Earth's Crust," *Nature*, **408**, 669–673.

Petsch, S. T., Eglinton, T. I., and Edwards, K. J. (2001). "^{14}C-Dead Living Biomass: Evidence for Microbial Assimilation of Ancient Organic Carbon During Shale Weathering," *Sci.*, **292**, 1127–1131.

Schopf, J. W. (ed). (1983). *Earth's Earliest Biosphere*, Princeton University Press, Princeton, NJ.

Taylor, S. R. (1999). "On the Difficulties of Making Earth-Life Planets," *Meteor. Planet. Sci.*, **34**, 317–329.

CHAPTER 12

Planetary Evolution II
The History of Mars

12.1 Introduction

Water defines our perceptions of Mars just as it does the Earth, but for completely different reasons. We think of the Earth as the water planet, the "pale blue dot," whose complexly intertwined history of life and geology revolve around the effects of its massive surface ocean and associated water reservoirs within and above the crust. Mars is, on the other hand, a comparatively dry planet, yet the evidence for the presence of water in the past is abundant and varied. Thus, most of the questions about the history of Mars—its formation and its current state—involve the origin and fate of its water: Where did Mars's water come from? How much was accreted? How much liquid water flowed across the surface and for how long? How was the surface water lost? How much water is present today in the crust of Mars? Does liquid water flow in a transient fashion today across limited parts of the Martian surface? Did life ever exist on Mars? If so, to what extent did it exist near or on the surface, and how far did it evolve? Did life find refuges against the drying up of the planet? Did Martian life and terrestrial life have a common origin, with primitive microbes transported from one planet to the other by impacts? Where are the likely sites for finding remnant liquid water, and life, on Mars?

Because Mars has been the subject of many technical and popular books (see the readings at the end of the chapter), at best what can be achieved here is a highly condensed summary. The basic properties of Mars are discussed first, along with a limited historical introduction. The chapter then summarizes what we know about past and present amounts of water on Mars and describes models for how Mars might have obtained its water and how it lost its water over time. The chapter ends with a discussion of the future of Mars exploration, although the details of the kinds of instrumentation that might be used for the search for life on Mars are deferred to Chapter 14.

12.2 Mars as a Planet Today

The Basic Parameters of Mars

Mars is a modest-sized planet—intermediate in size between the Earth's moon and Venus. Its mass is little more than one-tenth the mass of the Earth, its surface area is somewhat more than one-fourth Earth's surface area (about equivalent to the continental area on Earth), and the strength of Mars's gravitational acceleration is one-third that of the Earth. The 20 percent difference in bulk density between Mars and the Earth can be largely accounted for simply by a lesser degree of compression within the smaller planet. Thus, the bulk composition of Mars and its interior structure are like that of the Earth, with perhaps proportionately less metal in the core and more in the mantle.

The orbit of Mars is more eccentric than that of the Earth, with a mean distance from the sun of 1.5 AU. By virtue of this greater distance from the Sun, Mars receives less than half the amount of sunlight that the Earth does and was also formed in a part of the solar nebula colder, at any given time, than that of the Earth. Mars is also much closer to Jupiter than is the Earth, and Mars lacks the large Moon that the Earth possesses. For these reasons, the spin axis of Mars is much more profoundly affected by the periodic gravitational tugging of Jupiter than is Earth's. Our planet has an obliquity, or spin tilt, relative to the plane of its orbit of 23 degrees with little variation. The spin obliquity of Mars, on the other hand, varies from roughly 10 degrees to 50 degrees on timescales of millions of years. While the shift of the spin obliquity cannot be observed directly, because of the long timescale, repeating sets of ice and dust deposits imaged near the poles of Mars may have been formed as a consequence of the resulting oscillation in the extremes of the Martian seasons (Figure 12.1).

After the invention of the telescope, Mars has figured in the popular imagination most prominently among the planets as a possible abode for other life, despite the greater

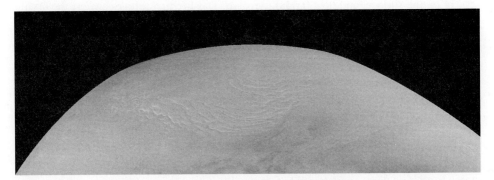

FIGURE 12.1 The layered terrain surrounding the north pole of Mars. Both the bright and dark layers contain dust, but the bright areas have a higher proportion of carbon dioxide. Water ice clouds and atmospheric dust create a haze above the features.

brightness of Venus in the sky. This greater interest is likely the result of the nonuniform nature of the Martian surface and its changes over time, both in contrast to Venus. Speculation about life on Mars rose with the observations by the Italian astronomer Giovanni Schiaparelli, who in the late 1870s made detailed maps of Mars from a small telescope in Milan. Schiaparelli identified a web of straight lines crisscrossing dark and light areas on Mars, referring to these as "canali"—which translates to English as channels (natural) or canals (artificial), depending on the context. Schiaparelli's maps were evocative, as were the names that he gave features, such as Amazonis and Elysium (Figure 12.2). These, in turn, figured prominently in a beautiful collection of Mars maps published in 1892 by Camille Flammarion, who embellished the maps with text describing the canali as having been constructed by a civilization superior to that on Earth. A copy of this book fell into the hands of the very wealthy Bostonian Percival Lowell, whose imagination caught fire at the prospect of a Martian civilization. With his considerable family resources, Lowell went about constructing a permanent, large observatory at Flagstaff, Arizona, on the high, pine-covered southern edge of the Colorado Plateau, to determine once and for all the nature of Mars.

Lowell's Mars bore even less resemblance to the real Mars than that of Schiaparelli, with more than 180 canals claimed. Of these, only a handful correspond directly to actual features imaged across the Martian surface, and these are all natural (e.g., Valles Marineris). Narrow artificial constructs such as canals require extreme optical quality and atmospheric transparency far beyond what Lowell's telescope and site could obtain.

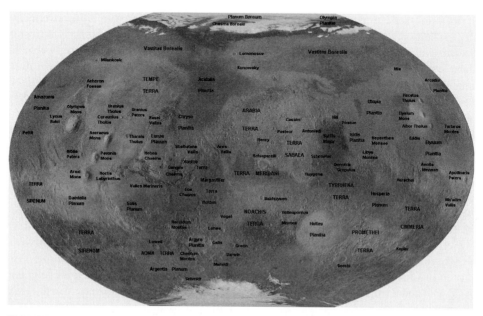

FIGURE 12.2 A map of Mars in which places discussed in the text are labeled.

Lowell knew this, and argued that what were seen were adjacent strips of verdant land much wider than the canals themselves. Lowell's persuasive manner, his high position in society, and the fervor with which he pursued his mission imbued in the American (and to a lesser extent European) popular culture the notion of an inhabited but dying Mars.

Lowell's vision influenced not only the popular culture, but also the scientific. For decades after the early 20th century, when the canal debate peaked, the issues raised by Lowell colored scientific observations of Mars. Periodic waves of darkening across the Martian surface, observed prior to Lowell, were interpreted as the seasonal oscillations of vegetation. This view persisted, with declining support, until the Mariner flybys of the 1960s, which imaged a rather lunarlike Mars (and, in fact, had accidentally missed many of the most interesting features known today by virtue of their trajectories across the planet). Spectroscopic observations of increasing fidelity, beginning in the middle of the 20th century, were at first interpreted incorrectly to indicate moderate temperatures and relatively high pressures. By the 1960s a consensus had emerged that the surface pressure was 80 mbar—12 times less than that on Earth but still conceivably allowing habitability at the surface.

The first robotic spacecraft to visit Mars, Mariner 4, flew by the planet in 1965. In addition to taking a few crude images of a lunarlike landscape, the spacecraft beamed a radio signal through the Martian atmosphere back toward Earth, where the attenuation of the signal due to refraction allowed the surface pressure to be revised downward an order of magnitude. Thus by the mid-1960s, the true Mars with its cold and tenuous atmosphere emerged from the data. Soon after, the waves of darkening were correctly interpreted as the waxing and waning of regional and global dust storms, and the Mars that we know today emerged just prior to the complete mapping by Mariner 9 and the Viking spacecraft in the 1970s.

It is too facile to pass judgment, in hindsight, on astronomers who either overinterpreted their optical observations of very faint features or who failed to properly interpret spectroscopic data indicative of a dry, cold, and almost airless planet. Science is always done in the context of a surrounding environment of expectations and prior prejudices. It is important, however, to note that the detractors of a habitable Mars existed as well, as far back as the time of Lowell. The British scientist Alfred Russel Wallace in 1907 (at age 83) applied straightforward physical reasoning to show that the properties of the canals, as inferred by Lowell, made no physical sense as conduits of polar water to the equator. And some astronomers in the mid-20th century correctly argued that the Martian atmosphere was very cold and very thin. Ultimately, additional and better data weighed in favor of these views. Finally, the construction of the Lowell Observatory has done vastly more than add intrigue to the question of the nature of Mars; among its discoveries are the distant planet Pluto (by Lowell astronomer C. Tombaugh) and the redshift in the spectra of galaxies (by V. Slipher). Today it is a world-class center of observational astronomy.

Perhaps the greatest value of the Mars debates of the 19th and 20th centuries is what they have to say about the search for habitable worlds beyond our solar system. As discussed in Chapter 15, the Terrestrial Planet Finder (TPF) and Darwin—respectively, the American and European versions of a spaceborne telescope designed to detect Earth-sized planets around other stars—will depend on spectroscopy to infer whether any detected terrestrial planets might be habitable. The lesson from Mars, and Venus for that matter, is that it is all too easy to allow our prior expectations to influence the interpretation of these difficult data sets. In some ways, the spectroscopic studies of Mars and Venus up through the mid-20th century are like those that will be attempted by TPF and Darwin for extrasolar planets in the next decade, making the history of their interpretation very relevant.

Surface Appearance of Mars

Mars presents a surface very different from that of the Earth, both from orbit (Figure 12.3, see Color Plates 12–13) and from the surface (Figure 12.4). A smooth, northern-hemisphere lowland relatively devoid of impact craters contrasts with a heavily cratered, southern-hemisphere highland. Interrupting the smooth northern plains on one hemisphere is the huge Tharsis plateau, which contains on its western end the solar system's largest basaltic volcano, Olympic Mons, as well as other smaller (but still giant) volcanoes further east. Like the northern hemisphere, it is lightly cratered. Extending out from the eastern side of the plateau is the solar system's largest canyon, Valles Marineris (an enormous fracture that contains within it side canyons bearing sedimentary deposits suggestive of water flow, Figure 12.5 on page 371), and at its western end begins in a network of apparently *fluvial* (water-caused) channels. The southern highlands, which contain the most ancient terrain, are also where most of the erosional features occur that are suggestive of the flow of liquid water, although younger features do exist on the slopes of some volcanoes. A giant impact basin has excavated a significant fraction of the southern highlands. In contrast to the Earth, where much smaller volcanoes follow ridges and troughs associated with plate tectonics, Mars's huge tectonic features are fewer and larger, and they are not organized into any evident pattern suggestive of plate tectonics.

Wind is a pervasive agent of erosion today on Mars, and on a global scale we can see both dune fields, in which larger particles are swept into repeating patterns (Figure 12.6, see page 371), and large-scale dust storms, which move both large and small debris around the planet (Figure 12.7, see page 372). Dust devils, which are vortices of wind resembling miniature tornadoes and are common in arid regions on the Earth, have also been seen in images from orbit. These may play a role in scouring the surface to create movable debris and to help loft that debris so as to initiate large-scale dust storms.

The two polar caps of Mars wax and wane with the seasons and, as described below, it is carbon dioxide moving from the gaseous to solid phase that causes the caps to change in this way (Figure 12.8, see page 373). Even as the so-called seasonal caps wane, there is residual dusty and icy material clearly marking the site of a permanent polar cap in each hemisphere. This permanent cap, discussed below, contains carbon dioxide ice but is largely water ice at both poles.

The surface of Mars seen from landers presents diverse views, although the need to pick relatively safe sites ensures that we have not yet seen the most extreme and beautiful scenery the Red Planet has to offer. The surfaces at the four landing sites of Vikings 1 and 2, Pathfinder, and Mars Exploration Rover 1 (Spirit) are littered with rocks of a range of sizes; the landing site of the second Rover (Opportunity) is devoid of scattered boulders and rocks but does have exposures of bedrock (Figure 12.4). The Pathfinder site contains very large boulders and background topography that are the result of an asteroidal or cometary

a

b

c

FIGURE 12.4 Three views of the Martian surface from (a) Mars Pathfinder at Ares Vallis, (b) Spirit in Gusev Crater, and (c) Opportunity on Meridiani Planum. The small Sojourner Rover is seen in the foreground of the Pathfinder lander image; the larger Spirit and Opportunity Rovers carried all the instruments and cameras themselves.

FIGURE 12.5 High-resolution camera system (HRCS) image from the Mars Express Orbiter of a portion of Vallis Marineris, showing features that appear to be caused by the erosive action of water.

FIGURE 12.6 Mars Global Surveyor Mars Orbiter Camera (MOC) image of a region of sand dunes, in the floor of Kaiser Crater in southeastern Noachis Terra. The steepest slopes on each dune point toward the east, indicating that the strongest winds that blow across the floor of Kaiser move sand in this direction. Wind features of different scales are visible in this image; the largest (the dunes) are moving across a hard surface (light tone) that is itself partially covered by large ripples. The image covers an area approximately 3 km (1.9 mi) across and is illuminated from the upper left.

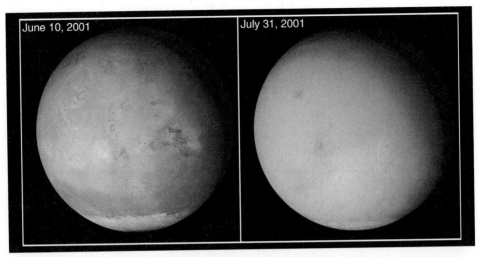

FIGURE 12.7 MOC global-scale images of Mars before and during the 2001 Great Dust Storm that swept over much of the planet. Such storms, as with this one, begin at the onset of southern-hemisphere spring. The clear Martian atmosphere in the left-hand image gives way to a dust obscured surface in the right-hand image.

impact, fluvial erosion, or both (Figure 12.9, see page 374). Erosion by wind is evident at all three sites, but, in particular, the Pathfinder lander with its meteorological package and imaging system camera appears to have actually captured a dust devil in action.

The Atmosphere of Mars

The Martian atmosphere is extraordinarily thin at the Martian surface by Earth standards. The 6-mbar average pressure, which fluctuates by tens of percent over the seasons of a Martian year, is equivalent to what we would experience at approximately 30 km above the Earth's surface—some three times higher than typical passenger jet cruise altitudes. Unlike Earth's mixed nitrogen–oxygen atmosphere, which has only an admixture of carbon dioxide, the Martian atmosphere is mostly carbon dioxide with an admixture of nitrogen. The thinness of the atmosphere precludes substantial greenhouse warming from absorption of upwelling infrared radiation, allowing huge temperature extremes—midsummer noon surface temperatures above 0°C contrast with polar temperatures of −80°C. Hence, water vapor is present but rare in the Martian atmosphere—about 10 parts water per million parts carbon dioxide.

That Mars has any atmosphere at all today is the result of its polar deposits of carbon dioxide, which are heated in each hemisphere as Mars swings alternately to its northern- and southern-hemisphere solstices. Both polar caps have permanent deposits of water ice, while carbon dioxide ice (dry ice) remains at the south pole throughout the southern summer. The seasonal caps are the dry ice deposits that sublimate off the summer cap into the atmosphere; at the winter cap the atmosphere snows out to build up an enlarged deposit of dry ice. The pressure in the Martian atmosphere is roughly what we would expect for a gas

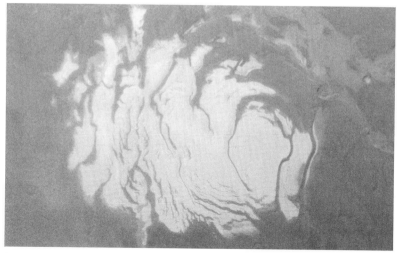

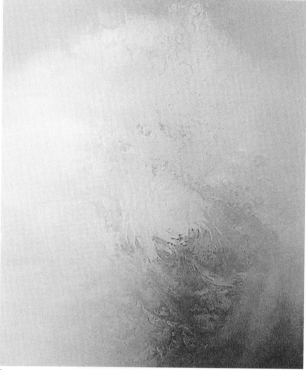

FIGURE 12.8 MOC images of the south polar cap during (a) summer and (b) spring in the southern hemisphere of Mars. The white areas surrounding the permanent cap in (b) are the seasonal carbon dioxide frost deposits, which are already sublimating away. Note that the scales in the two figures are not the same. The permanent cap is about 420 km in horizontal extent.

of carbon dioxide existing in steady contact with ice caps that have local temperatures of about $-80°C$. Thus, the cap temperatures are regulated by the sublimation and condensation of carbon dioxide vapor alternating from one pole to the other, from summer to winter.

FIGURE 12.9 "Yogi" is a meter-size rock about 5 meters northwest of the Mars Pathfinder lander and was the second rock visited by the Sojourner Rover, which performed simple chemical analyses. Pitting and other features possibly caused by wind erosion can be seen. Whether these and other large rocks were deposited by impact or by flood at the Pathfinder site remains undetermined.

This is a type of atmosphere very different from the Earth's: The bulk ingredient of our atmosphere, nitrogen, is nowhere close to condensing out on the surface as nitrogen ice. Indeed, Earth's atmosphere is warm enough that nitrogen simply cannot condense to a liquid or solid no matter what the pressure; it is significantly above its *critical point*. (Titan, one of the subjects of the next chapter, also has a nitrogen atmosphere that is below the critical point—and might at times in the past have suffered a rainout of this gas.) On Mars, the bulk atmosphere can—and does—condense out. Also on Mars, liquid water cannot exist stably under present-day surface conditions, and so to explain the existence of the erosional features described in Section 12.3, we must postulate both a denser and warmer atmosphere in the Martian past and transient conditions in the more recent past that allow for the temporary existence of liquid water.

A number of minor gases are present in the atmosphere as well. Most intriguing is the tentative detection in 2004 of methane in the atmosphere, with an abundance relative to carbon dioxide of 11 parts per billion (ppb). Spectrometers have detected one or more spectral lines (Chapter 3) of this gas, using instruments on the European Space Agency's

Mars Express orbiter and also on Earth-based telescopes. As discussed in Chapter 11, methane is unstable in an atmosphere like that of Mars or the Earth and should break down quickly—unless a source is present to resuppply the gas to the atmosphere. Sources proposed include volcanism and biological activity; methane reservoirs in the polar ices are a third possibility. The validity of the methane atmospheric signature and its implications for Mars remain unclear and may take some time to sort out.

12.3 Evidence for Past Epochs of Surface Liquid Water

Evidence for the action of water on Mars throughout the planet's history abounds. Impact craters appear to have melted ground ice; their peripheries show signs of extensive mudflows. Volcanic intrusions heat the ground and release water; a number of outflow channels reveal that water was melted and flowed through the eruptive heat. Most intriguing is evidence for an ancient warmer period on Mars contained in channels and canyons, as well as the geochemical evidence at one landing site for standing bodies of salty water in the ancient past.

Networks of dry channels and valleys are present on Mars. Three basic forms—outflow channels, valley networks, and runoff channels—can be identified (Figure 12.10). The outflow channels appear to have been formed by the very rapid release of large quantities of water or might have been carved by flows of debris (rocks, mud) mobilized by water. The flows in such channels were sufficiently energetic that they could have been sustained under virtually any atmospheric conditions, including the cold, dry climate existing now on Mars (under which slowly flowing water would quickly freeze and then sublime to water vapor). The wide variation in abundance of craters on surfaces in and around the channels indicates that the channels formed episodically over the history of Mars. The valley networks, on the other hand, have a form that suggests they were carved by more slowly flowing liquid water—not violent debris flows. The flow could have been above the surface, much like a terrestrial river, or just below. In the latter case, the networks reflect collapse of the surface (sapping) caused by underground flow. The possible sources of the water include melting of buried ice and expulsion to the surface, melting of surface ices, or even precipitation of snow or rain. The valley networks occur primarily, but not entirely, on surfaces that are very heavily cratered, and some of the impacts clearly occurred after the networks were formed. Most are therefore very ancient, dating to the end of the heavy bombardment some 3.8 to 4.0 billion years ago.

Because their formation requires conditions very different from those present today (much more restrictive than required for the outflow channels), valley networks seem to be a record of a time when the atmosphere was thicker and the climate warmer. A few younger valley networks, as well as runoff channels occurring on the slopes of some volcanoes, suggest that warm conditions (possibly localized) may have occurred multiple times in Martian history.

Some geologic features in various areas of Mars appear to have been carved by glacial action, that is, the movement of massive amounts of surface ices under their own weight. For such movement to occur, the ice must be relatively warm; ice too far below its melting

a

FIGURE 12.10 Three types of features on Mars thought to have been carved by water: (a) valley network in the southern highlands, shown in a Viking Orbiter image about 250 km across; (b) an outflow channel emerging from an area enclosed by cliffs, shown in a Viking image about 140 km across; and (c) runoff channels on the flanks of the Martian volcano Alba Patera, imaged by the Thermal Mapper and Imaging Spectrometer (THEMIS) on Mars Odyssey. The THEMIS image covers an area 58 km long by 27 km wide, showing detail as small as 20 meters.

point is rather brittle and immovable. The features include certain kinds of ridges and troughs that resemble terrestrial landforms carved by glaciers called moraines and eskers. Such interpretation is difficult to make without ambiguity; other nonglacial causes for the features might have been at work. If the glacial interpretation is correct, however, it implies surface conditions in which water ice was stable against rapid passage into the gas phase and hence thicker atmospheric conditions. Whether the glaciers were contemporaneous with the time of stable liquid water, followed the epoch of liquid water as Mars cooled, or were interspersed between warmer episodes for some period of time is unclear.

The amount of water required to explain all of the fluvial erosional features is at least 0.001 to 0.01 of an Earth's ocean—the equivalent of a layer 10 to 100 meters deep of water spread across the surface. Measurements discussed later in the chapter of high levels of heavy hydrogen—deuterium—in atmospheric water suggest that some or much of the hydrosphere was lost to space. Then values closer to 0.1 Earth's oceans, or even larger, must be entertained. One interpretation of the smooth northern-hemisphere basin is that it held a stable body of liquid water (that is, an "ocean"), and this provides another constraint on

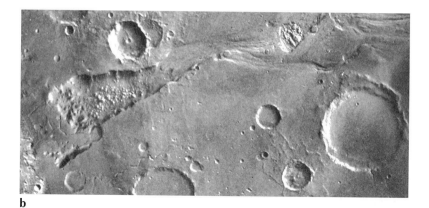
b

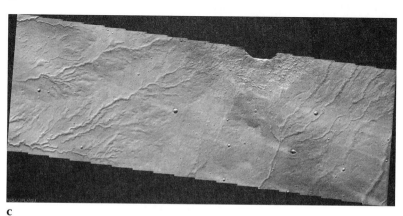
c

the past abundance of water. Based on the identification of a possible shoreline in data from the Mars Global Surveyor's laser altimeter (Mars Orbiter Laser Altimeter, or MOLA), which accurately measured Martian topography, the equivalent of roughly 1 percent of the Earth's ocean could have been contained in the basin. This is an upper limit, however, because others have argued that the basin could be filled with sediments derived from the erosion of highlands to the south and transported to the basin by the channels.

A determination for or against the hypothesis that the valley networks were formed on a planet where standing water existed for some time is crucially important in understanding whether Mars was habitable early in its history. The channels by themselves, though compelling, could have been formed during relatively brief releases of water from the subsurface of a cold Mars, perhaps associated with large impacts or a high flux of impactors. Geochemical evidence, which we now examine, suggests instead that liquid water persisted at or near the surface for some time.

The signature of mineral alteration by liquid water seen in data from orbiters and landers argues that there were times in the Martian past when liquid water persisted in the near-surface Martian crust, if not the surface itself. The Thermal Emission Spectrometer

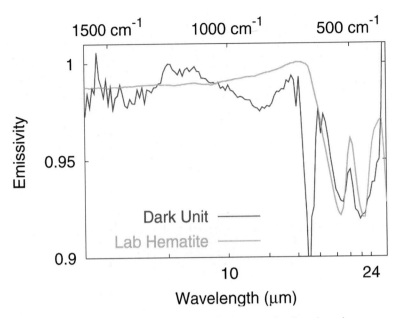

FIGURE 12.11 A spectrum of the soil at Opportunity's landing site using a miniature version of TES that is on the Opportunity Rover, showing the presence of grey hematite. The "W" shape on the right side of the spectrum is where the Martian soil signature (light grey) matches the laboratory signature for hematite (dark grey line). At other places in the spectrum, absorption by atmospheric CO_2 and other surface materials alters the spectrum from the pure magnetite.

(TES) aboard Mars Global Surveyor measures the infrared photons coming from the surface, and the spectrum of this light can reveal composition as described in Chapter 3. A few areas on Mars show the presence of grey hematite in the TES spectra, an iron oxide mineral that sometimes—but not always—forms in the presence of water. One area where this signature was seen is the flat plains of the Meridiani Planum, where landings could be safely conducted. Thus the area became the prime landing site for the second Mars Exploration Rover, Opportunity, which confirmed the spectral signature of the grey hematite (Figure 12.11) as well as direct geochemical evidence for past liquid water.

Opportunity landed in early of 2004 in a small crater within the vast dusty plain of Meridiani Planum. At Meridiani, Mars scientists got a lucky bounce—as Opportunity's cushioning but irregularly shaped airbags struck the surface, the spacecraft careered off in a direction different from its descent trajectory and came to rest inside the small crater, which is only 20 meters across. That crater—now named Eagle—preserves within it exposed areas, or outcrops, of the Martian bedrock. Such exposures are extremely important and so far rare in Martian surface exploration; rocks at other landing sites are only debris from impacts or ancient floods. Most extraordinary is that the Eagle outcrops hold geochemical evidence for past liquid water—the first quantitative demonstration that liquid water once was stable within the uppermost crust of Mars (Figure 12.12).

12.3 Evidence for Past Epochs of Surface Liquid Water

FIGURE 12.12 Eagle crater where Opportunity landed, viewed from the rover as it exited the crater in March 2004. The lighter colored areas are the outcrops that show strong chemical and physical evidence for the area having once hosted a salty body of liquid water. The carrier platform for the Rover during cruise, entry, descent, and landing is seen inside the crater.

The evidence at Meridiani can be divided into five separate areas:

1. Chemical analysis shows that the rocks in the outcrop formed in the presence of liquid water. The analysis involved two spectrometers on the rover. The "Mössbauer spectrometer" looks at the wavelength and intensity spectrum of gamma rays—photons of very short wavelength—from rocks against which the instrument is pressed; this spectrum is diagnostic of how elements in the rock are arrayed in mineral phases. This experiment found the mineral "jarosite," an iron-rich and sulfur-bearing mineral that is hydrated (has OH groups) and hence requires the presence of water to form. The alpha-particle X-ray spectrometer performs measurements at the somewhat longer wavelength photons called X-rays. In particular, it detected very large amounts of sulfur in the rock, as well as other elements, suggesting that 40 percent of the rock is made of salts, perhaps largely magnesium sulfate (commonly called Epsom salts). This discovery was corroborated by the thermal infrared spectrometer "mini-TES" on the Rover as well.

2. Other chemical elements such as bromine and chlorine are present, but their abundances relative to each other and to sulfur vary from the top to the bottom of the outcrop (Figure 12.13). Such a variation, together with the detailed geochemical data described in the item 1, suggest that the outcrop is an "evaporate sequence." Evaporites are deposits of mineral salts left behind by the evaporation of salty water. The significance of the variation from top to bottom of the outcrop is that the evaporation occurred very slowly, sufficiently slow that different minerals—bearing different concentrations of elements—were progressively deposited as the amount of water lessened and became saltier. The salt water environment could have been a lake, a sea, or a pore within a rock; the first two are favored based on the pattern of sediments discussed in item 5.

3. Tiny spheres, 2 or 3 mm in diameter, are abundant within the outcrop and, in color images, appear bluer than their surroundings. Microscopic images from the Rover (Figure 12.14) show that they are weathered (eroded) out of the surrounding rock. This militates against the possibility that they were molten blobs of rock from impact debris that happened to fall onto the site. Indeed, the thermal infrared spectrometer found these spherules to be rich in hematite, consistent with (but not

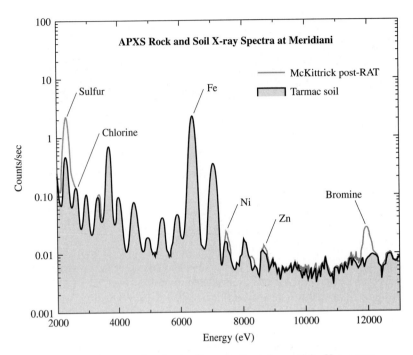

FIGURE 12.13 Spectra taken by the Opportunity alpha-particle, X-ray spectrometer showing that a rock dubbed "McKittrick" has higher concentrations of sulfur and bromine than a nearby patch of soil nicknamed "Tarmac." Only portions of the samples' spectra are shown to highlight the differences in elemental concentrations between McKittrick and Tarmac. The vertical axis is counts per second; the higher the counts, the more abundant the element in the sample. The horizontal axis is the energy of the X-ray photons, hence a proxy for wavelength, at which the diagnostic lines of different elements occur. A nearby rock named "Guadalupe" similarly has extremely high concentrations of sulfur but very little bromine. This variation in elements is typical of an evaporative sequence in which different salts are deposited as the increasingly briny water evaporates away. The rock abrasion tool, or RAT, was used to remove an outer layer of dust and weathering coating to analyze the minerals beneath.

requiring) their formation in a liquid water environment. The most consistent interpretation of the spherules is that they are "concretions," which form by the chemical precipitation of minerals and require the presence of mineral laden water where they grew.

4. Further evidence that the millimeter-sized spherules are not volcanic is that they are associated with unusually shaped voids, called "vugs," that have a very specific tabular shape. (Figure 12.15 see page 382). The simplest interpretation is that crystals of gypsum—the most common sulfate mineral on Earth—formed from the mineral-rich water and subsequently fell out of the rock due (perhaps) to wind erosion. The inference of such crystals from the voids they left behind are supportive

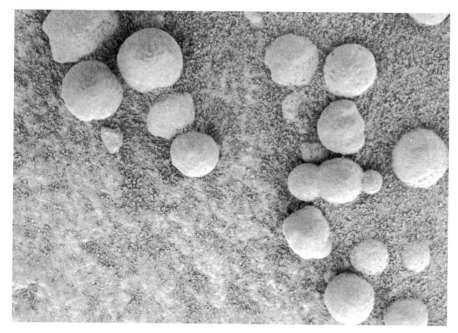

FIGURE 12.14 Microscopic image taken by Opportunity at the outcrop in Eagle Crater, showing the spherules that are abundant in the rock. The existence of the triplet seen in the image strongly indicated that the spherules grew in preexisting wet sediments rather than being glassy droplets of molten rock deposited after a volcanic eruption; these are unlikely to fuse along a line and form triplets. The image is about 3 cm across.

of the idea that the outcrop is sedimentary rock—laid down by water—rather than igneous rock left behind by volcanism.

5. Finally, the fine layers in the rock have a characteristic shape—they are angled relative to each other, truncated in places, and have curved shapes resembling "smiles" (Figure 12.16). These are characteristic of "cross bedding," a form of sedimentary layering that does not occur in deposition due to volcanism. Rather, it is indicative of deposition in slowly flowing water. While such an environment might occur in a subsurface aquifer, the most direct interpretation is that the sediments were laid down in a lake environment where water flowed gently for prolonged periods of time.

In sum, the geochemical evidence from this extraordinary outcrop is very strong that salty water existed on or within the surface of Mars. The visual ("geomorphological") evidence both supports this and leans in favor of a lake or sea environment where the evaporates formed, in the words of Cornell Professor Steven Squyres, the principal investigator for the Mars Exploration Rover payloads, "the shoreline of a salty sea on Mars."

In March 2004 Opportunity departed the Eagle crater to head toward the rim of an even larger one, called "Endurance," where the bedrock also is present and turns out to contain a much thicker sequence of layers. Along the way, additional outcroppings of rock similar

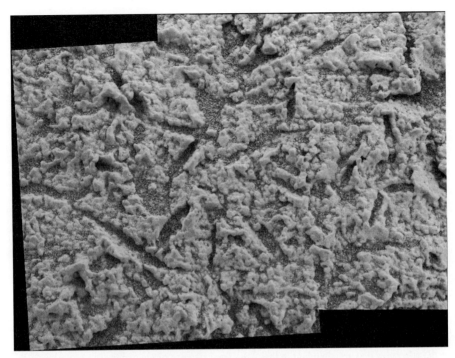

FIGURE 12.15 Microscopic image taken by Opportunity at the outcrop in Eagle Crater, showing the "vugs" that are believed to be locations where gypsum crystals formed in the sediments, then fell out as erosion occurred. Spherules are present in this image as well, which is about 3 cm across.

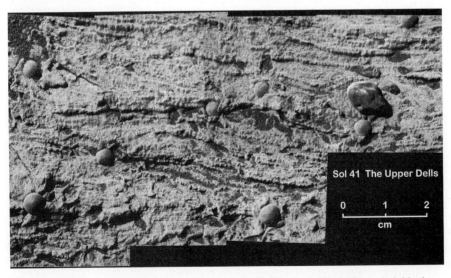

FIGURE 12.16 Microscopic image taken by Opportunity of a portion of a rock dubbed "Upper Dells" at the Eagle Crater outcrop site. Several aspects of the pattern of the fine layers are indicative of sedimentary layers that have been laid down in flowing water, as discussed in the text.

to that in Eagle crater have been encountered. If the geochemistry and layering proves to be similar in the Eagle and Endurance outcrops, it may well be that Opportunity is exploring what was once a large lake or sea. Based on the nature of the rock and geology seen from orbit, the sediments are likely quite old—billions of years in age—consistent with the view that the time of surface liquid water on Mars is long past. Opportunity's drive to the bigger crater was a successful race with the Sun; it landed during early autumn in the Martian southern hemisphere, and, as the autumn days wore on, the sinking Sun provided less and less light for the solar-powered Rover.

Spirit landed in January 2004 in a crater called Gusev, lying at the end of a 900-km-long valley network called the Ma'adim Vallis, which breaches one edge of the crater (Figure 12.17). It had been thought that the material filling the crater might be fluvial (river) sediments from the valley network. However, Spirit has found very little evidence of water at its Gusev crater landing site, and most of this is circumstantial. At the time of this writing Spirit had reached the so-called Columbia Hills, several kilometers from its landing site, where scientists found exposures of bedrock (Figure 12.18).

FIGURE 12.17 Image from orbit of Ma'adim Vallis and Gusev crater, where Spirit landed. The crater is 170 km across.

FIGURE 12.18 Columbia Hills, ultimate goal of the Spirit Rover in its search for evidence of flooding of the crater by the adjacent Ma'adim Vallis, are seen in the background of the crater Bonneville, which proved too steep for safe entry by Spirit.

12.4 Mechanism for a Warm Climate on Early Mars

The existence of stable liquid water, running in valley networks, ponded in canyons, or residing in large scale basins, requires that the ancient Martian environment be very different than that of today. It had to have been warmer and possessed of a larger atmospheric pressure so that surface conditions within the stability field of liquid water could have existed, at least for limited times and perhaps limited places.

The traditional proposal, developed a quarter century ago at the time of Viking, hypothesized that the ancient Martian atmosphere had much larger amounts of carbon dioxide, enhancing the atmospheric pressure and providing a greenhouse effect to warm the surface temperature beyond that of today, in turn providing a positive feedback through evaporation of surface waters. Very substantial amounts of carbon dioxide were required in this scenario—and not just because Mars is more distant from the Sun than the Earth. Adding to the carbon dioxide requirement is that the Martian climate was warmer at precisely the time when the solar luminosity was only 70 to 80 percent of the present value. Hence, carbon dioxide pressures of several bars were invoked in such models; this is an atmosphere with several times the total pressure and perhaps 10^4 times the pressure of carbon dioxide that characterizes Earth's atmosphere. As we discussed in Chapter 11, this is enough carbon dioxide to generate cloud formation in the atmosphere and possibly a negative feedback on the greenhouse effect; other greenhouse gases (sulfur-bearing for example) may need to be invoked to aid the heating. There is little agreement on just how strong the greenhouse warming was, that is, whether liquid water on very ancient Mars was continuously and globally stable on Mars or a transient regional feature. However, an interesting indication that an ancient, thick atmosphere was present is seen in the fact that the $^{15}N/^{14}N$ value in the Martian atmosphere is enriched relative to terrestrial by only 1.6. This enrichment can be achieved through atmospheric loss in less than one-third the age of the solar system. So why isn't nitrogen even more fractionated? One way to decrease the efficiency of fractionation is to dilute the nitrogen in an early dense CO_2 atmosphere, consistent with (but not requiring) the picture of a warm and wet early Mars.

Even more challenging to models is accounting for later episodes in which glaciation, ponding, or sedimentary layering associated with liquid water may have occurred. The principal problem is that the an important mechanism invoked for loss of carbon dioxide and water is atmospheric escape. Once lost to space, the greenhouse gases were no longer available to renew a warmer climate under a thicker atmosphere. The extent to which the Martian climate had to warm to produce glaciers, and the timescale and timing of such warmings, are very contentious issues. The existence of gullies in young terrains, discussed in Section 12.6, may indicate that the temporary presence of liquid water in recent times does not require global rewarming, only locally special considitions.

The outflow channels, which do not require *global* rewarming of the Martian climate, do imply that crustal reservoirs of liquid water are present throughout Martian history, and they, along with the gullies and other photographic evidence for erosion involving water, suggest that liquid water has appeared in certain times and places over the last half-billion years of Martian history. It is therefore crucial to obtain a map of crustal water sources—in whatever phase (ice, mineral, . . .)—and this was a primary goal of orbiting spacecraft that has already borne fruit. Then, to connect the ancient Martian water inventory to that of the recent past and present requires considering mechanisms of loss and sequestration, which are reviewed next.

12.5 The Drying and Freezing of Mars: Causes and Timing

The geologic evidence is strong that Mars underwent a rapid evolution sometime in its history from a relatively warm and wet surface environment to a much colder and dryer one. Erosion rates were larger earlier in Martian history than at later times. The largest craters and the impact basins, representing the oldest impact events, show heavy erosion. Many are missing central peaks or rims, and ejecta blankets are incomplete or missing. While the Martian erosion rates even at their peak are nowhere close to the average values for the Earth today, they do approach that in some very dry places like the Atacoma desert. On terrains younger than those represented by the heaviest cratered southern highlands, the geologic evidence suggests a steep drop in erosion rates, perhaps by a thousand times, signaling the end of the wet early climate regime. Some of the younger, water-carved features exhibit at their downstream ends U-shaped valleys more indicative of erosion by glaciers than by liquid water. Not long after the earliest period in Martian history—the time of the valley networks—the surface environment of Mars dried and cooled quickly and dramatically.

How was the water lost? Atmospheric chemistry suggests that loss of water may have been largely, but not exclusively, into space. The atmospheric ratio of deuterium to hydrogen (D/H) is five times higher than the value measured in terrestrial ocean water, and more significantly, is three times higher than the value found in very ancient meteorites thought to have come from Mars (Section 12.6). If the meteorite value represents the starting value for Mars, as we discuss later in the chapter, then the atmospheric value today suggests an enrichment process occurred that was associated with the loss of water. As atoms and molecules escape from an atmosphere, the heavier ones are preferentially left behind. The heavier isotope of a given element, then, will tend to become enriched in an atmosphere that is escaping. The extent of the enrichment depends on the mechanism and rate of escape—a large impact that catastrophically blows part of an atmosphere away from a planet will not distinguish among the isotopes, while a more gradual escape of atoms associated with photochemical breakup of molecules or heating will lead to greater enrichment of the heavy isotope.

On Mars the enrichment of deuterium is much less than on Venus; hence the loss process was either less efficient or much less water was lost from Mars than from Venus. Because of Mars's low surface gravity, impacts of asteroidal or cometary bodies would have been efficient at removing water both in the atmosphere and vaporized from the crust. Such impact stripping would have been indiscriminant with respect to the heavy versus light isotope of hydrogen, but some modest fractionation might have occurred. That another escape process was in play is suggested by the ratio of the heavy to light isotope of nitrogen in the atmosphere—$^{15}N/^{14}N$—which requires that the molecular nitrogen be dissociated high in the atmosphere by solar wind particles leading to preferential removal of the lighter isotope.

It is likely that both impact-generated loss of water and solar wind stripping of the background atmosphere played roles in the evolution of Mars toward a cold and dry climate. The loss of atmosphere would have reduced the greenhouse warming of the atmosphere and contributed to the freezeout of water. Thus, Martian water was lost both off the top of the atmosphere and through crustal trapping. The relative amounts remain controversial, but if large amounts were stored in the crust of Mars as the climate cooled, then both carbonates (Chapter 11) and sulfates (sulfur-bearing minerals that are often formed in the presence of water) ought to be seen in locales where the widespread windblown dust is not present. Carbonates have not been detected by orbiting spacecraft, but the infrared spectrometer aboard the Mars Express orbiter has detected sulfates in the Vallis Marineris canyon complex. Thus, the relative importance of the two loss processes—crustal sequestration versus loss to space—remains unclear. But Mars's vulnerability to climate change in either case was largely a result of its small size. The low gravity aided impact loss of water and carbon dioxide, while the rapid cooling of the small planet would have made plate tectonics difficult or impossible, hence preventing recycling of carbon dioxide and water from the crust to the surface.

Because much of the surface contains impact craters, it is possible to work out a relative chronology for various kinds of geologic activity on Mars based on the terrains in which they are located and the numbers of craters that are superimposed on, or cut through, by the features. As discussed in Chapter 6, the most heavily cratered regions are the oldest, that is, have been modified at the most distant time in the past. On Mars, the heavily cratered southern uplands probably represent the oldest regions, and their age has been given the name "Noachian," suggestive of a time of abundant water. More lightly cratered plains and plateaus correspond to the so-called Hesperian epoch, and the least cratered areas to the Amazonian epoch (Table 12.1 on pages 388–389). Because we have no systematic sample collection of the Martian crust (the only samples are meteorites that fall from Mars as described later), we cannot pin down absolute ages of various cratered terrains on the Martian surface as is possible for Earth's Moon via the rocks collected by the Apollo missions from 1969 to 1972. However, the earliest Noachian probably corresponds to the end of the formation of Mars, some 4.5 billion years ago. Some geologists divide the Noachian into two periods, namely, Early versus Middle-Late, with the dividing line being the end of the formation of the large southern impact basin. By analogy with the Moon, this would have occurred approximately 4 billion years ago. With that tie-point in the chronology, the Hesperian and Amazonian would have begun 3.7 billion and 1 billion years ago, respectively, although a minority of planetologists put the beginning of the Amazonian as far back as 3.1 billion years ago.

The valley networks tend to follow slopes that are associated with the Tharsis plateau, the major volcano-bearing platform. Thus, the valley networks, at least those seen today on the surface, must be younger than Tharsis. Because they appear to be Noachian, Tharsis must be Noachian in age as well. The massive volcanoes on Tharsis may well have injected large amounts of carbon dioxide and water from the interior into the atmosphere. But the association of the valley networks with Tharsis is most likely the result of volcanic heating of the subsurface leading to melting and runoff or (in a very wet early Mars) the topographic effect of the volcanoes on rainfall. There is evidence for buried valley networks derived from tracking of the Mars Global Surveyor spacecraft, which revealed variations in the gravitational pull of Mars over certain regions. These buried networks indicate transport of water from the southern highlands to the northern basin, and this supports the notion that the smooth northern basin once contained a shallow sea.

Table 12.1 lists major events in the history of Mars placed in the context of the crude chronology described earlier. These are based on examining a wide range of geological features from orbit-resolving features down to tens to hundreds of meters in size, selected areas at meters resolution, and close examination of sites visited by landers. Imaging is supplemented by chemical analysis at landing sites, analysis of Martian meteorites (see Section 12.6), and analogy with features observed on the Earth. The story as painted by this table is tentative, controversial in some aspects, and will certainly change as we learn more about Mars. However, the table is useful in portraying the complexity of the story relative to the simpler one for the Moon. What is also striking is the unidirectional nature of the evolution from relatively wet to dry. It is the details and causes of the transition from a potentially habitable Mars to the present-day cold and dessicated world that is one of the major unsolved challenges in planetary science.

12.6 Martian Crustal and Surface Water Today

There is a particularly extensive literature reviewing imaging and other evidence for the presence of ice and water in the crust of Mars today (see the Suggested Readings at the end of this chapter for an entry into the literature). Martian water reservoirs include water vapor in the atmosphere, water ice in the atmosphere, seasonal water ice deposits at the surface, permanent water ice deposits at the polar caps, and high-latitude permafrosts. The total mass of the permanent water ice and carbon dioxide caps has been determined using basic physics. Mars Global Surveyor, which has orbited the planet since 1997, carried with it a laser and high-speed electronic timing devices to enable the duration of travel of laser pulses from the spacecraft to the surface and back to be measured. This device, the Mars Orbiting Laser Altimeter, allows the topography of the planet to be determined to an accuracy corresponding to the knowledge of the spacecraft's orbital position relative to Mars—about a meter or so. This is enough to measure the general depression of the polar regions due to the weight of the ice lying on top of them. The effect is well known on the Earth and can be used to gauge the extent of already retreated ice sheets from the last glaciation, since the crust does not rebound from such a weight instantaneously. It is standard (and, from Earth studies, appropriate) to consider the crust of the planet (Earth or Mars) like an elastic plate, which sags under the weight of the ice. For Mars, MOLA topographic profiles indicate that the ice in the north and south polar caps is several kilometers thick, and, if spread uniformly over the planet, it would correspond to a shallow sea 20 to 30 meters thick. (The analysis assumes that most of the ice cap is water, as supported by other types of orbiter observations).

Another way, less direct, to estimate the amount of polar ice is to use the ratios of certain noble gases (argon, krypton and xenon) measured in the atmosphere of Mars. The ratios of xenon to argon and xenon to krypton are very different from that seen in carbonaceous chondrites, which is regarded (by assumption) as a fair representation of the starting composition of Mars. One possible cause of the alteration of these ratios is the selective trapping of the heavier noble gas xenon in the Martian ice caps relative to argon and krypton.

TABLE 12.1 Mars: Major Events in Geological History

	Volcanism	Tectonism	Fluvial Events	Cratering	Erosion and Surficial Processes	Number of craters per 10^6 km² >2 km
Amazonian	• Late flows in southern Elysium Planitia. • Decreased volcanism in northern plains. • Most recent flows from Olympus Mons.		• Channeling in southern Elysium Planitia.		• Emplacement of polar dunes and mantle.	— 20
					• Development of polar deposits?	— 30
						— 40
					• Formation of ridged lobate deposits on large shield volcanoes.	— 50
	• Emplacement of massive materials at S. edge of Tilysium Planitia. • Waning volcanism in Tharsis region.		• Late period of channel formation.		• Emplacement of massive materials at S. edge of Elysium Planitia.	— 60 — 70 — 80 — 90
					• Local degradation and resurfacing of northern plains.	— 100
	• Waning volcanism in Elysium region. • Widespread flows around Elysium Mons.	• Cessation of Tharsis tectonism. • Formation of Elysium Fossae. • Initial formation of Olympus Mons aureoles.	• Formation of channels NW of Elysium Mons.		• Erosion in northern plains. • Deep erosion of layered deposits in Valles Marinaris. • Development of ridges, grooves, and knobs on northern plains.	— 200 >5 km — 300 — 50 — 60

388

Hesperian	• Volcanism at Syrtis Major. • Formation of highland paterae. • Volcanism at Tempe Terra. • Major volcanism in Elysium and Tharsis region. • Emplacement of ridged plains (IIr).	• Formation of Noctis Labyrinthus. • Formation of Valles Marinaris. • Formation of wrinkle ridge systems. • Memnonia and Sirenium Fossae, fractures around Isidis.		• Degradation of northern plains materials. • Dorsa Argentia formation at South Pole. • Resurfacing of northern plains.
Noachian	• Formation of intercrater plains. • Decreasing highland volcanism. • Beginning of widesread highland volcanism.	• Ceranius, Tempe, and Noctis Fossae. • Tectonism south of Itellias. • Archeron Fossae. • Claritas Fossae.	• Formation of extensive valley networks.	• Warning impact flux. • Intense bombardment • Argyre impact. • Hellas and Isidis impacts. • Formation of oldest exposed rocks.	• Extensive dessication and etching of highland rocks. • Formation and erosion of heavily cratered plateau surface. • Deep erosion of basement rocks.

— 400
— 500
— 600
— 700
— 800
— 900
— 1000

— 70
— 80
— 90
— 100

— 200

— 300

— 400
— 500
— 600

>16 km
— 100
— 200
— 300
— 400

All three noble gases can be scrubbed from the atmosphere and near-surface crust by water ice, but xenon is trapped preferentially relative to the other two. Thus, the volume of ice required for the trapping can be derived. The answer is consistent with that derived from MOLA—a permanent cap kilometers thick. But, because the trapping is primarily accomplished in water ice—and not carbon dioxide ice—this exercise provides additional independent support for the notion that the bulk of the permanent caps is water ice.

Beyond the polar reservoir of ice, there is strong evidence for subsurface water ice in the form of permafrost. Models predict that water should be present today within the top few meters of the surface at latitudes as low as 20 to 30 degrees from the equator in favorable locations, depending on porosity. The gamma-ray spectrometer and neutron detector (XGRS) package on the Mars Odyssey orbiter sensed the signature of hydrogen in the Martian crust in two different ways; the hydrogen is almost certainly an indicator of water ice. Both detection techniques depend on the thinness of the Martian atmosphere, which allows cosmic rays to strike the surface and release high-energy photons (gamma rays) and particle radiation, which are detected by XGRS. The pattern of energies of the cosmic rays and particle radiation reveal the abundances of the elements in the surface, including hydrogen. An XGRS map of the hydrogen in the Martian crust matches almost exactly the theoretical map of water ice stability in the near-surface crust based purely on thermodynamic considerations. Whether this layer is meters or kilometers thick cannot be determined by the XGRS, as it senses only the upper meter or so of the Martian surface. Potentially, however, the subsurface layers of water ice—the Martian permafrost—could exceed all other present-day reservoirs of water on Mars. Ground penetrating radar on the Mars Express orbiter should be able to detect the ice and constrain its depth if the layer is thin; more powerful radars that could penetrate kilometers into the crust are planned for future missions.

If the ground ice had once covered the entire planet, the remnant of an even earlier time when liquid water was stable globally, it would have gradually migrated from the equatorial and midlatitude regions at depths where it is not stable toward the high latitudes. The estimated volume of the polar, layered terrain observed today is comparable to what would have been removed from equatorial regions over several billion years. This volume would have required for its storage a very porous crust, or *megaregolith*, some several hundred meters deep at least. Whether the Martian crust has such a volume of porous rock cannot be determined by current observations, although future planned orbiting radars will provide some constraints.

A number of meteorites whose origin is almost certainly Martian also provide constraints on the amount of water (in whatever form) that may remain in the crust of Mars today, or that was present in the past. These meteorites are characterized by the presence, in some samples, of trapped gases that have a composition almost identical to that of the Martian atmosphere. (The initial three classes of meteorites determined to be from Mars—Shergotty, Nahkla, and Chassigny—led to the acronym SNC for these objects, a term that is used progressively less to describe Martian meteorites as a whole but that the reader may find in the older literature.) Therefore, they almost certainly originated on Mars, having been ejected in large impacts, and then were intercepted by the Earth on their interplanetary trajectories. Computer simulations show that such interplanetary delivery is not common but plausibly has happened multiple times, given the large number of impacts over the history of the solar system. One key Martian meteorite, Allan Hills 84001, is considered further in the discussion of the search for life on Mars in Chapter 14. Some of the Martian meteorites display evidence in their mineralogy that they were exposed to limited amounts of liquid water at some time, or repeated times, in the past.

Constraints on the amount of exchangeable water with the atmosphere (i.e., near-surface crustal water) have been made on the basis of the D/H ratio in the SNC meteorite Zagami, which is approximately 180 million years old. Because the D/H ratio in the hydrated minerals in Zagami is comparable to that in the atmosphere, there must be a lower limit to the extent of the water ice reservoir with which the atmosphere presently exchanges deuterium. Based on detailed calculations, this reservoir is the equivalent of a layer of ice between 50 and 250 meters in thickness averaged over the Martian surface. Because data from the Mars Odyssey XGRS show that the distribution of ground ice is highly nonuniform, the actual subsurface ice deposits are likely much thicker; the radars aboard the Mars Express and future orbiters can test the extent of these layers. The deuterium data in Zagami also constrain the original reservoir of water on or near the surface of the planet to have been between 200 and 2000 meters in equivalent depth. These latter numbers overlap with other estimates for the amount of water on early Mars but should be regarded as upper limits because other minerals in the rock matrix of Zagami and its parent Martian crust may have contributed deuterium besides the hydrated phases.

Whether liquid water exists today in reservoirs accessible to life remains a controversial issue. The near-surface crust of Mars probably contains ground ice at high latitudes. The temperature increases moving downward from the surface, less steeply than for the Earth but still enough to allow the stability of liquid water kilometers down. However, that does not mean that liquid water actually exists there, today. Evidence from the Mars Orbiter Camera (MOC) on the Mars Global Surveyor of gullies on very youthful terrain scattered around Mars has raised the possibility that liquid water flows transiently today, or at various times in the recent past, across parts of the Martian surface (Figure 12.19). The presence of recent or ongoing liquid water activity on Mars is somewhat surprising, given the current conditions in which liquid water would freeze or boil away. However, some scientists have suggested plausible mechanisms for transiently melting subsurface water, stimulated by the change of seasons or the change in tilt of the Martian spin axis on long timescales, which would then be expelled onto the surface to flow briefly and rapidly downward along cliff faces to form gullies. There are alternative explanations for the gullies that cannot as yet be completely ruled out at this time, including the buildup and release of carbon dioxide or landslides of dry debris. However, the largest gulley system, that in the 55-km-diameter "Mohave" impact crater on Mars, exhibits young alluvial fans (water-driven sedimentary deposits near mountains) that compellingly seem to require significant amounts of liquid water for their formation.

The existence and ages of liquid-water-driven features on Mars remains an area of continuing controversy. But it is a key one: If liquid water can be demonstrated to be the working fluid for the gullies, then the potential conditions for life to exist today on Mars would have been demonstrated. Radar sampling that could detect the subsurface presence of ices, along with repeat imaging and spectroscopy of features on very small (meter or less) scales to identify frosts of water or carbon dioxide, are essential to identifying whether liquid water is present in, and driving, the gulley systems.

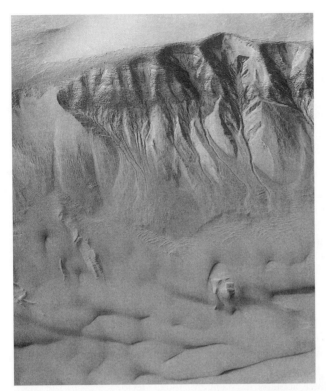

FIGURE 12.19 Mars Global Surveyor MOC images of very recent gullies on the Martian surface: (a) gullies on the edge of a cliff face, several hundred meters high, (b) alluvial fans and gullies on the edge of Mohave Crater, Mars.

12.7 The Origin of Martian Water

The origin of the Earth's oceans is a contentious issue because estimates for the amount of water in our planet's interior vary so widely. Some isotopic evidence, along with the relative compositions of the Earth and the Moon, lean in favor of the hypothesis that Earth accreted its water from local planetesimals in which water was present and bound to the rocks in the form of hydrated minerals. Under this hypothesis, in which nebular temperatures were sufficiently low to allow water bound to the minerals to exist in silicates, Earth ought to have accreted abundant water and, at least prior to the Moon-forming impact, been suffused with water throughout its crust and mantle. This hypothesis predicts an even wetter origin for Mars, since that planet accreted beyond the Earth at lower nebular temperatures, perhaps low enough even to have allowed some water ice to be directly added during formation. How much water Mars retained depends on whether that planet suffered very large impacts during its growth by accretion, akin to that which formed the Earth's moon (Chapter 6) because such impacts could have removed large amounts of water. Dynamical simulations suggest that the small size of Mars is the result of accretion accomplished in the absence of giant impacts, in which case it might have more efficiently retained the water delivered to it than did the Earth. The hypothesis of local accretion of water by Earth and Mars might therefore imply that Mars began its history with a large amount of water, that is, of order or greater than an Earth ocean's worth.

The local model of accretion of water has been criticized on the grounds that it requires the disk from which the planets formed to have been very cold in the terrestrial planet region, and it is inconsistent with the water content of meteorites whose parent bodies are thought to have formed in the inner part of the asteroid belt. The alternative hypothesis for the origin of the Earth's ocean, and by extension Martian water, is accretion of bodies from more distant orbits—at or beyond the asteroid belt. This hypothesis has been described in detail in Chapter 6. Its main conclusion—that Earth accreted most of its water from lunar-to-Mars-sized bodies in the asteroid belt—cannot apply to Mars. On the grounds of its size—comparable to the mass at which bodies were isolated from each other in the early stages of planet growth—Mars must have escaped impact with planet-sized bodies. Absent these giant impacts, Mars accreted its water from a mixture of small asteroids and comets, leading to a starting inventory of water between $\frac{1}{20}$ and $\frac{1}{4}$ the mass of the Earth's ocean, with a D/H value of twice SMOW.

The starting inventory of Martian water in this model is equivalent to a layer of water, if placed on the surface, 500 to 2,500 meters thick, which is well within the range of estimates based on the observed geology. However, not all of this water may have outgassed (escaped) from the interior; we return to this issue in the next paragraph. The D/H value derived, 1.6 times SMOW, is close to but slightly below the value of twice SMOW measured in an SNC meteorite QUE94201 that is thought to have been derived from a very ancient piece of the Martian crust. Given the uncertainties in both the laboratory and theoretical analyses, these numbers may be regarded as essentially the same. That is, the SNC value of D/H could well reflect a primordial Martian value derived from accretion of a mixture of asteroidal and cometary water. Alternatively, the D/H value in the SNC meteorite could be enriched from the terrestrial (SMOW) value through a very early period of massive escape of water, and the original source of water would then be undetermined.

Unfortunately, therefore, we cannot decide between the two hypotheses for the origin of Martian water, local versus nonlocal accretion of planetesimals, on the basis of the deuterium alone—both models are consistent with this isotopic indicator.

The heavy isotope of the noble gas argon, ^{40}Ar, as a radioactive product of ^{40}K (the potassium isotope of mass 40), provides constraints on the amount of outgassing that has occurred over the history of Mars. The ratio of ^{40}Ar to stable ^{36}Ar as measured in the atmosphere by the Viking landers implies that between one-fourth and one-half of the water in the Martian interior has been outgassed to the surface since the time of the heavy bombardment. This assumes a particular history of volcanism but is considered a reasonable outgassing fraction from other considerations as well. Thus, from our original estimate of the amount of water Mars acquired in the non-local model, we might argue that the water that made it to the surface was equivalent to a layer as little as (of order) 100 meters or as much as 1,000 meters thick. This is, again, within the geologically derived estimates for the amount of water that was at the surface and within the near-surface crust of Mars early in the history of the planet.

The different hypotheses for accretion of water predict a wide range of starting amounts of water for Mars—but consistent with the important role water played at the surface of Mars early in the planet's history. The nonlocal model implies that the amount of water Earth versus Mars received early in their histories depended very much on the particular details of the early solar system: the location of Jupiter, the abundance of rocky and icy solids, and other factors.

12.8 Future Exploration of Mars

The United States and Europe have ambitious plans to continue exploration of the Martian surface from landed spacecraft and from orbit. The goal of these explorations is to understand the following:

- How much water existed on Mars in the past, and how much was liquid?
- What was the climate like on Mars during the time of the formation of the valley networks?
- Did life form on Mars in the past?
- How did Mars evolve toward its current state of cold dryness?
- Are there sources of liquid water present today in the Martian crust?
- Does life exist today on Mars?

Exploration plans change often on the basis of both technical and financial considerations and new discoveries. To maintain its flexibility, NASA has constructed its plans for the exploration of Mars around four different pathways, each of which focuses on a slightly different emphasis in understanding the history of Mars. A chart illustrating the

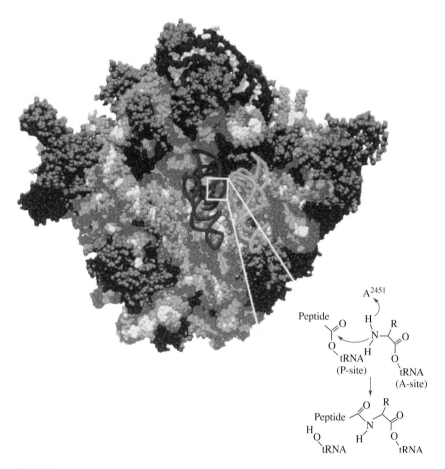

FIGURE 8.2 Structure of a major portion of a ribosome, illustrating the importance of RNA. Color code: proteins in purple, rRNA groups in orange and white, and burgundy and white, tRNA in green and red. Inset is the peptidyl transfer mechanism catalyzed by RNA.

Color Plate 10

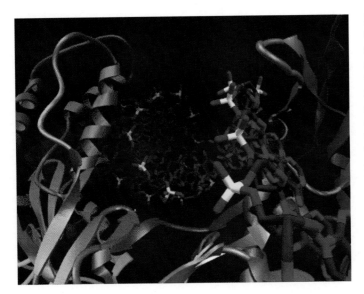

FIGURE 8.5 Structure of the active site of the enzyme reverse transcriptase, showing the various secondary structures that in turn make up the tertiary structure of the catalytic site of the protein (Chapter 4). There are four different regions within this site. Nucleotide templates and products are shown within the catalytic site as yellow-red-black and blue-black segments.

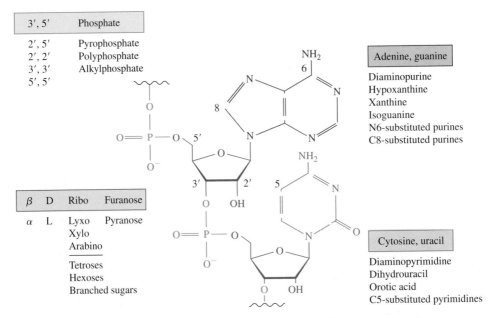

FIGURE 8.7 Basic components of the RNA molecule and related but alternative molecules that could have been incorporated instead to make a different polymer. The molecules actually used in RNA are boxed with border colors keyed to the chemical structure diagram. Ribofuranose is ribose; furanose refers to the fact that it is a five-member heterocyclic ring sugar (Chapter 4). Pyranose sugars have six-member heterocyclic rings.

Color Plate 11

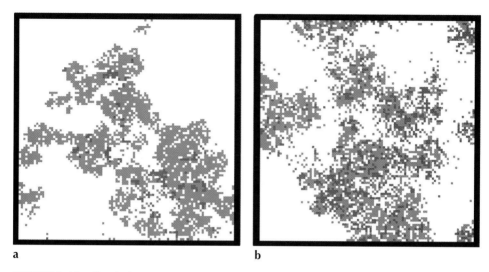

FIGURE 8.11 Simulation of the bootstrapping of a system of catalytic replicating molecules toward larger and more complex polymers. Polymers composed of at least 25 units ("altruists") are shown in blue, with smaller "parasites" depicted in purple. (a) Panel shows the situation early in the evolution, (b) panel much later.

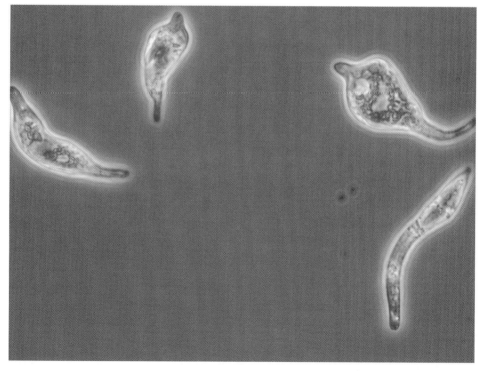

FIGURE 10.3 Extremophile organisms alive in the Rio Tinto in Spain, portions of which have a pH of 0.

Color Plate 12

a

FIGURE 12.3 The topography of Mars based on the Mars Global Surveyor (MGS) MOLA instrument, shown to 2 km spatial resolution. Three hemispheres are shown, emphasizing (a) the northern plains, (b) the southern highlands, and (c) the Tharsis plateau. The color shading shows altitude ranges.

Color Plate 13

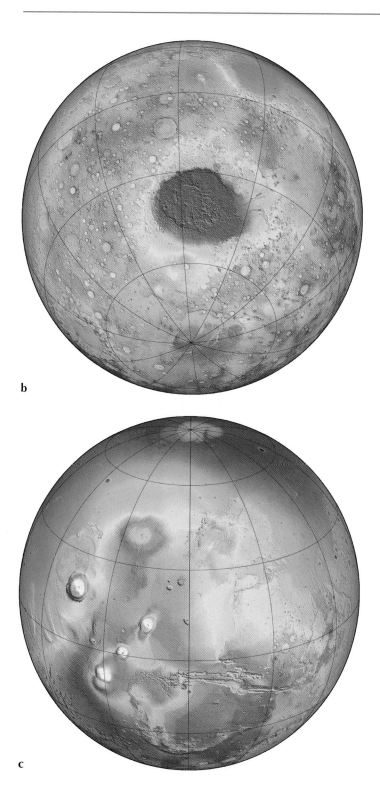

b

c

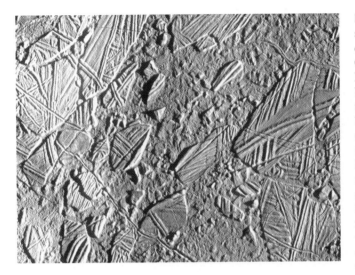

FIGURE 13.2 Galileo spacecraft image of thin, disrupted ice crust in the Conamara region of Europa. The white and blue colors outline areas that have been blanketed by a fine dust of ice particles ejected at the time of formation of the crater Pwyll some 1,000 km to the south. The unblanketed surface has a reddish brown color, possibly from mineral contaminants exuded from a subsurface ocean. The smallest detail that can be seen in the image is 100 meters across, and the terrain imaged is 70 km by 30 km in area.

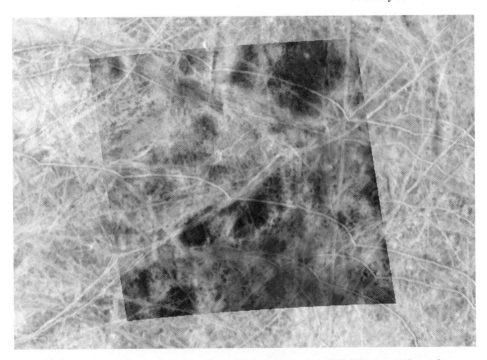

FIGURE 13.8 Data from the near-infrared mapping spectrometer (NIMS) presented as color-coded compositional differences across one of the most geologically active ("youthful") regions of Europa. The area with the NIMS color data superimposed is 400 km by 400 km in size. Based on analysis of the spectra that provided the color differences, blue is coded as clean ice, and red is material other than water ice, either sulfur compounds or mineral salts (the NIMS spectra cannot distinguish between the two). Intermediate colors represent a mixtures of these two endmembers.

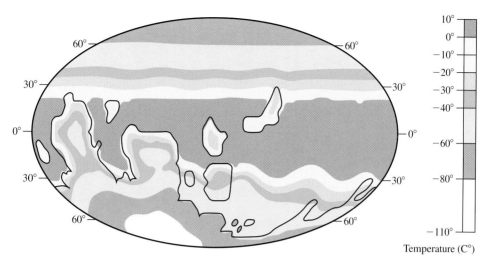

FIGURE 16.7 Result of a climate simulation for the Neoproterozoic Snowball Earth, showing the annually averaged temperature over the Earth. Continental boundaries are drawn. Note the region near the equator, which would remain liquid for most of the year.

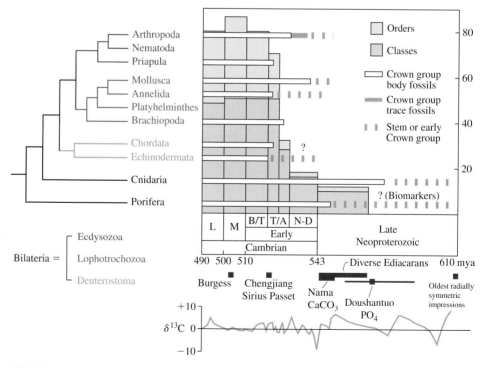

FIGURE 16.8 The timing of the sudden increase in animal diversity relative to climate events. The carbon isotopic record, showing dips at cold times, is at bottom. The sediments in which key animal fossils have been found are given in red. The first appearances of representatives (crown groups) of the modern phyla are shown, white bars representing body fossils and gray trace fossils. The phylogeny of basic body types is shown at left.

Color Plate 16

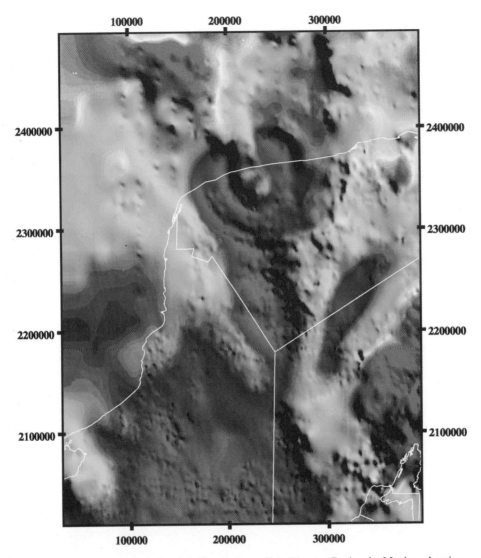

FIGURE 16.9 Map of gravitational field variations off the Yucatan Peninsula, Mexico, showing the circular feature identified as the Chicxulub crater. Blue and purple areas are lower in elevation than the mean; red areas are higher.

TABLE 12.2   Four Pathways for the Exploration of Mars

Pathway	Lines of Scientific Enquiry

Search for Evidence of Past Life	• If science from current missions indicates ancient Mars was wet and warm for extended period(s) of time, then
	  – Locating and analyzing water-lain sedimentary rock is primary goal.
	  – Pathway includes search for evidence of past life.

Explore Hydrothermal Habitats	• If exploration discovers hydrothermal deposits (active or fossil), then
	  – Probability of hydrothermal regions being discovered is potentially high.
	  – Hydrothermal habitats are focus of second decade of Mars exploration.
	  – Potential for discovery of evidence of past and present life is greatly improved.

Search for Present Life	• If potentially habitable sites are discovered on Mars, then
	  – Search for life at active hydrothermal deposits or polar margins (wet zones?).
	  – Path would be taken *only* following a discovery that revolutionizes our understanding of the potential of Mars to harbor present life.
	  – Mars Science Laboratory with mobility is included as the most reliable, means of detecting life under most circumstances.

Explore Evolution of Mars	• If current exploration does not find evidence of past or present liquid water environments (at least long-lived), then
	  – Determine the loss mechanisms and sinks for water and carbon dioxide over time.
	  – Determine why the terrestrial planets evolved differently, much more so than we had thought (i.e., how to explain layered sequences on Mars without water).
	  – Determining whether the initial conditions on Venus, Earth, and Mars were similar or very different.

four pathways is reproduced as Table 12.2. The exploration plans may shift from one pathway to another, depending on the nature of the discoveries of ongoing missions.

Regardless of which pathway is selected, a mixture of orbiters and landers will carry a range of experiments to Mars. Techniques and corresponding instruments for the identification and study of past and present life on Mars will be described in Chapter 14. The types of experiments that will likely be on orbiters include the following:

1. *Orbiting radar.* The global mapping of ground ice and, if it exists, deeper subsurface liquid water will be a primary input into models of the amount of water existing today at the surface of Mars. Analysis of polar ice deposits with orbiting radar will

help test the idea that polar ices trap a variety of other materials. The extent and nature of regions where water is sequestered in the crust will serve as an important indicator of how Mars lost its water over time.

2. *Very high resolution orbital imaging.* Very high resolution images showing features less than a meter across will be used in conjunction with spectroscopy to identify promising sites from orbit for landers. The capability to assist landers was demonstrated by the Mars Orbiter Camera aboard Mars Global Surveyor, which returned images that provided guidance on where the outcrops were located in Opportunity's roving area in Meridiani Planum. High-resolution imaging can be used to more closely study the gulley systems to ascertain whether they are presently active, as well as to search for frost patches at cliff edges and other clues to how they form. Combined with thermal infrared data from orbiters that indicate local surface temperatures, the nature of the gullies and whether they are carved by water—as well as whether they operate on obliquity or seasonal timescales—can be constrained. Furthermore, the high-resolution maps of regions where valley networks occur, where they flow into the northern basin, and their morphologies will better constrain their history as Mars lost its water.

3. *Advanced, high-resolution spectrometers.* The search for minerals and elements that indicate the past presence of liquid water remains a primary part of the orbital reconnaissance necessary to place landers in interesting locales. The detection of sulfur-bearing compounds in the Vallis Marineris canyon system of Mars by the near-infrared spectrometer on the Mars Express Orbiter and the thermal infrared detection of hematite at the surface by Mars Odyssey illustrate the utility of these techniques. The ability to look at locales on the surface in finer detail is essential to using this information to pick future landing sites but requires returning much more data to the Earth than has previously been possible. The Mars 2005 orbiter will be the first to have this new generation of spectrometers and imagers, along with a powerful communication system capable of returning more data from these experiments than has been returned from all other planetary probes to date.

4. *Atmospheric infrared spectroscopy.* The tracking of water seasonally across Mars as it moves from surface to atmospheric reservoirs contributes to an understanding of how much remains in the upper crust of Mars. An advanced infrared spectrometer designed to sample the Martian atmosphere finally flew in 2003, on the European Mars Express Orbiter, after two others were lost on failed U.S. missions in 1992 and 1999. Such observations will likely be important to pursue over several missions; Hubble Space Telescope and Earth-based telescope tracking of Mars suggests that there are changes in the Martian climate state over decades.

5. *Ultraviolet spectrometers and charged particle detectors.* Loss of water from the Martian atmosphere over its history can be quantified by looking at the interaction today of the upper atmosphere of Mars with the solar wind. Devices to measure the charged particles directly, as well as spectrometers to measure atoms and ions, can be flown on orbiting spacecraft to accomplish this task. The Japanese mission Nozomi was designed to accomplish some of this science, but unfortunately the spacecraft was lost in 2003 before it could begin its mission. Future orbiters will be needed to monitor the upper atmosphere.

On the surface of Mars, a very wide variety of instruments—some not yet conceived of or developed—may be deployed for chemical, geological, and biological analyses. How these are used depends on whether the lander is fixed, mobile, or equipped to return samples to Earth:

1. *Rovers*. Roving packages can perform a range of tasks, from close-up geological investigation of a site to the search for life on site and through retrieval of a sample for return to Earth. The Mars Exploration Rovers Spirit and Opportunity covered about 10 km^2 each of ground around their landing sites. Future larger rovers will need to be capable of covering at least 10 times that area in order to thoroughly explore the geology of a particular region or to have a good chance of finding special places where past or present water has created a habitable environment. While Opportunity was lucky in landing very close to key outcrops, Spirit was less so—it had to drive kilometers to find outcrops. Identification of mineral alteration by water at landing sites other than that of Opportunity will allow a global map to be drawn of where sources of standing water occurred on Mars and for how long. The next generation of rovers, capable of traveling more than 10 kilometers across the surface and carrying large scientific payloads of diverse instruments, is called the Mars Science Laboratory and will first be landed on the Martian surface in 2010.

2. *Polar landers*. The high latitudes of Mars are the places where water and possibly organics trapped therein are most accessible. Because the obliquity of Mars has varied in a chaotic fashion over time (Chapter 11), near-polar sites may have sustained habitable environments of liquid water in the recent past. If volcanism is not the source of the atmospheric methane hinted at by the Mars Express and ground-based spectrometers, it could have its source in the water ice. Methane is trapped very effectively in what is called "clathrate hydrate," a phase of water ice that is widespread in the shallow seafloors of the Earth where organisms generate methane in contact with seawater. The high latitudes are a part of Mars very different from elsewhere in ways that affect the design of landers, but they are crucially important to explore after the discovery by the Mars Odyssey Gamma Ray Spectrometer of vast fields of permafrost. The first surface exploration of the permafrost regions of the northern hemisphere will be undertaken by NASA's Project Phoenix, which will land in a high-latitude site in 2008 and drill into the surface for water and organics. The identification by Project Phoenix of water, methane, and other species trapped in the ice will be important for understanding how water is distributed today across the surface, as well as a potential indication (albeit weak and indirect) of past or present life.

3. *Sample return*. Even the Mars Science Laboratories, as powerful as they are envisioned to be, will not be able to carry all of the tools necessary to fully analyze rock samples for extant or extinct life—nor will they be able to fully characterize the age and mineral structure of rock samples. Both these tasks require returning rocks to the Earth. Although we have meteorite samples that originated from Mars, a fully documented sample from a preselected site is of much higher value than a rock that came randomly from an unknown part of Mars. One of the biggest challenges of sample return is the issue of possibly contaminating the Earth with Martian life—a

concern even though Earth and Mars have been exchanging meteorites (and mostly one way, from Mars to Earth) for billions of years. Both this "reverse" contamination problem and the "forward" contamination concern of bringing Earth microbes to Mars will be with us at Mars and beyond as we search the solar system for life.

QUESTIONS AND PROBLEMS

1. Compare images of the Martian valley networks with those of typical river systems (including tributaries) on the Earth. What are the primary differences between the two?

2. If the Grand Canyon of the Colorado River were placed on Mars, how would it compare in size and length with the various side canyons in Vallis Marineris and to the valley networks? Which of the types of fluvial features on Mars best describes the Grand Canyon, if any?

3. Use images from Magellan and Mars Global Surveyor to compare erosion on Venus and Mars. What is absent from Venus as a primary erosive agent? How does that manifest itself in the images of Venus? Now find aerial photos of desert mountains on the Earth (e.g., those in the southwestern United States). Looking at the image of Mohave Crater on Mars (Figure 12.19), in what ways do the erosional features there resemble those in the desert mountains on Earth? How do they differ?

4. Compare the global topographic maps of Earth, Venus (Chapter 11), and Mars (this chapter). Do all three planets exhibit a global asymmetry in their geology (e.g., from one hemisphere to another)? Is this asymmetry in latitude (north–south) or longitude (east–west)? Speculate on why planets might or might not have such asymmetries due to external causes or purely random effects. Now look at a map of the Moon, and speculate on the origin of its evident asymmetry between the sides facing toward and away from Earth. What might be the cause?

5. Is there any apparent evidence for global-scale tectonic features on Mars that might be due to plate tectonics? Argue for and against the notion that Vallis Marineris is akin to an ocean trench and that Tharsis is a continental platform.

6. Do a Web or literature search for cross bedding in sedimentary rocks on Earth. How do the fine layers compare with those in the Eagle crater site on Mars? What were the terrestrial environments in which the cross beds you found were laid down? Do any of these patterns occur in layers of volcanic dust laid down during eruptions?

7. Do a literature search to find a list of the typical minerals that occur in an evaporate sequence on the Earth. What is the usual variation in elements associated with these salts from the top to the bottom of such a sequence?

8. Radar systems to be used in this decade on Mars are akin to those used to explore the driest areas of the Earth, where the radio signals can penetrate a great distance below the surface. (In wet regions the water attenuates the radio waves before they can get very far below the surface.) Find studies of the Sahara and other deserts using ground penetrating radar. What kinds of geological features have been found?

9. What must the air pressure be at the Martian surface for liquid water to be stable today, without boiling away? Is this larger or smaller than the present-day surface

pressure of carbon dioxide? Can liquid carbon dioxide exist at the surface of Mars today? (Hint: What pressure is required at the triple point of carbon dioxide?)

10. The spiral troughs in the polar regions represent a beautiful manifestation of the deposition of dust and frost, modulated by the seasons and the changing obliquity of the Martian spin axis. Do a literature search on mechanisms for the formation of these polar spirals. Compare the various mechanisms offered. What do the different mechanisms imply about the obliquity tilt? Do the mechanisms offered, or the spiral patterns themselves, suggest particular strategies for sampling ices in the polar regions and adjacent high-latitude regions of Mars?

SUGGESTED READINGS

A classic reference, accessible to undergraduates, is M. H. Carr, *Water on Mars* (1996, Oxford University Press, New York). More comprehensive, but going out of date, is H. H. Kieffer, B. M. Jakosky, C. W. Snyder, and M. S. Matthews, eds., *Mars* (1992, University of Arizona Press, Tucson). A set of review articles written for nonspecialist scientists on aspects of the current understanding of Mars is in *Nature*, **412**, 207–253 (2001). A detailed, specialist set of articles on the results on the results of the extremely successful Mars Global Surveyor mission is given in the *Journal of Geophysical Research*, **106**, 23289–23929 (2001).

REFERENCES

Abe, Y., Drake, M., Ohtani, E., Okuchi, T., and Righter, K. (2001). "Water in the Early Earth." In *Origin of the Earth and the Moon*, K. Righter and R. Canup, eds., University of Arizona Press, Tucson, pp. 413–433.

Baker, V. R. (2001). "Water and the Martian Landscape," *Nature*, **412**, 228–236.

Bandfield, J. L., Hamilton, V. E., and Christensen, P. R. (2000). "A Global View of Martian Surface Compositions from MGS-TES," *Sci.*, **287**, 1626–1630.

Boynton, W.V., and Colleagues. (2002). "Distribution of Hydrogen in the Near-Surface of Mars: Evidence for Subsurface Ice Deposits," *Sci.*, **297**, 81–85.

Carr, M. H. (1996). *Water on Mars*, Oxford University Press, New York.

Feldman, W. C., and Colleagues. (2002). "Global Distribution of Neutrons from Mars: Results from Mars Odyssey," *Sci.*, **297**, 75–78.

Jakosky, B. M., and Phillips, R. J. (2001). "Mars' Volatile and Climate History," *Nature*, **412**, 237–244.

Kasting, J. F. (1991). "CO_2 Condensation and the Climate of Early Mars," *Icarus*, **94**, 1–13.

Kerr, R. A. (2004). "A Wet Early Mars Seen in Salty Deposits," *Sci.*, **303**, 1450.

Kieffer, H. H., Jakosky, B. M., Snyder, C. W., Matthews, M. S., and eds. (1992). *Mars*, University of Arizona Press, Tucson.

Leshin, L. A. (2000). "Insights into Martian Water Reservoirs from Analyses of Martian Meteorite QUE94201," *Geophys. Res. Lett.*, **27,** 2017–2020.

Lunine, J. I., Chambers, J., Morbidelli, A. and Leshin, L. A. (2003). "The Origin of Water on Mars," *Icarus*, **165,** 1–8.

Malin, M. C., Carr, M. H., Danielson, G. E., Davies, M. E., Hartmann, W. K., Ingersoll, A. P., James, P. B., Masursky, H., McEwen, A. S., Soderblom, L. A., Thomas, P., Veverka, J., Caplinger, M. A., Ravine, M. A., Soulanille, T. A., and Warren, J. L. (1998). "Early Views of the Martian Surface from the Mars Orbiter Camera of Mars Global Surveyor," *Sci.*, **279,** 1681–1685.

Segura, T. L., Toon, O. B., Colaprete, A., and Zahnle, K. (2002). "Environmental Effects of Large Impacts on Mars," *Sci.*, **298,** 1977–1980.

Smith, D. E., Zuber, M. T., Solomon, S. C., Phillips, R. J., Head, J. W., Garvin, J. B., Banerdt, W. B., Muhleman, D. O., Pettingill, G. H., Neumann, G. A., Lemoine, F. G., Abeshire, J. B., Aharonson, O., Brown, C. D., Hauck, S. A., Ivaniv, A. B., McGovern, P. J., Zwally, H. J., and Duxbury, T. C. (1999). "The Global Topography of Mars and Implications for Surface Evolution," *Sci.*, **284,** 1495–1503.

Zahnle, K. (2001). "Decline and Fall of the Martian Empire," *Nature*, **412,** 209.

Zuber, M. T., Smith, D. E., Solomon, S. C., Abshire, J. B., Afzal, R. S., Aharonson, O., Fishbaugh, K., Ford, P. G., Garvin, J. B., Head, J. W., Ivanov, A. B., Johnson, C. L., Muhleman, D. O., Neumann, G. A., Pettengill, G. H., Phillips, R. J., Sun, X., Zwally, H. J., Banerdy, W. B., and Duxbury, T. C. (1998). "Observations of the North Polar Region of Mars from the Mars Orbiter Laser Altimeter," *Sci.*, **282,** 2053–2060.

CHAPTER 13

Planetary Evolution III
The Significance of Europa and Titan

13.1 Introduction

Beyond Mars, two worlds in the outer solar system beckon as places where life might exist today or where the organic reactions leading to life might be replayed and preserved many times. Jupiter's smallest Galilean satellite, Europa, is suspected to harbor a liquid water ocean under its icy crust, and it is possible that life adapted to a deep ocean is present there. Saturn's largest moon, Titan, is enshrouded in an organic haze produced by the action of sunlight on methane, and, as these haze particles drift downward through the dense atmosphere, some melt, producing mixed liquid–solid surface deposits. Further chemistry occurs on the surface and is of keen astrobiological interest to the extent that it could inform us of the steps leading from simple organic chemistry to autocatalytic reactions and self-organizing structures. In this chapter these two worlds are examined, their astrobiological potential discussed, and near-term exploration plans described. This sets the stage for the next chapter in which detection techniques are considered in detail. Before moving to detailed descriptions of Europa and Titan, some general considerations regarding outer solar system conditions are discussed.

13.2 Setting: The Outer Solar System

The outer solar system is defined as that region beyond the asteroid belt—roughly, 4 AU and outward. Vastly larger than the inner solar system, the outer solar system hosts both the largest and the smallest planets, the largest moons, and debris left from the formation of the planets. The stability of water ice may determine the fundamental differences between the inner and outer solar system. The vapor pressure of water over its ice is such that, during the formation of the planets, condensation of the ice from the gases of the protoplanetary disk (Chapter 6) was possible at and beyond approximately 4 to 5 AU. Early in the history of the disk, temperatures were very high and the stability line of water ice was pushed outward to a larger distance, but as the disk aged and the rate of addition of new material ("accretion") declined, the ice boundary moved within the present orbit of Jupiter.

Inward of this boundary (which is sometime referred to in the planetary literature as the "snowline"), water could only exist as a vapor, bound chemically to the rocks as so-called water of hydration or affixed to (adsorbed onto) dust. Outward of the boundary, water ice is stable. Examination of the list of elements given in Table 3.3 shows that oxygen is about 20 times more abundant by number than the primary rock-forming element, silicon, and 10 times more abundant by mass. Therefore, where water ice condensed out in the early solar system, it greatly enriched the amount of solid material and, hence, increased the speed of planet building and perhaps the size of the objects formed. More details on this process were provided in Chapter 6, particularly with regard to the formation of the giant planets.

The same conditions that made water ice stable in the outer solar system (namely, distance from the Sun and low temperatures) renders habitable environments there a rare commodity. Many solid bodies in the form of planetary moons exist in the outer solar system, but their surfaces are all so cold as to rule out stable liquid water. Even atmospheres are a rare occurrence among the solid bodies of the outer solar system. For example, while Triton and Pluto both have nitrogen atmospheres, these are tenuous layers of gas with pressures 100,000 times lower than sea-level pressure on Earth, that are simply in vapor pressure equilibrium (Chapter 7) with nitrogen ice on the surface; only Titan has a dense atmosphere.

Three additional heat sources besides sunlight potentially are available to warm the interiors and surfaces of the solid bodies of the outer solar system. The first is accretional heating—that gained from impacting bodies as the moons grew. Like the Earth, the larger moons of the solar system are held together and made spherical by the mutual gravitational pull of every bit of rocky and icy material of which they are composed. This is a negative potential energy state (Chapter 2) relative to a zero point obtained by distributing the material, in the form of fine particles, across an arbitrarily large volume of space—a good approximation to the original state of the debris in the protoplanetary disk. Where, then, does the potential energy go as material is gathered in on itself during the growth of a large moon? It must go into the kinetic energy of the infalling material. The infalling motion is, at least initially, relatively ordered. However, as debris impacts the growing moon, the energy of motion is degraded into the random microscopic motion of the material in the outer layers of the object; hence, the thermal energy of the interior of the body increases over time. Some of the added energy is quickly radiated outward as infrared photons. How much is retained in the interior depends on how much of the accreting material is in the form of large bodies (which deposit their kinetic energy on average deep below the surface) and how much is in the form of small bodies (which deposit energy on the surface, much of which is radiated away).

The maximum accretional energy per unit mass of material that a body could acquire is given by the gravitational potential energy $3GM/5R$, where G is the universal gravitational constant, M the mass of the Moon, and R its radius. (The factor of $\frac{3}{5}$ comes from having to sum up each increment of energy added as the body grows and properly accounting for the changing volume relative to the surface area.) If we equate this "specific" (i.e., per unit mass) energy to the gain in specific thermal energy, $c_p \Delta T$, where c_p is the specific heat at constant pressure of the material, then the temperature rise can be hundreds of degrees Kelvin for large bodies such as Europa and Titan. The maximum possible temperature rise is often not realized during accretion because some of the energy is immediately reradiated. But enough is buried in the interior to have been sufficient to melt the water ice early on—though perhaps not maintain it in a liquid state against the slow cooling over time that would cause it to freeze. The accretional heating of Titan almost certainly melted liquid water in the upper layers, and some liquid may yet remain in the interior.

The second heat source is that of radioactive isotopes. In abundances of typical chondritic meteorites (Chapter 5), the most abundant natural radioactivities are those of the isotopes ^{40}Ar, ^{232}Th, ^{235}U, and ^{238}U. These decay with half-lives ranging from 1 billion to 10 billion years, releasing thermal energy, which further heats the rocky bodies that contain them. Assuming compositions for Io and Europa akin to those of carbonaceous meteorites (an assumption for which there is as yet no more specific substitute), concentrations by mass of 780 ppm (potassium), 0.04 ppm (thorium), and 0.013 ppm (uranium) are predicted at present. However, the abundances were larger and the relative abundances different 4.5 billion years ago based upon the decay rates for the isotopes. Integrating over the abundances and the decay rates gives the total thermal energy released by radioactivity over the age of the solar system. Because this energy is released slowly, its effect on the internal temperatures depends strongly on the size of the body (hence the surface area-to-volume ratio, hence the cooling rate) and the composition. Radioactive heating is sufficient to keep ammonia–water compositions partly melted in the deep interiors of the icy-rocky bodies, but probably not sufficient, in the absence of other heat sources, to maintain a liquid water ocean under the ice crust of Europa. (In this chapter, because the term *crust* implies rocky material to most geologists, we refer to the rigid outer layer of these bodies as the *ice crust*.)

The third energy source is tidal heating. The force of gravity, as described in Chapter 2, has a strength that depends on the square of the distance between two bodies. Therefore, the portion of the Earth closest to the Moon gets pulled more by the lunar gravity than does the center of the Earth, and the farthest point of the Earth from the Moon gets pulled the least. This gradient in the strength of the Moon's pull on the Earth (and the equivalent for the gradient in the Earth's pull on the Moon) is called the *tidal force*. There is also a tidal force on the Earth due to the Sun, which is much farther away but also much more massive than the Moon. The tidal forces lead to a distortion of the shape of the Earth: A bulge tends to move around the Earth as the Moon revolves in its orbit and the Earth spins on its own axis. The Earth's oceans in particular slosh back and forth in response to the tidal force and the diurnal rotation of the Earth on its axis, converting the potential energy of the gravitational force of the Moon and the Sun into the kinetic energy of the heaving tides, and then, via friction, down to the microscopic thermal energy of the stirred waters. The conversion of orbital energy to heat is accompanied by a transfer of angular momentum, which moves the Moon outward in its orbit while slowing the spin of the Earth. This is called *tidal dissipation*.

For the large moons of the outer solar system, the tidal dissipation occurs mostly within their own interiors rather than in their parent planets, which are largely gaseous. The massive gravitational fields of the giant planets make for enormous effects on the moons themselves—which, in the case of Io, has transformed that moon into a volcanic wonderland.

The difference in gravity acting on the edge of each moon relative to its center raises a bulge in the solid ice or rock of the moon itself—many meters in the case of Europa and more for Io. The bulge faces the giant planet as the moon orbits, but those orbits are eccentric, not circular. Hence, the moon orbits faster when it is closer to the planet and slower farther away. The bulge does not face directly toward the planet, and it moves back and force as friction slows its response to the changing speed of the moon in its orbit. The frictional movement within the solid moon and sloshing of surface or interior fluids creates heat and dissipates the energy of the orbit.

Left to itself, the frictional dissipation of the gravitational potential energy would make the orbit circular, and tidal heating would cease. Indeed, for Titan, which stands alone as Saturn's massive moon, the small orbit eccentricity sets a strong limit on just how much sloshing there might be in surface or interior liquids. But the Galilean satellites are a family of four massive moons, and the inner three are bound together by eccentric orbits whose periods are simple multiples of each other—Europa's period is just twice Io's, and Ganymede's is twice Europa's. The repeated mutual tugs at given places in the orbits, as the moons make their close approaches to each other, reinforce and maintain the eccentricities—and the tidal heating. The source of energy (the gravitational pull of Jupiter) is enormous, and this dissipative dance likely has played out for billions of years.

The tidal heating is a "higher-order" effect of the gravity of Jupiter, that is, it depends on the difference in gravitational pull from one end of a moon to the other. Hence this effect, a derivative of the main force of gravity, diminishes with distance far more steeply than r^{-2}, where r is the distance from Jupiter to the particular moon. The strength of the tidal force varies as the gradient (change with distance) of the force, hence as r^{-3}. But the heating is derived from the differential movement of the bulge on the moon and, indeed, more generally on the general bulging and twisting of the moon as it moves around its orbit. This effect dies away much more steeply than the cube of the distance and depends on where in the interior the heating takes place. Io is intensely heated to the point where the silicates themselves melt, generating a profusion of volcanic eruptions continuously across the surface. Ganymede seems hardly heated at all. And Europa, the subject of Section 13.3, might have sufficient tidal heating to ensure maintenance of a liquid water layer beneath its crust. How thick or thin is the ice crust depends, in part, on the magnitude of the tidal heating. In turn, the tidal bulge (i.e., the periodic distortion of Europa's shape as it moves around Jupiter) will turn out to be a crucial measurement to find out just how close to the icy surface such a putative ocean lies.

13.3 Europa

Galileo Galilei discovered Europa with the newly invented telescope in 1610. Little else was known about this moon, especially about its differences from the other Galilean satellites, for more than three and a half centuries. While some Earth-based observations (radar, spectroscopy) hinted at the unusual nature of Europa's surface, in particular showing that large areas are clean water ice, the most dramatic indications that Europa might be an active world came from the Voyager 1 and 2 flybys of the Jovian system in 1979. Europa's extraordinarily bright surface seemed to be mostly smooth, apparently devoid of craters. This simple "cue ball" geology was largely the result of a lack of high-resolution

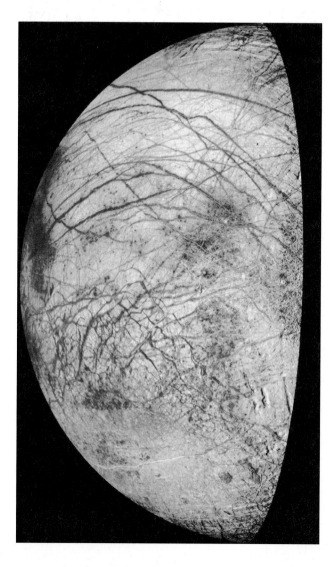

FIGURE 13.1 Global-scale view of Europa from the Galileo orbiter. The smallest details in this image are 4 km across.

images; when the Galileo orbiting spacecraft—named after the famous Italian scientist—began imaging Europa in the mid-1990s, a much more complicated surface emerged (Figure 13.1). However, the detailed Galileo images suggested even more strongly an active, youthful surface and hinted at the presence of a liquid water ocean existing today or in the recent past beneath the surface (Figure 13.2, see Color Plate 14).

Surface Features

A complicated surface of ridges and fractures, strange patterns of blocks that look like ice rafts once adrift in a now-frozen sea, craters and dark pits, and domes that might be sites of eruptions characterize the surface of Europa. Systems of fractures, many composed of multiple ridges and troughs, extend over the surface of Europa. They overlay each other

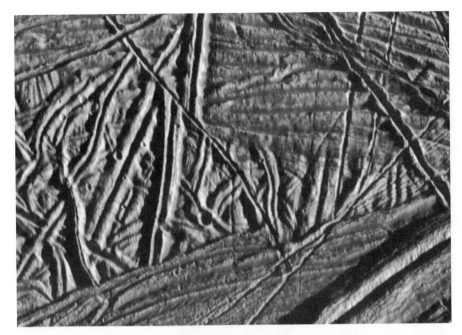

FIGURE 13.3 Galileo spacecraft image of a complex sequence of ridges in a region about 12 km by 15 km in size on Europa. Different episodes of fracturing can be seen as ridges overlay each other, and some ridges have been split and slide apart, indicating the crust has moved. The smallest detail in the image is 40 meters in size.

many times, and some fractures have slid apart associated with movement of the Europan crust (Figure 13.3). A large pattern of semicircular (or "cycloidal") fractures (Figure 13.4) appear to be a response of a relatively thin crust to the tidal stresses exerted by Jupiter, tearing the crust apart and then bringing it back together as the stress is released. Other areas look like jigsaw puzzle pieces of ice moved apart, ready to be put back together by the viewer's eye (Figure 13.2). Such geology suggests, if not requires, that a liquid layer of water underlies the solid ice crust of Europa (the only other alternative being a warm, very pliable or ductile layer of ice itself).

Were such features formed yesterday, and are they in the process of forming today? The answer is ambiguous. Europa does have craters—a very few large (Figure 13.5) but most quite small. The rarity of craters implies a relatively youthful age for the surface of this satellite compared with the heavily cratered neighboring moons Ganymede and Callisto. The Europan surface must be active enough to erase craters by fracturing and flooding the craters with liquid water. How often this must happen to account for the paucity of craters is uncertain because the modern-day cratering rate is poorly known. Some geologists have argued that the cycloidal ridges imply that the crust is opening and closing on the timescale of Europa's orbit around the Sun; others point out that a surface with few craters could still be many millions of years old.

Active resurfacing is not seen on Europa, but unusual domes and pits (Figure 13.6, see page 408) suggest that warm water has tried (and in some cases succeeded) in pushing

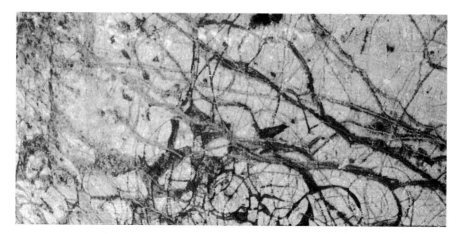

FIGURE 13.4 Galileo spacecraft image of a wide swath of Europan terrain, showing circular features that appear to be fractures in the crust formed by the stresses of Jupiter's tides. The smallest features visible are 1.6 km in size.

FIGURE 13.5 Pwyll crater on Europa as photographed by the Galileo spacecraft. The crater is 26 km in diameter, shows a collection of central peaks raised by the impact, and may be darker than the surroundings due to dark material excavated from deeper in the crust or from the impactor itself. The smallest feature discernable in the image is 1 km across.

through the relative rigid, icy crust. Volcanoes on Earth are the result of silicate melts, or magmas, extruded to the surface as lavas. On Europa, the "working fluid" of this geologic activity is water ice. Its movement in analogy to terrestrial volcanism is called "cryovolcanism." The origin of this activity lies in Europa's interior, which the Galileo orbiter was able to investigate indirectly.

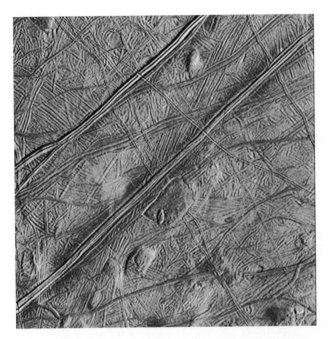

FIGURE 13.6 This complex Europan terrain, covering 80 km by 95 km, is illuminated by the Sun from the upper left and shows domes. The origin of the domes is unclear, but one mechanism is the upwelling of warm water from a liquid layer into the cold icy crust. The smallest features visible are 360 meters across.

The Interior of Europa: Structure and Magnetic Field

The Earth's gravity field is not uniform, despite our perception that it always pulls us toward the center of our home planet. But hang a weight on a string (a "plumb bob") and take it to the foothills of a massive mountain range. Careful measurement will reveal that the plumb bob is not pointed directly down at the ground but is shifted ever so slightly toward the mountains because of the gravitational pull of the huge granite masses piled high into the local sky. Indeed, such measurements have been made, and now satellites orbiting the Earth can be accurately tracked so that variations in their orbital path reveal variations in density associated with the topography of the Earth. The same is true for spacecraft orbiting the Earth's moon; early robotic orbiters sped up and then slowed down in their orbits in response to the presence below of dense masses of basaltic lavas filling the lunar "mare" basins.

Flyby spacecraft have a difficult time compared to orbiters detecting smaller features like mountains and basaltic basins, but they can infer the extent to which denser material is concentrated toward the center of a planet or moon versus being uniformly distributed throughout the volume of the object. How material of various densities (metal, rock, ice) is arranged in the interior affects the shape of the gravitational field the Galileo spacecraft experienced as it flew by Europa. The spacecraft had multiple opportunities to perform this measurement and produced data showing Europa to be strongly "differentiated," with the densest material at the center and least dense at the top, as shown in Figure 13.7.

From the earlier Voyager flybys it was possible to measure a mass for Europa; then knowing the size of the moon from images, a density of 2.94 g/cm^3 was derived; that of

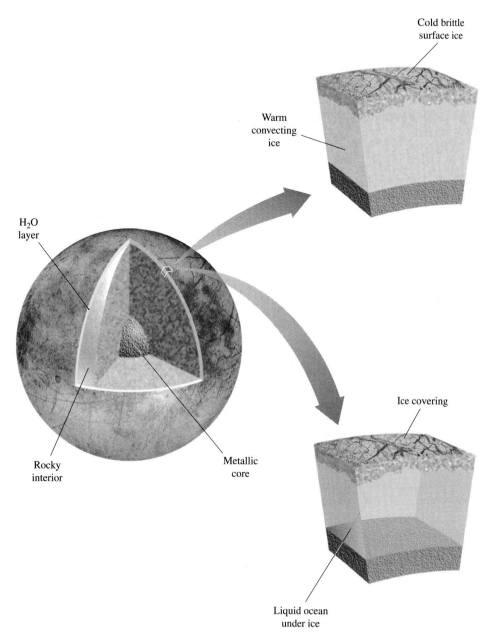

FIGURE 13.7 Possible internal structures for Europa. The left side shows a cutaway of Europa, revealing the thin outer layer of water (ice or liquid) underlain by a thick mantle of silicates and a core of iron–nickel alloyed with other elements. This structure was revealed by measurements of the Europan gravitational field made by the Galileo spacecraft. What is not known is how much of the water layer is today ice or liquid. On the right are two extreme alternatives for Europa's water crust—all ice or a thin ice layer underlain by a deep ocean.

Earth's Moon is 3.4 g/cm^3. It was thus known that the icy surface of Europa is the top of a water region hundreds of kilometers thick—but with no knowledge of how much (if any) might be liquid. The bulk of Europa, conversely, is like Earth's moon—a rocky body with some admixture of iron (indeed, Europa's rocky interior is not much smaller in size than our Moon). The shape of the gravitational field determined by the Galileo flybys required that Europa's interior be further subdivided into a metallic core (probably mostly iron, which is abundant in the cosmos, with an admixture of the also-abundant nickel) and a rocky mantle. This structure, well ordered according to density (that of ice is 0.92 to 1 g/cm^3, silicates 2.5 to 3.5 g/cm^3, and iron 5 to 8 g/cm^3, depending on impurities) suggests that Europa was once hot enough for bound water and iron to separate from the silicates and move up and down, respectively.

The Galileo spacecraft was also equipped with a magnetometer, a device capable of measuring the strength and, through multiple flybys, the direction of a magnetic field around planets and moons. In its tour of the Jupiter system, the spacecraft was always within a region where the magnetic field of Jupiter is strong, and the field's structure and strength was measured extensively. When the spacecraft made close approaches to Europa, it did not detect an intrinsic magnetic field around that moon but rather an induced magnetic field that can be thought of as a local alteration of Jupiter's field. The nature of the perturbation is very similar to what happens when a large, electrically conducting body is immersed in an external magnetic field; electric currents generate "induced" magnetic fields that partially neutralize the ambient field. Flybys at different angles relative to the pole of Europa itself allowed the data from the magnetometer to "map" the geometry of this perturbation and determine that the conducting region was the outer layer of Europa. The required conductivity could not be met by water ice; liquid water with dissolved salts, akin to our own terrestrial ocean, is required. Because the response of Jupiter's magnetic field to the conducting fluid is immediate, and would die away if the liquid water layer froze, the measurement provides a window into a liquid water layer within Europa existing *today*—not a fossil of millions of years ago.

These observations were not done by the magnetometer alone—it was necessary to use other instruments to measure the background fluxes of electrons, protons, and other charged particles in the magnetosphere (region of magnetic influence) of Jupiter to properly calibrate the field variations. The result is extraordinary—a real-time sensing of a layer of liquid water beneath the ice of Europa. Salts are a plausible solute from a chemical standpoint, as the base of the liquid water layer is in direct contact with the silicate mantle. Pressures at the base are not quite high enough to allow for ultradense layers of ice to form there. (Such phases are important in Titan, which we turn to in Section 13.4.)

While exciting, the magnetic results do not allow us to examine the ocean in any direct sense. Might there be places on the surface where the ocean has leaked out onto the surface? Mapping of the composition of Europa's surface provides some hints.

Interior Contaminants at Europa's Surface

The instrument complement on the Galileo spacecraft included the near-infrared mapping spectrometer (NIMS) capable of taking spectra in an imaging mode, where each picture element is a spectrum. The wavelength region of the near-infrared provides diagnostic spectral features of many compounds. Much of the surface mapped by NIMS is very clean water ice, but in some places, other compounds are present (Figure 13.8, see Color Plate 14). Examination of the spectral bands (Chapter 3) of water ice on Europa reveals that they are distorted relative to the shape of pure, water ice bands. That distortion could be the result of the presence of hydrated minerals (Chapter 12) in place of pure water ice. Such minerals include evaporite salts or clays. Another spectrometer on Galileo, working in the ultraviolet, found that the geologically most active areas of Europa are actually dark and then lighten as radiation from the Jovian magnetosphere bombards the surface. This behavior is consistent with the presence of sulfur or sulfur dioxide.

What are we to make of this menagerie of compounds? The sulfur or sulfur dioxide could be the result of contamination of Europa from nearby Io, which has large areas of sulfur compounds and active volcanoes powered by strong tidal heating. The salts, however, are more difficult to argue as a contaminant from the space environment. Instead, they are most plausibly produced by the interaction of liquid water below the Europan ice crust with the silicate mantle that lies beneath the water layer. The squeezing and cracking of the surface of Europa by tidal stresses might, at times and in places, cause water to be extruded to the surface, accompanied by salts and carbonates. How long such materials might survive unblemished on a surface bombarded by enormous quantities of particle radiation, which break chemical bonds and alter the spectral signatures of the fresh substances, is an active area of laboratory research today.

The correlation of salts and carbonates with geologically active areas of the surface is suggestive of a resurfacing process that has occurred recently—and perhaps is ongoing to the present day. The Galileo NIMS instrument was limited in its ability to resolve small features because, in each picture element, an entire spectrum was returned to Earth. The limitations of data that could be stored on Galileo and then returned to Earth thus prevented the NIMS data from having the detail that the camera provided; hence, it cannot be determined from the NIMS data whether individual fractures on the surface are the source of the salts and the carbonates. That will have to await other missions. Also important to note, from an astrobiological point of view, is that NIMS failed to detect organic molecules on the surface. This does not necessarily mean that organics do not exist in the liquid layer beneath, but rather that they are fragile: Any that reach the surface would have their carbon-hydrogen bonds broken readily by the particle radiation and the hydrogen would escape.

Given the inferred presence of a liquid water layer and the suggestive evidence that ocean material has been or is being extruded onto the surface, it is tempting to suggest that the next mission to Europa should land and drill. However, contradictory geologic evidence for the thickness of the Europan ice crust provides a cautionary note.

Thickness of the Europan Ice Crust

Gravity and magnetic field data from Galileo do not constrain the thickness of the ice crust of Europa. Nor do the calculations of tidal dissipation. Although tidal dissipation is a less powerful source of thermal energy than radiogenic, tidal dissipation is localized. The thermal energy input due to the flexing of Europa is preferentially localized to areas of large flexure, such as the interface between a liquid water ocean and an ice crust above it. Thus, if accretional and radiogenic heating initially melted a portion of the icy crust of Europa early in its history, tidal dissipation could maintain it. However, the details of the dissipation and resulting deposition of thermal energy could be so variable—depending on initial conditions of the ice crust, the variations in its thickness, and its material properties—that we cannot simply predict the thickness at which the ice crust is maintained. This may seem surprising to readers who quite reasonably expect that the material properties of water ice (such as its stiffness and shear strength) should be known, but uncertainties arise from the low temperatures, the presence of impurities, and the microscopic properties such as crystal size and presence of cracks. A consensus after two decades of modeling is that tidal dissipation depends too heavily on small details to allow us to establish the ice thickness by theory.

It is therefore necessary to rely on geologic data. The ice raft features of Figure 13.2 constrain the thickness of the rigid ice crust, at the time the blocks broke free, to be less than or equal to the dimensional scale of the blocks themselves. The crustal thickness could have been virtually zero if the blocks were floating in liquid with only a thin ice crust, but it could not have been many kilometers thick; otherwise, the blocks could not have broken free.

A stronger constraint on the thickness of the ice crust comes from the Pwyll crater and its central peaks seen in Figure 13.5. The formation of craters such as Pwyll involves a large impactor gouging out a volume of the ice crust. The enormous energies released in the impact create sufficient shock that the crust behaves temporarily as a liquid, and waves are set up that spread and then rebound back toward the crater center. The resulting central peak is a well-known phenomenon on the Earth's Moon. The fact that Pwyll has one is not surprising, but its persistence is what is of interest. To support such mountains during and immediately after formation requires that the solid ice crust into which the impact occurred be of the same scale as the excavated crater diameter—otherwise the topographic features typical of such a crater, including the central peak, would relax quickly since the liquid water beneath cannot support such a form. Hence, in the Pwyll region the solid ice crust must be kilometers thick.

On the other hand, the cycloidal features of Figure 13.4 are best explained as a result of tidal flexing along the orbit of Europa. Then the ice crust in those regions must be quite thin, perhaps tens of meters or less. The difference between kilometers and meters of ice is the difference between a plausible and implausible robotic mission to drill through the ice or to seek recently extruded samples of the Europan ocean. Further, the most interesting places to drill might not be the safest landing sites (Figure 13.9).

A Mission to Orbit Europa

The solution to the problem of determining the thickness of the ice crust is to deploy a spacecraft into orbit around Europa to make the necessary geophysical measurements. Key among those measurements is the changing shape of Europa as it makes its noncircular orbit around Jupiter. As Europa is subjected to the varying tidal force along this path, its shape changes. How much it changes depends on the stiffness of the weakest part—its ice crust and liquid water layer. A thin ice crust overlying a deep ocean will lead to a tidal bulge of 30 meters; a completely solid ice layer will distort only 1 meter. Furthermore, if

FIGURE 13.9 This oblique view of Europa obtained by the Galileo orbiter during a very close pass in 1997 shows a region 2 km across, with details visible at the bottom of the image as small as 12 meters across. These "badlands" of Europa are part rugged and part "painted" dark and light by varying surface composition. Where to put a lander with a drill in such alien and unfamiliar terrain is a daunting question.

the ice crust varies in thickness (putatively, from kilometers in the Pwyll region to meters at the cycloid features), the size of the bulge will vary over the surface and, hence, with time along the orbit of Europa.

Is it possible to detect time-dependent bulging of the ice crust of Europa to within a meter? A very precise way to determine distance from a spacecraft to the local surface beneath is with a laser altimeter, which emits, and then receives back, shaped pulses whose travel time can be carefully determined. This, combined with accurate tracking of the spacecraft in orbit around Europa using Doppler frequency shifts in the signal received on Earth from the spacecraft, allows the position of the spacecraft relative to surface features to be known within meters. This feat was demonstrated at Mars with the laser altimeter on, and precise radio tracking of, the Mars Global Surveyor spacecraft. To distinguish between a tidal bulge and a surface feature on Europa requires making many passes over given regions of the moon over time, and doing so from different angles so that a grid of crossover points on the surface is created over many weeks of spacecraft tracking (Figure 13.10). As Europa flexes periodically, and the orbiting spacecraft passes again and again over large parts of the surface as it interrogates those regions with the laser altimeter, a map of the size of the time-varying bulge over different parts of the surface (and hence the variation in crustal thickness) can be determined.

A mission so equipped to orbit Europa should also carry cameras to complete the mapping of the surface of Europa that was begun by the Galileo orbiter. An advanced spectrometer designed to return compositional information regarding smaller features on the

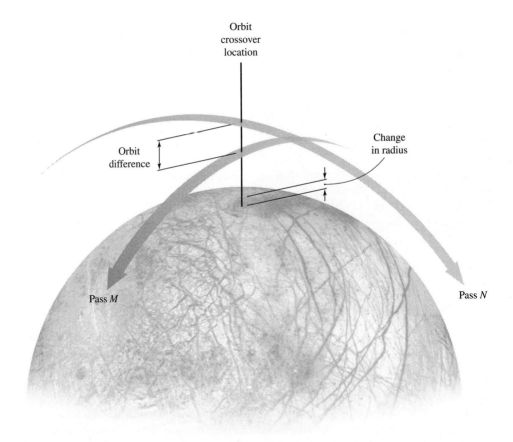

FIGURE 13.10 Gravity and altimetry measurements with a Europa orbiter. By making passes over the surface of Europa at many different orbit angles or inclinations (passes M and N being two arbitrary ones), measuring distance to the surface continuously with its laser altimeter, Europa Orbiter will establish the changing shape of Europa as it orbits Jupiter. Details in that changing shape will allow the variations in crustal thickness to be determined.

surface would also allow examination of the areas of thinnest crust for evidence of extrusion of salts, carbonates, and other potential solutes from the liquid water ocean. Such a mission cannot last long because the spacecraft—trapped in orbit around Europa, which is deep in the Jovian magnetosphere (in contrast to Galileo, which dipped in and out of the near-Jupiter region on a highly elliptical path)—will be damaged quickly by radiation. Practical shielding around such a spacecraft might allow "only" 10 Mrad of radiation to be absorbed by the internal electronics of the spacecraft over a month-long mission—an extreme challenge even for radiation-hardened electronics. But a month of geodetic tracking and laser interrogation of the surface would be enough to determine crustal thickness and its variation, and pave the way for the even-more-difficult surface sampling missions that are required if we are to search for evidence of life in the subsurface ocean of Europa.

Life in the Europan Ocean

Liquid water beneath an Europan ice crust is shielded from the horrendous radiation environment of the surface, allowing contemplation of the notion that the ocean may play host to a biosphere. Essential ingredients for life include sources of organics and of free energy. As noted earlier in the section on possible interior contaminants we do not know the abundance of organic molecules in the interior of Europa. Were Europa to have been composed of the most primitive meteorites known (carbonaceous chondrites), then perhaps a few percent of the moon (minus the ice and the metal) could be carbon, with the other elements key to life on Earth also present. Maybe Europa was bombarded early on with comets that brought even more organic compounds into the Europan interior. However, all of this remains speculation until direct, or compelling indirect, evidence for organic molecules in the Europan ocean becomes available.

The question of energy sources is only slightly more quantifiable. Should there be places in the Europan crust that are of order 10 meters thick or less, enough visible sunlight could be scattered into the upper Europan ocean to make photosynthesis possible, while at the same time shielding the liquid from particle radiation to a reasonable extent. However such a photosynthetic ecosystem would have to have a start with some other energy sources, and the prebiotic synthesis of requisite organic molecules cannot be hosted near the surface since much thinner ice layers and consequently far higher radiation dosages would have to be envisaged. On the other hand, the charged particle chemistry in the ice generates H_2O_2 (peroxide), and, if this is mixed into the water, it could serve as a source of oxygen to oxidize reductants such as hydrogen, hydrogen sulfide, and other compounds—and hence provide chemical-free energy for biological processes.

Tidal dissipation at the ice–water boundary provides a source of heat to warm the ice and maintain the liquid layer against the loss of heat through the crust to space—but little else. Radiogenic heat at the ocean base, which is the top of the silicate mantle, is a much more plausible heat source for biology, and organisms might feed directly off the thermal energy or the emission of reduced species to be combined with oxygen. Tidal currents themselves might serve as a source of kinetic energy, which could be harvested by organisms with movable "hairs," whose forced opening by the currents allows the flow of ions or other chemical energy sources into a cell against compositional or electrical gradients that would push the organism toward equilibrium. The induced magnetic fields in the ocean associated with passage of Europa across the magnetic field lines of Jupiter is another possible energy source. Magnetic minerals that might orient along the field lines could generate potential fields and free energy.

Compared with the Earth, any contemplated biosphere on Europa would be a poor producer of biomass—one quantitative estimate being no more than 10^{12} g/yr biomass production with present energy sources, orders of magnitude less than on Earth. Making some assumptions about the ocean volume and mixing timescales, roughly one cell or less per cubic centimeter might be found in the ocean—making life detection extremely difficult. Another issue is whether an ocean in contact with a chondritic silicate mantle is suitably buffered in pH for the prebiotic assembly, as opposed to breakdown, of polymers such as peptides. Some calculations suggest not, leading to the question of how life could get started on Europa in the first place.

The difficulty with all of the arguments presented is that we have little information on the chemical and physical state of affairs within the Europan ocean, on its organic inventory, or on its history. We are thus in an extraordinarily poor position to speculate on the possibility for life. Also of concern is that the biomass concentration would be so small in the ocean that detection would be extremely difficult. Sending a submersible a hundred kilometers or more to the ocean base to find richer colonies and then communicating back to the distant Earth, strains an imagination tempered by the reality of space program budgets now and in the future. Above all else, it is essential to understand that the precursor to any such search must be a mission to determine if the ice crust of Europa is thin anywhere and, if so, where. This entrée to the astrobiological exploration of Europa is planned for launch, as of this writing, no earlier than 2015.

13.4 Titan

Formation

Titan is the second largest moon in the solar system after Jupiter's Ganymede. It dwarfs the other moons in the Saturnian system, the next most massive (Rhea) being less than 2 percent the mass of Titan. Yet, Titan does not stand out in size or mass with respect to the Galilean satellites of Jupiter. Place Titan with Ganymede and Callisto around Jupiter and you have a remarkably matched set—three moons within 10 percent of each other in radii and 30 percent in mass. What process might generate such similar bodies in two different giant planet satellite systems?

The maximum thermal energy gained from accretion is approximately the latent heat of sublimation of water ice; that is, $\frac{3}{5}GM/r \approx L$. The potential significance of this correspondence for limiting the size of icy satellites can be appreciated by imagining what happens during accretion. Small bodies (planetesimals) of rock and ice fall into the growing satellites with ever-increasing velocity. As the satellites near their present mass and the accretional energy per unit mass approaches the latent heat of sublimation, water ice is preferentially sublimated rather than accreted, while the silicates continue to accrete. Thus, the satellite becomes progressively rock-rich. This appears to describe Titan well, because the satellite's bulk density, equivalent to half-rock half-ice, is significantly larger than that of its much smaller and more ice-rich neighbors in the Saturn system.

What distinguishes the appearance and evolution of Titan from that of Ganymede and Callisto is Titan's dense atmosphere. Titan's interest as an astrobiological target stems from this atmosphere. It is to the atmosphere, therefore, that we turn next.

The Basics of Titan's Atmosphere

Although the existence of Titan's atmosphere was inferred from the Earth, through spectroscopic observations of methane and other hydrocarbons beginning in the 1940s, the real breakthrough in understanding its chemical and physical nature came during the Voyager 1 close flyby of Titan in 1980. Voyager 1 passed within 4,400 km of the surface of Titan and returned many, nearly featureless, images, confirming that at optical wavelengths we cannot see through this body's global haze to the surface (Figure 13.11).

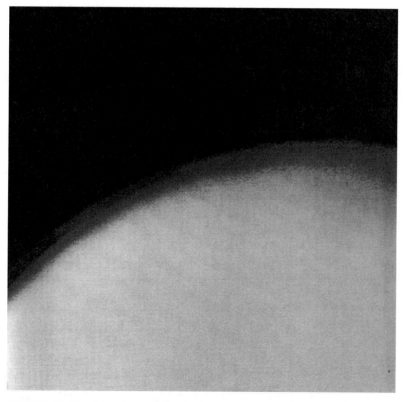

FIGURE 13.11 A quadrant of Titan as images by the Voyager 1 spacecraft in 1980. The surface is not visible; rather, a haze of hydrocarbon particles that exists hundreds of kilometers above the surface is seen in the image, including some detached layers near the limb. All of the haze seen here is orange in appearance to the human eye, with the detached layers slightly greyer or more neutral in color.

Had the Voyager spacecraft been equipped only with imagers, we would have learned little about Titan from the mission. But three other experiments provided a thorough characterization of the atmosphere in a nice example of multitechnique science. The radioscience subsystem on the spacecraft was capable of sending radio waves at two highly stable frequencies when the data transponder was turned off. As Voyager 1 passed behind Titan, as seen from the Earth, radio signals at these two frequencies were transmitted by the spacecraft through Titan's atmosphere. The gas molecules are poor absorbers of the radio signals, but they do alter the velocity and hence the path of the signals; that is, they "refract" them. Measurement of the weakening of the signals by refraction yields the atmospheric number density versus altitude.

To go further requires some basic physics. Because the atmosphere is stable—not collapsing or escaping—then Newton's law (Force = Mass × Acceleration) tells us that the change of pressure as we move upward in altitude is just the density of the air multiplied by the gravity. (Because the pressure gradient is a force per unit volume and the density is a mass per unit volume, this is equivalent to Newton's law.) Now the atmosphere has one other property: The density is low enough that the air behaves approximately as a so-called ideal or perfect gas (Chapter 7), wherein the pressure is proportional to the density multiplied by the temperature and divided by the average mass of the molecules that make up the atmosphere. This average mass per molecule is called the molecular weight of the atmosphere.

A little thought by the reader regarding these two relationships should make clear the following: If the density at every altitude is measured, then the pressure and the temperature divided by the molecular weight at each level in the atmosphere are determined—if we know the gravitational acceleration, which depends on the mass of the planet and was determined by the bending of the spacecraft trajectory as Voyager flew past Titan.

Where the radio signal was cut off completely, the spacecraft had found Titan's surface, and from the preceding analysis, the number density of the atmosphere at Titan's surface was found to be four times that at sea level on the Earth. This is the second-highest atmospheric density among the four solid bodies with atmospheres—Venus, Titan, Earth, and Mars—in descending order of atmospheric density. To proceed further in understanding Titan's atmosphere required data from other instruments.

The infrared spectrometer on Voyager determined temperature at several altitudes by measuring the flux of photons emitted from Titan's atmosphere at different wavelengths. The shape of the temperature profile is crudely similar to that of the Earth, with a declining temperature from the surface upward and then an increase in temperature above a certain altitude. On Earth this "tropopause" level is roughly 20 km; on Titan it is 50 km. The infrared spectrometer indicated a minimum temperature of 70 K; this must correspond to the tropopause at 50 km. Using this value for the temperature leads to a molecular mass of the atmosphere (expressed in units of the proton mass and hence as molecular weight) of roughly 28 atomic mass units (amu), equal to that of nitrogen (N_2) or carbon monoxide (CO) and well above that of methane. The surface temperature from the infrared data is 95 K; hence, from the ideal gas law, the surface pressure is 1.5 bars (1.5×10^5 newtons/m^2). This high pressure and low temperature also eliminate methane, with its low vapor pressure, as the primary gas in Titan's atmosphere. However, the Voyager infrared spectrometer saw characteristic lines of methane in the mid-infrared, so it must be present. By modeling the shape of the infrared spectrum, we get "best-fit" composition in which methane is between 2 and 10 percent of the atmosphere near the surface, and a heavy gas—perhaps argon—is about 6 percent or less of the atmosphere, to balance the low molecular mass of methane.

But still undetermined by the infrared spectrometer was whether the primary gas was nitrogen, carbon monoxide, or something else. Another instrument on Voyager 1 resolved the compositional ambiguity. The ultraviolet spectrometer measured atomic emission lines in the upper atmosphere, as well as ultraviolet absorption lines created as Voyager observed the Sun setting through Titan's atmosphere. The predominant ions were of nitrogen; oxygen (part of CO) was found to be present only in trace amounts. Hence, the dominant gas in Titan's atmosphere is molecular nitrogen, with methane next, not more than 6 per cent argon, 0.1 percent molecular hydrogen, 10 ppm carbon monoxide, and a very small admixture of carbon dioxide (at the parts per billion level, discovered in the infrared spectra). In the stratosphere, there is a menagerie of hydrocarbons and nitriles detected by the infrared spectrometer on Voyager and later by the Infrared Space Observatory (Figure 13.12). The ultraviolet spectrometer also detected a stream of atomic hydrogen escaping from Titan, a telltale signature of the breakup of methane by ultraviolet light from the Sun.

The similarity in shape of the atmospheric temperature profile, and the identity of the primary gas, between Earth and Titan is remarkable (Figure 13.13). Titan's atmosphere

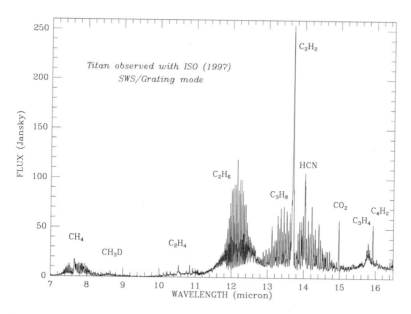

FIGURE 13.12 Spectrum of Titan taken by the European Infrared Space Observatory in 1997, from a vantage point in orbit around the Earth. Although more distant from Titan than was Voyager, the larger telescope and more advanced detectors allowed a spectrum with greater sensitivity to be obtained. The number of photons received per second from Titan is plotted versus wavelength of the photons. The emission lines of hydrocarbon species and one nitrile (hydrogen cyanide, HCN) are labeled. The lines are in emission because all species are seen in the stratosphere, where they are warm, superimposed on the colder lower atmosphere.

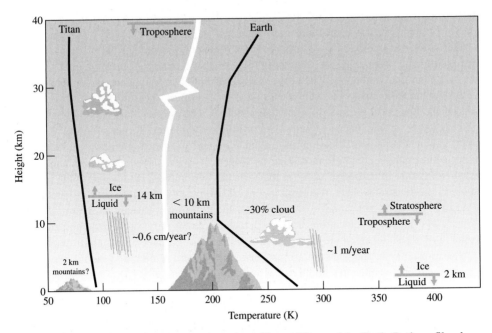

FIGURE 13.13 Comparison of the temperature profiles on Titan and the Earth. Both profiles decline with altitude to the tropopause and then increase again, but Titan's atmosphere is much more extended than the Earth's (a consequence of the lower gravity). Positions of clouds and rain are shown, with estimated rainfall rates. On Earth the rain is water; on Titan it is a mixture of methane and nitrogen.

has a classic greenhouse thermal balance that warms the surface from 85 K, the temperature in equilibrium with sunlight at 9.5 AU, to 95 K. Because of the low temperatures, water vapor is almost nonexistent in the atmosphere and carbon dioxide extremely scarce. In their place, absorption of mid- to far-infrared thermal radiation by nitrogen, methane, and hydrogen predominates. In the region above the tropopause, absorption of solar ultraviolet by hydrocarbons and solar blue light by haze particles create a warm stratosphere and filters out all but 10 percent of the sunlight before it reaches the lower atmosphere. To walk on Titan's surface would be to experience a red twilight at high noon, and no stars at night.

Hydrocarbon Chemistry and Oceans on Titan

The rich suite of hydrocarbons (carbon–hydrogen compounds) and nitriles (carbon–nitrogen–hydrogen species) seen in Titan's upper atmosphere are a testament to the action of sunlight on methane and nitrogen. The photochemical cycle begins hundreds of kilometers above the surface, where ultraviolet light breaks the carbon–hydrogen bonds in CH_4, methane. Acetyelene (C_2H_2) is the second-most abundant product of Titan photolysis, with ethane (C_2H_6) first based on the infrared spectra of the stratosphere from Voyager

TABLE 13.1 Abundances of the Hydrocarbons and Nitriles in the Stratosphere of Titan

Molecule	Abundance (ISO data)
C_2H_2	$5.5 \pm 0.5 \times 10^{-6}$
C_2H_4	$1.2 \pm 0.3 \times 10^{-7}$
C_2H_6	$2.0 \pm 0.8 \times 10^{-5}$
C_3H_4	$1.2 \pm 0.4 \times 10^{-8}$
C_3H_8	$2.0 \pm 1.0 \times 10^{-7}$
C_4H_2	$2.0 \pm 0.5 \times 10^{-9}$
C_6H_6	$4.0 \pm 3.0 \times 10^{-10}$
HCN	$3.0 \pm 0.5 \times 10^{-7}$
CO_2	$2.0 \pm 0.2 \times 10^{-8}$
H_2O (400 km)	$\sim 8 \times 10^{-9}$
HC_3N	$5.0 \pm 3.5 \times 10^{-10}$
C_2N_2 (Voyager)	$<10^{-9}$

Note: Abundances (number relative to N_2) in the region of the atmosphere above 50 km in Titan's atmosphere (except for water, which is 400 km and above). Uncertainties (plus or minus) are given. These abundances are updated from Voyager by the European Infrared Space Observatory, which orbited Earth with a very sensitive, cooled telescope and spectrometer.

(From Coustenis et al., 2003).

and the European Space Agency's Infrared Space Observatory. This is somewhat surprising because the dissociation of methane more readily makes acetylene than ethane. Table 13.1 lists the abundances of some of the products of methane photochemistry in Titan's atmosphere, as well as the abundances of CO_2 and water (which is derived from small, icy meteoroids).

The ultraviolet breakup of acetylene may well be the origin of the ethane. The carbon–hydrogen bonds in acetylene are more easily broken than in methane, and hence longer-wavelength ultraviolet light, penetrating deeper into the stratosphere, can break acetylene into reactive radicals. These "sensitize" the production of ethane when they react with methane by pulling a hydrogen off the methane to make CH_3 from the methane and C_2H_2 (acetylene) from the original C_2H radical. Then the collision of two CH_3 radicals with a third molecule (usually nitrogen) yields C_2H_6, ethane.

In the creation of ethane and acetylene from methane, hydrogen is liberated from the carbon. This hydrogen escapes rapidly from the atmosphere, making the conversion of methane to higher hydrocarbon and nitriles irreversible. Because essentially all of the photochemical products are much less volatile than methane, their production is followed almost immediately by the condensation of much of the product in the form of aerosols. Some of the material remaining in the gas phase is processed further to very heavy hydrocarbons and nitriles, which do not have individual spectral signatures but appear to be the orange haze obscuring Titan's surface. This material, referred to by the name *tholin* (first coined by Carl Sagan and colleagues) serves as nucleating seeds for the aerosols and thus falls to the surface as well.

Aerosols, perhaps layered with tholin cores, acetylene/ethane/nitrile mantles, and outer crusts of liquid or solid methane condensed in the troposphere, drop to Titan's surface. The acetylene remains solid, but ethane and methane melt at 91 K and hence exist on Titan's surface as liquid. If photochemistry has occurred in steady state over the 4.5-billion-year history of the solar system, Titan possesses a layer of liquid hydrocarbon a few hundred meters thick and a solid acetylene and nitrile layer 100 meters thick. Because the solubility of the solid material in the liquid is limited, it is expected that the solid organic sediments underlie the lower density liquid.

This simple picture of a progressive loss of hydrogen from the atmosphere over geologic time, and hence irreversible conversion of methane to higher hydrocarbons, requires that a source of methane other than the atmosphere be present. The amount of methane inferred to be in the atmosphere from Voyager data will be depleted in a timescale of a few tens of millions of years, or 1 percent of the age of the solar system. Therefore, either methane is resupplied to the atmosphere or the methane-bearing atmosphere we see today is a rare occurrence over the whole history of Titan. Resupply of methane from impact or volcanism is possible, but any such process will tend to "overfill" the atmosphere to saturation in methane and produce lakes or even seas of methane. Such seas would mix almost ideally with the ethane to produce a mixed ethane–methane ocean. The co-existing methane vapor pressure would be less than that for pure methane at the same temperature and thus is consistent with the Voyager-derived limits on methane relative humidity near Titan's surface. Nitrogen would dissolve in such an ocean, making a ternary mixture.

Such an ethane–methane–nitrogen surface ocean seems to be a straightforward way of explaining steady-state methane photolysis over the age of the solar system: The ocean is the source (methane) and sink (ethane) of photolysis and is drawn down in volume as methane is converted to ethane and acetylene with loss of hydrogen. Such an ocean, though, would have observable consequences on the shape of Titan's orbit, which is slightly eccentric. Unlike the large satellites of Jupiter, which mutually maintain their orbital eccentricities against dissipation by tidal interactions with Jupiter, Titan has no such buffer. Any strong tidal currents on the surface would dissipate orbital energy and make Titan's orbit circular on short timescales, in a process analogous to the effect of Earth's tides in transferring angular momentum to the Moon and slowing down our planet's rotation rate. If Titan's ocean exists, it is either very deep (submerging all topography completely) or it is confined to individual crater basins.

Remote-sensing data from the Earth after Voyager seem to rule out deep global oceans of methane. It is possible to see through the photochemical haze at infrared wavelengths, and both Hubble Space Telescope and ground-based telescopes equipped with adaptive optics systems to correct for blurring caused by our own atmosphere have imaged Titan's surface (Figure 13.14). These images show a mottled appearance, with one large (Australia-sized), bright area that is likely to be water ice, perhaps mixed with ammonia. The darker mottlings could be liquid or solid hydrocarbons. Radar bounced off the surface of Titan from the Arecibo Radio Observatory in Puerto Rico show that 50 to 75 percent of the surface of Titan reflects like a mirror when directly under the beam but otherwise is very dark. This is the signature of either liquid—hydrocarbons—or clean

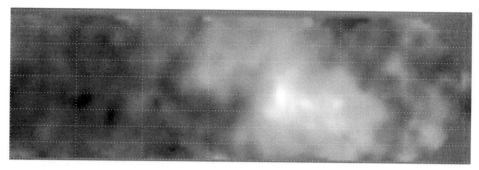

FIGURE 13.14 Map of the midlatitude region of Titan's surface (north and south of the equator) from the Hubble Space Telescope Wide Field and Planetary Camera. The bright region is about the size of Australia. The smallest features visible are roughly 200 to 300 km in size.

sheets of ice. But the infrared spectra of the surface rule out much of the surface being water ice. Thus, some but not all of the surface may be covered with liquid hydrocarbons. If liquid methane–ethane regions exist on Titan, they are restricted in size. Whether enough liquid methane and ethane exist on the surface to render plausible the idea that methane photochemistry has been ongoing for billions of years is a question that must be answered by the Cassini-Huygens mission to Titan.

The Exploration of Titan with the Cassini-Huygens Mission

Cassini-Huygens, launched in 1997, is a joint $3 billion mission among three partners—the United States, through NASA; Europe, through the European Space Agency (ESA); and Italy, through the Italian Space Agency (ASI). (Italy also has a share through ESA.) The Orbiter and Probe together represent one of the most heavily instrumented planetary missions ever, exceeding that of its Jovian counterpart Galileo. The Orbiter weighed 5,500 kg at launch and stands more than 6 meters high (Figure 13.15). The Orbiter payload can be divided roughly into four classes of instruments as described in the following sections.

Plasma Instruments A suite of experiments measures the magnetic and charged-particle environment of the Saturn system in different ways, from mapping the electromagnetic fields to direct sampling of particles and particle imaging of the entire magnetosphere. Although Titan does not have a magnetic field, it interacts with the magnetic field and magnetosphere of Saturn through which it moves. Because the solar wind pushes the saturn magnetosphere with strength that varies on an 11-year cycle, Titan is in an unusual position of being sometimes in the magnetosphere and sometimes in the solar wind. This provides an unusual opportunity to map the response of an atmosphere to two very different charged-particle environments. One of the devices to directly sample material, an ion and neutral mass spectrometer (INMS), is used when the orbiter dips within the atmosphere of Titan, at an altitude of 950 km above Titan's surface. This direct sampling of the upper atmosphere provides important comparisons of the major elements carbon, nitrogen, and hydrogen at high altitudes with that in the lower atmosphere.

FIGURE 13.15 Artist's concept of the Cassini Orbiter and Huygens Probe near Titan.

Remote Sensing Pallet A second group of instruments rides on the Orbiter's "remote sensing pallet." These are devices to sense photons from the ultraviolet out to the mid-infrared, with imaging and spectroscopic capability. They are an ultraviolet spectrometer, two CCD (charge-coupled device) imaging cameras for optical and near-infrared wavelengths, a near-infrared imager that produces spectra at each pixel, a polarimeter (to measure the polarization of light reflected from bodies, and a mid-infrared spectrometer that functions as a radiometer to accurately measure temperatures. These instruments observe Titan on a large number of the 40-odd close flybys that the orbiter is planned to make during the mission.

Radio System A third set of instruments works in the radio part of the spectrum. A radar system transmits and receives through the main antenna, producing images with resolution up to several hundred meters, and can determine altitudes of features as well as how strongly they radiate at radio wavelengths. The radio transmitters used for communication can, as on Voyager, measure atmospheric refractivity and hence generate pressure–temperature profiles. They can also be used to track the path of the spacecraft's orbit during close flybys to measure the distribution of mass within the moon's interior, that is, to determine whether the rock and the ice have separated from each other.

The Huygens Probe The Huygens probe is designed to enter the atmosphere of Titan in early 2005 at hypersonic velocity, slow to transonic speed via a heat shield, and then collect and relay to the Orbiter scientific data while descending on several parachutes from a 170-km altitude down to the surface. The probe has six instruments. The Doppler wind experiment uses an ultrastable oscillator to create a precise radio frequency, whose Doppler shift will be tracked by ground stations and the Orbiter as the probe descends; shifts from the nominal trajectory provide a measure of wind speed. The Huygens atmospheric structure experiment measures temperature, pressure, electrical potential, and other physical parameters as the probe descends. It shares a set of accelerometers with the surface science package that, upon impact, provide data on how soft or hard the surface is. The surface science package makes various measurements of the surface and is designed specifically for direct measurement of surface liquids. The descent imager and spectral radiometer is a multipurpose instrument that makes near-infrared panoramic images of the surface during descent, measures cloud properties through observations of optical phenomena, and, before impact, records a near-infrared spectrum of the surface using artificial illumination.

The final two instruments on Huygens form a crucial set to analyze in situ the nature of Titan's atmosphere. The gas chromatograph and mass spectrometer (GCMS) is the first such combined device to be deployed on a planet since the Viking landings on Mars in 1976. The gas chromatograph works by separating compounds according to how readily they are trapped by materials within several long but tiny tubes. The mass spectrometer strips electrons from molecules and then uses electric fields to spread the charged particles in a beam according to mass; this spread can be analyzed to determine composition. Operated together, the two types of chemical analyzers can avoid ambiguities in identifying certain important species in the atmosphere, most notably nitrogen versus carbon monoxide.

The GCMS will determine abundances versus altitude of various organic species; noble gases that otherwise are spectroscopically inactive and hence difficult to detect; and the isotopes of several major elements, including carbon, nitrogen, and hydrogen. Should the probe survive its landing on the surface, the warm inlet of the GCMS may vaporize surface hydrocarbons and permit one surface chemical analysis before the radio link is lost with the Orbiter as the latter speeds away from Titan.

This single chemical assay through the atmosphere from stratosphere to surface is supplemented with collection of the uppermost atmosphere by a mass spectrometer on the Orbiter and by the remote sensing instruments. Together these represent a detailed assessment of the bulk and organic composition of Titan's atmosphere, which, in turn, will allow a number of key questions about Titan's history to be addressed. These include how much methane is present in the surface and atmospheric inventory of Titan, how much other organic material is present, the origin of Titan's atmosphere, and the evolution of the atmosphere over time. The details of how Cassini will address these questions are too numerous to include more than two examples here.

How have the surface and atmosphere changed with time? Dramatic evidence of secular changes in planetary evolution can be seen on Mars, with its drying and cooling, and on Earth, where the atmosphere became rich in free oxygen billions of years ago. Cassini-Huygens could provide a record of the history of Titan's atmosphere through identification of surface features that represent past episodes of fluvial modification (which, under current conditions, would most likely be caused by methane) and through the distribution of impact craters.

The signature of impact cratering can be seen on the airless bodies of the Saturnian system, just as it can be seen throughout the rest of the solar system. Although there are variations in the pattern of size versus abundance of craters (size–frequency distribution) on objects in different giant planet satellite systems, the general trend is that the number of craters steeply declines with size on the airless bodies. On a body with a thick atmosphere, the crater size–frequency distribution should be distorted by the screening effect of the atmosphere. For incoming projectiles that are relatively weak, and hence shatter under the effect of the dynamic pressure of the atmosphere, an impactor of a given size and density will break apart when it has encountered a mass of atmospheric gas exceeding its own mass. The larger the atmospheric density, or the more extended the atmosphere, the larger the threshold impactor size that breaks up or is slowed. This effect is seen clearly on Venus, where the relative abundance of small craters is depleted relative to large ones, compared with that for the Moon or Mars; most of the smaller craters on Venus are secondary impacts of debris tossed up in the initial impact of a large body. The diameter of the crater formed in a hypervelocity impact depends on a number of factors but is roughly 10 times the diameter of the impactor itself.

Titan's large atmospheric density and slow dropoff in density with altitude should make it a candidate for a screening effect similar to that on Venus, and calculations predict that craters less than a few kilometers in diameter should be rare on Titan compared with what is seen on its neighboring, airless satellites. Cassini's instruments are easily capable of seeing craters below a kilometer in size. If Titan's atmosphere were much thinner in the past, it would have allowed survival to the surface of hypervelocity impactors of a size much smaller than those screened out by the present-day atmosphere. The record of this epoch might exist in places on the surface in the form of a size–frequency distribution that includes a large number of craters below a kilometer radius.

One cause of a thinner atmosphere at times in the past could be the depletion of methane from the atmosphere. If Titan's atmosphere does not have access to a steady-state source of methane from the surface or interior, then there may have been epochs in the past, perhaps of long duration, when photochemistry depleted the atmosphere of methane. This would remove as well the molecular hydrogen from the atmosphere on a timescale comparable to the removal of the methane. With two of the three major greenhouse gases gone, the atmosphere might cool. On the other hand, the photochemical haze, which prevents 90 percent of the sunlight from reaching Titan, would be removed as well, and this could encourage a warmer surface.

Modeling of the energy balance in Titan's atmosphere reveals that the net effect is a cooling of the atmosphere. But the extent of cooling is difficult to predict. There is the potential for a large "collapse" of the atmosphere if the nitrogen gas in the atmosphere were cooled below its saturation vapor point, which, for the current 1.5-bar surface pressure, is 82 K. Interestingly, whether this threshold ever was crossed depends on the history of solar illumination of Titan. For the current solar luminosity, removal of all the methane (with its consequences for hydrogen and the aerosols) would lower the temperature only modestly, to somewhere between 85 and 90 K, with a resulting atmospheric density similar to that observed today. No change in the crater size–frequency distribution is predicted for this case. On the other hand, if the Sun has indeed increased in brightness roughly linearly by 30 percent over the past 4.5 billion years (Chapter 11), then the situation up to 2 billion years ago would have been quite different. Loss of methane prior to that point would have lowered the surface temperature below 82 K, precipitating rainout of the atmosphere.

Thus, Titan's atmospheric history in the absence of a steady-state supply of methane for photochemistry could have been a roller-coaster ride of cold and warm epochs, up to approximately 2 or 3 billion years ago when the Sun brightened sufficiently to stabilize a pure nitrogen atmosphere (Figure 13.16). Identification of an early epoch, or multiple epochs, of a thin atmosphere on Titan could be the only planetary indication of the faint early sun other than the evidence for an early global ice age on Earth (Chapter 11); Mars shows no evidence of a less-luminous Sun early in the history of the solar system (Chapter 12).

What and where are the organic surface deposits? Cassini Orbiter data, combined with Huygens probe atmospheric and serendipitous surface measurements, will provide a map of the presence of organic sediments on the surface and some indication of the uniformity of their composition. The photochemical production of hydrocarbons and nitriles in the upper atmosphere should produce sediments of approximately uniform composition over a range of latitudes. Hence, except near the poles—where deposits of binary mixtures of liquid methane and nitrogen might occur—the organic deposits ought to vary little. However, further processing of the organics on the surface, particularly in regions where heating from the interior of Titan has melted the water ice crust or impacts have occurred, might yield different compositions. Most exciting is the possibility that transient heating events have exposed the organics to liquid water, albeit briefly, and hence have generated compounds of interest to the origin of life. This possibility, whose investigation will require going beyond Cassini-Huygens, is one of the keys to astrobiological interest in Titan.

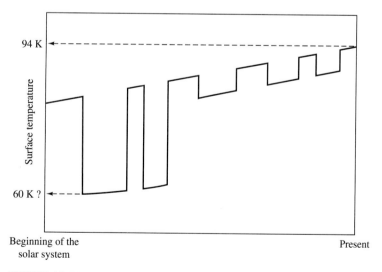

FIGURE 13.16 Schematic history of the surface temperature, and hence thickness of the atmosphere, on Titan. The temperature excursions are more extreme early on when the Sun's luminosity was less than 80 percent of its present-day value.

The Astrobiological Exploration of Titan

Although Titan's surface is too cold to support stable liquid water, even water–ammonia mixtures with a minimum liquid temperature of 173 K, other aspects of Titan's surface and atmosphere system make it attractive as an astrobiological target. Titan has a dense atmosphere that is not strongly reducing, unlike the giant planets, because of the relative scarcity of atomic and molecular hydrogen. Were oxygen to be added to the hydrocarbons and nitriles deposited onto the surface from Titan's atmosphere—for example, in reactions with liquid water—amino acids could be produced. Sources of free energy on Titan, while much less than on Earth, are there nonetheless. While cosmic rays and solar ultraviolet radiation dominate in the upper atmosphere, impacts and possible volcanism are potential sources of energy at the surface. Energy stored in the bonds of certain unsaturated hydrocarbons such as acetylene, generated in the stratosphere, may also be released much later at the surface during formation of polymers. This energy, whose release in an anoxic environment depends on the presence of initiators such as metals, could drive other chemistry on the surface. Because acetylene is the most abundant solid product on the surface of methane photolysis, the availability and magnitude of this source is not negligible. Titan is a large body that has existed for billions of years; organic chemistry ongoing in the absence of life at Titan's surface operates on temporal and spatial scales unachievable in a terrestrial laboratory, and, with a varied surface of liquids and solids, the potential exists for interesting surface-catalyzed or templated chemistry.

Might liquid water exist for short periods on the surface of Titan, despite the cold 95 K mean temperature? Regions where water and ammonia have flowed out of the interior as liquids are possible given the relatively large size of this satellite, but the timing and extent are hard to predict. A stochastic yet more predictable mechanism for generating liquid water at the surface is impact of kilometer-sized or larger bodies into Titan's icy crust. Such an impact forms a crater, ejects material, and leaves a few percent of liquid water at the base of the crater. Numerical simulations of the impact process show that, for oblique impacts, a lightly shocked tongue of organics and water ice from the lee side of the impact drops into the liquid water (Figure 13.17). A rind of water ice forms quickly over the liquid water and organics, but complete freezing is delayed for 10^2 to 10^3 years, depending on the amount of antifreeze (ammonia or methanol) in the water.

The exciting prospect associated with transient pools of liquid water is that the organic sediments on Titan's surface, generated in the stratosphere in a very oxygen-poor environment, are suddenly able to react with liquid water. Oxygen-bearing compounds such as amino acids should form quickly. Whether enough time, volume, and surfaces are available to go further is uncertain. Might some chiral amino acids, or chiral sugars, undergo amplification of initially small enantiomeric excesses? Could simple peptides form in the time available? All of these possibilities are speculation because it is extremely difficult—in spite of sophisticated numerical models of chemical reactions—to predict a priori what the results will be of exposing a complex mixture of organics to an aqueous medium for timescales much longer than are obtained in the laboratory. But whatever is produced will be well preserved as the last of the aqueous medium freezes over. Drilling several meters into the crater bottom to access and analyze the organics may be required. Should Cassini-Huygens identify places on the surface where organic deposits are associated with geolog-

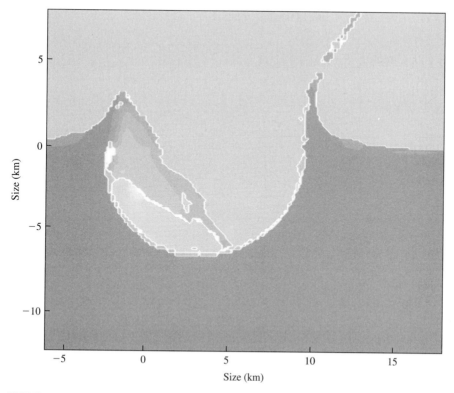

FIGURE 13.17 Computer simulation of the impact of a comet on Titan, performed by N. Artemieva (Russian Academy of Sciences, Moscow) and the author. The frame represents the situation 15 seconds after an impact of a kilometer-size comet with an initial velocity of 7 km/s and an angle of impact of 45 degrees. Liquid water is present in the crater volume, from ice melted as a result of the impact. Organics are contained within the tongue of water ice at the lee side of the impact. Sizes on the axes are in kilometers.

ical features, follow-on missions to examine these regions in detail would be of keen interest to astrobiology.

A possible strategy for identifying and sampling interesting organics on Titan's surface would use a blimp to examine potentially interesting sites at close range and then land or drop a small package for analysis of the organics. Titan's low-gravity and high-density atmosphere is ideal for exploration with small aerial vehicles. The weak sunlight and small temperature differences across latitudes implies a very weak surface wind of meters per second, making for easy maneuvering. But at very high altitudes, winds are measured from telescopic studies to increase to 100 meters per second, allowing blimps to rise and catch the strong current to head eastward across the planet. A blimp could be deployed to a site identified from Cassini-Huygens data to be of particular interest. A remote-sensing package on the blimp, consisting of visible imaging and high-resolution ultraviolet and infrared spectroscopy, would examine the site at close range. The goal is to identify organic samples that appear to have been altered by liquid water. Once a plausible set of deposits is identified during this phase, the blimp is set down on, or drops a package on, that site. Determination of the total oxygen content, the pattern of abundance of organic molecules as a function of carbon number, polymer structure, the presence of chiral molecules, and the ratio of left- to right-handed forms would be analyzed with schemes discussed in Chapter 14. If the sample looks promising, additional analyses and even drilling (which is difficult!) might be undertaken; otherwise, the blimp is raised from the site to begin a new reconnaissance or to navigate immediately to a backup site for analysis.

In summary, Titan offers three types of opportunities for study of organics systems of relevance to astrobiology; all three require access to the surface:

1. *Chemistry in nonaqueous organic systems.* Even in the absence of water, organic chemical systems may evolve over long timescales, fueled by the released enthalpy of polymerization, cryovolcanism, or impacts. Polymer formation could, in some cases (such as that of polyacrylonitrile) exhibit self-organizing properties in terms of the orientation of the monomers along the chain, a property called *tacticity*. Going one step further, we cannot rule out the possibility than a form of life, organic but very different from Earthly life, is able to use the liquid hydrocarbons as a solvent for living processes in place of liquid water. Exactly how this would work is unclear, because the nonpolar nature of the liquid hydrocarbons precludes defining the inside and outside of polymers in the way that, for water, arises from hydrophobic and hydrophilic sides of a molecule, but it cannot be ruled out.

2. *Chemistry in aqueous organic systems.* As discussed, it is possible that organic molecules on Titan have been exposed to liquid water for modest periods in the basins of newly formed craters. The cratering rate for the Saturnian system, integrated over the age of the solar system, suggests that a significant fraction of the surface of Titan has already been struck with large impactors. Only the freshest craters are likely to yield accessible organics altered by contact with liquid water, the older craters having been tectonically modified, mechanically stirred by secondary crater formation from nearby large impacts, and filled in by many hundreds of meters of hydrocarbon sediments over billions of years. It is difficult to predict a priori what organic compounds might be found in such places should it be possible to locate a fresh crater, identify organic sediments, and drill through the upper ice layer to access the deep-frozen products.

3. *Life in Titan's interior.* Finally, life might exist in the deep interior of Titan where ammonia–water could still be liquid, heated by radioactive elements in the rocky core. This liquid mantle might contain both the raw materials and perhaps the energy necessary for life. The kilobar pressures in this deep layer appear not to be a problem either, based on laboratory experiments at the Carnegie Institute of Washington, in which terrestrial organisms were subjected to such pressures. However, even if life of some kind were to exist there, the thickness of the ice layered above it (tens of kilometers) rules out direct sampling. Possible isotopic signatures of life outgassed to the atmosphere would be subtle and difficult to detect or distinguish from

abiotic processes. For the detection of subterranean biota, Europa is a better (but still very difficult) target. Titan, farther from the Sun but with a surface more benign and easier to land on than that of Europa, is of the greatest interest with regard to the organic chemistry leading toward the origin of life.

QUESTIONS AND PROBLEMS

1. The pressure scale height gives the altitude over which the atmospheric pressure drops by roughly half (actually, by $1/e$, where e is the base of the natural logarithm, or 2.3) and is kT/mg, where T is temperature, g gravity, k Boltzman's constant, and m molecular weight of the gas. It is a good measure of how extended an atmosphere is. What are the scale heights near the surface for the atmosphere of the Earth and the atmosphere of Titan?

2. Using the perfect gas law, justify the claim that, even though the surface pressure on Titan is 1.5 bars (50 percent higher than Earth's), the density of the air at Titan's surface is almost four times that of the air at sea level on Earth.

3. Do a literature search to determine the biomass production rate for the Earth's oceans and for the entire Earth's biosphere. Compare it with the theoretical value for a Europan ocean cited in the chapter.

4. The tidal dissipation occurring between the Earth and Moon leads to the slowing of the Earth's rotation rate as the Moon's orbit spirals outward. Angular momentum is conserved in this process even though orbital energy is going into (largely) friction associated with ocean tides on the Earth. Do a literature search to find estimates for how short an Earth day was 2 or 3 billion years in the past. Calculate (or quote from the same source if the number is available) how far from the Earth the Moon was at that time, and compare with the present value.

5. How orbital energy gets dissipated in Earth's oceans is a surprisingly complex problem and includes not only tidal rising and falling in narrow straights (e.g., the Bay of Fundy) but also waves generated in the oceans themselves. Do a literature search to find models for how the orbital and spin energies of the Moon and Earth are dissipated in the Earth's oceans. What do these mechanisms imply for the dissipation of orbital energy in shallow hydrocarbon seas on Titan or a subcrustal ocean on Europa?

6. The chapter highlights a potential future mission to Europa that emphasizes the geophysics of the satellite. Make some arguments in favor or against an alternative idea, which is to go directly to the surface of Europa and try to access an ocean beneath the crust. What do we know now that would enable such a direct mission, and what are the current uncertainties?

7. Occasional impacts on Titan, or expulsion from the interior of liquid ammonia–water solutions, will expose the surface organics—largely devoid of oxygen—to aqueous chemistry. Consult a chemistry textbook to take the argument in the text for amino acids a bit farther. What kinds of organic compounds might be made from the list of materials in Table 13.1 upon mixing with liquid water or water–ammonia solutions?

8. With a pair of scissors, or an electronic image-editing program, take a portion of Figure 13.2 and cut out the individual block or "rafts." Can you reassemble them, jigsawlike, in a way that convinces you that they drifted apart from an initial smaller set of "superrafts" at some time in the past?

9. Molecular nitrogen (N_2) is very hard to detect using optical and infrared spectroscopy on Titan because it is a symmetric molecule with very weak spectral bands. Go back to the original discovery paper of methane on Titan (cited in the References) and consider the problems Kuiper had in determining whether methane was a secondary gas or the primary gas in an atmosphere around Titan. Do a literature search for subsequent papers in the 1960s and 1970s by others that continue the debate, to be resolved by Voyager 1 in 1980.

10. Turtle and Pierazzo (cited in the References) make arguments for a thick ice crust on Europa based on the appearance of a large crater on the surface. Read this paper to understand their argument. How could both this argument for a thick crust and the arguments presented in the chapter for a thin crust be reconciled? Can you find any other studies in the literature, based on Galileo spacecraft data, that support the notion that the crust is kilometers thick?

SUGGESTED READINGS

Two recent and excellent books on Titan will give the reader a complete overview of current knowledge of Titan: A. Coustenis and F. Taylor, *Titan: The Earthlike Moon* (1999, World Scientific, Singapore); and R. Lorenz and J. Mitton, *Lifting Titan's Veil: Exploring the Giant Moon of Saturn* (2002, Cambridge University Press, Cambridge). Unfortunately, no equivalent book exists for Europa. The National Research Council's Space Studies Board has published a very readable report on current knowledge of Europa and plans for future exploration entitled *A Science Strategy for the Exploration of Europa* (1999, National Academy Press, Washington). R. Shapiro, *Planetary Dreams: The Quest to Discover Life Beyond Earth* (1999, John Wiley and Sons, New York), speculates about life on Titan and Europa.

REFERENCES

Europa

Chyba, C. F. and Phillips, C. B. (2001). "Possible Ecosystems and the Search for Life on Europa," *PNAS*, **98,** 801–803.

Domingue, D. L., Lockwood, G. W., and Thompson, D. T. (1995). "Surface Textural Properties of Icy Satellites: A Comparison Between Europa and Rhea," *Icarus*, **115,** 228–249.

Geissler, P. E., Greenberg, R., Hoppa, G., McEwen, A., Tufts, R., Phillips, C., Clark, B., Ockert-Bell, M., Helfenstein, P., Burns, J., Veverka, J., Sullivan, R., Greeley, R., Pappalardo, R. T., Head, J. W. I., Belton, M. J. S., and Denk, T. (1998). "Evolution of Lineaments on Europa: Clues from Galileo Multispectral Imaging Observations," *Icarus*, **135,** 107–126.

Greeley, R. (1997). "Geology of Europa: Galileo Update." In *The Three Galileos: The Man, the Spacecraft, the Telescope*, C. Barbieri, J. H. Rahe, T. V. Johnson, and A. M. Sohus, eds., Kluwer Academic, Dordrecht, pp. 191–200.

Greenberg, R., Geissler, P., Hoppa, G., Tufts, B. R., Durda, D. D., Pappalardo, R. T., Head, J. W., Greeley, R., Sullivan, R., and Carr, M. H. (1998). "Tectonic Processes on Europa: Tidal Stresses, Mechanical Response, and Visible Features," *Icarus*, **135**, 64–78.

Irwin, L. N. and Schulze-Makuch, D. (2003). "Strategy for Modeling Putative Multilevel Ecosystems on Europa," *Astrobiol.*, **3**, 813–821.

Johnson, R. E., Quickenden, T. L., Cooper, P. D., McKinley, A. J., and Freeman, C. G. (2003). "The Production of Oxidants in Europa's Surface," *Astrobiol.*, **3**, 823–850.

Kargel, J. S. (1998). "The Salt of Europa," *Sci.,* **280**, 1211–1212.

Kivelson, M. G., Khurana, K. K., Russell, C. T., Volwerk, M., Walker, R. J., and Zimmer, C. (2000). "Galileo Magnetometer Measurements: A Stronger Case for a Subsurface Ocean at Europa," *Sci.*, **289**, 1340–1341.

McCord, T. B., Hansen, G. B., Matson, D. L., Johnson, T. V., Crowley, J. K., Fanale, F. P., Carlson, R. W., Smythe, W. D., Martin, P. D., Hibbitts, C. A., Granahan, J. C., and Ocampo, A. (1999). "Hydrated Salt Minerals on Europa's Surface from the Galileo Near-Infrared Mapping Spectrometer (NIMS) Investigation," *J. Geophys. Res.*, **104**, 11827–11851.

Reynolds, R. T., McKay, C. P., and Kasting, J. F. (1987). "Europa, Tidally Heated Oceans, and Habitable Zones Around Giant Planets," *Adv. Space Res.*, **7**, 125.

Turtle, E. P., and Pierazzo, E. (2001). "Thickness of a Europan Ice Shell from Impact Crater Simulations," *Scie.*, **294**, 1326–1328.

Zolotov, M. Y. and Shock, E.L. (2003). "Energy for Biologic Sulfate Reduction in a Hydrothermally Formed Ocean on Europa," *J. Geophys. Res.*, **108 (E4)**, article 5022.

Titan

Bird, M. K., Dutta-Roy, R., Asmar, S. W., and Rebold, T. A. (1997). "Detection of Titan's Ionosphere from Voyager 1 Radio Occulation Observations," *Icarus*, **130**, 426–436.

Campbell, D. B., Black, G. J., Carter, L. M., and Ostro, S. J. (2003). "Radar Evidence for Liquid Surfaces on Titan," *Sci.*, **302**, 431–434.

Coustenis, A., Salama, A., Schulz, B., Ott, S., Lellouch, E., Encrenaz, Th., Gautier, D., Feuchtgruber, H. (2003). "Titan's Atmosphere from ISO Mid-Infrared Spectroscopy," *Icarus*, **161**, 383–403.

Dermott, S. F. and Sagan, C. (1995). "Tidal Effects of Disconnected Hydrocarbon Seas on Titan," *Nature*, **374**, 238–240.

Fortes, A. D. (2000). "Exobiological Implications of a Possible Ammonia–Water Ocean Inside Titan," *Icarus*, **146**, 152.

Gibbard, S. G., Macintosh, B., Gavel, D., Max, C. E., de Pater, I., Ghez, A. M., Young, E. F., and McKay, C. P. (1999). "Titan: High-Resolution Speckle Images from the Keck Telescope," *Icarus*, **139**, 189–201.

Grasset, O., Sotin, C., and Deschamps, F. (2000). "On the Internal Structure and Dynamics of Titan," *Planet. Space Sci.*, **48**, 617–636.

Griffith, C. A., Hall, J. L., and Geballe, T. R. (2000). "Detection of Daily Clouds on Titan," *Sci.*, **290**, 509–513.

Khare, B. N., Sagan, C., Thompson, W. R., Arakawa, E. T., Suits, F., Callcott, T. A., Williams, M. W., Shrader, S., Ogino, H., Willingham, M. W., and Nagy, B. (1984). "The Organic Aerosols of Titan," *Adva. Space Res.*, **4**, 59–68.

Kuiper, G. P. (1944). "Titan: A Satellite with an Atmosphere," *Astrophys. J.*, **100**, 378–383.

Lellouch, E., Schmitt, B., Coustenis, A., and Cuby, J.-G. (2003). "Titan's 5-Micron Lightcurve," *Icarus*, **168**, 209–214.

Lindal, G. F., Wood, G. E., Hotz, H. B., Sweetnam, D. N., Eshleman, V. R., and Tyler, G. L. (1983). "The Atmosphere of Titan—An Analysis of the Voyager 1 Radio Occulation Measurements," *Icarus*, **53**, 348–363.

Lorenz, R. D., Lunine, J. I., and McKay, C. P. (1997). "Titan Under a Red Giant Sun: A New Kind of "Habitable" Moon," *Geophys. Res. Lett.*, **24**, 2905–2908.

Lorenz, R. D. and Mitton, J. (2002). *Lifting Titan's Veil*, Cambridge University Press, Cambridge.

Lunine, J. I., Lorenz, R. D., and Hartmann, W. K. (1998). "Some Speculations on Titan's Past, Present and Future," *Planet. Space Sci.*, **46**, 1099–1117.

McKay, C. P., Lorenz, R. D., and Lunine, J. I. (1999). "Analytic Solutions for the Antigreenhouse Effect: Titan and the Early Earth," *Icarus*, **137**, 56–61.

Owen, T. C. (1982). "The Composition and Origin of Titan's Atmosphere," *Planet. Space Sci.*, **30**, 833–838.

Stevenson, D. J. (1992). "The Interior of Titan." In *Proc. Symp. Titan, ESA, SP-338*, 29–33.

Welch, C. J., and Lunine, J. I. (2001). "Challenges and Approaches to the Robotic Detection of Enantioenrichment on Saturn's Moon, Titan," *Enantiom.*, **6**, 69–81.

Yung, Y. L., Allen, M. A., and Pinto, J. P. (1984). "Photochemistry of the Atmosphere of Titan: Comparison Between Model and Observations," *Astrophy. J. Supp. Ser.*, **55**, 465–506.

CHAPTER 14

Life Elsewhere I
Direct Detection of Life and Its Remnants Within the Solar System

14.1 Introduction

The search for life on other planets within our solar system is perhaps the most difficult goal of planetary exploration, largely because we do not know what to expect in terms of the basic polymers of life elsewhere, the nature of their metabolisms, or even size and morphology. The most sensitive techniques for detecting life are based on the amplification of nucleic acids (DNA and RNA), yet these are so specific to terrestrial life that it is unlikely that they would work on extraterrestrial life with even small differences from terrestrial. The most general techniques, involving imaging of structures that life might leave behind, are open to ambiguity because some such structures might have nonbiogenic origins as well. This is the dilemma that faces those who would design instruments and protocols (detailed procedures) for life detection.

This chapter discusses previous attempts to detect extraterrestrial life, from the Viking missions of the 1970s and the studies in the 1990s of the martian meteorite ALH84001. We then survey a range of techniques that could be applied today or in the near future in the search for life. The focus will be on Mars because efforts to detect life have already taken place there, and a substantial effort is being mounted in the next decade for a more thorough search. Europa is very much an unknown in terms of what we would look for, and we briefly survey why. Finally, Titan is a locale for the detection and characterization of advanced organic chemical evolution that might be on the road to life, and we discuss possible approaches to studying such chemistry, with the analysis of organic phases in meteorites as the background.

14.2 What Are We Searching For?

We must know what it is that we are searching for, if the search is to have any chance of success. In this regard, the search for "life" is particularly difficult. In previous chapters we avoided the issue of a general definition of life in two ways. First, in the chapters on chemistry and biology, we limited the discussion to the biochemistry of the life that we know, that is, life on Earth. Other chapters tacitly assumed that we know or will know what it is we are searching for when we are ready to search for it. These are not merely word games, because as yet there is no general definition of life. We might imagine that the diversity of life on Earth—from the most exotic microscopic Archea to elephants, giraffes, and desert creosote bushes—enables us to formulate a general definition of life. Such would be the case were we to include only life that operated strictly by terrestrial rules, namely, using DNA with the appropriate four nucleic acid bases, encoding for the set of 20-plus amino acids used by earthly life, and so forth. We could even extrapolate to slightly more exotic living systems, using different nucleic acid bases, a different *number* of nucleic acid bases, different amino acids, etc. But we quickly begin to lose a sense of what is plausible or possible as an alternative biochemistry as we move farther and farther away from what is known.

The problem in providing a general definition of life is well articulated using water as an example—an analogy explored by philosopher Carol Clelland of the University of Colorado. What is water? Prior to the development of atomic theory, water could be defined only by its properties. Its liquid form is fluid over a broad range of temperatures; generally colorless except in very deep, still waters; odorless; and nourishing (essential) to the body. Indeed, liquid water is so ubiquitous on our planet that the ancient Greeks identified it as one of the four fundamental "elements" of nature. Can we generalize this definition of water to other "kinds" of water—substances that have some of the same properties? How about hydrogen peroxide, which appears much like liquid water under normal conditions yet is a poison? Is it a different and poisonous form of water, or is it something else? The problem of how to uniquely define water went away when the atomic nature of matter was understood and the distinct types of atoms—the chemical elements—were discovered. Water, it turns out, is one oxygen atom bound to two hydrogen atoms—H_2O. Whether in gaseous, liquid, or solid form, this is water. Hydrogen peroxide—H_2O_2, with its different ratio of oxygen to hydrogen, is not water, even though it shares some of its properties. Purely and simply, when we say water we mean H_2O.

There is no such equivalently unambiguous definition of life. Many of the definitions that depend on observable properties apply in some form to things that we would generally agree are not "alive." Crystals grow in well-ordered shapes. Complex chemical systems can "evolve," in the sense that new catalysts can be invented as the products of previous reaction cycles, and these in turn can change the mix of products and reactants in a unidirectional fashion. Galaxies evolve with time: The internal composition of succeeding generations of stars become enriched in heavy elements, leading potentially to the production of more planetary systems, or perhaps more massive planets, with implications for the habitability of those planets that we cannot as yet predict. Perhaps we are at the stage in understanding life equivalent to the "pre-atomic" stage of chemistry—we are missing some key underlying principle that would yield an unambiguous definition of life. Or, perhaps, the notion of "life" or a "living thing" is intrinsically ambiguous.

Could we not simply define life as adhering to the operational principles of earthly life? That is, could we not define it as organization into cells with various membranes; DNA composed of the four nucleic acid bases guanine, adenine, cytosine, and thymine; RNA the same except with uracil replacing thymine; and so forth? In such a case, life would be well defined. What then, if we were to discover a form of life elsewhere that substituted a different nucleic acid base for thymine? We would be required to broaden our definition. What if we were to stumble upon a moving, self-replicating creature whose fundamental structural units looked nothing like cells or contained an information-bearing molecule that looked nothing like DNA? We would be forced to broaden the definition even further or to deny that such a creature is living. Finally, nonchemical forms of life present a different challenge: Might there be self-organization, even awareness (as difficult as that is to define), in systems that are nonmolecular, nonatomic, or even nonbaryonic?[1]

Within our own solar system, the search for life will be performed at the microscopic level, seeking chemical and morphological clues listed in Table 14.1. The techniques for doing so are discussed in the remainder of this chapter. The search for life around other stars, either by taking spectra of newly discovered planets or by listening for radio transmissions from other civilizations, will be covered in Chapter 15.

TABLE 14.1 Biosignatures and Their Specificity for Life Detection and Applicability for Detecting Extant and Extinct Life and Terrestrial Contamination of Spacecraft

Signatures of Life	Application for Life Detection	Specificity	Fossil and Non-Terrestrial Life Detection
I. *Morphology (Micro- and Macroscopic)*			
• Shape, size, replication structures (buds, chains of cells, septa, fruiting bodies and spores) • Some biominerals • Macrostructures such as biofilms and stromatolite-like structures	Detect extant terrestrial life, fossils, indication of active cells; application to spacecraft contamination	Shape and size not definitive; terrestrial life is >100 nm in diameter; replication structures are definitive indicators of life; can identify eukaryotes; biofilms and stromatolite-like structures could be definitive.	Replication structures can be definitive; size, shape, and numbers of identical morphotypes such as seen in biofilms or laminated-structures observed in stromatolites may or may not be definitive for life; additional chemical and isotopic analyses are necessary

(continued)

---

[1] Humans have reported interactions of all sorts with supernatural entities (elves, extraterrestrials). Such encounters are fictitious; they are generated within the mind of the reporting individual. Care must be taken in dismissing these, however, because they can well be defined self-consistently as encounters with a completely parasitic intelligent life that lives—and dies—with the host interlocutor.

TABLE 14.1   continued

Signatures of Life	Application for Life Detection	Specificity	Fossil and Non-Terrestrial Life Detection
II. *Organic Chemistry and Biochemistry*			
• Cell walls (variety of biopolymers) • Membranes (fatty acids) • Nucleic acids (DNA, RNA) • Proteins • Hydrocarbons, steroids, hopanes • Amino acids • Organic metal and phosphate compounds • Porphyrins, flavins, etc. • Carbohydrates	The first four items detect extant terrestrial and extraterrestrial life; nucleic acids with same genetic code as terrestrial life would likely indicate terrestrial contamination; steroids generally indicate eukaryotes; hopanes found in cyanobacteria; chirality and presence of the 20 key amino acids associated with terrestrial life indicate universality or terrestrial spacecraft contamination	Bacteria, archaea, and eukaryotes have specific cell-wall chemical structures; chirality, enantiomeric excess, and repeating structural units such as C_5, C_6 of sugars, C_2 of polymethylenic lipids, C_5 of polyisoprenoids, α-substitution of proteins amino acids, and L-amino acids and D-sugars are canonical for terrestrial life; nucleic acids, proteins and phosphates, and organic-phosphate compounds could be indicative of recent or extant life	Hydrocarbons, steroids, and hopanes have been observed in the fossil record; other macromolecules (nucliec acids, proteins, and carbohydrates) are extremely labile; nothing is known about the long-term stability of cell-wall polymers of archaea and their chemical transformations during fossilization
III. *Inorganic Chemistry*			
• Iron minerals (e.g., magnitite) • Sulfur compounds • Carbonates • Silicates • Other biologically important metals (e.g., Cu, Mo, Ni, W, etc.) • Nitrogen compounds • Phosphorus compounds • Ratios of biologically important elements (CHONPS) • Disequilibrium in biologically important oxidation/reduction couples	Best application as additional information in conjunction with microscopic, isotopic, and organic chemical analyses for fossils and possibly for detecting presence of living extraterrestrial organisms; probably not applicable for detecting spacecraft contamination	C, N, and S can be highly specific for terrestrial life when appearing in conjunction with stable isotope or organic analyses; some bacteria form iron compounds with highly specific structures (e.g., magnetosomes and the ferruginous ribbons formed by the bacterium *Gallionella spp*); other microbes deposit silicates and carbonates and elemental sulfur as metabolites	Some crystal structures of magnitite are thought to be produced only by organisms; C, S, and N isotopes are essential additional analyses for inferring past life; oxygen isotope ratios associated with phosphates may be indicative of life; the heterogeneous distribution of biologically important minerals (Cu, Mo, Ni, W) and disequilibrium in the chemistry of rock samples could be supporting evidence for past life

Signatures of Life	Application for Life Detection	Specificity	Fossil and Non-Terrestrial Life Detection
IV. Isotopic Analyses			
• C, N, S, O, and possibly heavy metals	Best application to confirm extant and fossil life where there may not be evidence of intact cells; new methods also allow stable isotopic analyses of individual organic molecules and iron; probably not applicable for detecting spacecraft contamination	Stable C, N, and S isotopes can be definitive indicators of different metabolisms; best used to detect CO_2 reduction by photosynthesis, chemosynthesis, methanogenesis, and sulfate reduction	Vital analyses to help confirm biogenic origin of minerals or cell-like structures observed microscopically on rock or soil samples; used to understand the nature of C, N, and S cycles in ancient environments
V. Environmental Measurements			
• Global atmosphere measurements (spectral identification of volatiles such as ozone, hydrogen, methane, oxygen, water) • Macroscopic life-forms (imaging systems)	Identify metabolic indicators of extant life, potential for habitability, and sites of high concentrations of volatiles and water; visual indication of vegetation or other indications of life	Other measurements necessary to confirm presence of living organisms and ecosystems	Not applicable for detecting extinct life unless there were some visual indication of past vegetation such as stromatolites

Source: John Baross, University of Washington.

14.3 The Basic Characteristics of Life That Define What We Are Looking For

To proceed further in a search of our solar system for life, we must be more specific and recognize that some definitions or restrictions must be put on the characteristics of living ("extant") or once-living ("extinct") organisms. An excellent starting point is a list of characteristics of chemically based, cellular, self-replicating, and evolving life that can be rationalized as being common. They are arranged in order from characteristics that are more likely to those that are least likely to be general:

1. *Life is based on the element carbon.* The two fundamental types of polymers upon which Earthly life is based, proteins and nucleic acids, depend on carbon for their

bonding and hence chemical properties. From a practical standpoint, that of the chemical detection of extraterrestrial life, it is almost imperative to assume that life elsewhere is made of carbon, because the chemical detection techniques rely on the characteristics of organic compounds and bonding mechanisms. It might be possible to design detection techniques for life based on other elements, such as silicon, but the potential for missing the key chemical clues (isotopic enrichments, products of metabolism, etc.) is much greater. Silicon is often cited as an alternative to carbon because it is just one row below and in the same column as carbon in the periodic table, but silicon is not completely equivalent in its bonding properties. The variety of polymers that can be built from silicon is far less than for carbon, and the correspondence between the temperatures over which liquid water is stable—and where carbon bonds are stable yet reactions are rapid enough—is not replicated for silicon. It is possible that early forms of life used mineral surfaces, such as clays, as templates or even catalysts for sequences of reactions, but the reactions would still center on carbon.

2. *Life requires liquid water.* All life on Earth requires liquid water. Water is essential for the transport of nutrients and waste products, for the catalytic control of reactions, and for the structure of proteins and nucleic acids (which shape themselves in large measure based on repulsion or attraction with water). Psychrophiles, that is, extremophiles that live in extremely cold environments (Chapter 10), rely on thin layers of liquid water within water ice, and the interiors of their cells are, of course, liquid.

Liquid water is a particularly attractive medium for life. The temperature range of stability of the liquid is large and suitable for both the stability of organic bonds and moderate rates of reactions that break them. The high latent heat and specific heat of water tend to buffer perturbations associated with temperature changes in the environment. And water is quite polar, which enables charge-based mechanisms for controlling the transport of molecules across membranes and cellular structures.

Molten sulfur or silicates can be ruled out as the liquid media for organic life on the basis of the excessively high temperatures that would denature proteins and in general sever organic bonds. Liquid ammonia, or ammonia–water mixtures, have properties resembling liquid water but are stable to as low as $-100°C$. Such liquids are good candidates for supporting life in planetary environments colder than that of the Earth. Ammonia–water mixtures might make possible some form of life in the moons of the outer solar system. Nonpolar organic liquids seem less likely candidates as the solvent of life because of the lack of polarizability and consequent mechanisms for transport of nutrients across membranes. But enzymes will function in some organic solvents, and the addition of impurities to pure hydrocarbon liquids can provide some polarity. The hydrocarbon heptane is a liquid at room temperature but is unlikely to compete with liquid water. Ethane and methane may exist as liquids on the surface of Titan, at about $-180°C$, but here organic reactions are so sluggish that the coordinated networks of reactions required to sustain some form of life might not be possible, even with powerful catalysts.

This particular "requirement" for life has motivated the search beyond the Earth for environments containing liquid water as an indicator of where life might have existed or could exist today. Nonetheless, wherever there is liquid of some composition, there will be interest in examining the possibility of living or protoliving processes. Some have even suggested the possibility of "dry life"—existing without a liquid medium—though this would be so exotic that it is not clear how or where we would search for such a phenomenon.

3. *The number of fundamental chemical processes underlying metabolism are finite and small.* Sometimes called "metabolic unity," this is the notion that there are a limited number of ways in the universe to utilize free energy from the environment (defined in the thermodynamic sense of Chapter 7). There are two fundamentally different ways that organisms can do so: by converting light to chemical energy or through oxidation-reduction ("redox") reactions. Therefore, it should be possible to look for key indicators of metabolic processes in the geologic record, provided that they are stable enough to be preserved. The challenge, however, is to recognize that biochemistry is potentially capable of tapping thermodynamic free energy from a system in any of the ways that abiotic chemistry can—since it utilizes chemical processes to do so. This is becoming more and more evident as extremophiles are discovered (e.g., within the Earth's crust) that utilize energy via novel (and sometimes very simple) chemical processes (Chapter 10). Thus, the key in identifying life is to find a chemical or isotopic signature associated with a metabolic process that would not be seen in the underlying chemical reactions were life absent.

   Redox reactions involving inorganic compounds and metals can produce crystals with very specific characteristics, and one example, that of magnetite, is discussed below. Some metabolic processes may not leave stable crystals or other long-lived evidence behind and may be difficult or impossible to detect. One potentially general characteristic of metabolic processes is the generation of isotopic anomalies, particularly enrichment of lighter isotopes when this is energetically favored in organic reactions. Another is the preferential consumption of organic compounds of one or the other handedness, since the buildup of functional polymeric structures often requires the asymmetry that use of a single handedness brings. A problem with this type of evidence of metabolic processes is that it degrades with time since, for example, left-handed amino acids can spontaneously convert to right handed, leading over long periods of time to roughly equal numbers of left- and right-handed forms (a "racemic" mixture). Hence, the challenge in using metabolic processes as biomarkers is to identify those that are widespread, unique to living (as opposed to nonliving) processes, and long lasting (or degrade over time in a predictable way).

4. *The energy for biological processes is stored and manipulated primarily through phosphate bonds.* Phosphate bonds in ADP and ATP are fundamental to the ability of cells to store and release energy derived from metabolic and photosynthetic processes (Chapter 4). Phosphates are also the key to structural polymers such as DNA and RNA, phospholipids, and others. Would phosphate bonds be a common theme among life-forms throughout the universe? Certainly as a relatively abundant element from the cosmochemical point of view, phosphorous ought to be readily available on habitable worlds. Living systems based on carbon and liquid water would likely use phosphorous and phosphate bonds for at least some of the same roles, as does Earthly life. Should a microscopic ecosystem be detected, measurement of the total phosphate would then be a useful indicator of the total biomass of the system. Further, should ATP be used as a fundamental molecule of energy storage, the ratio of ATP to ADP

and AMP would be a measure of whether microbes are active or dormant, but largely by analogy to terrestrial microbes.

5. *The unity of biochemistry in all Earth life suggests that life on another planet will have its own form of biochemical unity.* All life on Earth is biochemically the same. While there are a number of different kinds of metabolic strategies, they (as noted earlier) are variations on the themes of extracting energy from redox reactions or from visible light (sunlight). The information carrying and replicating molecules (DNA and RNA) are universal in Earthly life, as is the construction of proteins from a small subset of the chemically possible amino acids. Presumably there was a time in the earliest development of life when other "kinds" of living systems existed that used only RNA, or something entirely different for replication, or no replicating molecule at all (life based solely on metabolism). It is possible as well that catalytic or structural molecules were built on a foundation of a broader range of amino acids or on a combination of amino acids, nucleic acids, and possibly other chemical compounds (e.g., other carboxylic acids). However, the biochemistry that we know today was advantageous enough that all other versions of living systems fell by the wayside. The assumption then is that a similar victory would be won on other worlds—once life develops (or is introduced by impact) on some planet, the most robust biochemistry will take over at some point. From the point of view of life-detection techniques, it would be most convenient if this were always something like terrestrial biochemistry, but there is no guarantee that this would be the case.

6. *A robust molecule for genetic storage and replication, namely DNA and RNA, is a foundational requirement for life to be sustainable over geologic time.* There is, in principle, no barrier to living systems being based solely on metabolism. However, the work of Dyson discussed in Chapter 9 suggests that "metabolism-only" life is restricted in the number and complexity of polymers that can be employed in the living system. Metabolism-only life would be very simple biochemical systems that we might consider only to be at the very threshold of life. To go beyond these requires that a large number of long-chain polymers be synthesized repeatedly and reliably in the cell and that provision be made for preserving and replicating the instructions leading to such polymers. DNA and RNA are fine-tuned to perform these replication and transcription functions in modern cells, but different (perhaps simpler) replicating molecules might sustain a biosphere on another planet. We do not know how DNA and RNA developed, which molecule was first (but likely RNA), or what were the possible precursor or alternative molecules of replication and transcription.

7. *The earliest cells relied on a single replicating and catalyzing polymer (e.g., RNA).* Chapter 4 discusses in detail the elaborately intertwined system of DNA, RNA, and proteins that modern cells use to store genetic information, replicate it during cell division, and transcribe it for the production of proteins. DNA, as the molecule of genome storage and replication, relies on proteins for the enzymatic (i.e., catalytic) actions needed for its repair and duplication, and it relies on RNA for transcription, transport, and implementation of the genetic code. In turn, of course, proteins and RNA rely on DNA for the information to produce them. This self-referential system could not have started out so tightly interwoven, and of the three biopolymers cited (DNA, RNA, and proteins as a class), RNA was likely the first, because it seems

capable of catalytic function as well as of information storage and transcription. If very primitive life on Earth relied on RNA in place of DNA and some or all proteins, cellular structure and function might have been very different. The search for life on other worlds must take account of the possibility that extant life elsewhere might exist in a "single biopolymer" state or might have been extinguished in such a state, leaving only traces of its very simple biochemistry. What such a biochemistry would look like and how it would affect metabolism and the signatures of metabolic processes (e.g., alteration of isotopic ratios) is unknown but could perhaps be amenable to simulations in model laboratory systems.

8. *All life must be surrounded by a membrane and hence be self-contained.* The fundamental topological principle that underlies cells is the selective exclusion of the environment at large from the region of biochemical activity. More simply, life must define an "inside" and an "outside" in order to set up free energy gradients for polymer syntheses and to control the rates of reagent input and product output associated with such syntheses. While it is conceivable that life might be sustained outside such a membrane (e.g., as a string of polymers secured to a submerged rock), selective exclusion of the external environment is most chemically straightforward with a membrane. Thus, the separation of an "inside" from an "outside" ought to be a very early (and repeated) step in the chemical evolution toward life. We would look then for shapes or structures associated with biological containers and their membranes. It is far more difficult to resolve the question of whether very primitive membranes or enclosures could be recognized as such, and to what extent they would leave behind recognizable traces.

9. *The first cells were small.* There are a number of lines of argument for the first cells on Earth being smaller than what exists today as bacteria. First, the overall evolution of the genetic code has been toward increasing numbers of base pairs, and hence longer polymers. Thus, the earliest information carrying polymers occupied less volume. A workshop conducted in 1999 by the National Research Council of the United States concluded that the most primitive nonparasitic organisms potentially had, at most, a few hundred genes with a consequent size (radius across the primitive cell) of tens of nanometers. Self-folding, so-called amphiphilic vesicles described in Chapter 8 are approximately 10 times smaller than bacterial cells (Figure 14.1). Lest we miss signs of such extraordinarily small organisms, the search for evidence of life in extraterrestrial samples must include the capability to detect cells with approximately 1,000 times less volume than the smallest Earthly bacterium. Whether a system even smaller than this could be living depends on the plausibility of the "metabolism only" approach to life (Chapter 9) and the minimum packing size for the catalytic polymers involved in such a system.

10. *Earthly life defines the limits of life.* The most conservative assumption we can make is that the forms of life we are looking for are no more exotic than the most extreme forms of life we know of on Earth today. While this leads to an extremely narrow approach to the problem of the search for life, it is at the same time very practical. The most sensitive techniques that we discuss in Section 14.4, based on the polymerase chain reaction, will likely work only on known Earthly forms of life or modest variations of such life.

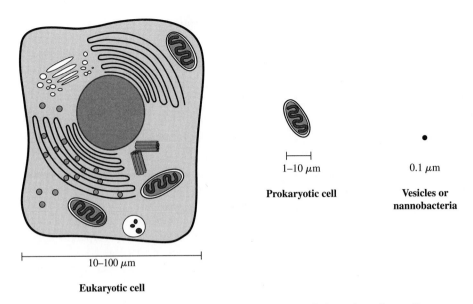

FIGURE 14.1 Size progression from eukaryotes to prokaryotic bacteria to the smallest feasible cells bearing DNA.

14.4 Modern Techniques to Search for Life on Other Worlds

The resurgence of efforts to detect life on Mars and other solar system targets (a key element of astrobiology) follows on the heels of a revolution in the field of microbiology, detection of life in extreme environments, and extensive efforts to measure small levels of contaminating terrestrial microorganisms in spacecraft prior to launch. In the past decade, thousands of new species of microorganisms have been discovered and genetically sequenced. Phylogenetic trees are being redrawn and restructured, and genomics—the listing of an organism's entire genetic code—is now economical for most microbial ecologists to undertake. The important consequences for life detection are the ability to better gauge the range of genomes represented by terrestrial microorganisms, the standard against which life detection must occur. Laboratory tools for analyses of extraterrestrial samples have become vastly more sensitive since the 1970s when certain kinds of techniques were simply not available. Table 14.2 lists a number of such techniques for detecting past and present life. Some of them are also used on Earth to determine whether spacecraft leaving Earth are essentially free of microorganisms that might contaminate other planets as well as to certify that samples returned to the Earth are free of microorganisms similar enough to terrestrial that they could pose an infectious threat. This section briefly discusses a few example techniques in more detail, moving roughly from techniques that are more specific to those that are more general.

TABLE 14.2 Specific Life Detection Techniques, Sensitivities and Present Limitations

Biosignature and Technique	Measurement for Life Detection	Sensitivity	Limitations and Developments Needed
Global			
• Spectroscopy (see Chapter 15)	Spectral lines in planetary atmospheres of extrasolar planets	Oxygen: 1% of Earth atm; other gases TBD	Imaging interferometry needs technical development
• Macroscopic imaging systems	Morphology of macroscopic life and ecosystems	Spatial resolution <10 meters	Solar system objects only
Morphological			
• Light microscopy	Structure; evidence for viabliity (motility; biofilms); noninvasive	200-nm spatial resolution	Morphology only; no chemistry
• Electron microscopy (environmental scanning electron microscopy)	High-resolution morphology and chemical composition (ESEM is noninvasive)	1–10 nm, 0.2 KeV, 1% relative abundance	Need contamination-free microscopes
• Electron microscopy	Structure, redox state, minerology; invasive	1 nm, 0.2 KeV, 1% relative abundance	Invasive sample preparation
• X-ray microscopy	Measures electronic state of molecules	Roughly nm resolution	Requires sectioning
• Fluorescence microscopy	Structure; detect very small entities, macro-molecules (nucleic acids and proteins); identify organisms (^{16}S and ^{18}S rRNA) and possibly viability (number of ribosomes; mRNA)	Can be used to enumerate viruses (30–50 nm)	Better preparation methods needed with rock and soil samples; present time needed to dislodge organisms for fluoresent in situ hybridization (FISH)
• CAT scan/imaging	Image of internal structure; noninvasive	Millimeter to centimeter scales	Higher spatial resolution needed
Mineralogical			
• Scanning electrom microscopy	Structural and abundance of elements	1–10 nm; 0.2 KeV, 1% relative abundance	Reduce diameter of EDX beam (<200 nm)

(continued)

TABLE 14.2 continued

Biosignature and Technique	Measurement for Life Detection	Sensitivity	Limitations and Developments Needed
• Ion and electron microprobes	Chemical and isotopic composition	Single organic molecules	
• Light microscopy, optical broadband spectroscopy	Chemical composition; noninvasive	0.2-μm spatial resolution	Limitations in spatial resolution
• Infrared spectroscopy	Structure and composition	1-μm spatial resolution	Improve signal to noise
• X-ray diffraction/ fluorescence	X-ray structure; minerals and elements; non-invasive	200-μm spatial resolution	Develop higher spatial resolution
• Mossbauer spectroscopy	Fe valence	Bulk sample	Measures Fe only
Organic Chemistry			
• Gas chromatography and mass spectrometry	Chemical composition, enantiomeric excess, diasteriomer specificity, structural isomer perference, and repeating structural units; lipid biomarkers; isotopes	Mass resolution 1:60,000; 10^{-15}–10^{-18} mol	Development of ionization techniques
• Chromatography for chirality (capillary zone electrophoresis)	Structure and chirality; enantiomer excess and repeating structural units	pmol	Disadvantage is need to derivatize sample
• Laser desorption/ laser ionization	Measure intact biomolecules	10^{-20} moles	Measures molecular weights only; need to develop better lasers and improve sample preparation
• Raman (IR spectroscopy and UV fluorescence)	Presence of organic compounds, pigments, biomineralization; noninvasive	1 μm	Need to develop low-noise detectors
• GC-isotope ratio mass spectrometry (CHONS isotope analysis)	Measure all biogenic elements and isotopic composition	nmol to pmol	Improve chromatography and ion source

Biosignature and Technique	Measurement for Life Detection	Sensitivity	Limitations and Developments Needed
• Chip chromatography-μ array antibody binding	Measure single organic molecule	single molecule	Need further development of sensors, detectors and arrays
• Liquid chromatography	Suitable of detecting enantiomer excess, diasteriomer specificity and repeating structural units	μmol	Improve resolution
Molecular and Biochemical			
• Polymerase chain reaction (PCR) and sequencing	Detect and sequence DNA and RNA; identify specific taxa	Theoretically can detect a single cell	Develop methods for in vivo PCR on single cells and fossilized cells
• Nanopores	Size and some structural information about biopolymers	Single molecules	Nanopore technology in development phase; linear molecules only
• Protein-chip/ chromatography/ stable isotopes	Molecular weight and structure of macro-molecules; isotopic composition		Need further development of sensors and detectors
• Metabolic analysis	Detect biological activity including metabolic pathways, and bioenergetic and biocatalytic activities	Depends severely on the metabolic process	Methods in the early development stage
Isotopic Analyses			
(See above for GC/MS and infrared spectroscopy)	Done with gas chromatography, mass spectrometry, and infrared spectroscopy		Methods in common use to test for life in the terrestrial geologic record

Source: John Baross, University of Washington.

PCR Amplification

The polymerase chain reaction, or PCR, is a way of making a large number of copies of a portion of a DNA molecule, taking advantage of the catalyzed biochemical reactions that permit the rapid replication of DNA during cell division. Each gene consists of a double strand of DNA, with the two strands running in opposite directions. As described in Chapter 4, the two strands have complementary base pairs and thus contain the same information. In the PCR amplification technique (Figure 14.2), the DNA strands are separated from each other by heating or chemical means; an enzyme called DNA polymerase is then used to make a copy of a portion of one of the two strands in an aqueous solution of free nucleotides. In addition to the enzyme, short pieces of DNA, called primers and corresponding to the starting sequence desired on each of the complementary strands, must be provided to initiate the replication. The solution is usually cycled in temperature, with several different peaks that allow separation of the DNA strand, binding of the primer to the exposed strands, and then finally polymerase catalysis of the construction of a new strand beginning at the primer and continuing along the remainder of the exposed strand. The technique can also be used with ribosomal RNA, although here an enzyme called *reverse transcriptase* is used to generate DNA from the RNA; this yields the highest sensitivity to detecting microorganisms.

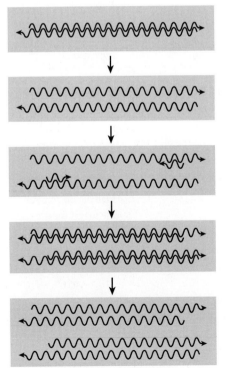

FIGURE 14.2 Schematic processing involved in the PCR technique. First, the complementary DNA strands are separated from each other by heating or chemical means. Next, primers initiate the replication catalyzed by the polymerase. The process is repeated again, now with four strands.

With an adequate supply of the raw materials, this doubling of the DNA strands can be repeated many times in rapid succession, leading to amplification of the DNA by factors of hundreds of millions to billions. Thus, detection of a single viable cell, or one recently dead, is possible. The technique is excellent for identifying potential terrestrial contaminants in spacecraft because very small amounts of DNA are detectable. It is less useful in searching for life-forms whose genetic code we do not know. Do we assume the same nucleic acid bases? What is the appropriate starting sequence? How long should the primer be? Because the number of possible sequences is enormously larger than what is found in the genes of organisms on Earth, simple guessing of the starting sequence is not sufficient to ensure a high likelihood of success, and the longer the primer the more specific the guess. However, shorter primers may be less effective in initiating the amplification process.

Were a PCR life-detection experiment to yield a positive result on another planetary surface (e.g., amplifying ribosomal RNA genes in a sample), it would almost certainly indicate a common origin with terrestrial life. The problem then would be to assess whether the genetic material detected was from recent terrestrial contamination of the spacecraft (prior to launch) or indigenous life with an Earthly heritage from a long time ago.

PCR amplification is unsuitable for detecting traces of extinct life, since both DNA and RNA degrade rapidly into nucleic acid bases and residues of sugars, from which amplification is not possible. However, it is difficult to obtain a false positive from the technique, that is, indicating life where life does not exist. The possibility of a false negative—missing the genetic signature of life present in the sample—is very high.

Tests for Metabolic Activity

These tests were the basis of the life-detection experiments conducted on the Martian surface by the Viking landers in the 1970s. Modern versions of this approach have the potential for detecting as few as 10^3 to 10^6 cells and require much less knowledge about the organisms sought after than does PCR amplification. Any form of biology is potentially detectable so long as the correct metabolic processes have been selected—or, at least, the right food. The approach is subject to false positives because abiotic chemical processes can mimic metabolism (e.g., redox reactions in the strongly oxidized Martian soil). Multiple tests can separate true metabolic processes from chemical imposters, though the Viking experience—discussed in Section 14.5—illustrates that controversy can still remain even in carefully designed protocols.

Enantiomeric Excesses

The detection of chiral molecules and the assessment of the population of left- versus right-handed forms is potentially a very general approach to identifying organic material from living or once-alive organisms. The essential enantiometric purity (the occurrence of a particular molecule exclusively in its left- or right-handed forms) of the amino acids used in proteins and the sugars used as DNA and RNA backbones may reflect a nearly universal characteristic of life. Enantiomeric purity imposes structural control and specificity of action that seems to be a central requirement of life (Chapter 4). The problems in detecting and characterizing chirality are several. First, techniques are not very sensitive, and 10^{10} molecules equivalent to 10^6 or more cells are often required to detect chiral excess. Second, some amount of specification of the molecules under consideration is required because particular chemical tests generally detect only certain classes such as amino acids and sugars. Third, enantiomeric excesses might exist in abiotic organic chemistry that has evolved to a certain level. That is, if enantiomeric selection is required for life, it must have evolved in at least some organic chemical systems prior to the origin of life itself. Of course, where the threshold between life and its precursors should be defined remains an open question (Chapter 9), so the detection of enantiometic purity by itself is exciting. Finally, after biological or protobiological processes cease, enantiomeric excess declines with time as molecules spontaneously convert to the oppositely handed form, at a rate that is dependent on temperature and availability of water. Hence, detection of small or modest enantiomeric excesses could be interpreted either as a sign of abiotic (but interesting) organic chemistry or as the remains of biological systems that ceased at some time in the past.

Techniques for detecting chiral excess vary. In commercial applications where the molecules are known ahead of time and where it is desired to establish the enantiomeric purity of the known substances, high-pressure liquid chromatography (HPLC) is a popular technique of choice. It is similar in basic principle to gas chromatography (GC), though intended for compounds of higher molecular weight and lower volatility than those suited for GC (including pharmaceuticals). Both techniques involve selective separation of different molecular species using a chemical or physical adsorptive coating in long columns, in which different compounds take different amounts of time to pass through the system and hence are separated. In chiral separations, one column may have a coating made of a chiral substance in which only one of the two enantiomers is present; in the second column the coating is racemic; that is, it possesses no enantiomeric excess. Chiral molecules that are passed through the columns will interact with the substrates such that two peaks are obtained in the racemic column while only one (or two with extremely different heights) is obtained in the column with the enantiomerically pure substrate. Figure 14.3 shows an example.

Any technique that generates a preferential trapping or sticking of one enantiomer versus another can be used to detect and measure chirality and the resulting enantiomeric excesses. Even on a very small scale, that of a monolayer of chiral sample sticking to a plate only a few millimeters in dimension, detection can be accomplished. One means to doing so is called a *quartz crystal microbalance* and consists of two plates, each of which is a sandwich of quartz underlying a copper conductor which in turn underlies the chiral adsorber. The two plates are each connected mechanically by a rod to a block, so they can oscillate, and to an electronic circuit that measures the frequency of oscillation. As the two plates absorb different amounts of enantiomer, corresponding to the excess of one enantiomer over another in the environment, their frequencies of oscillation change. By electronically mixing the two oscillatory signals, very small frequency shifts can be detected.

In a complex mixture of different amino acids and other chiral molecules, chiral discrimination may be difficult because the various compounds may or may not interact with the column substrates, with the coatings on the tiny plates, or with whatever other substrate for selectively trapping enantiomers is chosen. In fact, amino acids usually have to be derivatized (reacted with other substances to make the amino acid more easily identifi-

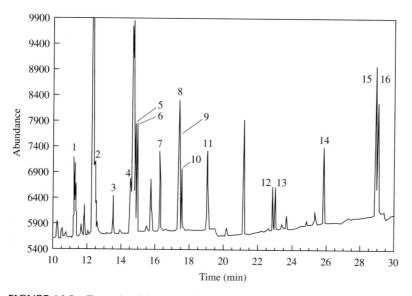

FIGURE 14.3 Example of the output from a chiral gas chromatographic column, such as that to be used in the analysis of the nucleus of a comet by the European Space Agency's Rosetta Spacecraft, launched in 2004. The different amino acids are well spaced from each other with one or two exceptions: The closely spaced peaks are samples of the same amino acid, but there is a different ratio of right- versus left-handed forms. Here, D- or L- indicates a nearly pure right- or left-handed enantiomer of the amino acid; where L- or D- are not indicated, the amino acid is in a racemic mixture: (1) α-amino-isobutyric acid, (2) alanine, (3) glycine, (4) β-alanine, (5) D-valine, (6) L-Valine, (7) norvaline, (8) D-isoleucine, (9) L-isoleucine + leucine, (10) 4-aminobutyric acid, (11) norleucine, (12) D-aspartic acid, (13) L-aspartic acid, (14) glutamic acid, (15) D-phenylalanine, and (16) L-phenylaniline.

able) before insertion into a chiral column, which may be too complex a procedure for an automated laboratory package. An initial separation of amino acids according to their electric charge is possible using a technique called *capillary electrophoresis*. The net result of such an initial separation would be a much higher sensitivity for amino acid detection, separation, and chiral characterization.

An entirely different approach to chiral detection rests on the fact that chiral molecules rotate the direction of the circularly polarized light with which they interact. Indeed, chiral compounds were (and less often, still are) referred to as "optically active compounds" in the chemical literature because of this property. Left- versus right-handed enantiomers generally rotate the direction of the light (i.e., the position of the electric field vector) in opposite directions. A device called a polarimeter can detect the absolute directions of circularly polarized light, and, hence, with suitable filters or other differencing techniques, the ratio of left- versus right-handed enantiomers can be determined. A compact, portable version of such a device has been developed in the United Kingdom for use on planetary missions. Where this technique again runs into ambiguity is in analyzing a complex mixture of different chiral compounds, with potentially different enantiomeric excesses (some perhaps favoring left-handed compounds, others favoring right-handed compounds), so that the net rotation of the polarized light may be difficult to interpret or even detect.

Generalized Gas Chromatographic and Mass Spectrometric Analyses

Experience with GC and mass spectrometry in various space applications, as well as broad applicability, make both of these techniques prime candidates for any life detection scheme. These were described in Chapters 11 and 13; an example of how the two techniques can be used to remove ambiguities in the analysis of organic compounds is shown in Figure 14.4.

Gas chromatography with mass spectrometry by itself can be used to determine whether biological or abiotic processes might have generated a particular sample of organic molecules, even in the absence of chiral or other special measurements. Because enzymes strongly alter the outcome of organic chemical reactions, preferentially generating some compounds while destroying or suppressing others, a plot of the abundance of organic compounds versus molecular weight will look very different for biologically mediated samples than for those that are abiotic. Figure 14.5 illustrates this effect. A sample

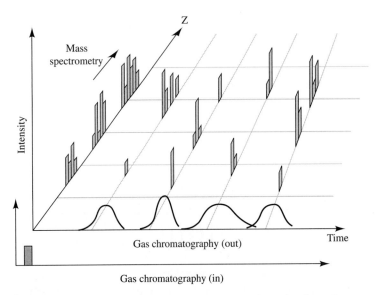

FIGURE 14.4 Connecting a mass spectrometer to the ends of gas chromatographic columns allows molecular species to be identified that otherwise could not be from either technique alone. A mass spectrometric analysis is shown on the z-axis as being replete with overlapping peaks leading to ambiguities in the identification of species. Preseparating classes of organic compounds using a gas chromatograph, as shown on the time axis, allows the individual mass spectrometric analysis of each separation to be much less ambiguous, both in terms of overlapping peaks and eventual reconstruction of the original composition of the gas.

14.4 Modern Techniques to Search for Life on Other Worlds

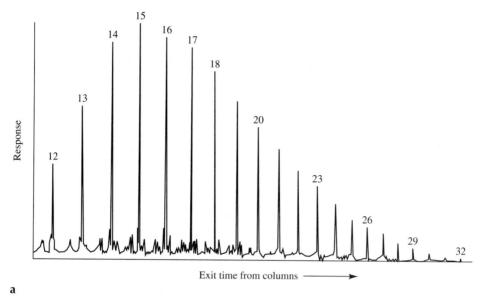

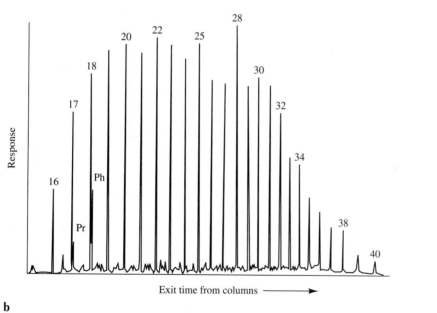

FIGURE 14.5 Comparison of gas chromatograms of organic systems formed by (a) abiotic and (b) biological processes. Panel (a) is the abundance of hydrocarbons produced in a set of reactions performed on formic acid. Panel (b) shows hydrocarbons produced by organisms.

of formic acid (HCOOH), the simplest carboxylic acid, undergoes reaction with water and some other simple organic and inorganic compounds in the absence of enzymes. It produces a suite of organic compounds, whose abundance declines in a smooth envelope according to molecular weight or, as in this figure, according to the number of carbon atoms. A sample of organic compounds derived from marine organisms, on the other hand, shows special preferences for compounds with a certain number of carbon atoms, and in particular (as is often the case) for compounds with an odd number of carbon atoms.

This simple technique of determining the pattern of molecular weights or carbon number has been used to argue that the organic phases found in meteorites were produced abiotically. The nonexistent (or, in the case of the Allende meteorite, weak) preferences of left- versus right-handed amino acids support the conclusion. Great care must be taken, however, in assessing the significant of such measurements, because organic remains of long-dead organisms might degrade in such a way as to mimic a purely abiotically produced sample.

Isotopic Enrichments

Carbon occurs in three isotopic forms (Chapter 3). The most abundant is ^{12}C, possessing six neutrons and six protons. By adding a neutron, we obtain ^{13}C, which is also stable, but addition of another neutron makes radioactive ^{14}C with a short half-life of 5,700 years. While the last of these isotopes is crucial for dating recent events in Earth's history (over the last few tens of thousands of years), ^{13}C on the other hand provides a tracer of the presence of past life in the geologic record of the Earth. The heavier ^{13}C more tightly bonds to carbon dioxide (CO_2) than does ^{12}C, and this difference in bonding energy translates to a slower reaction rate for the conversion of carbon dioxide to organic molecules, for example, via photosynthesis. The net result is that the lighter isotope, ^{12}C, is preferentially incorporated in the carbohydrates manufactured from the carbon dioxide and water.

Plants in the oceans and on the continents show a deficit of ^{13}C relative to ^{12}C in their organic phases of 2 to 3 percent when compared with the same ratio in carbonate sediments. Carbonate sediments should reflect reasonably well the isotopic ratio in carbon dioxide in the atmosphere at the time of precipitation; other inorganic carbon phases might show an enrichment in the value of $^{13}C/^{12}C$ if they were part of a closed system from which the lighter hydrocarbon was preferentially extracted by biological processes. Other biochemical processes that convert inorganic to organic carbon will show the same direction of fractionation, though to differing extents. The isotopic fractionations are generally well preserved in the sedimentary record for very long timescales, upward of billions of years. Therefore, a reduced value of $^{13}C/^{12}C$ in organic sediments, relative to the inorganic carbon phases found there, could be an indication of biological processes occurring at one time in the sediments (Figure 14.6). Measurements of fractionations of a few percent are easily done on organic samples by mass spectrometry.

The measurement of a distinctive fractionation in carbon alone is not a definitive test of the biological origin ("biogenicity") of an organic sample. Abiotic processes can also produce the same fractionation, albeit with less efficiency. On a planet such as Mars, where organic phases may be difficult to find in conjunction with contemporary inorganic phases, gauging the baseline value of the isotopic ratio, and hence the extent of fractiona-

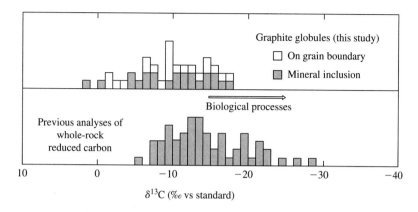

FIGURE 14.6 Values of $^{13}C/^{12}C$ for various organic and inorganic phases of carbon, expressed as $\delta^{13}C$, which is $\{(^{13}C/^{12}C)_{sample}/(^{13}C/^{12}C)_{standard} - 1\} \times 1000$. (Thus a 1 percent deficit is written here as 10‰, or 10 "per mil"). The organic (reduced) carbon is depleted in the heavy isotope.

tion, may be difficult. Therefore, it is advantageous to identify isotopic ratios of several elements that together more strongly argue for the past presence of biological processes.

Iron is an element that occurs often in biomolecules of various kinds, including proteins and co-factors. The stable isotopes have masses of 54, 56, 57, and 58, the second of these being most abundant. Iron phases are generally more robust against alteration in the geologic record than are carbon phases, and there are biological processes that fractionate the isotopes of iron one against the other in a recognizable pattern. Other elements of potentially similar utility in identifying biological processes include, but are not limited to, copper and zinc. Not all organic phases found in sediments will have associated with them multiple elements that fractionate in definite patterns associated with biological processes. However, for those that do, the prospect is very high for identifying the presence of past life, or ruling out a biogenic origin, with confidence on the basis of cross comparing isotopic fractions in multiple elements.

Mineralogical Biosignatures

The composition and appearance of minerals that may be the product of, or have been altered by, living processes, provide potentially powerful tools for the detection of past or present life. The measurement of isotopic fractionation patterns in elements associated with minerals, such as iron, represents one such technique that we covered in more detail in the previous section. The number of other approaches is so numerous that they are only listed and briefly described here. Banfield and colleagues (given in the References) provide much more detail.

1. *"Unexpected" phases implying a biogenic origin.* A phase is a region that is thermodynamically uniform (Chapter 7). Normally the ambient pressure and temperature

history of a mineral sample is sufficient to define the phase, namely, the crystal structure, in which it will form. However, some biological processes can generate phases out of equilibrium with the inferred temperature-pressure history of the mineral during its formation. $CaCO_3$ normally occurs as calcite during precipitation from ocean waters containing bicarbonate ions (Chapter 11). However, aragonite, which is a more compact crystal structure of the same chemical formula, is sometimes found even though the inferred pressures of the mineral-forming environment were too low to make this phase stable. Some biological processes, in fact, can generate aragonite in preference to the thermodynamically stable form calcite. As with other such examples, a limited number of abiotic processes can also produce "disequilibrium" mineral structures, so this test by itself is not definitive.

2. *Patterns of abundance of the elements can indicate a biogenic origin.* As with the isotopes, patterns of elemental compositions themselves can indicate that minerals had a biogenic origin. The beautiful mineral dolomite, which is $CaMg(CO_3)_2$ with certain particular structural ordering of the crystals, is generally a biogenic product. Some minerals, with bulk abundances that do not necessarily indicate a biogenic origin, may have minor or trace elements, whose abundances relative to the major elements (or relative to each other) are suggestive of biological synthesis. Finally, minerals produced abiotically may have their surfaces subsequently altered by the action of organisms. Bacteria that oxidize sulfur may obtain the element from the surfaces of metal sulfide minerals that are in the process of dissolving in water, leading to a dearth there of elemental sulfur that would normally be expected to accumulate because of the dissolution process. Organisms may themselves generate very small mineral grains or "granules" of highly unusual composition within their own cells; once the organisms die, the granules remain as a compositional oddity. Sulfur-oxidizing bacteria, for example, will sequester toxic selenium into sulfur granules.

3. *The surface morphology of minerals, and their particle sizes, may be altered by organisms.* The presence of amino acids from microorganisms can lead to particular, pitted textures at the surfaces of dissolving minerals. During the converse process, namely, precipitation of minerals, the mediation of microorganisms can alter the size range of mineral crystals. Large crystals generally require that the medium within which they grow be just above the thermodynamic threshold for their formation, that is, just above saturation. Enzymes can act to suddenly lower the free energy of the mineral phase, effectively supersaturating the medium and forcing rapid precipitation. This, in turn, favors the appearance of a large number of small crystals rather than a few large ones as in the abiotic case.

4. *The spatial arrangement of crystals may indicate a biogenic origin.* Minerals can precipitate out of solution directly on, and adhere, to the surfaces of cells and cellular structures such as bacterial strands or sheaths. Iron and manganese oxides are examples of minerals that do this preferentially. When the organism that served as the "mold" for the mineral precipitates dies, its signature in the sediments may disappear while the more durable mineral cast remains. If the cast retains the shape of the original organism, the resulting durable product is a microfossil (Figure 14.7).

FIGURE 14.7 Electron micrograph images of shapes produced by the "biomineralization" products of bacteria. In panels (a) and (b), bacterial sheaths are coated with iron hydrite minerals. In panel (c) on page 458, a biopolymer produced by a bacterial cell has been coated with the iron hydrite mineral akaganeite.

a

b

Searching for microfossils with microscopic imaging systems is a plausible approach to searching for life. Indeed, as discussed in Section 14.6, one of the early arguments for the detection of life in the Martian meteorite ALH84001 was a high magnification view of very small structures that "looked" like cell parts. However, abiotic mineral growth on surfaces can mimic such structures (Figure 14.8, see page 459); hence, identification of the past existence of life from mineral morphology is fraught with uncertainty. It must be combined with other, different tests to make claims of life detection believable and verifiable.

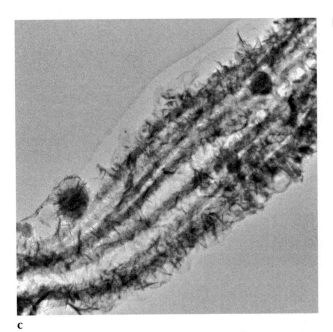

FIGURE 14.7 continued

c

5. *The details of the morphology of crystals may be diagnostic of biogenic origin.* Essentially all organisms require iron so that they may successfully complete key metabolic processes. Iron is not, however, very soluble in water when oxygen is present, as the process by which banded iron formations occur illustrates (BIFs; Chapter 11). Hence a number of microorganisms have developed ways to store iron in the form of crystals of the minerals ferrihydrite, magnetite, goethite, and lepidocrocite. These minerals exhibit magnetic properties, and organisms that generate such crystals are referred to as *magnetotactic*. Magnetotactic organisms produce crystals that usually are arrayed in the form of chains with characteristics that, in total, are diagnostic of a biogenic origin for the crystals (Figure 14.9, see page 460). These characteristics include crystals of unusually pure composition compared with abiotically generated ones, a crystal structure or "lattice" that has very few defects, particular volumes and aspect ratios of length to width that confer certain magnetic properties, and unusual shapes both in the overall crystal and the faceting of the ends of the crystal. The occurrence of such characteristics arises from the thermodynamic effect of organic ligands (active molecules that usually bind to a protein and create a specific chemical effect), which alter the relative stability of different crystal structures and hence force the growth of particular, and potentially diagnostic, characteristics.

After the bacteria or other microorganisms that produce such crystals die and disappear from the sedimentary record, the crystals and their special properties remain for geologically long times. Although the three-dimensional structure of the chains may collapse over time, enough remains recognizable that such chains and their

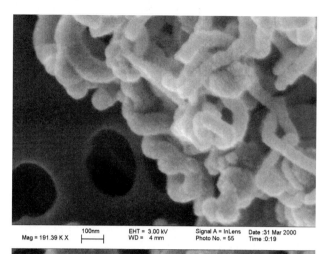

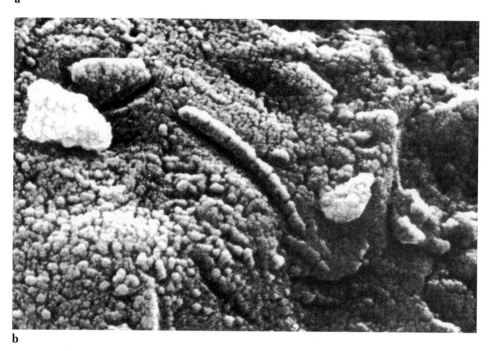

FIGURE 14.8 (a) Scanning electron micrograph images of filamentary forms in a terrestrial rock and aggregates of iron oxihydride mineral. (b) Scanning electron micrograph images of a portion of the Martian meteorite ALH84001, showing filamentary features. Scale bars for both images are shown.

crystal properties have been used to identify the past presence of magnetotactic bacteria in terrestrial soils. Use of such a biosignature to infer that life once occurred relies on the assertion that, while one or several of the characteristics of crystals generated by magnetotactic bacteria may occur in abiotically generated crystals, no such abiotically produced crystal has ever simultaneously shown all of the characteristics depicted in Figure 14.9.

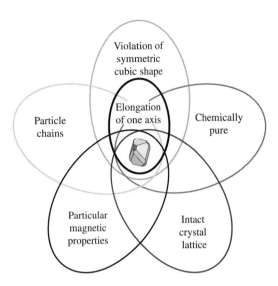

FIGURE 14.9 A Venn diagram showing the various characteristics of magnetite crystals produced by magnetotactic bacteria. If most or all such characteristics occur, the minerals are likely to be of biogenic origin.

6. *Certain groups of minerals may not form together in the absence of life.* Because assemblages (collections) of different minerals that form simultaneously must obey certain phase relationships (Chapter 7), the presence of unusual assemblages might indicate the action of biology. Metal sulfides in fine particulate form are sometimes found in contact with carbon-rich sediments, which should have prevented the formation of the sulfides. The small size of the grains is crucial to inferring the biogenic origin of the minerals, because it ensures effective contact between the sulfides and the carbon-rich sediments. The difficulty in using such assemblages to infer the past presence of biology, particularly in ancient samples, is that the two phases might have formed abiotically at different times and later come into contact with each other in a relationship that only mimics simultaneous, biologically mediated, formation.

14.5 Case Study in the Search for Life on Mars I: Viking

In 1976, the United States successfully landed two spacecraft on the surface of Mars. Unlike other planetary missions up to that time, the focus of the Viking landers was on the search for life as well as for evidence for organic molecules. (The mission also included two orbiters, which mapped the geology and the surface cycles of carbon dioxide and trace water on Mars.) Each spacecraft was equipped with a scoop for collecting samples of the upper few centimeters of soil; soil was placed in hoppers for analysis by the various instruments.

Two of the instruments are of particular interest here: a composite set of life-detection experiments and a gas chromatograph and mass spectrometer (GCMS) combination. Soil from the sampling arm was delivered to three separate experiments. The first, called the pyrolitic release (PR) experiment, was designed to detect photosynthesis occurring in the soil by measuring the uptake of carbon dioxide and carbon monoxide by Martian microbes as the raw materials with which to make (with water) organic compounds. The carbon provided by the experiment was relatively rich in radioactive ^{14}C, which served as the tracer for the uptake of carbon by biological processes. The soil was exposed to visible light equivalent in level to that experienced on Mars but without the damaging ultraviolet component that bombards the surface of the red planet. After an exposure period, the chamber was heated to a temperature sufficient to pyrolyze, or decompose, the organic compounds. The organic vapors were then trapped and the inorganic gases swept out; detectors then measured the level of ^{14}C in the organic phases. No detection of the radioactive carbon was made, suggesting that no photosynthesis actually occurred during the experiment.

The second and third devices, called the labeled release (LR) and gas exchange (GEX) experiments, looked for small amounts of gases released by organisms during metabolism. The first of the two used nutrients with radioactive tracers present, and then looked for the radioactive tracers to appear in the gas phase after metabolism was under way. The second used a GC to directly measure the gases released during metabolism by microorganisms.

All three devices "activated" their soil samples by adding liquid water, which is a nonexistent quantity on most or all of the surface of Mars today. No evidence of photosynthesis was seen, as noted, and the LR experiment was negative as well. However, the GEX showed evidence of reactions in the soil, through release of large amounts of oxygen, as soon as water was added. Here, rather than revealing the unambiguous evidence of life, the experiments illustrated their own ambiguities. The activity began suddenly and died away very quickly as well, which is not characteristic of the behavior of microorganisms. Other peculiarities in the response seen in the experiment led all but a handful of scientists to argue that life had not been detected. Instead, the prevailing opinion was that the soil itself was strongly chemically reactive and, in particular, contained highly oxygen-rich iron oxides. When the water (H_2O) contacted the soil, the iron oxides donated oxygen to make peroxide, H_2O_2, in a reaction that mimicked, very crudely and briefly, metabolism.

The GCMs results weighed in very strongly against the presence of life in the soil. Down at the level of parts per billion, no organic molecules were detected. Whether this is a result of the complete absence of life in the Martian crust or simply the destruction of organics in the uppermost layer of the highly oxidized soil remains unresolved. It is generally accepted, though, that the Viking experiments detected abiotic chemical reactions in soil that, for the first time in millions or billions of years, had been placed in contact with liquid water. The Viking experiments illustrated the dangers of relying on metabolic tests to reveal the presence of life, absent other techniques to study at the molecular or macroscopic level the structures of life itself.

14.6 Case Study in the Search for Life on Mars II: ALH84001

In August 1996 researchers brought the search for Martian life to the attention of millions of people around the world with the astonishing assertion that they found evidence for relict biological activity in a meteorite that came from Mars. Meteorite Allan Hills (ALH)

84001, found in Antarctica in 1984, is one of a number of meteorites delivered from Mars by impacts that put them on a collision course with Earth. Some of these meteorites contain trapped gases that, in their chemical and isotopic signatures, are identical to the present atmosphere of Mars as sampled by the Viking mission. Other Martian meteorites do not contain gas, but because their compositions and textures are similar to the gas-bearing rocks, they too are very likely to be Martian rocks. Work over the past decade by a number of researchers has shown that it is plausible for a number of large impacts to have gouged out portions of the Martian crust and sent a portion to Earth to account for the quantity of discovered Martian meteorites.

ALH84001 is an old igneous rock. Radioisotopic dating provides a history of the rock. The isotopes samarium and neodymium yield ages as old as 4.56 billion years and as young as 4.36 billion, based on analyses by two different groups. Therefore, ALH84001 is a rock that is representative of the earliest Martian crust. The rock, when embedded in the Martian crust, was heated again about 4.0 billion years ago by one or more strong shocks, based on dates from ^{40}Ar (argon) and ^{39}Ar, as well as the rubidium–strontium isotopic system (Chapter 6). The shock (or shocks) almost certainly was from a large, nearby impact, since it was in this period that the inner solar system was being intensely bombarded. Sometime after this, as early as 3.9 billion or as late as 1.4 billion years ago, fluids of some unknown composition rich in carbon dioxide flowed through the rock and produced globules of carbonates. These suggest (but do not require, because carbonates can form without water under certain circumstances) the presence of liquid water flowing through the rock.

The age of the carbonate formation in the rock quoted above, as determined by potassium–argon and rubidium–strontium systems, is highly uncertain. However, even the youngest age of the carbonate is much older than the estimated date when the rock was blown off Mars. The blow-off age is constrained by examining tracks made by cosmic rays on the surface of the rock exposed to space. The abundance of unusual isotopes of noble gases made by the cosmic-ray strikes gives a residence time of ALH84001 in space of between 10 and 20 million years. Once on Earth, isotopes such as ^{14}C and others of boron and chlorine, also made by cosmic-ray hits, begin to decay; their abundances indicate that ALH84001 has been on Earth only 13,000 years. Therefore, the impact and delivery of this rock to Earth were far more recent than the formation of the carbonates, indicating that the carbonates were produced when the rock was in the Martian crust.

David McKay of NASA's Johnson Space Center and colleagues went a step further in 1996 to delineate three properties of ALH84001 that are consistent with (but do not require) biological activity. First, ring-shaped, carbon-bearing molecules called *polycyclic aromatic hydrocarbons* (PAHs) were found near the carbonate globules. Although PAHs can be formed abiotically (e.g., in interstellar space, where their spectra are clearly detected by ground-based telescopes), there are subtle differences in structure between biological and abiotic PAH molecules. The ALH84001 PAHs more closely resemble those produced biologically. However, no one has demonstrated that it is *required* that such PAH structures form by biological processes, either, and so they provide only a consistency argument for biology.

The second line of evidence comes from the association of two mineral types, magnetite and greigite, in the carbonate globules. The formation of these two mineral types in contact with each other, and at the same time, requires very alkaline water, an unusual geological situation, or the mediation of biological activity. Further, the texture of the mineral grains resemble those produced by bacterial processes on Earth. While these seem to point strongly toward biology, there are some ambiguities: (1) The identification of greigite is tentative; (2) the unusual textures are consistent with, but do not necessarily require, biology; and (3) mineral assemblages that otherwise appear in thermodynamic disequilibrium (as do these) can be constructed by forming the minerals separately and bringing them together, or forming one first and then later the other, so that limited thermodynamic contact between the crystals is obtained. In this way, a disequilibrium situation suggestive of biological mediation can be simulated.

Most controversial are images, constructed by bombarding electrons into the sample, of very small structures that look like pieces of bacteria. They occur near the globules, and they are about 10 to 100 times smaller than terrestrial bacterial. Because they "look like" microbial forms, McKay and colleagues argue that they could be evidence of life. Their invocation as support for the Martian "microbe" interpretation has raised more controversy than resolution. Because the structures in ALH84001 are, in most cases, smaller than the putative nannobacteria, they raise the issue of just how small DNA-based life can be. Studies of extremely primitive organisms and of viruses suggest that a self-replicating cell can contain no fewer than several hundred genes (as noted in Section 14.3) and be perhaps 100 times smaller than normal terrestrial bacteria. Of course, Martian organisms might not have relied on DNA and RNA for information encoding, or been so primitive that they did not do information encoding at all (Chapter 9). UCLA paleontologist William Schopf points out that if that the forms in ALH84001 are cells, they should have cell walls, which cannot be seen in the images. Alternative explanations for the structures rely on the observations that certain kinds of minerals grow with very unusual surface structures, including some that look remarkably like the "cell" fragments of ALH84001 (Figure 14.8).

One final argument put forward in favor of biology was raised first by McKay and colleagues, then in more detail by Joe Kirschvink and colleagues at Caltech, and later by others, and it concerns the appearance of magnetite crystals found within ALH84001. One-fourth of the magnetite crystals in ALH84001 have an elongated, chemically pure hexagonal prism shape that is *identical* to that in certain bacteria on Earth. Magnetite grains with that shape have not been successfully produced abiotically. Some researchers have pointed out that many of the magnetite crystals in ALH84001 occur in chains, also consistent with biological production (Figure 14.10). However, others have disputed this interpretation. The origin of the Martian magnetite crystals remains ambiguous. The study by Kirschvink and colleagues stands as the most detailed and careful analysis of the crystals, and their interpretation of a biogenic origin for the crystals remains viable today.

McKay and colleagues argue that the history of ALH84001 is concordant with the popular view of an ancient Mars that was warm. If the meteorite's carbonates are more than 3 billion years old, then their formation via living processes is consistent with a picture in which life on Mars existed because liquid water was stable back then. But, as noted earlier, the carbonates could be younger. Further, analysis of organic carbon in the meteorite by two groups demonstrated that most—if not all—of the meteorite's carbon not in carbonates is terrestrial in origin. For example, analysis of the ratio of ^{14}C to ^{12}C indicates a large fraction of the organic carbon to be very recent and hence acquired after the meteorite fell

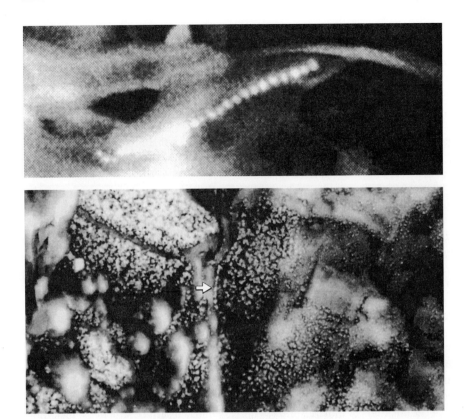

FIGURE 14.10 Chains of magnetite crystals imaged in ALH84001 (bottom, set off by arrows) seem to resemble those seen in bacteria on Earth (top).

on the Earth. In summary, *none* of the evidence present in ALH84001 requires interpretation in terms of Martian biological activity (except, arguably, the magnetite). The various lines of evidence would be consistent with Martian biological activity were there an additional "smoking gun" piece of evidence for life within that rock—but no such evidence exists. As the University of California paleontologist J. William Schopf has noted, "extraordinary claims require extraordinary evidence." Any claim of the detection of life beyond Earth is extraordinary!

The experience with the Martian meteorites illustrates the great challenge of finding definitive, irrefutable evidence for microbial life from subtle chemical clues and microscopic images of both mineral crystals and putative "microfossils." The strength of the techniques used to search for life in the meteorite is their generality, but that too is their weakness. It will be even more difficult to perform such analyses on the surface of Mars; the experience with ALH84001 illustrates that the best strategy is to locate promising sites on Mars where life may have existed and then return samples back to Earth for detailed laboratory analyses.

14.7 Future Searches for Life on Mars

The past half-decade of Mars exploration has yielded exciting results that suggest liquid water existed in many places and times over the history of the planet, and one landing site appears to have once been a salty lake or sea (Chapter 12). Orbiters later in the decade will survey much more of the surface at high spatial resolution and with powerful spectrometers to identify sites that might have liquid water for long periods of time, that might still hold organic molecules, and that would be targets for a mission to directly sample, test for organic molecules, and for the presence of life.

Such sites would be the goals of the Mars Science Laboratories (MSL) (Figure 14.11), described in Chapter 12 as going beyond the Mars Exploration Rovers in power available for instruments, roving distance, and computing power. These wheeled laboratories will have some decision-making capability. They would examine rocks microscopically, measure isotopic anomalies that could indicate life, search for organics, and possibly sequester some samples in a cache location for future pickup. Such landers will probably also have to drill as much as a meter or more below the surface to get below the destructive oxidizing layer found by the Viking life-detection experiments. Drilling is a difficult, time-consuming, and expensive operation that is especially tricky without the intervention of humans. Whether a significant drilling device can be included or not on the first of the MSLs is yet to be determined. Should the first MSL detect no positive indications of life, the exploration cycle might be repeated with more orbiting instruments and then with another MSL or two sent to new locations, for which drilling would become more imperative.

The search for organic molecules in the Martian crust will be taken to the high northern latitudes via the Project Phoenix Mars Lander, to arrive in 2008. By landing where there is elemental evidence of permanent, near-surface water ice (provided by the Mars Odyssey spacecraft), Phoenix can drill a short distance to determine what compounds—including organics—are present in the ice.

FIGURE 14.11 A mobile Mars Science Laboratory exploring Martian canyon country.

At the same time, or later, the technology for returning samples from Mars—much more difficult than from the Moon—would need to be validated in the Martian environment. Simple tests of rendezvous and docking might be conducted in Mars orbit while landed spacecraft might launch very small return capsules containing a "grab bag" sample of whatever is around. These tests would lead to the climactic drilling and collection of carefully selected samples that could be returned to the Earth or to the arrival of a mobile launch platform to pick up earlier cached samples from a site of interest.

Where to land remains one of the most crucial questions in astrobiology, to which there are no good answers. The Opportunity landing site (Chapter 12) is an excellent candidate. Canyons and chasms where sediments have built up, where water may have once flowed, and where there is evidence (in places) of standing water are highly attractive in the search for life but very difficult to reach safely with landers and to navigate with rovers. The edges of the shield volcanoes, where evidence exists of late heating of the crust and mobilization of liquid water, are also of high interest, especially given that Earthly life can exist in basalts and other volcanic or igneous rocks (Chapter 10). Almost certainly, though, gathering evidence for life in these regions (past or present) will require drilling below the subsurface. Finally, ancient evidence for life might exist around the perimeter of the large northern basin if that basin indeed at one time held an ocean. Places where channels apparently fed water at one time into the basin will be tempting sites, and their smoothness would make them relatively safe landing locales.

Sample return missions must be prepared with careful consideration of the risk of contamination. Both contamination of Mars by Earth organisms (forward contamination) and the contamination of the Earth by Martian organisms (back contamination) are of deep concern to planners of the Mars Exploration program. The Viking landers were heat sterilized, and no trace of life—terrestrial or otherwise—was found after arrival on Mars. Pathfinder and the Mars Exploration Rovers were also sterilized, though not to the extreme standards of Viking. International protocols now govern the sterilization, testing, and prelaunch isolation of Mars landers to avoid forward contamination. Examination of spacecraft under ultraviolet light to see fluorescence from organic molecules (metals do not fluoresce as effectively), swabbing of parts followed by PCR amplification, and introduction of molecules that would "tag" terrestrial hitchhikers are all under consideration. While the ultimate goal is to leave Earth with no terrestrial organisms, this is probably unrealizable, and sterilization efforts are geared to achieving cleanliness levels as high as are practical, without damaging the spacecraft itself (e.g., heat sterilization is particularly damaging to electronics).

Back contamination is a trickier issue because of the potential public concern that the return of Martian organisms might lead to new plagues. This concern echoes that for the Moon in the late 1960s, when a quarantine facility for returned lunar samples and astronauts was set up in Houston, Texas, an effort that proved unnecessary. In the case of Mars, if we seek evidence for life then we must acknowledge that returned samples could contain organisms never before introduced on the Earth. While Mars and Earth have never really been isolated, since rocks from both planets have been swapped by impacts and interplanetary transfer survivable by many bacterial organisms, we cannot rule out the possibility of introducing something that has never before been in contact with Earthly life.

Plans for handling returned Martian samples involve taking the sealed capsule immediately to a quarantine facility built to the highest standard based on facilities in use today to handle dangerous pathogens. A report of the National Research Council recommends dividing the sample into four pieces. One piece would be radiation and heat sterilized and removed from the lab for geochemical and geophysical analyses of the Martian crust. The second sample would be chemically treated and then heated to break apart proteins and nucleic acids and vaporize their components, and then these now-safe organic vapors would be transported to outside laboratories for analyses of their composition. The third component would remain in the facility and be tested for life with a variety of different approaches, including PCR amplification, metabolic and isotopic tests, and other techniques. A fourth component would remain in the facility, untouched, awaiting new technologies that might be used to test for life in the future. The rationale for analyzing portions of the returned sample outside the quarantine facility is that the accommodation of the totality of laboratory analytic devices within a single quarantine facility is impractical. Once sterilized, or once polymeric organic compounds are broken down to simpler components, the most effective analysis approach (fully safe) is to bring the sample to qualified laboratories around the world.

When will all of this happen? Current NASA plans put an advanced Mars Science Laboratory on the surface in 2010, with return of samples to Earth by the middle of the next decade (e.g., 2015) or thereafter. Results of the Mars Exploration Rovers and the continuing successes of the MGS, Mars Express, and Odyssey orbiters (Chapter 12) have created momentum to maintain an aggressive Mars exploration program, but the history of Mars exploration is one of unpredictable starts and stops, and so it is difficult to be confident about when a particular capability will be deployed in orbit or on the surface.

14.8 Searching for Life in the Subsurface of Europa

With the probability very high from Galileo magnetic field measurements that Europa possesses an internal global ocean, interest in sampling for life is now keen. However, to simply send a spacecraft to land somewhere on Europa, and then drill, is to accept a high level of risk that the mission would not reach liquid water. Currently we do not know whether the average thickness of the Europan crust is kilometers or meters, and if the former is the case, whether there might be isolated places where the crust is very thin (Chapter 13). Robotic drilling through hundreds of meters of cold ice is impractical, given the difficulties encountered and need for constant human intervention in the effort at Lake Vostok in Antarctica. Thus, an orbiter to measure the variations in the thickness of the Europan crust, described in detail in the previous chapter as the next step for Europa exploration, may be a key stepping stone for the search for life under the crust of Europa.

Should the orbiter identify places where the crust is very thin, the appropriate response would be to target those locations for a lander. Direct penetration through the crust with a projectile is one approach to breaching the ice, but it is risky. Alternatively, drilling or melting might be considered, but with the ice at a typical temperature of 120 K, melting will consume large amounts of energy. The intense radiation environment, instantly deadly to humans and destructive to electronics on short timescales, necessitates a robotic mission of limited duration. If the ocean can be reached, sampling for evidence and composition of organic compounds, isotopic analyses to test for patterns indicative of life, and both large-scale and microscopic imaging (with artificial illumination) to detect organisms will be part of the package required to explore the depths. Analytic tools could include mass spectrometers and gas or liquid chromatographs (the latter analyzing liquid phases in a way analogous to the functioning of the gas chromatograph). An advantage of selecting a portion of the ocean that is overlain by the thinnest crust is that photosynthetic life, if it exists, will likely reside there and could be detected through the spectral signature of chlorophyll or other light-capturing pigments. Direct exploration of the Europan ocean will be limited in spatial coverage and depth, given the operational difficulties of working and communicating back to Earth from below the Europan crust.

It has been proposed that regions of the surface of Europa open and close on a regular, tidal schedule (Chapter 13), exposing the ocean to space and depositing oceanic organics on the surface. A landed mission without drilling might take advantage of such regions by timing its arrival for the opening of a particular crack region or set of cracks where organics could be extruded onto the surface. Even were the vehicle not to attempt to pass through the open crack, radiation processing of the organics on Europa is so rapid that the analysis package would need to be on site very quickly after extrusion of any organics to enable analyses of their composition and structure as they existed in the protected environment below the surface. The timing required to arrive at a crack as it opened would at least require multiple observations from orbit and more likely would require the lander to first go into orbit around Europa so as to target a particular crack and arrive at the right part of the tidal cycle. Of course, before targeting a mission to take advantage of tides in this way, the hypothesis would need to be tested by orbital science that indeed there are cracks that open and close on orbital timescales.

In spite of the intense radiation environment surrounding Europa, panels of the National Research Council have recommended strict sterilization of spacecraft bound for Europa, including orbiters that might accidentally crash on the surface or will eventually crash after completing their mission as the gravity of nearby Jupiter changes the shape of the spacecraft's orbit around Europa. The Galileo Orbiter, after fulfilling its science on multiple Europa flybys, was directed into a series of orbits that led to its deliberate destruction within the atmosphere of Jupiter—eliminating any possibility of its contaminating Europa. The best example of a very radiation-resistant organism is the bacterium *Deinococcus radiodurans,* which exists in radioactive waste dumps on Earth and can survive somewhat more than the total dose within the aluminum walls of an orbiter during a month-long mission around Europa. Lower temperature and dessication make the organism even more resistant, and, in a hydrated medium, the irradiate cell can repair itself and then go on to reproduce. It would be tragic, scientific or otherwise, were we to send a spacecraft to penetrate the Europan ocean and there find—as a survivor of a "shipwrecked" Europa Orbiter from years before—*Deinococcus radiodurans* prospering in the alien ocean.

14.9 Exploration of Titan for Clues to the Origin of Life

Titan's contribution to astrobiology is in potentially providing us with samples of organic materials that have evolved beyond what is possible in terrestrial laboratories, but not yet to the living stage (Chapter 13). Sampling of the organic phases at the surface will require careful pre-selection of sites via the Cassini Orbiter, which has the capability to remotely sense, via radar, near-infrared imaging and spectroscopy over a range of wavelengths. Regions of the surface that exhibit young craters, or relatively recent ice–ammonia flows, and that show evidence for nearby organic deposits will be of particular interest. Should such areas manifest themselves, then astrobiologists will be keenly interested in fielding a Titan Organics Explorer to Saturn's largest moon.

Titan's high atmospheric density and low gravity, along with the benign radiation environment deep in the atmosphere, make this the ideal world for exploration via aerial vehicles. One strategy for identifying and sampling the significant prebiotic organics, using a blimp, was described in Chapter 13. A protocol for sampling the organics to search for signs of life is shown in Figure 14.12. Gas chromatography or liquid chromatography

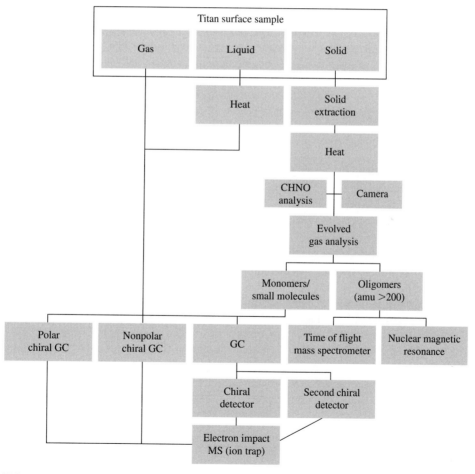

FIGURE 14.12 Protocol for the analysis of organic samples in the surface of Titan, progressing from simple elemental analyses to determination of enantiomeric composition and polymeric structure.

coupled to mass spectrometry would provide compositional data. Identification of chiral molecules and any predominance of left- versus right-handed forms is crucial to understanding how far toward life's origin the sample has proceeded (Chapter 8). This can be done using chiral chromatographic columns or other, separate, devices as described in Section 14.4. For the polymers, determinations of the total oxygen content, abundance as a function of carbon number, the presence of chiral molecules, and the enantiomeric excess are done after the sample is broken down into monomers. The intact polymers would be analyzed prior to this to assess their overall structure and to determine if there are any unusual preferences in the orientations of the monomers along the polymer chains. If the sample looks promising, more detailed analyses would be scheduled. If organic molecules were embedded in the water ice, a core would have to be extracted via drilling in the cold, hard water ice.

One area that might not be chosen for sampling is the Huygens landing site. The extreme cold and dryness of Titan led to the decision in the 1990s not to sterilize the Huygens probe. While almost certainly any hitchhiking life would be dormant or destroyed by the cold, no one would want to risk sampling terrestrial organic molecules. There is no need to move far from Huygens, however, to be assured of a clean sample site, since the pioneering Titan lander will touch down softly and likely remain intact.

14.10 Coda to the Search for Life

Go most places on Earth and life will jump up at you; only in the driest and coldest sites on Earth is life hard (though not impossible) to find. But try and conduct remote collection, handling, and analysis of samples, even on Earth, and you will find that the detection and characterization of the tiniest organisms, or of the incomplete record of organisms long extinct, is not easy. The real dilemma of Mars, and of Europa, is that life *could* be there but will be very hard to find or to access. Were these to have revealed themselves as worlds that were always dead and that did not then or do not now have liquid water, we could have passed on the search for life within our own solar system. But the strong evidence for a Mars that was warm and wet in the past, the hints of liquid water there today, and the vast ocean of water lying under the Europan crust beckon us. We will need the best of today's technology, and some of tomorrow's, to conduct these searches. And—together with the organic exploration of Titan—they will represent some of the most difficult frontiers of solar system exploration. Beyond such searches, we look to the nearest stars for hints there of Earthlike planets. That search, to be conducted by remote sensing over vast distances, is the subject of the next chapter.

QUESTIONS AND PROBLEMS

1. Imagine that you are charged with assembling a payload to search for extant life on Mars. Use Table 14.2 to select (a) up to three instruments, (b) up to six instru-

ments, and (c) up to 10 instruments to search for existing life in a recently discovered hydrothermal system on Mars. Justify your choice of instruments, and create a flowchart that describes the order in which you would use them on a sample. (Do not worry about mass and power on the spacecraft—assume you are doing this on Mars, with a highly powerful rover/driller that does not limit you in the listed commodities.)

2. Perform the same analysis as in question (1) but for a site such as the Eagle Crater outcrops on Mars (Chapter 12), where you are searching for traces of extinct life.

3. Perform the same analyses in question (1), but now for an ocean environment beneath the crust of Europa.

4. Take Figure 14.12 and imagine that you are limited by spacecraft power to three instruments to analyze organic phases on Titan's surface for evidence of life or organic chemistry on the threshold of life. Select the three instruments, justify why you chose them, and draw a flowchart indicating how you would use them. (Remember that you cannot assume knowledge of the organics that you would have obtained from other instruments you have left off the list!)

5. Do a literature search to find images of magnetite crystals produced by terrestrial bacteria and magnetite produced both by abiotic geological, and industrial, processes. Describe the differences.

6. Suppose life were discovered on Mars that used DNA, but the base sequences were virtually nothing like those in the terrestrial organisms. Could you argue for a separate origin from the Earth, or only an early divergence (last impact-generated exchange being, say, 4 billion years ago). Is there a way to tell the difference?

7. In the extraordinarily dry Atacama Desert of Chile, bacterial colonies are isolated enough that they show regional variations in their genetic type—suggesting that such differences develop quickly. Do a literature search on the sampling of the Atacama for bacterial colonies, and speculate as to why such colonies are so easily isolated one from the other, even in the Earth's environment.

8. Write a short and well-considered essay on the pros and cons of bringing back Martian samples that might contain life. What are the risks of contaminating Earth, based on National Academy reports (see the Suggested Readings), and are they worth the scientific payoff? What are the trade-offs in doing analyses directly on Mars?

9. Do a literature search on past proposals that worldwide plagues were generated not by extant terrestrial organisms but by material transported from space. Why are such arguments compelling or not compelling?

10. How might discovery on Mars of organisms that use a completely different scheme for gene encoding affect our perception of the commonality of life in the cosmos? Could you argue that it, like the life on Earth today, began on Earth and was transported to Mars during the heavy bombardment early in Earth's history—and was extinguished from the Earth at the same time? What is wrong with or awkward about such an argument, or the converse argument that both types of life began on Mars but only the DNA-based life was transported to Earth by impacts?

SUGGESTED READINGS

Several reports of the National Academies National Research Council provide readable and up-to-date discussion of the search for life, including techniques and approaches to exploration: Committee on the Origin and Evolution of Life, *Signs of Life: Report of a Workshop on Life Detection* (2002, National Academy Press, Washington, D.C.); Space Studies Board, *Size Limits of Very Small Microorganisms: Proceedings of a Workshop* (1999, National Academy Press, Washington, D.C.); Space Studies Board, *The Quarantine and Certification of Martian Samples* (2002, National Academy Press, Washington, D.C.); and Space Studies Board, *Assessment of Mars Science and Mission Priorities* (2002, National Academy Press, Washington, D.C.).

A popular discussion of the problems and promises in the search for life is that of P. C. W. Davies, *The Fifth Miracle* (1999, Simon and Schuster, New York). The same author has published a thought-provoking and speculative paper with ramifications for the search for Martian life: "Does Life's Rapid Appearance Imply a Martian Origin?" *Astrobiol.* **3,** 673–679.

REFERENCES

Banfield, J. F., Moreau, J. W., Chan, C. S., Welch, S. A., and Little, B. (2001). "Mineralogical Biosignatures and the Search for Life on Mars," *Astrobiol.,* **1,** 447–465.

Baross, J., Bada, J. L., Ferry, J. G., Fogel, M., Grange, J., Lambert, J. B., Mustin, C., and Pardee, A. B. (1991). Unpublished Report of Subpanel 1B, *Mars Sample Handling Protocol Workshop Number 4: Comprehensive Protocol Evaluation,* June 5–7, 2001, Arlington, VA.

Chyba, C. F. and Hand, K. P. (2001). "Life without Photosynthesis," *Sci.,* **292,** 2026–2027.

Chyba, C. F., and Phillips, C. B. (2001). "Possible Ecosystems and the Search for Life on Europa," *Proc. Natl. Acad. Sci.,* **98,** 801–804.

Committee on the Origin and Evolution of Life. (2002). "Signs of Life: Report of a Workshop on Life Detection," National Academy Press, Washington, D.C.

Figuerendo, P. H., Greeley, R., Neuer, S., Irwin, L. and Schulze-Makuch, D. (2003). "Locating Potential Biosignatures on Europa from Surface Geology Observations," *Astrobiol.,* **3,** 851–861.

Friedmann, E. I., Wierzchos, J., Ascaso, C., and Winklhofer, M. (2001). "Chains of Magnetite Crystals in the Meteorite ALH84001: Evidence of Biological Origin," *Proc. Natl. Acad. Sci.,* **98,** 2176–2181.

Haynie, D. T. (2001). *Biological Thermodynamics,* Cambridge University Press, Cambridge.

Jull, A. J. T., Courtney, C., Jeffrey, D. A., and Beck, J. W. (1998). "Isotopic Evidence for a Terrestrial Source of Organic Carbon in Martian Meteorites Allan Hills 84001 and Elephant Moraine 79001," *Sci.,* **279,** 366–369.

Kerr, R. (2002). "Reversals Reveal Pitfalls in Spotting Ancient and E.T. Life," *Sci.,* **296,** 1384–1385.

Kring, D. A., Swindle, T. D., Gleason, J. D., and Grier, J. A. (1998). "Formation and Relative Ages of Makelynite and Carbonate in ALH84001," *Geochimica et Cosmochimica Acta,* **62,** 2155–2166.

Marion, G. M., Fritsen, C., H., Eicken, H., and Payne, M. C. (2003). "The Search for Life on Europa: Limiting Environmental Factors, Potential Habitats, and Earth Analogues," *Astrobiol.,* **3,** 785–811.

McKay, D. S., et al. (1996). "Search for Past Life on Mars: Possible Relic Biogenic Activity in Martian Meteorite ALH84001," *Sci.,* **273,** 924–930.

Niemann, H. B., Atreya, S. K., Biemann, K., Block, B., Carignan, G., Donahue, T. M., Frost, L., Gautier, D., Harpold, D. M., Hunten, D. M., Israel, G., Lunine, J., Mauersberger, K., Owen, T., Raulin, F., Richards, J., and Way, S. (1997). "The Gas Chromatograph Mass Spectrometer Aboard Huygens," in *Huygens: Science, Payload and Mission,* ESA SP-1177, Noordwijk, pp. 85–107.

Pace, N. R. (1997). "A Molecular View of Microbial Diversity and the Biosphere," *Sci.,* **276,** 734–740.

Pace, N. R. (2001). "The Universal Nature of Biochemistry," *Proc. Natl. Acad. Sci.,* **98,** 805–808.

Pasteris, J. D., Wopenka, B. (2003). "Necessary, but Not Sufficient: Raman Identification of Disordered Carbon as a Signature of Ancient Life," *Astrobiol.,* **3,** 727–738.

Rodier, C., Laurent, C., Szopa, C., Sternberg, R., and Raulin, F. (2001). "Chirality and the Origin of Life: In Situ Enantiomeric Separation for Future Space Missions," *Chirality,* **14,** 527–532.

Schulze-Makuch, D. and Irwin, L. N. (2002). "Energy Cycling and Hypothetical Organisms in Europa's Ocean," *Astrobiol.,* **2,** 105–121.

Space Studies Board. (1999). *Size Limits of Very Small Microorganisms: Proceedings of a Workshop,* National Academy Press, Washington, D.C.

Space Studies Board. (2002). *The Quarantine and Certification of Martian Samples,* National Academy Press, Washington D.C.

Ueno, Y., Yurimoto, H., Yoshioka, H., Komiya, T., and Maruyama, S. (2002). "Ion Microprobe Analysis of Graphite from ca. 3.8 Ga Metasediments, Isua Supracrustal Belt, West Greenland: Relationship between Metamorphism and Carbon Isotopic Composition," *Geochemica et Cosmochimica Acta,* **66,** 1257–1268.

CHAPTER 15

Life Elsewhere II
The Discovery of Extrasolar Planetary Systems and the Search for Habitable and Inhabited Worlds

15.1 Introduction

The notion that the universe is filled with planets like our own Earth, and hence teeming with life, is widely held. The expectation that Earthlike planets should exist around other stars is, to many, a natural outgrowth of the Copernican revolution. But it is also a paradox with respect to Copernicanism: On the one hand, it displaces us from a unique astronomical and biological center, but on the other, it presumes that the processes of cosmic evolution must inevitably lead to something like the situation we find in our own solar system. This presumption need not be correct; until 1996 it was possible to challenge even the view that planets might be common in the cosmos, on the basis that none had yet been discovered.[1]

This chapter describes how planets have been detected around nearby stars up until now, as well as techniques that will permit detection and study of planets in the future, even those down to terrestrial mass. To date, all the planets detected are of the mass of Jupiter or larger, but even these objects provide us with important clues regarding the habitability of planetary systems. The chapter also outlines the efforts made to detect signals from extraterrestrial civilizations that might exist elsewhere in the galaxy.

---

[1]The exception was the discovery in the early 1990s of several planetary-mass bodies in orbit around a neutron star, the end product of a supernova explosion. The planets could not be relics of the neutron star's preceding existence as a normal star because the loss of mass associated with the explosion would have caused the planets to escape from the neutron star. How a neutron star could produce planets remains very poorly understood, but the significance of this singular example for the ubiquity of planets is arguably less than the discovery soon thereafter of planets around stars like the Sun.

15.2 The Systematic Discovery of Planets Using Doppler Spectroscopy

In 1995 Swiss astronomers Michel Mayor and Didier Queloz announced the first detection of a planetary companion to a normal, main-sequence star like the Sun, beyond our own solar system. However, the detection was not a direct one, that is, the "extrasolar" (beyond our solar system) planet itself was not seen. Instead, the wobble of the star about the gravitational center of mass or "barycenter" of the star–planet system was detected through variations in the star's radial motion, using the phenomenon of the Doppler shift (Figure 15.1). Because all bodies with mass possess a gravitational force, it is not correct to imagine planets orbiting around a star; instead, the star and the planet orbit about their common center of mass. For a star the mass of the Sun, a planet the mass of Jupiter will have one-thousandth the mass of the star. The sizes of the orbits are inversely proportional to the masses; therefore, the star will be a 1,000 times closer to the center of mass than the planet. But the time to go around the center of mass once for the star and planet must be the same, and hence the orbital velocities of star and planet are also inversely proportional to mass. Suppose the planet to be at 1 AU from the solar mass star, simply by way of example. The resulting year-long orbit at 150 million kilometers corresponds to the planet moving at about 30 kilometers per second around the star, so the velocity of the star in its orbit around the barycenter is 30 *meters* per second.

As described in Chapter 5, the Doppler shift in frequency is just proportional to the radial velocity (the motion toward or away from the observer), divided by the speed of light, roughly 3×10^8 m/s. So, to detect the barycentric wobble of the star, or its motion around the center of mass, requires seeing a frequency change of $30/(3 \times 10^8) = 10^{-7}$ for the 30 m/s stellar orbital velocity just computed. Such a tiny shift in a single line cannot be reliably measured, given various sources of error in the measurements. However, one can achieve the required precision by looking at large numbers of sharp lines in the spectrum of a star's photosphere (surface) using a gas cell that produces a fixed standard spectrum in the path of the light coming through the telescope. Because the star and the planet are in orbit around each other, the Doppler shift in the stellar lines must be periodic—first toward higher frequency then lower, with a complete period representing the time (period) of the orbit of the bodies around each other. A circular orbit will be represented by a curve that is a sine wave, because the velocity is a constant around the orbit. If the orbit of the planet is noncircular, then the star's motion speeds up and slows down in response to that of the planet, yielding a curve that is periodic but not sinusoidal (Figure 15.1).

While the Doppler spectroscopic approach has proved to be the most fruitful in detecting the mass of planets' bodies (indeed, there is, as of this writing, no planetary-mass body detected by other techniques), it has significant limitations. Because the planet is not directly detected, its properties cannot be studied (e.g., size or atmospheric composition). Further, because the detection of the barycenter wobble is via the radial motion of the star toward or away from the Earth, the magnitude of the effect depends on the orientation of the planet's orbit plane relative to the star–Earth line of sight. Hence, unless the orbit is exactly coplanar with the star–Earth line, the apparent mass of the planetary perturber is less than the actual, or physical, mass (Figure 15.1). The apparent mass is in fact $M \sin i$ where

a

FIGURE 15.1 The Doppler spectroscopic technique for detecting planets: (a) Basic geometry of the barycentric wobble and corresponding frequency shift of a spectral line is shown. The unseen planet causes the parent star to orbit around the center of mass of the system. As it moves away from the observer, the light from the star is reddened (wavelength increased) and made bluer (wavelength shortened) as it moves toward the observer, both due to the Doppler effect. (b) Examples of data for planets in a circular (left) and eccentric (e = 0.3; right panel) orbit, plotted as velocity versus orbit phase or versus time. (c) The inclination angle effect that makes the Doppler spectroscopic mass a lower limit to the physical mass is illustrated by two extremes—(top) the orbit plane perpendicular to the line of sight and (bottom) parallel to the line of sight. In the latter case, the derived Doppler spectroscopic mass is the physical mass, while in the former no Doppler signal is seen regardless of the planet mass.

M is the physical mass and i the angle between the orbit plane and the star–Earth line of sight. The latter is a particularly serious limitation because we cannot be certain of the mass of any given radial velocity candidate unless additional constraints are available; for a small fraction of cases, transits (see Section 15.4) provide the key constraint. However, for a random distribution of orbits in the sky as seen from Earth, roughly 90 percent of the planets will have masses determined by radial velocity within a factor of two of the physical mass. Finally, the technique is most sensitive to planets in close orbits around stars, because for a given mass a planet induces increasingly rapid stellar motion around the barycenter for progressively smaller orbits.

As of this writing, about 120 planets with masses from several times that of Jupiter downward have been detected by Doppler spectroscopy, and if the argument regarding system inclination is correct, most of these have masses within a factor of two of their measured values. The observed abundance of planets rises rapidly toward lower masses, to the minimum detected of one-fifth Jupiter's mass (less than the mass of Saturn). The semimajor axes of the orbits of the detected planets are distributed roughly uniformly; there is no preference for very distant or very close in giant planets. Many of the orbits are elliptical, suggesting that the formation of giant planets or their subsequent interactions with

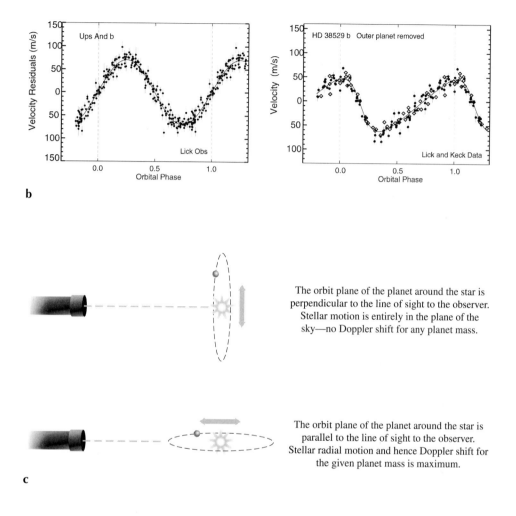

each other often lead to noncircular orbits. Table 15.1 lists orbital and mass information on the known extrasolar planets, including those located in multiple planet systems (i.e., a star with more than one planet).

It is becoming clear now that, far from being the only two giant planets known (two of four if we regard the "ice giants" Uranus and Neptune in the same class), Jupiter and Saturn are two members of a class of bodies. Of the solar-type stars surveyed, about 10 percent have detectable giant planets. Planets on orbits with periods much longer than five years must await discovery over the next decade, when the temporal baseline for radial velocity detections extend long enough to reliably detect such periods. Extrapolating from the current results (assuming a uniform distribution of orbits for giant planets out to 10 AU from the parent star corresponding to the orbit of Saturn around our own Sun), suggests that perhaps 15 percent of solar-type stars possess giant planets.

TABLE 15.1 Known Extrasolar Planets as of April 2004

Name of Star (planet 1 = b; planet 2 = c, etc.)	Mass Times Sine of Inclination (in Jupiter Masses)	Semimajor Axis (AU)	Orbit Period (days)	Orbit Eccentricity	Orbit Inclination[1]
OGLE-TR-56 b	1.45	0.0225	1.2	—	81.0
OGLE-TR-113 b	1.35	0.0228	1.43	—	—
OGLE-TR-132 b	1.01	0.0306	1.69	—	—
HD 73256 b	1.85	0.037	2.54863	0.038	—
HD 83443 b	0.41	0.04	2.985	0.08	—
HD 46375 b	0.249	0.041	3.024	0.04	—
HD 179949 b	0.84	0.045	3.093	0.05	—
HD 187123 b	0.52	0.042	3.097	0.03	—
Tau Boo b	3.87	0.0462	3.3128	0.018	—
HD 330075 b	0.76	0.043	3.369	0.0	—
BD-10_3166 b	0.48	0.046	3.487	0.	<84.3
HD 75289 b	0.42	0.046	3.51	0.054	—
HD 209458 b	0.69	0.045	3.524738	0.0	86.1
HD 76700 b	0.197	0.049	3.971	0.0	—
51 Peg b	0.468	0.052	4.23077	0.0	—
Ups And b	0.69	0.059	4.6170	0.012	—
c	1.19	0.829	241.5	0.28	—
d	3.75	2.53	1284	0.27	—
HD 49674 b	0.12	0.0568	4.948	0.	—
HD 68988 b	1.90	0.071	6.276	0.14	—
HD 168746 b	0.23	0.065	6.403	0.081	—
HD 217107 b	1.28	0.07	7.11	0.14	—
HD 162020 b	13.75	0.072	8.428198	0.277	—
HD 130322 b	1.08	0.088	10.724	0.048	—
HD 108147 b	0.41	0.104	10.901	0.498	—
HD 38529 b	0.78	0.129	14.309	0.29	—
c	12.70	3.68	2174.3	0.36	—
55 Cnc b	0.84	0.11	14.65	0.02	25?
c	0.21?	0.24?	44.28?	0.34?	25?
d	4.05	5.9	5360.	0.16	25?
Gl 86 b	4	0.11	15.78	0.046	—
HD 195019 b	3.43	0.14	18.3	0.05	—

Name of Star (planet 1 = b; planet 2 = c, etc.)	Mass Times Sine of Inclination (in Jupiter Masses)	Semimajor Axis (AU)	Orbit Period (days)	Orbit Eccentricity	Orbit Inclination[1]
HD 6434 b	0.48	0.15	22.09	0.30	—
HD 192263 b	0.72	0.15	24.348	0.0	—
Gliese 876 c	0.56	0.13	30.1	0.12	?
b	1.98	0.21	61.02	0.27	84
rho CrB b	1.04	0.22	39.845	0.04	—
HD 74156 b	1.86	0.294	51.643	0.636	—
c	>6.17	3.40	2025.	0.583	—
HD 168443 b	7.7	0.29	58.116	0.529	—
c	16.9	2.85	1739.50	0.228	—
HD 3651 b	0.2	0.284	62.23	0.63	—
HD 121504 b	0.89	0.32	64.6	0.13	—
HD 178911 B b	6.292	0.32	71.487	0.1243	—
HD 16141 b	0.23	0.35	75.560	0.28	—
HD 114762 b	11	0.3	84.03	0.334	Low?
HD 80606 b	3.41	0.439	111.78	0.927	—
HD 219542B b	0.30	0.46	112.1	0.32	—
70 Vir b	7.44	0.48	116.689	0.4	—
HD 216770 b	0.65	0.46	118.45	0.37	—
HD 52265 b	1.13	0.49	118.96	0.29	—
GJ 3021 b	3.21	0.49	133.82	0.505	—
HD 37124 b	0.75	0.54	152.4	0.10	—
c	1.2	2.5	1495.	0.69	—
HD 219449 b	2.9	~0.3	182.	—	—
HD 73526 b	3.0	0.66	190.5	0.34	—
HD 104985 b	6.3	0.78	198.2	0.03	—
HD 82943 b	0.88	0.73	221.6	0.54	—
c	1.63	1.16	444.6	0.41	—
HD 169830 b	2.88	0.81	225.62	0.31	—
c	4.04	3.60	2102.	0.33	—
HD 8574 b	2.23	0.76	228.8	0.40	—
HD 89744 b	7.2	0.88	256.	0.54	—
HD 134987 b	1.58	0.78	260.	0.25	—

(continued)

TABLE 15.1 continued

Name of Star (planet 1 = b; planet 2 = c, etc.)	Mass Times Sine of Inclination (in Jupiter Masses)	Semimajor Axis (AU)	Orbit Period (days)	Orbit Eccentricity	Orbit Inclination[1]
HD 40979 b	3.32	0.811	267.2	0.25	—
HD 12661 b	2.30	0.83	263.6	0.096	—
c	1.57	2.56	1444.5	<0.1	—
HD 150706 b	1.0	0.82	264.9	0.38	—
HD 59686 b	6.5	~0.8	303.	—	—
HR 810 b	2.26	0.925	320.1	0.161	—
HD 142 b	1.36	0.980	338.0	0.37	—
HD 92788 b	3.8	0.94	340.	0.36	—
HD 28185 b	5.6	1.0	385.	0.06	—
HD 142415 b	1.62	1.05	386.3	0.5	—
HD 177830 b	1.28	1.00	391.	0.43	—
HD 108874 b	1.65	1.07	401.	0.20	—
HD 4203 b	1.65	1.09	400.944	0.46	—
HD 128311 b	2.63	1.06	414.	0.21	—
HD 27442 b	1.28	1.18	423.841	0.07	—
HD 210277 b	1.28	1.097	437.	0.45	—
HD 19994 b	2.0	1.3	454.	0.2	—
HD 20367 b	1.07	1.25	500.	0.23	—
HD 114783 b	0.9	1.20	501.0	0.1	—
HD 147513 b	1	1.26	540.4	0.52	—
HIP 75458 b	8.64	1.34	550.651	0.71	—
HD 222582 b	5.11	1.35	572.0	0.71	—
HD 65216 b	1.21	1.37	613.1	0.41	—
HD 160691 b	1.7	1.5	638.	0.31	—
c?	1?	2.3?	1300?	0.8?	—
HD 141937 b	9.7	1.52	653.22	0.41	—
HD 41004A b	2.3	1.31	655.	0.39	—
HD 47536 b	4.96–9.67	1.61–2.25	712.13	0.20	—
HD 23079 b	2.61	1.65	738.459	0.10	—
16 CygB b	1.69	1.67	798.938	0.67	—

15.2 The Systematic Discovery of Planets Using Doppler Spectroscopy

Name of Star (planet 1 = b; planet 2 = c, etc.)	Mass Times Sine of Inclination (in Jupiter Masses)	Semimajor Axis (AU)	Orbit Period (days)	Orbit Eccentricity	Orbit Inclination[1]
HD 4208 b	0.80	1.67	812.197	0.05	—
HD 114386 b	0.99	1.62	872.	0.28	—
gamma Cephei b	1.59	2.03	902.96	0.2	—
HD 213240 b	4.5	2.03	951.	0.45	—
HD 10647 b	0.91	2.10	1040.	0.18	—
HD 10697 b	6.12	2.13	1077.906	0.11	—
47 Uma b	2.41	2.10	1095.	0.096	—
c	0.76	3.73	2594.	<0.1	—
HD 190228 b	4.99	2.31	1127.	0.43	—
HD 114729 b	0.82	2.08	1131.478	0.31	—
HD 111232 b	6.8	1.97	1143.	0.20	—
HD 2039 b	4.85	2.19	1192.582	0.68	—
HD 136118 b	11.9	2.335	1209.6	0.366	—
HD 50554 b	4.9	2.38	1279.0	0.42	—
HD 196050 b	3.0	2.5	1289	0.28	—
HD 216437 b	2.1	2.7	1294	0.34	—
HD 216435 b	1.49	2.7	1442.919	0.34	—
HD 106252 b	6.81	2.61	1500	0.54	—
HD 23596 b	7.19	2.72	1558	0.314	—
14 Her b	4.74	2.80	1796.4	0.338	—
OGLE-235/MOA-53	1.5–2.5	2.8–3	?	?	—
HD 39091 b	10.35	3.29	2063.818	0.62	—
HD 72659 b	2.55	3.24	2185	0.18	—
HD 70642 b	2.0	3.3	2231	0.1	—
HD 33636 b	9.28	3.56	2447.292	0.53	—
Epsilon Eridani b	0.86	3.3	2502.1	0.608	46?
c	0.1??	40??	280 years??	0.3??	??
HD 30177 b	9.17	3.86	2819.654	0.30	—
Gl 777A b	1.33	4.8	2902	0.48	—

[1] degrees, 90° = in plane of line-of-sight to observer

What is the lowermost mass limit that can be detected by Doppler spectroscopy? We might imagine that there is no fundamental limit except the widths of the lines whose cyclical shift is observed. Such lines could be so narrow that perhaps even Earth-mass planets could be detected. But there may, in fact, be a much coarser limit. The surface of any star, including the Sun, is a turbulent mass of hot gas with buoyant hot cells rising to release their heat to cold space, with the resulting cold gas falling. This turbulence is translated into line widths that are broader than their theoretical limits dictated by quantum mechanics. This turbulent broadening severely limits the accuracy with which the velocity of the barycentric wobble can be detected, and a practical limit may be 1 m/s or perhaps worse. For the tightest orbits imaginable, a few hundredths of an astronomical unit from the parent star (inward of that, gas is torn away from the planet by the star), this limit corresponds to detection of a planet perhaps 10 times the mass of the Earth. Hence, unless some statistical approach could remove the effects of turbulent line broadening, Doppler spectroscopy will be a search tool limited to finding giant planets, and perhaps "super-Earths" 10 times our own planet's mass. But Earths themselves will elude detection. Table 15.2 compares the threshold sensitivities required for detecting Jupiters and Earths around

TABLE 15.2 Observables, Measurement Precision, and Present-Day Capability for Extrasolar Planet Detection

Observable	Jupiter	Earth	Possible to Detect Today?
Direct detection: angular separation in sky	1 arcsec	0.2 arcsec	No (Jupiter); no (Earth)
Direct detection: brightness ratio planet/Sun	10^{-9} (vis); 10^{-4} (IR)	10^{-10} (vis); 10^{-6} (IR)	
Radial velocity	13 m/s	3 cm/s	Yes (Jupiter); no (Earth)
Astrometric wobble	1 marcsec	0.6 μarcsec	
Transit: photometric precision	1%	0.01%	Yes (Jupiter); no (Earth)
Transit duration	25 hours	11 hours	
Microlens amplification at 4 kpc	$\leq$10% (vis)	$\leq$1%	Yes (Jupiter); no (Earth)
Microlens duration	$\leq$3 days	$\leq$4 hrs	

Table gives parameters for a solar system twin, some 5 pc from our solar system. One arc second is 1/3600 degrees on the sky; the full moon subtends 0.6 degree. The brightness ratios for Jupiter-twin and Earth-twin relative to the solar-twin are given in the peaks of the optical and infrared parts of the planets' spectra. Photometric precision refers to the measurement of fluctuations in light from the star; 1 percent means that changes in brightness of 1 percent must be seen. The microlensing amplification is the percent increase in brightness of the background star at 4 kpc caused by the effect of a planet around the lensing star.

Source: Based on Tarter, J. (2001). "The Search for Extraterrestrial Intelligence (SETI)" *Ann. Rev. Astron. Astrophys.*, **39**, 511–548.

nearby stars and makes it clear that techniques other than Doppler spectroscopy will be required for detecting the latter.

15.3 Astrometry: Another Technique for Indirect Detection of Planets

The complement of Doppler spectroscopy is *astrometry*. A star executing a barycentric wobble associated with the presence of a planet will have side-to-side, or *transverse*, motions against the sky. This is true regardless of the inclination of the planet's orbit with respect to the line of sight with the Earth; hence, unlike Doppler spectroscopy, there is no $\sin i$ factor in the mass associated with orbit inclination. Measure the side-to-side shift of the star's position accurately enough, and you have the mass of the planet relative to that of the star. Transverse motion cannot be detected by Doppler shift and must instead be detected by perceiving the progressive shift in position of the wobbling star against the background sky. The need to measure tiny shifts in the position of a star against the background of more distant stars and galaxies was first encountered in 19th-century astronomy when distances to stars were measured by detecting their parallax shifts (Chapter 5). In the case of planet finding, the shifts are not perspective changes as the observer's position changes but real wobbles of the stellar position in the sky. And the wobbles are very small compared with what has traditionally been measured for stellar parallax.

Imagine trying to detect a planet the mass of Jupiter (one-thousandth the Sun's mass) orbiting a solar mass star just a parsec from our solar system. The star's parallax motion is, by definition, 1 arc second, (arc sec) which as defined in Chapter 5 is 1/60 of a minute of arc or 1/3600 of a degree (where the bowl of the sky is 180 degrees). This shift is readily detectable. What about the back-and-forth wobble as the star orbits the barycenter of motion? As before, the size of the star's orbit is proportional to that of the planet, multiplied by the mass ratio of planet to star. Assume the planet orbits at 1 AU from the star (or more strictly the center of mass of the system). Then the star orbits at a distance of 1/1000 of an AU from the center of mass. The parsec is defined in Chapter 5 by taking the base leg of the motion to be that of the Earth across its own orbit, 1 AU. Although in this case, the base leg is at the star rather than here at Earth's orbit, we can simply take the ratio of the two orbital base legs (Figure 15.2), and hence the star's barycentric wobble corresponds to 1/1000 of an arc second, or a milliarcsecond (marcsecond).

The best astrometry done so far is that from the orbiting satellite Hipparcos, which achieved a precision somewhat better than 2 marcsec, a bit too large to see the strawman case just described. However, the situation is actually worse, since only one star system (the multiple star system alpha Centauri) lies a parsec (pc) from our solar system. To have a reasonable sampling of stars for the planet search one must go out to 10 pc, which means the angular shift seen on Earth due to our strawman star–planet case drops a factor of 10—to 0.1 marc sec. If we wish to see an Earth-mass planet orbiting such a star—1/300 the mass of Jupiter—we must then be able to detect shifts of (0.1 marcsec)/300, or 0.3 "microarcseconds" (μarcsec). A μicroarcsecond, one-millionth of an arc second, requires measuring stellar positions to accuracies not yet possible with current ground or

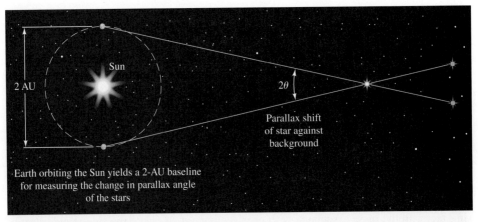

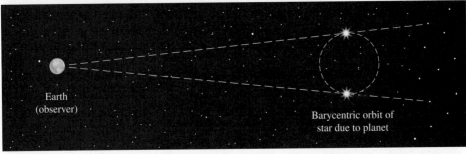

FIGURE 15.2 Geometry for figuring out the amount of angular shift in the sky of a solar-mass star 1 pc from the Earth, being pulled around a system center of mass by a Jupiter-mass planet at 1-AU orbit distance. Panel (a) recapitulates from Chapter 5 the geometry of the parallax shift in the sky of the star due to the Earth's motion around the Sun. Panel (b) shows what happens if the star itself is moving—the baseline is now the star's orbit around the center of mass of the star-and-planet system.

spaceborne telescope systems. Could we travel to the nearest stars and look back, our own Sun would be seen to execute a complex set of movements against the background as seen by a distant observer (Figure 15.3); this complex pattern is the result of superposing the gravitational tugs of all the planets but predominantly Jupiter and Saturn.

To date, no planets have been detected using astrometry, and this illustrates the difficulty in applying the technique even to Jupiter-mass planets. The intrinsic advantage of the technique, namely, the derivation of a direct physical mass (unaffected by the orbit angle relative to the line of sight to Earth), makes it attractive as a technique. It is very much complementary to Doppler spectroscopic searches: Planets in more distant orbits around a given star give a slower barycentric wobble to the star (more difficult for Doppler spectroscopy), but the wobble has a larger amplitude (better for astrometry).

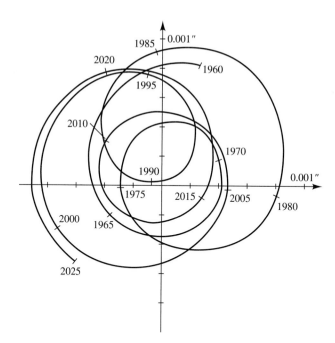

FIGURE 15.3 Movement of the Sun against the stellar background, resulting from the gravitational pull of the planets, as an observer 10 pcs from the solar system would see it, between A.D. 1985 and 2025. The edge of the axis is 1 milliarcsec in each direction.

Figure 15.4 illustrates the complementary nature of the techniques. The astrometric challenge is conceptually the same as in Doppler spectroscopy: to find a standard of reference that is fixed so as to detect very tiny motions of stars that might have planets. Unlike Doppler spectroscopy, for which a compact spectometer system containing iodine as a standard was developed by American astronomers Geoff Marcy and Paul Butler, no ingenious solution has been found to do astrometry on small ground-based telescopes. To provide the sensitivity and accuracy needed to detect planets like Jupiter, in orbits similar to those in our solar system appears to require large and expensive ground-based telescopes or spaceborne platforms. The American space agency, NASA, and the European Space Agency, ESA, have ambitious plans for spaceborne astrometry missions. One precursor, space-based system, Hipparcos, has already been flown by ESA and has precisely measured the distances to various stars while constraining the amount of astrometric shift they might undergo associated with the presence of planets.

The most ambitious astrometry mission planned, the Space Interferometry Mission (SIM), will fly around 2009 or 2010. A spacecraft designed to measure exquisitely small angular shifts of stars in the sky, it uses interferometry (Section 15.5) to detect the wobble of planets approaching the mass of the Earth to be measured around stars the mass of the Sun or smaller (Figure 15.4). SIM pushes the technology of astrometry to extreme lengths, requiring great control of optical paths and spacecraft stability. These will be required not only for detecting Earth-mass planets by astrometry (Table 15.2) but, ultimately, studying them with even more advanced spaceborne interferometers that are described in Section 15.5.

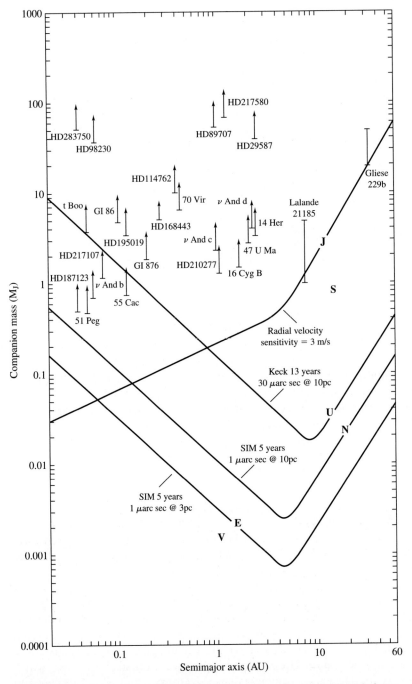

FIGURE 15.4 Sensitivities of Doppler spectroscopic and astrometric techniques for indirect detection of planets plotted as mass of the planet (in Jupiter masses) versus semimajor axis in astronomical units. The sample case shown is a twin of our solar system, with the planets labeled by their initials (V = Venus, etc.).

FIGURE 15.4 (continued) The sensitivity limits for the radial velocity technique and for astrometry are labeled. Although planets of increasing separation from the parent star create a larger astrometric wobble, the time required to observe enough of an orbit to make a detection increases with lengthening semimajor axis and hence orbit period. An assumption is made on the lifetimes of the SIM mission and that of the Keck ground-based astrometry program so that, beyond a certain orbit period, the sensitivity curve reverses slope and climbs steeply because insufficient time is available to make a detection. For radial velocity, the sensitivity drops with increasing semimajor axis; the break in the curve where the slope steepens corresponds to orbit periods longer than the observing campaign itself (assumed from 1995 through 2003, when the figure was prepared). The masses (as lower limits) and semimajor axes of a subset of the extrasolar planets detected by Doppler spectroscopy are shown. SIM sensitivities are shown for solar system twins at 3 and 10 pcs; for the former, SIM might barely be able to detect an Earth.

15.4 Planets Almost Seen: Detection of Planets by Transits, Phases, and Microlensing

Since the time of Giordano Bruno more than four centuries ago, the great difficulty in imaging extrasolar planets directly has been recognized; Section 15.5 describes the challenge in detail. However, there are two techniques that form a kind of bridge between the indirect detection of planets outlined in Section 15.3 and their direct observation by telescopes. These techniques depend on two phenomena—transiting of a star and microlensing of a star—and are referred to by those names. They are less technologically challenging than direct detection and provide complementary kinds of information. The first technique is intuitive—a planet passes in front of a star and blocks some of the light—but the latter is a more subtle effect arising from general relativity that leads to a kind of stellar twinkling.

Transits

The essence of the *transit* technique is shown in Figure 15.5. If the plane of the orbit of an extrasolar planet lies close to the line of sight to our Earth, then on each orbit the extrasolar planet will pass in front of the star. Because the planet is not self-luminous in the optical, and much dimmer than the star in the near-infrared, it acts to block the starlight at these wavelengths. Hence, by looking for a periodic dip in the light of a star, it is possible to detect the presence of the planet.

The observation of a transit immediately provides three pieces of information not obtained by Doppler spectroscopy. For a transit to occur, the orbit plane of the planet must be very close to the line of sight to us. Hence, the *sine* of the orbit inclination is nearly unity, and the Doppler spectroscopic mass of the planet *is* very nearly the physical mass. The fractional darkening of the star's light can be converted directly into the size of the planet (Figure 15.5). Also, knowing the period of the orbit of the planet, we know how far the planet is from the star from the timing of the transits. Hence, observations of transits yield the physical mass, size, and orbit of the extrasolar planet. The vast majority of transits will occur for planets in tight orbits around their stars, because the larger the orbit, the closer

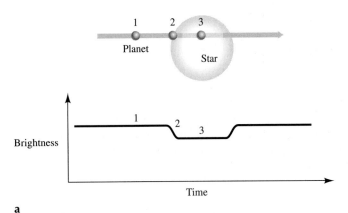

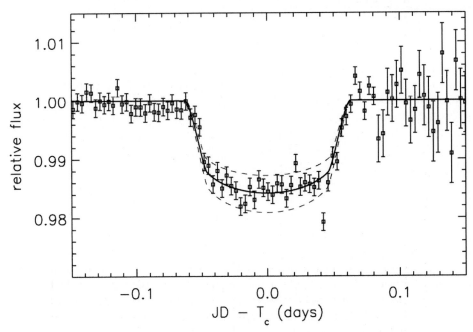

FIGURE 15.5
(a) Schematic of the transit technique. (b) Light curve of the star HD209458 as its planetary companion transited in front of it. Two transit events separated by a week are combined together. Time is given in days. The points with error bars are the data. The three lines correspond to the data for three different-sized planets; from top to bottom: 1.15, 1.27, and 1.40 times Jupiter's radius.

the orbit plane must lie along the line of sight to us for a transit to occur. For the close-in planets, only about 1 out of 10 systems randomly oriented in space will exhibit transits; for planets orbiting at 1 AU, the number drops to one in a few hundred. To date, indeed, only one of the planets detected so far by Doppler spectroscopy has been observed to transit its star. Three other giant planets in very tight orbits around their parent stars have been detected only by transits, the star–planet systems being too far from us for the Doppler spectroscopic technique to be usable.

The required accuracy in observing the brightness of the parent star in order to detect a planet by transit is readily computable. Because the distance to the star–planet system is vastly greater than the orbital distance between the star and the planet itself, the fractional amount of starlight blocked is simply the ratio of the area of the planet to that of the star. A planet the size of Jupiter has a surface area 1 percent that of the star, so, to detect the dimming, we must measure the star's brightness to an accuracy of at least 1 part in 100, and in practice somewhat better. This is achievable from carefully conducted experiments on ground-based telescopes, and is readily accomplished from the Hubble Space Telescope. To detect an Earth 1/100 the projected area of Jupiter requires an accuracy of at least 1 part in 10,000. Because of the fluctuating blurring effects of our own atmosphere, this is unlikely to be achieved from the ground. Even from space it requires more advanced optics than Hubble's, and special spaceborne telescopes must be designed for the task.

A breakthrough in the study of extrasolar planets came in the successful observation of a transit of one such planet across the disk of its parent star. The system, HD209458, consists of a star roughly the age of the Sun and just slightly more massive, along with a planetary companion at least 0.7 times the mass of Jupiter, based on Doppler spectroscopy, orbiting just 0.047 AU from the parent. This information, along with the circular nature of the orbit, was essentially all that could be derived from observation of the central star and its periodic wobble with Doppler spectroscopy. As with the other 100-plus radial velocity candidates, we could only wonder what the companion looked like. Was it a gas giant, like Jupiter? What effect did the extreme proximity of the parent star have on its size?

Surprisingly, these questions could be answered through the transit observations coupled with detailed modeling of the structure of a giant planet. Figure 15.5 shows the light curve from two transit events in the HD209458 system. The decrease in light as the planet blocks part of the stellar disk is best fitted by a planet with radius between 1.2 and 1.4 that of Jupiter, as shown in the figure. The fact that transits occur in this system immediately sets a tight limit on the orbital inclination of the planet–star system as seen from Earth. The timing of the transit allows a precise numerical value to be put on the orbit inclination, which in this case is within 4 degrees of being co-planar with the line to Earth. Thus, the minimum mass of 0.7 times the mass of Jupiter derived from the radial velocity studies for the planet is essentially the physical mass. The derived radius of HD209458b (the "b" refers to the planet as opposed to the star) immediately rules out a primarily rocky or icy composition, since such a body would have a radius less than one-half that of Jupiter. The planet must be a gas giant like Jupiter. But why is it larger than Jupiter by 20 percent to 40 percent, with a correspondingly low density (one-half that of Saturn and one-fourth that of Jupiter)?

The facile answer is that the proximity to the parent star causes the gas-giant planet to expand. There is, however, an important subtlety here that is key to understanding the early history of this planet. The expansion cannot be a superficial effect of the outer atmosphere because high gravity of the planet overall keeps the outer layers of gas relatively compressed. Instead, the bulk of the planet itself must be in an expanded state relative to Jupiter, the cause of which is the prodigious stellar flux retarding the cooling of the planetary interior.

Formation of gas-giant planets from the collapse of the gas and dust of the protoplanetary disk (Chapter 6) leads to an initially hot, distended object that then cools and contracts as thermal energy is removed from the interior. According to theory, which nicely fits both young stars and the details of the giant planets of our own solar system, giant planets far enough from their parent stars to be unaffected by the stellar light would cool and shrink quickly. It took less than 1 million years for Jupiter to drop below two Jupiter radii. The rate of shrinkage is large because heat is removed from the interior by convection, which is triggered by the steep drop in temperature from the deep interior to the surficial layers at the top of the planet. Near the top, the temperature declines more gradually, and the atmosphere transports the heat radiatively. In this situation, the thicker the radiative zone, the slower the loss of heat to space.

Strong stellar irradiation flattens the atmospheric temperature profile, reducing the rate at which energy can be transported outward, and extends the radiative zone deeper into the planet. These effects, in turn, reduce the flow of energy out of the deep interior and retard the contraction of the planet with time. (The strong irradiation also puffs up the atmosphere so that the transit also detects a very bloated but tenuous upper atmosphere.) Computer models provide good agreement with HD209458b's observed radius (correcting for the bloating effect) at its current age of somewhere between 4 billion to 7 billion years (the age being derived from the properties of the star and stellar evolution theory). Further, the transiting planet shows slightly different diameters at different wavelengths because various gases in the planet's atmosphere absorb more effectively at certain wavelengths. From this it has been possible to measure the presence of sodium, potassium, and carbon monoxide in the atmosphere of HD209458b. Hence, from very basic information derived from transit and radial velocity data, we can constrain the essential nature of the planet.

Like radial velocity searches, transit detections of giant planets do not require large aperture—just high accuracy in measuring brightness. The transit of HD209458 was observed from 2-meter-class telescopes. The wide availability of such telescopes bodes well for continued surveying of nearby stars for Jovian companions for transit opportunities. Use of larger telescopes, or those in space, allows for discovery of transits by smaller planets or more detailed studies of giant planets, including their atmospheric composition, by measuring the wavelength-dependent transit radius. The 6.5-meter aperture mirror of the next-generation James Webb Space Telescope, coupled with its advanced instruments, will allow much more detailed study of the compositions of extrasolar giant planet atmospheres—both those seen in transit and those detected directly using a blocking "coronagraph" described in Section 15.5.

The search for transits by Earth-sized planets requires photometric accuracy, that is, the ability to measure light fluctuations, of 10 to 100 ppm (Table 15.2). Impossible to do from the ground with our blurring atmosphere, such a search will need to be conducted from space. A low-cost space mission under the NASA *Discovery* program will do this beginning later in this decade. Called Kepler, it would view many stars at once and look for telltale changes in brightness over months to years that might indicate the periodic passage of an Earth-sized planet (or, as well, something larger) in front of a star. Random orientations of planetary orbits in the sky dictate (averaging over various planetary orbits, stellar sizes, and so forth), that Kepler will miss at least 99 out of every 100 planets (those in orbits not aligned with the line of sight to Earth). But by viewing tens of thousands of stars, Kepler will have a good chance of observing a handful of Earths. Because it must observe many stars, the vast majority of Kepler's targets are more than 100 pc away, still in our local part of the Milky Way galaxy but too distant for direct examination of any discovered planets. Kepler will give us the rate of occurrence of Earth-sized planets, but possibly few or no targets for follow-on missions to directly study.

Akin to transit searches for planets is the search by so-called photometric phase variation. As a planet orbits a star, it exhibits different phases as does Venus as seen from Earth. Even if the plane of the orbit is not aligned toward the Earth, the phase variation around the orbit will generate a change in brightness—unless the orbit plane is close to perpendicular with the line of sight to Earth. The variations in brightness are extremely small and very difficult to see from the ground against twinkling effects caused by our own atmosphere. However, several spaceborne telescopes may have the capability to see the extremely small, periodic variations as extrasolar planets wax and wane in brightness during their orbital march around the Sun's nearest stellar neighbors.

Microlensing

Microlensing is a technique that is capable of detecting small planets at very large distances. It relies on a phenomenon associated with Einstein's theory of general relativity, namely, that any object that has mass bends the path of a beam of light. Massive objects in the distant cosmos, such as galaxies, are observed to bend the light from yet more distant objects in a way analogous to the refraction of light by a glass lens. However, there is an important distinction here—the glass bends the light because the speed of the photons is slowed down by the electromagnetic interactions with the glass. In the case of the galactic "lens," the distances between the stars of the lensing galaxy are large, the galaxy is mostly empty space, and thus, in the absence of a general relativistic effect, the light would pass straight through. But the total mass of the galaxy in fact curves space through its collective gravitational force, and it is this curvature, predicted by the theory of general relativity, that focuses the light. What we see on Earth from such a lens is a doubling of the object behind it or background objects distorted into filamentary shapes (Figure 15.6).

Individual stars can also lens background objects. The result is not a second image of the background, because a star's gravitational field is not great enough to produce a separate image of a background body as seen from the Earth. But the lensing star can cause a brief fluctuation in the brightness of the background star. This effect is called "microlensing." One way to think of this result of microlensing is as follows. When we look at the night sky from the surface of the Earth, we see the stars twinkle because the turbulent nature of our own atmosphere leads to fluctuations in the air density, which bends the starlight this way and that. In the vacuum of space, then, the stars should not twinkle. But in fact, they do (for the reason just described)—microlensing produces a very slow (days-long) twinkling of stars. As stars relatively near us move against the background of stars that are farther away, a chance geometric superposition will occur every so often in which the nearer star passes directly in front of a background star as seen from our vantage point on the Earth. As this crossing happens, the gravity of the intervening star bends the light of the background star according to the principles of general relativity, causing a brief focusing and hence brightening. Could the human eye see this effect unaided and individually observe the vast array of the background Milky Way, we would perceive an occasional,

FIGURE 15.6 Example of gravitational lensing. This Hubble Space Telescope image of a massive cluster of galaxies, called Abell 2218, shows several arc-shaped patterns, which are duplicate images of galaxies that lie 5 to 10 times farther away than the cluster itself.

slow, and subtle twinkling of the background stars. The intervening star, the one that causes the background to twinkle, is called the *lensing star.*

This effect, by itself, is not directly relevant to the detection of planets. But should the lensing star possess a planet at an appropriate separation from itself, the planet will act as a very small additional lens and create a secondary brightening in the background star, as shown in Figure 15.7. While the main brightening effect can last days or more as seen from a telescope at Earth, the smaller planetary lens effect will span only hours. And yet, for a sensitive telescopic system, hours translates to a large number of well-characterized data points. And the technique lends itself to the observation of many hundreds of thousands of stars at the same time—in fact, it depends on it.

The microlensing detection of planets is a statistical survey of the occurrence of Earthlike planets in the galaxy rather than a study and characterization of individual planets in our own neighborhood. Only a fraction of stars with Earthlike planets will have those planets in a position to create a secondary signal in the lensing twinkle, and any such lensing systems are too far away for spectroscopy to study the planets found. But the technique takes in a broad sweep of the galaxy, and the fraction of such planetary systems that should cause a microlensing twinkle can be accurately estimated from geometric considerations alone. Also, the strength of the microlensing signal is related to the mass of the planet causing the extra twinkle, so the mass of the planet can be constrained—although the strength of the signal depends also on the unknown separation on the sky of the planet from the star (Table 15.2). The microlensing technique is best suited to detecting planets in orbits (around Sunlike stars) with semi-major axes from 1 AU to several AU, the most interesting "zone of habitability" around such stars.

15.4 Planets Almost Seen 493

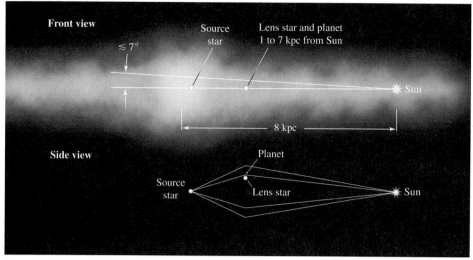

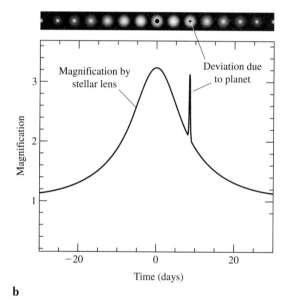

FIGURE 15.7 (a) Schematic of the microlensing technique, showing the effect of a planet on the bending of background starlight. The top view shows the Milky Way galaxy edge with a source star in the center, and a star/planet combination creating the lens some thousands of parsecs from our solar system. The bottom panel shows roughly the light paths that lead to the multiple image effect. (b) Illustration of what a light curve might look like that indicates the presence of an Earth-mass planet in an orbit with semimajor axis of 2 AU around a solar-type star.

Although microlensing observations can be done from the ground, a spaceborne telescope has the advantage of avoiding the blurring effects of the atmosphere that obscure detection of the microlensing signal. A spaceborne microlensing system need not be large but should have good photometric accuracy and be able to look at many stars at once. Such a system would search for stars that are in the process of being lensed, and the change of the brightness over weeks would be carefully examined for the much shorter "blips" of hours that could be evidence of a planet orbiting about the lensing star. Most stars that will serve as lenses are in the so-called bulge of our galaxy, some 10,000 pc distant. Concepts have been drawn up for a simple spaceborne telescope system that could detect microlensing events involving planets the mass of Earth or even smaller. Such an experiment, observing from Earth orbit over several years, could view tens of millions of stars and determine the fractional occurrence of Earth-mass planets around stars like the Sun or smaller, in orbits roughly corresponding to those ranging from our Earth's to that of our asteroid belt. The drawback is that we cannot return to the star–planet system doing the lensing to study the properties of the planet; these systems will, for the most part, be too distant to permit direct study. But obtaining the statistical occurrence of Earth-mass planets, from transits or microlensing, would address a key remaining question of the Copernican revolution: Are such planets common? It is a question that, should it be answered in the affirmative, will galvanize the development and launch of large optical interferometers to search the Sun's nearest neighbors for local examples of such earths, and to eventually characterize them. And it will provide a key constraint on how many nearby stars must be searched with such "direct" techniques before we expect to find another Earth.

15.5 Direct Detection

Development of techniques to detect planets directly, by their reflected or emitted light, remains a daunting problem. Imagine moving away from the solar system to some great distance—say, 5 pc—at which the Sun and its planets are no more than faint points of light. Jupiter is one-billionth the brightness of the Sun at optical wavelengths, peaking in the infrared at perhaps one ten-thousandth the Sun's brightness (Figure 15.8; Table 15.2). While this ratio is independent of the distance of the observer from the system, the ability to pluck Jupiter out from the glare of the neighboring Sun is not, as we shall now see.

Imagine that we have a perfect optical system—a telescope—with a mirror diameter of 2 meters: typical for modest-scale research telescopes today and just a bit less than the 2.4-meter-diameter mirror of the Hubble Space Telescope. Imagine further that this system is—like Hubble—deployed in space so that the images do not suffer from atmospheric blurring. Let us travel with this system 5 pc from our own solar system and look back on the Sun and planets. Here Jupiter would be seen as separated from the Sun by roughly an arcsecond of angle on the sky. (By way of comparison, an observer standing on Earth sees the full moon subtending 2,200 seconds of arc—approximately 0.6 degree.) Our Sun—now just a star in the sky—seems like a point of light when we look from our vantage point 5 pc away.

But in fact, what we see through our perfect telescope of the Sun is not an infinitesimal point of light, but a set of alternating bright and dark diffraction rings, the result of the well-known optical principle that even a perfect optical system cannot resolve objects below a certain size limit. (Our eyes produce the same effect, but they are not sensitive enough nor sufficiently optically accurately to see diffraction patterns around "point" sources of light.) This limit, called the *diffraction limit,* is just the wavelength of light divided by the mirror diameter. For visible light and our 2-meter telescope, the first dark ring is at about 0.05 arcsec beyond the center of the image of the central star (in our exam-

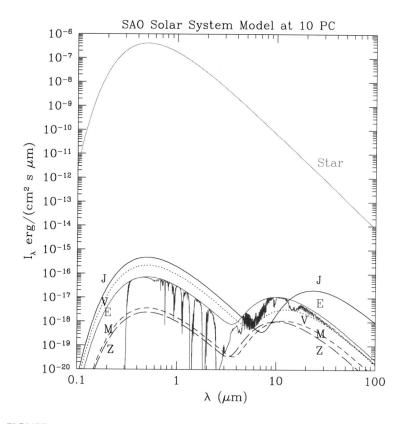

FIGURE 15.8 Intensity of light coming from the planets (labeled with initials) and Sun (labeled as "star") of our solar system, as seen from a 10-pc distance. Notice that the ratio of light from the Sun versus Earth or Jupiter is smallest in the infrared. All of the spectra assume that the planets behave as ideal blackbodies, with no discrete spectral lines except for Earth, where an actual spectrum is shown superposed on the blackbody curve to illustrate the effect of molecular absorptions.

ple, the Sun), and our putative Jupiter lies some 20 times farther away than this first ring. Although the Sun's image continues outward as an alternating pattern of dark and light rings, the brightness of the rings declines steeply as we move outward from the center. One arc second from the Sun, Jupiter is 10^{-5} as bright as the nearest diffraction ring—considerably better than the 10^{-9} ratio relative to the total solar brightness (Figure 15.9). Even so, this ratio—10 ppm—would be a challenge to detect even for the most advanced optical systems.

In the infrared, the ratio of planet to star brightness is much better—leading us to hope that Jupiter might be comparable in brightness to the diffraction ring nearest to it. But the diffraction limit is larger in the infrared than in the optical—for example, it is 10 times larger at a 5-μm wavelength than at a 0.5-μm wavelength. In effect, the diffraction ring

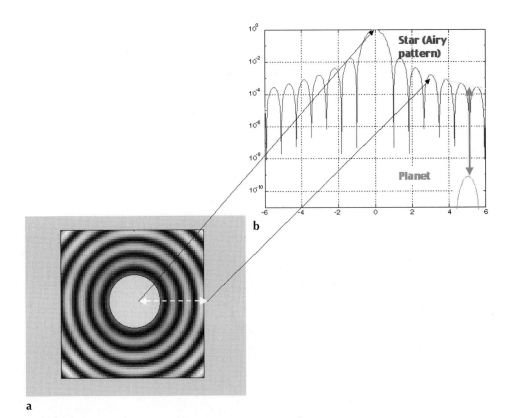

FIGURE 15.9 Diffraction rings from a point source will look something like the theoretical pattern in panel (a). These may correspond, for example, to the diffraction pattern shown in graphical form in panel (b) for a star some 5 pcs away seen at 5-μm wavelength in a 2-meter-class telescope. The planet at 5 AU, the size of Jupiter, will have its own diffraction pattern with the primary peak marked as shown, several orders of magnitude dimmer than even the secondary diffraction peak of the central star.

pattern is spread out (Figure 15.9) and now our putative Jupiter lies at just twice the distance of the first dark ring. The dropoff of stellar intensity is therefore much less impressive, and the detection of Jupiter remains very difficult.

Use of a larger telescope improves the situation as indicated by the formula given earlier: A 10-meter telescope (Keck in Hawaii) has a diffraction limit five times smaller than for a 2-meter telescope at the same wavelength. But such large telescopes (at present) are all located on the ground and must contend with the effects of atmospheric distortion. When combined with the imperfections of a real optical system, this makes direct detection of Jupiter even more challenging from the ground than from space (Figure 15.10). The atmospheric motions can be corrected to a large extent by allowing one of the mirrors in the telescope system to change shape in response to atmospheric distortions in the stellar image. Such corrections must be made fast, but modern computing power is up to the

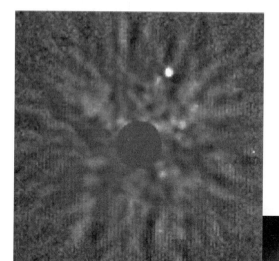

FIGURE 15.10 Coronagraphic image of a brown dwarf, an object about 60 to 80 times the mass of Jupiter, orbiting less than 20 AU from its parent star. The huge size of the star in the image shown below is an artifact of atmospheric blurring and optical imperfections in the telescopes used to make the image. The star is removed by image processing in the left panel to reveal the brown dwarf. Keck and Gemini telescope images.

task and such *adaptive optics* techniques are being refined today on a number of large telescope systems.

Even with adaptive optics, a 10-meter telescope by itself produces too broad a diffraction image to make planet detection practical, given the imperfections and consequent smearing of the diffraction rings in the telescope. Imagine, however, arraying two or more telescopes in a line and combining the light from the telescopes in such a way as to make a single, large telescope. To conceive of this possibility, let us imagine a telescope with a 10-meter-diameter mirror. It is difficult to build this from a single piece of glass, so assume it is made from a large number of smaller mirrors (say, 1 meter each in diameter), combined in a frame as a single large mirror. Indeed, this is the way the two Keck 10-meter telescopes were constructed. But to detect a planet around a star, we do not need to fill the 100 m^2 completely with meter-size mirrors—it is enough to have mirrors at each end of the 10-meter-diameter circle from which the light is combined. Less light is captured in this way, and hence less sensitivity, but the diffraction limit is still that of a 10-meter telescope. Then, what is to prevent us from taking our two mirrors and putting them 100 meters, rather than 10 meters, apart, to get a diffraction limited resolution equivalent to a telescope with a 100-meter diameter mirror?

To accomplish this requires combining the light from the two mirrors in a very particular way. Think of light at a single frequency as a simple plane wave, with crests and troughs. It is essential that the light from the two separated mirrors combines in such a way as to superpose the crest of the light wave coming through one mirror with the crest of the light coming from the other mirror. If this is achieved, then the diffraction limit formula applies to the total system of both mirrors, and the relevant diameter is the distance between the mirrors. Put another way, the light from the two mirrors must be made to combine or *interfere* in a way that preserves the "phase coherence" of the light beams. In our example, we considered light of one frequency. In reality, any beam of light, even put through a filter, will have a spread of frequencies, and perfect phase coherence is impossible to accomplish. Nonetheless, such *interferometers* have been shown using ground-based systems to dramatically increase resolving power over what can be done with single-mirror systems.

Successful operation of an interferometer depends on carefully controlling the lengths of the paths that the light travels from the two main mirrors. This is done with sets of auxiliary mirrors that are adjusted so that the light from the two mirrors is combined in phase. Such a technology has been available for radio telescopes since the mid-20th century, but it is much easier to combine electromagnetic waves that are centimeters or meters in wavelength than those that are a million times shorter. Hence, optical interferometry is a challenging new technology relying on high-speed computing, advanced optical systems, and clever techniques for measuring path lengths of light (Figure 15.11).

With optical interferometry we can, in theory, get extremely high spatial resolution in a system by making the diffraction limit very small. Then, for example, Jupiter sitting an arc second from the Sun would be so many "rings" in the fringe pattern away from the bulk of the starlight that it could be readily visible. Put such a system in space and the distorting effects of the atmosphere are eliminated without the use of adaptive optics. However, we must still contend with the imperfections of the optical elements themselves (the mirrors) and potential reflections off the mounting surfaces. If not tightly controlled, the starlight would be smeared over large angular distances and make the tiny diffraction limit irrelevant.

Suppose we now take advantage in a different way of the interferometer's capability (indeed, necessity) of controlling the phase of the light coming from the two mirrors. A standard interferometer would be set to constructively superpose the two beams of light—to ensure that a crest is merged with a crest and a trough with a trough so as to gain the sharpest possible stellar image. But what if we shift the optical elements to create *destructive* interference of the stellar light, that is, arrange for a crest of the light wave from one mirror to fall on the trough of the other? That would wipe out the light of the central star. Indeed, if this were done precisely, the central star would become invisible. But this is exactly what we want, because such a cancellation or *nulling* of the light from the central star would greatly reduce the problem of the scattering or smearing of that light over a larger area so as to obscure the planet.

This technique, called *nulling interferometry,* was proposed by Stanford University engineering professor Ronald Bracewell in the 1970s as a way of detecting planets from space. It is being practiced and refined today at various large, ground-based telescopic

a

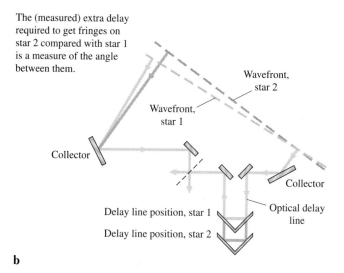

The (measured) extra delay required to get fringes on star 2 compared with star 1 is a measure of the angle between them.

b

FIGURE 15.11 The Palomar optical interferometer, developed by NASA's Jet Propulsion Laboratory. (a) View from the air, looking at the system located atop Mt. Palomar, California. (b) Schematic showing how the interferometer can measure precisely the angular difference between the positions of two stars. The delay lines change the pathlengths of the light for a given star in each of the collector mirrors. By adjusting the paths so that the light is in phase for each of two stars, and then measuring the difference in the delay line for each star, the relative positions can be precisely measured. Exactly the same approach can be used for a single star that is wobbling due to the presence of a planet—the star is observed over time and its changing angular position in the sky is determined.

facilities around the world (Figure 15.12). Done in space, outside of the interfering effects of the atmosphere, nulling interferometry could provide an unobscured view of the immediate regions around nearby stars, where planets might be lurking. Bracewell and his successors in this effort (e.g., J. R. P. Angel and colleagues at the University of Arizona) found that to maximize the nulling effect on the star requires four rather than two mirrors. The goal is to reduce the starlight by a factor of a billion in the optical, in order to clearly see, for example, a Jupiter-sized planet an arc second away from its parent star. But such a technique, if properly implemented, could go even further—to detect planets the size of Earth in orbits like the Earth around our Sun.

In thinking about this technique, we might wonder how we could pluck a planet out of the general pattern of dark and bright—constructive and destructive—fringes. What if we were unlucky and the Earth that was sought lay accidentally in a destructive fringe? Although we

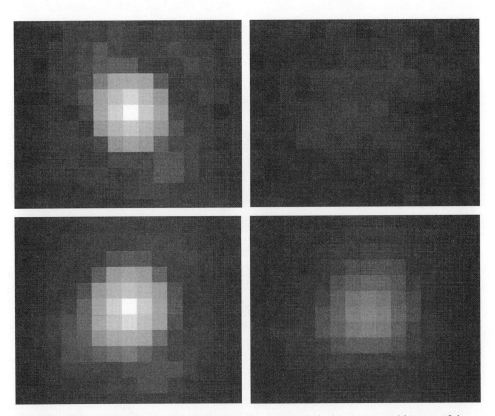

FIGURE 15.12 Examples of nulling interferometry. On the left side are normal images of the stars Aldeberan and Betelgeuse. The interferometer beams are arranged to null out the starlight in each right panel, allowing a search for planets or a disk around each star. Observations were performed on the 6.5-meter MMT facility on Mt. Hopkins, Arizona, operated jointly by Harvard Smithsonian Astrophysical Observatory and the University of Arizona.

have used the term "rings" to describe the interference fringes, so as to draw an analogy with the diffraction rings of the single mirror scope, in fact a system of mirrors can be arranged so as to produce bright and dark fringes that do not have circular shapes within the portion of sky being viewed. One of Bracewell's key insights was this: If the interferometer were rotated in space along an axis corresponding to the line of sight from the telescope to the star, then planets around the star would pass alternately into dark fringes and then light ones. This would be true regardless of the planet's angular separation from the star. If the starlight were properly nulled out, the signal from the planet would be a periodic brightening and then darkening as the interferometer spun in space (Figure 15.13).

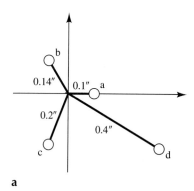

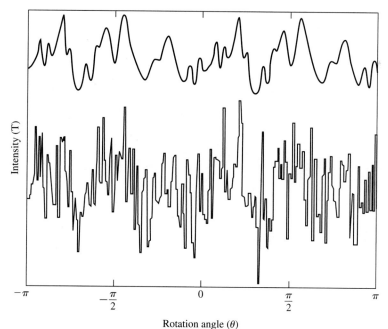

FIGURE 15.13 A four-mirror nulling interferometer can be used to detect the presence of planets around a nearby star. The positions of the planets and their separation from the parent star, in arc seconds as seen from the Earth (10 pc distant), is shown in panel (a). The interferometer projects a pattern of interference fringes on the sky. If the interferometer is rotated, like a pinwheel, centered on the parent star that is arranged to be nulled out, the planets produce a periodic intensity signal as a function of the angle of the interferometer during its rotation, shown in the upper curve of panel (b). Assuming noise from various sources, the planetary signals might be somewhat more difficult to detect as shown in the lower curve of panel (b).

The Terrestrial Planet Finder (TPF) and Darwin are, respectively, NASA and European Space Agency (ESA) concepts for doing nulling interferometry in space, to detect Earth's around nearby stars. By staring at a star with several properly spaced mirrors, the starlight can be nulled out by many orders of magnitude. As the mirror configuration is physically rotated on the sky, or electronically modulated, planets in orbit around the star periodically fall into the bright fringes between the nulls, during which time they are particularly visible. This modulation in signal could, for a sufficiently sensitive system, allow an Earth orbiting a Sunlike star to be seen from 10 pc away. The mirrors might be arranged on a metal structure—or mounted on free-flying spacecraft that maintain their relative positions to a very tight tolerance. While technologically daunting, either approach is conceptually powerful enough that serious studies are under way on the architecture of such a system in the United States and in Europe.

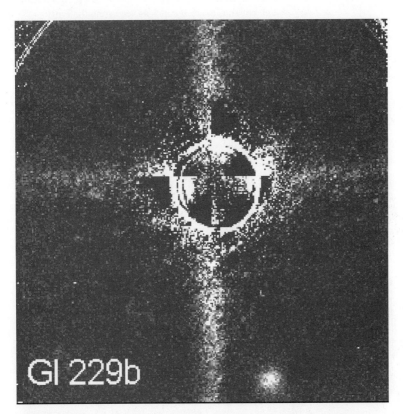

FIGURE 15.14 Example of a coronagraphic image made from a ground-based telescope. The brown dwarf Gl229b is seen in infrared wavelengths at the NASA Infrared Telescope Facility atop Mauna Kea, Hawaii, using a coronagraph with a specially designed soft edge. Additional removal of starlight was accomplished by constructing an image of the star itself and then subtracting this from the telescopic image. The brown dwarf is seen in the lower right; the cross-hairs are from the support structure that holds the mirror in the telescope.

Nulling interferometry is not the only way to make a giant "Earth detector" in the sky. Although it is a promising technology, it has not yet demonstrated the enormous degree of cancellation of starlight required for Darwin or TPF. An alternative means of removing the glaring light from the central star is called *coronagraphy*. It is a physical blocking of the star's light in the optical system of the telescope itself. The light is either blocked by a plate or using special pupils (holes through which the starlight is admitted) and associated optics. Coronagraphs have been used for decades on ground-based telescopes, yielding images of dust disks and brown dwarfs around nearby stars (Figure 15.14). However, the standard coronagraphs in use today are not perfect devices for blocking the light from stars. After all, careful readers of the preceding material on diffraction might realize that the edges of the blocking devices themselves must cause diffraction fringes to appear on the images, and, in searching the space very close to the star itself, such fringes (scattered by the rest of the optical system) would obscure planets. Clever schemes have been devised recently to minimize this effect. For example, we might shape the blocking device to "throw" the diffracted energy of the starlight into particular angles around the star, freeing the other angles for planet searching. If such a system were deployed in space, and rotated, the diffraction-free regions of space would sweep around the star, and planets would alternately appear and disappear with the period of sweeping. One clever design is shown in Figure 15.15. As of the printing of this book, NASA and ESA are working on both concepts—coronagraphy and interferometry—for finding terrestrial planets around nearby stars, and it is possible that both techniques will be put in space before the end of the second decade of the 21st century.

Class 1 – Classical coronagraph: Gaussian-type image-plane occulter

Class 2 – Optimized pupil: Spergel-Kasdin class of shaped-pupil designs

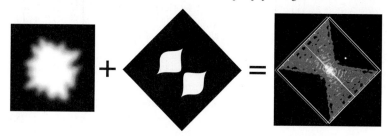

FIGURE 15.15 Diagram showing how a special coronagraphic shape, the Spergel-Kasdin pupil, can provide clear areas near a star for planet detection. In the first box, the planet is shown on top of the star just to indicate position—without the coronagraph, it is lost in the glare of the parent star.

15.6 Learning About Other Earths

The direct imaging of a planet around another star would yield a large amount of information. The size of the orbit and its inclination to Earth could be determined by tracking the position of the planet as it moved around the parent star. For a 1-AU orbit around a Sunlike star, this would require a year or less. Observations of the planet's brightness in the optical and infrared parts of the spectrum would permit us to determine the size of the planet and its reflectiveness, or *albedo*, given that the distance of the planet from its parent star is determined. We can do this because conservation of energy requires that the amount of energy in the form of starlight hitting the planet, which depends on the planet's size, be equal to the light reflected in the optical plus that reemitted in the infrared. If the star could be tracked by astrometry (requiring at least the full sensitivity of the planned Space Interferometry Mission), then the mass could be derived; knowing the size and the mass yields the planet's density. If the density were like that of the Earth, Venus, or Mars, then this would be a sure sign that we are observing a terrestrial-type planet. The Earth is one of the most variable objects in our solar system, when different hemispheres are observed, because of the strong color and brightness contrasts among clouds, oceans, and land. Observations of the changing brightness of the detected extrasolar planet, along with its color, would suggest whether it is cloud covered like Venus, dusty red like Mars, or blue and highly variable like Earth.

If an Earth-sized planet could be detected, and its orbit determined, the initial assessment of its habitability will be made based on that orbit. From our experiences with Mars and Venus, we might generously define a zone from 0.7 to 1.5 AU around a G-type (Sunlike) star as the *habitable zone:* a region in which liquid water may have existed at some point in the history of a planet with a nearly circular orbit, whose semimajor axis is between these bounds. This habitable zone would have to be scaled to the luminosity of the parent star—both the inner and outer distance will shrink or expand according to the square root of the luminosity of the star relative to that of the Sun. Hence, for a star with half the luminosity of the Sun, the habitable zone would extend from 0.49 out to 1.1 AU.

More difficult to define is the so-called continuously habitable zone, the location in orbital space for which a planet could sustain life over billions of years. Our understanding of the histories of Earth and Mars (Chapters 11, 12) suggest both the delicacy of maintaining an environment like that of Earth, and the robust persistence of liquid water at certain times and certain places on an otherwise dessicated world like Mars. Might the continuously habitable zone end just a bit beyond 1 AU for a Sunlike star, or do we extend it out to 1.5 AU to account for possible subcrustal life on Mars today or the possibility that an Earth-sized body at the orbit of Mars might have retained a substantial surface hydrosphere? How far inward do we extend this zone, given that Earth might lose its crustal water a billion years from now in the way that Venus apparently did some billions of years ago (Chapter 11)? The continuously habitable zone—an imaginary disk that could be broad or narrow compared with the overall extent of a terrestrial planet system—is very poorly understood and is not an unambiguous guide either to where to search for Earthlike planets around other stars or to which planets might be habitable.

But should enough light be available from the nulling interferometer or coronagraph of TPF, it is possible to take a spectrum of the reflected light from the planet and analyze it for signs of life. Because the contrast between star and planet is two to three orders of magnitude lower in the infrared than it is in the optical, spectra might most easily be taken in the 5- to 10-μm region. There, for an atmosphere like that of the present Earth, water, carbon dioxide, and ozone would dominate. Ozone, as a proxy for molecular oxygen, is particularly exciting because it could be an indicator of photosynthesizing life. If spectra in the optical were taken, molecular oxygen itself would be visible, along with a change in brightness in the far red, caused by the so-called red edge of chlorophyll (Chapter 4) in vegetation (Figure 15.16). The variability of the planet's spectra at optical wavelengths as it orbits its parent star might be tracked to see if it is as large as that of the Earth, where clouds, oceans, and land modulate the signatures of atmospheric gases and chlorophyll. Finally, the temperature of the lower atmosphere can be determined by observing the infrared spectrum. So, potentially TPF or Darwin could not only detect "other Earths" but determine their habitability as well.

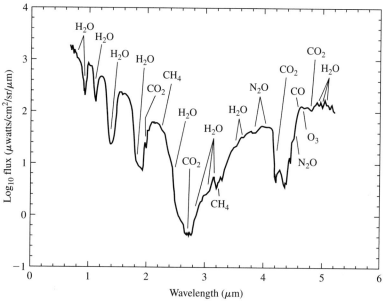

a

FIGURE 15.16 (a) A near-infrared spectrum of the Earth taken by the Galileo spacecraft on its way to Jupiter, with signatures of the atmosphere (carbon dioxide), the hydrosphere (water vapor) and the biosphere (methane and ozone) indicated. Shown in the upper part of panel (b) on page 506 is a spectrum at optical wavelengths of the "Earth-light" on the dark part of the Moon. This type of astronomical observation uses the Moon to average reflected light from different parts of the side of the Earth facing the Moon, and simulates the spectrum we would see from a distant planet around another star. Lines of ozone and molecular oxygen can be seen, as well as a dramatic change in brightness caused by the red edge of photosynthesizing vegetation. The lower part of the panel shows such lines in a computer model of the Earth's atmosphere, as well as signatures of the ocean and vegetation, for comparison.

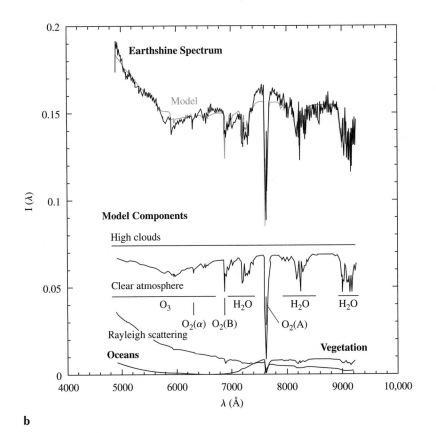

b

FIGURE 15.16 continued

The excitement of this prospect is diminished somewhat when the complications in the spectral signatures are considered. For example, Venus currently has virtually no water, but in the first 100 million years or so of its existence, water might have been escaping from the surface at a prodigious rate. During the time of rapid photolysis and escape of water, ozone derived from the water would have been present in measurable amounts. A spectrum of young Venus from TPF would then reveal carbon dioxide, water, and ozone—but from a planet that was as yet uninhabited and on its way to becoming uninhabitable. Further, prior to the time free oxygen was abundant on the Earth (see Chapter 11), the ozone signature would have been absent, even though life existed back then. (Figure 15.17). We might propose to look for other chemical species not in chemical equilibrium with a carbon dioxide atmosphere and hence indicative of life—for example, methane. Small amounts of methane (e.g., 10 ppm) could be abiotically produced in planetary environments, as may be happening on Mars (Chapter 12), but the presence of 0.1 percent or more methane could be a strong indicator of biological activity, according to Jim Kasting and colleagues at Pennsylvania State University.

15.6 Learning About Other Earth's 507

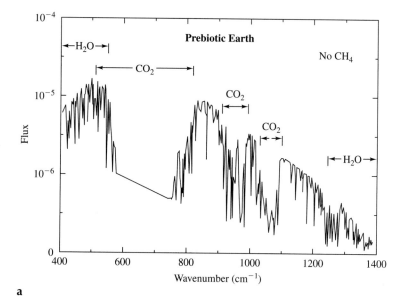

a

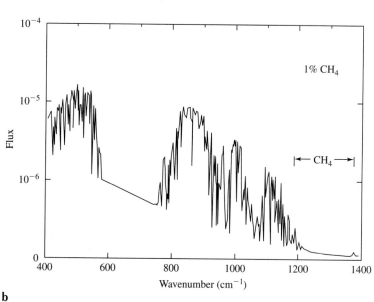

b

FIGURE 15.17 A model spectrum, in the infrared, of the Earth prior to the origin of life (a). This spectrum is little different from that of a primitive biosphere in which the lack of photosynthesis has as yet failed to generate significant amounts of oxygen, hence ozone. Therefore, it is almost impossible to tell whether a planet with this spectrum has primitive life or is sterile. The presence of large amounts of methane (b) might signify primitive life or, alternatively, an atmosphere that is so youthful that primordial methane is outgassing too rapidly from the interior to be consumed.

Regardless of the difficulties of designing an unimpeachable protocol for detecting life on another planet through spectroscopic analysis of the atmosphere, study of extrasolar Earth-sized planets would be a revolution in astronomy, planetology, and climatology. Given the unexpected diversity among the large natural satellites of our own solar system, and the apparently complex set of processes conspiring to make Earth (but not Mars or Venus) continuously habitable over 4 billion years, we should expect surprises in the characteristics of the planets that we find. The Danish Prince Hamlet, in Shakespeare's play of that name, admonished his friend Horatio: "There are more things in Heaven and Earth, Horatio, than are dreamt of in your philosophy." Today we might instead warn ourselves of the likelihood that there are more kinds of Earths in the heavens than are dreamt of in astrobiology.

Any mission to detect and spectroscopically characterize terrestrial planets around other stars must be designed so that it can characterize diverse types of terrestrial planets with a useful outcome. Also, because such telescopes must be optimized to have a certain field of view, they will likely be tuned to cover the habitable zones around stars as defined by the ability of a planet to sustain liquid water on its surface (Chapters 11 and 12). This would, for example, exclude the search for objects with deep subsurface liquid water only, equivalent to Europa in our own solar system (Chapter 13).

Despite the shortcomings, it is most remarkable that we can even conceive of a device that could be built within our lifetimes to discover and analyze Earth-sized planets and their atmospheres. TPF or Darwin could be put into space in the time frame around 2020. Although a generation would have passed between the first definitive detection of extrasolar planets in 1995 and the launch of TPF/Darwin to find other Earths, it would be less than 10 percent of the duration of the Copernican revolution itself (Chapter 1). If TPF/Darwin is successful, our view of the cosmos will change. Human beings in one or both hemispheres of our own world will be able to look up at the night sky, point to one or two or a handful of stars, and tell their children that there—around those stars—exist planets like the Earth. No previous generation has been able to do so, and arguably such an accomplishment will represent the penultimate step in completing the Copernican revolution. The final step may depend on something even more difficult—detecting signals produced by sentient beings elsewhere in the Galaxy.

15.7 The Search for Extraterrestrial Intelligence

The search for intelligent life on other worlds beyond the Earth represents the most romantic and publicly accessible aspect of the search for life, yet it is perhaps the most problematic. There is only one well-understood remote-sensing technique that can practically detect the presence of intelligent life-forms, namely, listening for electromagnetic forms of communication leaking from or deliberately sent from another world. Alternatives rely on short-lived or speculative signatures: Human civilization has put chlorofluorocarbons in the stratosphere of our own planet, which are demonstrably the product of a technological civilization but require much higher spectral resolution for their detection than is planned with TPF. Further, the damaging effects on the stratospheric ozone layer of these compounds implies that a civilization would do this only for a miniscule amount of time (several decades), making detection of such compounds on other worlds very unlikely. Thus, looking for the atmospheric signatures of a technological civilization is difficult and unlikely to succeed. Some astronomers have proposed that truly advanced civilizations might modify the architecture of their home planetary systems on a scale detectable from Earth, but this would require a civilization so advanced as to be very speculative.

Early attempts in the 1950s and 1960s to detect radio signals from extraterrestrial civilization yielded no positive results. The biggest challenges in making such searches were to pick the right radio frequencies and to guess the narrowness of the frequency bands ("bandwidths") with which extraterrestrials might communicate. The bandwidths of transmitters on our interplanetary spacecraft are a good starting point, and these are such as to permit enormous numbers of possible "channels" that must be listened to. Proponents of the search understood the need to build tunable radio receivers that could scan enormous numbers (literally, billions) of very narrow frequency bands (i.e., channels) at high speed, but the ability to do so was technologically limited until the 1980s, when inexpensive computing capabilities and receiver technologies matured. In the early 1990s NASA involved itself in the search through the high-resolution microwave survey (HRMS), which planned to utilize two complementary search modes: a cursory survey of the whole sky and a detailed ("targeted") search of selected areas, such as the inner part of the Milky Way galaxy. Both elements involved state-of-the-art digital spectrum analyzers and signal processing equipment to carry out the observations and data processing activities automatically and the use of existing large radio telescopes such as those in the NASA Deep Space Network, the Arecibo Radio Observatory, and others.

In 1993 the federal government terminated support for HRMS. Scientists involved in the teams, no longer funded by NASA, regrouped through a private organization, the SETI (Search for Extraterrestrial Intelligence) Institute founded in 1985 and based in the Silicon Valley of Northern California. The SETI Institute established the Phoenix Project in 1995 to implement the targeted search using equipment leased from NASA. Then the Institute raised private funds to complete the expansion of the system and to conduct the planned observations.

Phoenix does not scan the whole sky. Rather, it scrutinizes the vicinities of approximately 1,000 nearby (within 200 light-years) Sunlike stars most likely to sustain life. Phoenix looks for signals between 1,000- and 3,000-MHz frequencies. Exceedingly narrow band signals are assumed to be the "signature" of an intelligent transmission. The spectrum searched by Phoenix is broken into very narrow 1-hertz-wide channels, so 2 billion channels are examined for each target star. An automatic detection algorithm, one of the major software inventions of the SETI Institute, enables long-term observation with only occasional human intervention. Observations have been made during 2 three-week sessions each year using the 1,000-foot radio telescope at Arecibo, in Puerto Rico. With about half the target list stars searched as of this writing, no positive results have been found. However, the volume of the galaxy searched by Phoenix is vanishingly small, especially when compared with, for example, what microlensing surveys will cover in the search for planets the mass of the Earth.

A new project, the Allen Telescope Array, will cover a much larger volume of space and a much larger frequency domain. It is a telescope system designed and built specifically for the search for extraterrestrial intelligence, to be operated jointly with the University of California at a high-altitude site in northern California and funded with private donations.

In its current initial phase of operation, 32 dishes—each just a bit more than 6 meters across—are arrayed together as a radio-wavelength interferometer. The final phase of operation will see 350 such telescopes operating as a giant ear listening to the cosmos for signals from intelligent life.

The likelihood of finding an intelligent, communicating extraterrestrial civilization depends on a host of factors ranging from commonality of planets to the persistence of technological civilizations. We defer a discussion of these factors, folded into what is called the "Drake equation," until Chapter 17, where the origin of human consciousness and intelligence are considered.

Some scientists have argued that searching for extraterrestrial life is not worth pursuing because it is not a scientifically well-designed experiment—a null result does not tell us that intelligent life does not exist elsewhere, only that we (for some reason) have been unable to make contact. There is no way to set up the "control" experiment in which we know the detection system is capable of detecting other intelligence but, in fact, fails to do so. Lack of detection could mean lack of technologically sophisticated life elsewhere or the presence of civilizations so advanced that they use much narrower bandwidths or other parts of the electromagnetic spectrum. There may be intelligence without the curiosity or need to communicate, or we might be quarantined. Finally, we may simply not recognize such signals even though they are out there. The imponderables of the null results up to now have led to interesting philosophical discussions, but these fail to inform us how to change the experiment to ensure success or how to reformulate the hypothesis about the nature and existence of other-worldly life.

In this author's view, the criticisms of SETI are ill-founded because the underlying popular motivation for the search for life elsewhere is the aspiration to some day communicate with other forms of intelligence as much like or unlike us as we care to imagine. It would be dissembling, to say the least, to discourage a search (especially one enabled by private funding) at the same time that astrobiology as a whole taps into such basic emotions and aspirations to excite the public about the general search for life's origins, evolution, and cosmic ubiquity. Therefore, SETI should continue for a long time to come as a visible, technological expression of humankind's desire to meet other sentient beings among the stars.

QUESTIONS AND PROBLEMS

1. Consulting Table 15.2, suppose we replaced Earth at 1 AU with Jupiter. What then would the required radial velocity sensitivity be? What would the number be for Earth at Jupiter's orbit, 5 AU? How do these numbers depend on the semimajor axis of the orbit?

2. What would be the required sensitivity in the angular displacement (in milliarcseconds) for an Earth-mass planet at 5 AU from a Sunlike star 5 pc from our solar system?

3. Explain why—for a given mass of planet and star—the radial velocity signature of a planet decreases while that of the astrometric wobble (i.e., in the plane of the sky) increases as the star–planet separation increases.

4. For a young giant planet twice the radius of Jupiter, what would the required photometric precision be in order to observe a transit of this planet across the face of its parent star?

5. Assume for the Sun a habitable zone that extends from 0.7 to 1.5 AU in semimajor axis. Now (using a table of astronomical data) plot the positions of the inner and outer edges of the habitable zone as a function of stellar spectral type on the main sequence, from B stars down to M dwarf stars. Using the fact that the planet Mercury is tidally locked to the Sun (it does not have a 1:1 relationship between its spin and its year; rather it is a 3:2 relationship, but still one that is tidally locked), draw a horizontal line representing very approximately the tidal locking radius. Using your knowledge of stellar main sequence lifetimes, draw a vertical line representing the spectral type at which the main sequence lifetime is less than the time it plausibly took for life to form on the Earth (use Chapter 9 as a guide). Finally, plot the continuously habitable zone as a belt centered around 1 AU that extends ± 10 percent in semimajor axis outward and inward, for the Sun, and extend this according to the luminosity scaling for the main sequence stars from B to M.

6. Using the graph plotted in problem 5, describe a strategy for choosing which types of stars should be emphasized in the search for habitable planets by TPF and which ought to have lower priority. Draw on the arguments in Chapters 11 and 12 in making your plan.

7. Suppose terrestrial planets form uniformly in the semimajor axis space from 0.4 to 2 AU from a star like the Sun. What is the probability that a planet will end up in the habitable zone? What is the probability it will end up in the continuously habitable zone? Given this last probability, and assuming that half of the stars like the Sun have terrestrial planets somewhere in the 0.4- to 2-AU region, how many stars must TPF search to have a high likelihood of finding one terrestrial planet in the continuously habitable zone?

8. Read the paper by Raymond et al. in the References. What does this paper imply for the TPF search for Earthlike planets in terms of

    a. Probability of a terrestrial planet in the continuously habitable zone?

    b. Probability of a terrestrial planet in the habitable zone?

    c. Probability of such a planet having enough water to create an ocean?

    Discuss the assumptions and potential weaknesses of the approach in the paper. Should it be a guide to designing the TPF search strategy?

9. The next-generation, James Webb space telescope has a 6.5-meter-diameter mirror; Hubble Space Telescope has a 2.5-meter-diameter mirror. At what wavelength is the diffraction limit of the James Webb Space Telescope equal to that of the Hubble Space Telescope? By doing research on the Web, indicate whether this wavelength is one at which the James Webb Telescope can operate and whether Hubble can operate there as well. What is the essential difference between the two space telescopes, besides size, that affects the wavelength ranges of operation?

10. Consider the challenges of selecting the right frequencies, frequency widths, and other parameters in listening to the sky for signs of intelligent communication. Do a

literature search to identify favorable and unfavorable wavelengths for listening. Why would intelligent civilizations choose radio wavelengths? Would they choose optical communication or an exotic communications technique such as the release of gravitational waves from the collapse or collision of objects massive on an astrophysical scale, or something else? Why or why not?

SUGGESTED READINGS

Two popular books on techniques for detecting planets are D. Goldsmith, *Worlds Unnumbered: The Search for Extraterrestrial Planets* (1997, University Science Books, Sausalito, CA), and A. Boss, *Looking for Earths: The Race to Find New Solar Systems* (2000, Wiley and Sons, New York). Much more detail on planned space missions can be found in two NASA documents, more technical than the two books cited but still readable: *The Terrestrial Planet Finder* (1999, NASA Jet Propulsion Lab Report 99-3); and *SIM, Space Interferometry Mission: Taking the Measure of the Universe* (NASA Jet Propulsion Laboratory Publication 400-811). Both are downloadable from the JPL planet detection website http://planetquest.jpl.nasa.gov/.

REFERENCES

Angel, J. R. P., and Woolf, N. (1997). "An Imaging Nulling Interferometer to Study Extrasolar Planets," *Astrophys. J.,* **475,** 373–379.

Beichman, C., Woolf, N. J., and Lindensmith, C. A. (1999). *The Terrestrial Planet Finder,* NASA Jet Propulsion Laboratory Publication 99-3, Pasadena, CA, http://tpf.jpl.nasa.gov/.

Bracewell, R. N. (1978). "Detecting Nonsolar Planets by Spinning Infrared Interferometer," *Nature,* **274,** 780–781.

Burrows, A. S., Sudarsky, D., and Hubbard, W. B. (2003). "A Theory for the Radius of the Transiting Planet HD209458b," *Astrophys. J.,* **594,** 545–551.

Carter, B. D., Butler, R. P., Tinney, C. G., Jones, H. R. A., Marcy, G. W., McCarthy, C., Fischer, D. A., and Penny, A. J. (2003). "A Planet in a Circular Orbit with a 6 Year Period," *Astrophys. J. Lett.,* **593,** L43–L46.

Charbonneau, D., Brown, T. M., Latham, D. W., and Mayor, M. (2000). "Detection of Planetary Transits Across a Sun-Like Star," *Astrophys. J. Lett.,* **529,** L45–L48.

COMPLEX (Committee on Planetary and Lunar Exploration), (1990). *Strategy for the Detection and Study of Other Planetary Systems and Extra-Solar Planetary Materials (1990–2000).* National Academy Press, Washington, D.C.

Danner, R. and Unwin, S. (2000). *SIM, Space Interferometry Mission: Taking the Measure of the Universe.* NASA Jet Propulsion Laboratory Publication 400–811, Pasadeva, CA, http://sim.jpl.nasa.gov.

Des Marais, D. J., Harwit, M., Jucks, K., Kasting, J., Lin, D., Lunine, J., Schneider, J., Seager, S., Traub, W., and Woolf, N. (2002). "Remote Sensing of Planetary Properties and Biosignatures on Extrasolar Terrestrial Planets," *Astrobiol.,* **2,** 153–181.

Drossart, P., Rosenqvist, J., Encrenaz, T., Lellouch, E., Carlson, R. W., Baines, K. H., Weissman, P. R., Smythe, W. D., Kamp, L. W., and Taylor, F. W. (1993). "Earth Global Mosaic Observations with NIMS—Galileo," *Planet. Space Sci.,* **41,** 551–561.

Fischer, D. A., Butler, R. P., Marcy, G. W., Vogt, S. S., and Henry, G. W. (2003). "A Sub-Saturn Mass Planet Orbiting HD 3651," *Astrophys. J.,* **590,** 1081–1087.

Fischer, D. A., Marcy, G. W., Butler, R. P., Vogt, S. S., Henry, G. W., Pourbaix, D., Walp, B., Misch, A. A., and Wright, J. T. (2003). "A Planetary Companion to HD40979 and Additional Planets Orbiting HD12661 and HD38529," *Astrophys. J.,* **586,** 1394–1408.

Ford, E. B., Seager, S., and Turner, E. L. (2001). "Characterization of Extrasolar Terrestrial Planets from Diurnal Photometric Variability," *Nature,* **412,** 885–887.

Kuchner, M. J. and Traub, W. A. "A Coronagraph with a Band-Limited Mask for Finding Terrestrial Planets," *Astrophys. J.,* **570,** 900–908.

Lunine, J. I. (2001). "The Occurrence of Jovian Planets and the Habitability of Planetary Systems," *Proc. Nati. Acad. Sci.,* **98,** 809–814.

Mannings, V., Boss, A. P., and Russel, S. S., eds. *Protostars and Planets IV,* University of Arizona Press, Tucson, AZ.

Mayor, M. and Queloz, D. (1995). "A Jupiter-Mass Companion to a Solar-Type Star," *Nature,* **378,** 355–357.

Raymond, S., Quinn, T., Lunine, J. I. (2004). "Making Other Earths: Dynamical Simulations of Terrestrial Planet Formation and Water Delivery," *Icarus,* **168,** 1–17.

Schindler, T. and Kasting, J. F. (2000). "Synthetic Spectra of Simulated Terrestrial Atmospheres Containing Possible Biomarker Gases," *Icarus,* **145,** 262–271.

Tarter, J. (2001). "The Search for Extraterrestrial Intelligence (SETI)," *Ann. Rev. Astron. Astrophys.,* **39,** 511–548.

Turnbull, M. C. and Tarter, J. C. (2003). "Target Selection for SETI. I. A Catalog of Nearby Habitable Stellar Systems," *Astrophys, J. Suppl. Ser.,* **145,** 181–198.

Udry, S., Mayor, M., and Santos, N. C. (2003). "Statistical Properties of Exoplanets. I. The period Distribution: Constraints for the Migration Scenario," *Astron. Astrophys.,* **407,** 369–376.

Woolf, N. J., Smith, P. S., Traub, W. and Jucks, K. (2002). "The Spectrum of Earthshine: A Pale Blue Dot Observed from the Ground," *Astrophys. J.,* **574,** 430–433.

CHAPTER 16

External and Internal Influences in the Evolution of Life

16.1 Introduction

The last half of the history of the Earth has no analog, even remotely, elsewhere in the solar system. With the onset of an aerobic atmosphere, and the origin and spread of oxygenic photosynthesis and respiration, the stage was set for the appearance of multicellular animals, plants, and fungi. How early this occurred remains unclear: Reports in 2001 of animal fossils in sediments 1.6 billion years old have since been discredited. The first fossils of animal forms do not appear in the geologic record until about 570 million years ago, 2 billion years or more after the appearance of the first eukaryotes and of an oxygen-rich atmosphere. Analysis of animal genomes suggest that distinct phyla existed as long ago as 1 billion years before present, yet these were not expressed in the form of complex multicellular animals until the so-called Cambrian explosion roughly 545 million years ago, when essentially all of the modern animal forms appear in the fossil record. Did environmental catastrophes intervene? A period of intermittent but intense glaciations occurred in the period between 800 million and 550 million years ago and may have set back the flowering of complex life whose nascent pioneers were, genetically speaking, ready to go.

This chapter examines the evolution of complex life and the environmental and genetic agents that shaped this evolution. These agents include catastrophic events such as impacts and volcanism, as well as periodic or regular forcing such as that associated with modulation of the orbit and spin axis of the Earth. These processes external to biology are then considered in the context of the long-term evolution of life, which is characterized by the appearance of organisms with greater numbers of genes and new regulatory genes in their genome, and with greater numbers of different types of differentiated cells comprising individual organisms. The chapter then brings together the geologic evidence on environmental effects with the molecular evidence regarding evolution to address the following questions:

- What were the effects of sudden and gradual environmental changes on the evolution of life?
- What determined the timing of the emergence of eukaryotes and then multicellular eukaryotic organisms?
- Is complex life likely to require an extended incubation period on all habitable worlds?

These considerations set the stage for the final chapter, a discussion of the longevities of species, the origin and meaning of intelligence, its relationship to technological development, and the persistence of technological civilizations.

16.2 Mechanisms of Evolution

That life has changed over geologic time, with the introduction of new organisms and the disappearance of others, is a fact that cannot be disputed when the fossil record is carefully considered. Evolution is, in the broadest sense of the word, a process in which progressive change occurs. Specific to biology, evolution is the process by which organisms change with time such that descendants differ progressively more from their ancestors. Evolution is possible because the genetic code, while remarkably error-correcting, is not error-free. Individual organisms may exhibit a greater or lesser suitability or adaptability to a particular environment, resulting in a variation in the extent to which the genomes of given organisms persist or are distributed to later generations of organisms. Gene duplication or alteration, while usually neutral with respect to the benefits that are conferred on the organism itself, can be amplified in isolated populations of organisms and persist in later generations.

There is little dispute that evolution is "Darwinian" in how it operates. That is, the traits possessed by particular organisms confer greater or lesser survival advantage in a given environment. In consequence those organisms more suited for a given environment will have a greater chance of surviving and passing their genes on to succeeding generations—whether they do it by direct cell division, exchange of DNA, or the structured mixing of DNA known as sexual reproduction. The alternative, "Lamarckian," view is that physiological or morphological changes in an organism caused by particular interactions with the environment are then passed on to offspring. The classic Lamarckian example is that of the origin of the giraffe's long neck as residing in the effort, generation after generation, of ancestral animals to reach ever taller tree branches for food by stretching their necks. This view is discredited because there is no genetic mechanism for inheritance of traits acquired through an individual's interaction with the environment around it, and no evidence that it actually occurs.

The genome interacts with the environment, but indirectly—the genetic makeup of an organism provides a blueprint of morphological and behavioral characteristics that largely define how the organism responds to and affects its environment. The outcome for a lineage of organisms will depend on a host of factors beyond the ambient conditions, including the intensity and frequency of fluctuations of the characteristics of the environment, as well as random occurrences. Darwinian evolution is extremely difficult to see in action, particularly for eukaryotic species, because of the long timescales involved, the low probability that a given creature will be preserved as a fossil, and the lack of completeness in the geographic and temporal coverage of the geologic record. Many changes, for example, the suppression or activation of the production of novel enzymes, may not show up at all in the fossil evidence. Where the fossil record is unusually good, the development of new eukaryotic species from old seems to occur in populations geographically isolated from the larger cohort of a species, and when it happens it does so quickly.

One example of *speciation* (formation of new species) is that of snails in Bermuda, most thoroughly analyzed by the late Harvard paleontologist Steven J. Gould in the 1960s with his then–dissertation advisor Niles Eldredge. From this study and another involving trilobites, they proposed the mechanism of *punctuated equilibrium*, in which eukaryotic species remain static for long periods of time, interrupted by brief episodes in which rapid speciation occurs among a small, isolated, subpopulation. A geographically isolated subpopulation has a smaller set of mating partners from which to choose, and random genetic changes have a greater chance of being amplified in the smaller cohort. The smaller the geographically isolated population, the more chance it has of becoming reproductively isolated from the main population—that is, unable eventually to successfully replicate with the main population. Because the definition of species for sexually reproducing eukaryotes is the cohort of organisms that can successfully reproduce by generating fertile offspring, the process of reproductive isolation marks the birth of a new species. Speciation occurs in the isolated group, while the main line carries on unchanged. A number of eukaryotic species reproduce asexually, in addition to reproducing sexually; this complicates a bit the definition of speciation, but the basic idea remains the same—that speciation follows isolation of a segment of a population. (The mechanisms of sexual reproduction—the formation of an organism from the paired union of special cells from two individuals—and the "Mendelian" rules associated with inheritance of traits from prior generations of sexually reproducing individuals—require an extended discussion that is beyond the scope of this book. Readers are urged to review a standard biology text, including that suggested in the References, to attain familiarity with these mechanisms).

Punctuated equilibrium is Darwinian evolution for eukaryotes with a more staccato pace than the stately measured changes envisioned in the traditional or "gradualist" model, in which a large interbreeding population acquires new characteristics as a particularly advantageous mutation spreads through the gene pool over a very large number of generations. Punctuated equilibrium recognizes the extreme difficulty of the gradualist process in which the number of interbreeding members of the species is large and the trend toward mixing out of extreme variations is strong. It also explains why, except in rare circumstances (such as those that allowed Eldridge and Gould their classic study of speciation in Bermuda snails), the fossil record does not show directly the development of new species: Speciation occurs quickly, after long periods of stasis, and only among a small, reproductively isolated cohort of an existing species.

On the species level, punctuated equilibrium remains the best model for how natural selection works among eukaryotes. It is much less useful for explaining the appearance of new orders, or new phyla, that is, innovations higher on the taxonomic classification scheme for eukaryotes. Among animals the basic phyla, or body plans, appear in the fossil record to have been established in a short time in the Cambrian period. Punctuated equilibrium also does not directly address the details of the genetic mechanisms for change, although the rules of Mendelian genetics among sexually reproducing eukaryotes are such that reproductive isolation will aid in the persistence of new traits (especially recessive ones) associated with mutations. It does provide a useful context for understanding the interactions of the environment with evolution during times of dramatic climate change or of catastrophic events that lead to the flowering of large numbers of new species. We will examine these interactions beginning in Section 16.5.

The evolution of prokaryotic organisms is a very different matter—genetic theories of evolution were invented "by sexually reproducing eukaryotes, for sexually reproducing eukaryotes" (Levin and Bergstrom, 2000). Bacteria (the prokaryotic domain for which the vast majority of work on evolution has been done) reproduce essentially by cloning—replicating the whole genome from a single parent cell. So-called recombination—the mixing of genes from different organisms—happens much less regularly in bacteria than it does in sexually reproducing eukaryotes, for which the process is nearly always obligatory. Lateral gene transfer between prokaryotic cells, followed by *recombination* of the DNA from the two individuals into a single, altered genetic code in the recipient, is becoming recognized as a major factor in microbial evolution and innovation, however. Such transfer occurs via direct exchange of "naked" DNA through thin tubes called *pili*, by uptake of very small chromosomes called *plasmids*, or by viruses—and, crucially, not all such transfers are fatal for the recipient cell. Prokaryotic generations are short so, in terms of time, bacterial evolution can be rapid.

When recombination occurs in bacteria, the exchange can be between closely related types or (much more rarely) over a broad range of the phylogenetic tree, in contrast to the species-specific eukaryotes. The lateral transfer of genes across bacterial organisms makes definitive determination of the topology of the prokaryotic part of the phylogenetic tree difficult (Chapter 9).

Many of the capabilities that define the properties of bacteria, such as antibiotic resistance the ability to ferment certain sources of carbon, detoxify metals, and control the expression of selected genes, are acquired by horizontal gene transfer and hence look similar across diverse bacterial species. A number of the genes that code for the toxins that make bacteria fatal to host eukaryotes are features that are not part of the stable genetic component of the microbial cell, gained (and as often lost) through lateral gene transfer.

The existence of lateral gene transfer between widely different microbes does not fit the traditional Darwinian concept of natural selection—caricatured as "survival of the fittest"—in which the role of the genome and that of the environment are carefully separated from each other. Even among the eukaryotes, interactions with the environment can extend to the genome in subtle ways. The expression of a gene in the production of proteins that have enzymatic or structural functions can be turned on or turned off ("regulated") by the action of certain enzymes. In this way, the expression of certain genetic traits is often specific to certain cells or even regions in a multicellular eukaryotic organism. Hence, there can be thousands of differences in gene expression between different cell types, and these differences are controlled by a very small number of regulatory enzymes. Although the production of these regulatory enzymes is itself controlled by the organism's genome, there is the potential for environmental effects to alter the production or activity of the enzymes and hence force changes in the expression of whole suites of genes. To what extent this process affects the nature and rate of speciation is not well understood. But in the long history of eukaryotes, particularly early on when the symbiotic events described in Section 16.3 created the prototypes for such organisms, it is hard to imagine that genomic-environmental interactions did not play a strong evolutionary role.

16.3 On the Origin of Eukaryotes

Geologic Evidence

The geologic record suggests that, for the first 1 billion to 1.5 billion years of the history of life, the sole cellular type was prokaryotic. The complex eukaryotic cell, with its essential organelles that contain both the nuclear genetic code and (for those eukaryotes that possess mitochondria) their own separate genomes and is capable of oxygenic respiration and chlorophyll-based photosynthesis, came later. How much later is unclear, and some have argued based on the topology of the phylogenetic tree that it was fairly early in the history of life. But the first hint of eukaryotes in the geologic record comes from Archean-eon shales exposed in northwestern Australia and dated, with some controversy, at about 2.7 billion years ago. Extracted from the rocks were *steranes*, sedimentary deposits that come from *sterols*. The particular sterols are of interest: cholestane and related molecules that contain 28 to 30 carbon atoms. No prokaryotes existing today are known to produce these relatively elaborate sterols, and hence their presence is suggestive of the existence of a eukaryotic attribute—the complex-sterol synthesis mechanism—at the time these rocks formed (Figure 16.1). Whether the cells that formed the sterols resembled modern

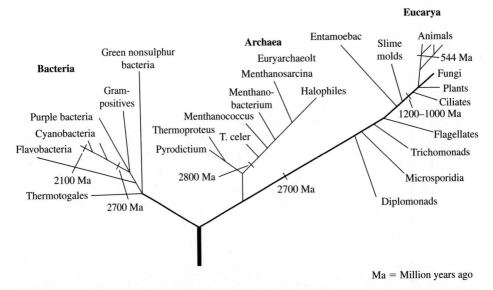

FIGURE 16.1 Phylogenetic tree of life as inferred from comparison of genetic sequences in ribosomal RNA. The dates indicate the *minimum* age of selected branches, based on fossil evidence and on chemical biomarkers. The age of eukaryotic origin based on the presence of steranes is indicated.

eukaryotes is unknown. More direct fossil evidence for cells of the size and appearance of eukaryotes does not occur until 1.7 billion to 1.9 billion years ago.

Hence, eukaryotes may be present in the geologic record around, but perhaps before, the time of the dramatic increase in the oxygen abundance of the Earth's atmosphere (Chapter 11). Although some eukaryotes lack mitochondria and do not undertake oxygenic respiration, the vast majority of eukaryotic cells do, and use oxygen as a powerful oxidizer to convert carbohydrates to stored energy in ATP (Chapter 4). (Whether this reflects the true balance between aerobic and and obligate-anaerobic eukaryotes, or is a sampling bias, is not known.) The 18-fold increase in production of ATP from ADP associated with mitochondrial respiration, relative to prokaryotic fermentation, for example, was one essential and enabling step toward the development of multicellular animals and the multistep planet–animal food chain. (There is a primitive predator–prey relationship among some prokaryotes but it is only a two-level chain in contrast to the extensive food chain evolved by the eukaryotic organisms.) In the existence of mitochondria lies a clue to eukaryotic origins.

The Origin of Organelles in Symbiosis

The two key innovations in the origin of the eukaryotic cell are the organelles and the cytoskeleton (Chapter 4). The organelles, including the mitochondria, the plastids (such as the photosynthesizing chloroplasts), and the membrane-bound nucleus that bears the central genetic code of the cell, all resemble bacteria embedded within the cell. But the mitochondria and the chloroplasts resemble them most of all. Mitochondria have their own DNA, messenger RNA, transfer RNA and ribosomes within a prokaryotic-like membrane. The DNA floats within the mitochondria as strands and is not bound in chromosomes. The ribosomes resemble bacterial ribosomes, and are sensitive to the same antibiotics (the eukaryotic cell wall, luckily, insulates the mitochondria from these drugs). Mitochondria divide separately from the rest of the eukaryotic cell, in the same simple fashion as do bacteria.

Although it was suspected more than a century ago that mitochondria might bear some relationship to bacteria, Boston University biochemist Lynn Margulis first formulated a quantitative hypothesis that the eukaryotic cell is the result of ancient symbiotic relationships between bacteria of various types. Sometime in the late Archean or early Proterozoic, aerobic metabolism became possible, forming mutually advantageous symbiotic relationships between aerobic bacteria, cyanobacteria, and other prokaryotes. These symbioses, undoubtedly consummated countless times among large numbers of various kinds of prokaryotes, with varying degrees of success and disaster, led in a few cases to chimeric organisms that survived and prospered. Horizontal gene transfer within proto-organelles of the nascent eukaryotes followed the symbiotic events themselves. Over time, the organelles became fully dependent on the host organism, losing their genetic independence and retaining only those gene sequences required for replication and existence as a structure fully dependent on the main cell. For example, the bacterium *Rickettsia prowazekii* is an obligate intracellular parasite with 834 genes in its genome (Table 16.1); by comparison, the mitochondrial genome can consist of as few as 60, and not more than 100, protein-coding genes. (*R. prowazekii* itself is not a candidate precursor to the mitochondrial organelle; it is simply an organism that possesses one of the smallest genomes, and it is a

TABLE 16.1 Increase in the Genome Size and Cell Types

Organism	Domain	Genome Size (genes)[1]	Number of Cell Types
Mitochondria	(organelle)	60–90	(1)
Mycoplasm genitalium	Bacteria	480	1
Rickettsia prowazekii	Bacteria	834	1
Escherichia coli	Bacteria	4,288	1
Aquifex aeolicus	Bacteria	1,512	1
Archaeoglobus fulgidu	Archaea	2,436	1
Methanococcus jannaschi	Archaea	1,738	1
Synechocystis sp.	Bacteria	3,168	1
Bacillus subtilis	Bacteria	4,100	2
Arabidopsis thaliana	Eukarya (plant)	24,000	30
C. elegans	Eukarya (animal)	18,424	50
Drosophila melanogaster	Eukarya (animal)	13,601	50
Zebrafish	Eukarya (animal)	80,000–100,000	120
Homo sapiens	Eukarya (animal)	80,000–100,000	120

[1]Protein-encoding genes only. For the small genome organism, this may underestimate the total gene count by 20 percent.

Source: Adapted from Carroll, S. B. (2001). "Chance and Necessity: The Evolution of Morphological Complexity and Diversity," *Nature,* **409,** 1102–1109.

model for the kind of intracellular behavior and reduction of the genome that may have occurred in the formation of organelles. It is the agent of a disease, typhus, carried by louses, which illustrates the point that the overwhelming majority of intracellular symbionts are, in fact, parasitic—and detrimental to the host.) For mitochondrial organelles to propagate from one cell to another during cell replication associated with organismal growth, they must reproduce; the missing genetic information to do so is contained in the eukaryotic nucleus as part of the nucleus-bearing, "nuclear," DNA. Most of the proteins needed to maintain mitochondrial function are constructed in the main cell, are specified by the nuclear DNA, and then transported to the mitochondria. In some cases, the transfer RNA (tRNA; Chapter 4) is built in the nucleus and transported to the mitochondrion, where protein construction is then conducted. Both approaches to supplying essential proteins illustrate that mitochondria are fully dependent subunits of the eukaryotic cell and are themselves incapable of independent existence. The DNA that they do possess, the so-called mtDNA, mainly codes for the proteins essential to carry out the respiratory chemical reactions that oxidize organic carbon and provide chemical free energy to be stored in ATP. Some mtDNA codes for a limited number of specific mitochondrial structures.

In the case of the mitochondria, it seems possible to actually identify the phylogenetic branch from which the ancestor to the organelle was derived. Genomic studies of mitochondrial DNA establish that the closest bacterial relatives of mitochondria, from a gene sequence point of view, are members of the so-called α-proteobacterial class of eubacteria (that is, of the bacterial phylogenetic domain). Some members of this class are obligate intracellular parasites, and phylogenetic analysis indicates that they may well be the closest extant relatives of the ancient proteobacterial forms that completely lost their identities to become eukaryotic organelles billions of years ago. Further, mtDNA is so similar among mitochondria in different eukaryotes that there appears to have been *one common ancestor* (or a very few, very closely related ancestors) to the modern mitochondria. This ancestor, a prokaryote capable of not only tolerating oxygen but also using it to oxidize organic matter to generate energy for biological processes, changed the course of life on Earth. Of course, that all mitochondria had this singular origin does not mean that other experiments in oxidative respiration did not lead toward some form of obligate symbiosis among other prokaryotic cells; this may have happened countless times.

Figure 16.2 illustrates schematically the incorporation of a mitochondrial ancestor into a host cell, which here is postulated to be an archaebacterium, that is, a prokaryotic cell from the domain Archaea. This is motivated by the phylogenetic tree analyses suggesting a

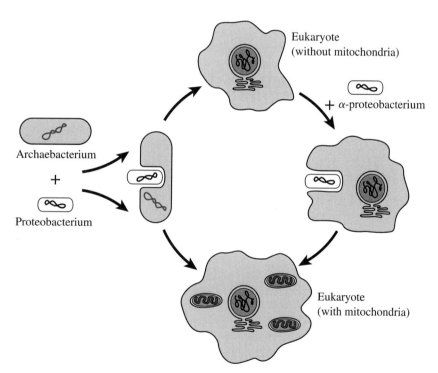

FIGURE 16.2 Schematic evolutionary paths toward the origin of the eukaryotic cell. Both scenarios sketched begin with the symbiotic fusion of two prokaryotes. Going counterclockwise, simultaneous creation of the eukaryotic nucleus and mitochondrion by fusion of a methanogenic Archaebacterium (host) with a hydrogen producing α-proteobacterium (symbiont). Moving clockwise in the figure is a two-step model initially involving formation of a eukaryotic cell by fusion of an Archaebacterium and a Proteobacterium followed by acquisition of the mitochondrion through endosymbiosis with an α-proteobacterium.

closer relationship between the eukaryotic and archaic than between the eukaryotic and bacterial genomes (see Figure 16.1). What stimulated such an odd alliance between archaea and bacteria? One hypothesis is that it was driven by hydrogen. The archaebacterium needed hydrogen, and the α-proteobacterium undertook a form of oxygenic respiration in which hydrogen or a reducing molecule (i.e., one that can donate hydrogen) was among the waste products. It, in turn, received some metabolic or other benefit from the host. While speculative, such metabolic "syntrophy" is a logical rationale, since it obviates the need for teleological hypotheses (e.g., the symbionts "knew" that eukaryotic existence would be advantageous) or for random symbiosis (which is difficult given the usually deleterious outcomes).

If the formation of the mitochondria was the defining evolutionary moment for the eukarya, it could not be the end of the evolution from symbiosis to organelle formation. As noted earlier, other structures in the eukaryotic cell were likely originally prokaryotic symbionts. Many different hypotheses for the progenitors of the nucleus, the chloroplast, and others have been offered, and space does not permit a detailed exposition of each of these. But the principle delineated by the story of the mitochondria—that obligate parasitism leads to symbiosis and then loss of identity (through loss of genes as in Spiegelman's experiments on viruses described in Chapter 8) as a single organism—presents a remarkable picture of evolution achieved through major modifications of organismal genomes. What came out of these events is a kind of cell that, paradoxically, is much more rigid in the rules of gene transfer than the prokaryotes from which it was derived—even though the eukaryotic nucleus itself has a strong propensity for gene transfer across its boundary. Eukaryotic sexual reproduction provides a structured set of rules for mixing genes that allows variation in every generation but, at the same time, limits the extent of such variation.

Oxygen and the Rise of Eukaryotes

Because the hallmark of most modern eukaryotes is oxygenic respiration, it is reasonable to postulate that the rise of oxygen was an important factor in triggering or allowing the diversification of the eukaryotic cell. Of course, respiring bacteria had to exist prior to this, because one of them was the ancestor of the mitochondria, but they likely occupied localized oxidizing niches prior to the onset of a fully aerobic atmosphere and ocean. With just a few exceptions, oxygen is not used in the chemical pathways synthesizing proteins and other biological molecules—it is just an oxidizer to generate free energy from organic molecules, which is then stored in phosphate bonds. Had abundant free molecular oxygen been available very early in the history of life, we might expect it to have played a more intimate role in biosynthesis and other cellular processes—or perhaps inhibited the formation of life by quickly breaking apart primitive biopolymers.

On the other hand, we must approach the relationship between the diversification of eukaryotes and the rise of oxygen with caution. Some eukaryotes that lack mitochondria probably did not evolve by losing mitochondria—they may never have had them in the first place. And the origin of eukaryotes in a tightly symbiotic relationship among prokaryotes does not address the origin of the other uniquely eukaryotic structure—the cytoskeleton. A more conservative picture would be one in which the number and types of bacteria had reached a point, sometime in the first half of the history of life, when symbiosis led to definite survival advantages over separate existences. Countless such experiments occurred and ended, leaving no trace. The compartmentalization of photosynthesis, respiration, and replication of the genome may have had particular survival advantages as the number of places on our planet in which oxidizing or aerobic conditions were the norm was increasing.

The successful acquisition of symbionts that exist within, and are wholly dependent on, the host cell, may have been just the one step in a number of innovations required to make a eukaryotic cell. These included losing the rigid cell wall in favor of a flexible cellular surface that could both increase the surface area for nutrient absorption and allow the nondestructive incorporation of other prokaryotes and the development of a cytoskeleton. This last, in particular, is perhaps the greatest challenge in understanding the origin of eukaryotes. The cytoskeleton allows the cell to maintain changes its shape, to distribute daughter chromosomes during reproduction, and move materials around the huge interior of the eukaryotic cell. The genes that encode for the production of a cytoskeleton are found neither in living bacteria nor archaea. It is, of course, tempting to then postulate a fourth domain of life, now extinct, that provided the prototypical cytoskeleton, or the genomic raw material for making one, but no other evidence for an extra domain exists in the phylogenetic tree of life. However the cytoskeleton may have been acquired, we can imagine many other chimeric cells that lacked one or more of the essential characteristics of eukaryotic cells and could not meet the challenge of the new oxygen-rich world that was arising. The eukaryotic cell type is perhaps not the sole product, but rather just the survivor, of a long and creative contest of symbiotic cellular evolution, occurring on a playing field increasingly tilted in favor of our eukaryotic ancestors.

16.4 Complexity, Diversity, and Evolution

Intuitively it is tempting to think of the evolution of life as proceeding from the simple to the complex, from the small to the large, and from the uniform to the diverse. But to quantify these notions is not straightforward. Size is easier, in the sense that the eukaryotic cell represents an increase over prokaryotic cells, and the development of multicellular eukaryotic life forms in the late Proterozoic represented a dramatic scale change for a "single" organism. However, all three domains of life have independently developed multicellular forms, though it can be argued that prokaryotic or archaeal colonies are not single organisms. In the eukaryotic domain, the mean size of organisms has not risen in a monotonic fashion. Some increases in size of related eukaryotic species along an evolutionary trend are "passive," in the sense that larger individuals represent part of a general increase in the variance, generation after generation, within a species or closely related groups of species; it might then be argued that smaller sizes are at a survival disadvantage because of predators, more extensive exposure to environmental hazards, and the like. However, larger organisms require more food; hence, it is questionable whether size alone confers a survival advantage. Some increases in size, though, do represent natural selection in action, the human brain being one example of a complex biological structure in which size (or at least surface area) matters.

Diversity can be defined in terms of shape and appearance (morphology), or in terms of the density and topology of the phylogenetic tree of life. New forms appear in the fossil record over time, but only very rarely do more dramatic occurrences, such as the symbiotic origin of eukaryotes, create major new pathways of evolution. The appearance of new forms does not necessarily mean the elimination of the old, since new species arise due to the reproductive isolation (usually geographically determined) of a subpopulation of preexisting species. However, ecosystems saturate in terms of the number of different kinds of species that can be supported: "Arms races" allowing novel organisms to take over a previously occupied niche may be suppressed until environmental changes alter the rules. Of the different designs of skeletons, external or internal, that animals have used since they appear in the fossil record, 80 percent appeared within the first 10 percent of the fossil record containing animals.

Has intrinsic post-Cambrian genetic variety played itself out? The answer is unclear, although (as discussed later in this section) much of the genetic potential for variation appears to remain unused by organisms. Diversity is initially quenched and subsequently encouraged by the geologic agents of "mass extinction events," in which large numbers of taxa are eliminated quickly based on evidence from the fossil record. Such extinctions provide empty ecosystems for new life-forms to occupy by wiping out large fractions of the previous inhabitants of those ecosystems. The magic of movie-making aside, we will never witness the remarkable animals called dinosaurs that occupied essentially all of the same ecosystems that mammals and birds occupy today; however, until the dinosaurs departed, both the mammalian and avian lineages were impoverished because the ecosystems they occupy today were already filled. Absent environmental changes, both periodic and catastrophic, the natural exploration of genetic phase space in the form of diversity both within species and between species would quickly have quenched in a kind of ecological stasis.

Whether complexity has increased over time is the proverbial hot potato. Complexity is a word so heavily used in science that it usually engenders confusion. In biology, complexity could refer to the size of the genome, which yields only potential complexity because the full information content of the code is apparently never expressed in extant or extinct organisms. A somewhat different definition of complexity, coming from information theory, is the amount of information an individual genome stores about the organism's environment. This is quantitative, and the measure increases with time in computer simulations of evolution but also requires assuming something about the adaptation of the organism to the environment (it is unclear what happens in a situation of rapid environmental change as might be caused by global catastrophies such as those discussed in Section 16.7.) More simply, complexity could refer to the number of parts, the ways in which such parts operate or interact, or the number of different cell types. Table 16.1 sugggests that both genome size and number of different cellular types have increased over time, but not in a gradual fashion. Genome size in particular seems more bimodal than progressive, especially when we consider that small-genome organisms (fewer than 1,000 genes) likely have lost a larger cohort of genes through parasitism or other dependent behaviors. Number of types of cells appears to have a closer and smoother correlation with evolutionary history within the eukaryotic domain. Finally, within certain closely related groups (species through phyla), there is often an evolutionary trend toward greater complexity in body parts. The classic example of this is the arthropods, whose trilobite representatives early in their history had a common basic limb, whereas today's arthropods have distinct limb shapes and uses from front to back (e.g., lobsters). We, too, bear the signs of a long history of gradual complexification, from our 120 different cell types (compared with 10 for sponges) to the impressive top-to-bottom variations in repeated structures, such as the variety of vertebrae, ribs, and our distinctive hands versus feet.

It seems remarkable that the intricate diversity in multicellular creatures is obtained with an increase in gene size of a factor of only 10 to 100 over unicellular bacteria and archaea. It is even more remarkable when we consider that the amount of potential genetic variation is far larger than the actual variation expressed, in any organism. With perhaps 100,000 genes, the number of potential variants in cell type, for example, is far larger than is actually realized in human beings. And much of the complexity we do see in advanced forms of life is achieved with extreme genetic parsimony through the modularity of biological structures—the repeated production of segments or units that, through the action of just a few regularity (*hox*) genes and regulatory proteins, develop in dramatically different ways. Suppress a particular *hox* gene in a housefly, and the two antennae develop instead as legs—because they are the same modular structure differentiated in a particular way. Our hands and feet, and for that matter legs and arms, are variations on a theme triggered by the *hox* genes. Likewise, a small number of regulatory proteins control the expression of genes within particular cells or regions of a body. This enables, for example, the structure programmed to be a "hand" to actually develop during embryonic growth into the hands we recognize as normal. Inhibition of the action of such regulatory proteins leads to drastic morphological changes such as flippers in place of hands. Such changes do not just occur from one species to another; pharmaceutical meddling with the action of regulatory proteins in fetal development can lead to severe developmental abnormalities in humans.

Evolutionary development of the *hox* genes appears to have occurred in only two bursts—first in the development of bilateral (as opposed to spiral or radial) body forms and then in the development of the vertebrates from the chordates. Likewise, essentially all of the body plans represented in animals today appeared in the Cambrian explosion of phyla. Like the development of the eukarya, these seminal events both enabled and limited the future diversity of complex life. That we do not have a grazing animal whose body plan is trilateral (three eyes, three ears, three limb points per segment) is a consequence of what was and was not invented in the Cambrian. Perhaps parsimony dictates that one symmetry axis is enough and two is too expensive in terms of complexity. More interesting than the Cambrian explosion itself, or the invention of eukarya, is what has *not* happened. The genomic potential for variation is enormously larger than what we see, yet it is hard even in the imagination of movie directors to generate imaginary creatures that do not possess essentially Cambrian body plans. Are these body plans then a kind of "eigenset" of the most likely or sensible plans from a geometric point of view, or are our imaginations biased by our Cambrian heritage?

In sum, the evolutionary trends from prokaryotic to eukaryotic life—and later toward multicellular organisms with highly differentiated, modularized parts—constitutes a temporal progression toward complexity, but at a level far below what could be achieved based solely on the size and potential information content of the genome. Much of the increase in complexity we recognize in "metazoans" is within phyla, and this increase has been achieved with no significant increase in the size of the genome after the invention of macroscopic animals; rather it has been accomplished through an expansion in the (small) number of regulatory genes and proteins. Why life has not explored much larger realms of complexity and variation may be tied both to environmental and genetic factors, as we explore next.

16.5 Environmental and Genetic Interactions in the Evolution of Life

The development and the evolution of the genome has occurred, not in isolation, but in interaction with the ever-changing environment of the Earth. Some effects of the environment on the genome are direct—for example, the variation in the amount and energy spectrum of ultraviolet light and cosmic rays that reach the Earth's surface translates into the variation in the rate of genetic damage in organisms at or immediately below the Earth's surface. Indeed, the wavelength distribution of the intensity of solar ultraviolet radiation has varied throughout the Sun's main sequence lifetime, as has the efficacy of the ozone shield that screens out most of this radiation. The flux of cosmic rays varies modestly with the solar cycle, and larger variations associated with the Sun's magnetic field and the occurrence of supernovae in the galactic neighborhood may have been experienced over the history of the Earth. These have, in turn, had effects on the viability of surface organisms and the rates of radiation-induced mutations, but they are difficult to quantify because there is no direct signature in the genomic or fossil record.

Another kind of environment-genetic interaction is an indirect one. The effect of the alteration of the genome, by whatever mechanism, can be magnified by environmental effects, local or otherwise, that isolate subgroups of species so as to prevent interbreeding. Indeed, hyperthermophilic microbes (Chapter 10) have acquired different regional characteristics by virtue of their isolation imposed by geographic separation of high temperature environments. Many kinds of environmental isolations occur all the time and, coupled with changes in the genome over time, provide opportunities for speciation, as we discuss in the remainder of this section.

The expansion of the genome requires that new genes be generated, and the predominant mechanism for doing this is gene duplication. *Gene duplication* is the result of a replication error that leads initially to a fully redundant pair of genes. From there, several outcomes are possible, just on the basis of enumeration. One copy mutates in such a way as to become a nonfunctioning ("silenced") gene, or it may acquire some beneficial function because of mutation and natural selection, or mutations could compromise both copies—but the two together retain the capacity of the single ancestral gene. It turns out that deleterious mutation seems to affect genes in an accelerated fashion after a duplication event, and the vast majority of duplicate gene pairs end up with one of the pair silenced. Such silencing might actually contribute to the emergence of new species: It allows duplication to recur without necessarily causing deleterious effects on the organism's genetic code.

The gene duplication process may be significant in the role it could play in reproductively isolating a cohort of a species, acting in concert with geographic isolation imposed by environmental changes (Figure 16.3). Geographic isolation prevents the intermixing of

16.5 Environmental and Genetic Interactions in the Evolution of Life

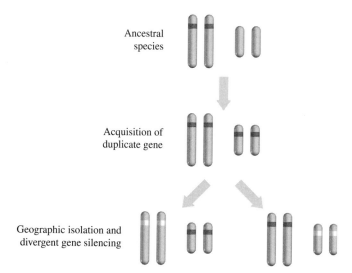

FIGURE 16.3 One example of the fate of duplicate genes is that the genes may be silenced—that is, no longer provide code for protein production. This diagram traces the initial history of chromosomes, beginning with a pair that possesses a duplicate of an essential gene (chromosomes with black bars). Descendants may then acquire the duplicate gene, and sometime later, in an unrelated environmental event, members of the species become geographically isolated. The genes may then be silenced (white bars) on different pairs of the chromosomes in the two different lines of the original species, leading to the potential for reproductive isolation in succeeding generations.

breeding populations when an unstable gene pair mutates and is silenced in succeeding generations. If alternate members of the gene pair happen to be silenced in two geographically isolated populations, then the potential is present for a reproductive barrier to be set up because the genomes are no longer compatible. Generally, the alteration of a single pair in this way will not isolate two populations reproductively, but, for example, 150 duplicate gene pairs will arise in a million years in the entire genome of a eukaryotic species with a genome size of 15,000 genes. Recurrent duplication of genes in certain regions of a chromosome can lead to rearrangement of the genes there, and hence (coupled with geographic isolation), reproductive isolation and the formation of a new species.

Here, then, is a type example of the interaction between genomic and environmental changes. The timescale for environmental changes is a function of the scale of the change. Localized or regional changes in ecosystems occurring on timescales as short as millennia or even centuries is in evidence in the late Pleistocene and early Holocene, when globally the Earth was moving from a glacial to an interglacial climate. Larger changes, on a global scale, are present with frequencies of tens of thousands to hundreds of thousands of years, representing, for example, the entire duration of global glacial and interglacial climates in the Pleistocene. On timescales of millions to tens of millions of years, shifts in the Earth's climate from relatively stable warm conditions to the rapid oscillation of glacial-interglacial cycles are seen in the climate record. On longer timescales, 10 million to

100 million years, climatic shifts and changes in volcanic activity due to plate tectonic motion (the supercontinent cycle) and the occurrence of globally catastrophic impacts come into play (Figure 16.4). Those environmental effects acting on million-year or shorter timescales, in isolating populations of a given species, provide the setting for subsequent reproductive isolation through gene duplication and mutation.

On longer timescales, there is the opportunity for wholesale changes in the eukaryotic genome. The gene duplication rate implies that 50 percent of the eukaryotic genome may be duplicated on a timescale of order 100 million years. Because the majority of these duplications lead to silenced genes, the net result is a larger amount of useless nucleic acid material in the organism's DNA rather than a monotonic increase in the number of genes. We do see dramatic flowering of genetic innovations in the fossil record, but only at special times, when mass extinctions have emptied the ecosystems of existing organisms, leaving the door open to new species, genera, orders, and the like. Because mass extinction events have occurred some half-dozen times since the Cambrian explosion, there have been a limited number of opportunities for the expression of major genetic innovations (Figure 16.5).

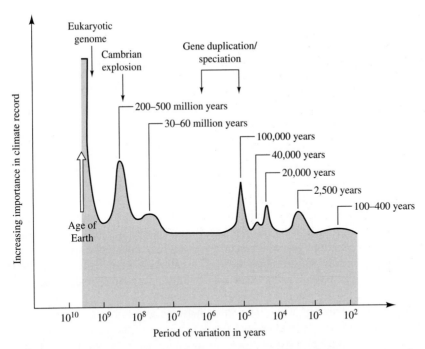

FIGURE 16.4 Timescales of various kinds of climate cycles and changes, expressed as the magnitude of climate fluctuations versus the period or extent of the variation. This is a "power spectrum" computed using a variety of data on climatic fluctuations over time; some timescales for variation are more important in the climate record, and they stand out. The time it takes for various key events (formation time of a genome of length typical for eukaryotes, the duration of the Cambrian explosion, the typical duration of species and time for gene duplication) are superimposed.

16.5 Environmental and Genetic Interactions in the Evolution of Life 529

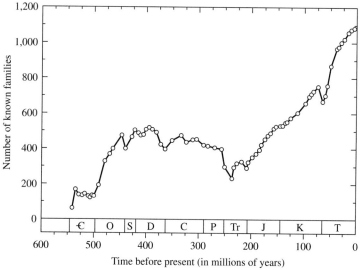

a

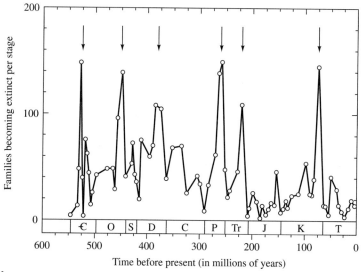

b

FIGURE 16.5 (a) Growth in the number of marine animal families alive from the Cambrian to the present plotted versus time before present and versus the Phanerozoic geologic periods (using the standard symbols for, left to right, Cambrian, Ordovician, Silurian, Devonian, Carboniferous, Permian, Triassic, Jurassic, Cretaceous, and the Tertiary). (b) Number of families becoming extinct as a function of time, with arrows indicating the times of major extinctions seen in the fossil record.

More gradual episodes of climate change than the catastrophic aftermath of large impacts (Section 16.7) might also permit increased diversity and complexity. In the last 60 million years, a sharp increase in the number of nonmarine animal families is seen in the fossil record and a gentler rise in the number of marine fauna. These stand in stark contrast to the general stagnation over the preceding approximately 480 million years after the Cambrian explosion. Whether this is the result of increased genomic complexity or of changes in the nature of the Earth's climate is unclear. Environmental change is a compelling cause; the last 60 million years have been characterized by the breakup of the supercontinent Pangea (Chapter 11) and an increase in the extent of climatic fluctuations on the continents—culminating in the extreme variations of the last 2 million years. The gradual opening of new oceans, bays, and seaways, as well as changes in ocean circulation patterns, may have provided more variable conditions for marine animals.

Under such circumstances, ecosystems change frequently, on timescales commensurate with, or shorter than, those for the formation of new species, and hence the opportunity for new variations in eukaryotic forms has perhaps never been better since the Cambrian. The last supercontinent breakup before Pangea was the breakup of Rodinia (after its second assembly) beginning in the mid-Cambrian, during a time when the number of faunal families also increased dramatically. (It is much more difficult to try to perform such environmental-genomic comparisons on the record of ancient prokaryotes. Biochemical, morphological, and behavioral factors determining speciation interact very differently in the prokaryotic versus eukaryotic domain. Indeed, the possibilities of lateral gene transfer by themselves render poorly defined such an exercise.)

In making the connection between environment and eukaryotic biological complexity, we should not argue that the supercontinent cycle by itself was the major stimulus for periods of innovation interleaved with periods of stagnation. The Earth's habitability is the result of a complex set of interacting processes, described in Chapter 11, among which the assembly and breakup of supercontinents provides one key driver for major climatic changes on a temporal scale of hundreds of millions of years. The intent is to set the stage for a discussion of the long eon *prior* to the Cambrian explosion when eukaryotes existed but multicellular plants and animals did not. Was the long run-up from the invention of eukaryotes to the end of the pre-Cambrian a time during which the requisite genetic complexity for large, complex animals had not yet been achieved? Was the atmosphere insufficiently oxygenated prior to the Cambrian to support multicellular organisms? Did environmental conditions extend the senescence period prior to the dramatic appearance of modern phyla? The details of the climate and the fossil records in the half billion years prior to the Cambrian hint at the answers to these questions.

16.6 Snowball Earth and the Cambrian Explosion

The first undisputed evidence of animals in the fossil record appears about 570 million years ago, in the Neoproterozoic, and most major groups appear shortly thereafter in the Cambrian explosion, beginning at 545 million years ago. However, estimates of divergence times for animals based on molecular clocks have indicated a much earlier period of animal evolution that is unrecorded in the fossil record and began perhaps a billion years ago. In particular, the rates of genetic divergence of animal phyla, calibrated using the appearance of various types of animals in the recent fossil record, suggest that the original bifurcation between vertebrate and invertebrate phyla occurred a half-billion years before the Cambrian explosion. Additional genes must be sequenced to check whether divergence rates might have been greatly accelerated in the Vendian and Cambrian relative to today, which would push forward the date of bifurcation.

If we accept for now the genomic-based conclusion that animal divergences occurred much earlier than the Cambrian, why do we not see evidence of such animals in the fossil record? It has been suggested that early animals were small and soft bodied, which would explain their absence from the fossil record. But why animals would have been sufficiently small, soft, and (perhaps) rare to escape fossilization for hundreds of millions of years, only to "burst out" in the Cambrian in a variety of hard-shelled and skeletal macroscopic forms, is a mystery. While it is possible that some genetic innovation was required to initiate the Cambrian explosion, geologic evidence hints that cold environmental conditions on the Earth from about 800 million to 550 million years may have played the key role.

Beginning sometime after 800 million years ago, and ending about 550 million years ago, evidence for widespread glaciation appears in the geologic record. In addition to geologic features associated with erosion caused by glaciers, the abundance in inorganic ^{13}C relative to ^{12}C decreases, consistent with a drop in biological productivity that would normally sequester the lighter carbon isotope in organic carbon and hence enhance the heavier isotope in the inorganic phases. Many of the glacial deposits contain carbonate debris, or have carbonate rocks on top of them, which would normally be found in low-latitude, warm-water environments. Furthermore, some of the glacial units contain sediments with banded iron formations, suggestive of fluctuating oxygen levels more than a billion years after the Earth's atmosphere should have stabilized in an aerobic state. Finally, sediments from western Australia, in which the local direction of the Earth's magnetic field at the time of sediment deposition (a function of geographic latitude) could be measured, suggest that glaciation occurred near the equator and not just at high latitude. Some of the sediments are also consistent with material deposited in marine delta environments near the edges of continents and not in high-altitude mountain ranges.

In total, the geologic data paint a picture of widespread glaciation (even down to low latitudes), a severe drop in biological productivity in the oceans, and possibly a return in restricted regions or (at particular times) to relatively oxygen poor conditions in the oceans consistent with a collapse in photosynthesis. What might have initiated the glaciation is unclear. The Sun was slightly fainter than it is today, by 8 percent, which in the face of other extenuating circumstances might have aided in precipitating glaciation. A single continent, Rodinia, was in the process of breaking up amid renewed tectonic activity (Figure 16.6). At the time of the most recent continental breakup, that of Pangea, the Earth's climate cooled as continental fragments separated, drifted toward higher latitudes, became partly ice covered, and hence reflected more sunlight. Increased erosion occurred because moving continental fragments collided to generate new, steep mountain ranges that would have produced regionally rainy conditions and thus scrubbed the atmosphere of carbon dioxide. All of this led, some 60 million years after the Pangea breakup, to the current

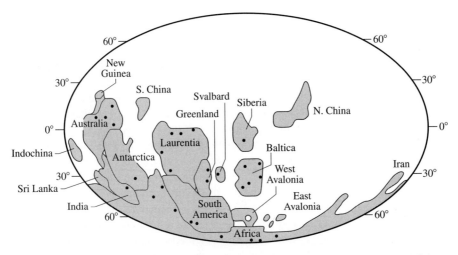

FIGURE 16.6 Reconstruction of the geography of the Earth at the time of the onset of the Neoproterozoic Snowball Earth. Dots show locations of the glacial deposits.

glacial-interglacial cycle we experience today. And 800 million years ago, under a fainter Sun, climate cooling engendered by Rodinia's breakup could have been even more severe, leading to glaciers extending not just over the continents but over the seas as well.

This "Snowball Earth" scenario for the late, or Neo-, Proterozoic is not universally accepted by geologists, but increasing numbers are seriously considering the notion that the Earth's climate was profoundly altered over the 250-million-year period ending some 550 million years ago. How profound an alteration remains a subject of contention. Some models suggest that, once ice covered the mid-latitudes, the resulting reduction in absorbed solar flux would have forced the entire Earth to become ice covered and possess a mean temperature of approximately 230 K. Under such conditions, life might have perished globally, except in a few refuges at mid-ocean ridge vents and perhaps at seasonal equatorial cracks in the ice. Detailed climate models, however, indicate that an equatorial ring of liquid water remains exposed to the atmosphere for the given conditions of the late Proterozoic Sun and carbon dioxide abundance somewhat above the present-day values (Figure 16.7, see Color Plate 15). Even with this ring, reduced productivity of the ocean environment would have produced elevated values of $^{13}C/^{12}C$ as seen in the geologic record and could have encouraged anoxic conditions in most of the ice-covered portions of the ocean.

The exit from the Snowball Earth glaciation may have happened as the extreme cold reduced precipitation overall and greatly increased the amount of snow versus rain. Consequently, erosion associated with rainfall and the associated conversion of carbon dioxide to carbonates were reduced. Continued volcanism associated with plate tectonic subduction zones would have continued to pump carbon dioxide into the atmosphere. Estimates vary, but the CO_2 abundance might have reached as much as 500 times the present-day value before precipitating, through intense greenhouse warming, the melting of the worldwide glaciers. Computer simulations of this process provide an explanation for the extensive carbonate layers seen above the glacial layers in the sedimentary record. As the ice begins to melt, and increasing amounts of sunlight are absorbed by the Earth's surface, a runaway warming melts essentially all of the glaciers. The resulting sudden exposure of the carbon dioxide–rich atmosphere to widespread liquid water at the ocean surface leads to rapid formation of carbonates—first in solution in the ocean, then rapidly precipitating out to form a thick layer of carbonates. The drop in carbon dioxide afforded by carbonate formation then regulates the climate to a value consistent with the ice-free conditions and the solar constant at the time. The geologic data suggest at least two deep glaciation and recovery episodes in the Neoproterozoic, one ending at 700 million years and the second at 580 million years. Unlike the current glacial-interglacial cycle, conditions may have been so cold within the deep glacials of the Neoproterozoic as to sustain glacial conditions for millions or tens of millions of years, although the time resolution of the geologic record is very limited compared with that of the Pleistocene.

Were it not for the geologic evidence, such Snowball Earth climate models should by themselves engender a skeptical attitude. For a century, modeling of the response of Earth's climate to ice formation has been plagued by the strong feedback between ice formation and reduced absorption of solar flux at the surface and the consequent positive feedback leading to a fully ice-covered Earth. The extreme response of the Earth to the addition of a small amount of ice has always been regarded as a defect of the models rather than a characteristic of the real Earth climate system. The most recent climate models include ocean circulation as a process to redistribute heat latitudinally and moderate the effect of the growth of ice, and they provide a more realistic response. Nonetheless, trying to model Earth's climate when the carbon dioxide flux changes by two or more orders of magnitude quickly, and when ice covers most if not all of the continents, is a tricky business whose results must be carefully tied to geologic evidence.

The effects of a Snowball Earth climate on life was profound, if the carbon isotopic data, and the occurrence of iron deposits, are being accurately interpreted. Under such conditions, it is plausible that the diversification of multicellular animals would have been suppressed. In particular, the development of large animals with thick muscles and well-developed internal organs would have required essentially modern levels of oxygen so that the decline in oxygen levels during the time of the Snowball Earth would have delayed the appearance of such animals. But this supposition really begs the larger question, echoed in Section 16.1, as to why such creatures did not appear even earlier in the fossil record, prior to the Neoproterozoic Snowball Earth.

An answer to this question is shown graphically in Figure 16.8 (see Color Plate 15). On the basis of biomarkers and trace fossils, animals corresponding to the phyla Porifera (e.g., sponges) and Cnidaria (e.g., jellyfish) do appear within the late Neoproterozoic. Whether animals extend back as far as a billion years ago, the number derived from molecular clocks, remains speculative. But, as pointed out by A. Knoll and S. B. Carroll, the bilateral animals, of which the arthropods and we are examples, required a set of genetic innovations such as *hox* and other regulatory genes not required in the simpler spiral and radial body plans of Porifera and Cnadaria. The basic enabling genome for multicellular animals with relatively simple body plans developed prior to the regulatory genes necessary for modularized complexity embodied in the bilateral animals. Perhaps members of the Porifera and Cnidaria phyla or (more likely) primitive precursors occurred throughout the Proterozoic, hidden in fossil traces that have yet to be discovered or properly interpreted. For all we know, bilaterans themselves could have made their appearance as early as, say, 700 million to 800 million years ago (we do not know the timescale over which gene duplication led to innovations such as *hox* genes) with their diversification delayed by the productivity and oxygen collapses associated with the Snowball Earth glaciations. Once oxygen levels rose to their current values, the environmental barrier to the success of the bilaterans was lifted, and their rapid diversification into many different forms in the Cambrian was enabled on an Earth whose ecosystems had been largely emptied of potential competitors by the unstable and unfavorable Snowball Earth climates. Knoll and Carroll draw a parallel between the Cambrian explosion—which was not the *origin* of animals but was instead the appearance of many new animal forms—with the diversification of long-extant mammals after the demise of the dinosaurs.

Geologic evidence points as well to an earlier snowball Earth epoch, between 2.2 and 2.4 billion years ago, at a time when the Sun's luminosity was only 85 percent of the current value. Glaciations at that time might have been even more profound than in the Neoproterozoic, but the greater depth of time to the early Proterozoic denies us the details needed to evaluate the earlier episode. That, too, is an intriguing time in history because the first chemical hint of the presence of eukaryotes occurs in the geologic record somewhat prior to that time, with a substantial gap until the eukarya appear definitively in the fossil record. We might speculate that the gap is not one of missing data but instead that the severe climate of an earlier Snowball Earth set back the spread of the eukarya, particularly if glaciation-induced productivity collapse suppressed the oxygen abundance at the time. Alternatively, the extreme cold of a Proterozoic glaciation might have encouraged the symbiosis that led to eukaryotes (although this would push the origin of eukaryotes several hundred million years later than the 2.7-billion-year number based on the chemical fossil evidence). Oxygen-respiring bacteria might have sought out the interior environment of the host cell as a warmer abode, while the thermal byproduct of oxygenic respiration could have aided the host cell in maintaining moderate internal temperatures in the face of the external cold.

Further discoveries and study of the geologic record may provide more definitive information on the role of Proterozoic climate fluctuations in the timing of the appearance, first, of eukaryotes and then of multi-celled animals. However it occurred, the subsequent explosive diversification of complex animals set the stage for a later history of mass extinction and rediversification continuing to the present day.

16.7 Impacts, Volcanism, and Major Extinctions

The causes invoked for sudden mass extinctions include the onset of deep global glaciation discussed in Section 16.6, large-scale volcanism, impacts of large comets or asteroids, supernova events, and passage of the solar system through a dense molecular cloud. We dispense with the latter two because little evidence exists for these in the geologic record, and the specifics of how they would influence life or the climate are speculative. We focus here on impacts. Many of the effects of large-scale volcanic eruptions are thought to be similar to those of impacts, and indeed battles over the interpretation of the geologic evidence associated with mass extinctions are usually divided between impacts and volcanism. Over the last several decades, the importance of impacts in terrestrial processes been increasingly recognized—a result in part of the reconnaissance of the solar system. The best-documented case for an impact-generated mass extinction is that at the close of the Cretaceous, some 65 million years ago. Indeed, of the seven major extinctions in the Phanerozoic, the last 600 million years of Earth history, only the one discussed here, at the close of the Cretaceous, shows strong positive evidence for impacts having a primary causal role.

The setting for the Cretaceous-Tertiary (K-T) extinction some 65 million years ago is an Earth with a wealth of flora and fauna on sea and on land. A single supercontinent, Pangea, was just beginning to break apart, and the clustering of land at low latitudes, coupled with a high carbon dioxide concentration, gave the Earth an overall much warmer climate than today. The spectacular dinosaurs diversified and spread after a previous mass extinction seen in the geologic record, that at the Permian-Triassic boundary. They occupied niches equivalent roughly to those occupied today by mammals and birds. Mammals had developed in the Triassic period but were unable to challenge the dinosaurs for the bulk of the ecological niches and were restricted to species of small rodentlike creatures. Remarkable for the enormity of their size and diversity of forms and habits, dinosaurs were the dominant large animal on land and sea up to 65 million years ago.

At the end of the Cretaceous, a dramatic demarcation in the fossil record occurs, wherein 15 percent of the shallow-water, marine families become extinct—including 80 percent of shallow-water, invertebrate species. The dinosaurs disappear too, though the massive size of their fossilized skeletons makes it difficult to determine how suddenly. Some have argued that the dinosaurs and other families were already stressed well prior to the K-T boundary itself. Above the Cretaceous sediments lies the Tertiary, a time when mammals finally diversified and occupied the niches left empty by the demise of the dinosaurs.

The apparently contemporaneous disappearance of the dinosaurs, other land species, and large numbers of marine organisms as recorded in the rock record is called the Cretaceous-Tertiary boundary extinction and is well dated in sediments worldwide at having occurred 65 million years ago. Most dramatic about the boundary is that it is demarcated in most sediments by a layer of clay inches in extent. In addition to the sudden disappearance of many small marine organisms just below the clay boundary, replaced in sediments above by new organisms, the clay contains peculiar abundance anomalies. The abundances in the layer of the platinum group elements (iridium, osmium, and gold) more closely resemble abundances in meteorites than in the crust of the Earth. Iridium and the other platinum group elements are rare in the crust because they have been extracted by the iron that sank to the Earth's center to form the core. In particular, at the K-T boundary, iridium is in excess of 100 times more abundant than in normal crustal rocks. Either the iridium is a product of the Earth's mantle where it is more abundant, or it was derived from impact debris which, whether asteroidal or cometary, is presumed to have had a much more chondritic composition than the Earth's crust.

Most of the properties of the thin, boundary clay layer and adjacent layers can be interpreted in terms of an impact. The impact of a large asteroidal or cometary fragment, which would be moving at many kilometers per second, produces distinctively shocked grains of quartz. These are seen in K-T boundary sediments around the world. Large impacts eject droplets of molten rock from the initial cavity, which cool rapidly and solidify during flight to form distinctive melt spherules. Their specific characteristics and occurrence over a large geographic area are diagnostic of an impact origin, and these are seen over a broad region around the impact site itself, which is buried under sediments in the Gulf of Mexico. Graphite and other evidence for intense burning are found at the K-T boundary in some parts of the world. Burning of an entire continent's worth of forests during an impact would be the result of large amount of debris lofted by the impact into suborbital trajectories above the bulk of the atmosphere. Reentry heating of this material, moving at many kilometers per second, then subjected the exposed hemisphere of the Earth to several times the equivalent photon flux of the Sun, igniting most forests. The location of the crater implies that the impact was into ocean, and hence huge tsunamis were generated; the destructive effects associated with such tidal waves, along with other effects of the direct blast, are seen in sedimentary layers from that time adjacent to the Gulf of Mexico.

The 1990s identification of a 150-km-sized crater, called Chicxulub, as being an impact structure 65 million years old, essentially clinched the K-T boundary impact hypothesis. A variety of geological analyses, including gravity mapping, magnetic field mapping, and direct sampling, delineate the crater off of the Yucatan Peninsula of Mexico, buried beneath large amounts of sediments (Figure 16.9, see Color Plate 16). The geology of the basin, which includes a large carbonate platform, is consistent with the chemistry of the impact target inferred from the boundary sediments worldwide and the age of the circular deformation defining the crater is that of the K-T boundary. Because most of the Earth's surface area is ocean floor, and because the lifetime of the ocean floor against subduction is roughly 100 million years, it was moderately lucky that this crater still exists; older oceanic impacts would have been obliterated by subduction, and even ancient continental craters are eventually destroyed by tectonic and erosive processes.

An asteroid or comet 10 to 15 km in size, striking the Earth at many kilometers per second, explains in detail the characteristics of the K-T boundary layer and adjacent material. That impactor size was sufficient to gouge out the Chicxulub crater and fling dust into the upper atmosphere that reduced the sunlight reaching the surface for months. The dust, composed of a mix of impactor and crustal material, fell onto continental and seafloor surfaces, carrying with it the chemical signature of the meteorite seen in the thin, K-T boundary layer. The energy released from such an impact is extraordinary. More than a million times more powerful than the Mount St. Helens eruption or this country's largest nuclear test detonation, such an impact has no rival with regard to anything experienced by humankind. The 1994 impact of the multiple fragments of comet Shoemaker-Levy 9 into Jupiter's atmosphere provide a guide, but each is much smaller than the estimated size of the K-T bolide (Figure 16.10). The lunar cratering record suggests that, on average in the Phanerozoic eon, a body the size of the K-T impactor strikes the Earth once every 100 million years. This number is consistent as well with the frequency of mass extinctions in the Phanerozoic, which, in turn, has led to the suggestion that most of the mass extinctions were impact driven, although as noted earlier only one of the extinction events shows very strong evidence for an impact.

In addition to the direct effects of the impact such as widespread forest fires and tidal waves, the plume of debris and fire smoke injected into the Earth's upper troposphere and

16.7 Impacts, Volcanism, and Major Extinctions 537

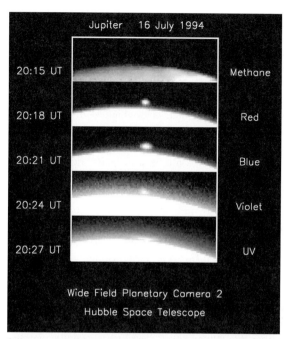

a

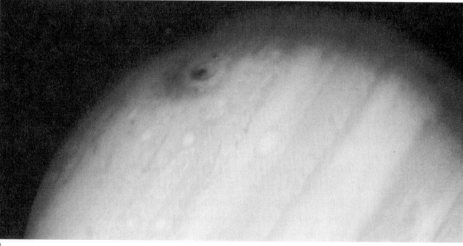

b

FIGURE 16.10 The Shoemaker-Levy 9 impacts into Jupiter, a reduced-scale analog to the K-T event: (a) impact plume from one of the fragments on the limb of Jupiter; (b) brown dust generated by impacts of the fragments; each is larger than the Earth.

stratosphere would have had a devastating effect on life through alteration of the climate. By physically blocking the rays of the sun, the dust would cause the lower atmosphere and surface of the Earth to cool suddenly, and would remain this way for weeks or months. Models suggest much of the continental area of the Earth would have had post-impact average daytime temperatures of only 10°C, much lower than the present average. In addition to the direct effects of cold on large animals and plants, the reduced sunlight would slow or shut down photosynthesis for up to a year, killing off large numbers of species dependent on various marine and continental food chains. The amount of carbon dioxide released from the carbonate target on impact, plus carbon released from burning forests, could have been very large, of order 10^4 gigatons (Gton), vastly more than the 5 Gton/yr of CO_2 that human activities are injecting into the atmosphere, and which would have led to substantial warming as the dust settled. Nonetheless, the bulk of the alteration of the atmospheric radiative balance may have been due to the injection of sulfur compounds from the impact site. Whichever, months of global winter would have been followed by years, decades, or more of global warming, further disrupting food chains and ecosystems.

Another important effect of the impactor was on the chemistry of the atmosphere. Nitrous oxides produced in the hot wake of the bolide would have altered the pH of poorly buffered lakes and bays, killing off whole families of marine organisms. Calcium-shelled species could have had their shells dissolved by a change in the acidity of lake waters; those types of organisms seemed to have suffered disproportionately in terms of species extinctions. Acid rain associated with sulfur compounds injected into the atmosphere may have defoliated vegetation and further altered water acidities over large areas. Large amounts of water injected into the atmosphere by the impact may have reduced the ozone abundance, allowing dangerous levels of solar ultraviolet light once the dust settled. Decay of large amounts of dead vegetation may have altered soil chemistry and generated locally anoxic conditions.

A large impact is not all that was happening at the close of the Cretaceous. Extrusion of massive amounts of volcanic flood basalts in the Deccan Traps in India may have released large amounts of sulfur into the atmosphere, as well as altering the carbon dioxide abundance (and perhaps that of oxygen as well). Indeed, this could have played a contributing role, stressing organisms that were then finally extinguished in the Chicxulub impact, but it is not clear that the sedimentary record supports this in terms of the timing of extinctions. The general properties of the K-T boundary and its extinctions appear to be best explained by an impact. Volcanism involving magma from the deep interior would produce an iridium enhancement, but the ratios of iridium to the elements gold and osmium are different from those in meteorites. Fire fountaining in volcanoes can produce spherules, but usually on a local scale, and they should be basaltic in composition; by contrast the K-T spherules are andesitic or dacitic (Chapter 11). The arguments against volcanism as the primary or even incidental cause of the K-T boundary phenomena do not rule out episodes of widespread volcanism as causing climate change or major extinctions at other times in the Phanerozoic or earlier. But the preponderance of the evidence strongly argues that the catastrophe at the close of the Cretaceous was caused by a large impact.

Although several other major extinction events, which mark boundaries in the geologic time of the rock record, have been tentatively proposed as being associated with impacts, no records with the clarity of the K-T boundary exist for them. Iridium enrichments of varying degree, shocked material, or other evidence appear sporadically in other sedimentary layers—but never as abundantly or clearly as at the K/T boundary. For example, a strong iridium anomaly is seen in sediments at the boundary between the Triassic and Jurassic periods of the Phanerozoic, which marks a dramatic change in flora and fauna.

But the other associated evidence available at the K-T boundary that argues for the iridium anomaly being impact generated, rather than volcanic in origin, is lacking here, perhaps because of the greater antiquity of the Triassic-Jurassic boundary. The Permian-Triassic mass extinction also lacks a "smoking gun" identifying its cause, in spite of being much more severe than later ones in terms of the number of families of organisms extinguished.

It may be that the quality of the impact evidence at K-T is an anomaly, associated both with the relative geologic youth of this event and its impact location. The characteristics of the Chicxulub impact site, with its carbon- and sulfur-laden marine sediments, probably ensured an unusually severe reaction on the part of the Earth's atmosphere and biosphere to that particular impact event. Episodes of widespread volcanism, which are likely triggered by major shifts of plate motion associated with continental collisions or disruptions, show up more clearly in the rock record than do impacts. Flood basalts typically cover an area 100 times larger than the area directly excavated in a 100-km-scale impact crater so that the direct evidence of a flood basalt is much more likely to be found. Flood basalts may erupt over a time span as large as several million years so that they are more likely to overlap with a mass extinction event, whether they cause it or not, than would a virtually instantaneous impact.

Turning now to the pre-Phanerozoic history of life, how might a "major extinction episode" among Proterozoic prokaryotic life-forms be identified, when the sedimentary fossil record itself was so sketchy and difficult to interpret? How much effect on the Archean biosphere could a K-T-type impactor have had, when the atmospheric carbon dioxide abundance was so much larger than today, and stands of burnable vegetation simply did not exist? A possible major extinction has been suggested associated with a well-documented, 580-million-year-old, 100-km-scale crater in southern Australia. Sharp changes in planktonic species, variations in isotopic carbon, and evidence for a major diversification in green algae correspond in rough time to the event, but they also correspond to the end of the major Neoproterozoic global glaciation. Perhaps there is yet another causal link here, but the half-billion-year horizon is already sufficiently more hazy than that at 65 million years that we despair of fully resolving the linkages between the impact itself and the global biological and climatological effects. Nonetheless, it should be clear to the reader that mass extinctions occurred throughout the history of the Earth, they had a number of different causes, and they were a dominant influence on the evolution of life.

16.8 Orbital and Spin-Axis Forcing of Climate Variations

The gradual cooling of the climate during the Tertiary period of Earth's history culminated, about 2 million years ago, in a set of climate swings that still occurs today. The Earth's climate swings between major episodes of glaciation that last of order 100,000 years, interspersed with warmer interglacials of about 10,000 years' duration. Neither the glacials nor the interglacials are as intense as the deep glaciations of the Neoproterozoic or the extended Cretaceous warmth. But the oscillatory nature of the climate may have played a role both in an overall decline in the diversity of mammals (which reached a peak some 10 million years ago) and—by emptying some ecosystems—in encouraging rapid changes and innovations in certain species, such as those of the genus Homo. Indeed, we may owe the origin of our uniquely intelligent species to the climatic oscillations of the past 2 million years.

The origin of these oscillations most likely lies in the nature of the Earth's spin and its orbit around the sun. Currently, the Earth's axis is tilted some 23 degrees from a line perpendicular to the plane of its orbit around the Sun, and the orbit itself is slightly elliptical. The closest approach of Earth to the Sun happens to occur when the southern hemisphere is tilted toward the Sun, that is, during southern summer and northern winter. Because most of the continental mass lies in the northern hemisphere, the current state is one where most of our planet's continental area does not experience the most summertime heating possible, because the Earth is slightly farther from the sun in July than in January. The difference in received sunlight from the closest to most distant point of the orbit is 7 percent, and the response of the climate system to a shift in orientation in which the northern hemisphere continents receive the strongest summer heating, versus the southern hemisphere ocean, can be significant.

As first introduced in Chapter 11, the Earth's axis precesses around a fixed point in space, much like a toy gyroscope can be made to do by pushing its axis once set in motion. The effect of this precession, for which a complete cycle takes 26,000 years, is to reorient the northern and southern hemisphere summers relative to the close and far points of the Earth in its orbit. Roughly 13,000 years ago, the onset of northern hemisphere summer occurred each year when the Earth was near the point of its orbit closest the Sun, opposite to the current state. Because plate tectonic motions are too slow to have shifted continental positions more than a few kilometers during that time, geography was the same, and the heating of the northern hemisphere continental masses was therefore 7 percent more severe at that time. It was then that the last Pleistocene glaciation episode ended (although glacial retreat poleward began 2,000 to 3,000 years earlier).

Other oscillations in the motions of the Earth are known to occur. The other planets of the solar system, in exerting very slight tugs on the Earth, not only cause the axial precession but also slightly alter the magnitude of the tilt by several degrees on 41,000-year cycles. The ellipse that is Earth's orbit drifts or rotates as well, having a net effect of shortening the precessional period from 26,000 to 22,000 years. Furthermore, these planetary tidal pulls also modulate the eccentricity of the Earth's orbit in a complicated way that approximates two cycles at 100,000- and 450,000-year periods. Serbian physicist M. Milankovitch, for whom the cycles are named, a century ago quantified the idea that all these cycles might affect climate. An approximate correlation can be found between the climate and several of these cycles in the recent geologic record. However, modeling of glacial and interglacial climates, as well as study of ice-core records, implies intricate feedbacks between ice cover, carbon dioxide abundance, ocean circulation, clouds, and sunlight that complicate the Earth's climate response.

It is in the last two interglacials that we find an interesting case of delayed development that faintly echoes those of the spread of eukaryotes in the early Proterozoic, the dramatic appearance of novel animal forms in the Cambrian, and the diversification of mammals in the Tertiary. Anatomically modern humans arose in Africa during or prior to the interglacial time, the *Eemian*, prior to this one (the *Holocene*). And there is evidence that these modern humans proceeded quickly from their birthplace in Africa to other continents. Yet, the development of agriculture as a way of life did not occur for another 100,000 years, until about 10,000 years ago or so. What prevented humans from settling and becoming agriculturalists, with the concomitant societal organization and development of cities, 110,000 to 130,00 years ago? Data from ice cores and oxygen isotopes in seafloor sediments suggest that the Eemian was a warmer, shorter, and less stable interglacial than the Holocene, with more abrupt and larger temperature swings. If this were true, it provides a natural explanation for the delayed onset of agriculture—the Eemian climate was simply too variable to support agriculture, frustrating any attempts by human to establish it at the time. This consigned modern humans to another 100,000 years of nomadic hunting and gathering, through the next glacial episode, through the first millennium of the Holocene (less stable than the rest of the Holocene), into the relative stability of a climate 10,000 years in duration that allowed a change in lifestyle that, in turn, changed the world. However, other interpretations of the data tend toward the view that the Eemian and the Holocene were comparable in variability. Should this have been the case, then the failure of humans to develop an Eemian agriculture must have some other cause—perhaps the population of the new hunter gatherers was as yet too small to require agriculture or perhaps, while physically modern, the *Homo sapiens* of the Eemian were not yet fully human in their mental processes (see Chapter 17). There is no argument, though, that the Eemian was a shorter interglacial, and further climate data might bring to light additional differences between the two interglacials that could have played a role in delaying agriculture.

16.9 Variations in the Pace of Biological Evolution on Habitable Planets

> *Non c'è differenza! Mostri e non-mostri sono sempre stati vicini!*
> *Ciò che non è stato continua ad essere.*[1]
> Italo Calvino, *The Origin of the Birds*.

The interactions between evolutionary innovations and climate history on the Earth appear to have affected, if not set, the timing of major innovations in life beginning with the origin of the Eukarya through to the agricultural lifestyle of modern human beings. While sudden climate changes and random catastrophes clear ecosystems, they pave the way for the appearance of new forms with new genetic innovations. Lengthy episodes of reduced habitability, such as deep ice ages, may, in turn, suppress or even squelch the expression of novel genetic innovations. The object lesson for life on other habitable planets is that the evolutionary pace may well be determined by the extent to which climate fluctuations and random catastrophes are exhibited to a greater or lesser extent than those on the Earth. A

---

[1]"There's no difference! Monsters and non-monsters have always been close together! Things that have not been, continue to be."

slowly rotating planet, with no large moon but with a nearby Jovian mass planet, for example, will experience wide obliquity swings that might frustrate the appearance of evolutionary innovations if the concomitant climate variations are too severe (Chapter 11). The extreme of marginal habitability may well be Mars, which was able to play host perhaps to prokaryotic forms of life, before irreversibly sinking into a cold, dry state that sterilized most if not the entire planet. What the other extreme would look like, that of a habitable planet with climatic swings more salubrious than Earth's to the development and diversification of evolutionary innovations, is limited today to the realm of guesswork.

One final speculation, and the entrée to the next chapter, concerns the ultimate limits of genomic innovation and complexity. The most complex life on Earth today does not express the full *potential* information content of the genome. As noted earlier, much of metazoan diversity is achieved through modularity. The pace of further innovation is dictated by the continued doubling of genes and the random accumulation over geologic time of genetic innovation such as regulatory genes. What might life look like if we go another 4 billion years ahead, or another 40 billion? If climate modelers are right, we will never know, because the Earth will lose its hydrosphere in a billion years. The pace of the genetic clock versus the climatological clock tells us there is little time left for the dramatic kind of innovation that the development of the eukarya or the bilaterans represented—unless we manipulate the genome (a subject of Chapter 17).

Absent that, the luminosity history of G-type dwarfs such as the Sun limits the timescale for genetic evolution, and to witness yet unimagined innovations might require examining ancient planets around the long-lived M-dwarfs. Here, unfortunately, we would likely be frustrated, because the most ancient stars today—Population II stars two or three times the age of the Sun (Chapter 5)—may be too impoverished in elements heavier than hydrogen and helium to support planets. Further, the habitable zone around M-dwarfs is narrower than around G-stars, making it less likely that a planet will happen to reside there, and is so close to the parent star that tidal locking of the planet to the star may compromise the planet's habitability. But M stars are enormously abundant, and so, in our imagination, we might look forward to a "someday"—when the Universe is 10 times older and M-dwarfs born in the Sun's time have stably outlasted our home star and, by luck, possess habitable worlds. Those rarest of planetary systems might foster Methuselan biospheres, whose most sophisticated organisms will carry genomes of extraordinary length and information content—and hence experience lives of complexity and richness unknown and unknowable here on Earth.

QUESTIONS AND PROBLEMS

1. Speculate on why the bacterial domain never developed differentiated, multicellular "creatures" like eukaryotic animals and plants. What is the stumbling block? Were a biologist to approach you and announce that, together, the domains archaea and bacteria had indeed successfully evolved into multicellular macroscopic creatures, would you refute or agree with his claim? Explain.

2. Calculate the kinetic energy released when a 10-km-sized asteroid (density that of silicate rock) strikes Earth at a speed of 10 km/s. Compare this number with the energy released in atomic and nuclear explosions (there are published tabulations of the energy release in nuclear tests) and large volcanic eruptions.

3. Calculate the velocity of an impact fragment that is lofted into a low-orbit around the Earth. (Hint: The potential energy of such an orbit is just one-half the energy needed to escape the Earth's gravitational field.)

4. Fungi are a Kingdom of multicelled eukaryotes that we have talked little about in this text. What are the distinguishing characteristics of fungal cells? What organelles are present in fungal cells? What common edible food is a fungus? Are all fungi multicellular?

5. It is claimed that, in fact, bacteria rule the Earth. In what quantitative way is this true? (Back your answer with numbers or a reference.)

6. Using the astronomical literature, look at the history of ultraviolet emission from M, G, and A main-sequence dwarfs. How do they compare? Given the fact that a habitable world must be farther from a parent A star than from a G star, is the intercepted ultraviolet flux at the top of the atmosphere of a habitable planet around an A star greater or less than that around a G star?

7. How does artificial selection by human agricultural industries differ from natural selection? Going back to Darwin's books, did he rely mostly on field observations of natural ecosystems or on artificial selection to make his arguments?

8. Take the organisms in Table 16.1 and, using the Tree of Life website (search for the title; the site has moved several times) or other sources, locate these organisms on the phylogenetic tree of life. Is there a relationship between where these organisms are rooted (e.g., relative to the position of LUCA) and their genome size? What is the significance of this correspondence or lack thereof?

9. Some have argued that species have life cycles—that species inevitably die out after becoming "old." Give arguments for or against this view, based on the references here and a general literature search.

10. How would you test the hypothesis that the genes for the eukaryotic cytoskeleton originated in a fourth domain of life? Why or why not could they have been present in an extinct branch of the archaeal or prokaryotic domains?

SUGGESTED READINGS

A superb discussion of biological complexity, evolution, and the genome can be found in the article by Sean B. Carroll, (2001), "Chance and Necessity: The Evolution of Morphological Complexity and Diversity," *Nature*, **409,** 1102–1109. A broad and deep insight into the nature of evolution, particularly eukaryotic, can be obtained by reading the books of the late Steven J. Gould, such as (but not limited to) *The Panda's Thumb* and *The Flamingo's Smile: Reflections in Natural History* (1980 and 1985, W. W. Norton, New York). An up-to-date discussion of the role of impacts in extinctions is given in the article by Kring (see References). The more recent history of climate change, Pleistocene and Holocene, is expertly explained in William James Burrows, *Climate Change* (2001, Cambridge University Press, Cambridge). A comprehensive discussion of sexual versus asexual reproduction, as well as the rules of inheritance developed by the 19th-century Austrian monk Gregor Mendel, can be found in W. K. Purves, D. Sadava. G. H. Orians, and H. C. Heller, *Life: The Science of Biology* (2001, W. H. Freeman/Sinauer Associates).

REFERENCES

Adami, C., Ofria, C., and Collier, T. C. (2000). "Evolution of Biological Complexity," *Proce. Natal. Acad. Sci.*, **97**, 4463–4468.

Bains, S., Corfield, R. M., and Norris, R. D. (1999). "Mechanisms of Climate Warming at the End of the Pleistocene," *Sci.*, **285**, 724–727.

Berger, A., and Loutre, M. F. (1994). "Precession, Eccentricity, Obliquity, Insolation and Paleoclimates. In *Long-Term Climatic Variations*, J.-C. Duplessy and M.-T. Spyridakis, eds. NATO ASI Series v. **I 22** Springer-Verlag, Berlin, pp. 107–151.

Brocks, J. J., Logan, G. A., Buick, R., and Summons, R. E. (1999). "Archaean Molecular Fossils and the Early Rise of Eukaryotes," *Sci.*, **285**, 1033–1036.

Carroll, S. B. (2001). "Chance and Necessity: The Evolution of Morphological Complexity and Diversity," *Nature*, **409**, 1102–1109.

EPICA Group (2004). "Eight Glacial Cycles from an Antarctic Ice Core," *Nature*, **429**, 623–628.

Gray, M. W., Burger, G., and Lang, B. F. (1999). "Mitochondrial Evolution," *Sci.*, **283**, 1476–1481.

Hoffman, P. F., Kaufman, A. J., Halverson, G. P., and Schrag, D. P. (1998). "A Neoproterozoic Snowball Earth," *Sci.*, **281**, 1342–1346.

Hyde, W. T., Crowley, T. J., Baum, S. K. and Peltier, W. R. (2000). "Neoproterozoic 'Snowball Earth' Simulations with a Coupled Climate/Ice-Sheet Model," *Nature*, **405**, 425–429.

Jenkins, G. S. (2000). "The 'Snowball Earth' and Precambrian Climate," *Sci.*, **288**, 975–976.

Knoll, A. H. (1999). "Paleontology: A New Molecular Window on Early Life," *Sci.*, **285**, 1025–1026.

Knoll, A. H., and Carroll, S. B. (1999). "Early Animal Evolution: Emerging Views from Comparative Biology and Geology," *Sci.*, **284**, 2129–2137.

Kring, D. A. (2003). "Environmental Consequences of Impact Cratering Events as a Function of Ambient Conditions on Earth," *Astrobiol.*, **3**, 133–152.

Kukla, G. J. (2000). "The Last Interglacial," *Sci.*, **287**, 987–988.

Levin, B. R., and Bergstrom, C. T. (2000). "Bacteria Are Different: Observations, Interpretations, Speculations, and Opinions about the Mechanisms of Adaptive Evolution in Prokaryotes," *Proc. Natl. Acad. Sci.*, **97**, 6981–6985.

Lynch, M., and Conery, J. S. (2000). "The Evolutionary Fate and Consequences of Duplicate Genes," *Sci.*, **290**, 1151–1155.

Lynch, M. (2002). "Gene Duplication and Evolution," *Sci.*, **297**, 945–947.

Olsen, P. E., Kent, D. V., Sues, H.-D., Koeberl, C., Huber, H., Montanari, A., Rainforth, E. C., Fowell, S. J., Szajna, M. J. and Hartline, B. W. (2002). "Ascent of Dinosaurs Linked to an Iridium Anomaly at the Triassic-Jurassic Boundary," *Sci.*, **296,** 1305–1307.

Peixoto, J. P. and Oort, A. H. (1992). *Physics of Climate.* AIP Press, New York.

Pierazzo, E. and Melosh, H. J. (1999). "Hydrocode Modelling of Chicxulub as an Oblique Impact Event," *Earth and Planet. Sci. L.,* **165,** 163–176.

Purves, W. K., Sadava, D. Orians, G. H., and Heller, H. C. (2001). *Life: The Science of Biology.* W. H. Freeman/Sinauer Associates.

Sepkoski, Jr., J. J. (1992). *A Compendium of Fossil Marine Animal Families. Milwaukee Public Museum Contributions in Bilogy and Geology 83*, 2nd ed. Milwaukee Public Museum, Milwaukee, WI.

Solè, R. V. and Newman, M. (2002). "Extinctions and Biodiversity in the Fossil Record." In *Encyclopedia of Global Environmental Change, Volume 2: The Earth System—Bilogical and Ecological Dimensions of Change.* H. A. Mooney and J. G. Canadell, eds. John Wiley and Sons, Chichester, U.K., pp. 297–301.

Whitaker, R. J., Grogan, D. W. and Taylor, J. W. (2003). "Geographic Barriers Isolate Endemic Populations of Hyperthermophilic Archaea," *Sci.*, **301,** 976–978.

Wray, G. A., Levinton, J. S., and Shapiro, L. H. (1996). "Molecular Evidence for Deep Precambrian Divergences among Metazoans," *Sci.*, **274,** 568–573.

CHAPTER 17

Evolution of Intelligence and the Persistence of Civilization

17.1 Introduction

The last 200,000 years of Earth's history include humankind—ourselves. Our predecessors, other members of the genus *Homo*, are in the fossil record at least as far back as 2 million years. Soon after the appearance of the earliest species of the genus *Homo* in Africa, its members were moving across the continent and into Asia, perhaps into Europe as well. After this first speciation event, others followed, and the final speciation led to modern humans, who spread across the entire land area of the Earth (including Antarctica, in the 20th century). The defining characteristic of humankind is intelligence, a quality that embraces self-awareness, analysis, retrospection, and complex structured expression in a variety of media from voice to electronically encoded words and images. We are, as best we can judge, the only form of life on Earth that consciously contemplates the universe and seeks the answer to a question of its own formulation: Are we alone in the cosmos?

This final chapter considers the conditions under which humans developed over the long Pleistocene ice age of glacials and interglacials, how and why we differ from other life-forms, the longevity of our species and its technological and cultural hallmarks, and the prospects for the further development of humankind into a multiplanet species. The chapter closes with a look at the likelihood that life, and technological civilizations, are common outcomes of the evolution of the cosmos.

17.2 Ecce *Homo*

The story of the origin of humankind is complex because of the wide variety of fossil types that have been found. This richness stands in stark contrast to the situation decades ago when a paucity of fossil species made even the concept of evolutionary development questionable and when the search for additional fossils was caricatured as the search for the "missing link." Today it is recognized that many different groups of primates existed throughout Africa, Asia, and South America in the millions of years preceding the rise of modern humans. Africa in particular (and perhaps Asia) hosted a number of genera and their species that were as closely, or more closely, related to our own genus *Homo* than are our closest primate relatives, the chimpanzee and the ape.

Human beings are members of the domain Eukarya, kingdom *Animalia*, phylum *chordata*, class *Mammalia*, order *Primates*, family *Hominidae*, genus *Homo*, species *sapiens*. While to some extent this taxonomic classification is a mental exercise, the structure it imposes can be overlaid with some important historical and genetic relationships. That humans are members of *Eukarya* means that our cells are large, have nuclei, and are an outgrowth of the symbiosis driven by the aerobic evolution of the atmosphere billions of years ago. Our heritage as a member of the animal kingdom is one of locomotion, oxygen-consuming respiration, sexual reproduction, and the capability for differentiation into large numbers of different cellular types—which defines our physiology. We owe the geometry of our body plan to membership in the chordate phylum, which is distinguished by its strongly centralized nervous system and a body plan with a high capacity for diversity driven by a sophisticated set of regulatory genes. As mammals, we have hair and we breast-feed our young, but that is less significant than the issue of timing. Mammals lost out to dinosaurs in the broad diversification of metazoans after the massive Permian-Triassic extinction, and it was not until the K-T extinction that mammals diversified and enlarged in body types and size. We can speculate as to whether humans would have evolved from mammals in the pre-Cretaceous world, had dinosaurs not dominated, given the very different geographic and climatological conditions then.

Among the big winners in the mammalian diversification was the order *primates*, whose considerable complexity of habitat and behavior, dexterity, and acute vision may have been the subtexts for the development of a large brain and, ultimately, intelligence. These found a broad epitome in the family *Hominidae*, whose current members include primates with very complex behavior and large body size, the great apes. What pushed the speciation of primates into large body types and high intelligence is unclear, and much of the phylogenetic record is lost because today's great apes are only a small remnant of the variety of similar creatures present in Africa and Asia some 5 million to 10 million years ago. *Sahelanthropus tchadensis* roamed the plains of Africa at the edge of what is now the Southern Sahara (Sahel region) some 6 million to 7 million years ago, and its recent discovery is significant because it could represent that part of the *Hominidae* family from which humans evolved. *Sahelanthropus tchadensis* is a relatively complete fossil skeleton, and it has a mixture of apelike and humanlike features that suggest an origin for both branches of *Hominidae*. But were *Sahelanthropus* to exist today, it is unlikely it would generate a great deal more interest than the pygmy chimps or gorillas featured in "wildlife parks" or on nature programs—its humanlike features would be taken for granted in the way we accept those of the great apes.

Sahelanthropus along with other *Hominidae* probably existed because the cooling, and increasingly fluctuating, climate prevented ecosystem stagnation and allowed the evolution of primates into larger and more complex creatures. Whether the speciation was passive or driven by particular natural selection pressures (see Chapter 16) is unknown and probably cannot be determined from the fossil record. But within the last 10 million years, as the Earth's ocean and atmosphere cooled to the point that the orbital Milankovitch cycles could amplify climate swings, first gently and then catastrophically in the glacial-interglacial Pleistocene oscillations, forests and savannas in Africa and southern Asia waxed and waned, and environmental pressures undoubtedly increased. Africa, with so much equatorial and southern hemisphere land, yet without massive high-elevation mountain ranges, would have been a salubrious place for natural experiments in increasing diversification associated with such swings—never so cold as to precipitate wholesale extinctions.

The fossil record, like an enormous jigsaw puzzle with plenty of torn and missing pieces, hints at a fascinating time through 2 million years of *Hominidae* genera (*Sahelanthropus, Ramapithecus, Paranthropus, Australopithecus*), whose members display increasingly human traits of large crania, upright mobility, and tool use. Eventually, about 2 million years ago, species arose whose physiological and behavioral (through toolkits) resemblance to modern humans was so great as to deserve membership in our genus, *Homo*. This is today a lonely club to be in: All of humanity belongs to the one extant species of *Homo*. Each of the latest models of *Homo* may have contributed to the demise of their predecessors, either incidentally or directly, in the ultimate phylogenetic version of the Greek tragedy *Oedipus*. That we do not have *Australopithecus afarensis, Paranthropus robustus, Homo erectus*, and so forth, roaming the African savanna and the forests of Africa and Asia today is a cruel trick of evolutionary biology that denies us the chance to face our origins and hence better understand ourselves. And it is a certain sign of the fragile nature of our self-awareness that we feel uncomfortable with the notion of an animal origin—symbolized in the Biblical creation story (Chapter 1) by Adam and Eve's immediate reaction to cover their *Hominidae* bodies, having just eaten the fruit that granted them self-awareness.

Integrated over time, though, the genus *Homo* is a rich one, as shown in the slice of the phylogenetic tree of Figure 17.1. An analysis such as this relies both on physical evidence from fossils (Figure 17.2, see page 550) as well genomic material from living species (in the case of *Homo neanderthalensis* there is also extant DNA in fossil remains). There is considerable disagreement whether two species, *habilis* and *rudolfensis*, are in fact in the genus *Homo* or in other genera that preceded and overlapped with ours. The phylogenetic relationships are also temporal relationships; our own species appeared last and is on the very end of the branch in a sequence that began (based on fossil dating and genomic clock estimates) several million years ago (the figure does not extend back to *Sahelanthropus* at 7 million years ago). In addition to the temporal progression, there seems to be a remarkable progression in brain size as shown in Table 17.1 on page 551. Brain size seems to jump in several increments corresponding to (and partly motivating the assignment of) progressive Hominidae genera. However, some caution must be exercised because when the brain size is normalized by body weight of a structural proxy for body size—the skull circumference squared for example—the progression is muted somewhat. Larger size means larger brain, given comparable complexity, but there is also a clear progression of behavioral complexity and tool-making capability in going from the earlier genera to *Homo*. Hence, increasing absolute brain size apparently enabled, perhaps through natural selection pressure, increased sophistication of activity.

If one must draw the line in the African sand at which humankind departs from its predecessors, between 2 million and 1.5 million years ago (Pliocene-Pleistocene transition) is

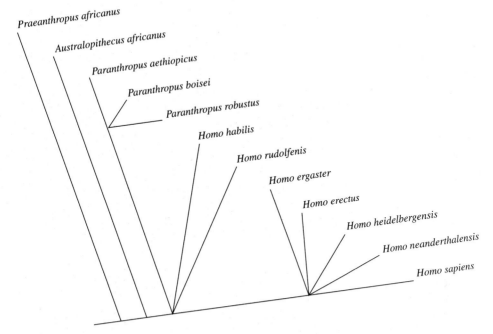

FIGURE 17.1 Phylogenetic relationships among humanlike precursors. Time moves roughly from the left, with modern humans at the right end of the graph.

as good a place as any. There, in the fossil record, we see the earliest definitively accepted members of the genus *Homo*, complete with tool kits. But more dramatically, it is at this time that the ancestors of the modern human species, and their relations within the genera, began voting with their feet. It is the genus *Homo* that is the great migratory one, epitomized by *Homo erectus* that left Africa and spread through Asia and Europe. The climatic swings deepened as the Pleistocene began, and the selection pressure for mobile and resourceful *Hominidae* to follow their food sources must have intensified. *Homo* was ready, with species that featured gracile bodies and sophisticated brain–hand capabilities for the unimaginable road ahead. Other members of the genus and the phylogenetic family were left behind, and the wanderers themselves would be supplanted in waves over hundreds of thousands of years by later members of the same genus better equipped in different ways. But the pattern of a physiologically generalized primate with extraordinary mental and manipulative powers migrating over large areas of the continents was established early in the Pleistocene.

The speciation of humankind, *Homo sapiens*, can be examined in some detail because of the genetic record available in all of us. Sexual reproduction mixes genes in the cellular nucleus from two separate individuals, obscuring the lineages of individual groups and making it difficult to determine whether, where, and when modern humans had a common origin. And the elaborate repair mechanisms in the genome of the nucleus ensure that mutation rates—that is, rates of change of the genome—are slow. In contrast, mitochondrial DNA (mtDNA) is inherited only from the mother and mutates at a faster rate than that in

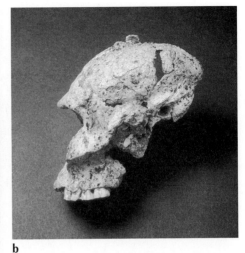
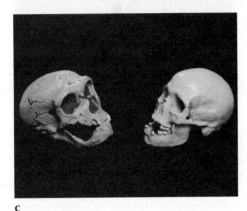

FIGURE 17.2 A gallery of skull casts and fossil remains from different species within the genus *Homo*, and some predecessors: (a) a representative of the genus *australopithecus*, (b) *Parantropus robustus*, and (c) *Homo Neanderthalensis* and *Homo sapiens* (right), side by side.

the nucleus. Therefore, by comparing the mtDNA from many individuals over large parts of the world, it is possible to derive a series of relationships of divergence of one group from another. We can apply a time stamp based on the mutation rate to make an estimate for the time when the last divergence of humans took place from a common ancestor. In the late 1980s, the first relatively complete such analysis was made and showed convincingly that all humans alive today descended from a single or small number of ancestors sometime between 250,000 years and 150,000 years ago. The divergence relationships among populations are such as to be consistent with the outward spreading of humans from the African continent, where fossil evidence also suggests the earliest modern humans appeared.

This so-called mitochondrial Eve picture is fully consistent with understanding of speciation from the point of view of the punctuated equilibrium model. A population of a precursor species of *Homo* (which, confusingly, carries the generic name of *archaic homo sapiens* because they are neither *erectus* nor modern-looking *sapiens*) was isolated from the larger population, perhaps by environmental pressures. This isolation, which undoubtedly happened to countless other *Homo* groups through the Pleistocene, was special in that

TABLE 17.1 Cranial Sizes of Humans and Other Hominidae

Taxon	Absolute Volume (cm^3)
Paranthropus aethiopicus	410
Paranthropus boisei	513
Australopithecus africanus	457
Homo (or *australopithecus*) *habilis*	552
Homo ergaster	854
Homo (or *australopithecus*) *rudolfensis*	752
Homo heidelbergensis	1,198
Homo erectus	1,016
Homo neanderthalensis	1,512
Homo sapiens	1,355

this isolated group underwent genome changes that then reproductively isolated them from the main population. This group, with morphological peculiarities that we would recognize as modern, represents the gene pool from which all present-day peoples arose. All races, all cultures, all men and women on the Earth today are descended from this small group of individuals in Africa who lived sometime between 250,000 and 150,000 years ago. Where they arose in Africa is very difficult to determine; while the genomic data hint at east Africa, there are fossil remains of the appropriate age in southern Africa. No further speciation has occurred since in spite of persistent isolation of groups because (1) there has not been time for the development of reproductively isolating gene changes in such groups; (2) what changes have occurred have not prevented reproduction with the larger population; and/or (3) speciation might have occurred at times, but the "rogue" population was subsequently exterminated. A key gene that may be part of the speciation event is one that controls the development of fine motor skills—and possibly cerebral capability—for speech. Studies of this gene indicate that it codes for a protein different in humans compared with that in apes, and the "apelike" version of the gene occurs today in a single related group of people who lack the ability to speak. On the basis of comparison among different human populations, the "speech" gene appears to be about 120,000 years old, which, given the uncertainties, may be consistent with the African speciation event in which modern humans appeared.

The mitochondrial Eve story has been challenged on the basis that mtDNA does, on occasion, acquire material from the male side of a mating pair or that the mutation rate is highly uncertain or that the phylogenetic relationships assembled from the DNA of living persons is topologically ambiguous. None of these objections has led to a fatal contradiction of the basic picture. The age may be off, the location of origin might be as well, but the notion that modern humans arose from a single group at a single time seems to be robust. The competing view is that modern humans represent a mélange of people who speciated separately from more archaic members of the *Homo* genus in different places on the

African, Asian, and European continents—with enough interbreeding to avoid reproductive isolation through time. This hypothesis has its basis in the existence of regionally distinct morphological characteristics that, some argue, persist from *Homo erectus* through to modern humans. There are difficulties, however, in maintaining enough reproductive isolation over millions of years to sustain regional (essentially, racial) differences while still maintaining reproductive viability worldwide. Where such viability fails, a new species arises, which is just what appears to have happened 150,000 years ago based on the mtDNA analysis. And there is morphological (fossil) evidence not only for plenty of earlier speciation events in Africa but in Asia and Europe as well—the speciation of the Neanderthals being an outstanding example.

Other genome analyses on modern humans further anchor the time of dispersal of modern humans from the African origin. Our genome contains a large amount of short, interspersed, repeated DNA that is peculiar to primates and can be more easily subjected to phylogenetic analyses than the remainder of the nuclear genome. This portion of the genome yields a dispersal about 140,000 years ago.

Humans have been in Australia for at least the last 40,000 years, and Australia holds a particular mystery with respect to the origin of modern humans. Mitochondrial DNA analysis of the living aboriginal people can be compared with an excellent record of human remains in Australia dating back into the last ice age. The skeletal remains are of anatomically modern *Homo sapiens*, but with a range of body types—suggesting diverse groups of people lived on that continent during the Pleistocene. The mtDNA analysis, when compared with that of other modern humans, suggests as deep a lineage in the aboriginal Australians as in the East Africans. Indeed, the DNA in the oldest of the Pleistocene remains implies that anatomically modern humans were in Australia *before* the African speciation event interpreted to be the origin of the modern human species. This contradictory result could imply that the model of a single origin of modern humans is wrong, but it could as well suggest some mixing of the Australian population with precursors to modern humans (presumably archaic *Homo sapiens*). Did that mixing occur in Australia? If so, it implies that members of archaic *Homo sapiens* were seafarers. Alternatively, could the population that migrated to Australia, very soon after the speciation event leading to modern *Homo sapiens*, have inherited a particular line of genes from an older, related species? The ambiguity illustrates that serious challenges remain in resolving ancient human history from analysis of the genome.

The situation in Europe, on the other hand, suggests direct replacement of archaic species with modern people migrating from Africa. *Homo erectus* was in Asia a million years ago, and speciation events took place there that led to other species, including some that entered Europe, such as *Homo heidelbergensis*. That species, in turn, is replaced in the fossil record by *Homo neanderthalensis*, whose overlap with modern humans in Europe over tens of thousands of years—up to 24,000 years ago—is the last and best-documented case of the coexistence of two species of *Homo*.

Homo neanderthalensis, or the "Neanderthals," first appear in the fossil record about 300,000 years ago, preceded by seemingly transitional specimens between Heidelbergensis and Neanderthal. Analysis of fragmentary DNA from three different Neanderthal specimens imply that the species has constituted a distinct lineage for at least a half million years, although the genetic clock is likely uncertain by a factor of two, and the lineage could include *heidelbergensis* for which no DNA has yet been found. Neanderthals lived primarily in

Europe, but fossil specimens have also been found in the Middle East, from 50,000 to 120,000 years ago, interleaved after 90,000 years ago with fossil remains of anatomically modern humans. Apparently Neanderthals moved south during the coldest glacial times of the Pleistocene, abandoning their European homeland, and may have retreated during warmer times when modern humans moved north to climates more temperate than interglacial Africa. Anatomically modern humans, of the cave-painting Cro-Magnon culture, were in Europe by 40,000 years ago, initiating a period of time in which the geographic range of Neanderthals was pushed westward and progressively reduced. The Iberian Peninsula appears to be the last refuge as the most recent Neanderthal remains occur there.

Analysis of DNA from Neanderthals and modern humans shows little or no inheritance of Neanderthal genes by us. Neanderthal was a separate species of the genus *Homo*, but one that left us a magnificent puzzle. Robustly muscular, with a mean body mass more than 20 kg larger than for modern humans, no chin, huge nose, and an extended skull (occipital bun) at the back of the head, Neanderthal has a distinctive face that resembles nothing possessed by the world's modern peoples. And yet, Neanderthal skeletons are modern in terms of upright mobility, dexterity, limb length, and brain size—the last averaging larger than modern humans on an absolute basis (Table 17.1). Neanderthal made sophisticated, if unimaginative, toolkits, for the vast majority of the time they existed as a species based on the fossil record. In the later stages of the coexistence of Neanderthals with modern humans in Europe, Neanderthals began to manufacture decent (but not perfect) imitations of the Cro-Magnon "master craftsman" sets. They valued trinkets, and they buried them with their dead. Although their brains were quite differently shaped (what did go on within those massively extended occipital buns?), they likely dreamed, they apparently grieved for those who left them, and they probably "knew themselves."

The mechanisms of genomic evolution and speciation created two species of quite different but certainly modern humans: Neanderthal, derived from an ancient *H. erectus* Asian stock that came long before from Africa, and us, derived from the *H. erectus* stock that had remained behind in Africa. When "we" met "them," perhaps in the Middle East 90,000 years ago but certainly 40,000 years ago in Europe, we faced in their eyes and minds a reality and an evolutionary trajectory that was separate from ours for a half million years—and evidently separate enough to ensure reproductive isolation and prevent the merging of the two species. These meetings, circumscribed by a breathtaking sweep of temporal quarantine, could have been no less strange and perhaps no less frustrating than our endlessly imagined meetings with intelligent, bipedal extraterrestrials. But, occurring as they did in the time before written history, the details are forever lost.

Had we stayed away from Europe 40,000 years ago, there is absolutely no reason there should not be living Neanderthals there today—the span of time since their extinction is less than 10 percent the duration of their species. In doing them in, however it may have happened, we lost the chance to contemplate (in historical times) different flavors of intelligence and consciousness, of self-awareness and reflective contemplation of the world, through a species close enough to us that intimate communication surely would have been possible. Instead, we are left with faint echoes of ancient encounters, expressed in the human propensity to imagine a host of magical creatures different from ourselves yet close enough in form and kinship to interact in meaningful ways. Are our most ancient legends faintly echoed tales of our encounters with the Neanderthals, of our departures from our archaic predecessors in Africa, of encounters with "strange ones" in Asia and Australia? If so, then they speak of meetings with other species, with *them*—never with others of *us*, because with *them* the consummation of kinship through fertile offspring was physically impossible, improbable, or forbidden by the deepest cultural taboos. Many modern tribes of humans traditionally call themselves—and only themselves—"the people," as if others are not. Is this a Pleistocene holdover?

If modern humans have (until the last 30,000 years) never been alone on this Earth, persistently encountering long-forgotten peripatetic predecessors in distant lands tens or hundreds of thousands of years after speciation events, why shouldn't we be programmed to imagine a universe teeming with intelligence, ready to receive us or do battle with us—regardless of the plausibility of our imaginings?

17.3 Human Intelligence as an Evolutionary Specialization and the Biology of Conscious Self-Awareness

Vishnu sleeps, and the universe is his dream; were he to awake it would cease to exist. This Hindu myth well describes how we can simultaneously sense physical reality, as do other animals, while at the same time being aware of it *and* being aware of being aware of it. Our brains process sensory input and create a perceptual theatre that allows the placement of a mentally constructed "I" within the perceived physical world. Because there is a single, nonexchangeable "I" for each human (with some pathological exceptions), consciousness represents a mechanism for allowing each human being to innovate, invent, and communicate with remarkable flexibility. Evidently consciousness is an emergent phenomenon that requires a particular functioning brain state involving coordinated activity in the cerebrum; when that is interrupted by nondreaming sleep, injury, or drugs, then consciousness (or remembered consciousness at least) ceases until cerebral function is restored.

Whether other extant species possess consciousness is unclear; the brain size and structural gap between the sophisticated pygmy chimp and ourselves is large. But reports that chimpanzees behave in front of mirrors as if they know it is they, as well as exercises in learning language, hint that chimps do have a form of self-awareness. Experiments with dolphins suggest that they too may have self-awareness. But if consciousness is an emergent phenomenon of specifically complex brain behavior, it is hard to imagine a kind of gradual reduction in the level of consciousness as we consider species with progressively less sophisticated brains. Perhaps, as the late Carl Sagan has lyrically suggested, to perceive the world as a dog is akin to (but much more sensory-rich than) how we perceive existence in dreams—they picture the world through their sharp senses but possess no reflective volition.

It is easy to "imagine" the selective evolutionary pressures to develop intelligence but much more difficult to do so for consciousness. Intelligence is a phenomenon of a big and complex brain, which enables sophisticated sets of manipulative operations that will yield tools effective for different jobs, clothes, fire where it is needed, and organized warfare against other tribes. Intelligence implies planning, which can exist without self-reflective awareness. As species representing the genus *Homo* progressively arose, their morphologies seem distinguished by bigger brains, more dexterity, and more elaborate toolkits. The advantages of more complex behavior and better tool making are clear for animals covering large distances in a perpetually changing world, where survival required a high probability of successfully hunting game in a variety of different kinds of environments. More neurons, organized in a relatively flexible network of connecting synapses, enable more complex behavior as surely as do more memory and a faster chip produce a more capable computer provided that the operating system is up to the task.

But consciousness is a different matter. What is it; does it arise from brain physiology, and if so, how; and is it a trait that parasitically arose out of a threshold number of neurons, or did it have survival value? Psychological theories of consciousness have been developed to address these questions. The traditional model is that consciousness is a manifestation of mental process that place the owner of that particular brain within a "theatre of the mind" that frames the external world and permits reflection and forethought. There are objections to this view promulgated by the 17th-century French philosopher and scientist Descartes and sometimes called the "Cartesian theatre." First, it localizes consciousness not only perceptually but also, of necessity, physiologically in a way that requires a special center of consciousness that does not seem to correspond to what is known about the rather distributed geography of brain function. Second, it requires the mental creation of an observer for which a direct neurological correspondence is lacking. While the model suggests a kind of evolutionary pressure for consciousness, in that the observer in the theatre is partially displaced from the action and hence has time to consider alternatives, there is no mechanism for creation of this observer. The sophistication of this homunculus within our heads simply displaces the question of the nature of consciousness inward one level, opening the door to a fallacious infinite regression without solving anything.

An alternative psychological model for consciousness is more in line with the neurological operation of the brain, as it is understood today. The so-called multiple-draft model, developed largely by D. C. Dennett of Tufts University, posits that information input to the brain (external sensory or internal neuron firings) stimulate multiple patterns of synaptic responses in a rapid-fire fashion. Hence, there is not a single stimulus–response pairing, but rather, a cascade of reactions that represent multiple responses to a set of evolving (internal and external) stimuli. Nothing is sent to a special area of the brain for interpretation—that is, to a Cartesian theatre of the observer—because there is no theatre. The repeated, changing, cascading responses of the brain to the stream of information, inducing responses that may include physical action or additional cortical stimuli (thinking), *is* consciousness. The observer in the theatre is an illusion promulgated by the multiple, repetitive neuronal responses to the stimuli.

A crude analogy, which gives the name to the model, is that of multiple drafts of a document. In the days of paper, typewriters, and ink, usually one draft of a paper at a time would be circulated to a group of readers. That draft—static, inert—would have none of the characteristics of a living, changing thing. But in the electronic age, a draft can be changed by the push of a button and sent around to a group of readers at a second push. Imagine a compulsive set of authors who are, without coordinating with each other, changing the draft of a paper and sending out altered electronic copies continuously—without informing the readers that changes are being made. This last point is important because there is no mechanism in the brain for labeling drafts—the brain simply reacts and reacts again to its own waves of synaptic firings. What the readers would perceive is a kind of living document, one that, when read, would never quite seem the same and, in its unpredictable alterations of words, changing shades of meaning, and so forth, would seem "to have a mind of its own." And that would be correct because the mind in this case is simply the changing patterns of words that are ongoing, unexpected, and beyond the control of the readers. Now accelerate the changes and the readings so they take place in tens of milliseconds or less, appropriate for travel times of neuronal signals. And let the readings be the responses of the various parts of the brain to external stimuli and to the stimuli of other neurons in the brain reacting to the stimuli. These waves of changing responses manifest themselves as a master interpreter sifting and considering the information—an "I" that is entirely illusory.

For this model, "the devil is in the details"—both literally and figuratively. Having no way to isolate humans from the operations of their brains, we cannot step outside the box and assess the correctness of the model. But the fact that we tend to animate inanimate stimuli just as in the phrase above hints at the correctness of the model. It also suggests a natural, mechanistic role for consciousness that confers upon the possessor a potential evolutionary advantage. As brains became bigger and behavior more complex, the possible actions our *Homo* predecessors could take became more complex—and more generalized. It is one thing to be a cheetah that spies its prey, springs off using powerful hindquarters, and brings that prey down with a set of mighty claws. It is quite another to find stones, grind and scrape them to make sharp points, tie them to appropriate sticks, head out with a gang to find a herd, make the herd run in a particular direction via coordinated distractions, and then insert a point into the jugular of one of the prey. The more complex and generalized the behavior, the easier it is to make a mistake. Simple stimulus–response processes would become more error prone as the complexity of the tasks increased, making the owner of the bigger brain simply a bigger fumbler. However, those whose brains operated in the multiple-draft mode possessed a layered set of responses that imperceptibly move from internal rehearsal to external rehearsal to the action itself—interleaved with sensory updates. Such creatures, much more engaged through neural feedback in the actions they are undertaking, might possess a survival advantage. And, while they are at it, their complex brains could generate multiple drafts of neuronal responses to a sunset (or to dancing or to painting), the colorful stimulus playing repeatedly tens or hundreds of times per second in changing synaptic patterns, multiple drafts, within the brain. Would not the neurons—the cells of the brain—existing for the sole purpose of sending signals or "firing," become addicted early to such stimuli, and is not that addiction what we articulate as the joy of beauty?

Some support for this view comes from the fact that the brain's "organ" of conscious thought, the cerebrum, overlies and postdates (from an evolutionary point of view) the cerebellum (Figure 17.3). There is no evidence of conscious activity in the cerebellum, even though it has nearly as many neuronal connections as does the cerebrum. The cerebellum controls complex activities (riding a bicycle, operating a car) that normally are not perceived as involving conscious thought. But while they are *learned*, conscious thought is heavily involved. We all remember our first time doing any complex mechanical task, and all of the mental effort required to work out what to do. Those aspects of the task that are a set of rote muscular responses then become controlled by the cerebellum and are lost from conscious thought. This frees the individual to perform other tasks that may require the cerebrum. Many tasks or aspects of a task may never devolve to the cerebellum because

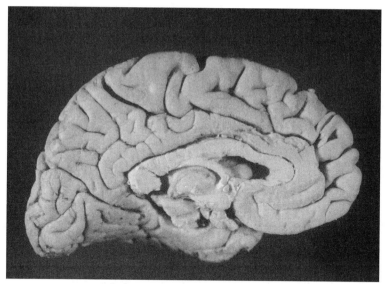

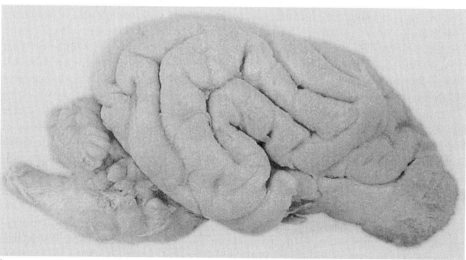

FIGURE 17.3 (a) The human brain in cross section, showing the large cerebrum arching over the rest of the organ. (b) Brain of a domestic dog, indicating the absence of a large cerebrum.

they require a complex, changing set of reactions to stimuli that cannot be encapsulated in a set of preprogrammed muscular movements. Thus, the cerebellum controls the act of throwing the spear, but deciding where to throw (and coordinating with others in the hunting party) requires conscious intervention.

Despite its attractiveness, the multiple-drafts model does not explain everything about consciousness. What, for example, is at the core of the perception of the passage of time? This cannot simply be a filing system for memory; individuals who have lost the capability to retain short-term memory still perceive the passage of time. Also, what are the specific neuronal processes that lead to the internal and external processes we describe as conscious volition and thought? The British mathematician Roger Penrose has developed an elaborate set of arguments to advance a view (which dates back to the German mathematician Goedel) that thoughts associated with mathematical problems cannot be broken down into a finite set of *algebraic* manipulations. Penrose goes further, arguing that this property can be generalized to other conscious processes and that it implies that computers, which rely on circuits that perform simple algebraic processes in repetitive algorithmic steps, cannot truly think—regardless of the complexity of their architecture. But what is going on then within the brain that allows this so-called noncomputational set of processes to take place? Penrose suggests that mental processes are, at their root, quantum mechanical in nature; that is, they involve structures within neuronal cells that are so small and complex in their connectivity that quantum processes play an important role. Unfortunately, it is difficult to test the model, and there is substantial skepticism among biologists that the particular structures he suggests—microtubules within the cellular cytoskeleton—are really key to conscious thought.

From the point of view of astrobiology, the importance of consciousness is that it is a mechanism for living creatures to contemplate the universe and to undertake activities to find life elsewhere. But is consciousness a universal accompaniment to intelligence? That is, taking multiple drafts as a straw-man mechanism for consciousness, is it necessary that the coordinating mass of cells for an organism operate in that or a similar fashion once a certain threshold of complexity of behavior is reached?

Examples of complex behavior in the absence of consciousness include the structures that insect colonies build—though the secret of their elaborateness lies in the repetition of underlying, simpler, building blocks. But even in the human realm, there is no universal agreement that consciousness has always worked in our own species the way we perceive it today. The Princeton psychologist Julian Jaynes argued a daring hypothesis more than a quarter century ago, based on the observation of the behavior of schizophrenic patients, that the modern human cerebrum has two potentially very different modes of operation. The "Cartesian theatre," which may be a poor model of a mechanism but not of our personal perception of consciousness, is, he argues, a *historically* recent innovation that supplanted the so-called *bicameral* mind. The owner of a bicameral mind does not have the conscious sense of self that we can universally describe to each other today but is instead an obligate responder to a sequence of instructions provided, in perceived voices or other stimuli, by a higher authority (Figure 17.4). The higher authority is the result of coordinated synaptic firings within the brain, as is the self of the Cartesian theatre, but is less well integrated with the sequence of neuronal firings that generate the response. Ergo, when Homer relates in the *Iliad* (Book 13) that "Zeus had brought Hector and the Trojans up to the ship," he is describing, in Jaynes's model, a mental process in which Hector's decision-making function—manifest as a separate being or deity in Jaynes's view—commands Hector to take the Trojans up to the ship. There is no formulation or contemplation of the action by a "Hector" of the Cartesian theatre, because his brain

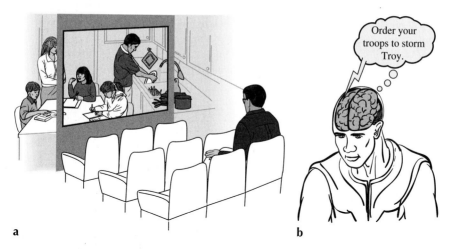

FIGURE 17.4 Two different forms of consciousness: (a) the Cartesian theatre and (b) the bicameral mind.

was not capable of operating that way. Nonetheless, Hector himself formulated the action—in a process that was perceived by him as separate from, and compelling, his subsequent action. Likewise, in Book 18, the deity Iris tells an armor-bereft Achilles, "We gods are well aware that your glorious armor has been taken. Nevertheless, go to the trench as you are and present yourself to the Trojans." And Achilles obeys, his appearance constituting a strategically important challenge to the Trojans that gives his own tired army much needed rest—a daring tactic formulated by and within the experienced warrior's bicameral brain.

Jaynes's assertion that these are not literary devices is part of a sequence of arguments in favor of an alternative form of mental processing of stimuli and consequent decision making. And, as bizarre as it sounds, it is actually difficult to disprove the argument. We individually assume there are no bicameral minds in the world today, except for schizophrenics. But Jaynes argues that we cannot a priori distinguish bicameral individuals via casual acquaintance. His view is that these different modes of operation represent a subtle change in how synaptic firings are organized, rather than a physiological difference (except for schizophrenics, whose disease-altered brains cannot function except bicamerally) and that we all retain residual bicameral functionality manifest, for example, in our religious yearnings.

Regardless of its applicability, or lack thereof, to humans, the bicameral model provides an intriguing alternative to how we imagine intelligent beings should act. That the physiology of the human brain might admit two such different modes of operation would not bode well for an interstellar "meeting of the minds" among complex organisms produced by billions of years of separate evolution. We might find ourselves in a universe teeming with life, including intelligent life, but horribly alone because there is no common "mental" basis for perceiving, reacting to, or contemplating the ensemble of matter and energy that is the universe.

17.4 Climate Change and the Timing of the Development of Human Civilization

The end of the last glaciation began about 15,000 to 16,000 years ago as European glaciers initiated a retreat, well in advance of the maximum northern hemisphere summer solstice sunshine at 11,000 years ago (when the axial tilt relative to perihelion was opposite that today). It is not understood whether fluctuating carbon dioxide levels drive or are driven by (or both) the glacial-interglacial swings, but CO_2 began an upswing with the glacial melting. An unstable climate epoch over a couple of thousand years, including the Younger Dryas cold snap, transpired as melting glaciers altered the salinity and heat balance in the oceans, thereby changing oceanic currents. These fluctuations and the onset afterward of the relatively stable Holocene interglacial, are well documented by isotopic data in ice core records (Figure 17.5). The ratio of deuterium to hydrogen in ice directly records the enhancement in the lighter isotope, associated with its preferential evaporation from the ocean relative to the heavier isotope, during colder climates when oceanic water is deposited on the continents in the form of glaciers.

At the end of the Pleistocene glaciation, modern humans had been the only representatives of the genus *Homo* on the Earth for perhaps 10,000 years. With the retreat of the ice, the spread of forests, and migrations of animals, human hunting groups moved into new regions, including the Americas, by both land and sea. In the Americas, the large mammals disappeared within a millennium. The rapid loss of large mammals was directly or indirectly the result of climate—the latter in the sense that human hunters took advantage of warming to spread into new territory where the large animals could not survive in the face of the new and growing human populations. A human-caused mass extinction of the

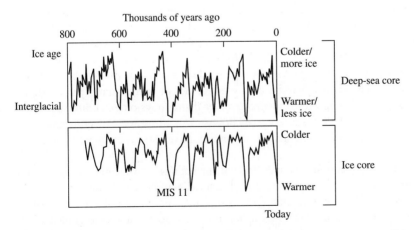

FIGURE 17.5 Pleistocene climate derived from the abundance of deuterium versus hydrogen in an Antarctic ice core, compared with the climate derived from oxygen isotopes in deep sea sediments. The record goes back 800,000 years, and shows among other features an interglacial at 400,000 years ("MIS-11") that lasted longer than the Holocene.

American large fauna is supported on a number of grounds, including the analogous disappearance of diverse Australia/New Guinea fauna 40,000 years ago coincident with the first migration of humans to those locations. The equivalent of a "mass extinction" in the fossil record continues over historical times as human activities lead to the demise of large numbers of species of animals and plants.

The climate then progressively warmed toward the Holocene "optimum" of 7,000 to 9,000 years ago. In this time, and later, very detailed data on climate in different regions are available from geologic evidence of glacial retreat and advance; studies of pollen deposited in lake sediments, and as well-preserved in packrat middens in very dry locations; the pattern of growth expressed by tree rings in ancient bristle cone remains; and other so-called climate proxy data. Hence, during the Holocene climate optimum, global temperatures were comparable to or somewhat higher than today, the region around the Nile was wetter than at present, but in North America the Laurentide ice sheet expanded and remained relatively far south (compared with today) until 6,000 years ago. Conversely, climate in the Middle East and adjacent areas of Asia remained relatively wet and mild until about 4,500 years ago, when rather dramatic desertification began.

These climate trends suggest that the region around the Middle East, a crossroad for migrations of humans and other species in the genus *Homo* for hundreds of thousands of years, was amenable to organized settlement and agriculture for many millennia. As populations grew, the innovation of a stable agricultural lifestyle became more desirable, if not essential. Modern humans, equipped with the ability to make elaborate tools, to plan, and to communicate with each other using elaborate sounds and symbols, may have undertaken agriculture earlier, but the combination of salubriously stable climate and increasing population may have made the time around 8,000 years ago a watershed in the agricultural history of humankind. North America, caught in the grip of sporadic but dramatic cooling during the same time period, would not have been as easy a locale for organized agriculture. There is far less information on early Holocene South America, but archaeological data suggest that the onset of organized agriculture and cities postdated that in Europe and Asia.

Once agriculture and the organization of cities began, almost everything that we recognize as a defining characteristic of civilization followed. The large-scale production of food and the division of labor enabled some people to undertake activities not immediately related to the survival of a particular culture. While art dates back to the earliest modern humans, suggesting that organized tribal groups also had some leisure time, the wholesale construction of monuments, art, literature, and scientific research were really enabled by the agricultural lifestyle developed during the remarkably stable set of millennia in the middle of the Holocene.

Over the last few thousand years, the Holocene climate has not been quite as salubrious as before. Climate fluctuations, desertification, and sudden cooling all show in the climate record. Most famous is the Little Ice Age, which (in the Europe of the 17th century) led to a number of decades of unreliable or failed harvests and associated decreases (small and large) in living standards. Around the rest of the world, evidence for dramatic climate change during that time is spotty, with some places showing evidence for climate changes associated with the onset of the Little Ice Age, others showing little. By the middle of the 19th century, temperatures had begun to rise in Europe, and the 20th century—especially the last half—evolved toward a climate of unusual warmth for the Holocene as a whole.

Much of this may be a result of the human introduction of substantial amounts of CO_2 into the atmosphere, about 30 percent more by the year 2000 than was present in the mid-20th century (see Section 17.5).

The cause of the unusual stability of the interglacial climate of the past 10,000 years, relative to the brevity and possible instability of the previous glacial, is not understood (Chapter 11). Nor is there general agreement among climatologists as to whether the Holocene climate is destined to become less stable as we approach the onset of another glaciation or, alternatively, as human input of greenhouse gases into the system changes the radiative forcing (heating due to infrared absorption) of the atmosphere. But were the climate to move into the more changeable state that characterized the last glacial and the earliest Holocene, agricultural productivity would be threatened as growing zones shift rapidly and unpredictably. This is but one of several challenges that 21st-century humankind faces, some eight millennia after the transition to agriculture produced a dramatic change in the way our species interacts with its environment.

17.5 Future Prospects for the Human Species and Its Civilization

Approximately 99 percent of all species that have ever existed on the Earth are now extinct, based on extrapolation of the fossil and genomic records available today. The typical lifetime of an animal species is somewhere in the range 1 million to 10 million years, and modern *Homo sapiens* has been around for at most 200,000 years. This would seem to presage a long future for humanity and its elaborate technological toolbox, which—in the form of agriculture and the rise of cities—has been around for only 10,000 of those 200,000 years, or about 5 percent of the lifetime of humanity. But the relatively rapid pace of evolution in the climatologically unstable Pleistocene put some species of *Homo* out of business after just a few hundred thousand years.

Since the Industrial Revolution began a few centuries ago, humankind has been able to command increasingly powerful sources of energy that range from fossil fuels to nuclear energy and to build machines that can move ever-larger amounts of goods at higher speeds with greater efficiency. This growth in capability to move goods has been motivated by a growth in agricultural food production, which, in turn, is a response to growing population. A significant fraction of humanity remains inadequately fed and protected from disease and drought, but there is no question that, in absolute numbers, more humans are being fed more effectively, and protected better against disease, than at any other time in human history (Figure 17.6).

Humankind faces challenges to its current industrialized civilization connected to the total world population (now in excess of 6 billion persons), the per-capita expenditure of energy, and the nature of the fuels that supply the energy. With respect to the last, the post–Industrial Revolution world economy is based on fossil fuels, which are the remains of organisms from earlier epochs in the history of the Earth. More than 90 percent of commercial energy generation worldwide is based on fossil fuels. Coal (25 percent of world energy production), oil (41 percent of world energy production), and natural gas are the

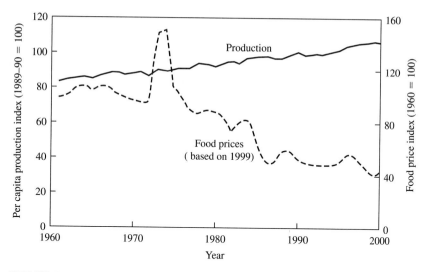

FIGURE 17.6 Production of food and its cost over the past few decades.

principal reservoirs of fossil fuels, the total abundance of which is estimated to be equivalent to less than 0.01 percent of the carbon stored as carbonates and kerogen in the crust of the Earth (Table 11.1). Different fossil fuels have different values of energy output from oxidation per amount of carbon dioxide produced—coal being the lowest and natural gas the highest in efficiency in this regard. The amount of carbon in fossil fuels is equivalent to 0.06 bar of carbon dioxide, or 150 times the current atmospheric abundance. And, indeed, burning of fossil fuel to generate energy is to first order the oxidation of the organic material, which forms CO_2 as the main waste product. Although there are sinks of carbon dioxide (e.g., the oceans), they can only partially buffer changes in the atmospheric abundance. Our use of fossil fuels inputs carbon dioxide into the atmosphere much faster than any natural sinks can take it up (Figure 17.7a), so that the atmospheric abundance of carbon dioxide has increased by 30 percent since the middle of the 20th century. Adding to the carbon dioxide input due to fossil fuels is that from forest burning as land is cleared for agricultural use.

Because carbon dioxide is a major greenhouse gas (Chapter 11), the increase in its atmospheric abundance associated with fossil fuel burning must lead to a commensurate rise in the surface temperature. For a constant amount of solar input, an increase in the infrared opacity due to an increase in the amount of carbon dioxide forces the radiating level—the altitude at which infrared photons are free to move outward without impediment—upward. At this higher altitude, the temperature is lower, and the atmosphere becomes less able to rid itself of infrared photons, that is, of heat. Therefore, the temperature at the radiating level increases, but in turn, this shifts the entire temperature profile in the troposphere (Figure 11.15) to warmer values because the temperature gradient itself cannot decrease without further impeding the flow of heat outward. The net effect is an increase in the global mean temperature. Because the Earth's climate system is extremely complex—with many different responses to a particular perturbation, such as changes in cloud cover,

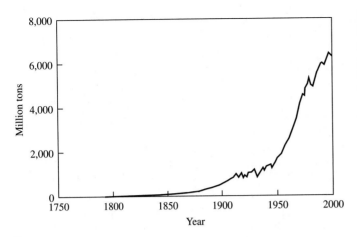

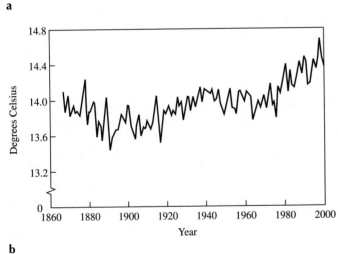

FIGURE 17.7 (a) Rise in annual global carbon emissions into the atmosphere since the 18th century. (b) Global mean temperature from the late 19th century through the 20th century.

rainfall, and oceanic transport of heat—it is difficult to predict with certainty specific climatic changes in particular areas due to the increase in carbon dioxide. But careful analysis of temperature data indicate that, particularly in the last two decades of the 20th century, global surface temperatures have been on the increase (Figure 17.7b), over and above a general 19th- to 20th-century warming out of the Little Ice Age of the past few centuries. Comparison of the current warm conditions with proxy climate data over the millennia of the Holocene, as well as the dramatic recession of glaciers and melting of arctic and Antarctic ice sheets, suggests that we are experiencing an anomalously warm period that may well be a response to the human-induced increase in carbon dioxide.

It is difficult to predict the degree to which agricultural and transportation systems will have to adapt to an extended bout of global warming—as well as how we might respond to sudden, unexpected shifts in the climate system associated with continuing input of carbon dioxide into the atmosphere. (To cite just one hypothetical example, should high-latitude rainfall increase substantially as a consequence of rising global temperatures, it is possible that oceanic salinity, and hence heat transport patterns, could be changed or interrupted. The oceanic transport of heat to the North Atlantic is, in part, salinity driven, and its interruption could drastically change the northern European climate, though models currently disagree on the direction and depth of these changes.) As a general approach to the question of global warming and its effects, the basic business principles of increasing efficiency and conserving where possible certainly apply.

There is little prospect today for an alternative energy source with a transportability and versatility akin to that of fossil fuels, at the price with which fossil fuels are recovered and processed. Solar energy, wind energy, and nuclear energy are potentially important in terms of their availability and comparative inexhaustibility (the last applying to fusion rather than fission sources of nuclear power), but they all remain expensive and difficult to adapt, especially to transportation technologies. The entire crustal supply of fossil fuels is probably not extractable economically in the foreseeable future, and no one wants to contemplate what our life on Earth would be like with 150 times more CO_2 in the air than is present today. Current reserves and estimated extractable future reserves of fossil fuels would, at projected trends in usage, last us for a century or more, perhaps several centuries; the regional differences between production and consumption of such fuels will likely continue for decades to dominate international politics (Figure 17.8). As the world's population increases, humankind will be challenged to produce increasing amounts of

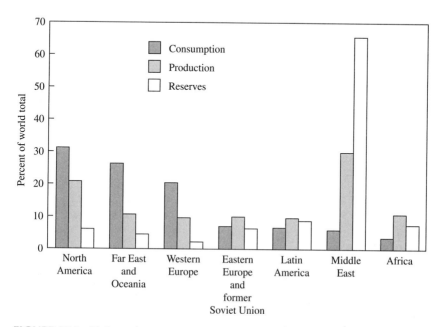

FIGURE 17.8 Estimated reserves, production, and consumption of oil in different parts of the world, expressed as percentages of the total.

food and to transport it to ever-larger numbers of people, while at the same time absorbing the impact of the very natural desire for billions of people to increase their standard of living, and hence energy consumption, well above current values.

The Malthusian interpretation of the future would bring to an end the remarkable trend of the 19th and 20th centuries in a population collapse engendered by starvation, disease, and war as resources fail finally to keep up with the growth in the number of humans—aided perhaps by a climate that changes with increasing unpredictability. Much of human history can be characterized in terms of such falls after periods of population increases, on a regional rather than a global basis. Today, we must think in terms of impacts on a global scale, and there is no refuge "beyond the next valley" when our technological civilization encompasses the entire world. Nonetheless, predictions of disaster are usually based on linear extrapolation from the current situation, without anticipation of novel developments that "change the rules of the game." Until the invention of the automobile, dire predictions about horse-drawn transportation systems bogging down in their own manure were made for the large cities of the 19th century. There is room for increased efficiency and innovation in the technologies we use today, and the future probably holds some dramatic technological surprises. Such surprises will not simply be given to us—there is no Prometheus waiting to hand humankind a better kind of fire. We must discover it for ourselves, if it is there. The challenge humanity faces today is to ensure that the last three centuries of remarkable technological progress are a prelude to a long-lasting, maturing civilization rather than to a global Malthusian collapse.

We cannot travel our way out of potential problems such as overpopulation or large-scale resource depletion. There is no planet in the solar system capable of hosting billions of humans; we would have to create an entirely new ecosystem on Mars and then deal with a planet whose surface area is less than that of the continents of the Earth. To loft all 6 billion of us into Earth orbit, with today's chemical-based rockets, would require approximately all of the fossil fuel estimated to be in the crust of the Earth in Table 11.1. While continued space exploration and expansion of humanity into the solar system may be a part of our future—this author fervently hopes it is—those who travel will continue to be a very few emissaries of a much larger population of humans who participate vicariously from the Earth.

Much more difficult to predict is the ultimate development of genomic technologies, and, specifically, genetic engineering, in the next century. The skyrocketing capabilities in analyzing and manipulating the genome form much of the basis for the laboratory part of astrobiology. How far will humankind take these technologies and use them? Much depends on ethical issues and how these evolve over the decades as the technology becomes progressively imbued in the cultural milieu of societies around the world. Today the most serious ethical question, in terms of the breadth of its impact, is the extent to which individuals should be given knowledge of their genetic propensity for diseases—and to what extent insurance companies should be tipped off as well. But other questions, tending more to the exotic from today's point of view, will arise. Should we manipulate large parts of the human genome to achieve changes in the growth and behavior of cells so that, for example, nerve cells can regenerate or disease-ravaged muscle can be rebuilt? Could we, or should we, invent new kinds of cells (e.g., other than muscle, nerve, skin cells, etc.) to enable humans to do things that they otherwise could not—for example, interface with silicon-based devices? To what extent should that technology be controlled to avoid individuals trying to become "smarter" or "stronger" with such technologies? Under what circumstances, if ever, should the human genome be manipulated in such a way that succeeding generations of people possess the genetic imprint of these manipulations? How should humans deal with, decades or centuries down the line, the urge to create new species out of humankind through large-scale manipulation of the genome, alteration or enhancement of regulatory genes, and so forth? Is it inevitable that the next human speciation event, centuries or millennia hence, will have as its seed the direct intervention of technology in the human genome?

The need to address the question of "designer humans" is decades away, maybe less. But the seeds to do such things have already been sown, and history shows that humans are incredibly inventive in using technologies both to solve problems and create new ones. Decades after the splitting of the atomic nucleus ushered in a new age of opportunity and dangers, it is the new understanding of a much different kind of nucleus—that of the eukaryotic cell—that promises things both wonderful and horrible to behold.

17.6 Epilogue: Is There Anyone to Talk to Elsewhere in the Cosmos?

Much of astrobiology as discussed in this book can be summed up in an equation first written by the astronomer Frank Drake, a pioneer in the radio search for extraterrestrial intelligence SETI (Chapter 15), over four decades ago. The following version of the equation expresses the number of technological civilizations that can communicate with us, in our galaxy:

$$N = N_{galaxy} f_{popI} f_{sun\text{-}mass} f_{planets} f_{habitable} f_{life} f_{complex} f_{intelligent} f_{techn} f_{outlook} \quad (17.1)$$

Here N_{galaxy} is the number of stars in our galaxy; f_{popI} is the fraction of those stars that are population I, that is, have a complement of elements heavier than hydrogen and helium similar to the Sun's; $f_{sun\text{-}mass}$ is the fraction of such stars within or below the mass range of the Sun, so that the main sequence properties are appropriate to life; $f_{planets}$ is the fraction of such stars with terrestrial planets; $f_{habitable}$ is the fraction of such planets that are continuously habitable over a time appropriate to the development of complex life; f_{life} is the fraction of such planets on which life does arise; $f_{complex}$ is the fraction of such biospheres within which complex (equivalent to eukaryotic) life develops; $f_{intelligent}$ is the fraction of such biospheres with complex life in which intelligence arises; f_{techn} is the fractional lifetime, relative to the age of the galaxy, over which the technological civilization exists; and $f_{outlook}$ is the fraction of such civilizations with the appropriate mental outlook to beam radio signals—deliberately or not. Let us examine each of these in turn, setting the first number, N_{galaxy}, equal to 2×10^{11} stars.

1. F_{popI}. Taking the disk and the halo of the Milky Way Galaxy, about half the stars are population I, and hence potentially formed with enough heavy elements to build rocky planets as well. Some have argued that there is a narrower range of metallicity over which sufficient rock-forming and biogenic elements will exist in a disk; thus,

50 percent might be an overestimate by a factor of several, so let us use 10 percent, or 0.1.

2. $f_{sun\text{-}mass}$. Stars 10 percent or more massive than the Sun do not exist on the main sequence long enough for complex, multicellular life to evolve, if our own planet's history is a good guide. The vast majority of main sequence stars are actually less massive than the Sun. The smallest stars, the M dwarfs, are half the mass of the Sun or less and put out so little light that the habitable zone is narrow and close to the star, where the spin of a planet would be tidally locked and stellar flares (more common and intense in M than G dwarfs) a large effect. While it is unclear that these would be a detriment to the evolution of life (both effects might only restrict a planet's zone of habitability to particular latitudes and longitudes), we will be very conservative and eliminate the M dwarfs, which are about 77 percent of the total number of stars. This leaves us with $f_{sun\text{-}mass} \sim 0.2$.

3. $f_{planets}$. Current planet searches, which are sensitive to giant planets, suggest that 10 percent of stars within the mass range considered in items 1 and 2 have giant planets during their main sequence; perhaps three times as many possess giant planets during formation (and the majority of such planets are then lost; Chapter 6). Terrestrial planet formation seems to be an easier process because massive amounts of gas need not be present. The presence of giant planets too close to the terrestrial planet zone would make the orbits of any terrestrial planets there unstable; based on computer simulations and the known distribution of extrasolar giant planets, this is unlikely to be a problem in more than half of the stars that produce giant planets. Binary systems, which comprise upward of 50 percent of Sunlike stars, may have terrestrial planets, but the range of possible stable orbits is smaller than for single stars. Thus, a reasonable but perhaps conservative estimate for the frequency of terrestrial planets around stars of roughly solar mass is 10 percent. To this we must add the possibility of moons of giant planets that orbit in the habitable zone—an entirely plausible possibility because giant planets have been discovered in such orbits around nearby Sunlike stars. This makes our value of 10 percent a lower limit.

4. $f_{habitable}$. Habitability here is defined as capable of sustaining liquid water on the surface over billions of years. In our own solar system, the Earth at 1 AU is habitable, Venus at 0.7 AU is not, and Mars at 1.5 AU was habitable for a time (Chapter 12) but now is not (at least at the surface). It is unknown whether an Earth-sized planet at 1.5 AU would have sustained habitability for billions of years; the severity of the "Snowball Earth" episodes on our own planet at 1 AU (Chapter 16) suggests not. Detailed numerical computations of planetary evolution can serve as a guide to whether liquid water is sustained for a given combination of orbital and planetary properties, but the complexity of processes associated with planet evolution make a priori mapping of the habitable zone an uncertain affair. Let the continuously habitable zone for a G-dwarf extend from 0.9 AU to 1.1 AU, a conservative width of 0.2 AU. The probability of terrestrial planets around Sunlike stars, computed in item 3, already included the consideration that giant planets within a few AU of the star would render such systems unstable. Hence the probability of a terrestrial planet occurring in the habitable zone is 0.2 AU/3 AU = 0.07, assuming a uniform probability of formation of terrestrial planets anywhere within the dynamically stable zone.

But habitability also requires that a planet be of reasonable size and have a supply of water and organics. In our own system, two of the four terrestrial planets are Earth sized, and computer simulations of terrestrial planet formation suggest that ingestion of lunar- to Mars-sized bodies to make Earth-sized planets is a common outcome. This process also seems to scatter water-laden and organic-bearing bodies throughout the terrestrial planet formation zone. Let us thus take half of our terrestrial planets to be Earth sized and possessed of water and organic molecules. Finally, for terrestrial planets that are within 5 AU of a giant planet, tidal effects could cause a slowly rotating terrestrial planet's obliquity to shift dramatically over time. The presence of a large Moon damps this, but, in the absence of such a moon, many terrestrial planets might maintain a sufficiently rapid rotation rate to avoid large shifts. Conservatively, we take another factor of two off $f_{habitable}$ for this effect, giving a final value of $0.07 \times 0.5 \times 0.5 = 0.02$. Habitable environments within subsurface oceans, which might exist, for example, beneath the surface of Jupiter's moon Europa, are missed by this requirement, as are other kinds of habitable moons around giant planets not represented by examples in our own solar system. Therefore, the value of 0.02 is almost certainly a lower limit.

At this point, let us take stock of the probabilities we have considered up to now, as beyond this point we will be considering factors for which far less information from data or from modeling is available. We have computed

$$N_{habitable\ planets} = N_{galaxy} f_{popI} f_{sun-mass} f_{planets} f_{habitable}$$

$$\sim 2 \times 10^{11} \times 0.1 \times 0.2 \times 0.1 \times 0.02 \sim 10^7 \text{ habitable planets} \quad (17.2)$$

which is the number of planets in the galaxy with sustained liquid water over billions of years.

5. f_{life}. Continuing into the biological factors, we immediately encounter a stumbling block in f_{life}. If Earth is the only planet on which we know life exists, then absent any theoretical guidelines to the probability that life will form on a habitable planet, we do not know whether f_{life} is closer to 0 or 1. Should we find evidence of past or present life on Europa or Mars, and then be able to establish that such life had an independent origin from that of the Earth, it would be appropriate to argue that $f_{life} \sim 1$, since two or more bodies in our own solar system developed life independently. Absent this, should indications of advanced organic chemistry, in the form of enantiomeric purity in chiral compounds or special arrangements in polymers, be found on Titan, we might be confident enough to give f_{life} a value close to 1—but arguably this is little more than a pure guess. Finding Earth life on Mars, or vice versa, does not boost f_{life}, because there is not a separate origin. Finally, the chances of interstellar transfer of life are vastly smaller than that of transfer between planets in a given planetary system. Absent any of these data, we venture into a realm that is only poorly constrained by Earthly laboratory experiments. Were the origin of life dependent on purely random assembly of nucleotides to make a gene, the probability would be essentially zero. But the concept of a metabolism-first autocatalytic system greatly improves the chances of life forming, because the system could co-opt molecules from the environment capable of storing information on the system (Chapter 9). But this simply pushes the uncertainty in f_{life} forward from the

origin of homeostatic systems to the formation of stably replicating systems capable of natural selection. Thus, let us retain f_{life} as unspecified, and we will examine the implications of it being close to 1 later.

6. $f_{complex}$. This is an immensely smaller conceptual jump than the origin of life itself, yet it apparently required more time on Earth than did life's origin. If ingestion and symbiosis are required to form complex cells with organelles, then surely this is a process whose likelihood is high in the billions of years available on planets around our specified long-lived stars within our estimated habitable zone. Essentially, it only requires that oxygenic photosynthesis begin, which harnesses so much free energy for biology that dramatic changes in cellular metabolisms are enabled. Whether fluctuating climate conditions are required for, or inhibit, obligate cellular symbiosis is not known. Absent other constraints, let $f_{complex}$ be 0.5. This is a number that may actually be testable through spectra of extrasolar terrestrial planets around other stars when large spaceborne coronagraphs or interferometers are flown (Chapter 15).

7. $f_{intelligent}$. The presence of a coordinated, central signaling (nervous) system in chordates led eventually, through natural selection, to a number of organisms with large brains and the capability to process external information in a complex fashion. It seems reasonable to suggest that once life of the complexity and interactive mobility of animals arises, intelligence is inevitable. But how much intelligence is required to eventually construct a civilization? Only humans evidently possess the requisite amount. But with perhaps 1 to 2 billion years left for the habitability of the Earth prior to moist convective loss of water (Chapter 11), a lot can happen—and maybe new, more intelligent species will arise within the mammals, or within an entirely new class or phylum (Figure 17.9). It is unclear whether deep episodes of glaciation in the Proterozoic set back the timetable for the flowering of complex animals or fostered it. Here, we are free to take either the warmer half of the habitable zone (or a bigger planet) to avoid such crises, or the colder half to exacerbate them and hence "stimulate" evolution. This gives us a factor of 0.5. Conservatives might point out that the 200-million-year reign of the dinosaurs, physiologically as sophisticated as us, produced little in the way of brainy creations—but the maximum size of dinosaur brains began to increase near the end of the Cretaceous period. So, of the two opportunities for the development of intelligence—the Mesozoic age of dinosaurs and the Cenozoic age of mammals—one panned out. Let us conservatively take another factor of 0.5 for that, for a total $f_{intelligent}$ of $0.5 \times 0.5 = 0.25$.

8. $f_{techn} \times f_{outlook}$. Given intelligence, is technology obligatory? Provided sufficient dexterity, the making of tools seems inevitable. Modern humans rocketed into the industrial revolution only 13,000 years after the last interglacial began—a time corresponding to 5 to 10 percent of the total duration of our species. But we did not create such a civilization during the last interglacial, 135,000 years ago, when, as a species, we were very young. So out of the two chances we had, we took one—a 0.5 probability. Neanderthals made the same style of tools for at least 100,000 years, and only when modern humans had been on their continent for 10,000 years did the Neanderthals change habits. Had they survived into the Holocene would

FIGURE 17.9 One of the extraterrestrials from the movie *Men in Black II*. Its body plan seems decidedly non-Cambrian, but it is still much too "familiar" to truly represent the forms extraterrestrial intelligent species might assume. As a stand-in for possible future eukaryotic life on Earth, though, one cannot rule it out.

they too have taken up the plow and started along the technological road? Let us say no and take another penalty of one-half for the failure of the "other" intelligent *homo* species to go technological. We must now multiply by the normalized lifetime of a technical civilization capable of communicating on interstellar terms. The normalization is the lifetime of the galaxy, 10 billion years. Humankind came close to destroying its 20th-century technological civilization three decades ago, and whether we can adapt either to the climate changes engendered by fossil fuel burning or to the depletion of fuels remains an open issue. If we cannot, a lower limit to the lifetime of this technological civilization is 10^3 years—the duration of the longest lived empires. (This is a lower limit because the demise of empires does not seem to result in more than a temporary loss of technological capability, which is inevitably recovered or restored by successor cultures.) If we can adapt, then the end is undetermined—but 10^8 years, an order of magnitude more than the lifetime of animal species, would be a fair upper limit. Hence f_{techn} ranges from $0.5 \times 0.5 \times 10^{-10} \times 10^3 = 2 \times 10^{-8}$ to $0.5 \times 0.5 \times 10^{-10} \times 10^8 = 2 \times 10^{-3}$. The second factor, $f_{outlook}$, recognizes the fact that not all technological civilizations might care to communicate. Would individuals whose minds are organized in a fully bicameral sense care to send signals outward, when all the deities they need are right inside their heads (or the equivalent)? Let us take a factor of $\frac{1}{2}$ for this to come up with the final number for this product of 10^{-8} to 10^{-3}.

Combining the biological factors with the result from equation (17.2) leads to

$$N = N_{habitable\ planets} \times f_{life}f_{complex}f_{intelligent}f_{techn}f_{outlook}$$
$$= 10^7 \times f_{life} \times 0.5 \times 0.25 \times (10^{-8} \text{ to } 10^{-3})$$
$$= (0.01 \text{ to } 10^3) \times f_{life} \text{ civilizations} \qquad (17.3)$$

For the numbers posited here, the existence of more than one communicating civilization in the galaxy would depend on f_{life} being greater than 10^{-3}. Even if it were unity, we could be alone in the galaxy. If this analysis were correct, it would address the question of why we have yet to be contacted by extraterrestrial civilizations. However, the numbers presented here are almost all lower limits, or extremely conservative guesses, and so the analysis is biased somewhat toward low numbers. More optimistically, the upper end of our estimate, $10^3 \times f_{life}$, would allow for hundreds of extant communicating civilizations if f_{life} is close to 1. And this, in turn, would bode well for the ongoing "SETI" search for signals from extrasolar technological civilizations in our Galaxy, described in Chapter 15.

There are tens of billions of galaxies throughout the universe, and hence if we ask what the value of N is for the cosmos as a whole—not just for our galaxy—the answer is that it could be huge. But the distances between galaxies are enormous—such that the gulf between putative intelligences is not just one of space but of time. A signal sent to Earth from the Andromeda galaxy would arrive 2 million years later, ten times the current age of our species, and would be a million times fainter than that sent by a planet a kiloparsec away from our solar system, that is, from a veritable cosmic neighbor.

Chapter 5 introduced the anthropic principle and the seeming suitability of the universe's physical laws and force constants for life. But for those who relish the prospect of communicating among the stars with other technological civilizations, this may be a cruel universe indeed. We may be trapped in a kind of galactic solitary confinement. If we contemplate instead, for our philosophical satisfaction, only the possibility of other intelligences without civilization, or even just of complex life itself, the chances of many such worlds existing in our own galaxy seem good *if* the formation of life itself is a common phenomenon. Go outside on a moonless night, away from the glare of the city, and look at the stars—our solar neighborhood in the Galaxy—and the faint band of light that traces out the disk of our Galaxy (Figure 17.10). Is life a natural physical outcome of the formation of planets? Whether you are staring at a stellar tapestry filled with life of unimaginable variety or at a galaxy sterile beyond the confines of this little Earth depends upon the answer to that question.

QUESTIONS AND PROBLEMS

1. Ice cores are being drilled deeper and deeper into Antarctic and Greenland ice sheets, providing a record of climate change through the last glacial time, the "Eemian" interglacial, and part of the glacial before that. Do a search of the most recent issue of *Science* and *Nature* (the weekly journals of science in the United States and the United Kingdom) to find the latest information on temperatures and carbon dioxide abundances determined by ice cores. How are the temperatures determined?

FIGURE 17.10 The stars of the solar neighborhood and the disk of the Milky Way Galaxy seen from the Pinaleños mountains of southern Arizona. The lights of Phoenix and Tucson are visible at the 10 o'clock and 8 o'clock positions on the horizon.

What is the latest information on the climate stability during the Eemian versus Holocene interglacials? What are the implications for the appearance, dispersal, and lifestyles of modern humans, who arose during or before the Eemian interglacial?

2. Mounting evidence for a human role in the increasing global temperatures has shifted the accompanying political debate from the question of the reality of the link to human activity to that of appropriate actions. Should industrial activity be curtailed, or should we simply mitigate the deleterious effects? Select articles in newspapers and magazines that discuss each of these sides, and write a few paragraphs on each of these views and their merits and demerits, in your opinion.

3. The possible effects on Europe of a reduction in the salt content of North Atlantic surface waters is hard to predict, and recent studies have ranged from little or no effect to the possibility of profound glaciation being induced. Search the literature for studies that deal with this issue. (Hint: You will find some of these articles under the

subject heading of the "Younger Dryas" cooling anomaly—a short period of time at the end of the last glaciation when melting glaciers flooded North American river systems and fresh water poured into the salty Atlantic.)

4. The Little Ice Age was a period of time in European history that followed the Medieval Warm Period and ended in the 19th century. Read some studies of the Little Ice Age, its possible causes, when it is thought to have ended, and what its effects were on European culture. (Hint: Look for books that deal with the links between climate and history.) During the Medieval Warm Period before that, what was going on in terms of colonization of Iceland and Greenland? What was happening to cultures in the desert Southwest of the United States?

5. The story of the origins of modern humans derived from physical evidence in bones and fossil remains is continually changing and being updated. Look in *Science* and *Nature* for the latest articles in this area. Do the newest finds support or contradict some of the story laid out in this chapter?

6. Compare the scientific ideas for the origin of humans to traditional stories from your culture or that of your ancestors. In what ways do the two contradict or support each other? How might you reconcile the scientific evidence with the cultural tradition, or are they irreconcilable?

7. The intelligence level of dolphins claimed in various studies ranges from that of dogs to that of ourselves. In particular, their large frontal lobes and complex communication capabilities seem unparalleled in the animal kingdom except for the great apes (and humans). Check the animal behavioral literature for the latest studies on dolphins. Given that dolphins might be quite complex creatures, why is our communication with them relatively limited (i.e., what is the barrier that prevents deep discussions of feelings and values)?

8. Write an essay in which you falsify the following hypothesis: A cohort of human beings live among us who have absolutely no conscious thought whatsoever—the so-called zombie hypothesis. Now try to falsify the hypothesis that a subset of humans living among us have the bicameral mode of consciousness described in the Julian Jaynes book listed in the Suggested Readings.

9. Pick one or several stories of abductions or visitations by extraterrestrials you might find in newspapers or books. Can you falsify the stories with the evidence provided, or by finding logical or scientific inconsistencies?

10. The numbers used in the Drake equation discussion are very much personal choices by the author, and scientists involved in the search have chosen other values. Using the review of SETI work listed in the References or another set of citations you might find, select a different set of numbers to come up with alternate values for the number of communicable civilizations in the Galaxy.

SUGGESTED READINGS

An excellent and moderately up-to-date summary of human evolution is given in a special issue of the journal *Science:* "Human Evolution: Migrations," *Sci.,* **291,** 1721–1753 (2001). Although not as recent as some others, Daniel C. Dennett, *Consciousness*

Explained (1991, Little, Brown and Co., Boston) is thoughtful and very clear. Regardless of your views toward the bicameral hypothesis of ancient human thought, Julian Jaynes, *The Origin of Consciousness in the Breakdown of the Bicameral Mind* (1976, Houghton Mifflin Co., Boston) is must reading, for its clarity of thought and sheer gall of the hypothesis. Writing later in his career, Carl Sagan's perceptive musings about the nature of humankind contain some important insights. Notable among these books is Carl Sagan and Ann Druyan, *Shadows of Forgotten Ancestors: A Search for Who We Are* (1992, Random House, New York). William Calvin's *A Brief History of the Mind* (2004, Oxford University Press) comprehensively covers the past, present, and future of human intelligence and consciousness. A brilliant synthesis of the 13,000-year Holocene history of human civilizations, which provides insightful guidance into why cultures behave as they do toward others, is Jared Diamond, *Guns, Germs and Steel* (1999, W. W. Norton, New York). The Worldwatch Institute of Washington, D.C., publishes an annual "State of the World" assessment with excellent charts of energy use, food production, and other vital data. A forward-looking collection of articles on the question of our rarity in the cosmos is B. Zuckerman and M. H. Hart, eds., *Extraterrestrials: Where Are They* (1995, Cambridge University Press, Cambridge), pp. 215–225. For historical reasons, read Iosef S. Shklovskii and Carl Sagan, *Intelligent Life in the Universe* (1966, Dell, New York).

REFERENCES

Broecker, W. S. (2001). "Was the Medieval Warm Period Global?" *Sci.*, **291**, 1497–1499.

Burroughs, W. J. (2001). *Climate Change: A Multidisciplinary Approach*, Cambridge University Press, Cambridge.

Campbell, J. (1988). *The Power of Myth*, Doubleday, New York.

Crowley, T. J. (2000). "Causes of Climate Change Over the Past 1000 Years," *Sci.*, **289**, 270–277.

Dennett, D. C. (1991). *Consciousness Explained*, Little, Brown and Co., Boston.

Ekers, R., Cullers, K., Billingham, J., and Scheffer, L., eds. (2002). *SETI 2020: A Roadmap for the Search for Extraterrestrial Intelligence*. SETI Institute Press, Mountain View, CA.

Enard, W., Khaitovich, P., Klose, J., Zöllner, S., Heissig, F., Giavalisco, P., Nieselt-Struwe, K., Muchmore, E., Varki, A., Ravid, R., Doxiadis, G. M., Bontrop, R. E., and Pääbo, S. (2002). "Intra- and Interspecific Variation in Primate Gene Expression Patterns," *Sci.*, **296**, 340–343.

Edelman, J., and Tonini, G. (2000). *Consciousness: How Matter Becomes Imagination*, Basic Books, New York.

Flavin, C., et al. (2002). *State of the World 2002*, W. W. Norton Co., New York.

Hart, M. H. (1995). "Atmospheric Evolution, The Drake Equation and DNA: Sparse Life in an Infinite Universe." In *Extraterrestrials: Where Are They?*, B. Zuckerman and M. H. Hart, eds., Cambridge University Press, Cambridge, pp. 215–225.

Homer. *The Iliad* (1966 translation by E.V. Rieu), Penguin, Baltimore.

Jaynes, J. (1976). *The Origin of Consciousness in the Breakdown of the Bicameral Mind*, Houghton Mifflin, Boston.

Katz, M. E., Pak, D. K., Dickens, G. R., and Miller, K. G. (1999). "The Source and Fate of Massive Carbon Input During the Latest Paleocene Thermal Maximum," *Sci.*, **286,** 1531–1533.

Kennet, J. P., Cannariato, K. G., Hendy, I. L., and Behl, R. J. (2000). "Carbon Isotopic Evidence for Methane Hydrate Instability During Quaternary Inerstadials," *Sci.*, **288,** 128–133.

Kleinberg, R. L., and Brewer, P. G. (2001). "Probing Gas Hydrate Deposits," *Am. Sci.*, **89,** no. 3, 244–251.

Lander, E. S., and Weinberg, R. A. (2000). "Pathways of Discovery. Genomics: Journey to the Center of Biology," *Sci.*, **287,** 1777–1782.

Levitus, S., Antonov, J. I., Boyer, T. P., and Stephens, C. (2000). "Warming of the World Ocean," *Sci.*, **287,** 2225–2229.

Levitus, S., Antonov, J. I., Wang, J., Delworth, T. L., Dixon, K. W., and Broccoli, A. J. (2001). "Anthropogenic Warming of Earth's Climate System," *Sci.*, **292,** 267–270.

Lunine, J. I. (1999). *Earth: Evolution of a Habitable World*, Cambridge University Press, Cambridge.

Martin, P. and Klein, R., eds. (1984). *Quaternary Extinctions*. University of Arizona Press, Tucson.

McGraw, D. J. (2000). "Andrew Ellicot Douglas and the Big Trees," *Am. Sci.*, **88,** 440–447.

McManus, J. F. (2004). "Paleoclimate: A Great Granddaddy of Ice Cores," *Nature*, **429,** 611–612.

Penrose, R. (1994). *Shadows of the Mind: A Search for the Missing Science of Consciousness.* Oxford University Press, Oxford.

Ponce de Léon, M., and Zollkofer, P. E. (2001). "Neanderthal Cranial Ontogeny and Its Implications for Late Hominid Diversity," *Nature*, **412,** 534–538.

Schurr, T. G. (2000). "Mitochondrial DNA and the Peopling of the New World," *Am. Sci.*, **88,** 246–253.

Shackleton, N. (2000). "The 100,000 Year Ice Age Cycle Identified and Found to Lag Temperature, Carbon Dioxide, and Orbital Eccentricity," *Sci.*, **289,** 1902.

Shapiro, R. (1999). *Planetary Dreams: The Quest to Discover Life Beyond Earth*, John Wiley and Sons, New York.

Stahle, D. W., Cook, E. R., Cleaveland, M. K., Grissino-Mayer, H., Watson, E., and Luckman, B. H. (2000). "Tree-Ring Data Document 16th Century Megadrought over North America," *EOS*, **81,** 121–125.

Stott, P. A., Tett, S. F. B., Jones, G. S., Allen, M. R., Mitchell, J. F. B., and Jenkins, G. J. (2000). "External Control of 20th Century Temperature by Natural and Anthropogenic Forcings," *Sci.*, **290,** 2133–2137.

Swetnam, T. W., and Betancourt, J. L. (1998). "Mesoscale Disturbance and Ecological Response to Decadal Climatic Variability in the American Southwest," *J.Climate*, **11,** 3128–3147.

Taylor, K. (1999). "Rapid Climate Change," *Am. Sci.*, **87,** 320–327.

Vostok Project Members. (1995). "International Effort Helps Decipher Mysteries of Paeloclimate from Antarctic Ice Cores," *EOS*, **76,** 169.

Ward, P. and Brownlee, D. (2000). *Rare Earth*, Copernicus Books, New York.

Ward, W. R. (1982). "Comments on the Long-Term Stability of the Earth's Obliquity," *Icarus*, **50,** 444–448.

Weiss, H. and Bradley, R. S. (2001). "What Drives Societal Collapse?" *Sci.*, **291,** 609–610.

Wong, K. (2000). "Who Were the Neanderthals?" *Sci. Am.*, **282,** no. 4, 98–107.

Wood, B., and Collard, M. (1999). "The Human Genus," *Sci.*, **284,** 65–71.

INDEX

Abiotic systems
  degrees of freedom, 236–237
  entropy reduction, 280–281
  GC, *453*
  precursor molecules, 253, *253*
  syntheses, 246–248, 250, 450
Absorption spectra, 92, *92*
Acceleration, 34, 35, *36,* 37
Accretion, 180, 183, *194*
Accretional energy, 402
Acetylene (C_2H_2), *87,* 88, 420, 421, 422, 428
Acid rain, 538
Acid—base reactions, 85–86
Acids, definition, 85
Adaptive optics, 144, 497
Adenine (A), *110,* 110–112
Adenosine diphosphate (ADP), 112, 125, 231, *247,* 441
Adenosine monophosphate (AMP), 113, *247,* 298–299
Adenosine triphosphate (ATP)
  energy storage, 109, 112, *112,* 213, 230, 441
  RNA and, 246
  structure, *247*
  synthesis, 125, 132, 250
Adenosine triphosphate synthase, 125
Adsorption, definition, 282
Aerobic respiration, 115
Alanine, 118
Albedo, definition, 504
Allan Hills 84001 meteorite. *See* Meteorite ALH84001
Allen Telescope Array, 509–510
Allende meteorite, 454
Alpha centauri C, 148
Altruists, molecular, 263
Amino acids, 104–105, *105, 106, 107, 134,* 135–136, 260
Aminoethylglycine, 255
Ammonia (NH_3), 228, 342–343, 440
Amphiphilic vesicles, 258–259, *266,* 297
Amylopullunase, 309
Andesite rocks, 348, 350
Andesitic volcanism, 346
Angular momentum, 53, 179–180
Antarctica, 315, 462
Anthropic principle, 168
Antiprotons, 155
Aragonite, 456
Archaea, 105, 116, 290–292, 309, 311, 522
Archean Earth, 348, *349,* 354, 355
Arecibo Radio Observatory, 422
Argon, 75, 84, 187–188, 387, 394, 419

Asteroids, 259, 323, 385, 536. *See also* Comets; Meteorites
Asthenosphere, 345
Astrometry, 483–487, *486*
Asymptotic giant branches (AGBs), 164
Atomic fusion, 99–100
Atomic number (Z), 73, 75
Atoms, 51, 52, *55,* 77. *See also* isotopes
Australopithecus, 550
Autocatalysis
  abiotic systems, 237, 278–280
  cycles, *281*
  evolving, 282–289
  prebiotic, 255, 260
  ribose synthesis, *251*
Autopoesis, 239–240

Bacillus subtilis, 26–27, 324
Bacteria. *See also* Prokaryotes; *specific* Bacteria
  anaerobic, 355
  extreme environments, 318
  magnetotactic, 459
  membranes, 124–125
  methane-producing, 342
  methanogenic, 355, *356*
  phylogeny, 290–292
  reproduction, 517
  sulfur reduction, 309
Bacteriopheophytin, 132
Banded iron formation (BIF), 335, 354, 458
Barycenter, 475
Basalts, 330, 331, 345, 348, 349, 538
Base 20, 12, *13*
Bases, definitions, 85
Belousov-Zhabotinsky (BZ) reactions, *237,* 237–239, 278, *280*
Beryllium, 158
Bible, Judeo-Christian, 11
Big Bang concept, 10, 154–157
Bilaterality, 525, 533
Biodiversity, 234, 274–301
Biogenic elements, 104, 157–165, 197–199, 455–460, 463–464
Biopolymers. *See* Polymers
Black holes, 163, 168
Blackbody radiation, 90, 152–153, *153*
Bolides, 536
Boltzmann constant (k), 222, *222*
Bonding mechanisms, 77, 81–86
Bosons, 56
Brain, 554–559, *558*
Brightness, *90,* 141–147, *147,* 152
Bromine, 379
Brown dwarfs, 166, *497, 502*

Butane, *23*

Calcium carbonate ($CaCO_3$), 345, 346, 456
Calcium silicate ($CaSiO_3$), 346
Calendrical systems, 12
Callisto, *196*
Calvin cycle, 132
Cambrian explosion, 514, 525, 530–534
Cameca ims 1270 probe, *331*
Capillary electrophoresis, 451
Carbohydrates, 113, 115, 130–132, 230–232
Carbon (C)
  bonds, 22, 86–88
  cycling, 345, 346–348, *349*
  on Earth, *344*
  global emissions, *564*
  isotopes, *86,* 190–192, 235, 334–335, *335,* 454, *455*
  kerogen compounds, 344, 352
  life based on, 439–440
  organic, 130
  reactivity, 75
Carbon dioxide (CO_2)
  on Earth, 341–342, *342,* 532
  Earth-like planets, 505
  fixation, *268*
  greenhouse theory, 337–340
  in HD209458 system, 490
  loss, 344
  on Mars, *343,* 344, 384
  production, 563
  solubility, 345
  on Titan, 419, 421
  on Venus, 341, 506
Carbon monoxide (CO), 419
Carbon-silicate cycle, 344–348
Carbonates, 344
Carnot efficiencies, *216*
Carotenoid pigments, 122
Carousels, 35–37, *36*
Cartesian theatre, 555, 558, *559*
Cassini-Huygens mission, 423–428, *424,* 469
Catalysts, 237, 261–265, 278–289, 283
Cathode ray tubes, 41, 96
Cell walls, 83, 117–118
Cells. *See also* Eukaryotes; Plants; Prokaryotes
  energy conversion by, 216
  essential functions, 116
  precursors, 258–261
  structure, 27, *28,* 443, *444, 521*
Celsius scale, 63
Centrifugal forces, 35–36, *36*
Cepheid variables, 151
Cerebrum, 556

578

Charged particle detectors, 396
Chemical periodicity, 68–70
Chemical systems. *See also* Abiotic systems
    autocatalysis, 237, *251*, 255, 260, 282–289
    complex behaviors in, 276–281
    oxidation-reduction, 309
    self-organizing, *296*
    transitions, 288
Chemisynthesis, 130–133, *131*, 232–233
Chemoautotrophic prokaryotes, 318
Chemolithoautotrophs, 325
Cherts, 332–333
Chicxulub event, 536
Chinese cosmologies, 12–13
Chiral compounds, 30, 297, 449–452, *451*
Chirality, 105, *107*, 252–253
Chlorine (Cl), 75, 83, 379
Chlorophyll, *115*, 122, 131, 132, 232, 505
Chloroplasts, 122, 132, 519
Cholestanes, 518
Chondrites, 192, 194–195, 198
Chondrules, 192
Chromatin, 121
Chromatography, *447*
Chromosomes, 27, 119
Citric acid cycle, 127, *128*
Civilization, 560–562, 567–571
Clevite, 21
Climate. *See also* Temperature
    cycles, *528*, 530
    early Mars, 384–385
    human development and, 560–562
    stabilization, 347
    variations, 539–541
Cloning, bacterial, 517
Cnidaria, 533
CNO cycle, 160–162
COBE satellite, *153*
Codons, 133–137
Columbia Hills, Mars, 383, *383*
Comets, 184, *184*, 198, 259, 323, *429*, 538
Complexity, evolution and, 524
Concordia curves, 190, *191*, 192–193
Cone nebula, *178*
Conformations, definition, 248
Consciousness, 556–559, *559*
Contamination, 323–325, 398, 466
Continents, 345, 348–352, *351*. *See also* Pangea
Continuously habitable zones, 504
Convective motions, 180, 339–340, 345
Copernican model, 11, 15–18

Coriolis acceleration, 34
Coronagraphs, 490, *502*, 503, *503*
Cosmic distance ladder, 147–154, 151
Cosmic rays, 96, 526
Cosmogony, 11
Cosmologies, 11–14, 141, 153
Cosmos, 17–18, 141–154, *155*
Covalent bonds, 82–83
Crab nebula, 13, 152
Cranial sizes, *551*
Creation stories, 153
Cro-Magnon man, 533
Cross bedding, description, 381
Crust, planetary, 345, 348
Cryovolcanism, 407
Crystal functions, 260
Crystal structures, 456, 458–460, *460*
Current, electric, 43–44
Cyanoacetylene (HO_3N), *254*
Cyanobacteria, 131, 132
Cyanogens $(CN)_2$, *254*
Cysts, 315
Cytochromes, 132
Cytoplasm, 119, 120, 315
Cytosine (C), *110*, 110–112
Cytoskeleton, 122, 519, 522–523
Cytosol, 119

Dark energy, 157, 167
Dark matter, 154, 157, 167–168
Darwin spaceborne telescope, 368, 502
Darwinian evolution, 295, 515
Dead Sea, 309–310
Deep-sea vents, 130, 239, 309, 349
Deinococcus radiodurans, 313, *314*, 468
Deoxyribonucleic acid. *See* DNA
Desorption, definition, 282
Destructive interference, 499
Deuterium (D), 76, 156, 385
Deuterium/hydrogen (D/H) ratios, 197–198, 360, 390, 393
Diamagnetic materials, 48
Diatoms, *316*
Diffraction
    definition, 108
    limits, 144, 494–495
    rings, *496*
    telescopes, 496
    wave, 57
Dinosaurs, 535
Dioritic rocks, 350
Dipole moments, *43*, *45*
Discovery program, 490–491
Disks, protoplanetary, 179, 180, *182*
Distances, 144, 147–154
Diversity, evolution of, 523–526
DNA, mitochondrial, 519–521, 549–550, 551–552

DNA (deoxyribonucleic acid)
    bases in, *110*
    cell functions and, 116
    copying RNA into, *249*
    denaturation, 309
    deoxyribose in, 109
    DNA world, 301
    eukaryotic cells, 121
    exchanges, 515
    formation, 248–250
    function, 27, 104
    genomic code and, 135
    high-radiation environments, 312
    life and, 442
    phosphates and, 441
    prokaryotic cells, 119
    protein synthesis and, 133–137
    recombination, 517
    replication, 108, 136, *245*, 246, 256
    structure, 83, 110–112, *111*
    thermodynamics, 231
DNA polymerases, 136, 244, 250, 448–449
DNA—RNA systems, 283–284
Dolomite $[CaMg(CO_3)_2]$, 456
Doppler shift, 145, *146*, 475
Doppler spectroscopy, 475–483, *476*, *477*, *486*
Dry ice clouds, 343
Dust, 150, 369, *372*, 538
Dynamical chaos, 278
Dynes, 37
Dyson model, 281–289, *285*, *287*

Eagle crater, 378, *379*, 381, *381*, *382*
Earth
    accretion timescale, *194*
    atmosphere, 337, *340*, 353–356, *356*
    axis, 539–540
    carbon inventory, *344*
    continents, 345, 348–352, *351*, 531
    core, 199–200, 349
    crust, 345
    diversity of life, 436
    early, 259, 297–299, 330–334
    formation, 184, 199–200, 329–364
    geologic timescale, 201–205
    gravitational pull on, 360
    gravity, 408
    hydrosphere, 359
    hydrothermal systems, 307
    infrared spectra, *507*
    magnetic field reversals, 199–200
    mantle, 345
    meteorites, 192–197, 324

580 Index

Earth *continued*
   molten outer layer, 185–186
   oceans, 197–199, 346, *346,* 348
   orbit, 539–540
   oscillations, 540–541
   plate tectonics, *347*
   rocks, *200*
   schematic cutaway, *201*
   Snowball Earth, 352, 530–534, *532*
   stability of, 200
   temperature profiles, *358, 420*
   tidal heating, 403
   tropopause, 357, 359
   water on, 197–199, 329, *333*
Earth—Moon distance, 17
Earthquakes, 345
Eccentricity, orbital, *16*
Eemian interglacial, 540–541
Electric currents, 43–44
Electric dipole moments, 43, *43*
Electric fields, *45*
Electrochemical gradients, 124, *124*
Electromagnetism, 37, 40–49, 89, 90–92, *91, 252*
Electron-volts (eV), 83, *84*
Electrons
   cloud shapes, 84
   description, 41
   gold foil experiments, 51
   orbitals, 54–55, *69*
   quantum angular momenta, *54*
   quantum mechanics and, 66–70
   radical formation, 85
   size of, 52
   slit interference experiments, *58–59,* 58–60, *61*
   subshell filling order, 68–69
   transfer, 230–231
   transport, 127–130
   wave—particle duality, 52, 57–62
Electrophoresis, capillary, 451
Elements. *See also specific* Elements
   accelerator produced, 75
   biogenic, 104, 157–165, 197–199
   creation, 194–196
   definition, 70
   discovery of, 20
   meteorites, 192–197
   periodic table, *21,* 72, *72*
   solar system abundance, 76, *78–81*
   spectroscopy, 21–22, 144
   transformation, 76–77, *166*
Emission spectra, 92, *92*
Enantiomers, 105, 251–252, 449–452
Endolithic organisms, 317, 318
Endoplasmic reticulum, 122

Energy. *See also* Kinetic energy; Metabolism; Potential energy; Thermodynamics
   conservation, 62–64, 66, 212
   orbital order by, 72–73, *74*
   phosphate bonds and, 441–442
   radiant, 89
   storage and transfer, 123–126
   subshells, *73*
Entropy
   Boltzmann constant and, 222, *222*
   cell specialization and, 267
   chirality and, 252–253
   definition, 214
   engine schematic, *215*
   evolution and, 263
   Maxwell's demon, 217–220
   order and, 224
   per energy content, *224*
   second law of thermodynamics and, 214–215
Environments
   evolution of life and, 526–530
   extreme, 306–307, 309–319, *444*
   genomic interactions with, 515–516
Enzymes, 104, 116, 121, 294, 312
Epicycles, 15
Equilibrium constants, 229
Equilibrium thermodynamics, 217
Error-correction mechanisms, 136
Escherichia coli, 294
Ethane (C_2H_6), 420, 421, 422
Eubacteria, 116, 521
Eukaryotes
   cells, 117, 120–123, 294, *444*
   geologic record, 519, 534
   hyperthermophilic, 307
   origin of, 518–523, *521*
   oxygen and, 522–523
   phylogeny, 290
   punctuated equilibrium, 516–517
Europa
   contamination risks, 324–325
   description, 401, 404–405
   ice crust thickness, 411–412
   interior contaminants, 410–411
   interior structure, 408–410, *409*
   isotope concentrations, 403
   magnetic field, 410, 467
   mass, 408, 410
   mission to, 412–415
   oceans, 415–416
   orbit, 404
   possibility of life on, 323, 415–416
   subsurface, 467–468
   surface, 405–408

   tides, 468
   views, 405, *405, 406, 407, 408, 413, 414*
   water on, 323, 329, *409,* 411–412
European Infrared Space Observatory, *419*
European Renaissance, 13
European Space Agency (ESA), 423–428, 485
Evaporites, 379, 411
Event horizons, 163
Evolution
   Darwinian, 295, 515
   early history, 325
   habitable planets, 541–542
   of intelligence, 546–578
   Lamarckian, 515
   of life, 514–545, 526–530
   mechanisms of, 515–518
   self-awareness and, 554–559
Exclusion principle, 71
Exobase, 358
Extinction events, mass, 524, 534–539, 561
Extremophiles, 25, 305–328, *306*

Fahrenheit scale, 63–64
Fats, energy storage, 113
Felsic rocks, *350*
Fermentation, 113, 126, 130
Fermions, 56, 68
Ferrihydrite, 458
Ferromagnetic materials, 48
Ferroplasma acidarmanus, 311, 312
Fimbriae, 119
Fine structure constants, 169
Fission, 98
Flagella, 118, 119
Fluorine, 71, 74
Fluvial channels, 369, 375, *376*
Flux, light, *90*
Food, 131, *563*
Forces, 34, 44, 49–50. *See also specific* Forces
Formaldehyde (H_2CO), 250, *251*
Formic acid, *453,* 454
Formose reaction, 250, 251–252
Fossil record, 336
   climate changes, 561–562
   geologic timescales and, 202–205, *203*
   geological sequence, *205*
   Homo sap., 552
   innovations, 528
   marine animal families, *529*
   microfossils, 456–457
   skulls, *550*
Four-dimensional space, 167
Fourth dimension, 40
Friction, motion and, 34

Fusion, hydrogen, 150–151
Fusion reactors, 100

Gabboric rocks, 350
Galactic habitability zones, 170
Galaxies
    distribution, 143–144, *144*
    formation, 163
    Hubble Space Telescope
        images, *143*
    molecular clouds, 174–176
    red shift, 145
    spiral arms, 175
Galilean transformation, 46–47
Galileo entry probe, 156
Galileo Orbiter, 405, *405, 406, 407,* 410, *413, 414,* 468
Gamma rays, 96, 98, 320
Gas chromatograph, 450, *451*
Gas chromatograph/mass
    spectrometer (GCMS), 425, *446, 452–453,* 452–454, 460–461
Gas exchange (GEX), 461
G1229b brown dwarf, *502*
Gene duplication, 294, 526–528, *527, 528*
Gene selection, 517
General relativity, 39
Genes, 27, 135, 290, 294
Genetic code, 133–137, 442, 449, 515
Genetic engineering, 566
Genomes
    cell sizes and, 267, *520*
    codes, 135
    complexity and, 524
    environmental interactions, 515–516
    evolution of, 294
    mutations, 262
    RNA precursors as, 289
    sizes of, *520*
Genomic analysis, 290–295
Genomic technologies, 566
Geographic isolation, 526–527
Geologic time, 76, 201–205, 290, 529
Germ cells, 28
Gibbs free energy (G), 225–227, *227,* 230, 236–237
Glaciation, 375–376, 384, 531–533, 539–541, 560
Global warming, 563–566, *564*
Glucose, energy storage, 125
Gluons, 49
Glyceraldehyde 3-phosphate, 127
Glycine, 105
Glycoaldehyde phosphate, 250, *251*
Glycocalyx, 117
Glycolysis, 127
Gneisses, 330
Goethite, 458

Gold foil experiments, 51, *52,* 535
Golgi apparatus, 122
Gram-negative bacteria, 118
Gram-positive bacteria, 118
Grand unification epoch, 155
Granites, 331, 348–352, 350
Graphite, 336, 536
Gravitational acceleration, *36,* 366
Gravitational lensing, *492*
Gravitational mass, 38
Graviton, 40, 50
Gravity, 34, 37–40, 176
Gray hermatite, 378, 379
Greenhouse gases, 340–344, 346, 563
Greenhouse theory, 337–340
Greigite, 462
Guanine (G), *110,* 110–112
Gusev crater, *383*

Hadrons, 52, 156
Halobacteria, 310
HD209458 system, 489–490
Heat. *See also* Thermodynamics
    accretional, 402
    definition, 212–213
    energy transfer, 229–230
    outer solar system sources, 402
    radioactive decay and, 345, 403
    tidal, 403
Heavy water (HDO), 76, 360
Heisenberg uncertainty principle, 60, 93
Heliobacteria, 131
Helium, 21
    bonding, 81–82
    electrons, 71
    fusion, 158, 162
    origin, 154–157
    production, 158–160, *159,* 340
Heme, *115,* 132
Hemin, 114–115, *115*
Hemocyanin, 115
Hemoglobins, 115, *115*
Hertzsprung-Russel diagrams, 149, *150*
Hesperian epoch, Mars, 386
Heterogenous nuclear RNA (hnRAN), 244
Heterotrophs, 325
High-pressure liquid chromatography (HPLC), 450
High-resolution microwave survey (HRMS), 509–510
Hipparchos satellite, 149, 483
HOBr abundance, *238*
Hominidae, 547–548, *551*
Homo erectus, 548, 552, 553
Homo habilis, 548
Homo heidelbergensis, 552
Homo neanderthalensis, 548, *550,* 552, 553

Homo rudolfensis, 548
Homo sapiens
    brain cross-section, *557*
    cranial size, *551*
    evolution, 540–541
    fossil skulls, *550*
    future prospects, 562–567
    learning, 556
    phylogeny, 540–541, 547, 549
    reproduction, 23–25
    self-awareness and, 554–559
Homunculi, *24,* 24–25
Horizontal gene transfer, 517, 519
Hormones, 114
Hox genes, 525, 533–534
Hubble constant, 152
Hubble Space Telescope, 39, *143,* 179, 489
Humans. *See Homo sapiens*
Huygens probe, 423–427, 470
Hyades, 151
Hydrocarbons, *97,* 419, 420–423, *421,* 453
Hydrogen cyanide (HCN), 254, *419*
Hydrogen (H)
    abundance, 21
    archaebacteria and, 522
    bonds, 83
    electrons, 42–43
    energy levels, *95*
    fusion, 150–151, 158, 160–161
    helium production, *159,* 340
    interstellar, 150
    on Mars, 322, 390
    orbitals, *70,* 71
    origin, 154–157
    reactivity, 74
    release by rocks, 318–319, *319*
    s orbitals, *82*
    spectroscopy, 21
Hydrogen sulfide (H_2S), 343
Hydronium ions, 85–86
Hydrophilicity, 83
Hydrophobicity, 83, 107
Hydrosphere, 359
Hydrothermal vents, 355
Hydroxyl groups (−OH), 85–86, 104
Hypercane storms, *349*
Hyperthermophilic organisms, 307, 526

Ice giants, 477
Igneous rocks, 186, 201, 203
Impact craters, 195, 375
Inductive forces, 84
Inertia, definition, 38
Inflation epoch, 155, 168
Information theory, 524
Infrared radiation, 89, 338–339, 495–496, 507

Index

Infrared Space Observatory, 145, 420
Infrared spectroscopy, 396
Inheritance, 27, 103–104
Insect reproduction, 24
Intelligence, 508–510, 546–578, 554–559, 571–572
Intensity, light, 89
Interbreeding, 526, 552
Interference patterns, 57, 68
Interferometry, 57, 485, *498*, 499, *500, 501*
Interplanetary transport, 323–325
Interstellar dust, 142–143
Intracytoplasmic membranes, 119
Introns, 136, 245
Io, 403, 404
Ionic bonds, 83
Ionization potentials, 96
Ions, 85, 123, 309–311, 310
Iridium, 535
Iron (Fe), 114, 335–336, 455, 458
Iron Mountain mine, California, 311
Isochron diagrams, 188
Isotopes
    analysis, *447*
    dating technique, 186–192
    decay, 98, 99, 186–187, 345, 403, 415
    definition, 75
    enrichment, 454–455
    neutron/proton ratios, *164*
    nuclear stability and, 98–100

James Webb Space Telescope, 164, 490
Jarosite, 379
Jupiter, 156, 183, 323–325, 329, 361, 366, 401, *537*. *See also* Europa

Kaiser Crater, *371*
Keck telescopes, *145,* 496
Kelvin scale, 63–64
Kepler spacecraft, 490–491
Keplerian motion, 181
Kepler's third law, 149
Kerogen compounds, 344, 352
Kilocalories, 307
Kinetic energy, 62–63, 66–67, 176–177, 212, 402, 415
Kitt Peak National Observatory, Arizona, *184*
Krill, 315, 317
Krypton, 71, 387
Kuiper Belt, 179, 184

L-process, 165
Labeled release (LR), 461
Lamarckian evolution, 515
Large Binocular Telescope, *145*

Laser desorption/laser ionization, *446*
Laser emission, 95
Lateral gene transfer, 290, 294, 517
Law of partial pressure of gases, 20
Lead, 188
Lepidocrocite, 458
Leptons, 52
Life
    basic characteristics, 439–444
    biosignature specificity, *437–439*
    definition of, 436–439
    detection of, 235, 435–473, *445–447*
    on Europa, 415–416, 467–468
    evolution, 514–545, 526–530
    existence of, 334–337
    extrasolar planetary systems, 474–513
    extraterrestrial intelligence, 508–510
    limits on, 443
    on Mars, 394
    origin of, 23–30, 236–240, 243–276, *275,* 295, *297*
    on Titan, 468–470
Light. *See also* Sunlight
    absorption, 89
    bending, 39, *39,* 491–494
    brightness, *90*
    diffraction limits, 494–495
    Earth-like planets, 505
    as electromagnetic radiation, 48
    emission, 89
    intensity, 89
    phase coherence, 499
    scattering, 89, 92
    spectroscopy, 21–22
    speed of, 44
    ultraviolet, 85
    wavelengths, 89
Linear momentum, 53
Lipids, energy storage, 113
Lithification, 201
Lithium, 75, *95,* 156, 158
Lithosphere, 345
Long-count calendar, Mayan, 12
Lorentz transformations, 47
Lowell Observatory, 368
LUCA (last common ancestor), 292–293, 294, 295, 301
Luminosity, 149, *150,* 340–341, 360
Lysosomes, 122

M dwarfs, 163
Ma'adim Vallis, 383, *383*
Mafic rocks, *350*
Magellan radar mission, *358*
Magnesium (Mg), 114
Magnetic fields, 44–48, *46,* 199–200, 410
Magnetism, 40–49

Magnetite, 48, 458, *460,* 462, 463, *464*
Magnetometer, 410
Magnetotactic, definition, 458
Main-sequence objects, 149
Mammals, evolution, 535
Mantle, planetary, 345, 349
Mariner missions, 368
Mars
    ALH84001, 390, 435, 457, *459,* 461–464
    atmosphere, *343,* 344, 372–375, 387
    basic parameters, 366–369
    buried networks, 386–387
    freezing and drying, 385–387
    future exploration, 394–398, 465–467
    habitability, 356–361
    history of, 365–400, *388–389*
    Kasei Vallis, *321*
    life on, 320–322, 460–461
    map of, *367*
    meteorites, 196
    orbit, 366
    parallax shifts, 149
    planetary motion, *15*
    soil, *378*
    surface appearance, 369–372, *370, 371, 372,* 391, *391–392,* 466
    terrain, *366*
    water on, 320, 357, 360, *366,* 375–383, *376,* 387–392, 393–394
Mars Exploration Rovers, 369
Mars Global Surveyor MOC, *371, 372, 373*
Mars Odyssey, 320
Mars Orbiters, *321, 371, 391*
Mars Orbiting Laser Altimeter (MOLA), 387
Mars Science Laboratories, 397, *465,* 465–466
Mass, 38, 62–64, 491–494
Mass number, 75
Massive vector bosons, 49
Maxwell's demon, 217–220, *218*
Megaregolith, 390
Membranes
    cell precursors, 258–261
    extreme environments, 315, 318
    life and, 443
    phospholipid bilayers, 113–114, *114*
    prokaryotic cells, 118, *118*
    proteins, 116
    temperature thresholds, 308
    voltage maintained by, 124–125
Mesophilic organisms, 305
Messenger RNA (mRNA), 134, 135–136, 244

Index

Metabolic analysis, *447*, 449
Metabolic unity, 441
Metabolism, 268, 269, 318
Metamorphic rocks, 186, 201, 203
Meteorite ALH84001, 390, 435, 457, *459*, 461–464
Meteorites, 192–197, *200*, 324, 390, 393
Methane (CH_4)
 carbon bonds, 87
 greenhouse effect, 355
 on Mars, 374–375
 orbitals, *87*
 on Titan, 418, 420, 421, 422
 on Venus, 506
Methanogenesis, 309, 355, *356*
Methyl cations, 87, *87*
2-methyl-propane, *23*
Micelles, 114, *114*
Microfossils, 456–457
Microlensing, 491–494, *492*
Microscopy, *445–446*, 457, *457*, *458*, *459*
Microwaves, 154
Milky Way galaxy, 9–10, 163, 165, 177, *573*
Minerals, biosignatures, 455–460
Mitochondria, 121, 127, 519, 521. *See also* DNA, mitochondrial
Modularity, 525
Moist convective runaway, 359
Molar concentrations, 85
Molecular cloud complexes, 174–176, *175*, 178, *181*
Molecular structure, 88–98
Molecules, terminology, 51
Moles, molecules in, 85–86
Moment of inertia, 180
Momentum, 34, 66–67
Moon
 accretion timescale, *194*
 effect on Earth, 200, 360
 formation, 184, 193, 199–200
 surface, *196*, *197*
 tidal heating, 403
Moons, tidal dissipation, 403–404
Motion. *See also* Planetary motion
 Galilean transformation, 46–47
 laws of, 34–35, 37–38
 uniform, 44
Multiverse concept, 168
Muscle contractions, 122
Mutations, 262, 290, 526
Myosin, 122

Nanobacteria, 266, *444*
Nanopores, *447*
NASA Astrobiology Institute (NAI), 2
National Academies of Science, Engineering and Medicine, 2
National Aeronautics and Space Administration (NASA), 1–2, 423–428, 485. *See also* specific missions
National Research Council, 2
Natural selection, 523
Neanderthals. *See Homo neanderthalensis*
Near-infrared mapping spectrometer (NIMS), 410–411, 505, *505*, *506*
Neon, 74, 75
Neptune, 183, 184
Neutrino particles, 157
Neutrinos, 49, 53
Neutron stars, 163
Neutrons, 41, 49, 75, *77*
Newtons (N), 37
Nitriles, 419, 420, *421*
Nitrogen (N)
 abiotic reduction, 312
 on Mars, 374, 384, 385
 on Pluto, 402
 on Titan, 419
 on Triton, 402
Nitrous oxides, 538
Noachian region, Mars, *371*, 386
Noble gases, 75
Nuclear forces, 49–50
Nuclear fusion, 158
Nuclear magnetic resonance (NMR), 93, 108–109
Nuclei, atomic, 52, 98–100
Nuclei, cellular, *28*, 121, 519
Nucleic acids, 103–104, 109–112, *253*, 253–255, 308
Nucleoids, 119
Nucleons, 99, 157–158
Nucleosyntheses, 164
Nucleosynthetic epoch, 156
Nucleotides, 261–262, 263
Nulling interferometry, 499, *500*, *501*

Oceans
 animal families, *529*
 on Earth, 197–199, 346, *346*, 348
 on Europa, 415–416
 oxygen isotope ratios, 333–334
 oxygen isotopes, *560*
 on Titan, 420–423
Oil drop apparatus, *42*
Oil reserves, 565
Olbers's paradox, 141–147, *142*
Olympic Mons, 369
Oort cloud, 179, 183
Opportunity Rover, *370*, 378, *378*, *380*
Optical interferometry, 499
Orbital angular momentum quantum numbers (l), 55
Orbital magnetic quantum numbers (m_l), 55–56
Orbitals, *55*, 68, *69*, *70*, 72–73, *74*, 81–83, *87*
Orbiting radar, 395–396
Orbits, electron, 54–55
Orbits, planetary. *See* Planetary motion
Organelles, 117, 120–123, 519–522
Organic chemistry, 22–23, 87
Organic samples, *469*
Organized behaviors, 276–281
Orion, *175*
Osmium, 535
Osmotic pressures, 310
Oxygen (O)
 on Earth, 353–356, 522–523
 Earth-like planets, 505
 electrons, 42, 71
 iron levels and, 335–336
 isotopes, 331, 332
 loss, *353*
 outer solar system abundance, 402
 production, *353*
 sinks, 353
 on Titan, 419
 use of, 307
Ozone, 505, 506, 526, 538

Palomar optical interferometer, *498*
Pangea, 531, 535
Panspermia models, 274
Paralax, 147–148, *148*
Paramagnetic materials, 48
Parantropus robustus, *550*
Parasites, molecular, 262–263
Parasitism, 522
Pathfinder sites, 369, *370*, 466
Pauli exclusion principle, 56, 68, 77
Peptide linkages, *106*
Peptide nucleic acid (PNA), 255–258
Peptidoglycan, 118
Periodic table, *21*, 51, 70–77, *72*
Permafrost, 390
pH, 85–86, 125, 311–312
Phagosomes, 122
Phanerosoic eons, 203
Phase coherence, 499
Phase space, 214
Phase transformations, 227–228
Phosphate bonds, 109, 112–113, 441–442
Phosphoglyceraldehydes, 132
Phospholipid bilayers, 113–114, *114*, 118, 259
Phospholipids, 113
Phosphorescence, 95

584 Index

Photolysis, 420, 422
Photometric phase variation, 491
Photons, 48, 89–92, *90*, 338–339, *340*
Photosphere, 338
Photosynthesis, 122, 130–133, *131*, 353, 531
Phylogeny
  definition, 290
  domain relationships, *292–293*
  Homo spp., 548–549, *549*, 551–553
  LUCA in, 276
  tree of life, *518*
Pili, 119, 517
Planck distribution, 90
Planck's constant (h), 60, 89, 90, 227
Planetary motion
  angular momentum, 53
  eccentric, *16*
  elliptical orbits, 16
  Kepler's law and, 149
  Mars, *15*, 366
  orbits, 504
  quantum angular momenta, *54*
  retrograde, 15
Planetisimals, 183, 184, 198
Planets
  angular shift, *484*
  assembly from fragments, *185*
  detection, 483–503
  direct detection, 494–503
  Doppler discovery of, 475–483, *476, 477*
  extrasolar, 474–513, *478–481, 482*
  formation, 174–186, 178–182, 182–186
  gas giants, 489–490
  habitable, 541–542, 561–572
  ice giants, 477
  light from, *495*
  microlensing, 491–494
  photometric phase variation, 491
Plankton, calcareous, 347–348
Plants, *121, 233*
Plasma instruments, 423
Plasma membranes, 118, 131
Plasmids, 517, 519
Plastids, 122
Plate tectonics, 344–348, 345, *347*, 351
Platinum group elements, 535
Pleiades, 151
Pluto, 368, 402
PNA, 298–300
Polar caps, 369, 372, *373*, 387, 395–396
Polar landers, 397
Polarimeters, 451

Polarity, 83–84, 85
Polycyclic aromatic hydrocarbons (PAHs), 462
Polymerase chain reactions (PCR), *447, 448*, 448–449
Polymers
  biopolymers, 103–115
  carbon-based life, 439–440
  chain termination, 233–234, *234*
  life and, 442–443
  molecular, 295
  prebiotic world, 298
  replicating, 442–443
Polypeptides, 106, *106*
Polysaccharides, 117
Porifera, 533
Porphyrin, 114
Potassium (K), 75, 187–188, 490
Potential energy, 37, 62–63, 70, 130, 185, 212
Primases, 244
Primers, 244
Principle of equivalence, 39
Principle quantum numbers (n), 55
Project Phoenix, 397, 465, 509
Prokaryotes
  cells, 116–120, *117, 266,* 444
  chemoautotrophic, 318
  domains, 290
  evolution, 517
  geologic record, 518
  membranes, 519
Protein chips, *447*
Proteins
  assembly, 104–109, 119, 290
  crystal structures, 108
  denaturation, 307
  folding, 83, 107, 307
  structure, 104–109
  synthesis, 133–137, 244
  X-ray diffraction, *109*
Protium, 76
Proto-organelles, 519
α-protobacteria, 521, *521*, 522
Protobiological world, 298–299
Protocells, 265–269, *268, 299*
Proton-proton (p-p) chains, 158
Protons, 41, 51, 75, 124
Pseudoforces, 35, 37
Psychrophiles, 315, 316, 320–321, *322,* 440
Punctuated equilibria, 516, 550
Purines, 109–110, *110, 254*
Purple sulfur bacteria, 131–132, *133*
Pwyll crater, Europa, *407,* 412
Pyrimidines, 109–110, *110, 254*
Pyrites (FeS$_2$), 260–261, 354
Pyrobolus fumarii, 308
Pyrolysis, 461

Pyrrolysine, 105
Pyruvate oxidation, 127

Qβ virus, 264–265, 267
Quanta, 41
Quantum angular momenta, *54*
Quantum mechanics, 53–62, 66–70
Quarks, 52, 155–156
Quartz crystal microbalances, 450
Quasistationary states, 284–289
Quinones, 132

r-process, 164
Racemic acid, 30
Radical formation, 85
Radio system, 424
Radioisotopes
  dating, 462
  decay, 98, 99, *99,* 186–187, *187,* 345, 403, 415
  half-lives, 186–187
  high-radiation environments, 312–315
  production, 75
  units of measurement, 313
Raman spectra, 98
Reaction centers, 131
Recombination, DNA, 517
Red giants, 162
Red shift, 144, 145–146
Red-shift-*versus*-distance relationship, 152–153
Redbeds, 354
Redox reactions, 441
Reference frames, 35–36, 47, *47*
Regeneration, limbs, 25
Religion, 10–11, 25, 30
Remote sensing pallet, 424
Replicases, 262–265
Replication, life without, 282–289
Replicator—catalyst systems, 261–265
Reproduction, 23–30, 27, 28, 260, 516, 552
Respiration, 121–122, 127, *129,* 353, 522–523
Retinas, light on, 48
Retroviruses, 249
Reverse transcriptases, 249, *249,* 448–449
Ribonucleic acid (RNA). *See also* Messenger RNA; Ribosomal RNA; Small nuclear RNA; Transfer RNA
  abiotic formation, 246–248
  amino acids, *110,* 134
  cell functions and, 116
  copying into DNA, *249*
  early roles of, 244–248
  evolutionary significance, 246
  function, 104

life and, 442
mitochondrial, 519
phosphates and, 441
PNA in construction of, 257
precursors, 250–261, *253,
   253*–255, 295
protocell sizes and, 265–269
replication, 260
ribose in, 109
ribosomal, 119
RNA worlds, 299–300
simulated evolution, 263–264
structure, 112
as template, 264
Ribose, 109, *251, 252*
Ribosomal RNA (rRNA), 134–135,
   244, *291*, 448–449, *518*
Ribosomes, 119, 120, 122, *135*,
   267, 519
Ribozymes, 245–246, 250, 262
Ribulose-1,5-bisphosphate
   carboxylase (rubisco), 132, 316
Rickettsia prowazekii, 519
Ridges, subduction zones and, 345,
   346
RNA polymerases, 135
RNA—DNA transitions, 269
Rocks. *See also* Meteorites;
   Specific rocks
   composition, *200*
   crustal, 331
   formation, *350*
   free energy sources, 318
   hydrogen release, 318–319, *319*
   isotope dating, 188
   plate tectonics and, 348
   types of, 201
Rotational energy. *See* Spin
Rotational spectral features, *94*,
   96–97
Rubidium (Rb), 188–189, *189*
Rubidium—strontium ratios, *193*
Rubisco (ribulose-1,5-bisphosphate
   carboxylase), 132, 316

s-process, 164
Saheloanthropus tchadensis,
   547–548
Salmonella entereca, 294
Salts, 309–311, *310*
Sample return, 397–398
Saracen's Head, Atlantic ocean,
   310
Saturn, 183. *See also* Titan
Schrödinger equation, 67
Seawater, 310
Second law of thermodynamics,
   214–215
Sedimentary rocks, 186, 201, 330,
   336
Selenocysteine, 105

Self-awareness, 554–559
Self-organization, 276–281,
   288–289
Semi-major axes, *16*
SETI (Search for Extraterrestrial
   Intelligence) project, 29,
   509–510, 567–572
Shoemaker-Levy 9 comet, 536, *537*
Silicon, 75, 88, 440
Silicon dioxide (SiO_2), 88, 332, 345
Skulls, hominid, *550, 551*
Slit interference experiments, *58*
Small nuclear RNA (snRNA), 136,
   244
Small subunit ribosomal RNA
   (SSU rRNA), 290
SNC meteorite, 390, 393
Snowball Earth, 352, 530–534
Sodium chloride (NaCl), 83, *84*
Sodium (Na), 21, 83, *95*, 490
Sojourner Rover, *374*
Solar system
   cratered surfaces of, *196*
   element abundance, 76, *78–81,
      162*
   life in, 320–322, 435–473
   outer region, 401–404
   planet formation, 178–179,
      182–186
Sound wave transmission, 144–145
Space, flatness, 154
Space Interferometry Mission
   (SIM), 485
Space walks, *36*
Space—time distortion, *39*, 40, 154
Space—time expansion, 152
Special theory of relativity, 44
Speciation, 516, 528, 552
Spectra
   description, 92
   Earth's atmosphere, *338*
   resolution, 93, 95–97
   shifts, 144
   terrestrial vegetation, *233*
   wavelength and, 89
Spectroscopy, 21–22, 88–98, 396,
   445–446
Spergel-Kasdin pupil, *503*
Spin, 48, 53–54, 56
Spin angular momentum. *See* Spin
Spin magnetic quantum number
   (m_s), 56
Spiral density waves, 180
Spiral galaxies, *10*
Spiral nubulae, 9
Spirit Rover, *370*, 383, *383*
Spitzer Space Observatory, 180
Spontaneous generation, 23–26,
   25–26, *26*, 29
Spores, 25, 307, 315, 320–321, 324
Standard candles, 150

Standard Mean Ocean Water
   (SMOW), 198
Stars
   aging, 150–151
   biogenic elements and, 157–165
   classification, 22, 149–151,
      151, 163
   collapse, 162
   distances, 17, 149
   effect of aging, 150–151
   formation, 175, 176–178
   lifetimes of, 143
   luminosity, 149
   microlensing, 491–494
   neutron, 163
   nurseries, 177
   parallax angles, 148–149
   photometric phase variation,
      491
   planet formation and, 174–186
   spectra, 21–22, *22*
   stable hydrogen fusion,
      160–161, *161*
   T-Tauri, 175–176
   transits, 487–491, *488*
   variable, 151
   wobble, 475
State functions, 214
Static electricity, 41–42, *43*
Stellar standard candles, 150
Steranes, 518
Sterols, 518
Stoma, 122
Stromatolites, 336
Strontium (Sr), 188–189, *189*
Subduction zones, 345, *346*, 352
Subsurface rock environments,
   317–319
Sugars, 109, 250–252
Sulfur dioxide (SO_2), 343, 411
Sulfur (S), 71, 309, 354, 411
Sun
   element abundances, *162*
   fusion reactions, 100, 158
   gravitational pull, 360
   luminosity, 90, 340–341
   movement, 485
   photon energy, 130
   photosphere, 192, 232, 338
   temperature, 90, 177, 338
Sunlight, 89, 230–231, 337–340,
   339, 357. *See also* Light
Supercontinent cycles, 348–352
Supernovae, 13, 151–152
Surface temperatures, 90, 92
Symbiosis, 519–522
Syntrophy, 522

T-Tauri stars, 175–176
Tacticity, 430
Taurus constellation, 174

Telescopes, 144, 496, 497. *See also* Specific telescopes
Temperature. *See also* Climate; Environments; *specific* Planets
 carbon emissions and, *564*
 climate cycles, *528*
 effective, 149
 greenhouse effect, 337–340, 563–564
 scales, 63–64
Terrestrial Planet Finder (TPF), 368, 502, 504
Tharis plateau, Mars, 386
Thermal Emission Spectrometer, 378
Thermal energy, 213, 234–235, 402, 415
Thermodynamics
 concepts, 212–217
 description, 211–212
 equilibrium, 217
 laws of, 212–215
 living systems, 211–242, 225–236
 photon redistribution, 339
 polymer chain termination, *234*
Thermophilic organisms, 292, 307–309, 325
Tholin, 422
Thorium, 194–195
Threose nucleic acids (TNAs), 253–257, 253–258
Thyalkoid membranes, 122
Thymine (T), *110*, 110–112
Tidal dissipation, 403, 415
Tidal heating, 403
Time, dimension, 40
Titan
 atmosphere, 85, 417–420, *419*, 423–428
 Cassini-Huygens mission, 423–428
 comet impact, *429*
 description, 401
 exploration, 428–431, 468–470
 formation, 416–417
 hydrocarbons, *97*, 259, 420–423
 oceans, 420–423
 orbit, 404
 organic samples, *469*
 stratosphere, *421*
 surface, 422–423, *423*
 temperature, 403, *420*, *427*
 water on, 329, 421, 428, 430
Titan Organics Explorer, 469
Tonalites, 330
Tool making, 548
Traits, inheritance, 27
Transcription, 135, *135*
Transfer RNA (tRNA), 134, 136
Transits, 487–491, *488*
Tree of life, phylogeny, *518*
Trilobites, *202*
Triton, 184, 402
Tropopause, 357, 359

Ultraviolet radiation, 85, 313, 343, 526
Ultraviolet spectrometers, 396
Uncertainty principle, 55, 67–68
Uniform motion, 44
Universal gravitational constant, 37
Universal tree of life, 290
Universe
 distribution of matter, *169*
 entropy and evolution of, 224–225
 expansion, 146–147, 167
 missing mass, 165–167
 open or closed, 167
 origins, 154
 shape, 165–167
Uracil (U), *110*, 110–112
Uraninite (UO_2), 354, 355
Uranium, 188, 190, *191*, 192–193, 194–195
Uranium-lead (U-Pb) isotopes, 330
Uranus, 54, 183

Vaccine development, 27
Vacuoles, 122
Vacuums, 44, 48, 324
Valence shells, 71, 86–87
Valles Marineris, 369
Van de Graaf generators, 41–42, *43*
Van der Waals attraction, 84
Vegetation, spectra, *233*
Velocity, definition, 34
Venn diagrams, *460*
Venus
 atmosphere, 341, 357
 formation, 184
 habitability, 356–361
 parallax shifts, 149
 physiography, *358*
 solar luminosity, 360
 sunlight, 357
 TPF spectra, 506
 water on, 506
Very high resolution orbital imaging, 396
Very Large Telescope, Chile, *145*
Vesicles, 258–261, 265, *444*
Vibrational spectral features, *94*, 96
Viking spacecraft, 368, 369, 460–461
Virgo cluster, 152
Virial theorem, 176–177
Viruses, 517
Vision, light and, 89
Vital spark, 29
Vitamins, 114
Volatility, definition, 182
Volcanic activity, 345, 346, 407

Voyager flybys, 404–405, *417*, 417–420, 420
Vugs, 380, *382*

Water. *See also specific* Moons; *specific* Planets
 altitude and, 359
 chemical potential, 227–228
 in chondrites, 197
 clouds, 343
 comets as source of, 198
 Earth-like planets, 505
 HDO, 76
 of hydration, 402
 life and, 440–441
 melting, 226, *227*
 outer solar system ice, 402
 recycling, 346, *347*, 349
 slit interference experiments, 58–59
 waves, *49*, 57
Watson-Crick pairing, 263, 298
Wavefunctions, 56–57, 62, 67, 77
Wavenumbers, 93
Wavepackets, 60–62
Wave—particle duality, 48, 52, 57–62
Waves
 diffraction, 57
 Doppler effect, 145, *146*
 electromagnetic, 48, *49*
 gravitational, 50
 interference patterns, 57
 slit interference experiments, 58–59
 transmission, 144–145
 water, *49*, 57
 wave—particle duality, 48, 52, 57–62
Weak nuclear forces, 49, 252
Weathering, 344–348, *345*, *347*
Weight, definition, 38
White dwarfs, 163
Wilkinson Microwave Anisotropy Probe, 154
Winds, 183, 369, 385
Work, 213, 225, 229–230

X-ray diffraction, 108, *109*
X-ray spectrometer, *380*
X-rays, 27, 96
Xenon, 71, 75, 387

Yellowstone National Park, 307, *308*

Zagami meteorite, 390
Zeroth law of thermodynamics, 213
Zircon (zirconium silicate, $ZrSiO_4$), 330, 331–332